map p. 309

Eoalpine map (high pressure) p. 355

Eo, Meso, Neo

GEOLOGICAL SOCIETY SPECIAL PUBLICATION NO 45

Alpine Tectonics

EDITED BY

M. P. COWARD
Department of Geology,
Imperial College,
London SW7 2BP, UK

D. DIETRICH
Geologisches Institut,
ETH-Zentrum,
CH-8092 Zurich,
Switzerland

R. G. PARK
Department of Geology,
University of Keele,
Staffordshire ST5 5BG, UK

1989

Published for
The Geological Society by
Blackwell Scientific Publications

OXFORD LONDON EDINBURGH
BOSTON MELBOURNE

Geological Society Special Publications

Series Editor K. COE

Published for
The Geological Society by
Blackwell Scientific Publications
Osney Mead, Oxford OX2 0EL
(*Orders*: Tel. 0865 240201)
8 John Street, London WC1N 2ES
23 Ainslie Place, Edinburgh EH3 6AJ
3 Cambridge Center, Suite 208, Cambridge
Massachusetts 02142, USA
107 Barry Street, Carlton, Victoria 3053,
Australia

First published 1989

Set by Setrite Typesetters, Hong Kong
Printed in Great Britain at the
University Press, Cambridge

DISTRIBUTORS

USA
Publishers' Business Services
PO Box 447
Brookline Village
Massachusetts 02147
(*Orders*: Tel. (617) 524 7678)

Canada
Oxford University Press
70 Wynford Drive
Don Mills
Ontario M3C IJ9
(*Orders*: Tel. (416) 441 2941)

Australia
Blackwell Scientific Publications
(Australia) Pty Ltd
107 Barry Street, Carlton,
Victoria 3053
(*Orders*: Tel. (03) 347 0300)

British Library
Cataloguing in Publication Data

Alpine tectonics.
1. Europe. Alpine regions. Tectonic features
I. Coward, M. P. II. Dietrich, D.
III. Park, R. G. (R. Graham) IV. Series 554

ISBN 0-632-02508-5

Library of Congress
Cataloging-in-Publication Data

Alpine tectonics/edited by M. P. Coward,
D. Dietrich, R. G. Park.
p. cm. — (Geological Society special
publication; no. 45)
Includes index.
ISBN (invalid) 0-632-02508-5
1. Plate tectonics — Alps Region. 2. Geology —
Alps Region.
I. Coward, M. P. II. Dietrich, D. (Dorothee)
III. Park, R. G. (R. Graham) IV. Geological
Society of London. V. Series.
QE511.4.A46 1989
551.1′36′094947 — dc19

Contents

Preface

This volume is the result of a Conference on Alpine Tectonics held at Burlington House, London, in May 1987 under the auspices of the Geological Society of London. The conference was the second in a new series of meetings organized by the Society on controversial and topical subjects, and included the 1987 Fermor Lecture given by Professor K. J. Hsü. The Fermor Lecturer was supported by 23 other contributors, 11 of whom were non-British geologists, giving the meeting a strong international flavour.

The volume contains 23 papers. The opening contribution by Coward and Dietrich is an overview, based partly on the results of the conference, aimed at introducing the tectonic setting for the more detailed papers to follow. These are grouped into three sections. The first, *Tectonic evolution of the external zones of the Alps*, includes papers by Ramsay and Dietrich and Casey on Helvetic structure and by Gratier *et al.* and Fry on the external zones of the Western Alps.

The second section deals with the tectonic evolution of the *Internal and southern zones of the Alps*. Papers by Merle *et al.*, Platt *et al.* and Ellis *et al.* on the structure of the Pennine zone and Schmid *et al.* on the role of the Periadriatic line, are followed by papers by Ridley on the Sesia–Lanzo zone, Brodie *et al.* on the Ivrea zone, Roeder and Laubscher on the southern Alps and by Ratsbacher and Neubauer on the Austro–Alpine cover nappes in the eastern Alps.

The third section is entitled *Models of the development of the Alpine chain* and commences with a paper by Dewey *et al.* on the Cenozoic evolution of the Western Mediterranean and Channell and Mareschal on the development of the Tyrrhenian rift. Mueller then discusses the seismic evidence for deep seated geodynamic processes. Indentation and rotation in the Western Alpine arc are dealt with by Vialon *et al.* and plate-scale kinematics of crustal shear zones by Lacassin. This group is followed by two papers on thermo-tectonic evolution, by Hunziker *et al.* and Hurford *et al.* and a review by Heller *et al.* of the palaeomagnetic and magnetic anisotropy results from the Alps. The final paper in the volume is the Fermor Lecture by Hsü, entitled 'Time and place in Alpine orogenesis'.

The volume aims to represent current views, many of which are controversial, on a fascinating but rapidly-changing topic. The book should not be regarded as a monument, but rather as a milestone!

The editors are grateful to the Geological Society for the financial support which enabled such an unusually large number of foreign speakers to attend the meeting, which also benefitted from support from the Fermor bequest. It was particularly pleasing to the organizers that Lady Fermor herself was able to attend the meeting.

MICHAEL P. COWARD
DOROTHEE DIETRICH
R. GRAHAM PARK

Acknowledgements

The editors would like to thank the following colleagues and friends who kindly helped with the reviewing of the articles of this volume:

Alastair Beach, Lochend by Beith, Scotland, UK
Jan Behrmann, Giessen, FRG
Mike Bickle, Cambridge, UK
Derek Blundell, Egham, UK
Kate Brodie, London, UK
Rob Butler, Milton Keynes, UK
Jean Michel Caron, Lyon, France
Martin Casey, Zurich, Switzerland
Jim Channell, Gainesville, Florida, USA
Bob Cliff, Leeds, UK
Peter Cobbold, Rennes, France
Mike Daly, London, UK
John Dewey, Oxford, UK
Martin Frey, Basel, Switzerland
Norman Fry, Swansea, UK
Rod Graham, London, UK
Rick Groshong, Alabama, USA
Ruth Haas, Innsbruck, Austria
Marc Helman, Oxford, UK
Johannes Hunziker, Bern, Switzerland
Mike Johnson, Edinburgh, Scotland, UK
Roy Kligfield, Boulder, Colorado, USA
Nick Kusznir, Liverpool, UK
Hans Laubscher, Basel, Switzerland
Cherry Lewis, Milton Keynes, UK
Alberto Malinverno, Palisades, NY, USA
Stephen Matthews, Llanrwst, Wales, UK
Elizabeth McClellend, Oxford, UK
Olivier Merle, Rennes, France
Alan Milnes, Zurich, Switzerland
Roland Oberhänsli, Bern, Switzerland
Renée Panozzo, Zürich, Switzerland
John Platt, Oxford, UK
Graham Potts, Colwyn Bay, UK
Lothar Ratschbacher, Tübingen, FRG
Pierre Rochette, Grenoble, France
Dietrich Roeder, Denver, Colorado, USA
David J. Sanderson, Belfast, Ireland
Albrecht Steck, Lausanne, Switzerland
Peter Treloar, Oxford, UK
Rudolf Trümpy, Zurich, Switzerland
Graham Williams, Keele, UK
André Zingg, Basel, Switzerland

Alpine tectonics – an overview

M. Coward & D. Dietrich

SUMMARY: This overview summarizes aspects of 150 years of research in Alpine tectonics and in particular introduces the tectonic setting for the more detailed papers in this volume.

The Alpine Mesozoic ocean, Tethys, formed as a large elongate pull-apart basin in the Jurassic, as a consequence of the opening of the Atlantic and of the movement of Africa towards the east relative to a fixed Europe. The NNE trending Tethys was bounded by WNW trending transforms, by the European/Iberian margin in the W and by the Adriatic promontory of Africa in the E, and its shape determined the present day configuration of the arcs of the Alpine chain. The closing of this ocean and the collision tectonics began during the Cretaceous, as Africa moved to the NE relative to Europe and as the N Atlantic gradually opened, to drive Iberia and the southern part of the European plate to the E. Subduction of oceanic crust and adjacent continental crust led to high pressure metamorphism of Cretaceous age. Ophiolites were obducted over the southern continental margin, but after collision the shear sense reversed so that the Austro–Alpine nappes of the African Adriatic promontory overthrust Europe in a WNW direction. During the main Tertiary deformation the overall anticlockwise rotation of Africa led to a change-over from N to NW and WNW-directed collisional structures. The E–W striking sector of the Alps in Switzerland and Austria is therefore a diffuse transpressive dextral shear belt, approximately reworking the northern transform boundary of Tethys, modifying it by compression related to the rotation of the African Adriatic promontory.

Approximately 250 km of European lithosphere were involved in the building of the western Alps. As Alpine nappes consist largely of rock material confined to the upper crust, a large amount of lower crust and lithospheric mantle of the two continental blocks must be duplicated and/or subducted during the Alpine collision history.

The Alpine belt marks the collision zone between the African and Eurasian plates. After the breakup of Pangea and the opening of the Atlantic during early Mesozoic times, the area between the African and Eurasian plates was occupied by oceanic crust extending from the western Mediterranean to the eastern Himalayas. This oceanic region has been known as 'Tethys' since Neumayr (1883) and Suess (1893, 1901). The Alpine chain of western Europe (Fig. 1) resulted essentially from the subduction of this Tethyan oceanic crust followed by continental collision between the African and the western part of the Eurasian plates. The chain consists therefore of slices of sediments and of continental crust scraped off from both the African and European margins, as well as of oceanic crust. These scrapings, the Alpine thrust sheets or *nappes* (e.g. Heim 1921), are each at most a few kilometres thick, yet stacked on top of each other they have produced pronounced crustal thickening.

This Special Publication deals with the Alpine chain in western Europe (France, Switzerland, Italy and Austria), as shown in Fig. 1, with particular emphasis on the arc of the western Alps. A vast number of papers in English, French, German and Italian are dedicated to the tectonics of this region. In the following, we do not want to present a systematic digest of all this literature, but an overview of Alpine evolution, with special emphasis on its still controversial aspects. Some themes in this introduction and overview are based on the papers included in this Special Publication, as well as on others read at the Geological Society of London meeting, from which the volume derives. We must note, however, that this introduction contains our own view and prejudices, sometimes supporting but occasionally at odds with other papers in the volume.

The main structure of the Alps

As shown in Fig. 1, the Alps consist of:

i) Internal units, where the structures are mainly of Cretaceous–Palaeogene age and the rocks, which include relics of the Tethyan ocean crust, are often metamorphosed, locally up to high temperatures and pressures.
ii) External units, where the structures are mainly of Oligo–Mio–Pliocene age and the units consist of thrust and folded but

From: Coward, M. P., Dietrich, D. & Park, R. G. (eds), 1989, *Alpine Tectonics*, Geological Society Special Publication No. 45, pp. 1–29.

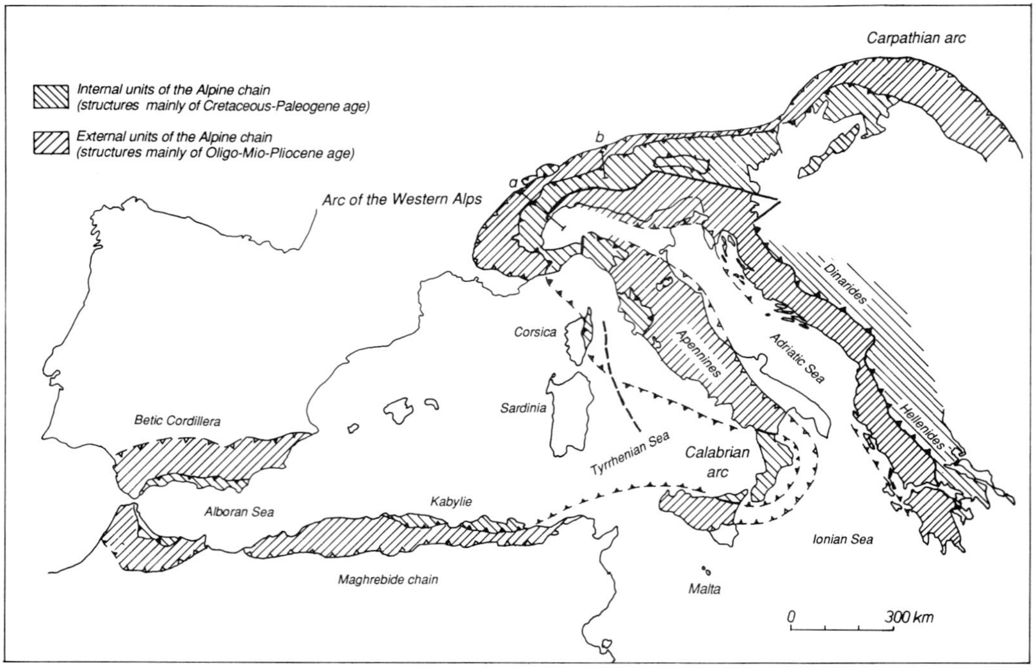

FIG. 1. A generalized map of the Alpine chain in western Europe, showing the tight arcs and changes in vergence of thrusting. The internal units comprise slivers of oceanic crust and basement nappes with associated sediments from both continental margins and are affected by Alpine metamorphism. The external units consist of sedimentary décollement nappes and were generally only slightly affected by Alpine metamorphism. The locations of the seismic sections of Fig. 4 are indicated.

weakly to non-metamorphosed basement and Mesozoic to Cainozoic cover of the European or African plates.

The main divisions of the western Alps are shown in Fig. 2, based on Ramsay (1963, 1981), Debelmas & Kerckhove (1980) and Trümpy (1980).

Interpretations of Alpine cross-sections have been controversial, as shown in the schematic sections in Fig. 3. In Fig. 3(a) units from the southern continental margin were thrust over the northern continent with strong deformation in a part of the northern margin, the internal Pennine zone. The external zone is made of largely autochthonous basement. In Fig. 3(b), the external units are represented as allochthonous thrust sheets because of assumptions made regarding section balancing. Both figures have shortcomings suggesting a 2-D model for a 3-D orogen and not considering or justifying volume balance of the lithosphere during the collision process.

The plate boundary is considered to lie at the base of the Austro–Alpine nappes, which overthrust the European plate for a minimum distance of 200 km to the NW, as shown from the presence of windows of European plate rocks beneath the Austro–Alpine nappes in the Engadin and Tauern regions of Switzerland and Austria. The overthrust kinematics are given from indicators such as stretching lineations (e.g. Choukroune *et al.* 1986, Ratschbacher & Neubauer this volume, Merle *et al.* this volume). In the southern part of the Swiss Alps, the Austro–Alpine and underlying thrust sheets are steepened and cut by an oblique to strike-slip shear zone known as the Insubric Line (e.g. Schmid *et al.* this volume). South of this line, in northern Italy, the Ivrea zone marks a part of the lower crust and upper mantle of the southern plate, deformed during Palaeozoic compressive events, but uplifted during Mesozoic extension and subsequent Alpine compressive tectonics (Brodie *et al.* this volume). The southern Alps, to the east, form a stack of S-vergent thrusts involving basement and Upper Palaeozoic to Mesozoic strata of the southern plate (Laubscher 1985). These structures and their relationships to other Alpine structures are described by Roeder (this volume).

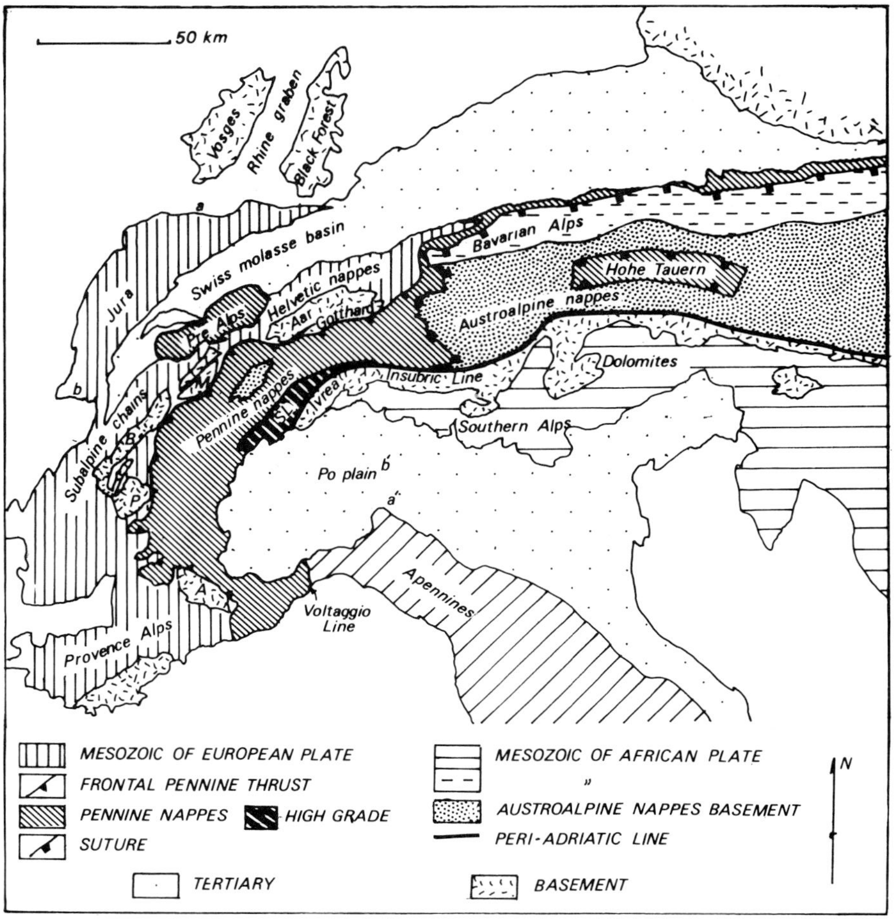

FIG. 2. Structural zones within the western Alps (after Ramsay 1963, Debelmas & Kerkhove 1980, Trümpy 1980). Section lines of Fig. 3 are indicated. SL = Sesia–Lanzo zone, M = Mont Blanc massif, B = Belledonne massif, P = Pelvoux massif, A = Argentera massif.

Along the suture zone between the two plates, there are slivers of ophiolitic material, representing Mesozoic oceanic crust, metamorphosed to high pressures and low temperatures. The underlying rocks of the European plate, forming the Pennine thrust and fold nappes of the internal zone, represent slices of basement with Mesozoic cover, which have been intensely deformed during Alpine compression. The lowest Alpine nappes emerge in the Ticino region of southern Switzerland, where they show high temperature regional metamorphism, due to burial of the Pennine nappe pile. The internal zone is limited in the W and NW by what is known as the Frontal Pennine Thrust, along which there must have been very large displacements, as Pennine rocks occur as klippen in the Pre-Alps.

The external Alps, forming the Helvetic Alps and the Jura mountains in Switzerland, the Chaînes Subalpines and Provençal fold and thrust arcs in France, consist dominantly of non to only weakly metamorphosed sediments of the European plate. On their internal margin there are large uplifted basement massifs: the Aar and Gotthard in Switzerland, the Aiguilles Rouges, Mont Blanc, Belledonne, Pelvoux and Argentera massifs in France (Fig. 2). These were originally considered to be uplifted but autochthonous basement (e.g. Ramsay 1963). Boyer & Elliott (1983) and Butler *et al.* (1986) reinterpreted them as far travelled basement thrust sheets, but more recent interpretations (e.g. Lemoine *et al.* 1986, Gillcrist *et al.* 1987) show them to be strongly deformed, originally Mesozoic extensional blocks, uplifted during Upper Cretaceous and Tertiary Alpine deformation. The frontal regions of the Alpine folds

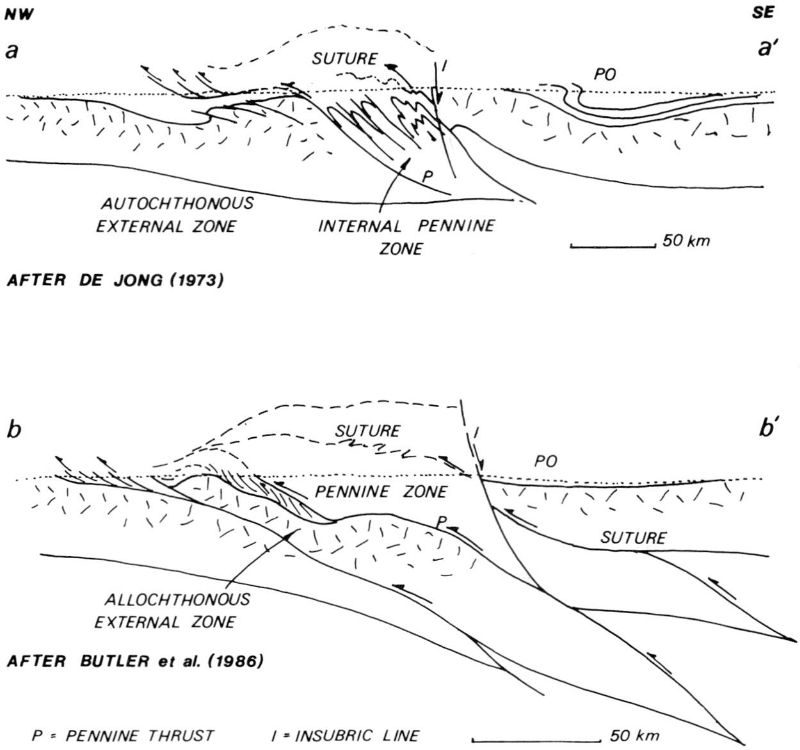

FIG. 3. Simplified NW–SE cross-sections through the western Alps, from (a) De Jong (1973), (b) Butler *et al.* (1986). Section lines are shown on Fig. 2: (a) is an unbalanced section with the external massifs shown as autochthonous or parautochthonous; (b) is a balanced section showing the Belledonne Massif as a major thrust sheet. Both sections show shortcomings; see text for discussion and Fig. 17 for comparison.

and thrusts form the Jura mountains, considered to be a folded sheet of Mesozoic rocks which detached above Triassic evaporites.

The external Alps show strongly divergent thrust directions: they are northerly directed in eastern Switzerland and Austria (e.g. Pfiffner 1981), change over from a northerly to a WNW direction in western Switzerland (Dietrich & Durney 1986, Dietrich & Casey this volume), they are NW-directed in the northern French Alps (Gratier *et al.* this volume) and SW-directed in the southern French Alps (e.g. Fry this volume). In southern France there is interference between thrust and fold structures related to a Palaeogene northerly directed system, linked to the Pyrenees, and these Neogene SW-directed structures.

The Alps are bordered by syn-orogenic sediments, infilling the molassic foreland basins of the Swiss and German plains, the Rhône valley in France and the Po basin in northern Italy. Earlier orogenic sediments form the Cretaceous and Palaeogene flysch deposits, deformed along with the Mesozoic sediments in the internal and external Alps. These deposits, together with the early molasse, allow some of the deformation to be dated and hence rates of underthrusting and fault propagation to be estimated (Homewood *et al.* 1986).

Alpine characteristics – Alpine problems

The Alpine chain can be characterized by the following main tectonic features, most of which still pose problems with interpretation:

(1) Closure of the Mesozoic Tethyan basins occurred at different times along the Alpine suture and involved different subduction kinematics. The timing and kinematics of subduction of Tethyan oceanic crust relative to the two major continental margins is still under discussion, with different models for different parts of the Alpine belt being proposed (e.g. Channell & Mareschal this volume; Hsü this volume; Hunziker *et al.*

this volume). Most authors agree that in the Alps *sensu stricto* the onset of compression between the two major plates was marked by subduction of oceanic crust beneath the African continental margin (e.g. Laubscher's 1983 synthesis), while along the continuation of the Alpine belt in Calabria, southern Italy, there is evidence for early subduction of oceanic crust beneath the European continental margin (e.g. Dewey *et al.* this volume).

(2) Much of the evidence for the earlier Alpine history is lost because of the overprinting character of the compressional history and because of localised middle Tertiary extension in the western Mediterranean, to open the Gulf of Lyons and the Ligurian Sea (Burrus *et al.* 1987), but particularly because of the fragmentation of the chain by the development of later extensional basins, such as the Pannonian and the Tyrrhenian basins. It is therefore still difficult to link kinematically the deformation histories postulated for individual sectors of the Alpine chain.

(3) The vergence of thrusting changes along the Alpine chain (Fig. 1). The main Palaeogene vergence, or direction of nappe emplacement in the Alps *sensu stricto* was on to the European continent, with partly superimposed Tertiary backthrusting, i.e. thrusting towards the African hinterland (Argand 1916, Laubscher 1970, 1985, Merle *et al.* this volume, Platt *et al.* this volume). The main vergence of Cretaceous to Plio-Pleistocene nappe transport in Calabria was on to the African plate (Dietrich 1988 and references therein). These changes in vergence can be related to the above mentioned changes in subduction kinematics.

(4) Opposite vergences of thrusting can be locally superimposed (e.g. in parts of the Alps and northern Apennines), and out of sequence thrusting is an important but yet poorly established fact of Alpine evolution. Simple palinspastic restorations, based on the assumption that the geometrically higher units were derived from palaeogeographic belts internal relative to the geometrically lower units, might therefore be locally impossible to construct.

(5) The Alpine chain is extremely arcuate with tight curvatures (e.g. Goguel 1963, Laubscher 1988) (Fig. 1). The sector of the chain discussed in this volume twice changes strike direction of nearly 180° through two major oroclines, the arc of the western Alps and the Calabrian arc. These tight curvatures might be explained by (a) complex palaeogeographic outlines of the original continental margins leading to important strike-slip movements between the colliding continents, and (b) rotational movements of the overthrusting margin relative to the underlying plate. Arcuate plate subduction (e.g. Frank 1968) and oroclinal bending (e.g. Ries & Shackleton 1976) are not considered as valid mechanisms for the formation of the arc of the western Alps.

An example of the complex palaeogeography of the original African continental margin is given by the Adriatic promontory (Channell & Horvath 1976), which indented the European plate. Such indentation in itself must lead to strike-slip motion along the plate boundaries and within the plate (e.g. Molnar & Tapponnier 1975).

(6) The continent–continent collision phase of the African and European margins has led to subduction of continental lithosphere. Many Alpine geologists came to this unorthodox conclusion because of concern regarding material balance (Ampferer 1906 and Laubscher 1969, 1970, for the first expression of this concept in plate tectonics terms). In the Alpine nappe structures only slices of the uppermost parts of continental lithosphere are conserved while huge volumes of continental lithosphere are missing. Alpine geologists also found it difficult to explain HP/LT metamorphism in tectonic units consisting of continental crust, if they did not assume subduction and later exhumation of these units (e.g. Dal Piaz *et al.* 1972). Subduction of continental crust could explain the impressive isostatic disequilibrium leading to the major uplift which is so apparent in today's Alpine topography (e.g. the uplift data by Hurford *et al.* this volume).

(7) Several sectors of the Alpine chain appear to have deep lithospheric roots, extending to over 200 km. These occur, according to Panza (1984) and Mueller (this volume) in the central and western Alps, the central Apennines and the Betic Cordillera. The subvertical roots beneath the central Alps have been interpreted following Laubscher's (1974) model of bivergent subduction, with participation of continental lithosphere from both continents in the subduction process (Panza & Mueller 1979, Mueller 1982, Finckh *et al.* 1987, Mueller

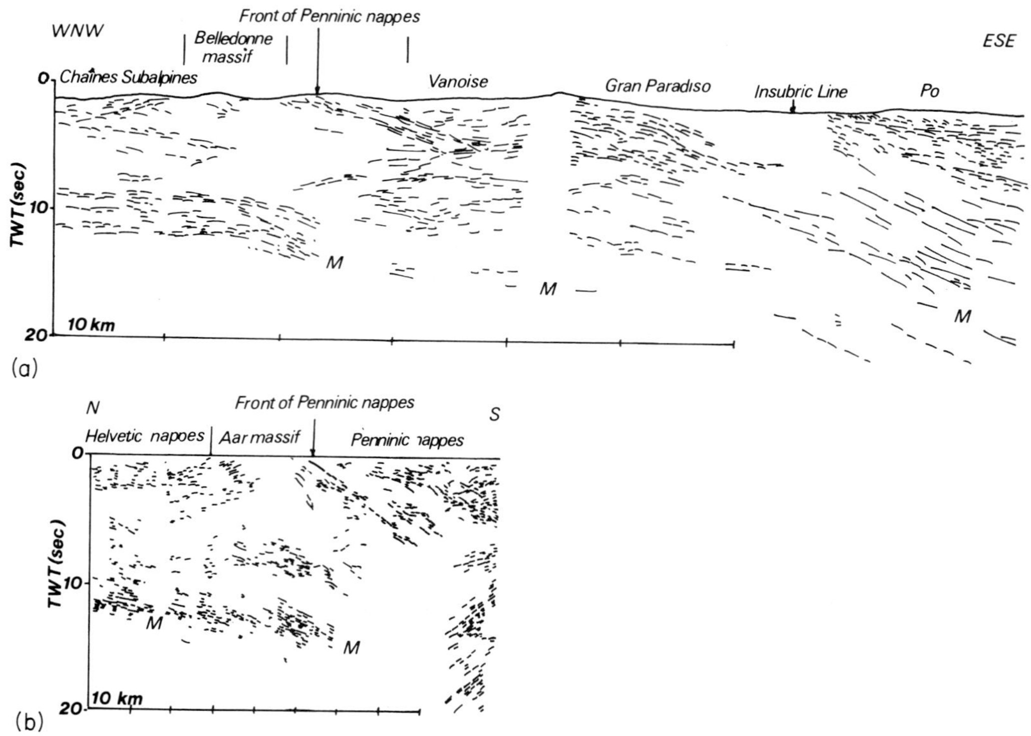

FIG. 4. The deep structures of the Alps, shown by seismic sections; the locations of the sections are indicated in Fig. 1. Note the two sections have a different horizontal scale. The principal reflection layers are represented by as yet provisional line drawings and are redrawn according to Bayer *et al.* (1987), section a, and Pfiffner *et al.* (1988), section b. M: Mohorovičić discontinuity, according to refraction data. TWT: two-way time-scale; 20 sec. corresponds approximately to 60 km. *Section a*: The ECORS–CROP line across the western Alps (from Bayer *et al.* 1987). The ESE-dipping reflectors suggest major lithospheric thrusting, with the African plate (s.l.) thrust over the European plate. *Section b*: The explosion-seismic reflection profile NFP-20 across eastern Switzerland (Mueller, this volume, Finckh *et al.* 1987, Pfiffner *et al.* 1988). A bivergent subduction zone with subduction of the decoupled lower part of the crust is assumed by Mueller (this volume) to be the best model to fit these data, though Roeder (this volume) interprets the S-verging thrusts of the southern Alps as the tip of a crustal flake back-thrust.

this volume). This interpretation is supported by the N-dipping reflectors seen deep in the southern plate on the Swiss explosion seismic reflection data (Fig. 4, section b, and Mueller this volume). However the deep structure of the western Alps is interpreted in a different way. The reflectors on the ECORS–CROP line (Fig. 4, section a, Bayer *et al.* 1987) generally dip towards the SE, and are interpreted by these authors as indicating lithospheric thrusting of the southern plate over the northern plate. This, as well as many other interpretations of the Alps in terms of crustal accretionary wedges pose problems of volume balancing regarding the lower parts of the northern and southern lithospheres.

(8) Alpine deformation was not confined to the Alpine chain, but affected much of NW Europe, as far N as the Shetland Platform and the Atlantic margin. The European plate did not behave as a rigid body, but deformed internally, reactivating Mesozoic extensional faults and earlier Palaeozoic steep strike-slip shear zones, to cause widespread tectonic basin inversion from the Middle Cretaceous to the Neogene (Ziegler 1982, 1987, Beach 1987, Coward & Trudgill in press). Alpine deformation chronology is sometimes easier studied in these inverted basins of the European craton than in the main Alpine chain, where much of the evidence for the earlier events has been lost during subsequent mountain building.

Mesozoic extension in the Alpine domain

Most authors who reconstruct the plate movements resulting from the breakup of Pangea, link the development of western Tethys to the gradual northward opening of the Atlantic (Smith 1971, Pitman & Talwani 1972, Dewey *et al.* 1973, Livermore & Smith 1985, Savostin *et al.* 1986, Dewey *et al.* this volume). Extension began in the Permo–Triassic, often reworking earlier Palaeozoic compressional and strike-slip structures, and developed basins on the future African margin and throughout much of NW Europe (Figs 5 and 6). Deep Triassic basins formed W of Britain on the hanging walls of Caledonian and Variscan thrusts (e.g. Cheadle *et al.* 1987, Gibbs 1984). Their extension direction was NW–SE, parallel to large steep tear faults, which acted as lateral extensional ramps to basins compartmentalising the extension and subsequent inversion (Fig. 6). These NW–SE structures were probably inherited from earlier Variscan steeply dipping strike-slip shears and in effect controlled the main Mesozoic extension direction and possibly the opening direction of the Atlantic by tramlining relative crustal movements (Arthaud & Matte 1977). Similar NW–SE trending shears occur in the Massif Central and the basement massifs of the external French Alps, such as the Pelvoux and Argentera.

On the northern continent, over much of western Europe, most of the Triassic deposits were continental, with the deposition of thick evaporites. On the future northern margin, locally thick carbonate platform deposits developed (e.g. in the Briançonnais domain), and locally basaltic volcanism occurred. On the future southern margin deep water basins were established with a very similar sedimentary facies development, e.g. the Imerese–Lagonegro basin in southern Italy, the Hallstatt basins in Austria, the Budva–Kotor–Pindos basin in Yugoslavia and Greece (Aubouin 1959, Broquet *et al.* 1966, Scandone 1967) (Fig. 5). All these basins show basalt intrusions and subsided below CCD in the Liassic (Scandone 1975). Lateral thick and very extensive carbonate platforms developed (e.g. Ippolito *et al.* 1975). Middle to Upper Jurassic ocean spreading was localised parallel to these Triassic rifts (Scandone 1975).

The Tethyan stretched crust was linked to the Atlantic by a major transform system, developed in the sinistral shear between Africa and Eurasia (Figs 5 and 7). The relative paths of Africa and Eurasia can be obtained from reconstructing and restoring the magnetic anomaly

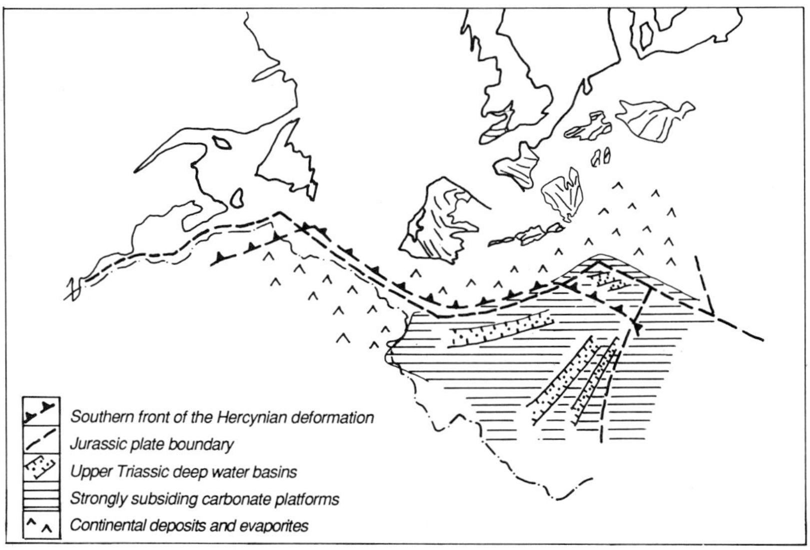

FIG. 5. The Upper Triassic palaeogeography and the Jurassic plate boundary. The plate boundary is roughly parallel to the southern front of the Hercynian deformation and does not follow closely the main sites of Middle Triassic rifting. The Germanic Triassic consists of continental deposits and evaporites (unornamented); the Alpine Triassic consists of strongly subsiding platforms and deep water basins. In the Jurassic the palaeogeography shows continuous extension of both margins (Fig. 9), and the basins and platforms of the southern margin sink partly below CCD. The restoration of the African Plate relative to the European Plate is based on the position of the continents at 175 Ma according to Dewey *et al.* (this volume).

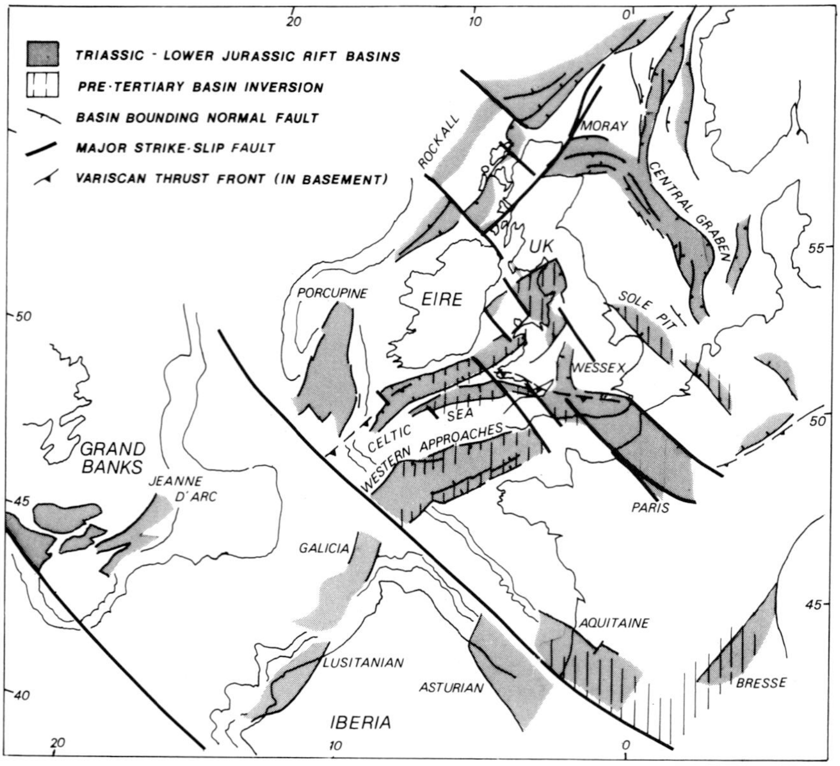

FIG. 6. Basin development in the NW foreland of the Alps; rift structures of Triassic–Liassic age. Regions with important Cretaceous basin inversion are shown. From Ziegler (1982) and Masson & Miles (1986), plus work by MPC.

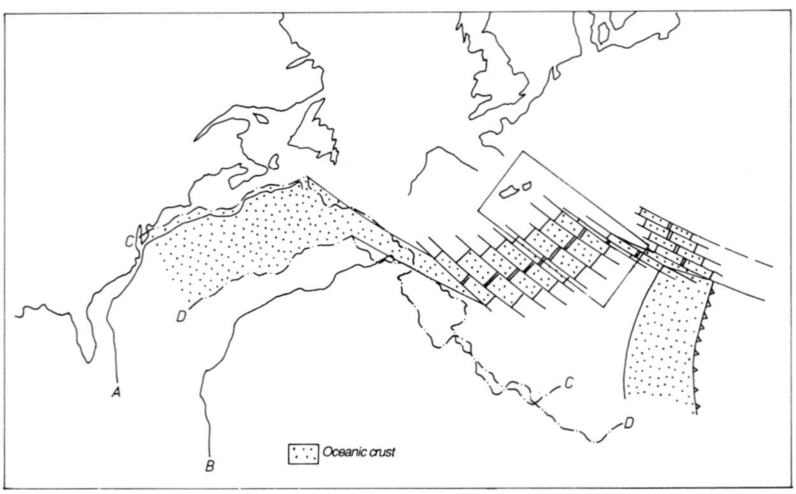

FIG. 7. The development of Tethys at 150 Ma (Upper Jurassic), interpreted as a transform belt between the African and European continents (Laubscher & Bernoulli 1977, Dercourt *et al.* 1986). The more recent reconstructions of the positions of the continents (e.g. Livermore & Smith 1985, Savostin *et al.* 1986. Dewey *et al.* this volume) agree qualitatively, but not quantitatively: A–B indicates the positions of Africa at 175 and 150 Ma (interpolated) according to Livermore & Smith (1985), C–D indicates the positions of Africa at 175 and 149.2 Ma according to Dewey *et al.* (this volume). The location of Fig. 9 is indicated (box).

data in the Atlantic; Fig. 8(b) shows the paths of Africa relative to stable Europe as suggested by Smith (1971), Dewey *et al.* (1973) and Livermore & Smith (1985). These paths differ from the latest paths presented by Dewey *et al.* (this volume) who suggest that the sinistral movement between the two continents was not followed by the pronounced dextral movement suggested by earlier models.

Ophiolites from the Tethyan ocean have been found in tectonic units all along the Alpine chain (e.g. the review by Knipper *et al.* 1986), from N Africa (e.g. Bouillin 1979), southern Italy (Dietrich & Scandone 1972, Haccard *et al.* 1972, De Roever *et al.* 1974, Amodio Morelli *et al.* 1976), from northern Italy (e.g. Boccaletti *et al.* 1971, Decandia & Elter 1972), from Corsica (Amaudric du Chaffaut *et al.* 1972, Ohnenstetter & Ohnenstetter 1980), from the western Alps (e.g. Debelmas & Lemoine 1970, Tricart & Lemoine 1983), from the central Alps (e.g. Steinmann 1906, Trümpy 1960, Bearth 1967, V. Dietrich 1969, 1980), from the eastern Alps (e.g. Bickle & Nisbet 1972, Tollman 1975), from the Carpathians (e.g. Savu 1980), from Yugoslavian Dinarides (e.g. Karamata *et al.* 1980) and from Greece (e.g. Moores 1969, Dercourt 1972, Bernoulli & Laubscher 1972). These ophiolites have generally been affected by Alpine metamorphism, but Jurassic ages of origin have been determined radiometrically (e.g. Bertrand & Delaloye 1976, Ohnenstetter *et al.* 1981). The Tethyan ophiolites are stratigraphically overlain by a characteristic sequence of Middle to Late Jurassic–Early Cretaceous age, consisting of radiolarian cherts and siliceous shales (the Radiolarite Formation), followed by pelagic limestones (the Calpionella Limestone Formation), overlain by alternations of siliceous shales, marls and pelagic limestones (the Argille a Palombini Formation of Decandia & Elter 1972). These pelagic deposits have been found in allochthonous units of the Alpine chain from eastern Switzerland to southern Italy.

Simple reconstructions such as that shown for the Kimmeridgian in Fig. 7 are based on the following considerations:

(1) The width of Tethys can be estimated kinematically from the relative motions of the African and European Plates. Figure 7 adopts the relative plate motion model of Dewey *et al.* (this volume) and results in a similar reconstruction to that of Laubscher & Bernoulli (1977). As the Jurassic–

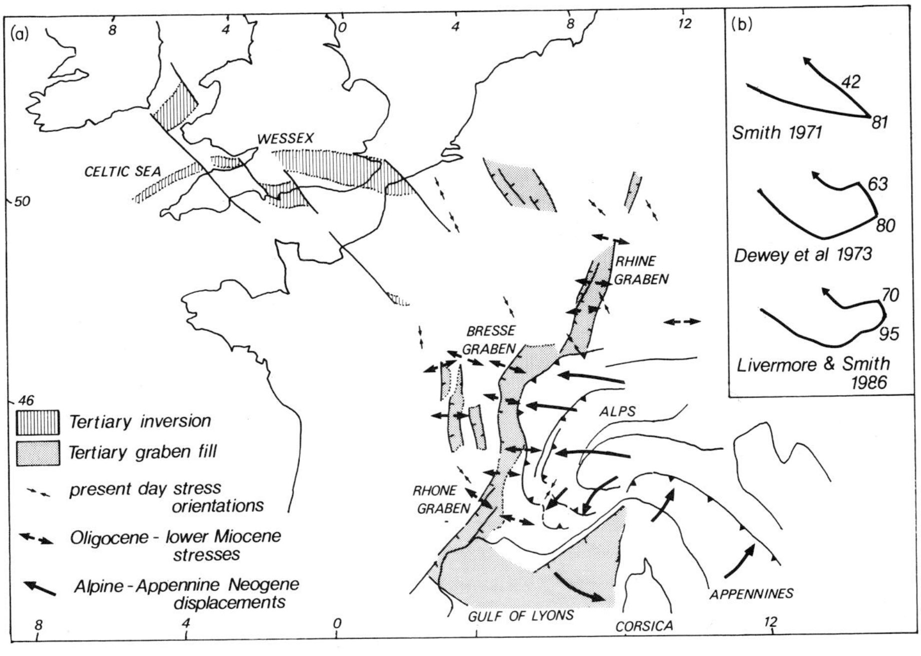

FIG. 8. (a) Map of western Europe showing the areas affected by middle Tertiary extension and middle Tertiary compression. The Alpine displacement directions for the Neogéne are shown.
(b) Paths of Africa relative to Eurasia based on reconstructions of Atlantic opening, from Smith (1971), Dewey *et al.* (1973), Livermore & Smith (1985).

Cretaceous path of the African plate relative to the European plate is visibly rotational, and as the Tethyan ocean is closer to the pole of rotation, the resulting width of Tethys is slightly smaller than the contemporaneous width of the central Atlantic. The Callovian Tethys as reconstructed by Dercourt *et al.* (1986, Plate 2) shows a much smaller width than does our reconstruction. Since only fragments of oceanic crust and associated sediments are preserved in the Alpine chain, there is no evidence from palinspastic reconstruction in favour of one or the other model. The exact width of Tethys however, would depend on how much displacement is taken up by basin development on the Tethyan margins.

(2) Tethys and the two continental margins were dissected by transform faults. Evidence for the existence of a transform pattern in the northern part of the Tethys domain has been presented by several authors (e.g. Lemoine 1980, Abbate *et al.* 1980, Weissert & Bernoulli 1985), where the ophiolite breccias and ophicalcites found commonly associated with the Tethyan pillow lavas and serpentinites can be compared with breccias from present-day oceanic fracture zones (Bonatti *et al.* 1974). The ophiolite suite from the southern part of Tethys is petrographically very similar to that of the northern part (De Roever 1972, Dietrich 1976, Lanzafame *et al.* 1979), so that we interpret the overall Tethys, including its continental margins, as a large pull-apart basin in a Jurassic–Cretaceous transform domain, in agreement with Abbate *et al.* (1980) and Dercourt *et al.* (1986). The northern and southern boundaries, which form the locations of the future Alpine and Calabrian Arcs, acted as the principal transform fault margins. Relative to present day geographical coordinates the transforms had a NW or WNW strike (Fig. 7), parallel to the tear faults which bound Triassic and Jurassic basins on-shore (Fig. 6) and which often reactivated earlier Palaeozoic strike-slip structures. We suggest therefore that the Alpine transform system may have been initially controlled by Variscan structures.

(3) The African Adriatic promontory (Argand 1924, Channell *et al.* 1979) was an extension of the African plate during the Mesozoic and early Tertiary. Based on palinspastic reconstructions of the African continental margin several authors suggested the continuity of continental crust from North Africa to Sicily and along the Italian peninsula to Yugoslavia and Greece, without an oceanic connection across the promontory from the western to the eastern part of the Tethys (Laubscher 1971, 1975, Scandone 1975, Laubscher & Bernoulli 1977, Channell *et al.* 1979). A comparison of the palaeomagnetic directions of Mesozoic and early Tertiary rocks from the undeformed part of the promontory (Gargano, Istria) with the corresponding directions from stable Africa confirms the view that the promontory moved relatively fixed to the African plate (Lowrie 1986, and references therein). Only in the post-early Tertiary did the promontory rotate counter-clockwise 27° around a pole near Malta (Van den Berg 1979) and was therefore decoupled from Africa.

Dercourt *et al.* (1986) selected a different model, where the promontory started to decouple from the African continent at 130 Ma and, according to these authors, represented an 'Apulian microcontinent.'

The Adriatic promontory however was not rigid, but suffered a considerable amount of crustal extension throughout the Mesozoic. Comparable extension is not recorded from the contemporaneous development of the North African crust (e.g. Wildi 1983) and thus active Mesozoic faults had to separate the two domains, or there was a major shear gradient between the two blocks.

(4) The width of the African–Adriatic promontory has been constrained from palinspastic restoration of the post-Mesozoic deformation (e.g. Bosellini 1973, Aubouin 1973, Bernoulli & Jenkyns 1974, Ippolito *et al.* 1975, Laubscher & Bernoulli 1977, Dewey *et al.* this volume). A width of about 1000 km is shown by all the more recent reconstructions.

The reconstruction done above meets some space problems. The restored Adriatic promontory and unrotated Spain tend to overlap. Dercourt *et al.* (1986) solve the problem by suggesting an extremely narrow Tethys, which was therefore not kinematically linked to the Atlantic. Laubscher & Bernoulli (1977, fig. 5) show the promontory shifted far eastwards to allow space for basin evolution on the European margin and from their interpretation, the promontory had not only to rotate during the Mesozoic, but also to shift westwards to get to its present position.

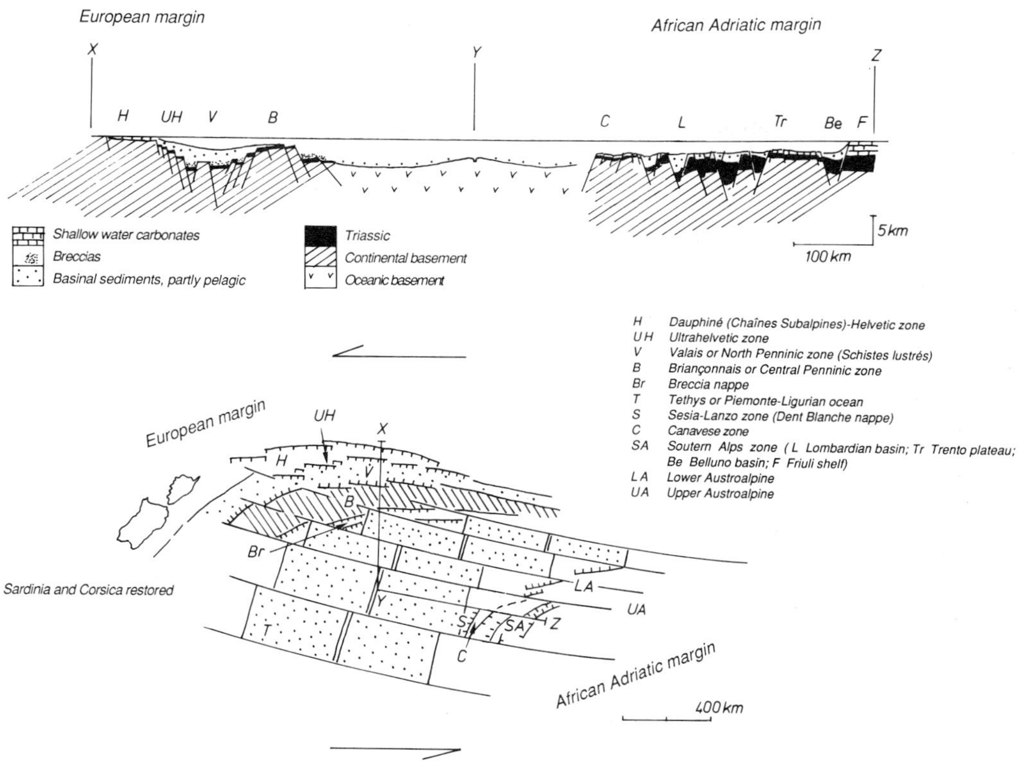

FIG. 9. The Alpine facies belts in the Upper Jurassic, near the major northern transform boundary of the Tethys, slightly modified from Laubscher & Bernoulli (1977), Bernoulli *et al.* (1979) and Weissert & Bernoulli (1985). The dominant palaeogeographic elements on the European margin are: the Valais trough, interpreted by various authors as being partly oceanic (e.g. Lemoine & Trümpy 1987, Platt *et al.* this volume) and the Briançonnais horst. A description is given in the text and, regarding the Briançonnais sequence, in Platt *et al.* (this volume). The African continental margin is dissected by extensional faults and dominated by shallow water sedimentation on the platforms and pelagic sedimentation in the basins. See e.g. Lemoine & Trümpy (1987) for an alternative palaeogeographic map of the Alpine realm with only a partly oceanic Tethys (oceanic lenses between thinned continental crust with individual microcontinents of normal crustal thickness). The simplified section illustrates the basin histories.

Figure 9 shows a rough palinspastic restoration of the northern part of the Tethys and its margins in the realm of the future Alps. The European margin was dissected by systems of early to middle Jurassic extensional faults, generally arranged in en echelon manner. Examples are the Cévennes, Nîmes and Durance fault systems in southern France (Fig. 10) and the faults in the Helvetic, Ultrahelvetic and Pennine realms (e.g. Lemoine & Trümpy 1987, and Fig. 9). Laubscher (1983) argues that such faults are responsible for the formation of a large trough, the north Penninic fracture zone, generally known as Valais trough (Fig. 9). This fault trough extended from Spain to Austria and represents a branch of the Pyrenean zone (Laubscher 1983). The Penninic fracture zone is therefore interpreted by Laubscher (1983) as a subplate boundary which will be of fundamental importance in determining the later subduction history of the European continent.

Most of the basement massifs of the western Alps were probably originally extensional fault blocks, which were later uplifted (e.g. Graciansky *et al.* 1979, Lemoine *et al.* 1981, Tricart & Lemoine 1986, Lemoine & Trümpy 1987). Furthermore, some authors (Lemoine *et al.* 1981, Lemoine *et al.* 1986) have emphasized the greater importance of extension relative to subsequent Alpine compression in determining Alpine tectonics and have compared the tilted fault blocks of the external Alps to structures found at the Atlantic margin, as in the Bay of Biscay (Montadert *et al.* 1979). Figure 10 comprises a region of the external Alps, centred at Pelvoux, where extensional faults are still well preserved; to the N, in the Mont Blanc region, most of the original extensional geometry has been destroyed by subsequent compression. In the Pelvoux, Grande Rousses and Belledonne original normal fault scarps have been re-eroded and preserved as cliffs on the sides of these basement massifs and the early half-graben fault geometry and sediment fill is still preserved. Figure 11 gives a diagrammatic section through the basins of the western Alps, based on Gillcrist *et al.* (1987), and shows the different subsidence histories. The extension direction, as determined by the extensional lateral ramps, was NW–SE, parallel to and possibly reworking Palaeozoic structures in the basement. Similar extensional fault geometries formed across much of NW Europe during Jurassic extension (Fig. 6).

The amount of extension in the external French Alps is uncertain. Gillcrist *et al.* (1987) argue that early normal faults were rotated to low dips during the Liassic, suggesting stretching factors greater than two. Downthrows on steeper late Liassic faults are in excess of 3 km, but the stretching factors were probably low. Parts of the Valais zone may have had greater extension, being stretched actively during the Cretaceous as well as the Jurassic, as volcanic rocks are interlayered with Lower to Middle Cretaceous deep-water sediments (Antoine 1971, 1972). Kelts (1981) and Loubat (1984) have compared the Cretaceous lithologies and structures of the Valais zone to those of the present-day Gulf of California pull-apart basins, characterised by the production of small, rapidly deepening basins, with thick sedimentary sequences intruded by basic sills.

Internal to the Dauphiné–Helvetic and Valais zones, the Briançonnais zone of the French Alps shows a very different subsidence history, with thick sediments of Triassic age, but no or an extremely reduced Jurassic sedimentary record. The Pennine zone of the Swiss Alps, however, shows initial rapid subsidence during the Jurassic, followed by a thermal phase of slow subsidence (Homewood & Lateltin 1988) (Fig. 11).

Deeper parts of extensional faults are preserved as ductile shear zones on the southern side of the Alpine suture, along the margin of the Ivrea zone (Brodie & Rutter 1987, Brodie *et al.* this volume). Kinematic shear indicators suggest an E–W to NW–SE extension direction of these ductile shear zones, in agreement with the direction of extensional faulting affecting the NW part of the African Adriatic promontory (Fig. 9). The contemporaneous extension direction in the Apenninic part of the promontory was NE–SW, relative to today's coordinate system. This suggests either a different extension direction on the southern part of the Adriatic promontory, relative to the Ivrea zone, or more likely, a subsequent anticlockwise rotation of the southern part of the Adriatic promontory during the Tertiary.

Early compression and early plate collision

The Eoalpine phase leading to a Cretaceous Alpine chain

Two principal sites of early collision are easily visible from the geometry of the blocks involved in transcurrent and rotational motion (Fig. 7). Jurassic compression occurred in the Vardar zone of the Hellenides–Dinarides, contemporaneous with ocean spreading in the western part of Tethys (Laubscher & Bernoulli 1977, Dercourt *et al.* 1986), and the component of anticlockwise rotation between the two continents led to consumption of the Tethyan oceanic crust followed by mid-Cretaceous compression between the northern and western front of the African–Adriatic promontory and the European continent. The following evidence points towards the existence of Mid-Cretaceous orogenic events:

In the northern calcareous Alps, formed of Permian and Mesozoic cover to the Upper Austro–Alpine nappes (Frank 1987), the limno-fluviatile to marine Gosau beds of Coniacian age unconformably transgress already formed thrust nappe structures. Most, if not all, of the Austro–Alpine basement nappes of the eastern Alps were affected by a Cretaceous metamorphic overprint (Frank *et al.* 1987a), reaching $\geq 600°$ and 6–7 kb in the Ötztal–Stubai complex (Frank *et al.* 1987b). From shear sense indicators, Ratschbacher & Neubauer (this volume and references therein) suggest W to NW-directed movement of the Austro–Alpine units during this phase.

In the eastern Alps there was also deformation of the European continental crust during this time. Part of the Penninic 'Bünderschiefer formation', composed of Triassic carbonates and other intensely deformed metasediments, were originally deposited on thinned European continental crust. These and metabasic and ultrabasic rocks, representing Tethyan ophiolites exposed in the Tauern window, below the Austro–Alpine units, have been affected by a blueschist metamorphic event, dated

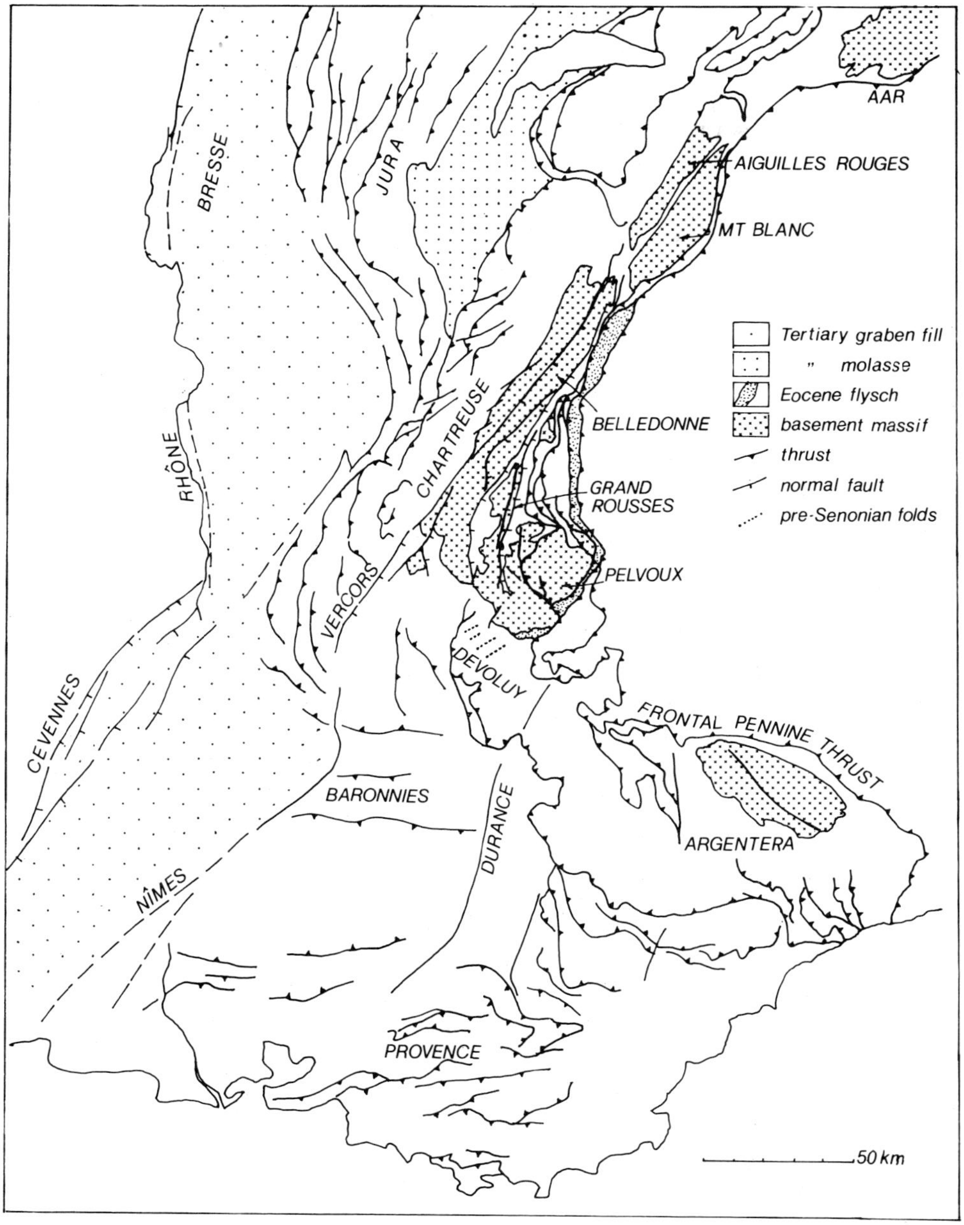

FIG. 10. Map showing the basement massifs of the western Alps, the regions where Mesozoic normal faults are preserved, and the interference between Pyrenean and Alpine thrusts.

at 90–60 Ma. This event overprints an earlier, but not dated, high pressure metamorphic event (Frank *et al.* 1987c). In summary, Cretaceous continental collision in the eastern Alps (Slapanski & Frank 1987) appears to have caused detachment of units from both continental margins and to have produced Eoalpine Europe-vergent nappes.

In the central and western Alps there is also widespread evidence for a Cretaceous orogeny. Williams & Compagnoni (1983) interpreted the early Alpine history of the eclogitic schists and 'Gneiss Minuti', in the lower part of the Sesia zone (see Fig. 9) as involving thrust tectonics of Austro–Alpine basement under conditions of increasing pressure. Peak conditions are recorded at >15 kb and 500–550° and the pressure peak is dated at about 110 Ma

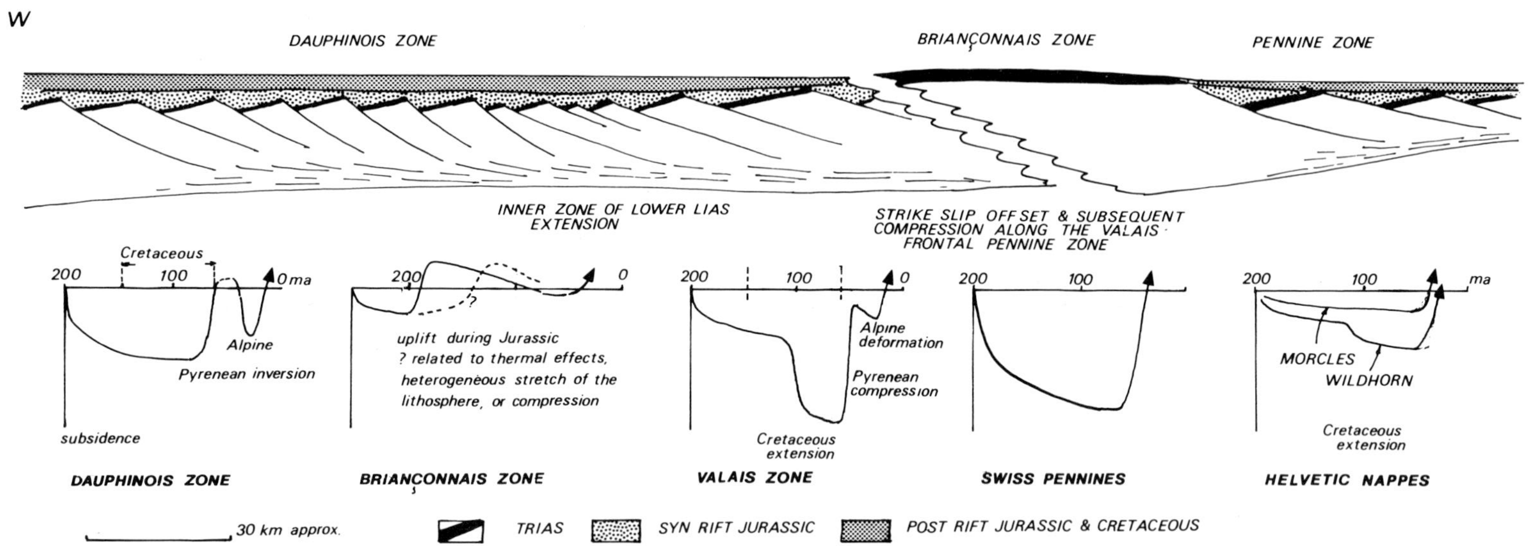

FIG. 11. Simplified cross-section through the western Alps showing the suggested basin structure and subsidence history; from Lemoine *et al.* (1986), Gillcrist *et al.* (1987) and Homewood & Lateltin (1988). The diagrams for the Valais zone, the Swiss Pennines and the Helvetic nappes refer to a section through the central Alps.

(Hunziker *et al.* this volume). Williams & Compagnoni (1983) suggest that the deformation under high pressure conditions occurred in an active subduction zone, located within the Austro–Alpine crust. Underplating of the Austro–Alpine crust by its own plate margin would conclude this deformation sequence, uplifting and folding the high grade rocks. The subduction zone would become inactive and a new subduction zone would develop further to the NW, within oceanic crust. Rubie (1984) presents a thermal model for such an underplating process.

Hunziker *et al.* (this volume) review the Cretaceous ages generally associated with high pressure metamorphism, which have been found in units of African, but also in units of European continental and of Tethyan oceanic crust. The peak P–T conditions were reached between 110 and 80 Ma (Hunziker *et al.* this volume, their Fig. 1). But it is still debated as to whether all these radiometric ages represent tectonic events; Hurford *et al.* (this volume) show that a well-defined Eoalpine Rb-Sr isochron from the Suretta nappe, considered to be of European central Penninic provenance, may be an artifact, i.e. could represent a 'possibly Hercynian system, partially disturbed during Alpine times' (Hurford *et al.* this volume). Assuming that the geochronological data indicate widespread Eoalpine tectonic events, and assuming that the palaeogeographical reconstruction of Fig. 9 is correct, it would appear that the Eoalpine compression had to involve subduction of both continental margins, to account for the high pressure metamorphism. Such a situation is difficult to conceive and therefore the classical palaeogeography has been questioned; Laubscher & Bernoulli (1982) discussed the possibility of a different palaeogeographical position of the Monte Rosa nappe, generally assumed to represent central Penninic basement, but showing a 110–100 Ma high pressure event (Hunziker *et al.* this volume); Hunziker *et al.* (this volume, their fig. 2D) further suggest the possibility that all the Cretaceous high pressure units originated from the Austro–Alpine margin of the Tethys. In this not generally accepted model, the European margin would end south of the Ultrahelvetic zone. According to these authors the Eoalpine orogeny would be related to subduction of the oceanic crust beneath a newly defined, tectonically thickened Austro–Alpine margin.

Laubscher (1970) and Laubscher & Bernoulli (1982) suggested a different model for the early convergence of the three plates. Bivergent subduction of the Tethys plate and of the Austro–Alpine margin of the African Adriatic promontory would lead to subduction of the ophiolites and as well to their obduction onto the Austro–Alpine margin, and would also account for the contemporaneous high pressure metamorphism in parts of this margin (Fig. 12). Cenomanian Cr-spinell and garnet-bearing flyschs of ophiolitic and Austro–Alpine provenance respectively are known to have been deposited on the southern margin of the Tethys (e.g. review by Wildi 1985). Such flyschs, testimonies of ophiolite obduction on the southern margin and of early erosion of Austro–Alpine crust, are interpreted by Laubscher (1970) as being deposited in the foredeep of the S-vergent Cretaceous orogen.

Cretaceous compression also affected the European continental margin in the central and western Alps, but according to Laubscher (1970) and Laubscher & Bernoulli (1982) predated continental collision. The movements were mainly concentrated along the north Penninic fracture zone (Fig. 9) and produced the north Penninic Cretaceous flyschs. Transpressive en échelon fault zones can develop characteristically double-vergent flower structures. Such structures rather than Cretaceous nappes are shown in Fig. 12 as a response to the Cretaceous compression on the western part of the European continent (e.g. Laubscher & Bernoulli 1982). In Dévoluy in the external French Alps, the deformation can be clearly dated as pre-Senonian, as there is a major unconformity of gently dipping Upper Cretaceous sediments resting on tight inclined NW-vergent folds involving Middle Cretaceous rocks (e.g. Mercier 1958) (Fig. 10). The origin of this deformation is not clear, since these compressive features cannot be linked easily to movements related to the North Penninic fracture zone.

It should be noted that an apparent discrepancy exists between the model of S-vergent Eoalpine ophiolite nappes and the model of W to NW-vergent Eoalpine thrusting of the Austro–Alpine units. South-vergent thrusting of ophiolite nappes would certainly predate continental collision, while W- to NW-vergent thrusting of the Austro–Alpine units of the eastern Alps would be related to collision. We therefore suggest that early ophiolite obduction onto the southern continental margin was followed by a diachronous continent–continent collision which occurred earlier in the eastern Alps than in the central and western Alps. The concept of diachronous deformation progressing from E towards W is of primary importance when analysing the Tertiary Alpine main phase structures.

As already mentioned, the Middle Cretaceous

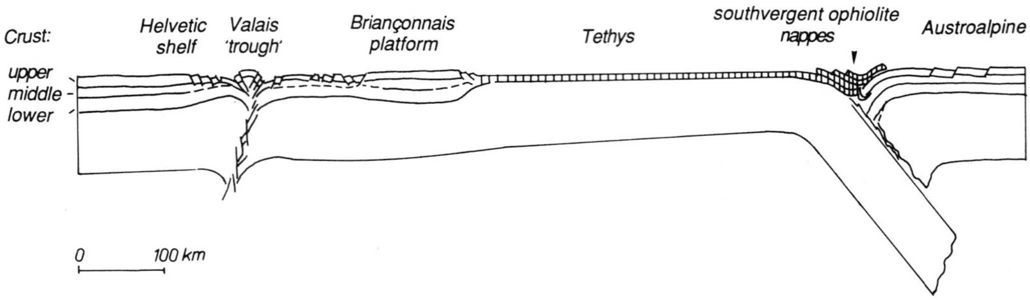

FIG. 12. The Cretaceous orogeny, according to Laubscher & Bernoulli (1982). The ophiolite nappes form a southvergent chain which is eventually thrust over the European plate during final closure of Tethys.

continental collision in the eastern Alps produced NW-directed overthrusts, suggesting dextral transpressional deformation between the African Adriatic promontory and the southern margin of the European continent (e.g. Ratschbacher & Neubauer this volume). Dextral transpression also played an important role during the Tertiary continental collision in the central Alps (Dietrich & Casey this volume, Merle *et al.* this volume, Schmid *et al.* this volume), as will be discussed later. Such a deformation regime can be easily deduced from the older plate models of Smith (1971) and Dewey *et al.* (1973), who postulated dextral shear between the African and European Plates as a consequence of the opening of the North Atlantic, but not from the new plate reconstruction by Dewey *et al.* (this volume), where during the Cretaceous, the motion of Africa was initially to the ESE relative to a fixed European plate, changing at about 90 Ma to an ENE to NE motion. There are several possible explanations for the apparent discrepancy between the observed deformation regime and the model by Dewey *et al.* (this volume):

(1) The recorded WNW directed motions of the Austro–Alpine nappes may not have been the original overthrust directions during the Cretaceous, but the whole thrust pile may have been rotated along with the Adriatic promontory during the Tertiary. This would require up to 90° of rotation of the nappe pile, a value greater than that recorded for the rotation of the promontory (e.g. Channell *et al.* 1979, Van den Berg 1979, Lowrie 1986).

(2) Several authors (e.g. Dewey *et al.* 1973, Dercourt *et al.* 1986) have assumed that the Adriatic promontory might have broken from the African plate to form a separate Apulian plate at this time, moving NW relative to Africa and to Europe. This movement could be linked with the sea floor spreading in the eastern Mediterranean as suggested by Dercourt *et al.* (1986) (see also Fig. 13a). Cretaceous extension does affect the platform carbonate sequences of the south Italian Apennines (e.g. Ippolito *et al.* 1975) but the observation that the facies belts of the southern Apennines continued through Sicily to North Africa (Scandone *et al.* 1974) as well as the palaeomagnetic data of Lowrie (1986) suggest that the promontory did not become a separate plate until the early Tertiary, and argue against the model of Fig. 13(a).

(3) During the Eoalpine period the N Atlantic began to open and the Iberian peninsula moved eastwards relative to the main European Plate (Fig. 13b). If the southern part of the European margin, south of the N Penninic fracture zone, was linked to the Iberian plate, then these movements would generate transtensional deformation in the Valais zone to the W, but transpression to the south-east, against the Adriatic promontory (Fig. 13b). This is a preferred model for the Eoalpine overthrust tectonics, as it requires no extra plate generation or motion, other than that which can be explained by the gradual opening of the Atlantic.

Cretaceous compression affected the European crust far to the north. There was important pre-Upper Cretaceous deformation of the Mesozoic basins of NW Europe, S and W of Britain and in parts of the North Sea (Ziegler 1982, 1987, and Fig. 6). This may have been caused partially by strike-slip movements on the large NW–SE trending basin bounding tear faults; for example, there was probably movement on the N Pyrenean fault at this time.

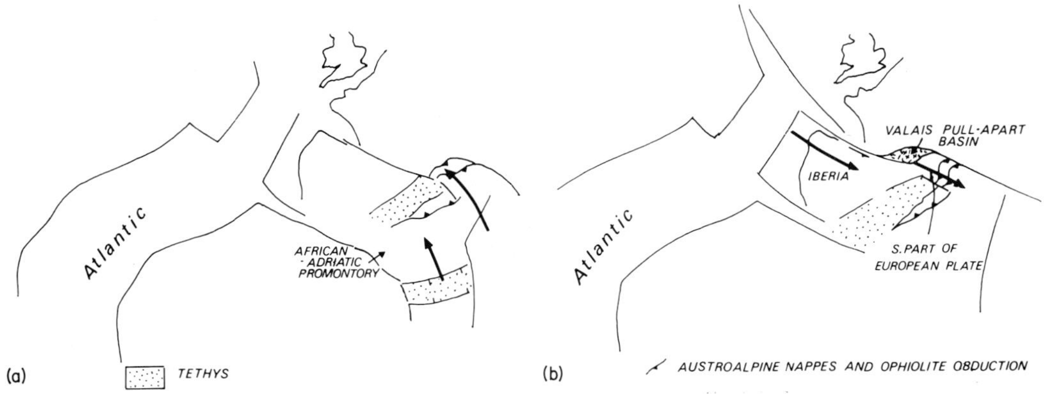

FIG. 13. Cartoons of the Eoalpine deformation:
(a) related to opening of an E Mediterranean basin/ocean, based partly on Dercourt *et al.* (1986), and
(b) related to the SE motion of the southern part of the European plate comprising Iberia. This motion is induced by the opening of the N Atlantic.

However, there was also regional compression and uplift of Mesozoic rocks, suggesting NW–SE compression across the basins (Fig. 6). We suggest that this intraplate deformation was related to the gradual opening of the Atlantic, and was formed by stresses generated at the spreading ridge or at the northward propagating tip of the ocean spreading zone.

The Cretaceous orogeny is also known in Calabria. In the Polia–Copanello unit (Amodio Morelli *et al.* 1976), a nappe consisting of pre-Alpine high grade rocks of European provenance (Dietrich 1988), biotite ages of between 150–120 Ma have been found (Borsi & Dubois 1968, Civetta *et al.* 1973). These are interpreted as the age of uplift of the high grade rocks from lower crust (Borsi *et al.* 1977). Similar high pressure metamorphism of probably early Alpine age also affected the Castagna nappe (Amodio Morelli *et al.* 1976) as well as the Calabrian ophiolites.

Laubscher & Bernoulli (1982) have emphasized that there already had to exist pronounced changes of the strike direction of the Cretaceous Alpine chain. From the palaeogeographic shape of the African Adriatic promontory, these authors suggest the existence of 'Cretaceous orogenic loops', i.e. early arc structures as precursors of the actual arcs of the chain.

The Pyrenean event

During the Upper Cretaceous and the lower Tertiary, the European sub-plate south of the N Penninic fracture zone became caught up in the N to NE directed motion of Africa relative to Europe, to close up its Lower Cretaceous basins and develop a region of pronounced crustal thickening from the Pyrenees to SE France. In southern France, north of the Penninic fracture zone, basement massifs were uplifted on generally northerly directed thrusts, with local back-thrust systems (Fig. 10). Some of these structures appear to have reactivated earlier extensional faults, related to Cretaceous stretching.

In the Pyrenees belt, there was probably no true oceanic crust except in the Parentis basin to the W of this belt, but deformation during the Pyrenean phase considerably shortened the previously stretched crust to produce a divergent thrust zone. On the N side, originally S to SSW dipping extensional faults were reactivated in the Aquitaine basin. The distal faults were fully inverted with syn-rift sediments pushed out of the half-graben, while the more proximal faults were less affected and many still show net extension. A foreland basin developed ahead of the thickened crust, infilled by Eocene molasse. The central portion of the Pyrenees, the axial zone, consists of Hercynian basement, uplifted on crustal scale thrusts (Seguret & Daignieres 1986) and also shortened and thickened to develop an upright schistosity. The southern margin of this axial zone forms an antiformal stack of thrusts involving basement and Mesozoic cover, with a transport direction generally to the SSW over the Ebro foreland basin (e.g. Hirst & Nichols 1986, Hossack *et al.* 1984). Thrust analyses (Hossack *et al.* 1984, Williams 1985, Munoz *et al.* 1986) suggest large displacements, involving probably over 100 km shortening of the Pyrenean basins and thickening of the crust to over 50 km. Southward thrusting over the Ebro basin developed from Lower Eocene to Miocene (Labaume *et al.* 1985). Deep seismic profiling across the Pyrenees (ECORS 1988) shows an apparent wedge of European crust driven over Spanish lower crust and mantle, with its southern tip somewhere beneath the axial zone. The N-vergent thrusts of the N Pyrenees and the Aquitaine basin appear to be linked to a S-dipping detachment at the top of the wedge, that is they form a backthrust to the main S-directed crustal scale overthrusting and probably formed along a reactivated earlier extensional detachment (Fig. 14). The ECORS section is very similar to the section proposed for the central and southern Alps by Roeder (this volume).

Pre-Eocene deformation affected the external basement massifs of the French Alps (Gidon 1979). Much of the uplift and closure of the Mesozoic half-grabens occurred before the deposition of the Eocene flysch, the 'Grès Nummulitique'. These synorogenic sediments were derived from thickened and uplifted crust to the S and SE, but along the southern and eastern margins of Pelvoux, they rest unconformably on a sequence ranging from basement gneiss and granite to Jurassic cover and overlie inclined folds and W-directed thrusts (A. Chambers pers. comm. 1988). Figure 15 shows a simplified map of the main regions of pre-Eocene deformation in the western Alps. The exact age of this deformation is unknown, though it is probably co-eval with the N to NNE directed Pyrenean thrusts seen further S.

The Alpine main phase

Nappe emplacement on to the European continent and anticlockwise rotation of the direction of overthrust shear

The main phase of Alpine deformation is here defined as the effect of continent–continent collision on the already deformed margins of the African Adriatic promontory and the European plate. It comprises thrusting of Austro–Alpine elements, i.e. nappes consisting of rock units of the deformed southern margin, onto the European continent, as well as metamorphism and thrusting of the European Pennine nappes. This was followed by the deformation of the Briançonnais domain, the formation of the major external structures, such as the Ultrahelvetic and Helvetic units in Switzerland and

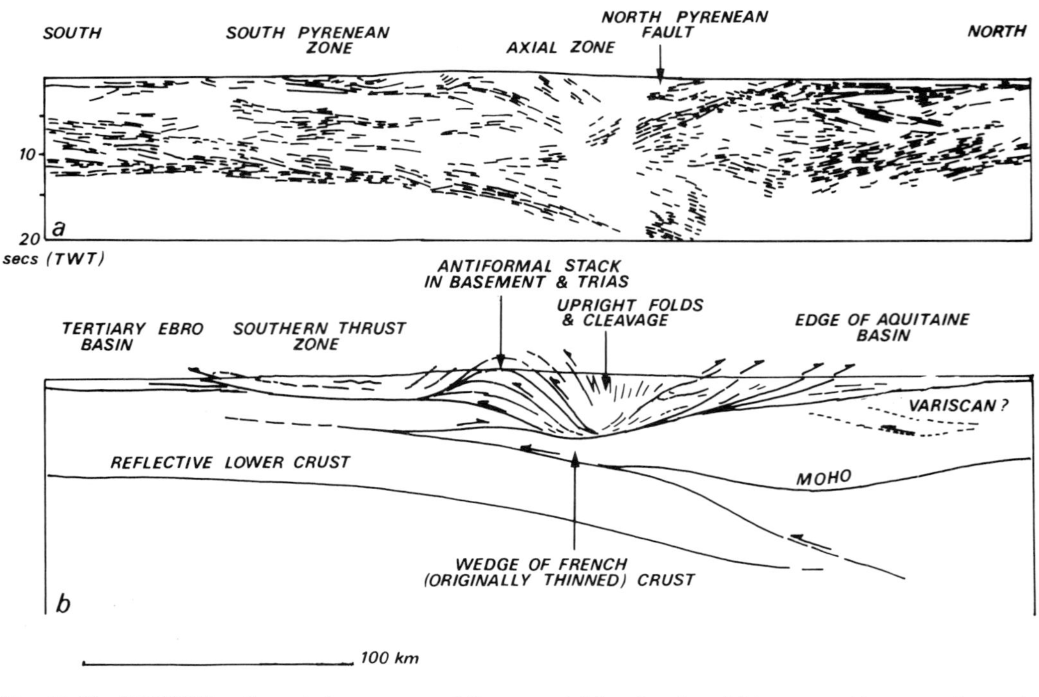

FIG. 14. The ECORS line through the east-central Pyrenees: (a) line drawing, (b) interpreted in terms of crustal scale wedge-type thrusting.

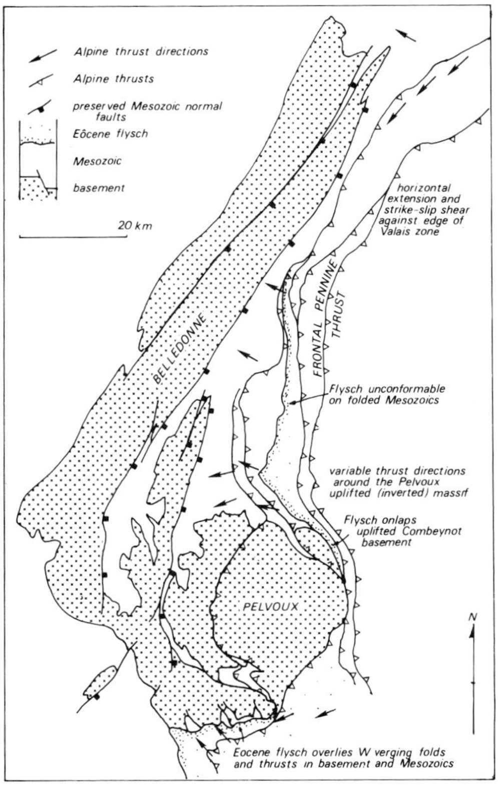

FIG. 15. Map of the Pelvoux area showing the locations of partially inverted half graben and pre-Eocene deformation, the Eocene unconformity and the subsequent Alpine tectonics (from Gillcrist *et al.* 1987, Chambers pers. comm. 1988, Spencer pers. comm. 1988, and work by MPC).

the Chaînes Subalpines in France and the Pliocene deformation of the Jura. The time of collision may be defined as the time at which all true ocean floor had disappeared. Later compression, forming many of the nappe structures in the European crust, can be considered as post-collisional, that is, due to intra-plate deformation.

The detachment horizon, from which the main phase nappes were stripped off the European lithosphere, rises from deep in the crust and from amphibolite facies rocks in the S, to close to the surface in the N, below the unmetamorphosed Jura. This upward progression of the detachment gives the Alps their characteristically asymmetric structure, verging towards Europe. Bally (1981) and Beach (1987) have suggested that there may be other decoupling horizons within the lower crust or lithospheric mantle, to transfer slip far into the European craton and cause uplift and inversion of Mesozoic basins in southern Britain. Alternatively, Gillcrist *et al.* (1987) link the basin inversion and deformation of the European craton with that of the Alps by steep strike-slip shear zones, rather than by flat detachments deeper in the lithosphere. Compressive stresses causing inversion in the NW European craton (Fig. 8a) may be generated within-plate or at the Atlantic extensional margin, rather than at the Alpine collisional zone.

Africa's anticlockwise rotation into collision appears as a key to the understanding of the Tertiary tectonic history of the Alps. Regarding Alpine deformation, there are two principal effects of an anticlockwise rotation of Africa relative to Europe:

(1) A diachronous collision, occurring earlier in the eastern Alps than in the central/western Alps and related diachronous deformation, progressing from E towards W.
(2) The re-occurrence of dextral transpression in subsequent zones of active deformation.

In the eastern Alps continent–continent collision is assumed to have occurred already in the early Late Cretaceous, according to Slapansky & Frank (1987). These authors suggested collision of the Austro–Alpine units with the Briançonnais domain of the eastern Alps at 90 Ma, causing large scale inversion of the frontal parts of the Austro–Alpine basement. On the contrary, in the western Alps the sedimentary cover of the Briançonnais zone ranges from Triassic to Eocene (e.g. Debelmas & Lemoine 1970). Laubscher (1970) assumes continent–continent collision in the central Alps as occurring during the Eocene.

The concepts of Alpine continental collision as suggested by the various authors depend entirely on their concepts of Alpine palaeogeography. If Tethys is viewed as a complex palaeogeographic domain consisting of small oceans and intervening stretched continental blocks (e.g. Lemoine & Trümpy 1987), the timing of continent–continent collision is less clearly defined than if relatively distinct continental margins are assumed, as, for example, in Fig. 9. Milnes (1978) suggested therefore for the central Alps mid-Cretaceous continental collision, not as a single subduction suture, but as a long period of convergence, leading to the deformation of the Pennine zone into a tectonic mega-mélange consisting of continental and oceanic crust and of sediments, followed by post-mid-Oligocene intra-plate compression. Trümpy (1980) expressed a very similar view. For Platt (1986) the Tethys domain consists of two, possibly three, oceanic zones separated by microcontinents. Platt *et al.* (this volume) therefore bracket the underthrusting of the Briançonnais domain of the western Alps beneath the Alpine accretionary wedge between possibly Paleocene and not later than Mid-Eocene; continental collision is assumed by these authors as Lower Oligocene. We prefer a simpler palaeogeographic reconstruction here (Fig. 9), with one Tethyan ocean which closed diachronously from E to W, though there were several Triassic to Cretaceous stretched basins on the continental margins. Following Laubscher (1970) and Laubscher & Bernoulli (1982) we explain the several oceanic slices found in the Alps as thrust sheets between units of continental crust formed by a complex deformation history, as, for example, Eoalpine S-vergent thrusts being incorporated in the main phase N-vergent chain, rather than by a complex original palaeogeographic configuration involving several small sutures.

The peak of metamorphism in the Pennine nappes of central Switzerland has been dated as ranging between 45–35 Ma (see, e.g. model A in fig. 20 of Hurford *et al.* this volume). The Pennine nappes consist of continental basement and Triassic–Jurassic sediments (Fig. 9); deep burial of this zone has to be assumed to account for the metamorphic peak temperatures of about 650° (Frey *et al.* 1980). The Pennine thrusting in the central Alps occurred under amphibolite facies conditions. The associated movement directions were towards NNW in the central part of the Pennine (the Lepontine) zone and towards NW in the western part (Merle *et al.* this volume, their fig. 4), relative to today's coordinate system. Convergence continuing during the Oligo–Miocene led to the formation of a N-dipping steep belt in the southern part of the Pennine zone (Milnes 1974); this deformation corresponds to the Insubric phase of Argand (1916, see also Schmid *et al.* 1987). In the early Miocene the Pennine zone was uplifted more than 9 km by backthrusting (thrusting towards south) along this steep belt, generating mylonites (Hurford *et al.* this volume). Ridley (this volume) considers that it is these shears that cause juxtaposition of rocks of different grade of high metamorphism in the Sesia–Lanzo zone.

Movements contemporary with this uplift and under greenschist facies conditions overprinted the earlier thrusting in the Pennine zone. These movements were parallel to the strike of the Alpine chain, i.e. towards W (Merle *et al.* this volume, their fig. 7). Uplift associated with westwards thrusting was responsible for the formation of the structural domes of the Lepontine area, the Ticino and Simplon domes (Merle *et al.* this volume). The characteristic onion-shell map pattern of the Lepontine metamorphic isograds (Niggli 1970) reflects this domal structure.

The deformation history of the Pennine zone therefore shows an overall anticlockwise rotation of the direction of crustal shortening (Steck 1987). Such a deformation pattern of foreland directed NNW to NW-wards thrusting, followed by steepening of the rear part of the nappes (the formation of a 'root zone'), and followed by massive uplift and backthrusting and by westwards 'lateral escape of rocks within the Pennine zone' (Merle *et al.* this volume) has been independently suggested for the subsequent deformation history of the Helvetic zone (e.g. Dietrich & Casey this volume). This rotational deformation sequence is therefore of regional importance as an expression of plate convergence associated with dextral shear. Anticlockwise rotation of overthrust shear led to westwards structural stacking and induced locally gravity collapse: Dietrich & Casey (this volume) interpret the Simplon normal fault (e.g. Mancktelow 1985, Steck 1987) as such a collapse structure, which again induced further westwards directed mass escape.

Anticlockwise rotation of the thrusting direction in time and in space on the scale of the arc of the western Alps is discussed by Vialon *et al.* (this volume). These authors compare the thrust directions around the arc of the western Alps to a ring shear model leading to radial compression directions in a dextral vertical simple shear band between a stable outer zone and a coaxially rotating inner cylinder.

Anticlockwise rotation of thrusting has been documented in detail for the external zones of the Alps, the Helvetic zone and the Chaînes Subalpines. Dietrich & Durney (1986) describe the rotation from northwards directed overthrust shear in the older, upper Helvetic nappes to NW or W directed shear in the youngest part of the Helvetic nappe pile, the inverted limb of the Morcles nappe. Gourlay (1986) describes a similar deformation sequence for basement and cover in the Mont Blanc area. Ramsay (this volume) shows how the Alpine structures developed in the external Aiguilles Rouges and Mont Blanc Massifs are anticlockwise obliquely superposed on the already emplaced Helvetic nappes. Dietrich & Casey (this volume) suggest that in the belt of the Helvetic nappes, the zone of active thrusting moved from east towards west, i.e. diachronous deformation with the youngest increment being observed only in western Switzerland. Gratier *et al.* (this volume) deduce the Alpine displacement of the external Chaînes Subalpines from a strain trajectory map, separating the displacements related to the Pyrenean event from the displacements related to Miocene–Pliocene compression; for the latter they find a regional anticlockwise rotation.

In western Pelvoux, there are pronounced local variations in thrust direction, often within individual thrust sheets (Beach 1981). Thus westwards along the trace of a thrust, the transport directions as observed from slickensides, shear bands, cut-off and lateral ramp data (Gillcrist *et al.* 1987, Gillcrist pers. comm. 1988), change from NW to W to SW directed. However, there seems to be very little discernible rotation of the Pelvoux Massif itself, as the pre-Alpine structures, such as the major NW–SE trending tear faults, have the same trend as on the Alpine foreland.

The divergence of the direction of thrusting as well as the non-plane strain nature of much of the deformation make coaxial section restoration across the Alps impossible (Fig. 8a). But it is easily visible from Alpine cross sections that the overall amount of displacement has to be large. Some indication of the amount and rate of shortening may be obtained from a study of the foreland basin sediments. The Swiss foreland basin sediments (e.g. Rigassi 1977, Mugnier & Vialon 1986, Homewood *et al.* 1986) are of Oligocene to Upper Miocene age. The earliest sediments are turbiditic and formed in basins ahead of the encroaching Alpine thrust sheets. These are followed by molassic sediments, from early wave-dominated shoreline sands to alluvial fans and floodplain and lake sediments (Homewood *et al.* 1986). From the balancing of sections and the migration of sedimentary pinch-outs on restored cross sections, Homewood *et al.* (1986) estimated an average thrust propagation rate of 7.35 mm/yr, with a shortening rate of 3.5 mm/yr and a basin migration rate of 5 mm/yr. Assuming therefore that Alpine deformation advanced at this 0.5 cm/yr time-averaged rate since final collision in the Eocene, there should have been about 250 km of European lithosphere involved in the building of the western Alps.

The importance of strike-slip movements during the Alpine main phase

It appears from the geometry of the pre-collisional plate boundaries (Fig. 7), that post-collisional plate convergence had to induce strike-slip movements in the region of the western Alps, to account kinematically for the N to NW-wards crustal shortening in the central and eastern Alps.

Ricou & Siddans (1986) reviewed the major N–S trending strike-slip movements from the internal and external zones of the western Alps. These movements were concentrated along faults with a sinistral sense of displacement. These authors showed how from the beginning of the Tertiary onwards the WSW–ENE trending palaeogeographic elements of the European margin were gradually sliced up by such faults. The resulting overall sinistral displacement of the palaeogeographical zones (Fig. 16a) lead to an Oligocene arc configuration.

On the other hand, the importance of dextral E–W strike-slip movements during the evolution of the central and eastern Alps has already been mentioned. Laubscher (1971) suggested from regional geological reasoning 300 km of dextral strike-slip movements between the European foreland and his Insubric plate, part of the Neogene African Adriatic promontory. These movements would account for the dominantly westwards thrusting in the N–S segment of the western Alps, and for the closure of the Ligurian part of the Tethys. The post-Oligocene westwards convergence of Laubscher's Insubric plate caused the following three major compressional events in the western Alps (Fig. 16b):

(1) The obduction of a slice of lower crust/upper mantle of the Insubric plate, the Ivrea body, from below into the internal part of the western Alps (the Ivrea body being defined through characteristic seismic and gravimetric data; see the impressive scaled section in fig. 8 of Schmid *et al.* this volume).
(2) The westwards thrusting of the Pennine and of the Briançonnais units and the particularly spectacular backthrusting of the latter (e.g. the sections by Platt *et al.* this volume).
(3) The décollement of the external units of the western Alps, the Chaînes Subalpines (e.g. the sections by Gratier *et al.* this volume).

Laubscher (1971, 1984) presented 3-D plate models to explore the evolution of the NW corner of the African Adriatic promontory and the European Plate on a lithospheric scale.

The strike-slip movements laterally induced by the westwards indenting Insubric plate were generally diffuse; they are represented by the dextral component of the Penninic Lepontine and Helvetic transpressive zones. A relatively minor part of these movements was concentrated along the Insubric Line. Schmid *et al.*

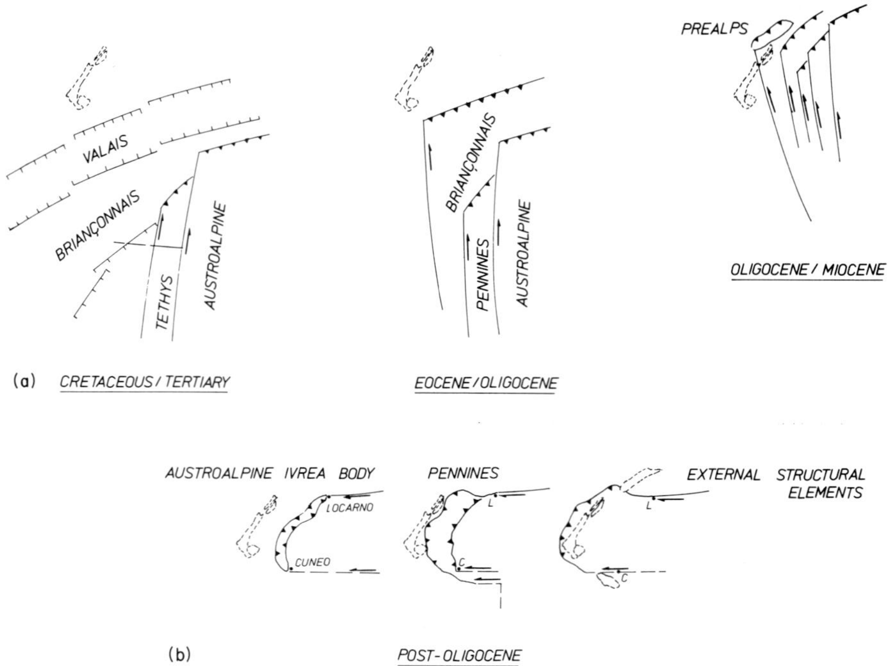

FIG. 16. The importance of strike-slip movements during the Alpine main phase. (a) Strike-slip deformation along the western edge of Tethys — according to Ricou & Siddans (1986), modified. (b) Strike-slip movements related to indentation of the Insubric plate — according to Laubscher (1971), modified. See text for discussion.

(this volume) discuss the deformation history of the rocks of the Insubric Line, the amount and timing of the displacements along it, and review the significance of this line and of its continuation into the eastern Alps, i.e. of the overall Periadriatic lineament, during Alpine evolution.

Strike-slip faults which partially acted as rails for plate movements are lithospheric structures. A particularly important strike-slip zone is the Valais or Penninic fracture zone as viewed by Laubscher (1983). During the Cretaceous the fracture zone was still far from the subduction zone (Fig. 12), but approached it in the Upper Eocene. Compression between this crustal scale buttress and the still mobile, i.e. buried, Pennine nappes caused the important backfolding of the Pennine front in the central Alps (e.g. Huber *et al.* 1980 for a section through the backfolded Gotthard massif). The obstacle was then during the Miocene 'scraped off by the N-moving orogenic lid' (Laubscher 1983), a process which led eventually to the tectonic extrusion of the Helvetic nappes (Dietrich & Casey this volume). The Miocene Alps can therefore be considered as a dextrally transpressive intracontinental mountain belt, comparable to the Pyrenees in the Eocene (Laubscher 1988).

The formation of the arc of the western Alps

The following model for the formation of the arc of the western Alps emerges:

(1) The structures of the Alps can be explained by movements generally related to the episodic northward propagation of the Atlantic rift. The Tethyan ocean formed in a sinistral strike-slip regime between Africa and Europe and can be considered as a large pull-apart basin. On both continental margins smaller basins developed; these were active at different times and had different subsidence histories.

(2) NE convergence of the African plate, relative to the European plate, together with eastward movement of the part of Europe S of the Pyrenean–North Penninic line, led to Cretaceous deformation on both continental blocks and to collision in the eastern Alps by NW to WNW directed overthrusting of the Austro–Alpine part of the Adriatic promontory onto the European plate. Cretaceous collision in the eastern Alps occurred, therefore, in a dextrally transpressive deformation regime, leading to an early arc structure (Fig. 13b).

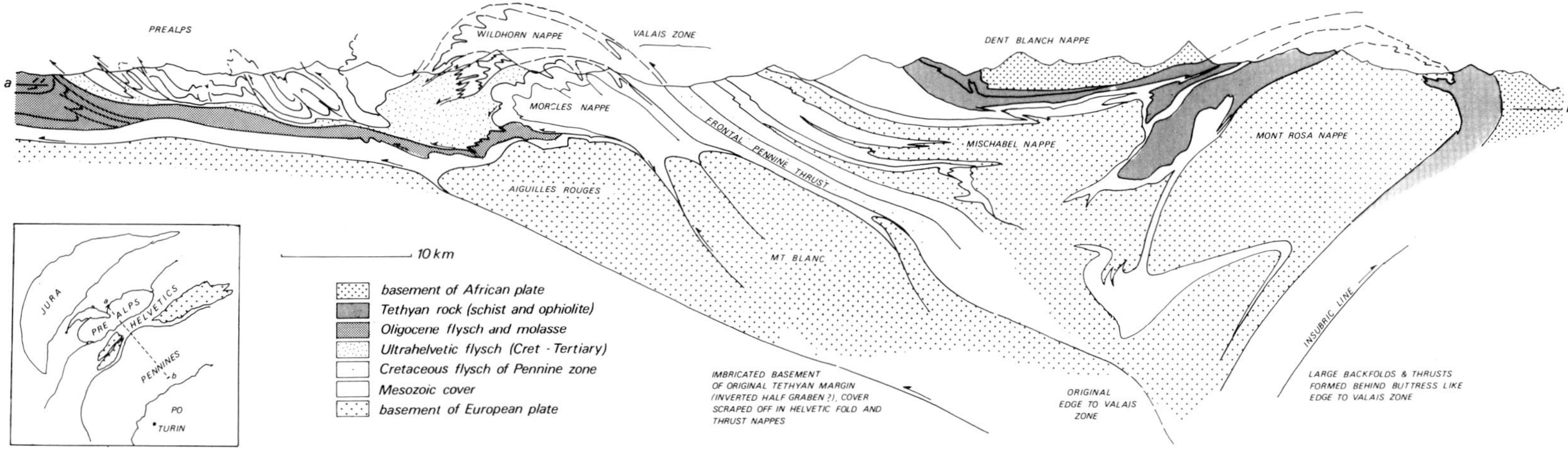

FIG. 17. Cross section through the Swiss Helvetic and Pennine Alps (section line shown on inset) after Escher *et al.* (1988) to show:
(i) The complex thrust and fold nappes of the Pennine zone (basement of European Plate), transported against the N Penninic fracture zone (Valais zone).
(ii) The interfolded ophiolite and Tethyan deep water sediments.
(iii) The Insubric Line backthrust and back-fold zone.
(iv) The Helvetic nappes tectonically extruded between the backfolded Gotthard Massif and the Mont Blanc by the overthrusting Pennine nappes. The basement slice SE of Mont Blanc represents the Mont Chétif, a massif in an analogous tectonic position as the Gotthand massif further east.
(v) The lower detachment beneath the Aiguilles Rouges Massif, climbing up into molasse to the NW, as imagined by MPC. For DD this detachment is steeper and not so deep.
(vi) The overall heterogeneity of deformation.

(3) After the consumption of Tethys, mostly by subduction, but partly by ophiolite obduction, the main phase of collision produced general thrusting of African units over Europe, as well as the northward thrusting of slices of European crust. Since the overall movement of the African plate relative to the European plate was strongly rotational, with a pole somewhere near the western edge of the Mediterranean (Dewey *et al.* this volume), and as there was, in the Tertiary, subsequent further rotation of Apulia relative to Europe, dextral shear movements between the continental blocks were again extremely important, especially along the NE–SW segment of the Alpine deformation zone, the eastern and central Alps. These transpressive movements interfered with the sinistral strike slip movements in the N–S sector of the Alps, the western Alps. An anticlockwise rotation of the direction of overthrust shear in the zone of interference produced the typical deformation pattern of the central-western Alps: an older, northwest- or northwards shear direction being overprinted by a younger fold-axis parallel stretching. The anticlockwise direction of overthrust shear is viewed also as an expression of the migration of deformation from east towards west, the westwards directed thrusting being younger than the northwards directed thrusting.

(4) Approximately 250 km of European lithosphere were involved in Alpine deformation, as deduced from foreland basin migration in the central-western Alps, with probably more in the eastern Alps. Shortening was associated with lower crustal and lithospheric mantle subduction, to form 'roots', beneath the Po plain for the western Alps, but possibly beneath the internal zones of the Alps further east. Net shortening exceeds net extension, as the Alpine basins have been fully inverted and the crust thickened to over 45–50 km. However, it is difficult to match extension with compression, as the overthrust plate also suffered rotation.

The overall tectonic framework given here represents only a first order approach to Alpine evolution. Detailed kinematics induced by plate movements were complex considering, for example, that during the Tertiary the African Adriatic promontory was partially decoupled from the African plate, and considering that the local kinematic indicators in fault zones do not always reflect the dominant plate motions (e.g. Baird & Dewey 1986). Complex was as well the response to deformation of the extremely heterogeneous rock material being subjected to long deformation histories under a wide range of physical conditions (Fig. 17).

Many of the points made in this overview will be discussed further in the papers given in this volume. These contributions illustrate the range of problems on which Alpine geologists are currently working, based on 150 years of intense research. Many problems remain for future generations. We suggest three important lines of research for the future:

(1) The sequential directions of extension and relative displacements are poorly known over large parts of the Alpine chain. These data are the basis for the establishment of a kinematically consistent model for individual sectors of the chain, from the eastern Alps to southern Italy, North Africa and Spain.

(2) The ages of some of the movements need to be found using stratigraphic and/or geochronological techniques; dating of well understood structures appears particularly important.

(3) Quantitative restoration of Alpine palaeogeographic belts by conventional methods and by unstraining techniques, particularly in 3-D, plus further knowledge of conditions of sedimentation and subsidence history of these belts, should allow better understanding of the crustal structure, the structural history and the collisional and post-collisional processes.

ACKNOWLEDGEMENTS: MPC is grateful for long and sometimes fruitful discussions with students and colleagues in Britain and Switzerland. DD enjoyed discussions with Daniel Bernoulli and Otto Kälin on this text and with Martin Casey on individual aspects of Alpine tectonics. Thanks are due to Carolyn Holloway for typing the manuscript. The topics discussed here involve some of the controversial discussions held after papers were presented during the Geological Society meeting. DD acknowledges financial support from Schweizerischer Nationalfonds, project No. 2.2140.86.

References

Abbate, E., Bortolotti, V. & Principi, G. 1980. Apennine ophiolites: a peculiar oceanic crust. *Ophioliti*, **5**, 59–96.

Amaudric Du Chaffaut, S., Caron, J. M., Delcey, R. & Lemoine, M. 1972. Données nouvelles sur la stratigraphie des schistes lustrés de Corse: la série de l'Inzecca. Comparaisons avec les Alpes occidentales et l'Apennin ligure. *Comptes-Rendus de l'Académie des Sciences de Paris D*, 2611–2614.

Amodio Morelli, L., Bonardi, G., Colonna, V., Dietrich, D., Giunta, G., Ippolito, F., Liguori, V., Lorenzoni, S., Paglionico, A., Perrone, V., Piccarreta, G., Russo, M., Scandone, P., Zanettin Lorenzoni, E. & Zuppetta, A. 1976. L'arco calabro–peloritano nell'orogene Appenninico–Maghrebide. *Memorie della Società Geologica Italiana*, **17**, 1–60.

Ampferer, O. 1906. Über das Bewegungsbild von Faltengebirgen. *Jahrbuch der kaiserlichen und königlichen Reichsanstalt (Wien)*, **56**, 539–622.

Antoine, P. 1971. La zone des brèches de Tarentaise entre Bourg St. Maurice (Vallée de l'Isère) et la frontière Italo-Suisse. *Mémoires Travaux du laboratoire de Géologie de l'Université de Grenoble*, **9**, 367 pp.

—— 1972. La domaine pennique externe entre Bourg St. Maurice (Savoie) et la frontière italo-suisse. *Géologie Alpine*, **48**, 5–40.

Argand, E. 1916. Sur l'arc des Alpes occidentales. *Eclogae Geologicae Helvetiae*, **14**, 145–91.

—— 1924. La tectonique de l'Asie. *Proceedings International Geological Congress*, **13**, 171–372.

Arthaud, F. & Matte, P. H. 1977. Late Paleozoic strike-slip faulting in southern Europe and northern Africa. *Geological Society of America Bulletin*, **88**, 1305–20.

Aubouin, J. 1959. Contribution a l'étude géologique de la Grèce septentrionale: les confins de l'Epire et la Thessalie. *Annales Géologiques des Pays Helléniques*, **10**, 1–403.

—— 1973. Des tectoniques superposées et de leur signification par rapport aux modèles géophysiques: l'example des Dinarides; paleotectonique, tectonique, tarditectonique, néotectonique. *Bulletin Societé Géologique de France*, **15**, 426–60.

Baird, A. W. & Dewey, J. F. 1986. Structural evolution in thrust belts and relative plate motion in the upper Pennine Piedmont zone of the internal Alps, SW Switzerland and NW Italy. *Tectonics*, **5**, 375–87.

Bally, A. W. 1981. Thoughts on the tectonics of folded belts. *In*: McClay, K. R. & Price, N. J. (eds). *Thrust and Nappe Tectonics*, Special Publication of the Geological Society of London No. 9, 13–32.

Bayer, R., Cazes, M., Dal Piaz, G. V., Damotte, B., Elter, G., Hirn, A., Lanza, R., Lombardo, B., Mugnier, J. L., Nicolas, A., Nicolich, R., Polino, R., Roure, F., Sacchi, R., Scarascia, S., Tabacco, I., Tapponnier, P., Tardy, M., Taylor, M., Thouvenot, F., Torreilles, G. & Villien, A. 1987. Premièrs résultats de la traversée des Alpes occidentales par sismique reflexion verticale (Programme ECORS–CROP). *Comptes-Rendus de l'Académie des Sciences de Paris*, **305**, 1461–70.

Beach, A. 1981. Thrust structures in the eastern Dauphinois zone (French Alps), north of the Pelvoux massif. *Journal of Structural Geology*, **3**, 299–308.

—— 1987. A regional tectonic model for linked tectonics in north-west Europe. *In*: Brooks, J. & Glennie, K. (eds). *Petroleum Geology of North West Europe*, 43–48. Graham & Trotman, London.

Bearth, P. 1967. Die Ophiolithe der Zone von Zermatt–Sass Fee. *Beiträge zur geologischen Karte der Schweiz* N. F., **132**.

Bernoulli, D., Caron, C., Homewood, P., Kälin, O. & Van Stuijvenberg, J. 1979. Evolution of continental margins in the Alps. *Schweizerische Mineralogische und Petrographische Mitteilungen*, **59**, 165–170.

—— & Laubscher, H. P. 1972. The palinspastic problem of the Hellenides. *Eclogae Geologicae Helvetiae*, **65**, 107–18.

—— & Jenkyns, H. C. 1974. Alpine, Mediterranean and central Atlantic Mesozoic facies in relation to the early evolution of Tethys. *In*: Dott, R. H. & Shaver, R. H. (eds). *Society of Economic Paleontologists and Mineralogists Special Publication*, **19**, 129–60.

Bertrand, J. & Delaloye, M. 1976. Datation par la methode K–Ar de diverses ophiolites du flysch des Gêtes (Haute-Savoie, France). *Eclogae Geologicae Helvetiae*, **69**, 335–41.

Bickle, M. J. & Nisbet, E. 1972. The oceanic affinities of some Alpine mafic rocks based on their Ti–Y–Zr contents. *Journal of the Geological Society of London*, **128**, 267–71.

Boccaletti, M., Elter, P. & Guazzone, G. 1971. Plate tectonics model for the development of the western Alps and northern Apennines. *Nature*, **234**, 108–11.

Bonatti, E., Emiliani, C., Ferrara, G., Honnorez, J. & Rydell, H. 1974. Ultramafic carbonate breccias from the equatorial Mid-Atlantic Ridge. *Marine Geology*, **16**, 83–192.

Borsi, S. & Dubois, R. 1968. Données géochronologiques sur l'histoire hercynienne et alpine de la Calabre centrale. *Comptes-Rendus de l'Académie des Sciences de Paris*, **266**, 72–5.

——, Hieke Merlin, O., Lorenzoni, S., Paglionico, A. & Zanettin Lorenzoni, E. 1977. Stilo unit and 'dioritic–kinzigitic' unit in Le Serre (Calabria, Italy). Geological, petrological, geochronological characters. *Bollettino della Società Geologica Italiana*, **95**, 219–44.

Bosellini, A. 1973. Modello geodinamico e paleotettonico dellé Alpi meridionali durante il Giurassico-Cretaciaco. In: *Moderne Vedute sulla geologia dell'Appennino*. Accademia Nazionale dei Lincei Roma, **183**, 163–205.

BOULLIN, J. P. 1979. La transversale de Collo et d'El Milia (Petite Kabylie): une region-clef pour l'interprétation de la tectonique alpine de la chaîne littorale d'Algerie. *Mémoires de la Société Géologique de France*, **57**, 1–84.

BOYER, S. E. & ELLIOTT, D. 1982. Thrust systems. *Bulletin of the American Association of Petroleum Geologists*, **66**, 1196–230.

BROCQUET, P., CAIRE, A. & MASCLE, G. 1966. Structure et évolution de la Sicilie occidentale (Madonies et Sicani). *Bulletin de la Société Géologique de France*, **7**, 994–1013.

BRODIE, K. H. & RUTTER, E. H. 1987. Deep crustal extensional faulting in the Ivrea Zone of Northern Italy. *Tectonophysics*, **140**, 193–212.

BURRUS, J., BESSIS, F. & DOLIGEZ, B. 1987. Heat flow, subsidence and crustal structure of the Gulf of Lyons (north-west Mediterranean): a quantitative discussion of the classic passive margin model. *In*: BEAUMONT, C. & TANKARD, A. J. (eds). *Sedimentary Basins and Basin-forming mechanisms*, Canadian Society of Petroleum Geologists Memoir, **12**, 1–15.

BUTLER, R. W. H., MATTHEWS, S. J. & PARISH, M. 1986. The NW external Alpine thrust belt and its implications for the geometry of the western Alpine orogen. *In*: COWARD, M. P. & RIES, A. C. (eds). *Collision Tectonics*. Special Publication of the Geological Society of London No. 19, 245–60.

CHANNELL, J. E. T. & HORVATH, F. 1976. The African–Adriatic promontory as a paleogeographical premise for Alpine orogeny and plate movements in the Carpatho–Balkan region. *Tectonophysics*, **35**, 71–110.

——, D'ARGENIO, B. & HORVATH, F. 1979. Adria, the African promontory, in Mesozoic Mediterranean paleogeography. *Earth Science Reviews*, **15**, 213–92.

CHEADLE, M. J., MCGEARY, S., WARNER, M. R. & MATTHEWS, D. H. 1987. Extensional structures on the western UK continental shelf: a review of evidence from deep seismic profiling. *In*: COWARD, M. P., DEWEY, J. F. & HANCOCK, P. L. (eds). *Continental Extensional Tectonics*, Special Publication of the Geological Society of London. **28**, 445–467.

CHOUKROUNE, P., BALLEVRE, M., COBBOLD, P., GAUTIER, Y., MERLE, O. & VUICHARD, J. -P. 1986. Deformation and motion in the western Alpine arc. *Tectonics*, **5**, 215–26.

CIVETTA, L., CORTINI, M. & GASPARINI, P. 1973. Interpretation of a discordant K/Ar age pattern (Capo Vaticano, Italy). *Earth and Planetary Science Letters*, **20**, 113–8.

COWARD, M. P. & TRUDGILL, B. in press. Basin development and basement structure, west of Britain. *Bulletin de la Société Géologique de France*.

DAL PIAZ, G. V., HUNZIKER, J. C. & MARTINOTTI, G. 1972. La zona Sesia-Lanzo e l'evoluzione tettonico-metamorphica delle Alpi nordoccidentali interne. *Memoire della Società Geologica Italiana*, **11**, 433–466.

DEBELMAS, J. & KERKHOVE, C. 1980. Les Alpes franco-italiennes. *Géologie Alpine*, **56**, 21–58.

—— & LEMOINE, M. 1970. The western Alps: Paleogeography and structure. *Earth Science Reviews*, **6**, 221–56.

DECANDIA, F. A. & ELTER, P. 1972. La zona ofiolitifera del Bracco nel settore compreso fra Levanto e la Val Graveglia (Appennino Ligure). *Memorie della Società Geologica Italiana*, **11**, 503–30.

DE JONG, K. A. 1973. Mountain building in the Mediterranean region. *In*: DE JONG, K. A. & SCHOLTON, R. (eds). *Gravity and Tectonics*, 125–39. Wiley, New York.

DERCOURT, J. 1972. The Canadian Cordillera, the Hellenides and the sea floor spreading theory. *Canadian Journal of Earth Sciences*, **9**, 709–43.

——, ZONENSHAIN, L. P., RICOU, L. E., KAZMIN, V. G., LE PICHON, X., KNIPPER, A. L., GRANDJACQUET, C., SBORTSHIKOV, I. M., GEYSSANT, J., LEPVRIER, C., PECHERSKY, D. H., BOULIN, J., SIBUET, J. C., SAVOSTIN, L. A., SOROKHTIN, O., WESTPHAL, M., BAZHENOV, M. L., LAUER, J. P. & BIJU-DUVAL, B. 1986. Geological evolution of the Tethys Belt from the Atlantic to the Pamirs since the Lias. *Tectonophysics,* **123**, 241–315, 9 plates.

DE ROEVER, E. W. F. 1972. Lawsonite–albite–facies metamorphism near Fuscaldo, Calabria (Southern Italy), its geological significance and petrological aspects. *GUA Papers of Geology*, **3**.

——, PICCARRETA, G., BEUNK, F. F. & KIEFT, C. 1974. Blue amphiboles from northwestern and central Calabria (Italy). *Periodico Mineralogia*, **43**, 1–37.

DEWEY, J. F., PITMAN, W. C. III, RYAN, W. B. F. & BONNIN, J. 1973. Plate tectonics and the evolution of the Alpine system. *Geological Society of America Bulletin*, **84**, 3137–80.

——, HELMAN, M. L., TURCO, E., HUTTON, D. H. W. & KNOTT, S. D. this volume. Kinematics of the western Mediterranean.

DIETRICH, D. 1976. La geologia della Catena Costiera Calabra tra Cetraro e Guardia Piemontese. *Memorie della Società Geologica Italiana*, **17**, 61–121.

—— 1988. Sense of overthrust shear in the Alpine nappes of Calabria. *Journal of Structural Geology*, **10**, 373–81.

—— & DURNEY, D. W. 1986. Change of direction of overthrust shear in the Helvetic nappes of western Switzerland. *Journal of Structural Geology*, **8**, 389–98.

—— & SCANDONE, P. 1972. The position of the basic and ultrabasic rocks in the tectonic units of the southern Apennines. *Atti Accademia Pontaniana Napoli*, **21**, 61–75.

DIETRICH, V. 1969. *Die Ophiolithe des Oberhalbsteins (Graubünden) und das Ophiolithmaterial der ostschweizerischen Molasseablagerungen*. Lang, Bern.

—— 1980. The distribution of ophiolites in the Alps. *Ofioliti*, **5**, 7–53.

ECORS PYRENEES TEAM 1988. The ECORS deep reflection seismic survey across the Pyrenees. *Nature*, **331**, 508–11.

ESCHER, A., MASSON, H. & STECK, A. 1988. Coupes géologiques des Alpes occidentales suisses.

Mémoires de Géologie de l'Université de Lausanne, **2**, 5–11.

FINCKH, P., FREI, W., FREEMAN, R., HEITZMANN, P., LEHNER, P., MUELLER, S., PfiFFNER, A. & VALASEK, P. 1987. Nationales Forschungsprogramm 20 'Geologische Tiefenstruktur der Schweiz–Problemstellung und erste Resultate. *Bulletin der Vereinigung Schweizerischer Petroleum-Geologen und Ingenieure*, **54**, 59–74.

FRANK, F. C. 1968. Curvature of island arcs. *Nature*, **220**, 363.

FRANK, W. 1987. Evolution of Austro–Alpine elements in the Creteaceous. *In*: FLÜGEL, H. W. & FAUPL, P. (eds). *Geodynamics of the Eastern Alps*. Deuticke, Vienna.

——, KRALIK, M., SCHARBERT, S. & THÖNI, M. 1987a. Geochronological data from the Eastern Alps. *In*: FLÜGEL, H. W. & FAUPL, P. (eds). *Geodynamics of the Eastern Alps*. Deuticke, Vienna.

——, HOINKES, G., PURTSCHELLER, F. & THÖNI, M. 1987b. The Austro-Alpine unit west of the Hohe Tauern: The Ötztal-Stubai Complex as an example for the Eoalpine metamorphic evolution. *In*: FLÜGEL, H. W. & FAUPL, P. (eds). *Geodynamics of the Eastern Alps*. Deuticke, Vienna.

——, HOCK, V. & MILLER, C. 1987c. Metamorphic and tectonic history of the central Tauern window. *In*: FLÜGEL, H. W. & FAUPL, P. (eds). *Geodynamics of the Eastern Alps*. Deuticke, Vienna.

FREY, M., BUCHER, K., FRANK, E. & MULLIS, M. 1980. Alpine metamorphism along the Geotraverse Basel–Chiasso – a review. *Eclogae Geologicae Helvetiae*, **73**, 527–46.

GIBBS, A. D. 1984. Structural evolution of extensional basin margins. *Journal of the Geological Society of London*, **141**, 609–620.

GIDON, M. 1979. Le rôle des étapes successives de déformation dans la tectonique Alpine du massif du Pelvoux (Alpes occidentales). *Comptes-Rendus de l'Académie des Sciences Paris*, **288**, 803–6.

GILLCRIST, R., COWARD, M. P. & MUGNIER, J. -L. 1987. Structural inversion and its controls: examples from the Alpine foreland and the French Alps. *Geodynamica Acta*, **1**, 5–34.

GOGUEL, J. 1963. L'interprétation de l'arc des Alpes occidentales. *Bulletin de la Société Géologique de France*, **5**, 20–33.

GOURLAY, P. 1986. La déformation du socle et des couvertures delphino–helvetiques dans la région du Mont Blanc (Alps occidentales). *Bulletin de la Société Géologique de France*, **8**, 159–69.

GRACIANSKY, P., BOURBON, M., CHENET, P. Y., DE CHARPAL, O. & LEMOINE, M. 1979. Genèse et évolution comparée de deux marges continentales passives: marge iberique de l'ocean Atlantique et marge européenne de la Tethys dans les Alpes occidentales. *Bulletin de la Société Géologique de France*, **21**, 663–74.

HACCARD, D., LORENZ, C. & GRANDJACQUET, C. 1972 Essai sur l'évolution téctogénetique de la liaison Alps–Apennins (de la Ligurie a la Calabre). *Memorie Società Geologica Italiana*, **11**, 309–41.

HEIM, A. 1921. *Geologie der Schweiz*. Tauchnitz, Leipzig, 476 pp.

HIRST, J. P. P. & NICHOLS, G. J. 1986. Thrust tectonic controls on Miocene alluvial distribution patterns, southern Pyrenees. *Special Publication of the International Association of Sedimentologists*, **8**, 247–58.

HOMEWOOD, P., ALLEN, P. A. & WILLIAMS, G. D. 1986. Dynamics of the molasse basin of western Switzerland. *Special Publication of the International Association of Sedimentologists*, **8**, 199–217.

—— & LATELTIN, O. 1988. Classic Swiss clastics (flysch and molasse). The Alpine connection. *Geodynamica Acta*, **2**, 1–11.

HOSSACK, J. R., DERAMOND, J. & GRAHAM, R. H. 1984. The geological structures and development of the Pyrenees. *Colloquium sur chevauchement et déformation, Toulouse, Abstract*, 46–47.

HUBER, M., RAMSAY, J. G. & SIMPSON, C. 1980. Deformation in the Maggia and Antigorio nappes, Lepontine Alps. *Eclogae Geologicae Helvetiae*, **73**, 593–606.

IPPOLITO, F., D'ARGENIO, B., PESCATORE, T. & SCANDONE, P. 1975. Structural-stratigraphic units and the tectonic framework of the southern Apennines. *In*: SQUYRES, C. (ed.). *Geology of Italy*, Earth Science Society of the Libyan Arab Republic, Tripoli, 317–28.

KARAMATA, S., MAJER, V. & PAMIC, J. 1980. Ophiolites of Yugoslavia. *Ofioliti*, **5**, 105–25.

KELTS, K. 1981. A comparison of some aspects of sedimentation and translational tectonics from the Gulf of California and Mesozoic Tethys, northern Penninic margin. *Eclogae Geologicae Helvetiae*, **74**, 317–38.

KNIPPER, A., RICOU, L. E. & DERCOURT, J. 1986. Ophiolites as indicators of the geodynamic evolution of the Tethyan ocean. *Tectonophysics*, **123**, 213–40.

LABAUME, P., SEGURET, M. & SEYVE, C. 1985. Evolution of a turbiditic foreland basin and analogy with an accretionary prism: example of the Eocene south-Pyrenean Basin. *Tectonics*, **4**, 661–85.

LANZAFAME, G., SPADEA, P. & TORTORICI, L. 1979. Mesozoic ophiolites of northern Calabria and Lucanian Apennine (S Italy). *Ofioliti*, **4**, 173–82.

LAUBSCHER, H. P. 1969. Mountain building. *Tectonophysics*, **7**, 551–63.

—— 1970. Bewegung und Wärme in der alpinen Orogenese. *Schweizerische Mineralogische und Petrographische Mitteilungen*, **50**, 565–596.

—— 1971. The large scale kinematics of the western Alps and the northern Apennines and its palinspastic implications. *American Journal of Science*, **271**, 193–226.

—— 1974. The tectonics of subduction in the Alpine system. *Memorie della Società Geologica Italiana*, **13**, 275–83.

—— 1975. Plate boundaries and microplates in Alpine history. *American Journal of Science*, **275**, 865–76.

—— 1983. Detachment, shear and compression in the central Alps. *Geological Society of America Memoir*, **158**, 19–211.

—— 1984. The tectonic problem of the Ivrea body and the Insubric line. *Annales Geophysicae*, **2**, 169–70.

LAUBSCHER, H. P. 1985. Large scale, thin skinned thrusting in the southern Alps: kinematic models. *Geological Society of America Bulletin*, **96**, 710–18.

—— 1988. The arcs of the western Alps and the northern Apennines: an updated view. *Tectonophysics*, **146**, 67–78.

—— & BERNOULLI, D. 1977. Mediterranean and Tethys. *In*: NAIRN, A. E. M., KANES, W. H. & STEHLI, F. G. (eds). *The Ocean Basins and Margins, 4A, The Eastern Mediterranean.* Plenum Press, New York.

—— & —— 1982. History and deformation in the Alps. *In*: HSÜ, K. J. (ed). *Mountain Building Processes*, 169–80. Academic Press, London.

LEMOINE, M. 1980. Serpentinites, gabbros and ophicalcites in the Piemont–Ligurian domain of the western Alps: possible indicators of oceanic fracture zones and of associated serpentinite protrusions in the Jurassic–Cretaceous Tethys. *Archives des Sciences, Genève*, **33**, 105–15.

——, BAS, T., ARNAUD-VANNEAU, A., ARNAUD, H., DUMONT, T., GIDON, M., BOURBON, M., DE GRACIANSKY, P. C., RUDKIEWICZ, J. L. & TRICART, P. 1986. The continental margin of Mesozoic Tethys in the western Alps. *Marine and Petroleum Geology*, **3**, 178–99.

——, GIDON, M. & BARFETY, J. C. 1981. Les massifs cristallins externes des Alpes occidentales: d'anciens blocs basculés nés au Lias lors du rifting téthysien. *Comptes-Rendus de l'Académie des Sciences de Paris*, **292**, 917–20.

—— & TRÜMPY, R. 1987. Pre-oceanic rifting in the Alps. *Tectonophysics*, **133**, 305–20.

LIVERMORE, R. A. & SMITH, A. G. 1985. Some boundary conditions for the evolution of the Mediterranean region. *In*: STANLEY, D. J. & WEZEL, F. C. (eds). *Geological Evolution of the Mediterranean Basin*, 84–98. Springer-Verlag, Berlin.

LOUBAT, H. 1984. Considerations préliminaires sur la configuration horizontale de l'édifice subvolcanique du Versoyen (Alpes Franco–Italiennes); analogue aux bassins en distension du type Golfe de Californie. *Géologie Alpine*, **60**, 37–44.

LOWRIE, W. 1986. Paleomagnetism and the Adriatic promontory: a reappraisal. *Tectonics*, **5**, 797–807.

MANCKTELOW, N. 1985. The Simplon line: a major displacement zone in the western Lepontine Alps. *Eclogae Geologicae Helvetiae*, **78**, 73–96.

MASSON, D. G. & MILES, P. R. 1986. Development and hydrocarbon potential of Mesozoic sedimentary basins around margins of North Atlantic. *Bulletin of the American Association of Petroleum Geologists*, **70**, 721–729.

MERCIER, J. 1958. Sur l'âge de la phase tectonique 'ante-senonienne' a l'ouest du Dévoluy (Drôme). *Bulletin de la Société Géologique de France*, **8**, 689–97.

MILNES, A. G. 1974. Structure of the Pennine zone (central Alps): a new working hypothesis. *Geological Society of America Bulletin*, **85**, 1727–32.

—— 1978. Structural zones and continental collision, central Alps. *Tectonophysics*, **47**, 369–92.

MOLNAR, P. & TAPPONNIER, P. 1975. Cenozoic tectonics of Asia; effects of a continental collision. *Science*, **189**, 419–26.

MONTADERT, L., ROBERTS, D. G. & DE CHARPAL, O. 1979. Rifting and subsidence of the northern continental margin of the bay of Biscay. *Initial report of the Deep Sea Drilling Project No 48.* US Government Printing Office, Washington, 1025–60.

MOORES, E. M. 1969. Petrology and structure of the Vourinos ophiolitic complex of Northern Greece. *Geological Society of America Special Paper*, **118**.

MUGNIER, J. L. & VIALON, P. 1986. Deformation and displacement of the Jura cover on its basement. *Journal of Structural Geology*, **8**, 373–88.

MUELLER, S. 1982. Deep structure and recent dynamics in the Alps. *In*: HSÜ, K. J. (ed). *Mountain Building Processes*, 189–99. Academic Press, London.

MUNOZ, J. A., MARTINEZ, A. & VERGES, J. 1986. Thrust sequences in the Spanish eastern Pyrenees. *Journal of Structural Geology*, **8**, 399–405.

NEUMAYR, M. 1883. Klimatische Zonen während der Jura und Kreidezeit. *Denkschrift der kaiserlichen Akademie der Wissenschaften Wien*, **47**, 1–34.

NIGGLI, E. 1970. Alpine Metamorphose und Alpine Gebirgsbildung. *Fortschritte der Mineralogie*, **47**, 16–26.

OHNENSTETTER, M. & OHNENSTETTER, D. 1980. Ophiolite sequences in Corsica. *Ofioliti*, **80**, 53–58.

——, ——, VIDAL, P., CORNICHET, J., HERMITTE, D. & MACE, J. 1981. Crystallisation and age of zircon from Corsican ophiolitic albitites: consequences for oceanic expansion in Jurassic times. *Earth and Planetary Science Letters*, **54**, 397–408.

PANZA, G. F. 1984. The deep structure of the Mediterranean–Alpine region and large shallow earthquakes. *Memorie della Societa Geologica Italiana*, **29**, 5–13.

—— & MUELLER, S. 1979. The plate boundary between Eurasia and Africa in the Alpine area. *Memorie de Scienze Geologiche (Padova)*, **33**, 43–50.

PFIFFNER, O. A. 1981. Fold and thrust tectonics in the Helvetic nappes (E. Switzerland). *In*: MCCLAY K. & PRICE, N. J. (eds). *Thrust and Nappe Tectonics*. Special Publication of the Geological Society of London No. 9, 319–27.

——, FREI, W., FINCKH, P. & VALASEK, P. 1988. Deep seismic reflection profiling in the Swiss Alps: Explosion seismology results for the line NFP 20 east. *Geology* (in press).

PITMAN, W. C. & TALWANI, M. 1972. Sea floor spreading in the North Atlantic. *Geological Society of America Bulletin*, **83**, 619–43.

PLATT, J. P. 1986. Dynamics of orogenic wedges and the uplift of high-pressure metamorphic rocks. *Geological Society of America Bulletin*, **97**, 1037–53.

RAMSAY, J. G. 1963. Stratigraphy, structure and metamorphism of the western Alps. *Proceedings of the Geological Association*, **74**, 357–91.

—— 1981. Tectonics of the Helvetic nappes. *In*: MCCLAY K. & PRICE, N. J. (eds). *Thrust and Nappe Tectonics*, Special Publication of the Geological Society of London No. 9, 293–309.

RICOU, L. E. & SIDDANS, A. W. B. 1986. Collision tectonics in the western Alps. *In*: COWARD, M. P. & RIES, A. C. (eds). *Collision Tectonics*, Special Publication of the Geological Society of London No. 19, 229–44.

RIES, A. C. & SHACKLETON, R. M. 1976. Patterns of strain variation in arcuate fold belts. *Philosophical Transactions of the Royal Society of London, A. Mathematical and Physical Sciences*, **238**, 281–288.

RIGASSI, D. 1977. Genèse tectonique du Jura: une nouvelle hypothèse. *Paleolaboratory News*, **2**, Terreaux du Temple, Geneva.

RUBIE, D. C. 1984. A thermal–tectonic model for the high pressure metamorphism and deformation in the Sesia zone, western Alps. *Journal of Geology*, **92**, 21–36.

SAVOSTIN, L. A., SIBUET, J. C., ZONENSHAIN, L. P., LE PICHON, X. & ROULET, M. J. 1986. Kinematic evolution of the Tethys belt from the Atlantic ocean to the Pamirs since the Triassic. *Tectonophysics*, **123**, 1–35.

SAVU, H. 1980. Ophiolitic rocks and other initial magmatites in the Carpathians. *Ofioliti*, **5**, 97–104.

SCANDONE, P. 1967. Studi di geologia lucana: la serie calcareo–silico–marnosa e i suoi rapporti con l'Appennino calcareo. *Bolletino della Società dei Naturalisti Napoli*, **76**, 3–175.

—— 1975. Triassic seaways and the Jurassic Tethys ocean in the Central Mediterranean area. *Nature*, **256**, 117–9.

——, GIUNTA, G. & LIGUORI, V. 1974. The connection between the Apulia and Sahara continental margins in the southern Apennines and in Sicily. *24 Congrès-Assemblée plenaire C.I.E.S.M.*, Comitée Géologique et Géophysique Marines, Monaco, 4–6. 12. 1974.

SCHMID, S. M., ZINGG, A. & HANDY, M. 1987. The kinematics of movements along the Insubric line and the emplacement of the Ivrea body. *Tectonophysics*, **135**, 47–66.

SEGURET, M. & DAIGNIERES, M. 1986. Crustal scale balanced cross-sections of the Pyrenees; Discussion. *Tectonophysics*, **129**, 303–18.

SLAPANSKY, P. & FRANK, W. 1987. Structural evolution and geochronology of the northern margin of the Austroalpine in the northwestern Schladming Crystalline (NE Radstätter Tauern). *In*: FLÜGEL, H. W. & FAUPL, P. (eds). *Geodynamics of the Eastern Alps*. Deuticke, Vienna.

SMITH, A. G. 1971. Alpine deformation and the oceanic areas of Tethys, Mediterranean and Atlantic. *Bulletin Geological Society of America*, **82**, 2039–70.

STECK, A. 1987. Le massif du Simplon – réflexions sur la cinématique des nappes de gneiss. *Schweizerische Mineralogische und Petrographische Mitteilungen*, **67**, 27–45.

STEINMANN, G. 1906. Geologische Beobachtungen in den Alpen II. Die Schardt'sche Uberfaltungstheorie und die geologische Bedeutung der ophiolitischen Massengesteine. *Beiträge Naturforschende Gesellschaft Freiburg im Breisgau*, **16**, 1–49.

SUESS, E. 1893. Are great ocean depths permanent? *Natural Sciences*, **2**, 180–7.

—— 1901. *Das Antlitz der Erde*. Temsky, Vienna.

TOLLMANN, A. 1975. Ozeanische Kruste im Pennin des Tauernfensters und die Neugliederung des Deckenbaus der Hohen Tauern. *Neues Jahrbuch Geologische Paläontologische Abhandlungen*, **148**, 286–319.

TRICART, P. & LEMOINE, M. 1986. From faulted blocks to megamullions and megaboudins: Tethyan heritage in the structure of the western Alps. *Tectonics*, **5**, 95–118.

TRÜMPY, R. 1960. Paleotectonic evolution of the central and western Alps. *Geological Society of America Bulletin*, **71**, 808–843.

—— 1980. *Geology of Switzerland, a Guide Book. Part A, an outline of the geology of Switzerland*. Wepf & Co., Basel.

VAN DEN BERG, J. 1979. Reconstruction of the western Mediterranean area for the Mesozoic and Tertiary timespan. *Geologie en Mijnbouw*, **58**, 153–60.

WEISSERT, H. J. & BERNOULLI, D. 1985. A transform margin in the Mesozoic Tethys: evidence from the Swiss Alps. *Geologische Rundschau*, **74**, 665–79.

WILDI, W. 1983. La chaîne tello-rifaine (Algérie, Maroc, Tunisie): structure, stratigraphie et évolution du Trias au Miocène. *Revue Géologie Dynamique Géographie Physique*, **24**, 201–97.

—— 1985. Heavy mineral distribution and dispersal pattern in Penninic and Ligurian flysch basins (Alps, northern Apennines). *Giornale di Geologia*, **47**, 77–99.

WILLIAMS, G. D. 1985. Thrust tectonics in the south central Pyrenees. *Journal of Structural Geology*, **7**, 11–7.

WILLIAMS, P. F. & COMPAGNONI, R. 1983. Deformation and metamorphism in the Bard area of the Sesia Lanzo zone, western Alps, during subduction and uplift. *Journal of Metamorphic Geology*, **1**, 117–40.

ZIEGLER, P. A. 1982. *Geological Atlas of Western and Central Europe*. Shell International Petroleum Maatschapp ij B. V., The Hague, 130 pp.

—— 1987. Evolution of the Western Approaches Trough, *Tectonophysics*, **137**, 335–40.

MIKE COWARD, Geology Department, Royal School of Mines, Imperial College, London SW7 2BP, UK.

DOROTHEE DIETRICH, Geologisches Institut, ETH-Zentrum, CH-8092 Zürich, Switzerland.

Tectonic Evolution of the External Zones of the Alps

Fold and fault geometry in the western Helvetic nappes of Switzerland and France and its implication for the evolution of the arc of the western Alps

J. G. Ramsay

SUMMARY: The main features of the fold and fault geometry of the Helvetic nappes and Chaînes Subalpines of Switzerland and France have been determined by the construction of tie lines between outcropping hinge points of folds and cut-off points of strata against thrusts. Longitudinal correlation of these structures is now possible. Folds are very persistent along their hinge lines and can often be followed for tens of kilometres. Fold hinges and thrust plane cut-off lines are extremely regularly oriented. The Alpine structures developed in the external massifs are obliquely superposed on previously emplaced nappes. The overall structural pattern is attributed to a progressive change of the relative displacement directions between the northern (European) plate and southern (Apulian) plate from N–S, through NW–SE, concluding with movements in an E–W sense. The well known structural culminations and depressions are interpreted as the result of the latest shortening in an E–W direction and not to complex ramp geometry or to duplex stacking.

The Swiss Helvetic nappe region is a classic one for the study of nappe tectonics. Although the concept of the horizontal translation of detached rock sheets by over thrusting was developed independently by workers in the Alps (Escher 1841), northwest Scotland (Callaway 1883, Peach & Horne 1888), in the Caledonides of Sweden (Törnebohm 1883, 1888) and in the western Alps of Provence (Bertrand 1884), the clarity of anomalous stratigraphic relationships produced by duplication of strata and stacking up of the thrust sheets in the Helvetic nappes have led geologists to regard this terrain as a showground for overthrusting phenomena (Heim 1921, Collet 1927, Trümpy 1980).

During the development of the Helvetic nappes the rocks were folded and subjected to secondary faulting of normal, strike-slip and thrust type. The rocks have also locally developed planar and linear fabrics as a result of their internal deformation. Previously published accounts of these structures give some detail of the fold geometry mostly in the form of map arrows representing the generalised directions of the fold axes and either cross-sections (e.g. Heim 1921, Boyer & Elliott 1982) or profile sections where the fold hinges plunge more than about 20 degrees (Durney 1972, Ramsay & Huber 1983, Huggenberger 1985, Langenberg *et al.* 1987, Dietrich & Casey this volume). There is practically no published information on the continuity of individual folds from section to section or on the changes in fold forms and in deformation style that occur laterally and vertically in the nappes. Very few authors have attempted to trace individual folds along their hinge lines in the manner developed by Albert Heim (1921, p. 361). In order to fill in these gaps in our knowledge a team of geologists at the ETH Zürich have remapped the Helvetic nappes at a scale of 1:25 000 and we have compiled this new information at a scale of 1:100 000. We have tried to establish the orientation features of all the major folds in terms of hinge lines and axial surfaces together with data of bed thickness variations and the fold style of the individual layers making up the folds. Other data important in the interpretation of fold kinematics that was also acquired during the course of the investigation were: small scale parasitic folds, finite and incremental strain states, type, orientation and intensity of rock cleavage and the significance of various types of linear deformation fabrics ('stretching' lineations, intersection fabrics and 'pencil' fabric). Presentation of this data in map form will appear in 1989. The aim of this present publication is to set out new information on the structure of the western Helvetic nappes of Switzerland, to relate this geometry to the structures mapped in the Chaînes Subalpines of France and to suggest a kinematic model to explain the features of the arcuate structure of the Alps of this region.

Principal features of the western Helvetic nappes

The Helvetic nappes consist mostly of overthrust sheets of Mesozoic and Tertiary sediments deposited in a carbonate shelf environment

From: Coward, M. P., Dietrich, D. & Park, R. G. (eds), 1989, *Alpine Tectonics*, Geological Society Special Publication No. 45, pp. 33–45.

located on the southern margin of the European continental mass bordering the Tethyan ocean. Some of the Tertiary sediments clearly show the influence of the closure of the ocean and the development of the Alpine collision zone in their development of synorogenic flysch facies. In the region which is the subject of this paper three major Helvetic nappes lie one above the other, the Morcles, Diablerets and Wildhorn nappes. These nappes override a so-called autochthon: a 'crystalline' base of metamorphic and plutonic rocks with infolded unconformable sediments of Permo-Carboniferous age exposed in the Mont Blanc and Aiguilles Rouges massifs (Figs 1 and 2) together with an unconformably overlying cover of Mesozoic and Tertiary sediments showing a more continent proximal environment of deposition than that seen in the overlying nappes. Although this basement and attached cover are termed 'autochthonous' in most previous publications, geophysical data suggest that they have been moved northward relative to the main European continent (Hsü 1979). In this present paper I will use the term 'parautochthon' for these tectonic units.

The Helvetic nappes are themselves overlain by other nappe elements (Fig. 1): the Ultrahelvetic nappes derived from a part of the European marine shelf situated to the S of the Helvetic depositional region, and the pre-Alpine nappes, derived from an even more southerly environment and tectonically connected with the Pennine nappes (Trümpy 1980). These tectonic units are best preserved today in the form of klippen (Romande pre-Alps, Chablais pre-Alps and the klippe of Les Annes, see Fig. 1) which have been isolated by erosion from their tectonic counterparts in the Pennine nappes and from the Ultrahelvetic units immediately underlying these nappes. The geometric relationships of the nappes overlying the Helvetic nappes varies from E to W. In the E the Ultrahelvetic nappes are exposed in the Rhône Valley as a type of 'root' zone while to the NW they are found as small klippen which occupy the cores of synformal folds on the top of the uppermost Helvetic nappe (Wildhorn nappe). Between the middle (Diablerets) and lowest (Morcles) nappes the Ultrahelvetic units occupy the core of a large ENE-plunging recumbent synform, a fold which has an amplitude of more than 12 km (Lugeon 1901, Heim 1921, Ramsay 1981, Dietrich & Casey this volume). The frontal parts of the Ultrahelvetic nappes are located as a more or less continuous rim around the main pre-Alpine klippen indicating that they form a continuous sheet beneath the main pre-Alpine units. In contrast to these relationships, the Ultrahelvetic and pre-Alpine nappes W of the Rhône discordantly cut across the folds developed in the frontal part and upper limb of the Morcles nappe (Fig. 2 in the regions around Cluses and Samoëns). These nappes are similarly unaffected by the large folds making up the Chaînes Subalpines Although the strata in the pre-Alpine nappes are highly folded the basal thrust surface appears to be only affected by open folds (e.g. Les Annes) with differently oriented, more northerly-trending, axes from those seen in the underlying Chaînes Subalpines. Although the nappe equivalents of the Diablerets and Wildhorn nappes cannot be identified for certain to the W of the Rhône valley, the relationships of the Ultrahelvetic elements with the folds of the Morcles nappe show either that the Morcles nappe folds were formed later in the E than in the W, or that the last movements of the overlying Ultrahelvetic and pre-Alpine nappes in the W were later than those in the E. In view of the geometric correlations of the individual folds within the Morcles nappe to be discussed below, the second of these hypotheses seems the most likely.

A well known feature of the Helvetic nappes is the systematic change that takes place in the amount of plunge of the hinge lines of individual folds and in the nappe sheets as a whole leading to regionally important culminations and depressions (Argand 1911, Arbenz 1912, 1913, Heim 1921, Ramsay 1981, Dietrich & Casey this volume). In the region shown in Fig. 1, two major depressions, the Wildstrübel depression in Switzerland and the Flaine depression in France, are separated by a culmination which more or less coincides with the Rhône valley between Martigny and St Maurice. As a result of the vertical displacements related to these structures the Helvetic nappes are not exposed at the surface in every NW–SE cross-section. The highest nappes have been completely eroded away alone the line of the Rhône culmination whereas the lowest nappes are hidden from view by a cover of higher nappes in the Wildstrubel depression. This means that not everywhere can surface connections between the exposed nappe elements be unambiguously established. In this paper the classic correlation of Doldenhorn with Morcles, and Gellihorn with Diablerets (Fig. 1) is accepted as providing the best structural linkage of homologous units. There are some differences in the internal stratigraphy of these correlated nappes but such differences are to be expected considering the distances involved between the sedimentation sites (Ferrazzini & Schuler 1979).

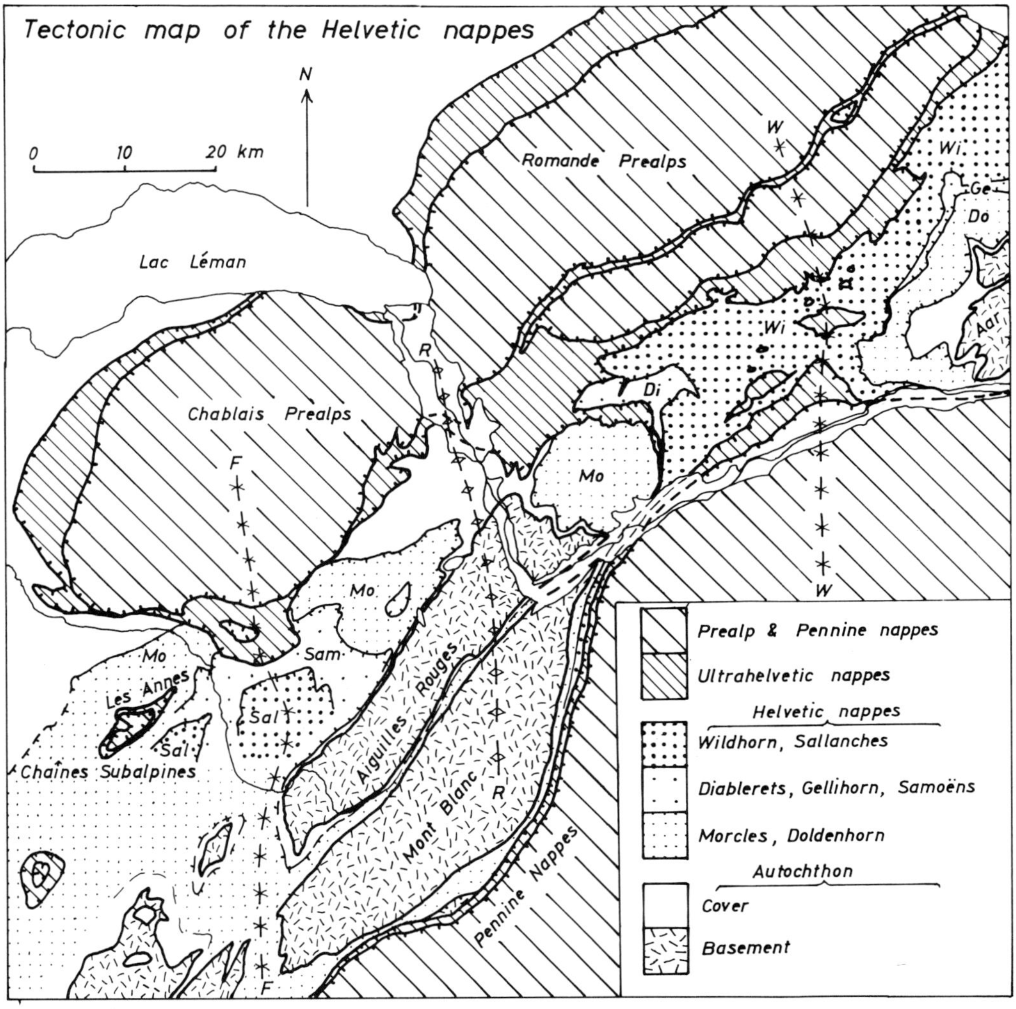

FIG. 1. Map of the Helvetic Alps of western Switzerland and eastern France indicating the main tectonic units with speculative nappe correlations. The N–S dashed lines F–F, R–R and W–W represent the traces of the Flaine depression, Rhône culmination and Wildstrubel depression respectively.

Fold geometry

The terms used to describe the geometrical features of folds in this paper are those defined in Ramsay & Huber (1987). Each folded surface contains one or more hinge lines and the joins of hinge lines through the fold defines the axial surface or axial plane. Because of the quite complex nature of the axial surfaces of the Helvetic nappe folds and the irregularity of the high mountain topography it is often difficult to directly measure these features in the field. However, it is possible to precisely identify the location of a hinge line where it passes through the land surface, that is a *hinge point* (Fig. 3). In order to make interpretative maps of the folds these points have been linked in two ways. First, by joining the hinge line points of successive adjacent bed surfaces, the fold *axial surface trace* is determined (Fig. 2), a line representing the intersection of the axial surface with the ground surface. The volume of rock situated between two adjacent fold traces (fold limb) has been ornamented in two ways: where an antiformal trace trace lies on the NW side of a synformal trace lies on the south east side of a synform the region has been stippled. The unornamented zones are those where strata are

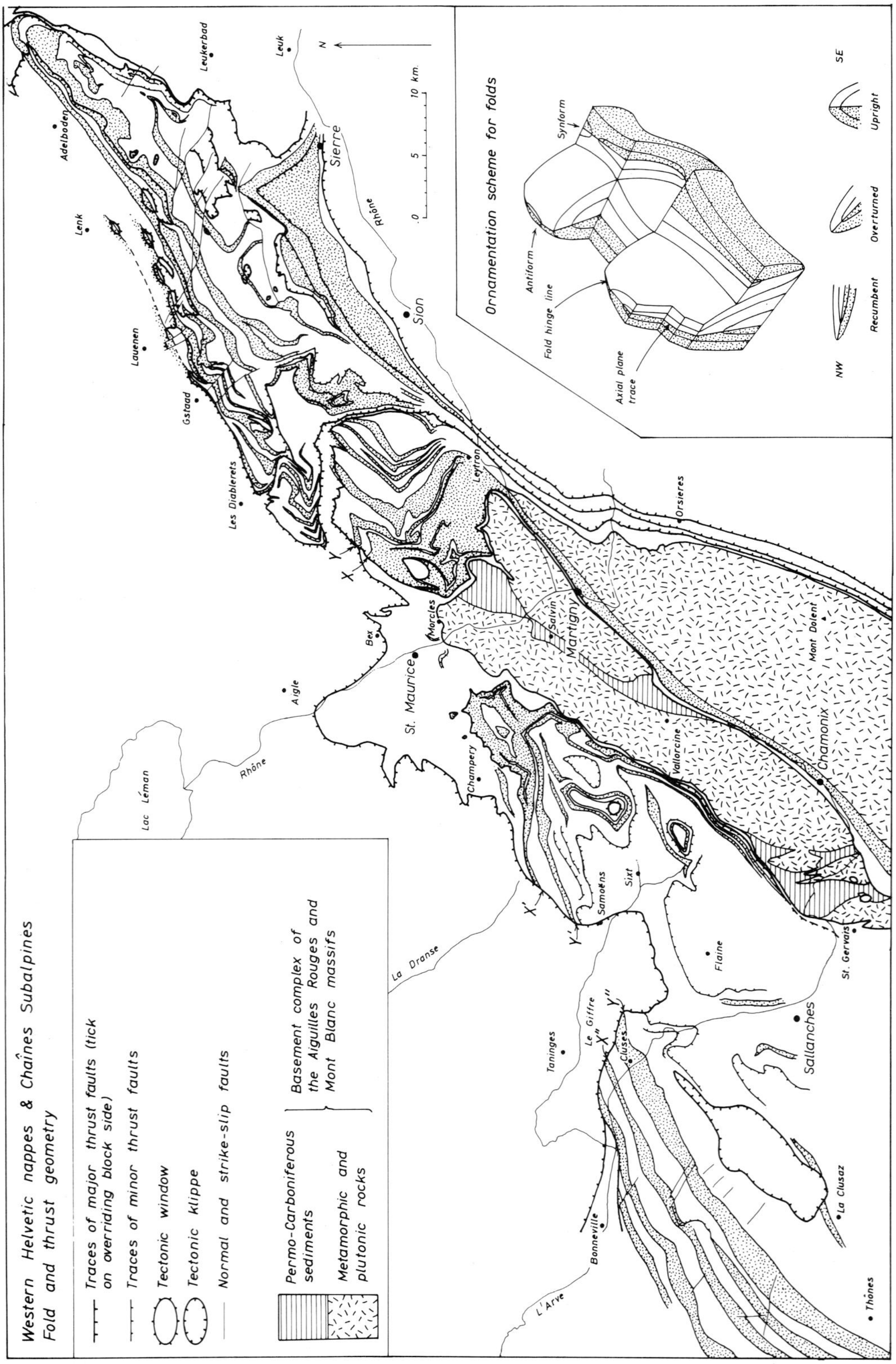

FIG. 2. Synoptic map of the folds in the Helvetic nappes indicating the positions of major fold limbs.

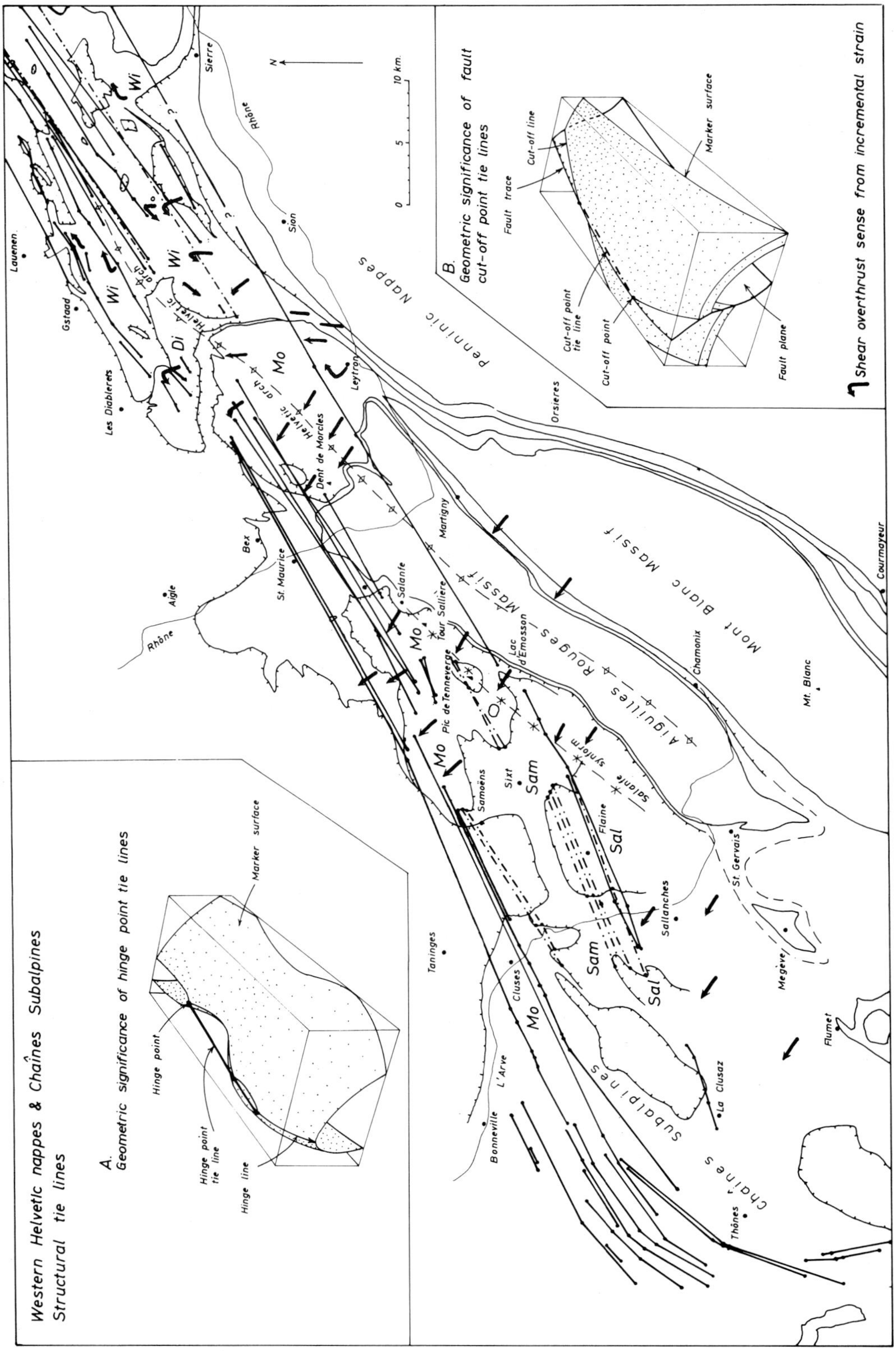

FIG. 3. Synoptic map of the Helvetic nappes showing the principal structural tie-lines. Mo-Morcles nappe, Di-Diablerets nappe, Wi-Wildhorn nappe, Sam-Samoëns thrust sheet, Sal-Salanches thrust sheet.

uninverted (normal fold limb), whereas in the stippled zones the rocks are in normal or reversed order depending upon the inclination of the fold axial surface and the interlimb angle (see schematic block diagram on the lower right hand side of Fig. 2). The importance of this method of representation is that the regional persistance of folds can be easily seen, fold correlations are established and no fold can be 'lost' (Ramsay & Huber 1987, p. 329). The second method of analysing hinge line points is shown in Fig. 3. The hinge line point at any one stratigraphic level is connected to the next point on the *same level* in the same fold by a straight line which is called the *hinge point tie line*. These lines do not always coincide with the actual hinge lines of the fold but give the overall direction of the hinge line between the two points. This construction provides extremely valuable information about the regional trends of fold hinges (Fig. 3).

The constructions made in Figs 2 and 3 clarify many features of the fold geometry and greatly assist the correlation of folds from locality to locality. The Upper Jurassic (Malm) to Upper Cretaceous strata in the core region of the Morcles nappe exposed on both sides of the Rhône show two major anticlinal components (Collet 1943, Badoux 1972, Ramsay 1981, Dietrich & Casey this volume). These have been located in Fig. 2 as X and Y east of the Rhône and X′ and Y′ west of the Rhône. They disappear under higher tectonic units N of Samoëns but reappear to the W as large asymmetric anticlines to the E and N of Cluses (Fig. 2, X″ and Y″). The hinge line point tie lines of these folds can also be correlated very well on either side of the Flaine depression, the sector between Morcles and Salanfe trending 60° and the sector W of Cluses trending 65°. These slight differences are interpreted as the result of a more forward (NW) advance of the Morcles nappe in its Swiss sector relative to that in the French sector, an interpretation coinciding with that deduced from thrust plane displacements to be described below. The tie line in the core of the Morcles nappe at the Lias–Dogger contact from W of Lac d'Emosson to that three km N of Leytron is absolutely parallel to the tie lines constructed on the frontal and upper limb folds of the Morcles nappe (Fig. 3). The hinge line point N of Leytron has also been joined with the equivalent point in the core of the Doldenhorn nappe to the E of Fig. 3 (tie line with a question mark in Fig. 3). Because the distance involved here exceeds 40 km this correlation might be open to question. However, it is a remarkable fact that this tie line continues the other in a direct line.

The tie lines in the Morcles nappe are parallel to those seen in the overlying Wildhorn nappe. However, W of the Rhône they show a marked discordance to the trends seen in the structurally underlying units of the Aiguilles Rouges and Mont Blanc massifs and in the intervening Martigny–Chamonix syncline. In fact the changes in trend leading to the fold arc bend in the Chaînes Subalpines only start to develop south of Bonneville (Fig. 3). A suggestion to account for this discordance in fold trends is that the structures developed in the Aiguilles Rouges and Mont Blanc massifs under previously folded rocks and that these later-formed tectonic features relate to more westerly oriented components of displacement than those responsible for the earlier foldings. This relationship is that proposed by Boyer & Elliott (1982): the time progressive propagation of upper crustal instabilities towards the orogenic foreland.

The continuity of fold structure seen in the Morcles nappe is also a feature of the Wildhorn nappe. Figure 2 shows that individual folds may be followed for distances of more than 35 km and, although not shown on these maps, correlatable folds continue for many tens of kilometres to the E with the same 60° hinge line trends. When this ETH research project started we were initially of the opinion that the fold pattern would be likely to show extremely complex geometric features with folds relaying each other in en-echelon pairs (Ramsay & Huber 1987, p. 313). We were greatly surprised to discover that this was not generally so and that most large folds persisted laterally in a manner more in accord with that seen in the Chaînes Subalpines and in the Jura fold belt. The lateral persistency is also generally associated in the Morcles, Diablerets and frontal zone of the Wildhorn nappes with a harmonic fold style; the axial surfaces of individual large folds can be followed with continuity through all rock strata irrespective of lithology. This style is particularly clear in the Morcles nappe where competent limestone layers lie close to each other and where a sideways deflection arising in one layer effectively transmitted into adjacent competent layers. In the Diablerets and Wildhorn nappes various lithological facies changes occur. These are principally a decrease in thickness of major competent layers (Schrattenkalk or Urgonian limestone, Valanginian Calcaire and Malm limestones) together with an increase in thickness of the interbedded incompetent shales and marls (Seewerschiefer, Drusbergschichten and Valanginian Schisteux) towards the S (see summaries in Heim 1921 and Collet 1927). These changes induce modifications in the fold style. Although the large scale folds represented in Fig. 2 remain harmonic, thinner competent units contained within the main incompetent groups become folded on two or more wavelengths (polyharmonic folding). In the most southern outcrops of the Wildhorn nappe the separation of the competent units is often so great that they fold with independent wavelengths in a disharmonic style (Ramsay 1981).

Returning to the lateral persistence of the folds, we must ask ourselves what this implies. Analysis of the thickness variations of the competent and incompetent beds in the folds strongly suggests an origin of the folds by buckling mechanisms, that is the sideways deflection of a mechanically unstable competent layer into neighbouring incompetent rock as a result of layer shortening. Only very rarely do folds show the features of fault bend folds enforced in a thrust sheet when it rides over an irregular ramp-flat morphology in its underlying thrust surface. The fold persistence and rather constant wavelengths (seen in Fig. 2 by the stippled-unstippled band widths) both suggest a certain degree of freedom of movement of the folding layers, and that the folds were not constrained by the forms of the nappe boundaries. The model which best fits all these features is that involving straining within the nappes at least in part by simple shear parallel to the nappe boundaries (Fig. 4a). This model accounts rather well for certain of the strain characteristics seen in the nappes as well as the changing fold forms (Ramsay *et al.* 1983, Ramsay & Huber 1987, p. 378). This model does not, however, account for the observed overall increase in strain within each nappe within its root zone, a feature best interpreted in terms of a wedge-like nappe geometry (Dietrich & Casey this volume). Certain modifications can be made of the simple model shown in Fig. 4 with respect to changes of dip of the shear zones or thrusts bounding the nappes. A ramp-like steepening of the thrust surface will induce a steeper than average dip in the transported ramp hanging wall sector and this orientation aligns the bed so that it is in a position to receive more than the average shortening along that layer (Fig. 4). The fold cascades seen in the frontal parts of the Diablerets and Wildhorn nappes south of the towns of Les Diablerets, Gstaad and Lauenen (Fig. 2) appear to be best interpreted in this way because they overlie thrust surfaces which are climbing more steeply than usual as they pass from relatively incompetent units (Dogger and Valanginian Schisteux) into more competent limestone-dominated parts of the succession (Kieselkalk, Valangian Calcaire, Schrattenkalk and Nummulitic limestones). The wavelengths of these cascade folds have nothing to do with the ramp length but appear to be controlled by rock lithology and bed spacing. The simple-shear dominated

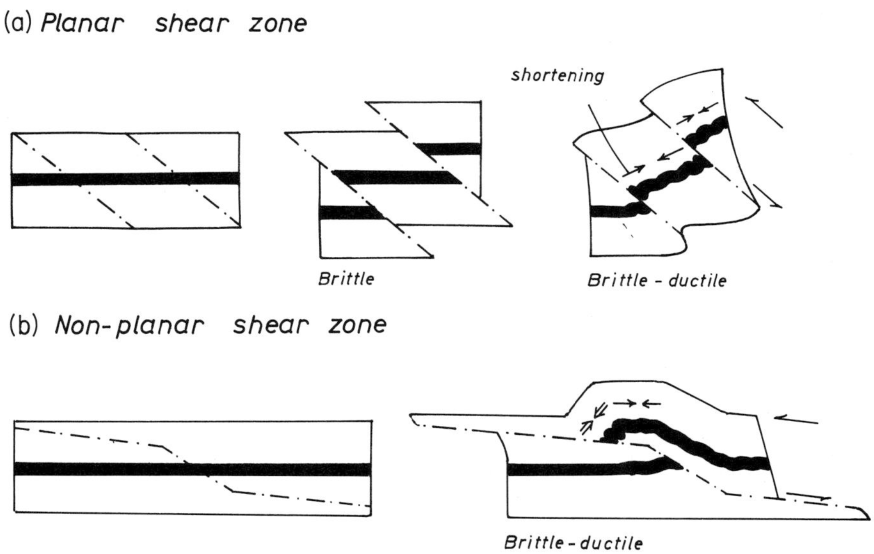

FIG. 4. Schematic diagram to illustrate the development of layer shortening and the formation of buckle folds as a result of shear strains within a nappe. A indicates a scheme where the bounding surfaces of the nappes are constantly oriented, B illustrates the differential shortening and fold formation where these boundaries are not planar.

strains in the models of Fig. 4 control the orientations of the fold axial surfaces. It should be clear why the axial surfaces are obliquely inclined to the nappe boundaries and why this inclination is least in the regions of maximum shear strain generally located along the nappe boundaries (Fig. 4).

Thrust geometry

All the nappe structures of the region of Fig. 1 are bounded by thrust faults which cut up section from SE to NW in the general direction of transport. Although the separation of the Diablerets and Wildhorn nappes is a thrust in the frontal part of the nappes, when this contact is followed to the SE to a point about 8 km NW of Sion the thrust is replaced by a synclinal fold, the Synclinal de Raccord (Lugeon 1914, Dietrich & Casey this volume) and the overlying reversed limb of this fold takes the position of the lower part of the Wildhorn nappe. The Morcles and Doldenhorn nappes are also rather special in that they show the structure of large recumbent folds with extensive inverted limbs lying directly on a basal thrust.

The nappes and their basal thrust planes are arched into an antiformal structure (Helvetic arch in Fig. 3), the crest of which coincides with the continuation of the crestal line of the Aiguilles Rouges dome. The Morcles nappe situated on the north west side of the Aiguilles Rouges massif shows downward facing folds. These folds become recumbent and then upward-facing when traced towards the NW, a variation which defines an open synform, the Salanfe synform. The trough line of this synform trends NE–SW parallel to the crestal line of the Aiguilles Rouges dome, making an angle of 25° with the hinge lines of the folds within the Morcles nappe.

The traces of the major thrusts are shown in Figs 2 and 3 together with other thrusts which, from the displacement of marker beds, have displacements of less than 4 km (e.g. the Ardon slice). The Morcles nappe between the Arve and Rhône can be divided into three distinct structural units. The lower part (Morcles in Fig. 1) appears to be a homologue of the main Morcles nappe in its type locality on the NE side of the Rhône. The upper limb of the great recumbent fold exposed in the cliff face of the Tour Sallière SW of Salanfe is overthrust by a well developed thrust sheet (the Samoëns thrust sheet–'Sam' in Fig. 3), a unit which also forms an impressive klippe on the summit of the Pic de Tenneverge (Fig. 3). NE of Samoëns the displacement on this thrust is 3 km. The Samoëns sheet is cut by the thrust at the base of the Ultrahelvetic nappes but reappears on the W side of the Flaine depression in the Arve valley SE of Cluses with a much reduced displacement of 0.5 km dying to zero in a westerly direction. The Samoëns sheet is itself overthrust by a higher element (the Sallanches thrust sheet, 'Sal' in Fig. 3) as it is very well exposed in the mountain sides NW and NE of that town. The displacement is about 1 km decreasing to zero westwards. The question that immediately arises is, what is the significance of these thrust sheets and what are their structural homologues, if any, on the NE side of the Rhône culmination? The Diablerets and Wildhorn nappes show much larger displacements than do the Samoëns and Sallanches thrust sheets (13 and 15 km respectively). However, if the increasing displacement gradient which can be measured on the Samoëns thrust from W to E is continued towards the NE to where the Diablerets nappe is exposed, a figure of 12 km displacement would be expected, a number very close to that actually observed in the Diablerets nappe. Although not too much emphasis should be placed on correlations of nappes based on stratigraphic resemblances (because of the unlikely prospect of exact parallelism of palinspastic facies lines and tectonic lines) it is striking that the very peculiar facies of the Eocene Taveyannaz sandstone, with its high content of basic volcanic debris, characterises both the Diablerets nappe and the Samoëns sheet to the N and NW of Sallanches. Because of the clear syntectonic origin of this deposit (late Eocene to early Oligocene: Trümpy 1980) there is, perhaps, a better likelihood of its isofacies lines coinciding with tectonic directions.

The thrust planes of the Helvetic nappes are generally discordant to the stratigraphy and cross-cut successive marker beds in cut-off lines. The separation distance between marker-equivalent cut-off lines in the hanging wall and foot wall of a thrust measured in the direction of the total transport movement vector gives the value of the total displacement across the fault (Dahlstrom 1969, Elliott & Johnson 1980). Unfortunately, like fold hinge lines, their exact three dimensional orientation is not always easy to establish. In the field it is possible to map or construct the point where the cut-off lines passes through the ground surface, which will be termed the cut-off point. Figure 3 shows an analysis of these points in the form of *cut-off point tie lines*. These lines represent the average trend of the cut-off lines between any two adjacent points (Fig. 3). A remarkable feature of these tie lines is that they are without exception, very closely aligned with the hinge point tie lines and extremely rectilinear. These features strongly suggest that the folds and thrusts are geometrically and kinematically (and therefore dynamically) related and that the development of the nappes was on frontal ramp situations. Nowhere did this analysis reveal oblique or lateral ramp geometry (Boyer & Elliott 1982, Ramsay & Huber 1987, p. 527), a point which is very important in the interpretation of the significance of the culminations and depressions to be made later. A rather remarkable coincidence of tie line in the cut-off of the upper limit of the Kieselkalk formation in the footwall of the Wildhorn thrust with the fold hinge tie lines in that nappe exists. Like the long distance, fold tie line correlation made in the Morcles and Doldenhorn nappes, this may be only a coincidence but, if correct, it supports the correlation of the Diablerets and Gellihorn nappes.

Tectonic synthesis

An analysis of the geometric data set out above leads to the following tectonic development sequence for this part of the Alpine arc (Fig. 5).

Stage 1 Emplacement of the Ultrahelvetic nappes on an unfolded and undisplaced European foreland now occupied by the Helvetic nappes and their underlying autochthon. Over most of the present area of the Wildhorn nappe and over all of the Diablerets and Morcles nappes and the autochthon the Ultrahelvetic nappes rest directly on Tertiary rocks, a feature which shows that the inclination of the basal thrust surface of the Ultrahelvetic nappes must have been subhorizontal over a transport-directed distance of at least 50 km. Such a relationship is more in accord with the structures seen today at accretionary plate margins than that of Rocky Mountain type foreland thrusting and is suggestive of a high fluid content in the overriding nappes. Many of the rather chaotic fault and boudinage structures within the Ultrahelvetic nappes seem to support this contention. The earliest formed crystal fibres around pyrite pressure shadows in these rocks are aligned NW–SE, while the geometry of imbricate thrust faults in these nappes situated within the large synform between the Morcles and Diablerets is consistent with a WNW direction of shear. However, it is not at all certain if these imbricate structures were contemporaneous with the main nappe transport and it seems probable that they were imprinted much later in the tectonic history when the Ultrahelvetic nappes had already arrived on their Helvetic base.

Stage 2 Beginning of the development of the Helvetic nappes. Folding began in the Wildhorn, Diablerets and normal limb of the Morcles nappe in previously unfolded strata probably along E–W hinge lines. Incremental strains measured from first formed crystal fibres around pyrite pressure shadows indicate a N to NNW directed transport direction sub-perpendicular to the fold hinges (Durney & Ramsay 1973, Dietrich & Durney 1986, Dietrich & Casey this volume). It seems probable that the frontal and upper limb folds of the Morcles nappe were initiated at this time, but that the NNW shortening was less than that seen in the Swiss sector. The Morcles equivalent folds in the Chaînes Subalpines are thought to have been initiated at this time as their trends are practically identical. Probably these folds terminated in unfolded strata somewhere south of Bonneville at the position where we now see the main 60° trends changing towards a more southerly trend.

Stage 3 Change in the transport direction of the Helvetic nappes to a more NW orientation together with the overthrusting of the Mont Blanc massif and formation of the lower limb of the Morcles nappe. A major ductile shear zone affecting the northwestern front of the Mont Blanc massif passes into the lower limb of the major recumbent fold of the Morcles nappe (Fig. 5, Stage 3A). The finite strains are extremely high in this zone and show regularly oriented SE-plunging directions of maximum elongation (Ramsay and Huber 1983, p. 204, Dietrich & Durney 1986, Dietrich & Casey this volume) which are not perpendicular to the hinge lines of the folds (Fig. 3). The form of the strain ellipsoids in this shear zone accord well with having been formed under conditions of plane strain simple shear without volume loss (with $(1+e_1)\ (1+e_3) = (1+e_2)^2$). Except for a small late extension parallel to the fold axes, all the strain increments have their maximum stretch directions oriented to the SE. Apart from some local deviations of this strain pattern caused by later sub-horizontal shear displacements (see stage 4) this strain pattern is also characteristic of the Mesozoic rocks within the Martigny–Chamonix syncline (Fig. 3). These data are interpreted as being the result of the northward and upward movement of the Mont Blanc massif under the already formed Diablerets and Wildhorn nappes and their previously emplaced Ultrahelvetic cover, together with the main NW transport of the Morcles nappe. In the French sector, particularly in the

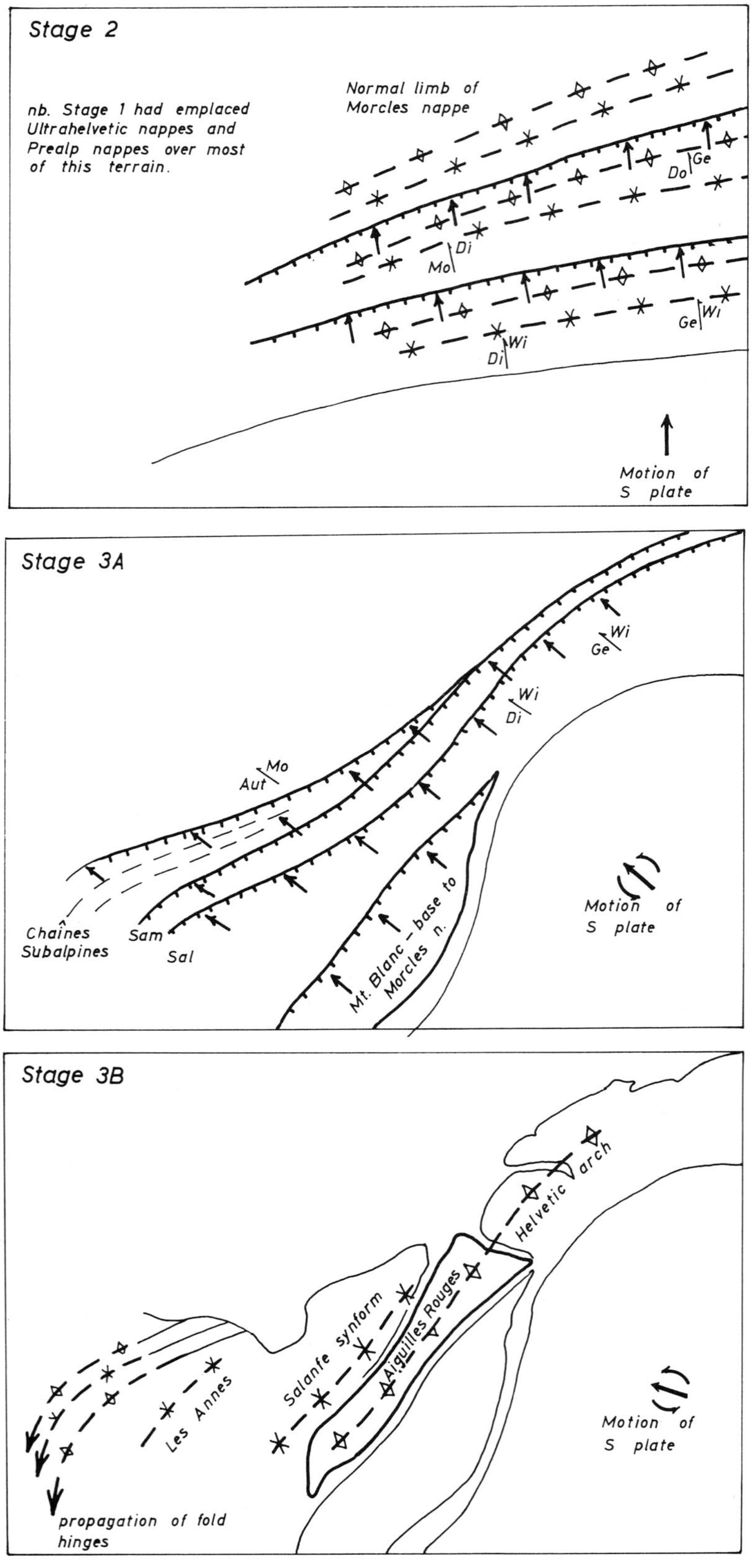
Stage 2
nb. Stage 1 had emplaced Ultrahelvetic nappes and Prealp nappes over most of this terrain.
Normal limb of Morcles nappe
Do
Ge
Mo
Di
Di
Wi
Ge
Wi
Motion of S plate
Stage 3A
Ge
Wi
Di
Wi
Aut
Mo
Chaînes Subalpines
Sam
Sal
Mt. Blanc - base to Morcles n.
Motion of S plate
Stage 3B
Helvetic arch
Aiguilles Rouges
Salante synform
Les Annes
propagation of fold hinges
Motion of S plate

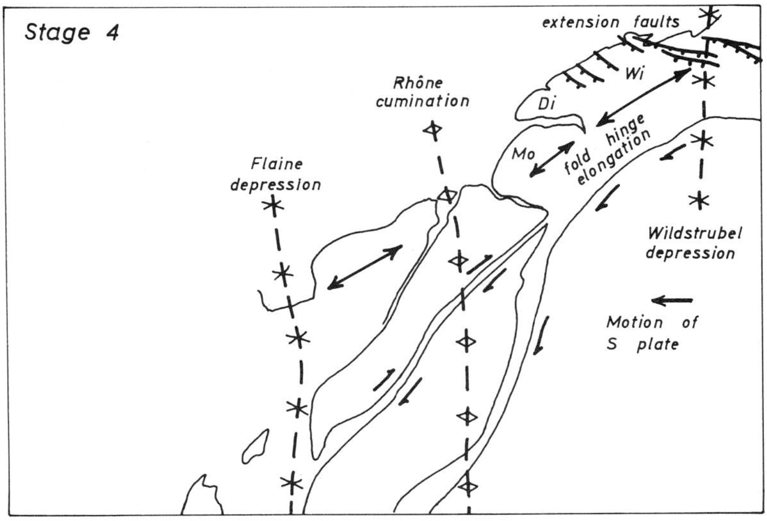

FIG. 5. Schematic diagrams illustrating some of the developmental stages during the formation of the folds and thrusts in the Helvetic nappes and Chaînes Subalpines. Aut — autochthon, Do — Doldenhorn, Di — Diablerets, Ge — Gellihorn, Mo — Morcles, Sal — Sallanches, Sam — Samoëns, and Wi — Wildhorn nappes respectively.

middle and lower Jurassic rocks in the southern part of the Chaînes Subalpines between Sallanches and Flumet, the strata also suffered a NW-directed shear strain. The strains set up during this deformation show similar features to those seen in the Morcles nappe in that the directions of maximum finite stretching are aligned at about 80° to the fold hinge lines and to the cut-off lines of the previously formed thrusts (Fig. 3). During this stage of development it seems highly probable that the terminated ends of the major folds of the Chaînes Subalpines propagated in a SW direction to begin to form the arcuate form seen today in the region NW of Thônes (Fig. 3).

Later during the development of the period of NW–SE shortening, the uplift of the Aiguilles Rouges massif probably occurred because the crestal trend of this dome structure coincides with that of the Mont Blanc massif (Fig. 5, Stage 3B). This uplift led to the formation of the Helvetic arch and the synformal form of the Morcles nappe seen in the Salanfe synform NW of the Aiguilles Rouges massif. The axial trend of the klippe of Les Annes also has a NE–SW orientation that might have formed at this time.

Stage 4 Movement of the SE Alpine block relative to the European foreland in a western direction. In the western Alps many of the fold and thrust trends appear to be related to an overall westward movement of the Penninic zone over the foreland. This displacement led to a strong right-handed strike-slip motion concentrated in the Helvetic root zones and along the Penninic front. Evidence for this motion can be seen in the form of sub-horizontal late stage striae and shear fibre lineations (Ramsay & Huber 1983, pp. 257–261) in the Helvetic root zone and in the Martigny–Chamonix syncline. These displacements led to an extension along the hinge lines of the folds in the Helvetic nappes (Durney & Ramsay 1973, Ramsay 1981, Dietrich & Durney 1986, Dietrich & Casey this volume). The right-handed shear motion also appears to have been responsible for the formation of the late stage southward-dipping extension faults which cut the Helvetic folds and which are especially well developed in the Wildstrubel depression between Lauenen and Leukerbad (Fig. 2).

It has been noted earlier that the major culminations and depressions of the nappe folds appear to affect all structural units, the individual units rising or falling together in the N–S oriented antiforms and synforms. The culminations and depressions neither appear to be related to the development of oblique and lateral ramps (Ramsay & Huber 1987, p. 527) nor to the stacking of duplexes beneath the major thrusts (Boyer & Elliott 1982). These structures seem best explained as being formed during an E–W shortening superposed across all the previously formed structures. As a speculation one might ask the question of whether there is any deep level correspondence between these structures seen in the frontal and central parts of the Alps with those seen in the foreland: are there correlations to be made between the high of the Rhône culmination and Vosges dome, the Wildstrübel depression and the Rhine graben and the Aar–Gotthard–Ticino culmination with the Schwarzwald?

Conclusions

It has been shown in this paper how the structure of the Helvetic nappes and Chaînes Subalpines can be analysed particularly effectively. Although some of the data shown in Figs 2 and 3 may be obtained from examination of published maps there are many localities where these constructions can only be carried out with new field data. Although these data are impartial and their interpretation is open to discussion, in this paper I have presented what I believed to be the most reasonable sequence of events, one which requires a progressively changing motion between the colliding continental plates from N through NW to conclude with a westward motion of the southern plate relative to the European continent.

ACKNOWLEDGEMENTS: The terrain chosen for this discussion was mapped by myself during the course of the ETH Helvetics research project (Nationalfonds Projekt 2.859–0.77). Although I alone am responsible for the interpretation set out here I would like to thank my colleagues Martin Casey, Dorothee Dietrich, David Dumey, Roy Kligfield, Neil Mancktelow, Adrian Pfiffner and Stefan Schmid for their close association with the work and for helpful discussion. I thank John Platt and an anonymous reviewer for their constructive comments. Support from Swiss Nationalfonds Project 2.214–0.86 and 4.963–0.85.20 is gratefully acknowledged.

References

ARBENZ, P. 1912. La structure des Alpes de la Suisse centrale. *Archives des Sciences Physiques et Naturelles*, **34**, 401–25.

ARBENZ, P. 1913. Die Faltenbogen der Zentral-und Ostschweiz. *Vierteljahresschrift der naturforschenden Geselleschaft Zürich*, **58**, 16–34.

ARGAND, E. 1911. Les nappes de recouvrement des Alpes occidentales. *Matériaux pour la carte géologique Suisse no.* **27** *et carte speciale no.* **64**.

BADOUX, H. 1972. Tectonique de la nappe de Morcles entre Rhône et Lizerne. *Matériaux pour la carte géologique de la Suisse, nouvelle Série*, **143**.

BERTRAND, M. 1884. Rapports de structure des Alpes de Glaris et du Bassin houiller du nord. *Bulletin de la Société Géológique de France*, **12**, 318–30.

BOYER, S. C. & Elliot, D. 1982. Thrust systems. *Bulletin of the American Association of Petroleum Geologists*, **66**, 1196–230.

CALLAWAY, C. 1883. The age of the newer gneissic rocks of the northern Highlands. *Quarterly Journal of the Geological Society of London*, **39**, 355–422.

COLLET, L. 1927. *The Structure of the Alps*. Arnold & Co., London.

COLLET, L. 1927. La nappe de Morcles entre Arve et Rhône. *Matériaux pour la Carte géologique de la Suisse*, **79**.

DAHLSTROM, C. D. A. 1969. Balanced cross sections. *Canadian Journal of Earth Sciences*, **6**, 743–7.

DIETRICH, D. & DURNEY, D. 1986. Change of direction of overthrust shear in the Helvetic nappes of western Switzerland. *Journal of Structural Geology*, **8**, 389–398.

DURNEY, D. 1972. *Deformation history of the western Helvetic nappes Valais. Switzerland.* Unpublished PhD thesis, University of London.

DURNEY, D. & RAMSAY, J. G. 1973. Incremental strains measured by syntectonic crystal growths. *In*: DEJONG, K. & SCHOLTEN, R. (eds). *Gravity and Tectonics*, 67–96, Wiley, New York.

ELLIOTT, D. & JOHNSON, M. R. W. 1980. Structural evolution in the northern part of the Moine thrust belt, NW Scotland. *Transactions of the Royal Society of Edinburgh*, **71**, 69–96.

ESCHER VON DER LINDT, A. 1841. Geologische Karte des Cantons Glarus und seiner Umgebungen, nebst Profilen. *Verhandlungen der Schweizerischen Naturforschenden Gesellschaft Zürich*.

FERRAZZINI, B. & SCHULER, P. 1979. Eine Abwicklungskarte des Helvetikums zwischen Rhone und Reuss. *Eclogae Geologicae Helvetiae*, **72**, 439–54.

HEIM, A. 1921. *Geologie der Schweiz*, **2/1**, 259–476. Tauchnitz, Leipzig.

HSÜ, K. 1979. Thin skinned plate tectonics during neo-Alpine orogenesis. *American Journal of Science*, **279**, 353–66.

HUGGENBERGER, P. 1985. *Faltenmodelle und Verformungsverteilung in Deckenstrukturen am Beispiel der Morcles Decke (Helvetikum der Westschweiz)*. Unpublished Dissertion, ETH-Zürich, **7861**, 1–326.

LANGENBERG, W., CHARLESWORTH, H. & LA RIVIERE, A. 1987. Computer constructed cross sections of the Morcles nappe. *Eclogae Geologicae Helvetiae* **80**, 655–67.

LUGEON, M. 1901. Les grandes nappes de recouvrement des Alpes du Chablais et de la Suisse. *Bulletin de la Société géologique de France*, **1**, 723–825.

LUGEON, N. 1914. Les Hautes Alpes calcaires entre la Lizerne et la Kander. *Matériaux pour la Carte géologique de la Suisse*, **30**.

PEACH, B. N. & HORNE, J. 1888. Report on the recent

work of the Geological Survey in the North West Highlands of Scotland. *Quarterly Journal of the Geological Society of London*, **44**, 378–441.

RAMSAY, J. G. 1981. Tectonics of the Helvetic Alps. *In*: MCCLAY, K. R. & PRICE, N. J. (eds). *Thrust and Nappe Tectonics*, 293–309, Special Publications of the Geological Society no. 9.

RAMSAY, J. G., CASEY, M. & KLIGFIELD, R. 1983. Role of shear in development of the Helvetic fold-thrust belt of Switzerland. *Geology*, **11**, 439–42.

RAMSAY, J. G. & HUBER, M. 1983. *The Techniques of Modern Structural Geology*, **1**, 1–307. Academic Press, London.

RAMSAY, J. G. & HUBER, M. 1987. *The Techniques of Modern Structural Geology*, **2**, 308–700. Academic Press, London.

TÖRNEBOHM, A. E. 1883. Om Dalformationens geologiska Alder. *Geologiska Föreningens Stockholm Förhandlingar*, **6**, 622.

—— 1888. Om Fjällproblemet. *Geologiska Föreningens Stockholm Förhandlingar,* **10**, 328.

TRÜMPY, R. 1980. *Geology of Switzerland, a Guide-Book. Part A, an outline of the Geology of Switzerland*. Wepf & Co., Basel.

J. G. RAMSAY, Geologisches Institut, ETH Zentrum, CH-8092 Zürich, Switzerland.

A new tectonic model for the Helvetic nappes

D. Dietrich & M. Casey

SUMMARY: A tectonic model must be based on geometrical information of both large and small scales. A new model of the Helvetic nappes of western Switzerland has been developed using constraints from:

(1) The geometric shape of the nappes. Accurate profiles of the Morcles, Diablerets and Wildhorn nappes show that all three nappes are wedge-shaped and that the axial planes of the major nappe internal folds converge towards the root zone.
(2) The strain state of the nappes. All the existing pressure shadow fibre data of the region are reviewed and reveal the existence of very steep gradients of increasing strain towards the root zone of each individual nappe.
(3) The metamorphism of the nappes. A compilation of illite crystallinity data indicates the existence of a vertical gradient of increasing metamorphism, but as well of a horizontal gradient of metamorphism increasing towards the root zone.

A geometrical and finite element compatibility model is presented which best accounts for these data. This model shows the following geometrical features:

(1) Combinations of simple shear and pure shear deformation to account for the horizontal strain- and metamorphic gradients.
(2) The heterogeneous simple shear part of the deformation is constant in magnitude from the root zone to the external part of the nappes.
(3) The pure shear part of the deformation decreases towards the external part of the nappes.
(4) Varying amounts of overthrust displacement are taken up by distributed ductile shear in the internal part of the nappes and by thrust faulting in the external part.

This model has kinematic implications. We envisage tectonic extrusion from the root zone to be responsible for the pure shear component of the deformation. The nappes formed sequentially being extruded between the underthrusting basement and the back-folded Gotthard massif.

Why a new tectonic model for the Helvetic nappes?

The Helvetic chain of Switzerland is made up by the Ultrahelvetic and Helvetic sedimentary nappes overlying the External Crystalline 'Massifs' and their 'autochthonous' cover. (The 'massifs' are formed by Hercynian basement strongly reactivated during the Alpine deformation, and their 'autochthonous' cover is generally very highly strained.)

The distribution of the internal deformation in a geological body is determined by the forces acting on this body. Every tectonic process occurs in the Earth's gravity field, but deformation patterns created by dominant gravitational body forces differ from patterns created by gravitational body forces which are small compared to boundary forces acting on the body under consideration, as, for example, forces causing overthrust shear.

For parts of the Helvetic chain, gravity sliding or gravity spreading models have been proposed. In eastern Switzerland Trümpy (1969, 1973) and Milnes & Pfiffner (1977) suggested gravity sliding for the emplacement of the south Helvetic and Ultrahelvetic units; in western Switzerland Lugeon (1943) and most later authors explained the 'mise en place' and related 'diverticulation' of the Ultrahelvetic units by gravity. Merle (1986) suggested a gravity spreading model for the Morcles nappe. On the other hand the totally opposed view of the Helvetic chain as a thin-skinned thrust belt has been put forward (Boyer & Elliott 1982, Groshong *et al.* 1984, Butler 1985). The latter authors assert that a top to bottom sequence of thrusts climbing over ramps would produce the geometric features seen in the individual nappes.

The importance of the facies distribution on the Helvetic shelf in determining the nappe development has been emphasized by Trümpy (1963). Already the early geologists had realised that there is a systematic facies change in the nappes with the upper nappes being of a more internal provenance than the lower ones. They also noticed that the isofacies lines are

From: COWARD, M. P., DIETRICH, D. & PARK, R. G. (eds), 1989, *Alpine Tectonics*, Geological Society Special Publication No. 45, pp. 47–63.

not parallel to the strike of the nappes (e.g. Arnold Heim in Albert Heim 1921). According to Trümpy (1963), in central and eastern Switzerland there are no anhydrite layers in the Triassic Aar Massif cover sediments, the Lower Jurassic is absent and the Middle Jurassic is very thin. In western Switzerland there are at least in some localities Triassic anhydrite layers, and in the Morcles nappe there exist well developed Lower and Middle Jurassic sediments. Trümpy proposed that these facies changes could determine the change in tectonic style of the Helvetic chain: the huge Morcles–Doldenhorn fold nappe, the lowermost nappe in western Switzerland, has no equivalent of similar dimensions in central and eastern Switzerland. In western Switzerland this nappe shows the same overall geometric shape on both sides of the Wildstrubel depression, i.e. over a distance of more than 60 km along strike (Arbenz in Heim 1921, and Fig. 1). Towards the E, in the Bernese Oberland, the fold nappe passes into a series of tight synclines pinched into the Gastern Massif, between the Gastern and the Aar Massif, and into the Aar Massif (e.g. the sections by Herb in Masson *et al.* 1980). This structural style is found eastwards all along the Aar Massif, with the local development of folds tectonically equivalent to the Morcles–Doldenhorn fold, but of a much smaller size (e.g. the Windgällen fold in central Switzerland, the Sandalp fold in eastern Switzerland, sections 3 and 4 by Heim 1921).

We would like to show in the following that the Helvetic structures can be best explained by overthrust shear. The internal deformation of the nappes cannot be reconciled with pure gravity sliding or spreading. Gravity sliding would only involve friction-induced internal strains, as, for example, locally developed layer-parallel extension and compression or internal deformation of layers close to the gliding surface. Gravity spreading would require an increase in net overthrust displacement from the internal to the external part of the nappes, which is clearly not observed. On the other hand, the Helvetic lineation- and fold-patterns are certainly not a consequence of thrust faulting alone. We restrict our detailed discussion to the Helvetic nappes of western Switzerland, Fig. 1, because for this region we have a large number of small scale structural observations at our disposal, but we think that our conclusions could as well be applied to the Helvetic nappes of eastern Switzerland. In contrast to many previous workers we would like to emphasize that small scale structural information is critical for the formulation of the characteristics of a tectonic model. Such data alone allow the choice of one out of many theoretically possible paths which could have led to the finite deformation. A scenario for the Helvetic overthrust shear deformation has been depicted by Laubscher (1983). The mechanically low-viscous Helvetic shelf sediments are deformed between a cold, rigid orogenic lid, represented by the already overthrust Pennine nappes, and the warm, but highly viscous, basement.

The disappearance of the Morcles–Doldenhorn nappe towards the E is not related to a facies change alone. Dietrich and Durney (1986) concluded that the inverted limb of this fold nappe was formed during a regional change of direction of overthrust shear. The overthrust shear directions for the upper nappes and for the normal limb of the Morcles–Doldenhorn nappe were towards N or NNW, whereas they were towards NW or WNW for the inverted limb of the Morcles–Doldenhorn nappe. This latest phase is not observed in central and eastern Switzerland, indicating that the zone of active thrusting moved westward during the Helvetic compression history. The change of direction of overthrust shear is in kinematic accordance with the movement of the Simplon hanging wall fault block determined, for example, by Mancktelow (1985). Gravity collapse of the Paleogene Alps and associated sidewards mass escape (change of direction of overthrust shear in the Helvetics) appear in this region to conclude the compression history.

In the following we discuss three basic sources of data for the tectonic model, the geometric shape of the nappes, the strain state of the layered rock sequence involved in the nappes, and the metamorphism. The first and the second items are linked, since the shape of the nappes is an expression of their state of strain; the strain state and metamorphism also appear to be linked.

The profile shape of the Helvetic nappes of western Switzerland

Axial depressions and culminations are a characteristic feature of the Helvetic chain. All the available data containing information on the plunge of the fold axes, that is bedding/cleavage intersection lineations measured regionally, and axial plunge values calculated from map patterns of folds, were projected on the average strike line of the Helvetic units. The individual plunge values were then integrated in an axial plunge diagram, Fig. 2. This diagram shows very

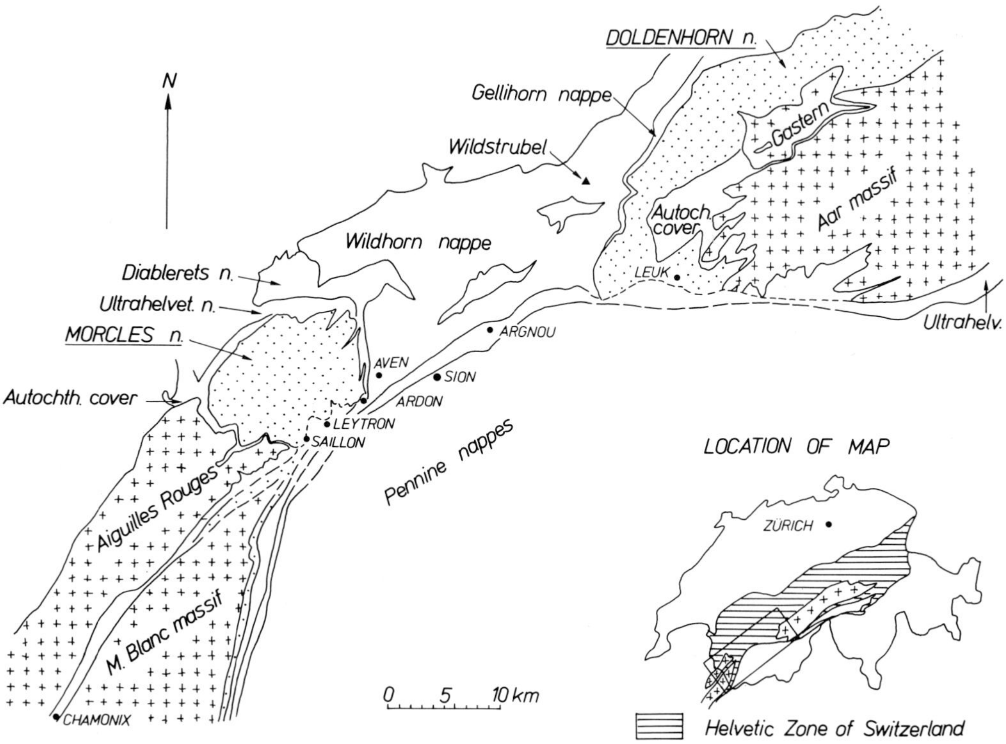

FIG. 1. A tectonic map of the Helvetic nappes of western Switzerland.

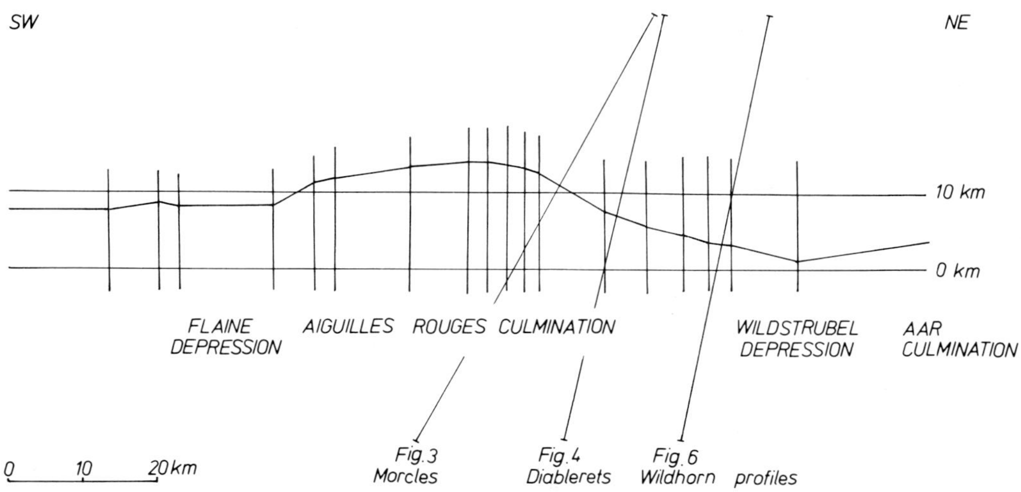

FIG. 2. An axial plunge diagram of the Helvetic chain of western Switzerland and of the French Chaînes Subalpines. The construction is based on over 200 data consisting of constructed trends/plunges of fold axes and of measured bedding/cleavage intersection lineations. The planes of projection of the Morcles, the Diablerets and the Wildhorn profiles are approximately indicated.

clearly, from W to E, the Flaine depression, the Aiguilles Rouges culmination, the Wildstrubel depression and the Aar culmination. The basement structures responsible for this configuration were formed late in the geological history; they deform the pre-existing folds and thrusts of the chain. This is seen from the absence of major sidewall ramps in the Helvetics, from the great lateral continuity of the individual fold structures and from the lateral continuity of the nappes (J. G. Ramsay, this volume). Lateral continuity of folds and thrusts would not be expected across e.g. a pre-existing Wildstrubel depression. The basement structures are, according to J. G. Ramsay (this volume), obliquely superimposed on the already emplaced nappes, in agreement with the change of direction of overthrust shear postulated by Dietrich & Durney (1986). The scale of the basement structures which give rise to the culminations is greater than the scale of the overlying nappes. The straightforward continuity of overthrust shear deformation passing in time from the cover to the basement has therefore to be questioned.

It can be seen from the high axial plunge values in Fig. 2, that the Helvetic nappes have to be represented in fold axis-perpendicular profiles instead of in vertical sections. The planes for the projection of the profiles representing the Morcles, the Diablerets and the Wildhorn nappes were chosen in principle perpendicular to the axial plunge diagram of Fig. 2.

For the *Morcles nappe*, Fig. 3, the strike of the profile plane lies between the northwards stretching direction of the normal limb and the WNW-wards stretching direction of the inverted limb. This compromise produced notwithstanding a reasonable section. The most prominent feature of the Morcles profile is the overall wedge shape of the fold nappe. The thickness of the nappe is less than 3 km at its most southern outcropping part, the so-called root zone, but more than 6 km at the northern front. The Upper Jurassic Malm limestones, the thickest of the competent layers which forms the skeleton of the nappe, shows a tremendous thickness variation from the front towards the root zone, but also from the normal limb towards the inverted limb. This thickness variation is much greater than the possible original stratigraphic thickness variation of the Morcles depositional realm pointed out by Huggenberger in Lemoine & Trümpy (1987, fig. 2). All the folds seen in the profile are first-phase folds caused by the main internal deformation during the emplacement of the nappe. The second-phase folds, very frequent on both limbs in the root zone and related to a late compressional event, are too small to be represented. The first-phase folds are tight on the upper limb, wider on the front and again extremely tight, and of opposed vergence, on the lower limb of the nappe. The fold axial planes converge towards the root zone. These geometric features are in accordance with the existence of high strain corresponding with the two limbs of the nappe. The particularly evident thickness variation of the Upper Jurassic limestones from the front to the root zone is therefore due to deformation.

In the *Diablerets nappe*, Fig. 4, quite strong variations of the fold axis trends and plunges were found. These variations are of local significance only, but they made it impossible to project the nappe on one plane without introducing in the section projection artefacts not observed in nature. The front, the central part and the root zone of the Diablerets nappe were projected according to the fold axis data, Fig. 5, on three different planes. The three projections had then to be fitted together manually. The profile of this nappe, like that of the Morcles nappe, appears wedge shaped. The Diablerets nappe is not a fold nappe like the Morcles nappe: the inverted limb is cut off by the thrust fault at the base of the nappe. The folds in the Middle Cretaceous Urgonian limestone are disharmonic relative to those in the Upper Jurassic Malm limestone. The fold axial planes converge towards the root zone. There exists a tectonically induced thickness variation in the Upper Jurassic Malm limestones from the central part of the nappe towards the root zone. The 'synclinal de raccord' between the Diablerets and the Wildhorn nappes shows an extremely elongate hinge zone in the competent Kieselkalk. The same geometry is seen in the next following anticline in the competent Urgonian limestone. These features point towards a high-strain zone close to the thrust contact of the overlying Wildhorn nappe. The shearing off of the inverted limb did occur in a high-strain zone at the base of the nappe. The two high-strain zones converge towards the S, concomitant with an increase in strain towards the root zone.

For the *Wildhorn nappe*, Fig. 6, two different projection planes had to be used for the frontal part and for the central and root zone parts respectively. A problem in constructing the profile were the normal faults which strike obliquely to the chosen projection planes. These faults appear late in the deformation history. The representation of the faults in the profile reflects the topography they meet. The thrust planes which delimit this nappe converge to-

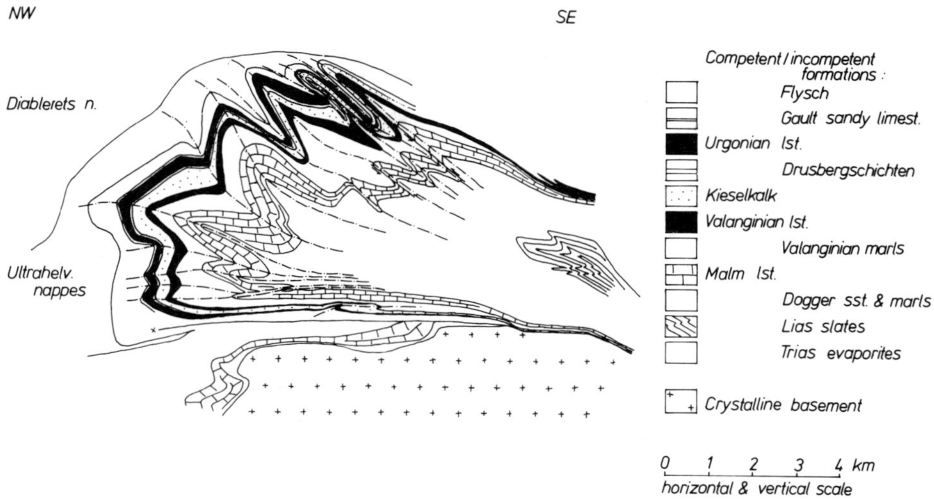

FIG. 3. A profile (combined down- and up-plunge projection) through the Morcles nappe. The plane of projection strikes N 330° and dips 60° SW. This profile is modified from Dietrich & Song (1984). Independently produced profiles of the Morcles nappe by Huggenberger (1985) and by Langenberg *et al.* (1987) show very similar features.

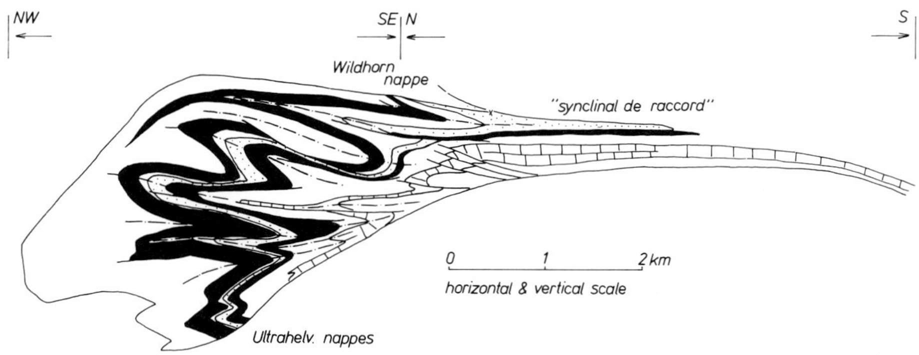

FIG. 4. A profile (combined down- and up-plunge projection) through the Diablerets nappe. According to the data of Fig. 5 three projection planes were used: for the front of the nappe a plane striking N 315/75° SW; for the central part a plane N 360/75° W; for the root zone a plane N 360/80° W. The symbols are as in Fig. 3.

wards the S. The Wildhorn nappe has an inverted limb, very clearly developed in the Upper Jurassic Malm limestones. The folding style of this nappe is highly disharmonic due to a gradual increase in thickness of the incompetent beds, the nappe being of a more internal provenance than the Morcles and the Diablerets nappes (J.G. Ramsay 1981, and this volume). Between the folds in the Upper Jurassic Malm limestones and the folds in the Cretaceous competent beds (Valanginian limestones, Kieselkalk and Urgonian limestones) lies a thick mass of incompetent Valanginian marls. The architecture of the nappe is complicated by the presence of several internal thrust planes in the Upper Jurassic Malm limestones and in the frontal folds. These thrusts cannot be followed through the Valanginian marls. The thinning of the competent beds towards the S, and the convergence of the fold axial planes, point towards the existence of a highly strained root zone, as seen for the Morcles and the Diablerets nappes.

The Ultrahelvetic sediments overlying the Wildhorn nappe show in the profile superimposed folds. These sediments were first folded independently of the Wildhorn deformation and then later refolded, being infolded between the Helvetic nappes.

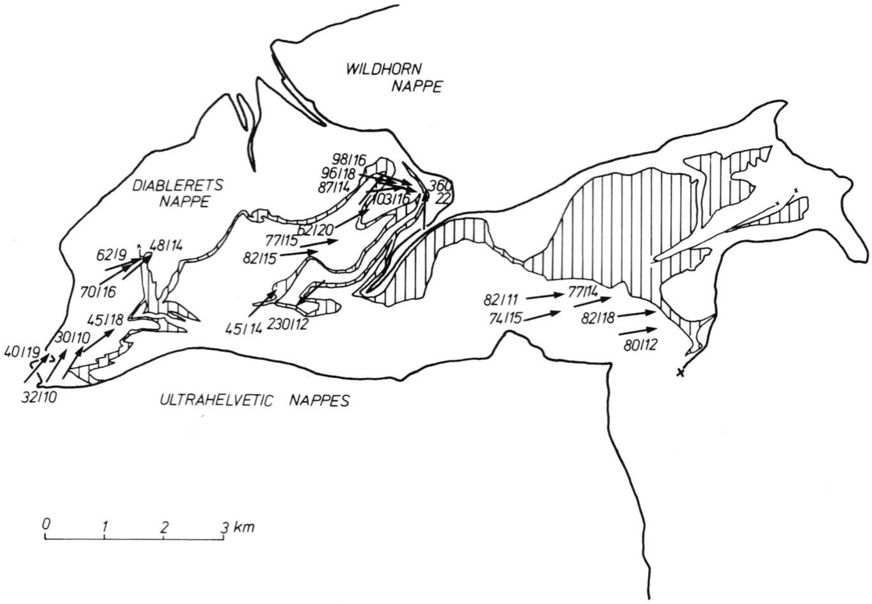

FIG 5. A map of the Diablerets nappe showing the variations in trend and plunge of the bedding/cleavage intersection lineations. The Urgonian limestone is indicated as a reference horizon.

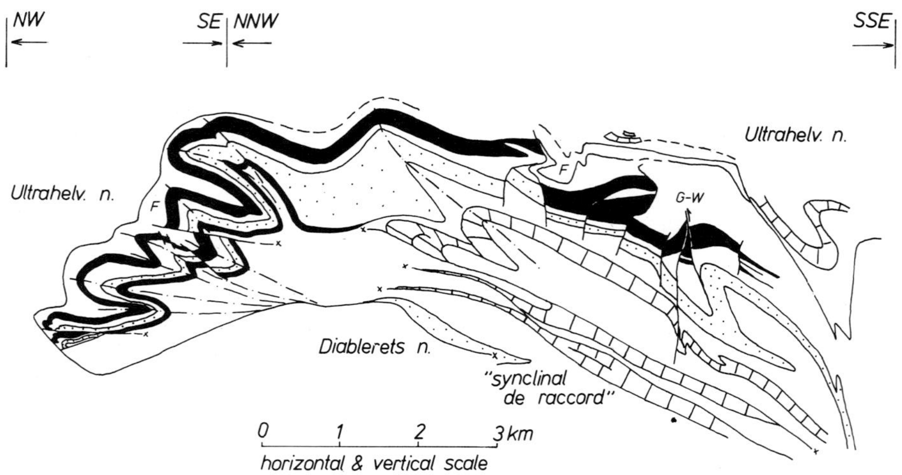

FIG. 6. A profile (essentially a down-plunge projection) through the Wildhorn nappe. Two projection planes were used: for the front of the nappe the plane N 340/78° SW; for the central part and the root zone the plane N 320/78° SW. F: flysch. G–W: Gault to Wangschichten. Other symbols as in Fig. 3.

The strain state of the Helvetic rocks

The Helvetic shelf sediments form a multilayer sequence. Competent layers, mostly limestones, alternate with incompetent layers, mostly shales and marls. Overthrust shear will deform such an anisotropic sequence very heterogeneously. A further complication of the deformation history of this region is represented by the change of direction of overthrust shear during the nappe formation (Dietrich & Durney 1986). The strain pattern is therefore expected to be complex.

There is a variety of strain markers in the Helvetic nappes. In western Switzerland the

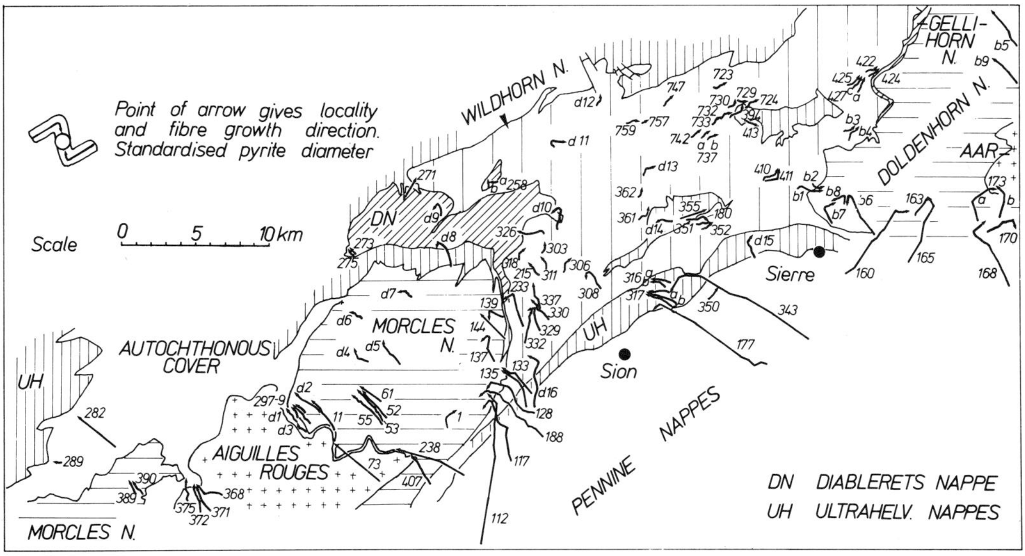

FIG. 7. The strain state of the Helvetic nappes, as indicated by pressure shadow fibres around pyrites. The fibres are represented in the XY finite strain plane (cleavage plane) without a dip correction. The lengths of the fibre increments are normalized relative to the standard pyrite size shown. There exists a horizontal gradient of increasing strain from the front of the nappes towards the root zone. The numbers refer to individual specimens. d: Data from Durney & Ramsay (1973). b: Data from Burkhard (1986(a)).

most evident ones are: stretched belemnites (Badoux 1963), deformed oolites (Badoux 1970, Durney 1972, Siddans 1983), deformed conglomerates (Siddans 1983), fibres in chocolate tablets (Durney 1972, Casey *et al.* 1983), and fibres in pressure shadows (Durney 1972, Durney & Ramsay 1973, Gray & Durney 1979, Ramsay & Huber 1983, Ramsay *et al.* 1985, Burkhard 1986a).

Fibres in pressure shadows are probably the most interesting deformation markers, because they contain information on the incremental deformation history. Since the work of Durney (1972) and Durney & Ramsay (1973) the Helvetic nappes are a classical region for the study of incremental deformation by syntectonic fibres. The discussion of the strain distribution in the region considered is therefore based on a pressure-shadow fibre map, Fig. 7. The finite strain calculated from pressure shadow fibres compares well with the finite strain calculated from oolites (Durney 1972).

The most prominent curvature of the pressure shadow fibres can generally be observed in the main cleavage plane, i.e. in the *XY* finite strain plane. Special attention was therefore given to the fibre geometry in the cleavage plane. Figure 7 is a compilation of all the available pressure-shadow fibre data in the cleavage plane. Care was taken to collect specimens containing pressure shadows as much as possible from competent beds and avoiding minor folds. Only fibres around framboidal pyrites were considered. Framboidal pyrite is a feature of early diagenesis of sediments and behaved rigidly during deformation. The fibres are assumed to record the full extension history (Durney & Ramsay 1973). The length and orientation of the submillimetric fibre increments was measured under the microscope. The fibre increments are normalised to a standard size pyrite radius, and represented in Fig. 7 on the cleavage plane without a correction for the cleavage dip.

The 112 fibre data of Fig. 7 are obviously insufficient to describe the deformation state of such a large region accurately, even if supplemented with all the other kinds of strain data mentioned earlier. Notwithstanding, the following strain distribution can be seen:

(1) A steep gradient of increasing strain from the front to the root zone of the nappes.
(2) An overall increase in strain from the central part of the Wildhorn nappe down to the central part of the Morcles nappe. In the front and in the root zone of the nappes such a downwards increase cannot be seen. The pressure shadow strains in the lowest

part of the nappe pile, along the inverted limb of the Morcles nappe, do not reach the pressure-shadow strain values in the root zone.

(3) A gradient of decreasing strain from the inverted limb of the Morcles nappe across the autochthonous cover to the Aiguilles Rouges basement. This gradient is especially clear from the data of Siddans (1983).

The highest strains in Fig. 7 are seen in the normal limb of the Morcles nappe (specimen 112, locality Ardon) and in the Ultrahelvetic nappes (specimen 343, locality Argnou). For both specimens the maximum finite extension amounts to $(1 + e_1) > 20$. The highest strain value found by D. Durney is from the root zone of the Wildhorn nappe (locality Aven), where $(1 + e_1) > 17$. These are extremely high strains, corresponding to a ratio $X{:}Z = 300$ to 400:1, if $Y = 1$. In the front of the Wildhorn nappe the fibre directions recorded cannot be related to overthrust shear; these directions are related to fold axis parallel extension. In the front of the Diablerets and of the Morcles nappe the strains amount to $(1 + e_1) = 1$ to 3.

A study of cathodoluminescence petrography of syntectonic quartz fibres (Dietrich & Grant 1985) has shown that there are systematic variations in luminescence measured between older and younger parts of the same pressure shadow fibre, and between fibres from different tectonic units. These variations are related by Dietrich & Grant (1985) first to a temperature increase during the formation of the quartz fibres, and second to deformation-induced differences in the microstructure of earlier formed parts of quartz fibres during later imposed strains.

The metamorphism of the Helvetic nappes

Regional Alpine metamorphism has for some time been known to have affected the Helvetic nappes (e.g. reviews by Niggli 1970, and Frey *et al.* 1980a). In the region considered, the metamorphic mineral associations observed in the Grès de Taveyanne formation belong to the Laumontite and to the Prehnite-Pumpellyite zones (Martini & Vuagnat 1965, Bussy & Epard 1984).

An increase in metamorphism from the front of the nappes towards the root zone and from the Wildhorn nappe towards the Morcles nappe was suggested for the first time explicitly by Durney (1972). The quantitative studies of Kübler *et al.* (1979), Mullis (1979), Frey *et al.* (1980b) and Burkhard (1986b) confirmed this view. Illite crystallinity data presented in the following show, in agreement with these studies, the existence of a horizontal and vertical gradient, ranging from diagenetic to epizonal metamorphic conditions. However, the distribution of the metamorphic zones in Fig. 8 is somewhat different from that shown by previous authors.

Illite crystallinity

The width of the first order illite basal reflexion at half height can be used for a zonation of very low grade metamorphism (Kübler 1968). The illite crystallinity data in Table 1 were obtained from X-ray diffractograms of smear slides of Helvetic shales, slates, marly limestones and marls. The specimens were ground for 30 seconds, then decarbonated with acetic acid, afterwards centrifuged to remove the calcium acetate, then disintegrated in an ultrasonic bath and finally the fraction $<2\ \mu$ was obtained by sedimentation. Air dried mounts of this fraction were prepared; tests by G. Kahr had shown that glycolated mounts did not produce essentially different results. X-ray operation conditions were 2°/min, the paper speed was 1200 mm/h. Measurement standards provided by Prof. Kübler were used, and the widths of the peaks in the resulting diffractograms are presented in Table 1 in °/2 Θ (e.g. Hunziker 1986). Even with several different measurements it was not possible to obtain good illite peaks from every specimen. The illite crystallinity of specimens with small and equivocal peaks was therefore determined by infrared spectrography (Flehmig 1973). The resulting infrared spectrography values have been converted to be comparable to the illite crystallinity values and are marked 'IR' in Table 1. *Note of caution*: The illite crystallinity values of specimens of different lithologies might not be simply comparable, and the influence of pre-Alpine mica clasts on the results cannot be totally excluded, even though the specimens were checked.

The metamorphic gradients

The illite crystallinity data of Table 1 have been projected onto a synthetic profile, Fig. 8. The following observations can be made: There exists a *vertical metamorphic gradient* apparently related to the tectonic overburden of the individual nappe. The profile in Fig. 8 shows a maximal thickness of the nappe pile of about 15 km without consideration of the now eroded,

TABLE 1. *Illite crystallinity data of Helvetic shales, slates and marly limestones, determined from the the fraction <2 μ of the ground specimen. 'IR' refers to infrared spectrography data. The coordinates of the specimen localities refer to the Swiss topographical map.*

Specimen Nr.	Tectonic Unit	Age and Formation Name	Illite Crystallinity °/2θ	Coordinates of of Specimen locality
191	Ardon tectonic sliver	Lower Cretaceous Valanginian Limestone	0.3 IR	585.450/117.875
192	Diablerets n.	Upper Jurassic Malm Limestone	0.4	585.475/117.975
193	Ardon tectonic sliver	Lower Cretaceous Valanginian Limestone	0.36 IR	585.475/117.925
196	Morcles n.	Lower Cretaceous Valanginian Limestone	0.2 IR	573.625/117.050
199	Ultrahelvetic n.	Lower Jurassic Lias Slates	0.4	596.650/123.700
203	Diablerets n.	Tertiary Grès Nummulitique	0.7	587.250/130.800
207	Wildhorn n.	Tertiary Globigerina Marls	1.0	587.675/134.250
208	Wildhorn n.	Lower Cretaceous Valanginian Marls	0.95	588.900/131.675
213	Diablerets n.	Upper Cretaceous Urgonian Limestone	0.7	589.050/130.300
221	Wildhorn n.	Middle Jurassic Dogger Slates	0.3 IR	590.525/122.150
222	Morcles n.	Lower Cretaceous Valanginian Marls	0.2 IR	585.000/117.150
227	Morcles n.	Lower Jurassic Lias Slates	0.23 IR	582.025/115.750
230	Morcles n.	Lower Cretaceous Valanginian Limestone	0.23 IR	585.500/122.150
240	Morcles n.	Lower Cretaceous Hauterivien limestone	0.18 IR	578.925/113.525
242	Morcles n.	Tertiary Flysch Shales	0.40	576.000/123.775
244	Ultrahelvetic n.	Eocene Flysch Shales	0.42	574.775/124.225
246	Autochthonous	Tertiary Flysch Slates	0.25	574.325/119.700
247	Morcles n.	Tertiary Flysch Slates	0.28	574.375/119.700
253	Diablerets n.	Tertiary Globigerina Shales	0.45	584.750/131.525
266	Wildhorn n.	Lower Cretaceous Barrémian Shales	0.85	582.750/132.475
268	Wildhorn n.	Tertiary Globigerina Shales	1.0	582.250/113.225
272	Diablerets n.	Tertiary Flysch Shales	0.6	575.225/126.975
281	Autochthonous	Tertiary Flysch Slates	0.38	557.375/116.125
291	Autochthonous	Lower Cretaceous Valanginian Marls	0.68	554.250/111.100
294	Autochthonous	Lower Cretaceous Valanginian Limestone	0.3	569.475/118.725
321	Wildhorn n.	Lower Cretaceous Valanginian Marls	0.58	586.900/127.300
323	Diablerets n.	Lower Cretaceous Barrémian slates	0.35	586.900/127.750
328	Wildhorn n.	Middle Jurassic Dogger Slates	0.4	587.775/123.450
345	Wildhorn n.	Middle Jurassic Dogger Slates	0.3	599.900/127.900
364	Wildhorn n.	Upper Cretaceous Wang Slates	0.51	595.475/129.575
365	Ultrahelvetic n.	Tertiary Flysch Slates	0.41	595.650/129.350
385	Morcles n.	Tertiary Flysch Slates	0.35	556.400/110.225
386	Autochthonous	Tertiary Flysch Slates	0.34	555.975/110.300
393	Ultrahelvetic n.	Lower Cretaceous Barrémian Shales	0.55	601.800/136.500
395	Wildhorn n.	Tertiary Hohgant Shales	0.85	601.425/136.900

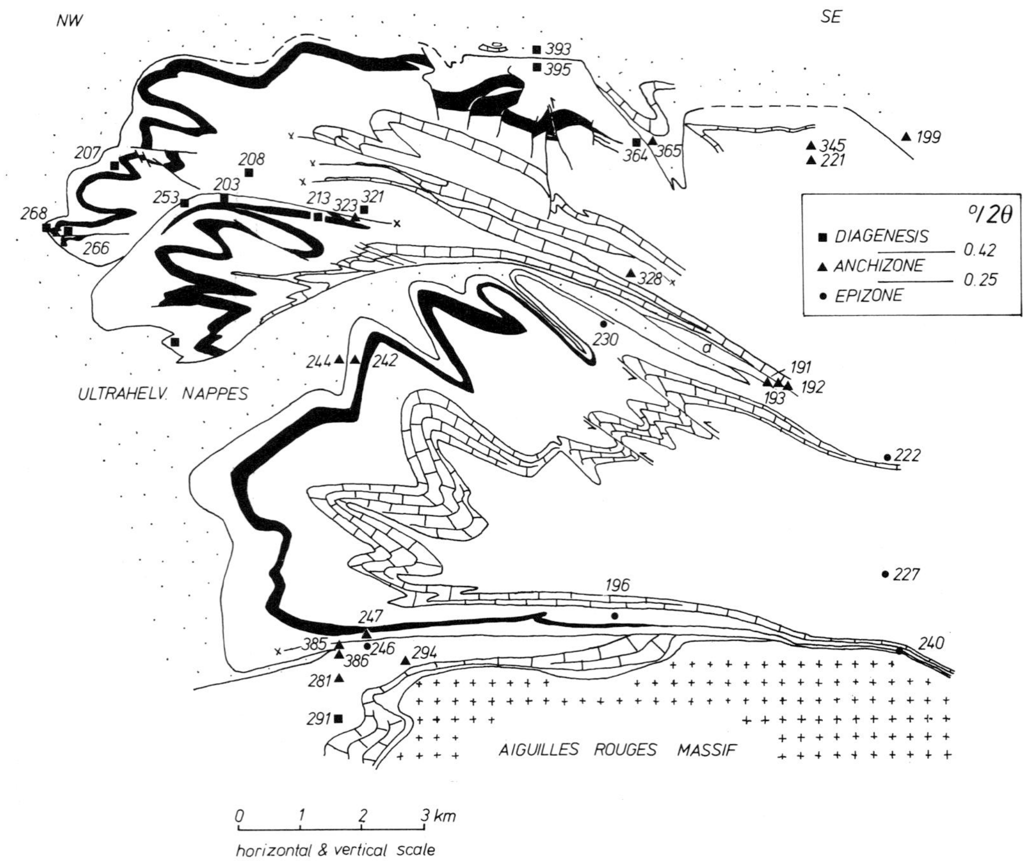

FIG. 8. The metamorphism of the Helvetic nappes, as indicated by illite crystallinity. The data are projected in a synthetic profile of the Helvetic chain. Details of this profile are slightly different from the profiles of the individual nappes, Figs 3, 4 and 6. For these profiles more than one projection plane had to be used resulting in a misfit between the individual projections in the synthetic profile. This misfit was manually minimized. The symbols are as in Fig. 3. a: Ardon tectonic sliver (Masson *et al.* 1980). The numbers refer to Table 1, where the specimen coordinates and the exact illite crystallinity values are listed. There exists a horizontal and a vertical gradient of increasing metamorphism. The horizontal metamorphic and strain gradients show a similar trend (cf. Fig. 7).

but once overlying Pennine units. This corresponds to a vertical thickness of about 14 km. In general, the metamorphism increases from diagenesis in the upper part of the Wildhorn nappe to epizonal conditions in the inverted limb of the Morcles nappe.

A *horizontal metamorphic gradient* exists in all the three nappes, and probably as well in the Ultrahelvetic nappes. The Wildhorn and the Diablerets nappes pass from diagenesis in the front of the nappes to anchizonal conditions in the root zone, the Morcles nappe from anchizonal conditions in the front to epizonal conditions in the root zone. A similar horizontal gradient exists in the Helvetic nappes east of the Wildstrubel depression (e.g. Frey *et al.* 1980b). Such a gradient cannot be explained by the thermal evolution as a consequence of the tectonic overburden alone.

Some data appear to indicate the existence of *inverted metamorphic gradients*: these are the illite crystallinity values between the Morcles nappe and the autochthonous cover of the Aiguilles Rouges Massif, and the crystallinity values between the Ultrahelvetic nappes and the Wildhorn nappe. The first is apparent too from the data by Kübler *et al.* (1979), the second from the data by Burkhard (1986b). In both cases the higher metamorphic unit appears to be transported over the lower metamorphic unit: the Ultrahelvetic nappes, already deformed and metamorphosed, were reinvolved in the deformation of the Helvetic nappes, and the Morcles nappe reached its actual position on the autochthonous cover of the Aiguilles Rouges Massif after its main internal deformation.

Discussion

A thin-skinned model for the Helvetic nappes, as proposed for example by Butler (1985), cannot explain the steep gradients of deformation and metamorphism. The strain distribution to be expected above a ramp (e.g. Ramsay & Huber 1987, p. 538) is totally different from that seen in the Helvetic nappes. Also the model of a nappe thrust over a ramp and successively sheared (Ramsay *et al.* 1983) cannot explain the observed strain and metamorphic gradients.

All the three Helvetic nappes have comparable internal fold structures. A tectonically comparable formation can therefore be assumed. The 'inversion' model suggested by Lemoine *et al.* (1986) and by Gillcrist *et al.* (1987) would hence require Mesozoic normal faults dividing the depositional realms of the three future nappes to represent the 'buttresses' required, according to these authors, during the folding of the multilayer sequences of the individual nappes. Whilst tectonic reactivation of synsedimentary Helvetic faults was observed long ago (Günzler-Seiffert 1941, 1952), these faults are not known to have delimited the nappe depositional realms. The clearly visible 'synclinal de raccord' between the Wildhorn and the Diablerets nappe and the Chamonix syncline between the Morcles nappe and the autochthonous cover of the Aiguilles Rouges Massif do not show any normal fault discontinuity. Moreover the 'buttressing effect' of such a fault would produce in the front of the nappes a significant pure shear component of the finite deformation which is not seen: the fold axial planes steepen up gradually, a feature particularly clear in the Morcles profile, Fig. 3. As a consequence of the 'buttressing effect' of a reactivated normal fault the lowermost axial planes of the nappe ought to steepen up more than the higher axial planes, and this is also not seen.

Casey & Huggenberger (1985) produced a mechanical model of folding in multilayer sequences which demonstrates that buckle folds can develop in an overthrust shear oblique to the bedding planes. That is, folds could form in the Helvetic nappes without the 'buttressing' considered necessary by Lemoine *et al.* (1986) and Gillcrist *et al.* (1987).

The geometric forms of the fold axial planes of the three nappes are comparable, converging towards the root zone in accordance with the strain gradient. The finite strain *XY* planes as defined by the strain markers are arranged in cleavage fans to the folds shown in the profiles of the nappes (Figs 3, 4 and 6). No gravitational nappe formation model can explain such a strain distribution. Casey & Huggenberger (1985) show that the fold enveloping surfaces and the fold axial planes within each nappe deform as passive markers, allowing them to be used as a marker grid in considering the internal deformation.

Deformation and metamorphism show in the Helvetic nappes a remarkably parallel development. At least the horizontal metamorphic gradient appears to be deformation-related, since in the same tectonic unit the diagenetic zone passes into the anchizone, or the anchizone into the epizone, over a horizontal distance of only a few kilometres. The sequence of tectonic events is illustrated by the two examples of transported metamorphism mentioned earlier: the Ultrahelvetic nappes, already deformed and metamorphosed, were reinvolved in the Helvetic deformation; the Morcles nappe, with its inverted limb strongly extended by distributed shear was thrust over a decreasingly deformed and metamorphosed autochthonous series. The nappes therefore formed in a top to bottom sequence between the already overlying Pennine units and the actively underthrusting basement.

The observed thinning of the nappes towards the root zone and the steepness of the deformational and metamorphic gradients leads us to propose a process of *tectonic extrusion* for their origin. 'Rear compression' as a possible deformation mechanism of the Helvetic root zone has been discussed for the first time by Durney (1982). At this point the deformation history of the Pennine units has to be considered. The last of the major Pennine deformation events, the backfolding event, affects also the Gotthard Massif: Pennine nappes and Gotthard Massif are folded together in huge km-size folds with steeply NW-wards plunging axial planes (Huber *et al.* 1980, fig. 7). The internal deformation of the Gotthard granitoids is overall flattening on subvertical planes (Choukroune & Gapais 1983, Marquer & Gapais 1985). The Gotthard Massif is classically considered to represent the basement of the Ultrahelvetic nappes (Trümpy 1963, 1980). The Ultrahelvetic nappes were sheared off during an earlier phase, before the metamorphism occurred which affected the remaining Gotthard cover sediments. The steepened-up fan-like Gotthard Massif, or, more in the west, its unexposed backfolded equivalent, represented a giant obstacle during the shear movements between the overlying Pennine units and the underthrusting basement. The Gotthard Massif was probably finally 'scraped off by the north-moving orogenic lid' (Laubscher 1983; especially his fig. 16c). This process had to influence the formation of the Helvetic nappes, leading eventually to their tectonic extrusion. In the root zone of the nappes extrusion is manifested in a pure shear component of the finite deformation; pure shear rather than flattening deformation as indicated by the parallelism of the pressure shadow fibres in XY sections. The wedge shape of the nappes and the steep deformational and metamorphic gradients can be related to this pure shear component.

From the evidence presented above it is clear that the models proposed by previous authors are too simple. A model of sequentially developing and southwards converging shear zones accounts best for the observed geological features. We present below such a model which is set up so as to reconcile the evident overthrust shear with the strong deformational and metamorphic gradients towards the root zone. Particular attention is given to the modelling of the transfer of overthrust displacement from a 'synclinal de raccord' into a thrust fault, this transfer representing one of the main characteristics of the Helvetic nappe geometry.

Tectonic models

Tectonic extrusion – superimposition of pure shear deformation on simple shear

We start this discussion modelling the large scale geometry with a model which meets the three principal conditions discussed earlier, the wedge shape of the nappes, their strain state and their metamorphism. The model is applied to the Diablerets nappe, Fig. 4. It could be applied to the other two nappes as well. The estimated pure shear part of the deformation is superimposed on the estimated simple shear part, according to

$$\begin{vmatrix} a & b \\ c & d \end{vmatrix} = \begin{vmatrix} R & 0 \\ 0 & 1/R \end{vmatrix} \cdot \begin{vmatrix} 1 & \gamma \\ 0 & 1 \end{vmatrix} \quad (1)$$

where a, b, c, d are the components of the deformation gradient matrix, R is the extension ratio $R = (1+e_1)/(1+e_2)$, and γ is the finite shear strain.

It should be mentioned that the order of the matrix multiplication, simple shear before pure shear, is only for mathematical convenience, and not the order in which the deformation processes are thought to have occurred. The two processes are thought to have occurred contemporaneously.

Estimation of the shear strain γ: The pure shear part of the deformation is removed, that is, the two converging thrust planes in Fig. 4 are restored to parallelism. The Urgonian limestone is unfolded between the two parallel planes, corrected for strain, and, as shown in Fig. 9,

$$\gamma = \frac{e}{f} \quad (2)$$

where e is the finite displacement and f is the shear zone width.

For the Diablerets nappe the shear strain obtained is $\gamma = 3$.

Estimation of the amount of pure shear: The parallel nappe boundary planes of Fig. 9 are deformed to the wedge shape seen in Fig. 4. In the outer part of the wedge shaped nappe the deformation produced is an extension in the vertical direction; in the inner part a compression. The nappe is therefore compared with a sector of a concentric fold with a neutral surface (Ramsay 1967, p. 398, Ramsay & Huber 1987, p. 620). The strains on the outer and

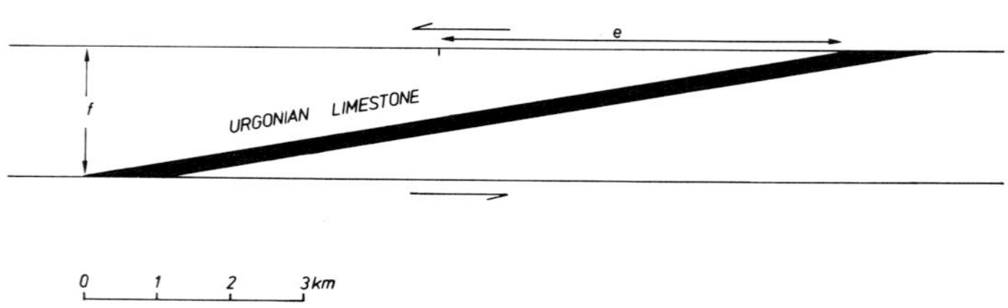

FIG. 9. Estimation of the amount of simple shear in the deformation of the Diablerets nappe. The Urgonian limestone is shown undeformed.

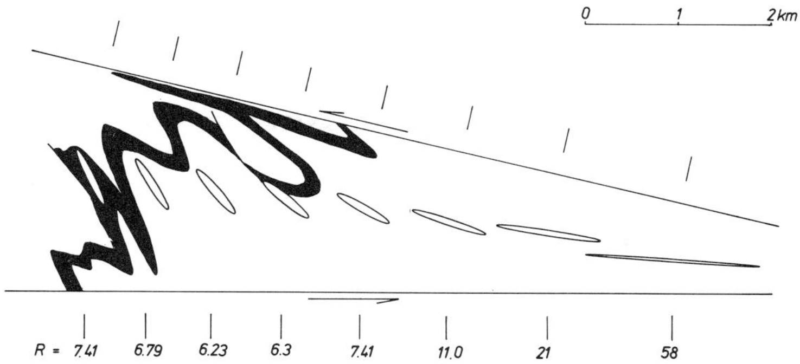

FIG. 10. Tectonic model for the Diablerets nappe. The Urgonian limestone is indicated as a structural reference horizon. The strain distribution has been obtained by superimposition of pure shear on simple shear. The pure shear part of the deformation has been calculated according to the concentric fold model of Ramsay (1967) and Ramsay & Huber (1987).

inner arc of the neutral surface were calculated and expressed in terms of the extension ratio R = $(1+e_1)/(1+e_2)$. The position of the neutral surface in the model was assumed to indicate the width of the shear zone when restoring the two shear planes to parallelism.

Strain compatibility is respected in this procedure. Figure 10 shows the strain distribution obtained in the Diablerets nappe after superimposition of pure shear on simple shear deformation, according to equation (1). The strain gradient from the external part of the nappe towards the root zone compares well with the strain distribution according to the pressure shadow data of Fig. 7. The finite strain orientation is parallel to the fold axial planes of the nappe. The unsatisfying point of this model is the increase in the finite strain values outwards from the neutral surface, which is in contradiction to the regional strain pattern.

The distribution of overthrust shear within the nappes

Once the above model is accepted, the structures within the nappes must be considered: the problem here is to model the transfer of overthrust displacement from continuous simple shear deformation in the internal parts of the nappes to discontinuous displacement along a thrust fault in the external parts. This was done by geometrical finite element modelling, the details of which will be published separately. The general features of the model are shown in Fig. 11. The model represents the deformation zone between two nappes. The upper and lower parts of the model are assumed rigid and the upper part is displaced parallel to x and to the left relative to the lower part. A fault runs from the left end of the model to about three-quarters of the model's length. The displacement on the fault is arbitrarily assumed to increase linearly, from zero at its right hand termination, to a value equal to the total relative movement of the blocks at the left end of the model. Deformation is assumed to be restricted to a zone in the centre of the model. The zone is narrower at the right of the model than at the left. Marker lines, vertical before deformation, were constructed. For example, a vertical line AB in the upper rigid part of the model intersects the edge of the deformed zone at point B. In the absence of fault displacements this line would intersect the centre line of the model at half the overthrust displacement to the right of B, point P. If P does not fall on the fault then the marker passes through it. If P does lie on the fault the total displacement is made up of a contribution from ductile deformation and a contribution from fault motion. The amount of the latter is a chosen function of distance, in this case linear, from the end of the fault and is given in Fig. 11. Thus the fault displacement is known and the actual point of intersection can be found.

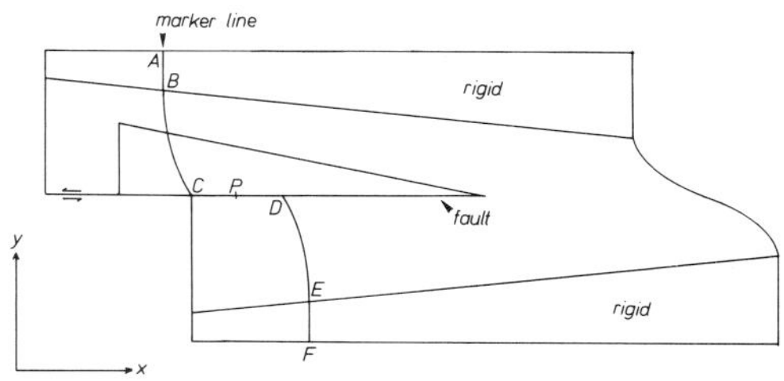

FIG. 11. The geometric model of an overthrust and the construction of marker lines defining the deformation.

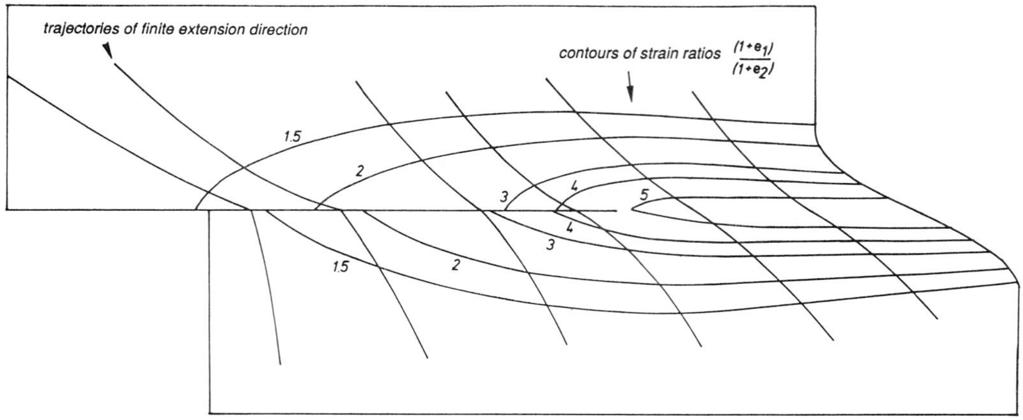

FIG. 12. Trajectories of the finite extension direction and contours of the finite strain ratios $R = (1+e_1)/(1+e_2)$ for the model of Fig. 11.

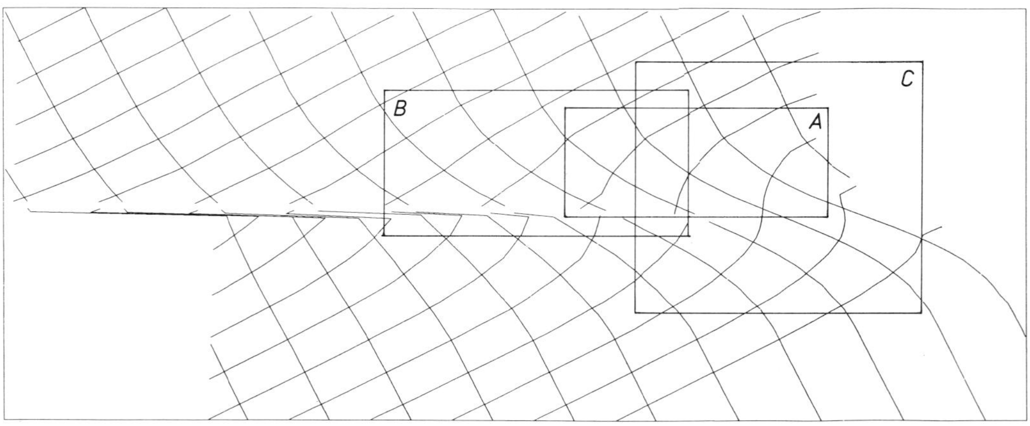

FIG. 13. A rectangular grid deformed according to the model of Fig. 11. The set of lines at 27° to the model boundary represents bedding planes before the deformation and enveloping surfaces of folds after the deformation. The set of lines at 63° represents fold axial planes. The boxes A, B and C indicate regions of the model chosen for detailed comparison to the Helvetic nappes in Figs 14–16.

The segment of the marker line between the edge of the rigid part and the centreline, BC, is made to be a quadratic function of height with coefficients such that the line passes through the endpoints of the segment and has no slope discontinuity at the boundary of the rigid zone. The line DEF can be drawn in a similar way.

To calculate the strains in the continuously deformed region the model is underformed according to a finite element compatibility procedure. The marker lines are restored to vertical using an area balance, which also allows the original x-coordinate positions to be calculated. The condition of constant area is used to set up a set of simultaneous linear equations which are solved for the y-coordinate values, thus defining the deformation. Strain ratios and the orientation of the principal extension have been calculated and are shown in Fig. 12. The deformed state of a grid that was rectilinear before the deformation is shown in Fig. 13. The grid lines made angles of 63° and 27° with the boundaries of the model.

The deformation distribution shown in Figs 12 and 13 can be compared to the deformations in the western Helvetic nappes. The finite strain ratios show gradients increasing from the external zones to the root zone and decreasing

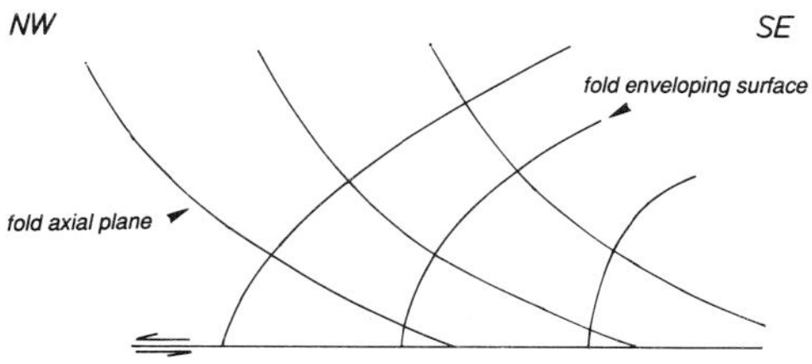

FIG. 14. Region A of the model of Fig. 13 chosen for comparison to the lower part of the Diablerets nappe, Fig. 4.

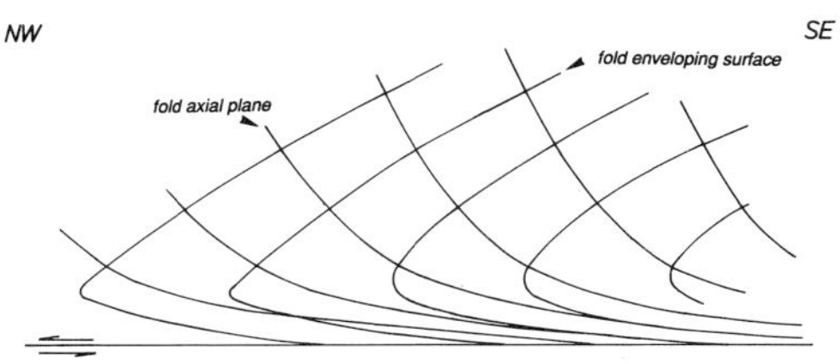

FIG. 15. Region B of the model of Fig. 13 chosen for comparison to the lower part of the Morcles nappe, Fig. 3. An additional component of heterogeneous simple shear has been added to form the inverted limb.

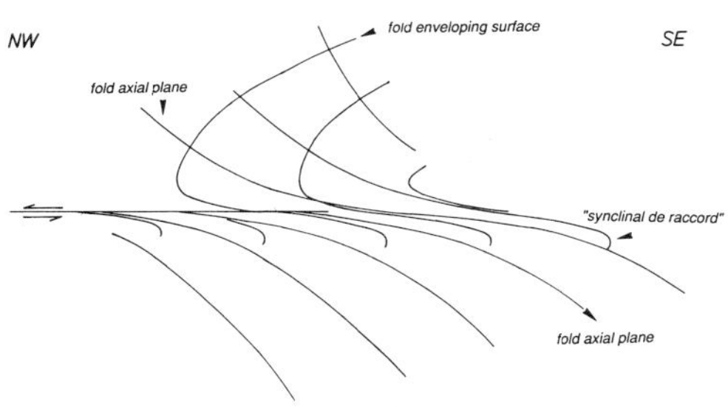

FIG. 16. Region C of the model of Fig. 13 chosen for comparison to the lower part of the Wildhorn nappe and to the upper part of the Diablerets nappe, Figs 6 and 8. A component of heterogeneous simple shear has been added to form the 'synclinal de raccord'.

upwards and downwards from the separation zone between the nappes. The finite strain trajectories may be compared to general cleavage trajectories. The deformed state of the rectilinear grid is comparable to the enveloping surface of folded layers (27° set) and to the orientations of axial planes (63° set).

Closer agreement between modelled geometry and observed deformations of the nappes can be obtained by selecting areas of the general model (Fig. 13) and in some cases adding further components of simple shear. This shear is localized near the centre of the model but constant in x so as to give no strain compatibility problems.

The outlined region A of Fig. 13, shown in Fig. 14, is in reasonable agreement with the Diablerets nappe, Fig. 4. The outlined region B of Fig. 13 combined with an additional component of heterogeneous simple shear, shown in Fig. 15, compares well with the Morcles nappe, Fig. 3. The steep gradient of strain from external to internal zones is explained by the super-imposition of plane strain as discussed above. The outlined region C of Fig. 13 plus heterogeneous simple shear, shown in Fig. 16, gives a good agreement with the Wildhorn nappe, Fig. 6. In the internal part the bedding planes pass unbroken down into the Diablerets nappe through the 'synclinal de raccord' whereas in the external part there is a thust contact.

Conclusions and kinematic implications

The finite strain, fold geometry and metamorphic data presented in this paper can only be adequately explained by tectonic models involving the interplay of several components:

(1) Superimposition of pure shear components on simple shear deformation. The magnitude of the pure shear components decreases towards the external parts of the nappes. The superimposition of pure shear on simple shear explains the steep horizontal strain- and metamorphic gradients observed in the three Helvetic nappes and is considered to be the result of tectonic extrusion from the root zone.

(2) Heterogeneous simple shear constant in magnitude from internal zones to external zones. This deformation component accounts for the ductile strains expressed in the fold geometry of the nappes.

(3) Varying amounts of overthrust displacement being taken up by distributed shear in the internal part of the nappes and by thrust faulting in the external part. This different response of rocks to deformation, leading in a distance of a few kilometres from a ductile 'synclinal de raccord' into a thrust fault, may be the effect of varying temperature. This temperature gradient cannot be explained by the deformation of a stratigraphic sequence containing bedding-parallel isotherms alone. The additional heat is provided by the juxtaposition of the root zone sediments with hotter basement as a part of the extrusion process.

ACKNOWLEDGEMENTS: John Ramsay generously allowed us to use his 1:25 000 geological field maps of the Helvetic chain. Without these solid maps full of structural information this paper could not have been written. John Ramsay is also thanked for critical reading of the manuscript. Eveline Herren helped and Günter Kahr advised with the illite crystallinity determinations. The reviews by Peter Cobbold, Rick Groshong and Roy Kligfield are appreciated. Financial support from Schweizerischer Nationalfonds, projects n. 2.2140.86, 2.0800.83 (MC) and 4.963.085.20 (DD), is gratefully acknowledged.

References

BADOUX, H. 1963. Les bélemnites tronconnées de Leytron (Valais). *Bulletin de la Societé Vaudoise des Sciences Naturelles*, **68**, 311, 1–7.

—— 1970. Les oolites déformées du Vélar (massif de Morcles). *Eclogae Geologicae Helvetiae*, **63**, 539–48.

BOYER, S. E. & ELLIOTT, D. 1982. Thrust systems. *Bulletin of the American Association of Petroleum Geologists*, **66**, 1196–230.

BURKHARD, M. 1986(a). Déformation des calcaires de l'Helvétique de la Suisse occidentale (Phénomènes, mécanismes et interprétations tectoniques). *Revue de Géologie Dynamique et de Géographie Physique*, **27**, 281–301.

—— 1986(b). *L'Helvétique de la bordure occidentale du massif de l'Aar.* Unpublished dissertation, University of Neuchâtel.

BUSSY, F. & EPARD, J. -L. 1984. Essai de zonéographie métamorphique entre les Diablerets et le Massif de l'Aar (Suisse occidentale), basée sur l'étude des Grès de Taveyanne. *Schweizerische Mineralogische Petrographische Mitteilungen*, **64**, 131–50.

BUTLER, R. W. H. 1985. The restoration of thrust systems and displacement continuity around the Mont Blanc Massif, NW external Alpine thrust belt. *Journal of Structural Geology*, **7**, 569–82.

CASEY, M., DIETRICH, D. & RAMSAY, J. G. 1983. Methods for determining deformation history for chocolate tablet boudinage with fibrous crystals. *Tectonophysics*, **92**, 211–39.

—— & HUGGENBERGER, P. 1985. Numerical modelling of finite amplitude similar folds developing under general deformation histories. *Journal of Structural Geology*, **7**, 103–14.

CHOUKROUNE, P. & GAPAIS, D. 1983. Strain pattern in the Aar Granite (Central Alps): orthogneiss developed by bulk inhomogeneous flattening. *Journal of Structural Geology*, **5**, 411–18.

DIETRICH, D. & DURNEY, D. W. 1986. Change of direction of overthrust shear in the Helvetic nappes of western Switzerland. *Journal of Structural Geology*, **8**, 389–98.

—— & GRANT, P. R. 1985. Cathodoluminescence petrography of syntectonic quartz fibres. *Journal of Structural Geology*, **7**, 541–53.

—— & SONG, H. 1984. Calcite fabrics in a natural shear environment, the Helvetic nappes of western Switzerland. *Journal of Structural Geology*, **6**, 19–32.

DURNEY, D. W. 1972. *Deformation history of the western Helvetic nappes, Valais, Switzerland.* Unpublished PhD thesis, Imperial College London.

—— 1982. Some observations concerning the question of gravity gliding and hinterland compression in the western Helvetic nappes. Abstract, International Conference on Planar and Linear Fabrics, Zurich. *Mitteilungen aus dem Geologischen Institut der Eidgenössischen Technischen Hochschule und der Universität Zürich,* **239a**, 92–4.

—— & RAMSAY, J. G. 1973. Incremental strains measured by syntectonic crystal growth. *In*: DEJONG, K. & SCHOLTEN, R. (eds). *Gravity and Tectonics*, Wiley, New York.

FLEHMIG W. 1973. Kristallinität und Infrarotspektroskopie natürlicher dioktaedrischer Illite. *Neues Jahrbuch Mineralogische Monatshefte*, 351–61.

FREY, M., BUCHER, K. FRANK, E. & MULLIS, M. 1980a. Alpine metamorphism along the Geotraverse Basel–Chiasso – a review. *Eclogae Geologicae Helvetiae*, **73**, 527–46.

——, TEICHMUELLER, M., TEICHMUELLER, R., MULLIS, J., KUENZI, B., BREITSCHMID, A., GRUNER, U. & SCHWIZER, B. 1980b. Very low-grade metamorphism in external parts of the central Alps: illite crystallinity, coal rank and fluid inclusion data. *Eclogae Geologicae Helvetiae*, **73**, 173–203.

GILLCRIST, R., COWARD, M. & MUGNIER, J. -L. 1987. Structural inversion and its controls: examples from the Alpine foreland and the French Alps. *Geodinamica Acta*, **1**, 5–34.

GRAY, D. R. & DURNEY D. W. 1979. Investigations on the mechanical significance of crenulation cleavage. *Tectonophysics*, **58**, 35–79.

GROSHONG, R. H., PFIFFNER, O. A. & PRINGLE, L. R. 1984. Strain partitioning in the Helvetic thrust belt, eastern Switzerland. *Journal of Structural Geology*, **6**, 5–18.

GUENZLER-SEIFFERT, H. 1941. Persistente Brüche im Jura der Wildhorn–Decke des Berner Oberlandes. *Eclogae Geologicae Helvetiae*, **34**, 164–72.

—— 1952. Alte Brüche im Kreide/Tertiär-Anteil der Wildhorn–Decke zwischen Rhône und Rhein. *Geologische Rundschau*, **40**, 211–39.

HEIM, A. 1921. *Geologie der Schweiz*. Band II/1. Tauchnitz, Leipzig.

HUBER, M., RAMSAY, J. G. & SIMPSON, C. 1980. Deformation in the Maggia and Antigorio nappes, Lepontine Alps. *Eclogae Geologicae*

Helvetiae, **73**, 593–606.

HUGGENBERGER, P. 1985. *Faltenmodelle und Verformungsverteilung in Deckenstrukturen am Beispiel der Morcles Decke (Helvetikum der Westschweiz)*. Unpublished thesis *7861*, ETH Zürich.

HUNZIKER, J. C. 1986. The evolution of illite to muscovite: an example of the behaviour of isotopes in low-grade metamorphic terranes. *Chemical Geology*, **57**, 31–40.

KUBLER, B. 1968. Evaluation quantitative du métamorphism par la cristallinité de l'illite. *Bulletin du Centre de Recherches de Pau, SNPA*, **2**, 385–97.

——, PITTION, J. -L., HEROUX, Y., CHAROLLAIS, J. & WEIDMANN, M. 1979. Sur le pouvour réflecteur de la vitrinite dans quelques roches du Jura, de la Molasse et des nappes préalpines, helvétiques et penniques (Suisse occidentale et Haute-Savoie). *Eclogae Geologicae Helvetiae*, **72**, 347–73.

LANGENBERG, W., CHARLESWORTH, H. & LA RIVIERE, A. 1987. Computer constructed cross sections of the Morcles nappe. *Eclogae Geologicae Helvetiae*, **80**, 655–67.

LAUBSCHER, H. P. 1983. Detachment, shear, and compression in the central Alps. *Geological Society of America Memoir*, **158**, 191–211.

LEMOINE, M., BAS, T., ARNAUD-VANNEAU, A., ARNAUD, H., DUMONT, T., GIDON, M., BOURBON, M., DE GRACIANSKY, P. -C., RUDKIEWICZ, J. -L., MEGARD-GALLI, J. & TRICART, P. 1986. The continental margin of the Mesozoic Tethys in the Western Alps. *Marine and Petroleum Geology*, **3**, 179–99.

—— & TRÜMPY, R. 1987. Pre-oceanic rifting in the Alps. *Tectonophysics*, **133**, 305–20.

LUGEON, M. 1943. Une nouvelle hypothèse tectonique: la diverticulation. *Bulletin de la Societé Vaudoise des Sciences Naturelles*, **62**, 301–3.

MANCKTELOW, N. 1985. The Simplon line: a major displacement zone in the western Lepontine Alps. *Eclogae Geologicae Helvetiae*, **78**, 73–96.

MARQUER, D. & GAPAIS, D. 1985. Les massifs cristallins externes sur une transversale Guttanen–Val Bedretto (Alpes centrales): structures et histoire cinématique. *Compte-Rendus de l'Académie des Sciences*, **301, II**, 543–6.

MARTINI, J. & VUAGNAT M. 1965. Présence du faciès à zéolites dans la formation des 'grès' de Taveyanne (Alpes franco–suisse). *Schweizerische Mineralogische Petrographische Mitteilungen*, **45**, 281–93.

MASSON, H., HERB, R. & STECK, A. 1980. Helvetic Alps of western Switzerland. Excursion I. *In*: TRÜMPY, R. (ed). *Geology of Switzerland, a Guide Book. Part B, Geological excursions*, 109–45. Wepf & Co, Basel.

MERLE, O. 1986. Patterns of stretch trajectories and strain rates within spreading-gliding nappes. *Tectonophysics*, **124**, 211–22.

MILNES, A. G. & PfiFFNER, O. A. 1977. Structural development of the Infrahelvetic complex, eastern Switzerland. *Eclogae Geologicae Helvetiae*, **70**, 83–95.

MULLIS, J. 1979. The system methane-water as a geologic thermometer and barometer from the external part of the central Alps. *Bulletin Minéralogie*, **102**, 526–36.

NIGGLI, E. 1970. Alpine Metamorphose und alpine Gebirgsbildung. *Fortschritte der Mineralogie*, **47**, 16–26.

RAMSAY, J. G. 1967. *Folding and Fracturing of Rocks*. McGraw-Hill, New York.

—— 1981. Tectonics of the Helvetic nappes. *In*: MCCLAY, K. R. & PRICE, N. J. (eds). *Thrust and Nappe Tectonics*, Special Publication of the Geological Society of London No. **9**. 293–09.

—— (this volume). Fold and fault geometry of the western Helvetic nappes of Switzerland and France and its implication on the evolution of the arc of the western Alps.

——, CASEY, M. & KLIGfiELD, R. 1983. Role of shear in development of Helvetic fold-thrust belt of Switzerland. *Geology*, **11**, 439–42.

——, DIETRICH, D. & CASEY, M. 1985. Field excursion C: structural transect of the Swiss Alps. Part 1: western Helvetic nappes: *In*: ALLEN, P., HOMEWOOD, P. & WILLIAMS, G. (eds). *International symposium on foreland basins, Fribourg, 2–4 September 1985, Excursion Guidebook*.

—— & HUBER, M. 1983. *The Techniques of Modern Structural Geology. Vol. 1: Strain analysis*. Academic Press, London.

—— & HUBER, M. 1987. *The Techniques of Modern Structural Geology. Vol. 2: Folds and Fractures*. Academic Press, London.

SIDDANS, A. B. W. 1983. Finite strain patterns in some Alpine nappes. *Journal of Structural Geology*, **5**, 441–8.

TRÜMPY, R. 1963. Sur les racines des nappes helvétiques. *Livre à la mémoire du Professeur Paul Fallot. Mémoire Societé Géologique de France, 1960–1963*, 419–28.

—— 1969. Die helvetischen Decken der Ostschweiz: Versuch einer palinspastischen Korrelation und Ansätze zu einer kinematischen Analyse. *Eclogae Geologicae Helvetiae*, **62**, 105–42.

—— 1973. The timing of orogenic events in the central Alps. *In*: DE JONG, K. & SCHOLTEN, R. (eds). *Gravity and Tectonics*, 229–51. Wiley, New York.

—— 1980. *Geology of Switzerland, a guide book. Part A: An outline of the Geology of Switzerland*. Wepf & Co., Basel.

D. DIETRICH & M. CASEY, Geologisches Institut, ETH Zentrum, CH-8092 Zürich, Switzerland.

Strain-displacement compatibility and restoration of the Chaînes Subalpines of the western Alps

J.-P. Gratier, G. Ménard & R. Arpin

SUMMARY: Finite strain values and trajectories of the sedimentary cover of the French Chaînes Subalpines (Chartreuse, Vercors, Diois, Baronnies) with respect to the stable platform west of these regions were deduced from the strain measurements of a folded and faulted competent layer (Tithonian). The removal of this finite deformation was obtained by a finite element analysis (84 elements of initial size of about 15×15 km) and the finite displacement field of this cover was calculated. Assuming a stable margin (one fixed element, and one line with a constant direction during the progressive deformation) the displacement of the eastern, southern and western limits of the Chaînes Subalpines were estimated. A displacement towards the west, associated with a counter-clockwise rotation (15–20°) was found for the Miocene–Present basement thrust of the Belledonne massif. The presence of local heterogeneities (such as Jurassic normal faults) deflect the general E–W displacement along dextral NE–SW and N–S strike-slip faults and induce a displacement of the Chaînes Subalpines towards the southwest. The comparison of the restored depth of the Tithonian with its estimated initial depth reveals an excess of elevation of the Chaînes Subalpines with a maximum value (1000–1500 m) for the Diois–Vercors which might be associated with an external basement thrust under the Chaînes Subalpines.

When considering the deformation of the external zone of the western Alps, three domains may be schematically distinguished on an E–W section (Fig. 1a).

The western part may be considered as a stable margin. The sedimentary cover overlying the eastern continuation of the basement of the Massif Central is horizontal. The basement/cover interface is also nearly horizontal, affected only by normal and strike-slip faults. The latest marine sediment (Miocene) lies only a few hundred metres above the level of its deposition in the Rhône Valley and in the Bas–Dauphiné plain (Fig. 2).

Within the central part (Fig. 1a), in general called the Chaînes Subalpines, the basement/cover interface also remains nearly horizontal and also is affected only by normal and strike-slip faults. The sedimentary cover, however, is folded, with a large décollement at the basement/cover interface, with several thrust faults within this cover, and with a high elevation of Miocene sediments (up to 2000 m above sea level).

Finally, the eastern part is the mobile margin, with large thrust faults in the basement carrying the Belledonne and Pelvoux massifs westwards over the Chaînes Subalpines (Ménard 1979), and with the basement/cover interface as much as 3000 m above sea level.

The aim of this study is to estimate the displacement of the mobile margin (Chaînes Subalpines) with respect to the stable area to the W. On a section parallel to the orientation of the principal strain (Fig. 1b), it is simple to estimate the displacement of a mobile margin. Following Ramsay (1969, 1976) and Hossack (1978), the displacement is obtained by the integration of the strain values along the strain trajectories. When considering a competent layer that has been folded (and faulted) without any change of thickness (such as the Tithonian limestone of the Upper Jurassic sequence), the displacement of the mobile margin (relative to the stable margin) is simply the difference between initial and final length of this layer. The problems in estimating this strain for the western Alps are first, that a part of the deformation is accommodated by strike-slip faulting, and second, that the finite strain trajectories are not straight, but curved. This clearly appears on a structural map (Fig. 2); the orientation of the fold axes changes from NNE–SSW in the Chartreuse to N–S in the Vercors, and in the Diois and Baronnies there are at least two orientations of Alpine folds (Goguel 1947). To remove the strain from such a heterogeneously deformed region, is not simple. Following Schwerdtner (1977) and Cobbold (1979), however, the removal of the finite heterogeneous deformation using strain trajectories may be accomplished by dividing the whole domain into several elements, such that the deformation of each element can be considered as homogeneous and continuous.

From: Coward, M. P., Dietrich, D. & Park, R. G. (eds), 1989, *Alpine Tectonics*, Geological Society Special Publication No. 45, pp. 65–81.

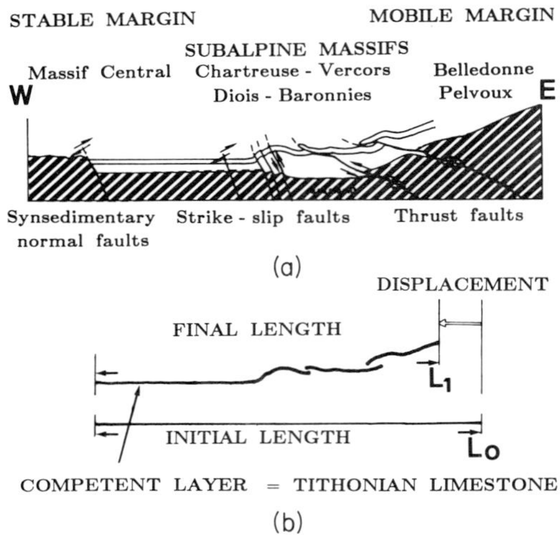

FIG. 1. (a) Schematic cross section through the Chaînes Subalpines from the stable margin (Massif Central) to the mobile margin (Belledonne–Pelvoux Massifs): hatched area = basement; arrows = movement on faults, either strike-slip or thrust. (b) Measurements of the displacement of the mobile margin, using the ratio of final length to initial length of a competent layer (folded and faulted without change of thickness), measured along a section parallel to the orientation of principal strain. The Upper Jurassic Tithonian limestone was used as a reference layer.

Method

The basis of the method used here was suggested by Schwerdtner (1977) and Cobbold (1979) and was clearly summarized by the latter author as follows: 'using the principal orientations, strain trajectories are constructed for the deformed state. The body is divided into finite elements bounded by these trajectories. Each element is then unstrained without changing its orientation or position. This process creates artificial voids and overlaps, which are minimized by imparting rigid translations and rotations to the elements. The result is the pattern of strain trajectories for the underformed state.' If we can choose both a line that did not undergo rotation during deformation and an element that did not undergo translation, then by comparing the strain trajectories before and after deformation, we can estimate the displacement of the mobile margins of the studied area with respect to this line and this element.

The first stage towards this aim was to determine the strain trajectories. This was done locally by measuring the horizontal strains (i.e. the ratio of the final, horizontal, length to the initial length of a folded competent layer) along three (or four) variously orientated cross sections across each area. (Fig. 3.) These regional strain measurements on cross sections allow us to calculate horizontal strain ellipses from place to place, with a slight overlap of the measured areas, for a greater accuracy of the strain orientations and magnitudes. To ensure a homogeneity of these strain values, the same value of 15 km was chosen as diameter of most unit circles (Fig. 3), in order to respect the assumptions of the method (i.e. homogeneous and continuous deformation of each element). This value is one order of magnitude greater than the displacement on the major thrust faults. For the eastern part of the Chartreuse, where large thrust displacement did occur, the diameter of the unit circle was 30 km.

The cross sections were drawn from 30 geological maps published at the scale of 1:50 000 (Bureau de Recherches Géologiques et Minières, France). For the Chartreuse and Vercors, recent seismic profiles were used to constrain the geometry of deep structures such as the décollement surfaces. The Tithonian limestone was chosen as the reference, or measured, layer, first, because it was folded without any change of thickness, and second, because it crops out throughout almost the entire Chaînes Subalpines. As often as possible the Urgonian limestone was also used to complete the estimation of the deformation from the Tithonian layer.

It was also necessary to take into account the effect of the strike-slip faults. This was accomplished for some regions, where the displacement along such faults is known from previous geological work (e.g. Goguel 1948, Gidon 1964, Arnaud 1981, François 1981), and a new strain ellipse was obtained as the product of the two different types of deformation. The deformation sequence is first folds and thrusts, then strike-slip faults.

With a knowledge of both the strain trajectories and the strain magnitude (treated as homogeneous and continuous within each measured area), the region can be divided into rectangular elements, with their sides parallel to the finite strain trajectories (Cobbold 1979). For a best possible adjustment of the elements restored to their initial unstrained state, the elements of the deformed state were drawn in order to have an initial size as close as possible to 15 km×15 km. Note, however, that if the grid in the strained state is chosen so that the sides of the elements are parallel to the finite strain trajectories, the shape and orientations of such elements in their initial unstrained state

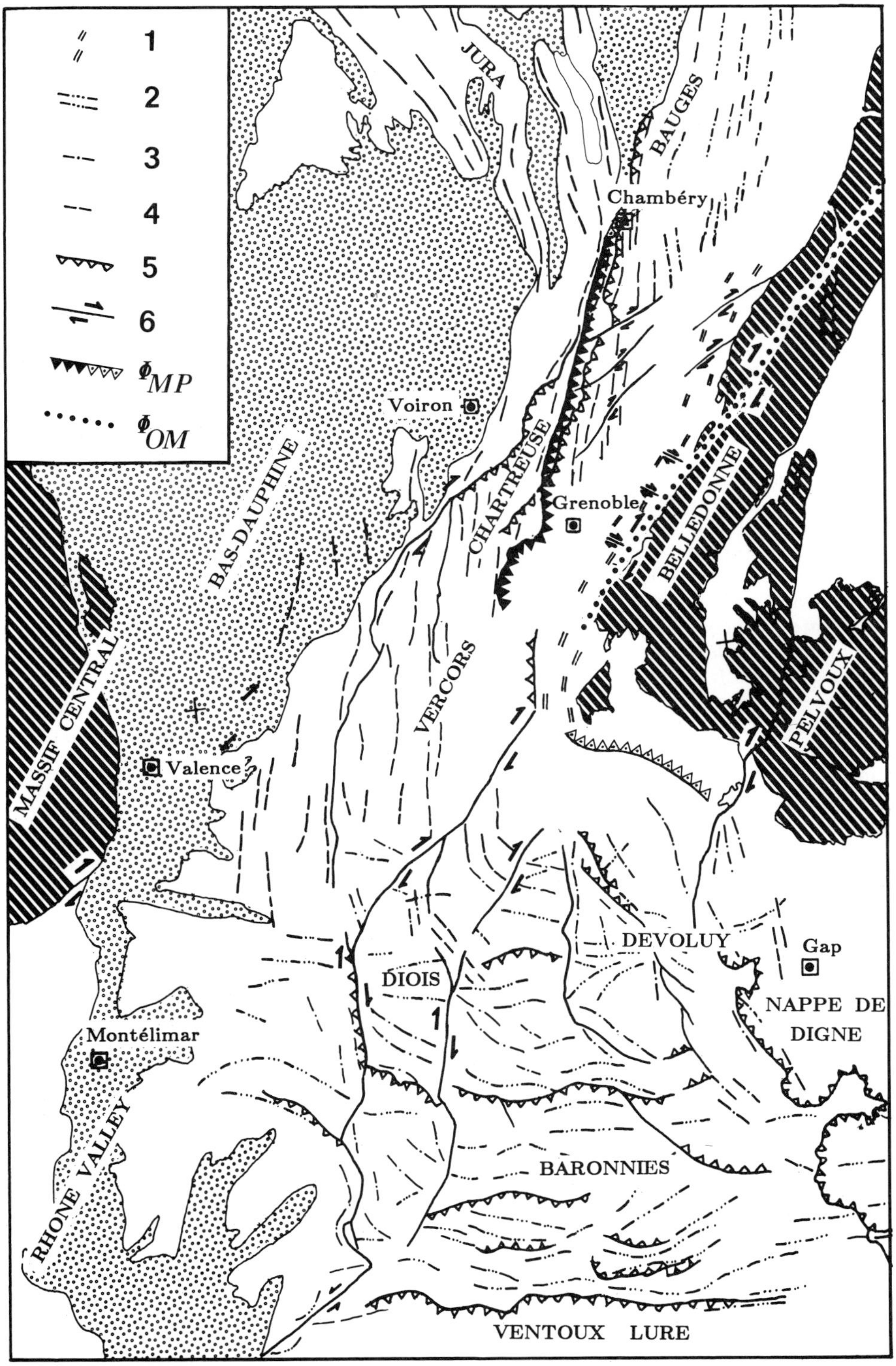

FIG. 2. Structural map of the Chaînes Subalpines showing the main structures of the region: 1 – spaced cleavage in the Jurassic slates of the cover of Belledonne; 2–3–4 – fold axes of different ages; 2 – Pyreneo–Provençal folds; 3 – Alpine folds (Oligo–Miocene); 4 – Alpine folds (Miocene–Present); 5 – main thrust faults; 6 – main strike-slip faults; hatched area – basement, shaded area – Tertiary; ϕ_{OM}–Oligo–Miocene basement thrust; ϕ_{MP}–Miocene–Present basement thrust (with black triangles = outcrop, with dotted triangles = presumed basement thrust concealed by the sedimentary cover).

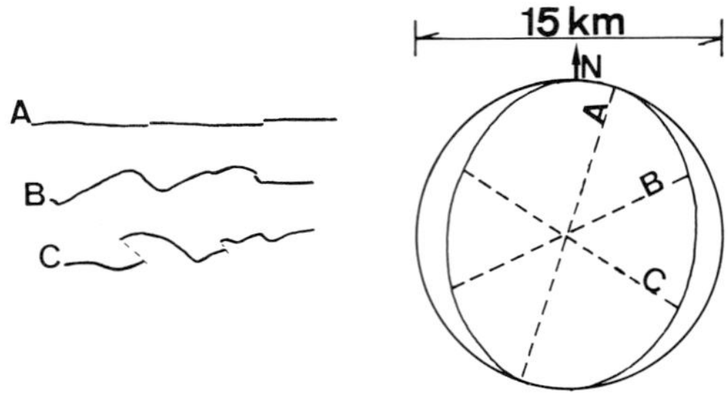

FIG. 3. Determination of strain trajectories and strain values by the measurement of the horizontal contraction (finite length L_1, initial length L_0) along three variously orientated cross sections A, B, C. The horizontal strain ellipses were calculated from place to place (with a slight overlap), for an initial length (L_0) of 15 km of the Tithonian limestone.

will not necessarily define a Cartesian grid (see Cobbold 1979). Therefore, since we cannot be sure that the initial strain trajectories conformed to a Cartesian grid (see Cobbold & Barbotin 1987), we cannot balance the magnitudes with the strain trajectories in order to have the same initial squares. To construct the deformed grid we simply used (independently) the strain trajectories deduced from the calculated strain ellipses (see above) and the magnitudes of strain measured along the sections parallel (or nearly parallel) to the principal strain directions. These continuous remeasurements of strain are necessary in order to avoid some gaps or overlaps of strain data along the strain trajectories.

An important source of error was the varying accuracy in the determination of the strain directions and magnitudes from place to place. In order to minimize these errors, the construction of the deformed grid was made by three successive stages, see Fig. 4(a):

(1) The first elements to be drawn were those with the most accurate strain magnitude data (regions with extensive outcrops of the Tithonian formation and with known displacements along the faults) and strain orientation data (those elements with only one principal direction of shortening).
(2) Second, we then drew the elements with accurate strain values (see above), but with relatively poor constraints on the orientations of principal strain (i.e. elements with either low strain values or with two principal orientations of shortening). For these elements, all of which lie between well constrained elements (Diois), the principal strain orientation was chosen to be compatible with that in the adjacent elements by assuming a smooth progressive variation in orientation as the strain values vary across the region. The problem of the western part of the studied area (Massif Central) will be discussed later (see discussion on the choice of the orientation of the undeformed grid). At this point, a first interesting observation is that it was possible to draw the best constrained elements in the deformed state such that the initial size of these elements was not far from a regular square of $(15\pm3)\times(15\pm3)$ km. This means that the strain trajectories transposed in the undeformed state were probably not very different from a Cartesian grid.
(3) To draw the elements with not very accurate strain data, complementary strain data were linearly interpolated from the adjacent elements in order to obtain an initial shape of these elements close to a square of $(15\pm1)\times(15\pm1)$ km. Of course, in the future this grid will have to be corrected after new strain measurements have been acquired for these elements. All the strain values used to construct the strain map are given in Table 1.

Another factor that can be useful for a tight control of the horizontal strain is the knowledge of the vertical strain. Both the depth of the décollement and the depth of the initial sedimentary reference layer (Tithonian limestone) are needed to estimate such vertical strains, when assuming constant volume in the deformation of each element. The depth of the décollement (which is often the basement/cover interface) is relatively well known in the western Alps, after the synthesis from seismic work and drill holes by Ménard (1979), and the results of new seismic profiles recently obtained through the Chartreuse (Thouvenot & Ménard 1989) and the Vercors (Arpin *et al.* 1988). On the contrary, an estimation of the depth of the Tithonian limestone before the sedimentary cover was folded is very difficult, because even over the region with horizontal sedimentary cover, the present depth of the top of the Tithonian formation varies from 0.5 km in the Bas–Dauphiné plain to 2.5 km in the Rhône valley (Fig. 2). In such areas it is clearly impossible to calculate the vertical strain, and therefore to balance these values with the horizontal strain. Nevertheless, using both the depth of the décollement and the horizontal strains (associated with folds and thrusts), it was possible to calculate the restored depth of the top of the Tithonian limestone (after the removal of the measured horizontal

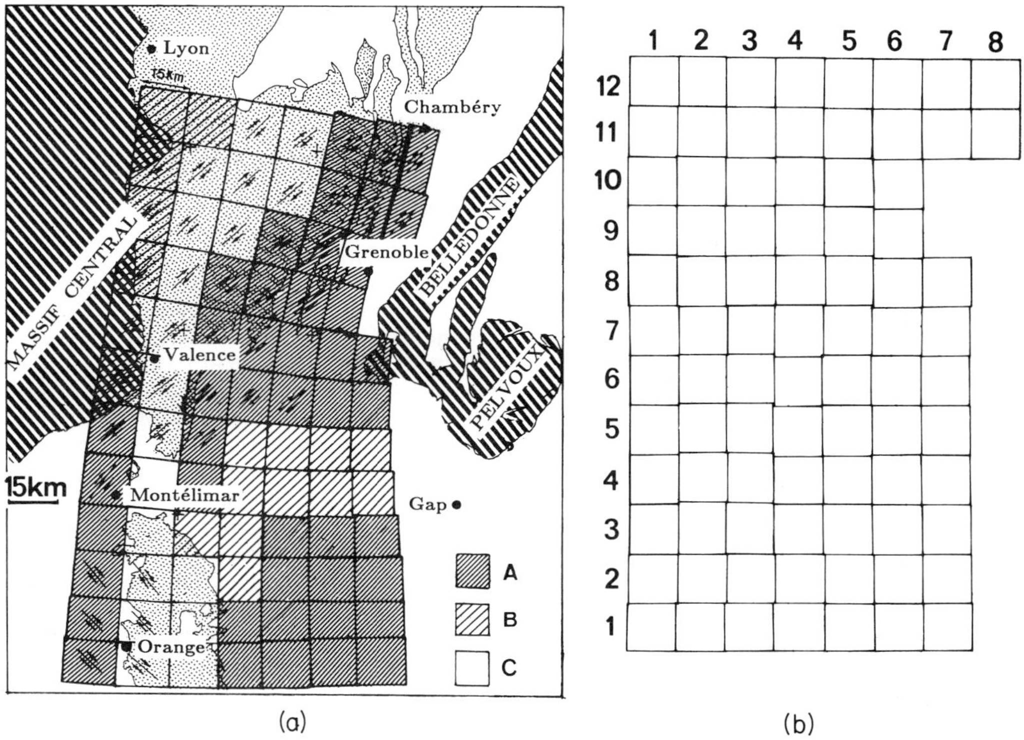

FIG. 4. (a) Deformation grid of the Chaînes Subalpines: using the strain values from measurements of the folded and faulted Tithonian formation (Fig. 3), the whole region was divided into 84 finite elements bounded by strain trajectories. The elements A, with both accurate estimation of strain values and principal directions, were drawn first. Then the elements B, also with good estimates of strain values but not with well constrained strain orientations, were drawn. Finally interpolations were made to draw the elements C, with few outcrops. The arrows within some elements indicate that strike-slip faults were taken into account to estimate the strain values for these elements. Hatched area: basement; shaded area: Tertiary. (b) Undeformed grid after the removal of the deformation of the elements and their adjustment. The figures of the rows and the columns give the reference of the elements in Table 1 where the strain values and strain orientation used for all the 84 elements are listed.

TABLE 1. *Strain values and principal horizontal strain orientations for the 84 elements of the finite deformation grid of the Chaînes Subalpines (Fig. 4a).*

1	– Reference number of the elements: horizontal row/vertical column of Fig. 4b
2 & 3	– Strain values associated with folds, thrusts and normal faults
4 & 5	– Strain values associated with strike-slip faults
6 & 7	– Finite strain values
8 & 9	– Finite strain directions

The elements 11/7–11/8 and 12/7–12/8 were grouped together in Fig. 4, since for these elements the displacement values on thrusts are larger than those for the other 80 elements.

strain). This depth must then be compared with the estimated initial minimum depth of the Tithonian formation necessary for the subsequent deposition of the entire marine sedimentary sequence after the Tithonian interval. The difference between the initial and the restored depths corresponds with the remainder of elevation after the strain has been removed, i.e. with the excess altitude of the Tithonian due to processes other than simple horizontal shortening of the cover above the basement (see discussion on vertical elevation).

TABLE 1. (*cont.*). *Elements of the finite deformation grid.*

1	2	3	4	5	6	7	8	9
1/1	1	.954	1.1	.909	1.1	.867	95	5
1/2	1	.971	1.1	.909	1.1	.883	91	1
1/3	1	.9	1	1	1	.9	88	178
1/4	.9	.933	1	1	.9	.933	88	178
1/5	1	.967	1	1	1	.950	88	178
1/6	1	.967	1	1	1	.967	88	178
1/7	1	.967	1	1	1	.967	88	178
2/1	1	.945	1.05	.952	1.05	.9	95	5
2/2	1	.911	1.05	.952	1.05	.867	90	0
2/3	.973	.850	1	1	.973	.850	87	177
2/4	.913	.863	1	1	.913	.863	87	177
2/5	.987	.820	1	1	.987	.820	87	177
2/6	.985	.757	1	1	.985	.757	87	177
2/7	1	.757	1	1	1	.757	87	177
3/1	1	.955	1.017	.983	1.017	.939	95	5
3/2	1	.955	1.017	.983	1.017	.939	90	0
3/3	.955	.980	1	1	.955	.980	89	179
3/4	.885	.922	1	1	.885	.922	89	179
3/5	.950	.894	1	1	.950	.894	88	178
3/6	.967	.9	1	1	.967	.9	88	178
3/7	.985	.937	1	1	.985	.937	85	175
4/1	1	.933	1	1	1	.933	95	5
4/2	1	.917	1	1	.990	.917	91	1
4/3	.937	.910	1	1	.937	.910	89	179
4/4	.885	.833	1	1	.885	.833	89	179
4/5	.934	.820	1	1	.934	.820	90	180
4/7	.917	.840	1	1	.917	.840	81	171
5/1	1	1	.968	1.033	.968	1.033	95	5
5/2	1	1	1	1	.985	.987	93	3
5/3	.935	.973	1	1	.935	.973	92	2
5/4	.867	.950	1	1	.867	.950	91	1
5/5	.906	.910	1	1	.906	.910	90	0
5/6	.870	.910	1	1	.870	.910	86	176
5/7	.877	.875	1	1	.877	.875	83	173
6/1	1	1	.968	1.033	.968	1.033	96	6
6/2	.983	1	.984	1.017	.967	1.017	95	5
6/3	.910	.917	.984	1.017	.895	.933	94	4
6/4	.867	.873	1	1	.867	.873	93	3
6/5	.910	.920	1	1	.910	.920	91	1
6/6	.885	.890	1	1	.885	.890	90	0
6/7	.833	.860	1	1	.833	.860	87	177
7/1	1	1	.923	1.083	.923	1.083	97	7
7/2	1	1	.923	1.083	.923	1.083	96	6
7/3	.924	1	.937	1.067	.866	1.067	96	6
7/4	.908	1	.937	1.067	.857	1.067	95	5
7/5	.913	1	.968	1.033	.884	1.033	94	4
7/6	.833	1	1	1	.833	1	93	3
7/7	.800	.933	1	1	.800	.933	90	0
8/1	1	1	.896	1.117	.896	1.117	97	7
8/2	1	1	.909	1.1	.909	1.1	97	7
8/3	.920	1	.937	1.067	.862	1.067	97	7
8/4	.850	1	.952	1.050	.809	1.050	97	7
8/5	.887	1	1	1	.887	1	97	7
8/6	.783	.967	1	1	.783	.967	96	6
8/7	.750	.950	1	1	.750	.950	92	2
9/1	1	1	.909	1.1	.909	1.13	97	7
9/2	1	1	.909	1.1	.909	1.1	98	8
9/3	1	1	.923	1.083	.923	1.083	99	9
9/4	.880	1	.923	1.083	.812	1.083	100	10
9/5	.887	1	.968	1.033	.859	1.033	102	12
9/6	.727	.933	1	1	.727	.933	102	12
10/1	1	1	.937	1.067	.937	1.067	97	7
10/2	1	1	.923	1.083	.923	1.083	98	8
10/3	1	1	.923	1.083	.923	1.083	100	10
10/4	.94	1	.937	1.067	.881	1.067	101	11
10/5	.87	1	.952	1.050	.829	1.050	103	13
10/6	.727	.983	1	1	.727	.983	104	14
11/1	1	1	.984	1.017	.984	1.017	96	6
11/2	1	1	.984	1.017	.984	1.017	96	6
11/3	1	1	.968	1.033	.968	1.033	98	8
11/4	1	1	.937	1.067	.937	1.067	100	10
11/5	.930	1	.909	1.1	.845	1.1	103	13
11/6	.75	1	.883	1.13	.662	1.13	105	15
11/7-8	.4	1	.870	1.15	.348	1.15	100	10
12/1	1	1	1	1	1	1	95	5
12/2	1	1	1	1	1	1	95	5
12/3	1	1	.968	1.033	.968	1.033	96	6
12/4	1	1	.937	1.067	.937	1.067	97	7
12/5	1	1	.896	1.117	.896	1.117	97	7
12/6	.8	1	.822	1.22	.658	1.22	98	8
12/7-8	.32	1	.789	1.267	.252	1.267	100	10

Strain calculation from cross sections

The strain data used to estimate the horizontal strain ellipses were deduced from numerous cross sections through the folded and faulted sedimentary cover. The geological data were obtained from the published 1:50 000 geological maps. The data on the deep structures were deduced from the synthesis by Ménard (1979) for the basement/cover interface (Diois–Baronnies), and from new seismic profiles (Chartreuse and Vercors).

A seismic profile from the Bauges to the Chartreuse (Thouvenot & Ménard 1989) revealed two interesting types of deep structures (Fig. 5, section 12a). The basement/cover interface was obtained both under the Chartreuse (parautochthonous and autochthonous basement) and under the Liassic cover of the Belledonne massif (overriding basement). A second interface within the sedimentary cover at a shallow depth (2–3 km) was also found to underlie the major part of the Chartreuse. This latter interface was interpreted as the top of a competent layer, and because no thick limestone layer is known to exist below the Tithonian limestone in the sedimentary sequence, this interface was considered to be the top of an autochthonous domain overlain by an allochthonous domain. After discussion (Mugnier *et al.* 1987), the interface was interpreted to represent the top of the Urgonian limestone instead of the Tithonian. With these geological and geophysical data, several balanced cross sections, which may be restored to their initial state, were drawn (Fig. 5, sections 6 to 12) (see Arpin *et al.* 1988 and Thouvenot & Ménard 1989).

The thrust sequence through the Chartreuse implies firstly the development of a large thrust (ϕ_1 = dashed line, Fig. 5) and secondly a later folding of this thrust associated with a décollement parallel to the Jurassic black shale (ϕ_2 = dotted line, Fig. 5). The ϕ_1 thrust surface is always complex with two levels of décollement (on an horizontal deep thrust surface within the crust, and along the Jurassic black shale) related to two ramps. During this thrust sequence the basement reached the top of the more external ramp, but did not climb over it (see sections 6 to 12, Fig. 5). The ϕ_2 thrusts occurred afterwards

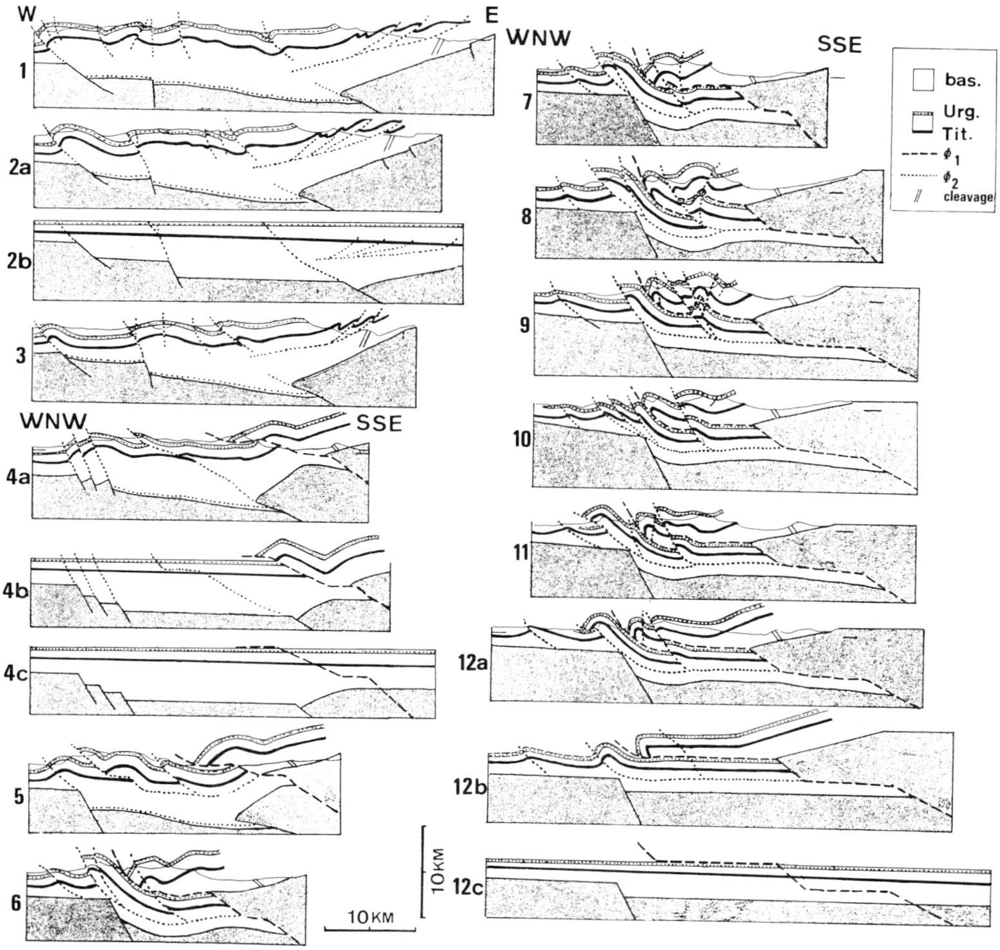

FIG. 5. Balanced cross sections through the Chartreuse and Vercors (after Arpin *et al.* 1988). Locations of sections are on Fig. 7. The restored state is given for three sections (2, 4, 12) with an intermediate stage between initial and finite deformation for sections 4 and 12. The geological surface data were obtained from the 1:50 000 published geological maps, from an unpublished 1:100 000 geological map of the Vercors (Arnaud 1981), and from local outcrop studies. The data on deep structures were deduced from seismic profiles (Arpin *et al.* 1988, Thouvenot & Ménard 1989). Shaded area: basement; the dotted double line is the Urgonian limestone (mid-Cretaceous formation); the thick line is the Tithonian limestone (Upper Jurassic); two types of thrust faults were distinguished: ϕ_1 = dashed line, ϕ_2 = dotted line.

because probably less work is needed to create new external thrust faults, parallel to the Jurassic black shales, than to cause large internal deformation of the basement by climbing over the second ramp. Note that the ϕ_2 surface may also be parallel to the basement/cover interface of the most external domain (see sections 6 to 12 in Fig. 5), where, because of the effect of normal faults, the depth of the basement/cover interface is almost the same as the Jurassic black shales under Chartreuse.

The ϕ_1 thrust can be followed up to the north of the Vercors (sections 4 and 5, Fig. 5, with an outcrop of the ϕ_1 surface on section 4a), but with a drastic decrease of the amount of the displacement towards the west. Consequently, towards the south of section 3 (Fig. 5), the geometry of the thrusts becomes different. The décollement surface is always found along the basement/cover interface. This discontinuity is deduced from seismic profiles, using explosions from working quarries near Grenoble (Arpin *et al.* 1988). Other structures appear from section 3 towards the south (sections 1, 2, 3 in Fig. 5); back thrusts developed above the overriding basement and partially accommodated the horizontal E–W shortening of the cover, south of the termination of the ϕ_1 thrust.

The cross sections through the Diois and Baronnies cannot be balanced because both principal horizontal strains indicate shortening. These sections were simply drawn using the widespread outcrops of Tithonian limestone, which is never very far from the topographic surface, and is well mapped on the 1:50 000 geological maps (Fig. 6).

In two regions there are no outcrops of the Tithonian limestone. In one region data from gravity studies reveal en échelon N–S folds along a NE–SW strike-slip fault (eastern limit of the Bas–Dauphiné plain, Orgeval & Rumeau 1955, Gamond & Odonne 1984). In the other region (Rhône valley), interpolations were made from adjacent regions (see method section).

Near the eastern limit of the Chartreuse, but east of the measured sections, a spaced cleavage affects the Jurassic slates of the Belledonne cover. Strain trajectories can be drawn assuming that this cleavage defines the X–Y strain plane (Ramsay 1967), (Fig. 2). A progressive change is observed in the orientation of the strain trajectories from west of Chartreuse (N10) to the western limit of Belledonne (N20–30), and will be discussed below.

Strain removal, displacement field

Strain removal and finite total displacement field

The first step was to remove the deformation in each element, then the elements in their initial shapes were displaced and rotated by hand in order to minimize the voids and the overlaps between neighbouring elements. The discrepancies between adjacent elements were distributed over the whole region by adjusting the lengths of the columns and of the rows in order to have the boundaries of the region as smooth as possible (Fig. 4b). The undeformed grid is not very different from a Cartesian grid, as anticipated previously when considering the small curvatures of the strain trajectories and the low finite strain gradient.

The next step was to choose the orientation of the Cartesian grid, superposed on the undeformed grid, in order to calculate first the transformation of the grid and then the displacement of the mobile margin. No sufficiently reliable data are available to infer the rigid rotation of a deformed region (see also Percevault & Cobbold 1982). Consequently, we used the following approach. As noted above, the continuation of the Massif Central (the Bas–Dauphiné Plain) acts as a rigid indentor, in front of which the folded sedimentary cover is heterogeneously deformed with a more or less symmetrical curvature of the strain trajectories between the Jura and the Vercors–Chartreuse Chaînes (Fig. 2). We chose the bisector of the triangular indentor as an unrotated line, in order to allow minor symmetrical shear zones on both the NE–SW (dextral) and on the NW–SE (sinistral) strike-slip faults along the limits of the indentor.

When points of the initial Cartesian grid are linked with those of the deformed grid, a schematic view of the relative finite displacement reveals a motion of the cover of the southern mobile zone towards the N and NW and a motion of the eastern mobile zone towards the W and SW (Fig. 8). Of course, this is not the true displacement path of the cover since several successive stages of deformation, with different orientations, occurred in the region. In order to discuss the displacement fields associated with each of these successive stages, the tectonic history and the major structural features of the Subalpine massifs must be considered.

Tectonic history and structural features of the Chaînes Subalpines

A simplified tectonic history of the Chaînes Subalpines may be summarized from the work of Goguel (1947), Flandrin (1966), Vialon (1974), Arnaud (1981), Gidon (1982), Tricart (1984) and Lemoine (1984) as follows.

Continental rifting of the European margin occurred during the deposition of the Lias–Dogger and is always associated with the tilting of large blocks bounded by variously orientated normal faults. These faults are often inherited from preceding structural events (i.e. the Massif Central structures). Within the area studied, synsedimentary normal faults were commonly reactivated as strike-slip or thrust faults (Fig. 2). At least four orientations of such normal faults have been revealed by stratigraphic studies: NE–SW faults (along the eastern limit of the Massif Central and Bas–Dauphiné Plain); N–S faults (through the Diois–Baronnies); E–W faults (between the Ventoux and the Baronnies); and NW–SE faults (along the south-west limit of the Belledonne–Pelvoux massifs). The development of these normal

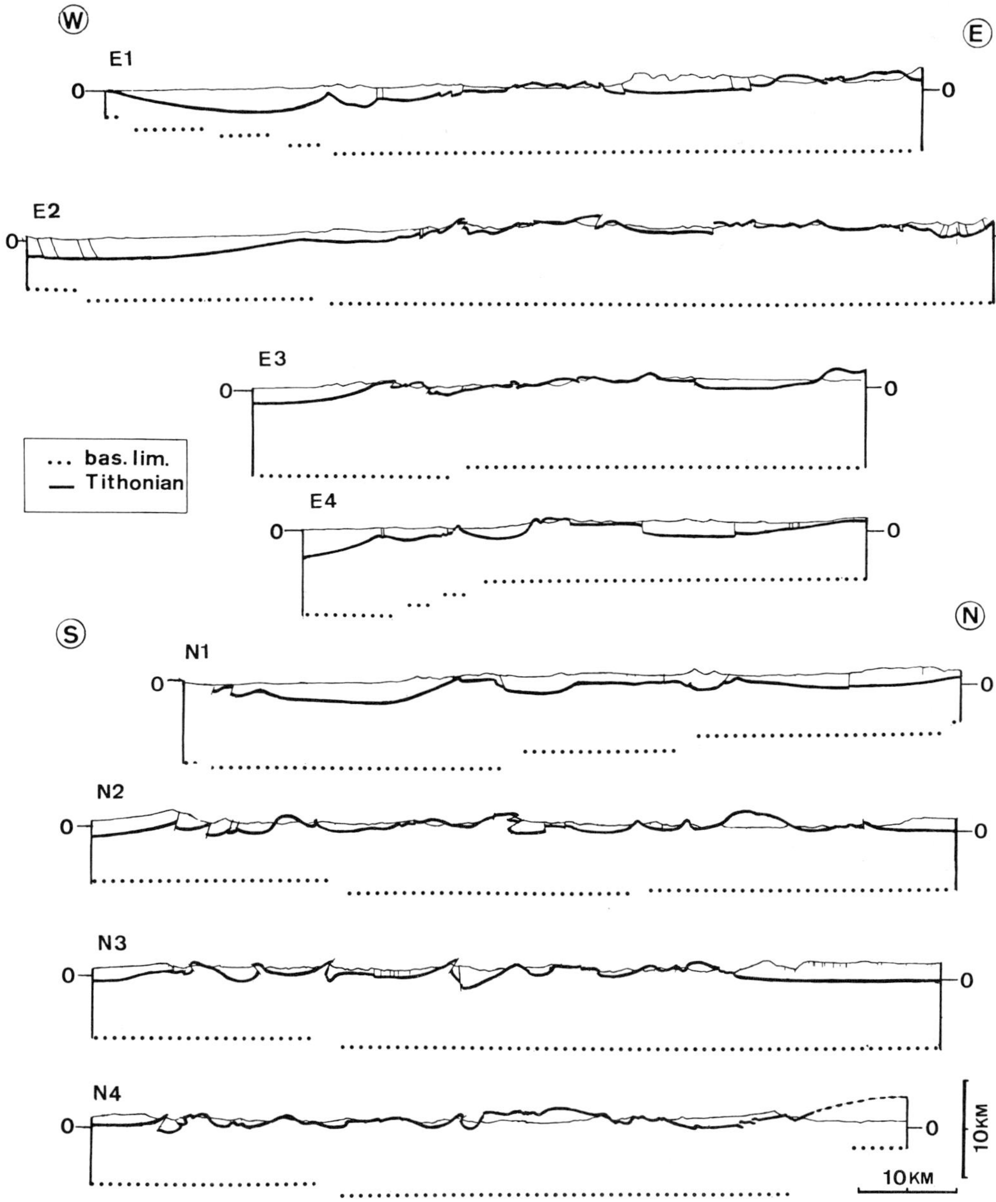

FIG. 6. Cross sections through the Diois and Baronnies (after Jouve, 1985). Locations of sections are on Fig. 7. The surface data were obtained from the 1:50 000 geological map, and the data on the basement/cover interface are from Ménard (1979). Dotted line: basement/cover interface; thick line: Tithonian limestone (Upper Jurassic).

faults implies extension with all horizontal orientations, but the E–W trending normal faults are rather localized in the southern part of the studied area. Nevertheless, the choice of the Tithonian limestone as reference layer allows us to assume that the effect of the Lias–Dogger horizontal extension was not registered by the strain measurements of the Tithonian limestone, because these phases of horizontal extension probably had ceased by the end of the Jurassic period. Nevertheless we shall see that these normal faults acted as heterogeneities and geometric discontinuities during the Alpine contraction (see discussion section).

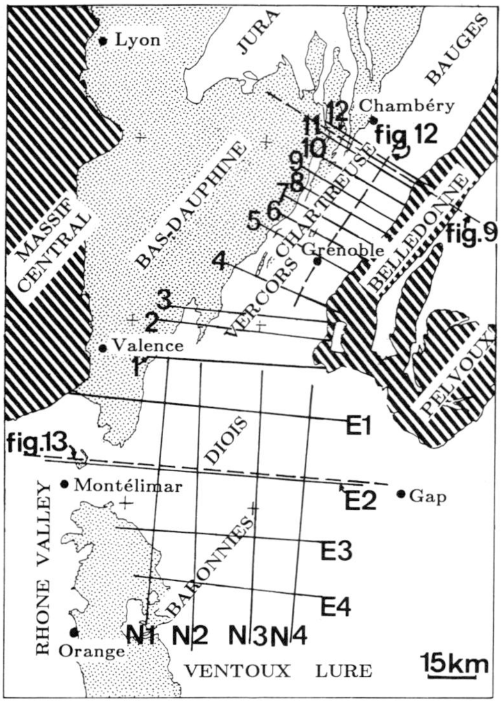

FIG. 7. Map showing locations of cross sections shown in Figs 5, 6, 9, 12 and 13a (1 to 12 on Fig. 5, E1 to E4 and N1 to N4 on Fig. 6). Dashed area: basement; shaded area: Tertiary.

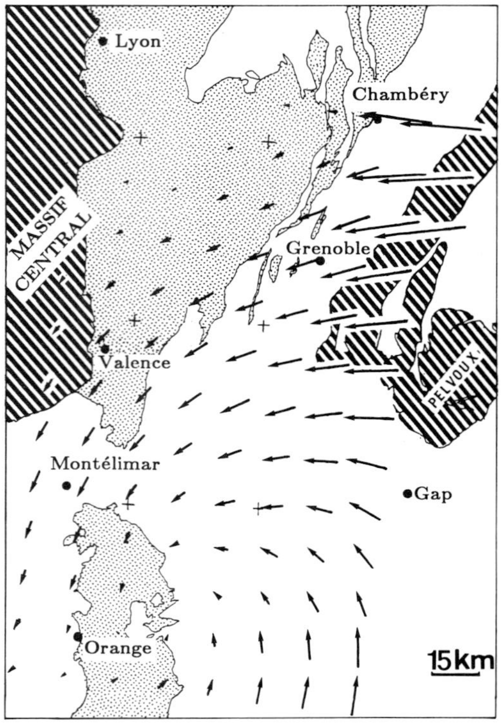

FIG. 8. Finite total displacement field: shortest path from points on a Cartesian grid (superposed on the undeformed grid) to points on a finite deformation grid (see Fig. 4b and a). Dashed area: basement; shaded area: Tertiary.

The contraction phases, which occurred afterwards, may be schematically divided into two stages: Pyreneo–Provençal and Alpine (Fig. 2).

The Pyreneo–Provençal stage began with the development of the pre-Senonian E–W folds of the Diois–Baronnies linked to a NNE displacement of the limit of the southern mobile belt. The N–S contraction increases from west to east and is associated with sinistral NE–SW strike-slip faults (Arnaud 1973). These E–W folds were locally amplified during the Cenozoic part of the Pyreneo–Provençal stage (up to the Miocene for the Ventoux thrust).

The Alpine stage, in the Chaînes Subalpines, probably began during the Oligocene, and is linked to the first large basement thrusts of the Belledonne massif, now sealed between the internal and the external domains of Belledonne (ϕ_{OM}, Figs 2 and 9). We named this first basement thrust 'Oligo-Miocene' (ϕ_{OM}). Its effect, largely seen in the deformation of the Bauges (Doudoux *et al.* 1982), decreases from north to south. Movement on this fault is probably not very important in the Chartreuse and is almost nil in the Vercors, but the spaced cleavage of the Jurassic slates between Chartreuse and Belledonne may be associated with this thrust. The major E–W contraction of the Chartreuse–Vercors–Diois instead was associated with the development of a second more external basement thrust of the Belledonne massif, active from the Miocene to the present, here named Miocene–Present thrust (ϕ_{MP}). The N–S fold axes (often along pre-existing N–S normal faults), the dextral strike-slip movement along these faults, and an amplification of the Pyreneo–Provençal E–W folds may be associated with this Alpine stage (Diois–Baronnies), see Goguel (1963). Other basement thrusts have been anticipated in the western part of the main ϕ_{MP} basement thrust (Fig. 5), but following Ménard & Thouvenot (1984), these thrusts do not continue below a shallow décollement surface within the basement (Fig. 9). Finally, the effect of the basement thrusts of the Pelvoux massif toward the west (Fig. 2) must also be taken into account to explain the E–W contraction of the Diois–Baronnies (see discussion section).

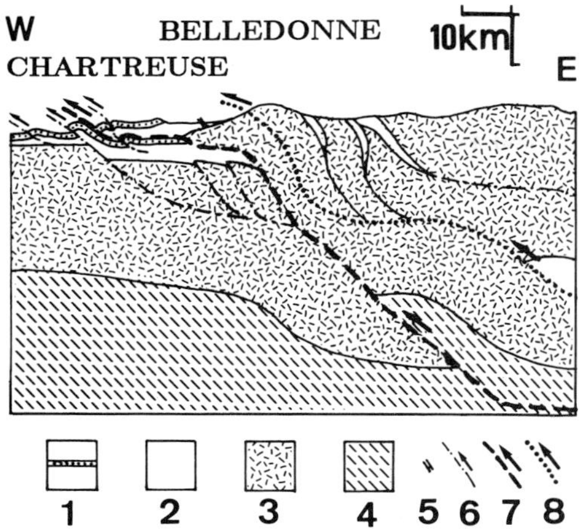

FIG. 9. Balanced crustal cross section, from Ménard & Thouvenot (1984), showing the two main basement thrusts: Miocene–Present external thrust dashed line (7) = ϕ_{MP}; and Oligo–Miocene thrust, dotted line (8) = ϕ_{OM}; (6) other thrusts. The spaced cleavage of the Jurassic slates of the Belledonne cover (5) may be related to the ϕ_{OM} thrust. (3) and (4), upper and lower crust; (2), sedimentary cover; (1), Tithonian to Urgonian formations.

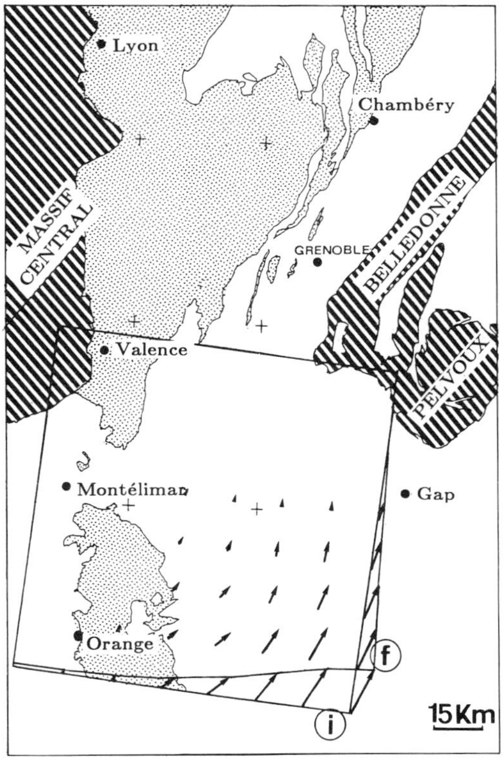

FIG. 10. Finite displacement field associated with the Pyreneo–Provençal stage of deformation: initial (i) to final (f) state; dashed area basement; shaded area Tertiary.

Finite Pyreneo–Provençal displacement field

The discrimination between the Pyreneo–Provençal and Alpine deformation is not easy, but an attempt was made to distinguish the displacement pattern for each stage. Following the geological evidence given above, a Pyreneo–Provençal deformation model was drawn using the N–S contraction values measured within the Diois–Baronnies, associated with the displacement towards the NNE. The amount of the NNE displacement decreases from east to west (from 15 to 1 km) and from S to N (up to an underformed domain near the limit between Vercors and Diois). The displacements associated with the E–W horizontal extension were taken to vary from zero along the northern limit of Diois to a maximum value of 5 km along the southern limit of Baronnies. The region with extension is limited by the Rhône valley (using the whole measured finite E–W horizontal extension, Table 1), and no significant extension is seen along the E–W Diois–Baronnies folds. The resulting displacement pattern is given in Fig. 10. The progressive changes of both the amount and the orientation of the displacements were derived from the measured strains and assumed boundary conditions.

Finite Alpine displacement field

The difference between the total finite deformation and the Pyreneo–Provençal deformation gives the Alpine deformation, with a simpler pattern of displacement (Fig. 11) than that of the total finite displacement (Fig. 8). The main characteristics of this Alpine displacement are that the limit of the eastern mobile zone is carried towards the W. The progressive increase of this E–W cover displacement, from the S (5 km) to the N (30 km) implies a counter-clockwise rotation of this eastern limit of the Tithonian outrops of about 15–20° (Fig. 11). The presence of a NE–SW stable domain along the western limit of the Vercors–Chartreuse massifs induces a lateral expulsion of the sedimentary cover towards the southwest, with a general counter-clockwise rotation of the Chaînes Subalpines.

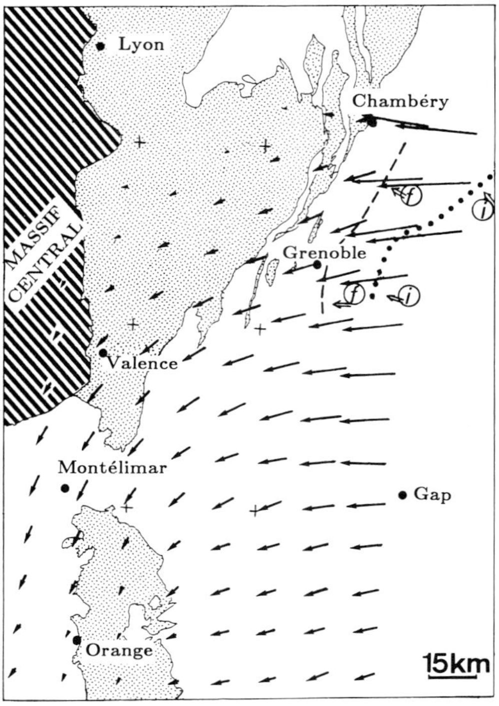

FIG. 11. Finite Alpine displacement field deduced from the difference between the total displacement field (Fig. 8) and the Pyreneo–Provençal stage (Fig. 10). The present western limit of the hanging wall ramp of the Miocene–Present thrust of the Belledonne Massif is drawn as a dashed line, its initial state (after the removal of the Alpine deformation) is drawn as a dotted line. The double arrows indicate the direction of displacement deduced from measurements of magnetism of samples in their present position (f) and in their reconstructed initial position (before the Miocene–Present thrust) (i); dashed area, basement; shaded area, Tertiary.

(a) Displacements of the mobile zone along the western stable limit

Along the NE–SW trending limit of the Vercors–Chartreuse (probably a Jurassic system of normal faults) the displacement of the sedimentary cover is clearly associated with a dextral strike-slip movement on these NE–SW trending faults, and also with the development of N–S trending folds along pre-existing N–S normal faults (with reversal in the sense of slip on these faults, see Gillcrist *et al.* 1987, and Fig. 5). Through the Diois–Baronnies, the displacement towards the SW is mainly deflected along two major N–S trending dextral strike-slip faults (Fig. 2), again Jurassic normal faults reactivated with a component of thrust displacement. The geological observations bear no clear evidence for relating these N–S trending faults with the faults on the NE–SW limit of the Vercors–Chartreuse. The Alpine dextral displacement, at the north termination of the N–S faults of the Diois instead is partitioned into several divergent branches (N–S to NE–SW dextral strike-slip, see Fig. 2). The general counter-clockwise rotation of the Chaînes Subalpines is limited in the W by two parallel fault systems that curve from NE–SW to N–S: one along the eastern part of the Massif Central–Bas Dauphiné, and the other along the western limit of the Diois–Baronnies. Such a system of lateral expulsion, associated with large amounts of counter-clockwise rotation, with curved X–Y finite strain trajectories within this deformed domain and limited by NE–SW dextral strike-slip faults, has already been described within the Bourg d'Oisans basins by Gratier & Vialon (1980). The same geometry of lateral termination of a mountain belt is also proposed by Molnar & Lyon-Caen (1989).

The horizontal displacement of the sedimentary cover, which progressively decreases from the mobile (eastern) to the stable (western) limit, has to be related to the mechanical and geometrical properties of the décollement zone. In the particular case of the Diois–Baronnies, the pre-existing Cretaceous E–W folds induced an anisotropic behavior of the cover, with a relative weakness along the N–S direction that facilitated the counter-clockwise rotation of these regions.

(b) Displacement along the eastern mobile limit: Belledonne massif

Another problem is the relation between the displacement of the cover and that of the basement, especially near the eastern mobile limit of the Chaînes Subalpines. The evidence of two successive basement thrusts in the Belledonne Massif (Fig. 9) requires an estimation of the part of the cover displacement associated with each of these basement thrusts. For example, the strains registered by the spaced cleavage of the Jurassic slates (east of the eastern limit of the shortening measurements of the Tithonian, Fig. 2), must not be used to estimate the displacement of the basement associated with the Miocene–Present basement thrust, because this cleavage is clearly above the ϕ_{MP} basement thrust (see Fig. 9). Probably, either this part of the cover was passively transported (piggy-back) by the Miocene–Present thrust with the cleavage being associated with the Oligo–Miocene basement thrust, or this cleavage developed during gravity sliding movements (in this case associated with the tilting of the basement acquired during the Miocene–Present thrust). In contrast, the deformation registered by the Tithonian folds and faults through the Chartreuse–Vercors may be associated entirely with the displacement of the Miocene–Present basement (including the back-thrusts between Vercors and Belledonne). This displacement may be visualized by drawing the western limit of the basement hanging wall ramp and by calculating the initial position of this line using the strain data of the Tithonian layer. This was done in Fig. 11, where a 15–20° counter-clockwise rotation was associated with the westwards displacement of the basement, together with an horizontal extension parallel to the Belledonne Massif (less than 20%). As in the Chartreuse, this NNE–SSW extension of the Belledonne Massif was probably accommodated by strike-slip faulting: dextral ENE–SSW faults and WNW–ESE sinistral faults (clearly visible all along the Belledonne Massif, Gratier *et al.* 1973). A part of the internal deformation of the Belledonne basement may also be associated with the Alpine deformation. The deformation history observed through the Belledonne schists closely matches the Alpine deformation history of the sedimentary cover of the Belledonne and Pelvoux Massifs (Gratier & Vialon 1980): asymmetric folds with slaty cleavage, reorientation of the fold axes from the Y into the X direction with increasing strain, crenulation cleavage, and Riedel faults. All these structures indicate thrust movement towards the west, on shear planes dipping towards the east.

(c) Displacement along the eastern mobile limit: Belledonne–Pelvoux transition

The problem of the SW end of the Belledonne basement thrust will now be discussed. The ϕ_{MP} basement thrust (Fig. 11) shows a decrease of displacement from N of Chartreuse (30 km) to the mid Vercors (15 km), but most of this decrease is acquired S of the Chartreuse and N of the Vercors. On a schematic NNE–SSW section (Fig. 12), it clearly appears that the decrease of the thrust displacement is associated with a lateral ramp. More exactly, the depth of the flat level of the autochthonous Urgonian unit (acting as the décollement interface) rapidly increases from its value N of Vercors to that S of Chartreuse, just at the latitude of Grenoble. It may be noted that after the removal of the Belledonne thrust the southwest limit of Belledonne is just the SE prolongation of the NW–SW part of the Isère valley (from Grenoble to Voiron, Fig. 2). This NW–SE trending limit could represent an old fault re-actived during the Alpine deformation firstly as sinistral strike-slip fault (to accommodate the different amounts of westward displacement between Chartreuse and Vercors), and secondly as fault with vertical movement (Fig. 12). When taking into account the geometric compatibility of the deformation of the cover associated with the counter-clockwise rotation (Fig. 4), these vertical movements must be associated with a thrust fault near Grenoble and a normal fault near Voiron. An explanation for the abnormal vertical elevation of the Vercors massif will be discussed later (see discussion on the partitioned vertical elevation): the NW–SE part of the Isère valley could be a lateral ramp of a recent external basement thrust under the Baronnies, Diois and Vercors (see Fig. 13a).

Finally, even if the ϕ_{MP} Belledonne thrust displacement decreases from north to south, there remain 15 km of horizontal E–W displacement SW of the end of the Belledonne Massif. From this zone to the actual end of the region with E–W contraction (south of the Baronnies) other basement thrusts through the Pelvoux Massif have to exist (Fig. 2). Understanding

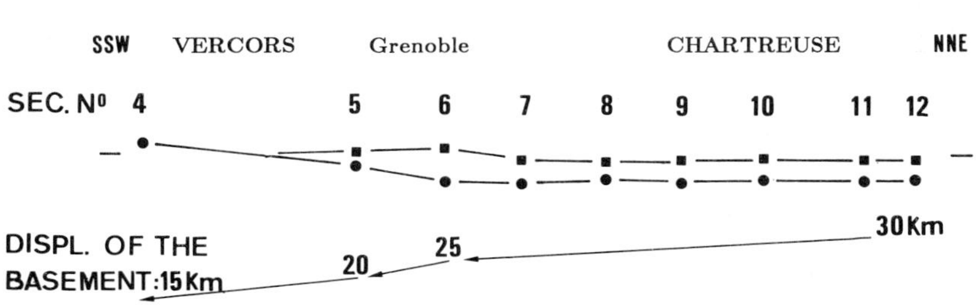

FIG. 12. Schematic longitudinal cross sections along the Chartreuse–Vercors, showing elevations of the autochthonous Urgonian limestone (black points), estimated along a flat level of décollement (ϕ_1), and showing the variation in elevation of the allochthonous Tithonian limestone (black squares). The position of this longitudinal cross section is indicated on Fig. 7. The values of the westwards displacement of the basement are given below the section (15–30 km).

this clearly requires complementary strain measurements through the Devoluy and the nappe de Digne structure. In any case, it is possible that the basement of the Pelvoux massif was thrust towards the SW, laterally bounded by NE–SW dextral strike-slip faults (Gidon *et al.* 1976). The early Oligo–Miocene thrust of Belledonne was reactived as a strike-slip fault during this Pelvoux thrusting movement (Ménard & Fréchet 1989), (Fig. 2). As this Belledonne strike-slip fault may continue through the Vercors–Diois as a NE–SW strike-slip fault (Fig. 2), the same movement is observed for the basement and the cover: a lateral expulsion towards the SW limited by strike-slip faults, and associated with a counter-clockwise rotation.

(d) Relation with the development of the western Alpine arc

The strain removal based on the Tithonian limestone indicates a progressive change of the orientation of the displacement from N95 (north of the Chartreuse) to N80 (mid-Vercors). The same change in orientation was recently revealed by the study of the magnetic anisotropy of cleaved Jurassic slates sampled between Chartreuse–Vercors and Belledonne Massif (Lamarche & Rochette 1987, Aubourg 1987): the displacement direction varies from N125 at the latitude of the north of Chartreuse to N90 at the mid-Vercors latitude (double arrows on Fig. 11). As noted above, the cleavage in the Jurassic slates was probably not directly associated with the Miocene–Present basement thrust. If this cleavage is associated with the Oligo–Miocene basement thrust (see Fig. 9), the removal of the Miocene–Present deformation implies a variation of the Oligo–Miocene thrust orientation from N145 (Chartreuse) to N110 (Vercors). A contemporaneous NNE–SSW extension parallel to the Belledonne Massif is registered by numerous tension gashes (up to 10% of horizontal extension, Gratier 1984) in the Jurassic slates of the Belledonne cover, and probably also by strike-slip faults (see above). These directions are relatively approximate depending on the mechanism of rotation of the basement (rigid rotation and/or internal distortion). In all cases, the same counter-clockwise movement could be associated with the westwards motion of the two successive basement thrusts of the Belledonne Massif. The same evolution is also suggested by Vialon *et al.* (this volume), in the context of the movement of the entire western Alps. A counter-clockwise rotation of the Belledonne Massif associated with its westwards movement is linked to the general displacement of the Apulian indentor (African Plate) towards the N and the NW.

Limits and accuracy of the results

As noted above, the method used to remove the finite deformation of a sedimentary cover is based upon several assumptions. Different assumptions could lead to a different displacement field. The strain map (Fig. 4) is clearly a first attempt to take into account all the geometric features of a deformed area, but, of course, corrected strain values may always be introduced. Here, the various possibilities of error in the strain data will not be discussed for each element, we discuss only the possible errors associated with the orientation of the initial Cartesian grid (see method section). If the same fixed element (e.g. 12/1, the nearest to Lyon in Fig. 4) is considered, but with a slightly different orientation (±5°) of the initial grid, the following differences in the displacement field are introduced.

With a different orientation of 5° clockwise of the initial grid, the estimated direction of the displacement towards the west for the eastern limit varies from N115 (north of the Chartreuse) to N135 (Vercors–Diois limit), with an associated change in the displacement of the eastern limit from 30 to 10 km. A N135 direction of displacement along the southwestern limit of the Pelvoux Massif seems unrealistic since most measured orientations of contraction through this region (Goguel 1963, Gidon 1982) are approximately NE–SW (from NNE–SSW to ENE–WSW). For the western limit, a displacement towards the east would be required by strain compatibility. This similarly unrealistic displacement can be reduced by changing the interpolated strain values along the Rhône valley (see Fig. 4), with a progressive increase southward of the E–W horizontal extension from Lyon (nil) to more than 15 km for the southern limit, and of course yet more towards the Mediterranean. This also seems to be incompatible with the geological evidence.

With a different orientation of 5° counter-clockwise, the displacement direction of the eastern limit varies from N75 (north of Chartreuse) to N70 (Vercors–Diois limit), the displacement values varying from 30 km (Chartreuse) up to 25 km (Vercors). In this case, the N75 direction of finite displacement is clearly difficult to accept for the Chartreuse with its NNE–SSW trending folds, and also N75 trending strike-slip faults (see Fig. 2). For the western limit, a large displacement toward the SW, with an increasing value from Lyon (nil) to for example 15 km (Montelimar), and yet more towards the S, is also difficult to accept (but it is not as unrealistic as are the directions of displacement).

These two examples show the limits and also the accuracy of the results. Taking into account all the geological constraints, it is possible to estimate that the directions of the finite displacement may vary by ±10°, and the displacement values by ±10–20%, due to the discrepancy in the orientation of the initial grid.

Partitioned vertical elevations

It remains now to discuss the vertical elevations associated with the measured horizontal contraction. As noted previously (see method section), it is not possible to balance the horizontal and the vertical strain since the initial level of the reference layer (Tithonian), just before the Alpine contraction, varies greatly from place to place. A restored elevation of the Tithonian was calculated, using the amount of horizontal contraction for groups of four elements, and assuming constant volume during the deformation of the cover above the décollement surface, which commonly lies at the basement/cover interface. The depth of this décollement is known from seismic profiles (Thouvenot & Ménard 1989, Arpin *et al.* 1988). The great size of the analysed blocks (30×30×10 km) allows us to assume constant volume deformation. This restored elevation was then matched with the estimated initial elevation of the Tithonian limestone (that is the minimum depth of the Tithonian limestone necessary to allow the deposition of the marine sedimentary sequence above this Tithonian limestone, Fig. 13a). The difference between the two values gives an excess of elevation of the Tithonian limestone due to processes other than the folding and thrusting of the cover above the basement. This excess altitude was plotted on a map (Fig. 13b). A maximum excess elevation of 1000 to 1500 m clearly appears for the south Vercors, the northern and eastern Diois, and the eastern Baronnies. These values decrease towards the west to less than 500 m in the Bas–Dauphiné plain and in the Rhône valley (southwards of a prolongation of the Diois). Applying the same model used to explain the elevation of the Belledonne basement (basement involved thrust faulting), the excess altitude of the Chaînes Subalpines might be due to the development of a new external basement thrust (see Fig. 13a), with a blind thrust now just below the Rhône valley. A very small displacement (of some kilometres, according to Molnar & Lyon-Caen 1988), is sufficient to explain the calculated excess altitude if the thrust affects the whole thickness of the crust as do the Belledonne and Pelvoux thrusts.

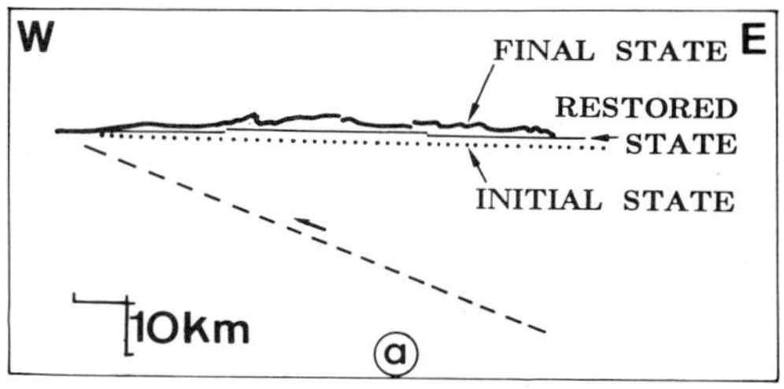

FIG. 13. (a) Comparison between the present elevation of the Tithonian limestone, the elevation after the removal of the finite deformation, and the elevation necessary for the deposition of the marine sequence covering the Tithonian. An external basement thrust (dashed line) may explain the excess of elevation of the Tithonian. The location of the section is shown on Fig. 7.

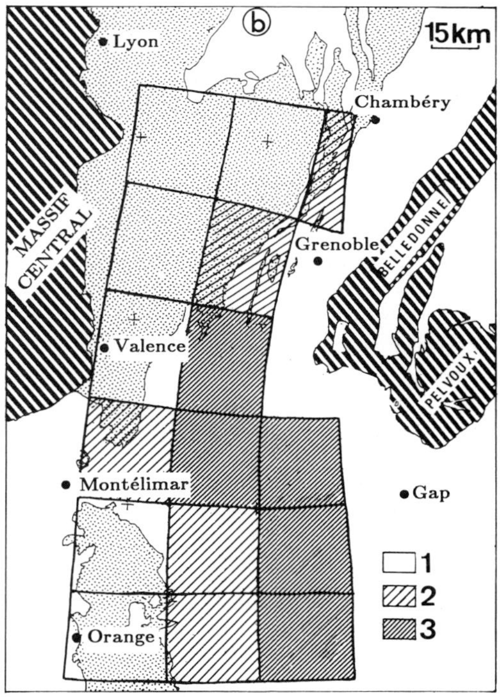

FIG. 13. (b) Distribution of the excess elevation of the Tithonian s.l. through the Chaînes Subalpines. Dashed area, basement; shaded area, Tertiary; excess altitude from 1000 to 1500 m (3), from 500 to 1000 m (2), and from 0 to 500 m (1).

Conclusions

In a large heterogeneously deformed region, the finite displacement field may be deduced from an analysis of regional finite strains by dividing the whole region into several homogeneous finite elements and by integrating these strain values along the strain trajectories.

From strain localised in the folds and thrusts of the Tithonian limestone of the Chaînes Subalpines, associated with the Miocene–Present basement thrust of the Belledonne Massif, the Alpine finite displacement directions and amounts indicate a westwards movement of the eastern limit of the Chaînes Subalpines, and an associated counter-clockwise rotation of 15–20° of this limit, with respect to the underformed stable part of the Massif Central.

NE–SW to N–S trending Jurassic normal faults along the stable margin (Massif Central) deflect the westwards displacement of the Chaînes Subalpines toward the SW and locally towards the S. A general counter-clockwise rotation of the Chaînes Subalpines is observed, limited along the stable margin by dextral strike-slip faults which reactivated the pre-existing Jurassic normal faults.

After the removal of the folding and thrusting of the Tithonian limestone, the restored depth of this layer was calculated (using the décollement surface deduced from seismic data) and matched with the estimated initial depth of the Tithonian (depth of the marine deposition of the overlying sedimentary sequence). An excess of altitude was found (up to 1500 m) especially for the southern Vercors, the Diois and the eastern Baronnies, which could be due to a present external basement thrust under the Chaînes Subalpines associated with the lateral expulsion of these regions.

Such a system of Alpine lateral expulsion, associated with large amounts of rotation, with arcuate X–Y finite strain trajectories within this deformed domain, and limited by curved strike-slip faults along the stable margin, is one of the possible geometries of lateral termination of a mountain belt.

Finally, this work is only a first attempt to take into account all the geological and geophysical data that constrain the amounts and orientations of deformation in a heterogeneously deformed region. The finite deformation grid may always be corrected, as soon as new data are acquired. In a general way this method of restoring a sedimentary cover which is heterogeneously deformed between stable and mobile margins could be applied to other regions, to study the geometry and the displacement field of lateral termination of mountain belts.

ACKNOWLEDGEMENTS: We thank D. Dietrich, P. Molnar and the anonymous reviewers for their careful revision of this paper.

References

ARNAUD, H. 1973. Mise en évidence d'un important décalage antémiocène de sens senestre le long de la faille de Presles (Vercors occidental). *Comptes-Rendus de l'Académie des Sciences Paris*, **276**, 2245–8.

—— 1981. De la plate-forme urgonienne au bassin vocontien: le Barremo–Bédoulien des Alpes occidentales entre Isère et Büech. *Géologie Alpine*, **12**. Unpublished Thèse d'Etat, Grenoble.

ARPIN, R., GRATIER, J. P. & THOUVENOT, F. 1988. Chevauchements en Vercors et Chartreuse déduits de l'équilibrage de données géologiques et géophysiques. *Comptes-Rendus de l'Académie des Sciences Paris*, **307**, 1779–89.

AUBOURG, C. 1987. *Etude de la fabrique magnétique des Terres Noires subalpines*. Unpublished Diplôme d'Etudes Approfondies, Université de Grenoble.

BUREAU DE RECHERCHES GEOLOGIQUES ET MINIÈRES: 1:50 000 geological maps of La Tour du Pin, Chambéry, Voiron, Montmélian, Serrière, Beaurepaire, Grenoble, Domène, Romans, Vif, Vizille, Valence, Charpey, La Chapelle en Vercors, La Mure, Crest, Die, Mens, St Bonnet, Dieulefit, Luc en Diois, Gap, Valreas, Nyons, Serres, Orange, Sederon, Avignon, Carpentras, Sault. Orléans, France.

COBBOLD, P. R. 1979. Removal of finite deformation using strain trajectories. *Journal of Structural Geology*, **1**, 67–72.

—— & BARBOTIN, E. 1988. Kinematic significance of fanning strain trajectories. *Journal of Structural Geology*, **10**, 811–8.

DOUDOUX, B., MERCIER DE LEPINAY, B. & TSARDY, M. 1982. Une interprétation nouvelle de la structure des massifs subalpins savoyards (Alpes occidentales): nappes de charriage oligocène et déformation superposées. *Comptes-Rendus de l'Académie des Sciences Paris*, **295**, 63–8.

FLANDRIN, F. 1966. Sur l'âge des principaux traits structuraux du Diois et des Baronnies. *Bulletin de la Société Géologique de France*, **7**, **VIII**, 376–86.

FRANCOIS, D. 1981. L'étude des accidents méridiens de Saillans–Mérindol et de Die-la-Motte–Chalançon (Diois–Baronnies). Unpublished rapport, Institut Français du Pétrole.

GAMOND, J. F. & ODONNE, F. 1984. Critères d'identification des plis induits par un décrochement profond: modélisation analogique et données de terrain. *Bulletin de la Société Géologique de France*, **7**, **XXVI**, 115–28.

GIDON, M. 1964. Nouvelle contribution à l'étude du massif de la Grande-Chartreuse et de ses relations avec les régions avoisinantes. *Travaux du Laboratoire de Géologie Alpine*, **40**, 187–205.

—— 1982. La reprise de failles anciennes par une tectonique compressive: Sa mise en évidence et son rôle dans les Chaînes Subalpines des Alpes occidentales. *Géologie Alpine*, **58**, 53–68.

——, PAIRIS, J. L. & APRAHANIAN, J. 1976. Le linéament d'Aspres-lès-Corps: sa signification dans le cadre de l'évolution structurale des Alpes occidentales externes. *Comptes-Rendus de l'Académie des Sciences Paris*, **282**, 271–4.

GILLCRIST, R., COWARD, M. & MUGNIER, J. L. 1987. Structural inversion and its controls: examples from the Alpine foreland and the French Alps. *Geodinamica Acta*, **1**, 5–34.

GOGUEL, J. 1947. Recherches sur la tectonique des Chaînes Subalpines entre le Ventoux et le Vercors. *Bulletin de la Service Carte Géologique de la France*, **223, XLV**, 533–78.

—— 1948. Sur le rôle des failles de décrochement en Chartreuse. *Bulletin de la Société Géologique de France*, **5, XVII**, 227–35.

—— 1963. L'interprétation de l'arc des Alpes occidentales. *Bulletin de la Société Géologique de la France*, **7, V**, 20–33.

GRATIER, J. P. 1984. *La déformation des roches par dissolution-cristallisation: aspects naturels et expérimentaux de ce fluage avec transfert de matière dans la croûte supérieure*. Unpublished Thèse d'Etat, Grenoble.

——, LEJEUNE, B. & VERGNE, J. L. 1973. *Etude des déformations de la couverture et des bordures sédimentaires des massifs cristallins externes de Belledonne, des Grandes Rousses et du Pelvoux*. Unpublished Thèse 3ème cycle, Université de Grenoble.

—— & VIALON, P. 1980. Deformation pattern in a heterogeneous material: folded and cleaved sedimentary cover immediately overlying a crystalline basement (Oisans, French Alpes). *Tectonophysics*, **65**, 151–80.

HOSSACK, J. R. 1978. The correction of stratigraphic sections for tectonic finite strain in the Bygdin area, Norway. *Journal of the Geological Society of London*, **135**, 229–241.

JOUVE, J. F. 1985. *Analyse quantitative de bassins plissés et reconstitution de leur état initial: exemple du Diois–Baronnies*. Unpublished, Diplôme d'Etudes Approfondies, Université de Grenoble.

LAMARCHE, G. & Rochette, P. 1987. Microstructural analysis and origin of lineation in the magnetic fabric of Alpine slates. *Tectonophysics*, **139**, 285–93.

LEMOINE, M. 1984. La masse occidentale de la Tethys ligure et les Alpes occidentales. *In*: BOILLOT, G. (ed). *Les marges continentales en mer et à terre autour de la France*, Masson, 155–248.

MENARD, G. 1979. *Relation entre structures profondes et structures superficielles dans le Sud-Est de la France. Essai d'utilisation de données géophysiques*. Unpublished Thèse 3ème cycle, Grenoble.

—— & THOUVENOT, F. 1984. Coupes équilibrées crustales: méthodologie et applications aux Alpes occidentales. *Geodinamica Acta*, **1**, 35–45.

—— & FRECHET, J. 1988. Mécanismes au foyer de séismes des Alpes occidentales et modèle de déformation actuelle de la Chaîne. *Earth and Planetary Sciences Letters*, in press.

MOLNAR, P. & LYON-CAEN, H. 1988. Some simple physical aspects of the support, structure and evolution of Mountain belts. *Geological Society of America*, Special publication no. 218.

—— & —— 1989. Fault plane solution of earthquakes and active tectonics of the northern and eastern parts of the Tibetan Plateau. *Philosophical Transaction of the Royal Society of London*, in press.

MUGNIER, J. L., ARPIN, R. & THOUVENOT, F. 1987. La construction de coupes équilibrées: méthodes et applications au nord du massif subalpin de la Chartreuse. *Geodinamica Acta*, **1**, 125–37.

ORGEVAL, M. & RUMEAU, J. L. 1955. Préreconnaissance du bassin de Valence. Bureau de Recherche du Pétrole. *Collection Institut Français du Pétrole*, **5054**.

PERCEVAULT, M. N. & COBBOLD, P. R. 1982. Mathematical removal of regional ductile strains in central Brittany: evidence for wrench tectonics. *Tectonophysics*, **82**, 317–28.

RAMSAY, J. G. 1967. *Folding and Fracturing of Rocks*. McGraw-Hill, New York.

—— 1969. The measurement of strain and displacement in orogenic belts. *In*: KENT, P. E., SATTERTHWAITE, G. E. & SPENCER, A. M. (eds). *Time and Place in Orogeny*, 43–79.

—— 1976. Displacement and strain. *Philosophical Transactions of the Royal Society*, **A283**, 3–25.

SCHWERDTNER, W. M. 1977. Geometric interpretation of regional strain analysis. *Tectonophysics*, **39**, 515–531.

THOUVENOT, F. & MENARD, G. 1989. Allochthony of the Chartreuse Subalpine massif: seismic constraints. *Journal of Structural Geology*, in press.

TRICART, P. 1984. From passive margin to continental collision: a tectonic scenario for the western Alps. *American Journal of Science*, **284**, 97–120.

VIALON, P. 1974. Les déformations synchisteuses superposées en Dauphiné. Leur place dans la collision des éléments du socle préalpin. *Bulletin Suisse de Minéralogie et Pétrographie*, **54**, 553–690.

J.-P. GRATIER, G. MÉNARD & R. APRIN, LGIT Observatoire de l'Université de Grenoble, IRIGM BP 68 38402 Saint-Martin d'Hères Cedex, France.

Southwestward thrusting and tectonics of the western Alps

N. Fry

SUMMARY: The gross geometry of the western Alps is the result of three processes operating since Eocene times: thrusting to WNW; thrusting to SW; arcuate lithospheric flexure due to loading. In the *external zones*, taken to include the SubBriançonnais domain, these processes are more substantial than earlier phases of Alpine tectonics. In the *internal zones*, these processes rework substantial earlier Alpine structures. A major feature of Alpine thrusting is the preservation of basement topography, which influences the geometry of the thrust system. This is demonstrated by the Digne thrust system W from Remollon and is implicated in both formation and emplacement sequence of the Lower Embrunais–Ubaye nappes, in the Embrunais–Ubaye arc, in the localization of WNW thrust outcrops S of the Pelvoux Massif and perhaps in the southerly derivation of Gres d'Annot sediments. A major (>100 km) phase of SW thrusting affects the Embrunais–Ubaye terrain (continuity of external and SubBriançonnais) before the development of the lower Embrunais–Ubaye nappes and the arcuate deformation of their external floor. All these Embrunais–Ubaye processes are consistent with maintenance of a single 230° transport direction, and with thrusts arising from depth, causing real crustal shortening. Later events with southwestward transport are the emplacement of the combined Parpaillon–Briançonnais thrust stack, the Argentera Massif 'pop-up', and the Digne thrust system. Despite circumstantial evidence, there is no substantial contribution to the regional tectonics of the western Alps from strike-slip tectonics, inversion tectonics, gravity tectonics, rotation during thrusting, radial thrusting or arc bending. The dominant features are the maintenance, spanning hundreds of kilometres and tens of millions of years, of the two principal directions of thrust transport (230°; 290°) and the regional continuity of displacements which this implies.

A comparison of any collection of recent papers on the geology of the western Alps, such as this volume, clearly shows a lack of any consensus as to the general character of the Alpine orogenic regime. Some recent papers assume only one dominant direction of orogenic contraction, although disagreeing whether the general style is of *thin-skinned thrusting* to WNW (e.g. Butler 1985, Butler *et al.* 1986) or *inversion tectonics*, in which structures produced by Mesozoic southeastward extension are reactivated in reverse sense by Alpine compression (e.g. Lemoine *et al.* 1986, Cooper & Williams, in press). Other recent authors recognize a spread of transport directions on Alpine structures and invoke a variety of hypotheses which permit divergent directions to develop simultaneously at different locations around the Alpine arc. Chief among these are *rotation* of thrust sheets (e.g. Gratier this volume), *radial thrusting* (e.g. Malavieille *et al.* 1984), which requires *lateral spreading* of an overthrust mass to occur on the scale of the western Alpine arc, or the radial pattern to result from later *arc bending*. These may easily be combined (e.g. Platt this volume).

The western Alps also display examples of structures classically interpreted as gravity nappes, notably the pre-Alps and the nappes of Embrunais–Ubaye (Debelmas & Kerckhove 1973). The same *gravity sliding* hypothesis has been extended by a number of authors to structures of the external zones (e.g. Graham 1981).

This paper will argue that none of the hypotheses mentioned above is satisfactory as a general characterization of the western Alpine tectonic regime. Such a characterization must include the following essential features:

(1) Major structures interpreted in the past as gravitational in origin, including the Embrunais–Ubaye nappes, have geometry indicative of thrusting from depth.
(2) There are two major directions of thrust transport in the external western Alps (and more in the internal zones).
(3) The transport direction of each thrust system remains constant throughout the western Alpine region and through geological time. Divergent directions are the result of superimposition of more than one system.

The most important single feature lacking from most Alpine interpretations is a southwesterly directed thrust-system of regional extent in the SW Alps, which is the main subject of this paper.

Historically, recognition of two main directions of thrusting of regional extent is not new.

From: Coward, M. P., Dietrich, D. & Park, R. G. (eds), 1989, *Alpine Tectonics*, Geological Society Special Publication No. 45, pp. 83–109.

Nor are most other current concepts applied to the western Alps. Goguel's view of the western Alpine arc (Goguel 1963) for example incorporates radial transport directions, rotating thrust sheets and lateral spreading, together with the two dominant directions of basement thrusting. He also recognized the constraints imposed by the need for regional continuity, for example relegating proposed radial transport patterns to superficial gravitational emplacements, not basement thrusts. Butler *et al.* (1986) mention the SW system, but then evade the real problems of continuity by relegating it (incorrectly) to being later, of small displacement and beneath the WNW system with which they are more familiar. To the extent that the views in this paper are new, they result from paying more detailed attention to the geometry of thrust surfaces, ramps and cut-offs, and to constraints imposed by restorability and continuity of displacements. These give improved estimates for directions and distances of thrust displacement and provide a distinction between gravity sliding surfaces and thrusts from depth.

The region within which southwestward thrusting can be discerned lies approximately between the Pelvoux Massif in the north and Provence in the south, between Sisteron in the west and the Dora Maira Massif in the east (Fig. 1). An attempt has been made (line TF of Fig. 1) to define a thrust front. In the south, this has been drawn to include the outside of the Castellane Arc, from which structures appear to trace northwards into the molasse basin of the Plateau de Valensole. Thrusts in the Mesozoic strata beneath the molasse are reported by Dubois & Curnelle (1978) and confirmed by drilling at Mirabeau, west of Digne. Thrusts within the molasse itself are inferred by Vann *et al.* (1986) from the structure of the eastern margin. These authors imply that such thrusts do not break surface in the main part of the basin, but they do crop out in its northern part. The external limit to thrust outcrops in Fig. 1 has been drawn on the assumption of a shallow floor which does not extend westwards of the molasse in the vicinity of Sisteron. It has then been traced northwestwards, to include slight repetitions in the Jurassic 'Terres noires' of Lower Bochaine and to link up with the thrusts of Col de Cabre, Grimone and eastern Diois. If thrust displacements extend west of the limit drawn in Fig. 1, the effects are restricted to backthrust reactivations of the early E–W folds of Provence and west of the Buech, and the uplift on the Durance line.

Southwestward thrusting becomes difficult to trace with certainty both to the SE, around Nice, and in the N, around the Pelvoux Massif, because of the superimposition of this thrust system with others of different transport direction. The question of a radial pattern of thrusting, whether real or apparent, on the scale of the western Alps as a whole will be reconsidered after presentation of evidence on individual thrust systems.

When considering crustal shortening, little distinction will be made between forethrusts and backthrusts sharing the same direction. However, there is a spatial separation as far as NE-SW shortening is concerned. Forethrusts, carrying the overthrust slab southwestwards over its floor, dominate structures in the external, Sub-Briançonnais and parts of the Briançonnais zones. Backthrusts, overthrusting to the NE, occur mainly as ductile shear zones in the Dora Maira Massif, Monviso ophiolite and the immediately overlying schistes lustrés, and may extend further in subcrop beneath the sediments of the upper Po Plain. These shear zones cut the dominant structures of these units, which indicate an independent and earlier system of ductile overthrusting to the W rather than early forethrusts of the same southwestward thrust system.

This paper will tackle separately four different aspects of the regional structure, and then try to draw conclusions together in a consideration of the nature of Alpine thrust tectonics.

Southwestward thrusting (mainly Neogene) of the Digne system near the Durance

The first area to be discussed spans the northward extension of the Digne thrust system and includes the northern end of a strip of country with shallow pre-Triassic basement (Menard 1979). As such, it is a convenient area for making some introductory remarks about mapping in thrust terrains and continuity of thrusts, and about the effects of both frontal and lateral steps in basement level on thrust networks. The history of the Digne system is complex, but its main southwestward transport occurred in Neogene (Aquitanian, Tabianian) times (Gidon *et al.* 1970, Gigot *et al.* 1974, Siddans 1979). The area extending from Remollon, on the Durance, for 20 km to the W, south of Gap, is represented schematically in Fig. 2. (Fig. 1 of Ehtechamzadeh Afchar & Gidon 1974 covers most of this area, which is centred on the letter 'E' on the regional structural map Fig. 1 of Siddans 1979.)

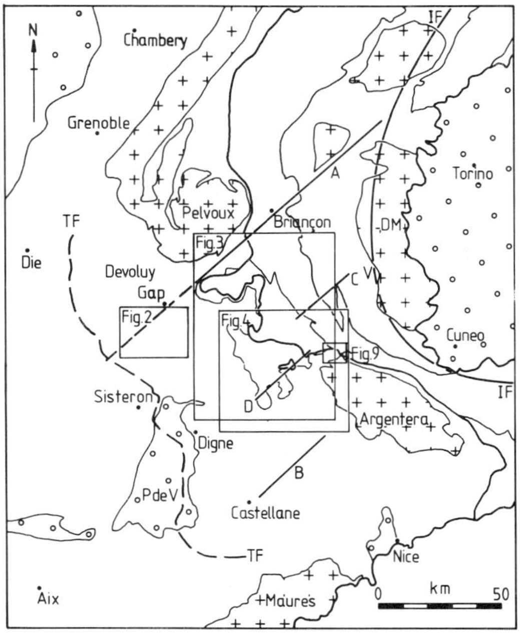

FIG. 1. The southwestern Alps, showing locations of areas referred to in more detail in later figures. Lines A–D are present or past monoclinal flexures which trend along the direction of thrust transport 050° to 230°. Line A (solid): monocline to SE, with (pecked) continuation as pre-Oligocene monocline to NW. Line B: persistent lateral line in Castellane Arc thrusts. Line C: monocline preceding later thrusts, now seen as an alignment of wedging out of Briançonnais units. Line D: monocline affecting Embrunais–Ubaye thrust surface geometries. IF: approximate front, at depth, of the seismic surface to the Ivrea Body. TF: present 'thrust front' for all SW thrusting; the southwesternmost extent of thrust repetitions recognized at the present erosion level. P de V: Plateau de Valensole (molasse basin). DM: Dora Maira (Internal) Massif. V: Monviso ophiolite.

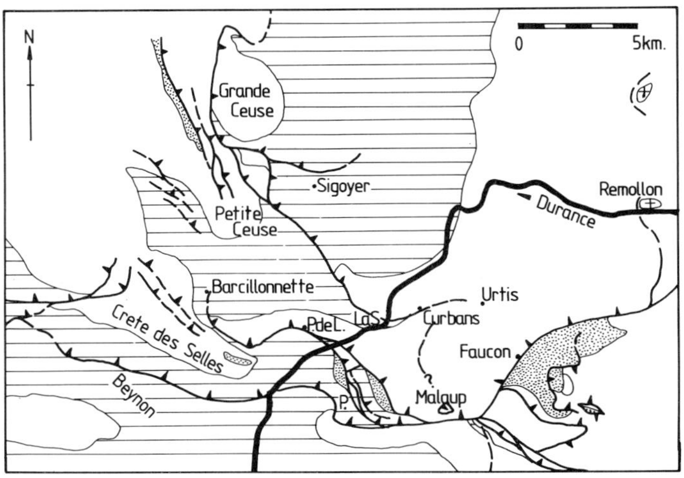

FIG. 2. Schematic map of thrusts of the Digne thrust system *s. l.* to the west of Remollon. Crosses denote pre-Triassic basement. Ruled lines indicate Middle and Upper Jurassic. Speckle denotes Oligocene, unconformable on Jurassic and Cretaceous formations. Note that thrusts are not continuously mappable, that they appear to climb from Lower Jurassic to Cretaceous from SE to NW, but that the Oligocene occurrences show this to be due to a pre-Oligocene northwesterly tilt P. = Peyrouard. P. de L. = Plan-de-Lardier. LaS. = La Saulce.

Several major thrusts, clearly displayed (Fig. 2) by stratigraphic repetition along parts of their outcrops, become untraceable where they lose both stratigraphic separation and angular discordance. (1) The Sigoyer thrust, carrying the Grande Ceuse structure, can be traced eastward past Sigoyer but becomes lost within Terres noires. (2) Stratigraphic repetition near Curbans and Urtis within Lower Jurassic calcareous shales east of the Durance appears to be a direct continuation of the Saulce thrust in the W (contrary to Ehtechamzadeh Afchar & Gidon 1974). It has not been traced further E and may trace up (or reappear on) the dip slope of the mountainside on the S, rising up towards the klippe of Malaup. The continuity of Liassic succession across the Durance in the thrust slice immediately beneath (denied by Ehtechamzadeh Afchar & Gidon 1974) imposes on the Saulce thrust a continuation somewhere to the E. (3) The thrust carrying the Lias just referred to, the Plan-de-Lardier thrust, is itself lost northwards in Terres noires near Barcillonnette.

Thrusting has in several areas given rise to an array of thrusts rather than a single thrust surface. Figure 2 indicates schematically the existence of two such arrays NW of Petite Ceuse. One of these traces to the N above, the other below the Montmaur thrust of Mesozoic on Oligocene, W of Grande Ceuse. The former is linked by a series of splays directly from the underlying mappable Montmaur thrust, whereas the latter does not link to a mappable thrust directly. The Peyrouard thrust-array links to its overlying Plan-de-Lardier thrust.

It should be clear from these few observations that, when mapping thrust terrain such as this, it would be unwise to erect named rock-masses, e.g. Plan de Lardier thrust sheet, and to assume that they can be extrapolated by mapping their bounding surfaces as unambiguous single and continuous thrusts. Residual ambiguity about the amount of the thrust displacement transferred on splays or on to invisible thrust flats without discernible stratigraphic separation is an element of any realistic thrust mapping.

The Remollon dome: persistent basement topography; limited inversion tectonics

Figure 2 includes the northwestern edge of a region of shallow basement which around its northern extremity gives the outcrops of pre-Triassic rocks at Remollon and 4 km to the north. It extends southwards and includes pre-Triassic outcrops near Clamensane, Nibles and Barles. The nature of the eastern margin of this Remollon–Barles–Clamensane high can be assessed from variations of Liassic stratigraphy now found within the area of Fig. 2. These allow us to comment on the debate between 'thrust tectonics' and 'inversion tectonics'.

Very thin Lias is seen within the complex Faucon–Turriers structural window which overlaps the SE corner of Fig. 2 (Arnaud *et al.* 1977, and refs.). Two kilometre-thick Lias forms the 'Remollon dome', over the gypsiferous Trias and basement outcropping at Remollon. The latter is recognized as allochthonous (Gidon 1975) and extends westward to near La Saulce, where the Lias of the Saulce thrust sheet is thicker and more shaly than that of the underlying Plan-de-Lardier sheet. From where do these sequences derive? It is not possible to be exact in the absence of precise data on distance, direction and timing of movements on all the thrusts. However, the Plan-de-Lardier Lias must restore approximately northeastwards by a distance which is at least the sum of displacements on the Beynon thrust, each of the thrusts in the Peyrouard duplex and the Plan-de-Lardier thrust. Repeated detachment within Jurassic shales as well as Triassic gypsum in the Peyrouard duplex suggests that this is likely to exceed by far the minimum of 6 km implied by the outcrop pattern north of the Durance, and even that implies an origin within the area of the Remollon dome. The thick Lias now over Remollon must derive from a location northeastwards from there by a greater distance, equal to the sum of that noted above plus all the displacement represented in the Petite Ceuse duplex and under the Saulce thrust. This unknown distance must exceed the 10 km needed to derive the crest of the Remollon dome from the area of deep basement in the Embrunais.

To summarize, thin Lias originated from an area where basement is shallow at the present day, while thick Lias originated within a present-day basin with a deep basement.

The above argument rests on generalizations, indicating preservation of the eastern margin of the Remollon–Barles–Clamensane high created during early Mesozoic extension (Arnaud *et al.* 1977, Lemoine *et al.* 1986). It does not preclude some modification by later Mesozoic or Alpine tectonics. Reversal of fault motion, causing local basin inversion, is locally evident 20 km further W, at Clamensane (Arnaud *et al.* 1977). In addition, the outcrops of basement at Remollon and Barles indicate local uplifts superimposed on the wider basement high, due to some reverse fault motion within basement. Transpressional effects on the eastern margin of the basement high could have occurred during northwestwards thrusting. Nevertheless, the general pattern of Liassic basement topography still exists.

Basin inversion in the context of sub-horizontal detachments is in vogue (Lemoine *et al.* 1986, Cooper & Williams, in press). Such inversion can mean that basement of an area first sank, the resulting basin was filled with sediment, and then the basement returned to near its previous level, so 'inverting' the basin. If the term 'basin inversion' has such connotations, it is misleading to apply it to the Remollon dome for three reasons:

(1) The basement which sank has remained relatively deep beneath the Embrunais. It has not returned, regionally, to its previous level. Basement topography is not 'inverted'.
(2) Thrust tectonics is the dominant factor. The Remollon dome is typical of thrust sheet geometries. It is a large anticline, the size of which results directly from the size of the ramp which it overlies. Its western limb is located by the hanging wall ramp, its eastern limb by the footwall ramp. The tectonic style is of thrusting, even if the ramp location is determined by the basement margin (that is, by the basement not being 'inverted').
(3) Only the margin of the basin has been affected by uplift. Most of the basin contents still lie on the deep basin floor, though now displaced and thickened by internal thrusting.

If that is 'basin inversion', the term must apply so generally to thrusted orogenic forelands as to be fairly meaningless. The implications for Alpine tectonics are clearly that while variations in basement depth and Mesozoic succession resulting from Mesozoic extension must be taken into account, 'inversion tectonics' is an inadequate conceptual framework with which to do so. The pattern of thrusts at the margins of the Remollon–Barles–Clamensane high will now be considered, to show how they reflect the variations in depth of basement.

Effects of basement margins on thrust patterns

The nature of the northwestern edge of the basement high is evident from the scattered Oligocene outcrops. Within Fig. 2, these lie on Jurassic shales south of the Durance, oldest Cretaceous across the Durance on the Crete des Selles, and on younger Cretaceous further N, W of Grande Ceuse. This is just part of a regional pattern from the S, where Oligocene locally cuts into Lias, northwards to Devoluy where it overlies Palaeogene limestones on top of thick Upper Cretaceous. The northwestern margin of the basement high is clearly due mainly to a regional tilt developed in Cretaceous to Palaeogene time.

Regional floorthrust ramp

The effect of the eastern margin of the basement high, approximately from Remollon to Barles, is to force thrusts propagating deep within the basin in the E to climb rapidly up a major ramp to near surface. In Fig. 2, this is evident in the way the anticline of thick Lias (overlain by thick Terres noires) has been carried over an array of thrusts, many of which are shallow enough to preserve Oligocene molasse in their footwall. This capacity of shallow basement to cause a regional rise in floorthrust level will be referred to again in another context.

Constant depth versus constant detachment

The effect of the northwestern margin to the shallow basement is also regionally significant. The shallow level of thrust propagation is continued from the area of shallow basement northwestwards by those thrusts immediately beneath the Saulce thrust. This is demonstrated by the way Oligocene molasse is repeated in the Peyrouard duplex and continues to occur from the Petite Ceuse duplex northwards in the footwall of the Montmaur and the Median Devoluy thrusts. It is seen as a northward climb through the tilted Mesozoic succession, as shown by the Saulce and Plan-de-Lardier thrusts in Fig. 2. It might be suspected that propagation of thrust surfaces would follow levels of easy detachment, such as the Terres noires. Although this can be demonstrated on a local scale, on a regional scale this is not the case. This area provides verification of the hypothesis that thrusts have greater tendency to constant depth than to constant stratigraphic level of detachment.

Displacement continuity in the Digne system

The pattern of laterally extensive shallow propagation described above becomes modified by further footwall collapse. This is able to achieve forelandward migration of deep detachments only NW of the basement high. The result is seen as the westward splay of thrust traces such as the Beynon thrust NW of the Durance in Fig. 2. From Digne northwards as far as the area in Fig. 2, the displacements of the Digne thrust system are carried by relatively few shallow thrusts which crop out in a narrow strip above the shallow basement. To the NW of the Durance, these displacements are shared by an increased number of thrusts spread over an increased outcrop area. Thrust repetitions are ubiquitous in SW Devoluy, in the country W of the Beoux river and N of Veynes (NW of Fig. 2). They demonstrate both that a large amount of thrust shortening is accommodated W of and beneath the Median Devoluy thrust and that such shortening is dispersed over a large number of thrust surfaces, many of which have only a few metres stratigraphic repetition. It is worth stressing the point that it is the increased number of thrusts which makes most thrusting intraformational and so not susceptible to normal geological mapping.

In view of the above it is misleading to talk of a 'Digne–Remollon nappe' and to equate its frontal structure in the S (the Digne thrust) with only one structure (the Median Devoluy thrust) in the north (Gidon *et al.* 1970). It is also not necessary to invoke any rotation (presumably clockwise) of the overthrusting sheet, or any lateral 'transfer faults' through the poorly exposed parts of Gapencais, to explain the northward decrease in displacement from more than 20 km near Digne to perhaps as little as 6 km in NE Devoluy. A large, though indeterminate, amount of this displacement has 'leaked' forwards to produce all the thrust structures westwards from the Median Devoluy thrust to the Col de Cabre thrust.

Palaeogene southwestward thrusting of Embrunais–Ubaye

From the comments in the previous section of this paper, it is clear that attention should be paid to intraformational thrusts of small stratigraphic repetition. This author is not aware of any study of small thrusts within the Digne system. However, a new study (Lawson 1987) undertaken further E makes good use of such information. One important new observation is that such thrusts have been folded with bedding around those folds which developed during the emplacement of the lower Embrunais–Ubaye nappes. Therefore, it is necessary to recognize a new early phase of thrusting in the region and to abandon any assumption that thrust repetitions correlate with nappe emplacement (Kerckhove *et al.* 1978). Before discussing the Embrunais–Ubaye structures further it is necessary to summarize their regional setting.

The area of deep basement on the E side of the Remollon dome includes the Embrunais and extends S across the lower Ubaye valley into the mountains at the heads of the Verdon, Var and Tinee (Fig. 3). It is here that a thick, shale-rich succession, ascribed variously to the Dauphinois or Vocontian domain of the external zones, is structurally overlain by thrust sheets (nappes) of quite different stratigraphy. The lower group of these Embrunais–Ubaye nappes consist of tectonically interleaved Autapie helminthoid flysch and elements of a Mesozoic–Palaeogene succession termed 'Sub-Briançonnais'. In the lower Ubaye valley it is evident that these lower nappes were emplaced and then folded and reverse-faulted with their external zone floor before being overridden by the later and higher composite nappe of mainly Parpaillon helminthoid flysch. (More complicated relationships between the Parpaillon nappe and its floor further E will be considered in detail later.) The geometric disposition, emplacement history and lithological contents and associations of the region, particularly the Embrunais–Ubaye nappes, are well documented, mainly by Kerckhove (1963, 1964, 1965, 1969) after Latreille (1961) and Schneegans (1938). The external zone is well documented by Pairis (1965), Plan (1964, 1968) and others, who document considerable stratigraphic variability, suggestive of a similar history of extensional faulting to that documented further NW (Lemoine *et al.* 1986). This provides lithological contrasts and particular structures which help to determine style and location of the later, Alpine structures. In addition to these local controls, there are three important regional features which constrain Alpine thrusting.

Regional constraints

(1) The area is bounded in the S by the Provençal platform, with a thinner, limestone-dominated Mesozoic succession over persistently shallow basement, which was emergent before the end of Cretaceous times.

(2) The area experienced progressive tilting to the S during Cretaceous–Palaeogene times. In the southern valleys, the Palaeogene 'Nummulitique' succession overlies a thick, tripartite Cretaceous of Aptian–Cenomanian shales sandwiched between older and younger limestones. Between the head of the Bleone river and the Ubaye, the Upper Cretaceous limestones are cut out by the Nummulitique, while themselves cutting

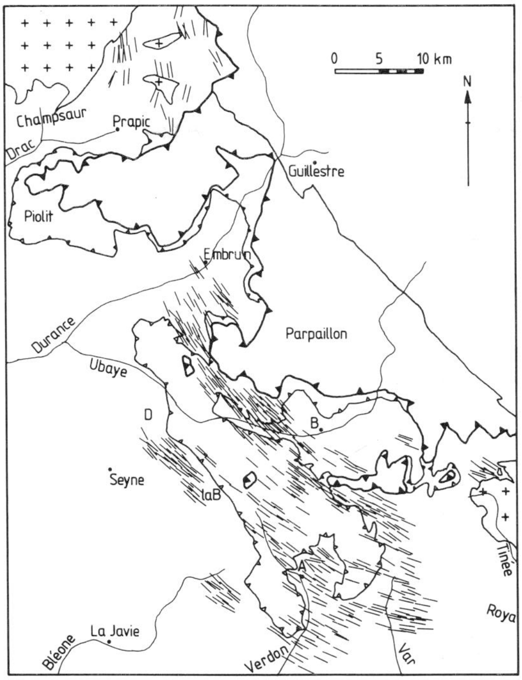

FIG. 3. The extended Embrunais–Ubaye region. B = Barcelonnette. D = Dormillouse. LaB = Montagne de la Blanche. Crosses denote areas of dominantly pre-Triassic basement. The surface separating the lower Embrunais–Ubaye nappes from the underlying external zones crops out along the line marked by open thrust tooth symbols. The Embrunais–Ubaye arc is created by deformation during emplacement of the lower nappes; cleavage trends are shown by line elements, taken from Lawson (1987) and Plotto (1977). The Parpaillon–Briançonnais thrust (solid tooth symbols) cuts through the Embrunais–Ubaye arc structures beneath.

down through the shales on to the older limestones east of Seyne and on to the Jurassic before the Ubaye. In the N, in the Drac valley, the 'Nummulitique' cuts down from Jurassic Terres noires, through older Mesozoic formations, on to the pre-Triassic rocks of the Pelvoux massif.

(3) At least from the Ubaye southwards, folds of E–W trend are ubiquitous. These are well known in Devoluy and in Provence to originate before the main Alpine structures. This can be demonstrated locally in this region either where E–W folds are overlain unconformably by later deposits (by Upper Cretaceous at Le Villard, by Nummulitique on Montagne de la Blanche (Kerckhove *et al.* 1978), by Nummulitique at Barreme), or where they are modified by the later Alpine structures (e.g. upper Var valley, Lawson 1987). These 'Pyrenean' or 'Provençal' folds sometimes show a maximum amplitude in the Upper Jurassic 'Tithonique' dying away upwards and downwards into thick shale formations. Although it is not possible to demonstrate whether most folds originated by buckling or above a thrust tip, the distinction is probably unimportant on a regional scale. Such folds indicate N–S shortening, probably above a regional detachment in evaporitic Trias. Local thrust splays rising from Trias beneath tip folds are seen at Norante, near Barreme and show sense of overthrust to the S. Although thrusts as well as folds undoubtedly started well back in Cretaceous times, individual thrusts continue to show southward displacement cutting Nummulitique (M. Hamiti & G. Apps pers. comm.). The reworking by late Alpine movements makes the shortening represented by these structures (many kilometres) indeterminable.

The lower nappe sequence

Kerckhove (1969) has shown that the part of the Embrunais–Ubaye edifice composed of Sub-Briançonnais and Autapie lithologies can be divided into several tectonic units. These have been separately emplaced on to their immediate substrate, but form such a closely related sequence that their emplacements should be considered the parts of a single continuous orogenic phase rather than independent emplacement events (Kerckhove 1969, Debelmas & Kerckhove 1973). The sequence is one of over-step, with addition of each successive unit on to the footwall, rather than the piggy-back accretion to the hangingwall which is now considered normal (Butler 1982). The internal structure of each unit is complex. Kerckhove (1965, 1969) provided evidence that the Autapie flysch had been emplaced on to Sub-Briançonnais terrain, with an intervening olistostrome, before the resulting assemblage was repeated by thrusts which are internal to the emplacement units. Lawson (1987) has developed this latter point by making a clearer distinction between earlier thrusts, which have geometry typical of a piggy-back sequence, and the later emplacing thrusts. His tectonic units (Fig. 4) differ slightly from those of Kerckhove (1969). Lawson (1987) shows that by restoring both his sets of thrusts it is possible to achieve one approximately planar and horizontal contact ('Basal Autapie thrust') between the Autapie flysch and underlying Sub-Briançonnais lithologies, as Kerckhove's model requires.

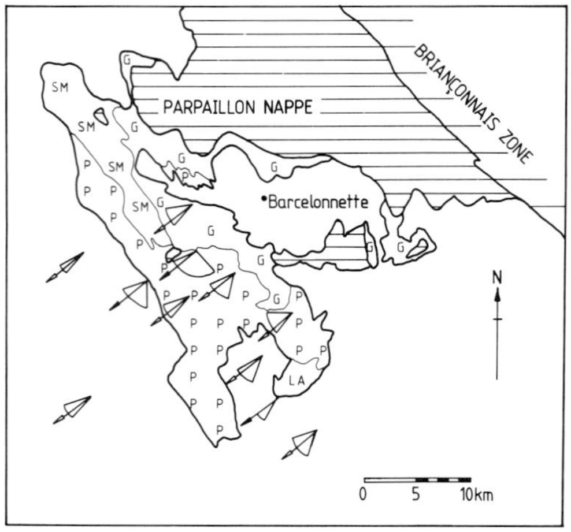

FIG. 4. Constraints on directions of thrust transport in the country S and W of Barcelonnette. Units as defined by Lawson (1987): LA = Lac d'Allos (Upper and Lower) Units; P = Pelat-Trois Eveches Unit; G = Gimette Unit; SM = Petite Seolane and Morgon Units. Each sector of a circle, with included arrow, indicates the range of direction of thrust transport permitted by a data set. Solid arrows: constrained by emplacing thrust surface geometry. Open arrows: constrained by ramp climb patterns. Directions between 220° and 230° are consistent with all data; all arrows here drawn towards 230°.

Minor thrust ramps in the external, Sub-Briançonnais and Autapie lithologies

A study of what by analogy with minor folds may be termed 'minor thrusts' is included in Lawson's (1987) study of regional structure. Faults which ramp through one or two beds, or sometimes as much as a few metres of bedding, are common in the well-bedded lithologies which dominate most of the successions represented in the Embrunais–Ubaye region. His most important observation in terms of tectonic history is that, as a rule, they are folded with bedding around structures previously considered the earliest post-Nummulitique effects in each particular unit. A major difficulty with this kind of work is knowing how to present data so that they may be scrutinized by the reader. Where the combined effect of thrusting is to produce an antiformal stack, it is possible to illustrate how the stack is cut by other structures. For example, the culmination in the external zone west of Barcelonnette is cut across by the base of the earliest nappe visible in the area. Within the Pelat-Trois Eveches unit the basal Autapie thrust cuts an earlier imbricate stack of the Sub-Briançonnais (Fig. 5). The total amount of shortening represented by these early thrusts is difficult to assess. Lawson (1987) estimates 50% in the external zone on the basis of section balancing and 50% in one part of the Pelat-Trois Eveches unit by restoration. It is worth pointing out that in units which have suffered several episodes of shortening, 50% at an early stage represents a far larger absolute amount of crustal shortening than does a similar percentage later in its history. It is difficult to assess the accuracy of such estimates, but Lawson (1987) is probably correct in his opinion that the largest component of crustal shortening in the Embrunais–Ubaye edifice is represented by this previously unrecognized episode of early thrusting. By implication, areas where thrusting has been recognized previously, such as Dormillouse–Montagne de la Blanche (Kerckhove *et al.* 1978, Pairis 1965) and Roya/Tinee (Fallot & Faure-Muret 1949, Faure-Muret 1955) on Fig. 3, are unusual not in that thrusts exist but in that their ramps climb through succession steeply enough to repeat formations.

An important use of minor ramps is to constrain the direction of thrust transport. Certain assumptions are necessary for the method to work. The reference orientation, assumed horizontal, against which ramps are measured is

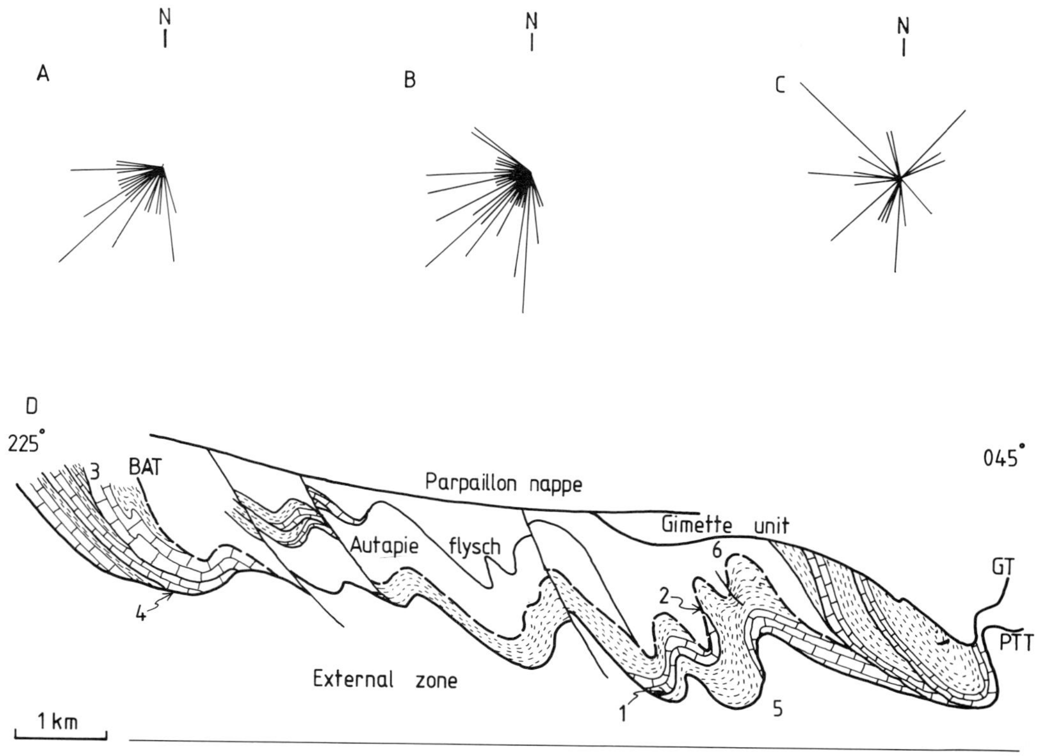

FIG. 5. From Lawson (1987), rose diagrams of directions of climb through bedding and a section illustrating the thrusting history of the Pelat–Trois Eveches unit. (A) Ramp orientations in external zone, Col des Champs, consistent with climb in the direction of thrust transport between 190° and 250°. (B) Aggregated data on ramp orientations from three areas of Pelat–Trois Eveches unit, consistent with transport direction between 220° and 245°. (C) Ramp orientations in Autapie Flysch of Pelat–Trois Eveches unit, not consistent with assumption of one direction in which thrusts climb through bedding forelandwards. (D) Simplified section after Lawson's Fig. 72, from Uvernet to les Trois Eveches, with higher relationships extrapolated from NW of section line. Early imbrication (1) of Sub-Briançonnais Mesozoic (brick symbol) and Cainozoic (tick symbol) is cut (2) by the Basal Autapie Thrust (BAT), which is domed by second imbrication in piggy-back style (3) before being cut (4) by the emplacing thrust (PTT) at the base of the Pelat–Trois Eveches unit. This unit is folded with its external zone floor (5) before being cut (6) by the emplacement thrust (GT) of the Gimette unit. All deformation in these lower units is cut by the later base of the Parpaillon nappe.

that of bedding, and it is assumed that thrusts climb in their direction of transport. Such a study therefore also constitutes a test of these assumptions. Rose diagrams from three areas are shown in Fig. 5. The diagram from the external zones shows the pattern expected from thrusts which share a single transport direction and consistently climb through bedding towards the foreland. This consistency requires that bedding orientation had not been significantly affected by structures earlier than the thrusts which were propagating. This confirms the early age of the thrusts relative to other Alpine structures. It also confirms field observations of piggy-back geometries because it is only in a piggy-back sequence that thrusts always propagate into the undeformed footwall, so retaining a simple ramp-climb pattern. The same is true of the diagram (Fig. 5) from part of the Sub-Briançonnais of Lawson's 'Pelat-Trois Eveches unit'. It is to be noted that consistent climb southwestwards towards the foreland, or, to put it the other way about, tracing consistently deeper in a NE, hindward, direction implies real NE–SW crustal shortening at depth. Superficial transport either by gravity sliding or by an over-riding mass plucking small scales of rock from its floor should produce some extensional faults cutting down in the direction of transport. Although long considered a classic example of gravitational emplacement, the lower Embrunais–Ubaye nappes do not have a suitable geometry for this in their early thrust pattern which is cut by the basal Autapie thrust, in the imbrication of this thrust, or in the emplacement structures. In contrast, the ramp pattern in the Autapie flysch of the same 'Pelat-Trois Eveches unit' is not consistent. Gravity sliding may have emplaced it on to the Sub-Briançonnais terrain.

Lawson (1987) produced ramp orientation diagrams for four areas in the external zones and four in Sub-Briançonnais elements of the lower nappes. Each area individually, and all the data together are consistent with a transport direction of between 220° and 240°. By comparison, the shapes of three of the later emplacement thrust surfaces which bound tectonic units are consistent with an emplacement direction of between 220° and 230° (Lawson 1987). The simplest hypothesis, according with all thrust data collected, is that the whole thrust history of both external zone and Sub-Briançonnais elements was in a single transport direction towards 230°. Limits to direction in Lawson's eight areas and three unit boundaries are shown on Fig. 4.

The Embrunais–Ubaye terrain and the status of the Sub–Briançonnais zone

Two aspects of Lawson's (1987) work prompt a reconsideration of the Sub-Briançonnais zone. One is the apparent lack of pervasive tectonism prior to southwestward thrusting. The other is the way in which redefining the lower Ubaye structural units leads to domino-like matching when they are restored, with the local external zone succession included in this pattern (Fig. 6). These features appear to provide a stark contrast with other internal zones while showing affinity with the external zones. On what is the distinction between 'Sub-Briançonnais' and 'external' based?

Considering first the structural situation, there is evidence of some pre-thrusting Alpine structure in the Sub-Briançonnais units. It is of limited development, but provides a substantial unconformity at the base of the 'Nummulitique'. Such a statement is equally applicable to much of the external zones. It contrasts with earlier pervasive tectonism over most of the internal zones. Is there a pre-thrusting structural boundary which can be discerned as separating Sub-Briançonnais from external domains? None is documented. It appears that if the distinction between Sub-Briançonnais and 'external' is structural in nature, it arises during the process of thrusting.

Considering the stratigraphic evidence, it is certainly true that the Sub-Briançonnais elements contain a less complete and less shale-dominated succession than the external substrate which immediately underlies them now. This substrate happens to be abnormally thick and shaly also by external zone standards. Both zones are highly variable. Within the external sequence of the Embrunais–Ubaye region there are marked changes in Mesozoic succession (Pairis 1965) and the Lias E of Barcelonnette is even considered to have Sub-Briançonnais affinities (Plan 1964, 1968).

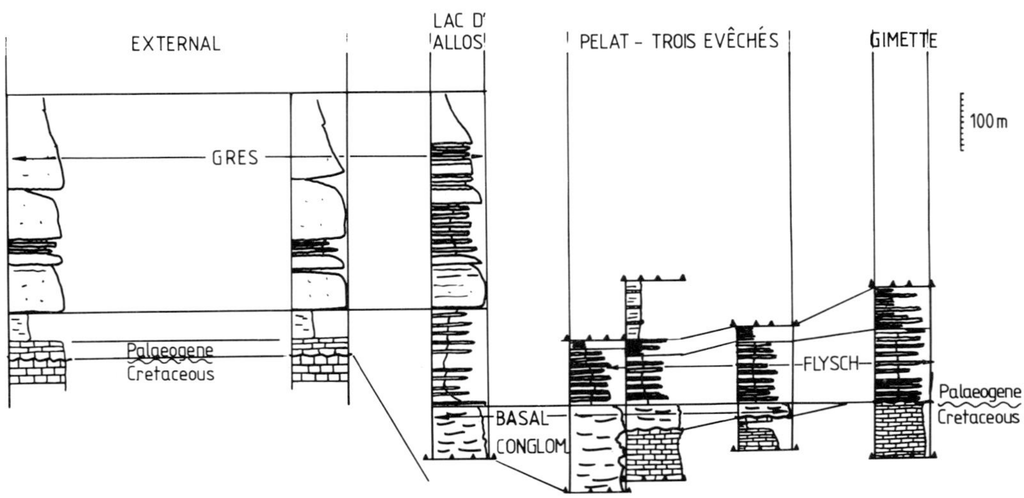

FIG. 6. Cainozoic successions, showing variation within units and matching between units, giving overlapping facies without any dividing line between external and Sub-Briançonnais elements of the lower Ubaye nappes. Unit definitions after Lawson (1987). Successions from Lawson, after Kerckhove, and from Kerckhove (1969).

Just as in the external zones, a highly variable Sub-Briançonnais Jurassic–Lower Cretaceous mosaic is covered by a more weakly and broadly variable Upper Cretaceous calcareous sequence. Kerckhove (1969) comments on 'the pre-Nummulitique mosaic of Mesozoic formations whose isopic trends look like a gross sketch of those of the south Sub-Alpine zone'. The Nummulitique (Fig. 6) provides no clear distinction, nor any location for one, separating Sub-Briançonnais and external domains.

The situation may be summarized as follows. The Sub-Briançonnais zone lacks features characteristic of the internal zones. Its Mesozoic history is locally variable, but not distinguishable in general from the external zones. Its Palaeogene history (unconformity, Nummulitique sedimentation, thin-skinned thrusting to the SW) is characteristic of the external zones, and in particular those parts now incorporated into the Embrunais–Ubaye region. The Embrunais–Ubaye 'external' and 'Sub-Briançonnais' are components of one continuous Embrunais–Ubaye terrain, best considered as a whole, and as essentially 'external'. Any further distinction is structural in nature, arising out of the SW thrusting history (see following and Fig. 7).

A summary of the thrusting history

The following summary history is an attempt at a synthesis of the information now known about the thrust and emplacement history of the lower nappes and their external zone substrate in the region of Embrunais–Ubaye.

(1) A shallow thrust system migrates in general in piggy-back sequence southwestwards, telescoping the whole Embrunais–Ubaye terrain.

(2) Autapie flysch is emplaced onto the rear part of this thrust-shortened terrain. The direction of emplacement and the mechanism, which could be gravity, are not known.

(3) The locus of the shortening at upper crustal levels takes a major step back. This sets off a second southwestwards migrating piggy-back system, imbricating Autapie flysch with its once thrust-shortened substrate.

(4) Footwall collapse fails part of the way through the terrain. The floorthrust climbs from deep in the rear portion to run at or near sediment surface nearer the foreland. The effect of the regional rise of floor-thrust level is to cause thrusting of twice-imbricated materials over an almost complete once-imbricated floor. This establishes a structural distinction between 'Sub-Briançonnais' and 'external'. We may conjecture that this resulted from a step in basement analogous to that described earlier in this paper. The thin lower Jurassic succession now found W of Embrun (Pairis 1965) is evidence of a source area of shallow basement to the E of present external zone outcrops.

(5) The external zone floor has an array of weaknesses on earlier faults and southward directed structures, above a Triassic detachment. Overthrusting of the Sub-Briançonnais-plus-Autapie wedge causes this floor to deform by a combination of ductile shortening, folding, cleavage formation and reverse faulting (Kerckhove 1969). The dip of axial planes, of cleavage and of reverse faults decreases downwards, in most cases to detachment in the Trias (Lawson, 1987). This detachment permits deformation to spread southwestwards to a thrust and fold front approximately through Barreme and eastern Champsaur. Such spread is blocked in the N, where Nummulitique cuts out Trias on to shallow basement, and in the S where E–W folded limestones cap the shallow basement of the Provençal platform, giving rise to the markedly arcuate pattern of fold axes and cleavage traces (the Embrunais–Ubaye arc of Lawson 1987) in Fig. 3.

Deformation of the floor causes readjustment of the thrust surface. Where anticlines are developing, they are 'scalped' (e.g. Bouchiers, Allos). Where relative subsidence is occurring, propagation of a more planar thrust surface leaves a slice of the overriding sheet attached to the footwall, below the re-established thrust level. Locally such slices are found in developing synclines, as at les Thuiles, west of Barcelonnette. Regionally, the centre of the cleavage/fold arc becomes a relative low because the stratal shortening and thickening are dispersed forward across a wider terrain than at the arc's edges. Hence, the centre of Embrunais–Ubaye is destined to become a relative low in which the Embrunais–Ubaye structural units have been progressively accreted from the overriding sheet in their distinctive overstepping sequence, lowest first, highest last.

(6) Erosion and the arrival of the Parpaillon nappe remove the Sub-Briançonnais-plus-Autapie overriding thrust sheet, except for the overstepping sequence of accreted slices

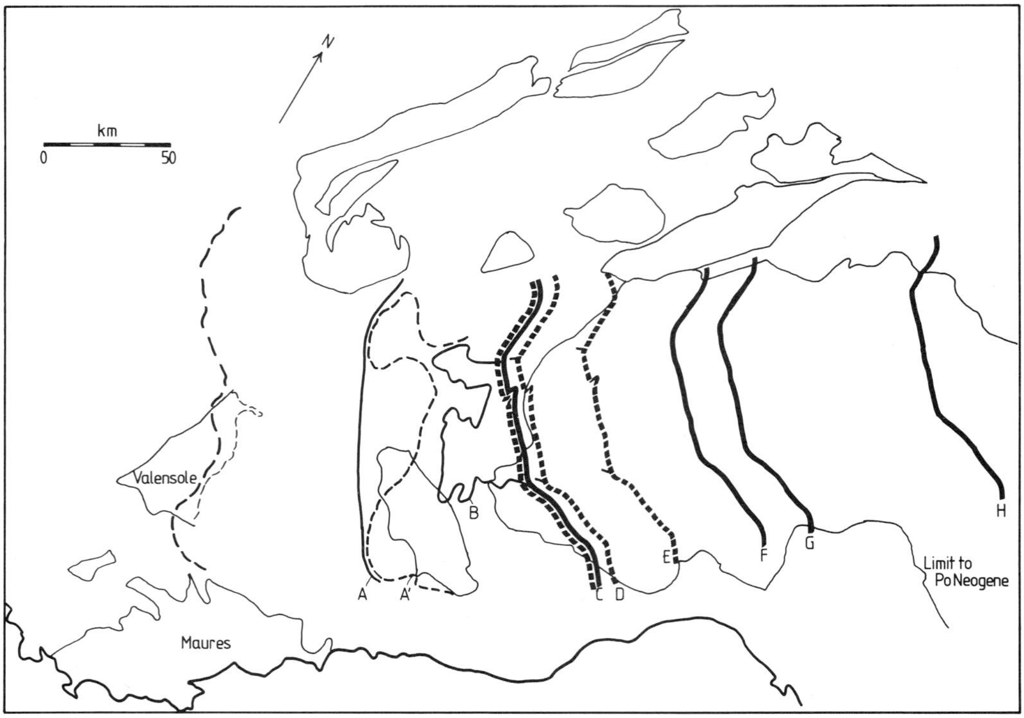

FIG. 7. Approximate restored location of the Embrunais–Ubaye terrain during Palaeogene thrusting. Valensole thrust front assumed as fixed foreland. Outcrop outlines of the coastline, the limit to post-Palaeogene sediments of the upper Po Plain and various Alpine units (Belledonne, Monte Rosa, Sesia, etc.) are present-day locations for reference; their existence in Palaeogene times is not implied.

Embrunais–Ubaye locations assume, for convenience of construction, that nominal 30 km of SW shortening on the Digne system and nominal 30 km of WNW shortening south of Pelvoux are entirely later than Embrunais–Ubaye thrusting. A: thrust front, aproximately Sannoisian. A': outcrop of floorthrust (A) at present erosion level. B: extent of lower Embrunais–Ubaye nappes at present erosion level. C: internal limit to present outcrops of external zones (pecked line) and Sub-Briançonnais (solid line).

Removal of 6 km of syn-emplacement shortening restores external zone limit to D. Removal of 26 km of pre-emplacement thrusting restores external zone to E.

Present lower nappes derive from behind E, requiring 66 km or more transport to prevent overlap, restoring their present internal limit (C) to F. Removal of syn-emplacement deformation with the external zone, plus overlaps between nappes restores the internal limit at least by the 20 km to G, giving a minimum width (E–G) of 55 km for the emplacing slab. This slab is produced by imbrication (2nd system) of Sub-Briançonnais and Autapie flysch, for which 17 km of shortening is seen in P.T.E. and Gimette units. Assuming zero (2nd system) imbrication in other Sub-Briançonnais units, and nominal 45% average shortening in the first system (50% in P.T.E.), the internal limit to Sub-Briançonnais now seen in Embrunais–Ubaye extends an additional 77 km to line H.

Known shortening under Valenzole, external zone cover extending beneath the Briançonnais zone, nappes now removed by erosion, and Sub-Briançonnais beneath the Briançonnais zone will make the true restored distance larger than implied in this diagram.

which remain preserved in Embrunais–Ubaye beneath the regional thrust level.

Age of southwestward thrusting

It is difficult to assess the reliability of the ages attributed to tectonic events in the literature. We need to distinguish between (1) N–S shortening, (2) thrusting to SW within the external zones, (3) overthrusting by the Sub-Briançonnais-plus-Autapie wedge, and (4) thrusting to WNW, as discussed in the next section of this paper. This has been discouraged by the concept that external zone thrusting, possibly in more than one direction, is a direct consequence of nappe emplacement. In the circumstances it is wise to refer tectonic episodes directly to biostratigraphic data on syntectonic deposits, without correlating different tectonic processes.

Apps (1987) argues that the first thrust displacements to the SW, as distinct from N–S shortening, affect the external zone in the area of the upper Verdon immediately before the marine transgression of Lutetian age (Bodelle 1971). Continuing development and modification of the Nummulitique 'Gres d'Annot' basin is explained by Apps (1987) as a direct effect of southwestward thrusting in the underlying Mesozoic succession. The top of the Gres d'Annot, which continues directly into an olistostrome containing blocks from the Sub-Briançonnais-plus-Autapie wedge (Kerckhove 1969), belongs to zones NP19–20 of the Priabonian (Muller, cited in Cavelier *et al.* 1984). Sub-aerial erosion of Sub-Briançonnais-plus-Autapie materials and fluviatile transport provide debris from this wedge to the Barreme basin during NP22 'Sannoisian' (Bodelle 1971, depo-units redefined by Evans 1987). The closeness of timing of the episode of external thrusting immediately followed by wedge emplacement supports the hypothesis that they are successive responses to essentially continuous southwestward shortening at depth.

Rates of shortening, thrust migration and superimposition of sub-systems

The southwestward thrust system exists, on the basis of the ages cited, for 15 Ma or less, corresponding approximately to the second half of Eocene times. Dates vary between approximately 42–30 Ma and approximately 45–35 Ma, according to the time-scale applied. The Nummulitique basin is diachronous (Bodelle 1971), and apparently structurally controlled, presumably migrating with the thrust system on a regional scale. Evans (1987) points out that the thrust system should more locally migrate faster than the basin. For example, if the shortening is on average 50%, material must be incorporated into the front of the system by thrust front migration at twice the shortening rate. Stability of an increasingly broad but shallow thrust wedge should break down sooner or later. The locus of shortening may then step back, starting a second sub-system, as in the Embrunais–Ubaye history. There should be no surprise at superimposition of thrust sub-systems with common transport direction as part of a continuous longer episode of crustal shortening. The surprise is that such a history has not been commonly recorded. If a sedimentary basin is caused by lithospheric loading, rather than local thrust geometries, it should migrate only at the shortening rate, being left behind by the first sub-system and then overridden by the second, as recorded in the Embrunais–Ubaye region.

Thrusting to WNW, within the external zones S of the Pelvoux Massif

Much discussion has taken place concerning the nature and extent of thrusting with transport towards WNW in the western Alps (e.g. Goguel 1963, Tricart *et al.* 1977, Butler 1983, 1984a, 1984b, 1985; Butler *et al.* 1986, Gilchrist in press, Platt 1984, this volume, Tricart 1984). Most of this literature refers to areas north of the Pelvoux Massif. Evidence for displacement directions between 270° and 360° are scarce further S. This author is aware of the following. Their locations are numbered 1 to 6 in Fig. 8.

Arnaud *et al.* (1977) describe thrusting towards WNW and overturning of folds to the W of N in the area of Faucon–Turriers, S of Remollon.

Gidon & Pairis (1980–81) record striations aligned 110°–290° on cleavage surfaces beneath the thrusts which emplace Mesozoic over Nummulitique, with ramps climbing to N and W, around Soleil Boeuf, S of Vieux Chaillol on the southern margin of the Pelvoux Massif.

One kilometre NE of Soleil Boeuf there is a duplex repeating basal Triassic strata (Harwood, pers. comm.). The sense of overthrust is to WNW. There are lineations on the thrusts within this structure with a trend 110°–290°.

Kerckhove *et al.* (1978) record the association of overturning approximately to the NW with thrusting, 13 km east of the last two locations mentioned, near Prapic in eastern Champsaur.

Butler *et al.* (1986) mention thrusts for which opposed lateral ramps constrain the transport direction to WNW observed 'W of Vallouise', N of the previous location.

Matthews (1984) reports that thrusts in the Rochers de L'Yret have a transport direction clockwise of W.

The striking features of these reports are their agreement about direction and timing of thrusting, where these can be determined. All the well constrained directions are recorded as 290° or WNW (which corresponds to 292½°) and three reports (Faucon–Turriers, Soleil

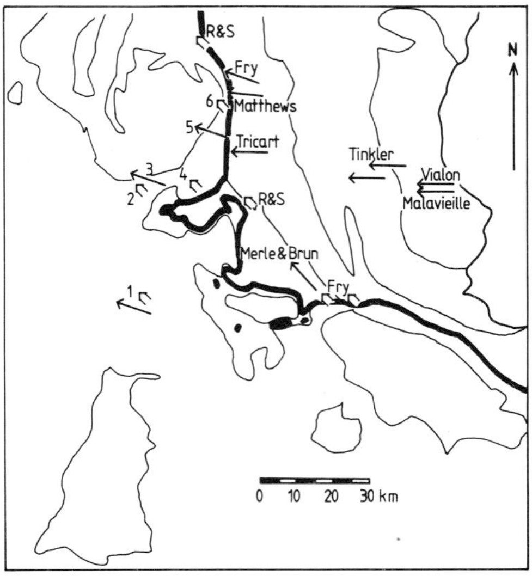

FIG. 8. Some reports of thrusting to directions between west (270°) and north (360°) in the SW Alps. Well constrained directions are shown by long arrows.

Locations 1 to 6 in the external zones are individually referred to in the text. All may be to WNW and of Priabonian–Sannoisian age. East of the Parpaillon-Briançonnais front (heavy line) the directions are varied and this author considers ages to be uncertain. 'Fry' indicates a new report, 'R & S' indicates reinterpretation in this paper of data from Ricou & Siddans. Other reports in the references named.

Boeuf and Prapic) are of movements synchronous with sedimentation, in a time range from end-Eocene into the early Oligocene.

The structural situation of all these locations is the same, in as much as they all lie on the NW margin of the Embrunais–Ubaye region and are repeating the sedimentary cover of a shallow crystalline basement. Presumably these thrusts are climbing towards their foreland from deep detachments beneath Embrunais–Ubaye, ahead of deeper detachments beneath the Maures Massif crystalline basement. It is at present an interesting point of speculation whether such displacements come over, or arise beneath the Triassic top surfaces of what are now the Dome de Barrot and the Argentera Massif.

N of the Pelvoux Massif, the age of WNW thrust structures is considered to range from late Oligocene through the Miocene (and Pliocene, out towards the Jura). Therefore the relative timing and location of the thrusts S of Pelvoux accord with their inclusion within the same overall WNW migrating system. Any earlier thrusts of this system probably precede the propagation of detachments beneath the Provençal platform into the Embrunais–Ubaye region.

The most critical relationships are those associated with structures around Prapic. According to Kerckhove *et al.* (1978) there is a direct link between the thrusting and overturning of the 'Nummulitique' of Plateau de Basset and deposition of the olistostrome 'complexe d'Orcieres'. Olistoliths in this complex include rock-types from the external Nummulitique and from Sub-Briançonnais and Autapie sources. Such an olistostrome, deposited locally at the top of the external succession, demonstrates that the overriding Sub-Briançonnais-plus-Autapie sheet was already a sediment source, but had not yet extended southwestwards over the Prapic area to Piolit. This confirms the relative timing, established regionally, between the SW Embrunais–Ubaye thrust systems and the WNW thrusting south of Pelvoux. The WNW thrusts start to carry the Embrunais–Ubaye edifice passively, just before the last stages of the Embrunais–Ubaye structure are completed.

Extrapolation of the interactions between the two systems to the regional scale gives a regional pattern which may be summarized as follows.

(1) From Lutetian until late Priabonian times (late Eocene), WNW thrusting is restricted to the far S, with thrusts banking up on to the shallow basement of the Provençal–Corsican platform and supplying sediment from the S into the Gres d'Annot basin, as it migrates on the SW thrust terrain.

(2) The region between the Provençal platform and Champsaur (approximately 'Embrunais–Ubaye'), is dominated by southwestward thrusting, except at its NW margin where thrusts show that it has been carried passively WNW since the end of the Eocene.

(3) From the internal margin of the Pelvoux Massif northwards, any Eo-Oligocene SW thrusting of the Embrunais–Ubaye systems is restricted to shallow external and Sub-Briançonnais levels near the contact with internal zones. If recorded at all, it has been thoroughly reworked during the later widespread WNW thrusting. This regional pattern quite clearly still dominates the Alpine structure W of the Briançonnais zone, despite late movements such as the Argentera uplift and on the Digne thrust system.

Relationship to more internal elements

With the exception of the Sub-Briançonnais of Embrunais–Ubaye, noted earlier, all the elements of the southwestern Alps considered to belong to the internal zones have complex Alpine histories, involving overthrusting or subhorizontal shearing towards more than one direction. These histories are not the same for different units. Forethrusts to SW, or backthrusts towards the NE, are normally the latest of the structures implying large tectonic displacements.

The least internal of these units comprise the frontal part of the Briançonnais zone. They show no evidence of even moderate tectonic burial. Their tectonic style is one of thin-skinned thrusting and their grade of diagenesis/metamorphism is lower than that of the external zone beneath (Davies 1983). The tectonic style changes and metamorphic grade of the Briançonnais zone increases rapidly eastwards due to post-metamorphic thrusts telescoping the regional metamorphic gradation. The internal Briançonnais shares with more internal units, such as the Monviso ophiolite, a history of early blueschist metamorphism and several, later superimposed greenschist fabrics.

Monviso

The deepest structural elements of the southwestern Alps are the Dora Maira Massif, the Monviso ophiolite and the units of Schistes Lustrés which lie within and immediately above the ophiolite (Fig. 1). Their metamorphic history is long, and in some cases spectacular (Chopin 1984, in press). They consist of many tectonic slices, and may have done so from early in their history. In low strain patches within Monviso, the earliest stretching lineation has a trend 090°–270° (Tinkler 1980), parallel to that in the Dora Maira Massif (Vialon 1967). This lineation consists in part of aligned needles of glaucophane. Tinkler (1980) recognizes three later episodes of penetrative ductile deformation, of which the last produced a subhorizontal shear fabric which is now warped over the late domal uplift of Dora Maira. In all units the trend of the associated stretching lineation is 050°–230°. Within the Dora Maira and Monviso meta-igneous rocks this last deformation is restricted and concentrated into zones of high shear strain. At the margin of one shear zone in ophiolite, Tinkler (1980) records actinolite needles which have grown aligned 050°–230°, overprinting a blueschist 090°–270° fabric. Within the Schistes Lustrés the late shear fabric is not restricted to shear zones, but increases upwards through the pile. At low structural levels the trend of earlier folds is 155°–335°. As deformation increases upwards, the fold hinges rotate out of this trend until at high levels they are aligned 050°–230° (± 10°).

Correlation of internal zone fabrics is dangerous. Should a 270° blueschist fabric in Monviso be correlated with either the 290° blueschist fabric, or the 270° greenschist fabric in Vanoise (Platt, this volume)? Should we assume parallelism, or a radial fabric pattern (Malavieille *et al.* 1984)? If it is radial, is that arrangement original or due to bending? Such questions are considered later. Here we should be alerted to the danger of subjectivity when selecting from amongst several superimposed lineations in a single tectonic unit the one to be compiled regionally with those chosen from other units in the same manner.

The one reasonably safe correlation would appear to be the array of shear zones, all of which are the last major deformation and which share a sub-horizontal original orientation and a 050°–230° stretching direction. Tentatively we can suggest that they originated during part of the same episode of crustal shortening as the Embrunais–Ubaye thrusting of the same transport direction. Radiometric ages would be useful. If these lie in the 45–38 Ma age range discerned further north (Chopin & Maluski 1980), they could be synchronous with the early Embrunais–Ubaye thrust system, and represent a deeper level of that system, now uplifted over the crest of the Ivrea body and accentuated by the Dora Maira dome. Between these most internal outcrops and the external zones are the structurally overlying Briançonnais zone and higher Schistes Lustrés. It is possible that large displacements (and variety of directions) recorded within these higher units derive across the top of Dora Maira and do not transfer on to structures giving displacement between Dora Maira basement and the basement of the Embrunais–Ubaye external zone.

The problem of the Parpaillon nappe

The upper part of the present Embrunais–Ubaye structure consists of a deformed flysch sequence, generally considered as a single unit, the Parpaillon nappe. Its structural history and association are in some dispute. The main points at issue are included in a description of its main features before referring in more detail to some recent work.

The Parpaillon nappe differs substantially from the lower Embrunais–Ubaye nappes in several ways. In lithological content, it is almost entirely composed of a sequence of Cretaceous to Palaeogene helminthoid flysch, whereas the lower nappes contain both helminthoid flysch and elements from the more normal Sub-Briançonnais, Mesozoic to Eocene succession. In structural association, the Parpaillon nappe is related more closely to the Briançonnais zone than are the lower nappes, in that it is the only nappe to have scales of Briançonnais materials decorating its basal contact, and its Parpaillon flysch is the only one of the Embrunais–Ubaye flysch types to occur in structural outliers on top of the Briançonnais zone. The Parpaillon nappe is less structurally associated than the lower nappes with external zone structures. In particular, the folds and cleavage of the Embrunais–Ubaye arc in both the external zones and the lower nappes are cut by the base of the Parpaillon nappe and do not occur within it (Fig. 3). This relationship leads Kerckhove (1969) to the opinion that the Parpaillon nappe is emplaced after the main structural development of the external zones and the lower nappes. In contrast, Tricart (1980) has shown that the Parpaillon nappe and underlying external zone rocks N and E of Embrun share in common a cleavage-forming deformation, associated with westward overshear of the Briançonnais zone. Merle & Brun (1984), recognizing that the Parpaillon nappe is not a single coherent sheet, propose that different parts have been emplaced at different times. In their view, a lower and more internal unit arrived before its deformation with the external zone, the other later and higher unit only arriving after external zone deformation had ceased. Such an explanation is based on the assumption that the deformation of the external zone beneath the overthrusting Briançonnais zone can be correlated with the Embrunais–Ubaye arc structures. This is not the case. The arcuate structures, cut by the Parpaillon nappe, dip steeply at high levels, but curve downwards to flat décollement surfaces in the external zone sedimentary sequence (Lawson 1987). In contrast, the cleavage shared by Briançonnais, Parpaillon and external rocks has an attitude close to that of bedding in the highest, Nummulitique formations of the external zone and in the Parpaillon flysch, suggesting shear associated with an overthrust mass (cf. Tricart 1980). This distinction can only be made easily in the south (Ubaye, Larche). It is here that a thick Mesozoic sequence exists above Triassic décollement, and the curved profile of the arc structures is developed. It is also possible here to trace the overshear cleavage, which is very strong near the Briançonnais contact S of Col de Larche, and see that it becomes weaker and dies out westward, SE of Barcelonnette. North of Embrun, where the Nummulitique formations lie directly on the shallow Dormillouse pre-Triassic basement, both deformations probably contributed to westward shear and are not separable. These structural correlation problems are reflected in disagreements about ages of emplacement. Correlation of the overshear cleavage with the arcuate structures (Tricart 1980) leads to an Oligocene age for Briançonnais overthrusting of Parpaillon and external units, hence also for emplacement of the Parpaillon nappe (Tricart 1980) or at least its earlier portion (Merle & Brun 1984). Because the present author denies the validity of that correlation, he has no reason to reject the previously established view that the whole Parpaillon nappe was emplaced across an early Miocene erosion surface (Kerckhove 1969).

Further observations on the Parpaillon nappe, and its relationships to Briançonnais and external zones, arise from a study of the area S of Col de Larche, which will be discussed in some detail. These will relate particularly to the concept of a 'Briançonnais front'. Can the western boundary to the Briançonnais zone have the same structural significance against the Parpaillon flysch as against other units, in view of the existence of outliers of Parpaillon flysch on the Briançonnais zone?

Parpaillon–Briançonnais relationships near Argentera

Tanner (1985) reported that three of us (N. Fry, V. M. Davies, M. A. Jones) had reinterpreted the area around the SE corner of the Parpaillon nappe. This area extends from Col de Larche southeastwards to Pontebernado and includes Argentera and Bersezio, near the head of the valley of the Stura di Demonte. It was described in some detail by Sturani & Kerckhove (1963). Figures 9 and 10(a) may be compared directly with their figs 4 and 1 respectively. The disposition of formations shown by Sturani & Kerckhove (1963) is accepted without modification, but the continuity of tectonic contacts is markedly changed. This, together with new observations on folds and cleavages, leads to the conclusion that the structural continuity between Parpaillon and Briançonnais elements is far greater than previously reported.

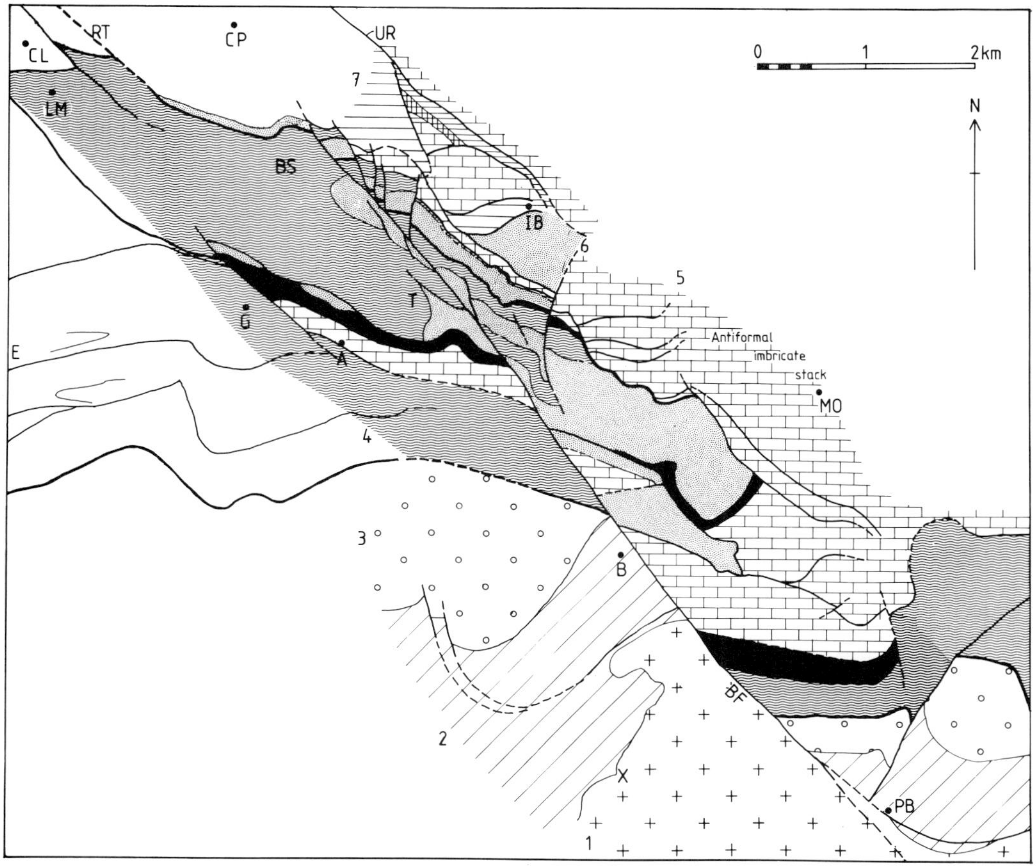

FIG. 9. The area between Col de Larche (Colle della Maddalena) and Pontebernardo, upper Stura di Demonte valley, based on mapping by N. Fry, V. M. Davies and M. A. Jones. 1: crystalline basement. 2: Mesozoic cover. 3: Cainozoic, mostly Gres d'Annot. 4: Incompetent bedded sequences of Sub-Briançonnais, Briançonnais and helminthoid flysch units (Cretaceous to Palaeogene) subject to intensive folding and multiple cleavage formation. 5: Briançonnais massive competent limestones and dolomites (Triassic), in places cut by thrusts into large wedges. 6: Bedded calcareous units (approximately = Jurassic) in the frontal Briançonnais, subject to decametric ramps and folds, with limited parasitic folding on bed-scale and with patchy cleavage development. 7: White crystalline limestones (Triassic–Jurassic) of higher units. Other symbols are black for gypsum formations, and close ruling to indicate Permo–Triassic sediments just SW of the Upper Roburent line. Lines extended west of the mapped area, delimiting the Parpaillon helminthoid flysch, are taken from Sturani & Kerckhove (1963).

A: Argentera. B: Bersezio. BF: Bersezio fault. BS: Becca della Sabiera. CL: Col de Larche. CP: Cima de Peyrassin. E: Enclause. G: Grangie. IB: Il Bric. LM: Lago della Maddalena. MO: Mont Oserot. PB: Pontebernardo. RT (and thick line): Rouchouze thrust. T: La Tinetta. UR: Upper Roburent Line. X: Location of NE-facing fold in basement surface.

Before considering continuity of structures, it is first necessary to establish an opinion on the nature of structures within the Briançonnais zone. Figure 10(c) illustrates the difference between the observations of this author and those of previous works, synthesized as the 1:50 000 geological map 'Larche' (BRGM). The upper part of Fig. 10(c) shows the style implied by previous work with (1) large tectonic slices between few thrusts; (2) continuity around unbroken folds; (3) substantial inversion of succession; (4) rapid variations in formation thickness. The reinterpreted section, in the lower part of Fig. 10(c) shows (1) many thrusts between slices of a wide range of sizes (metric to hectometric), with an implication that more may exist than are discerned at outcrop; (2) first-order folds truncated at thrust surfaces (their second-order parasitic folds are not); (3) predominance of right-way-up sequences;

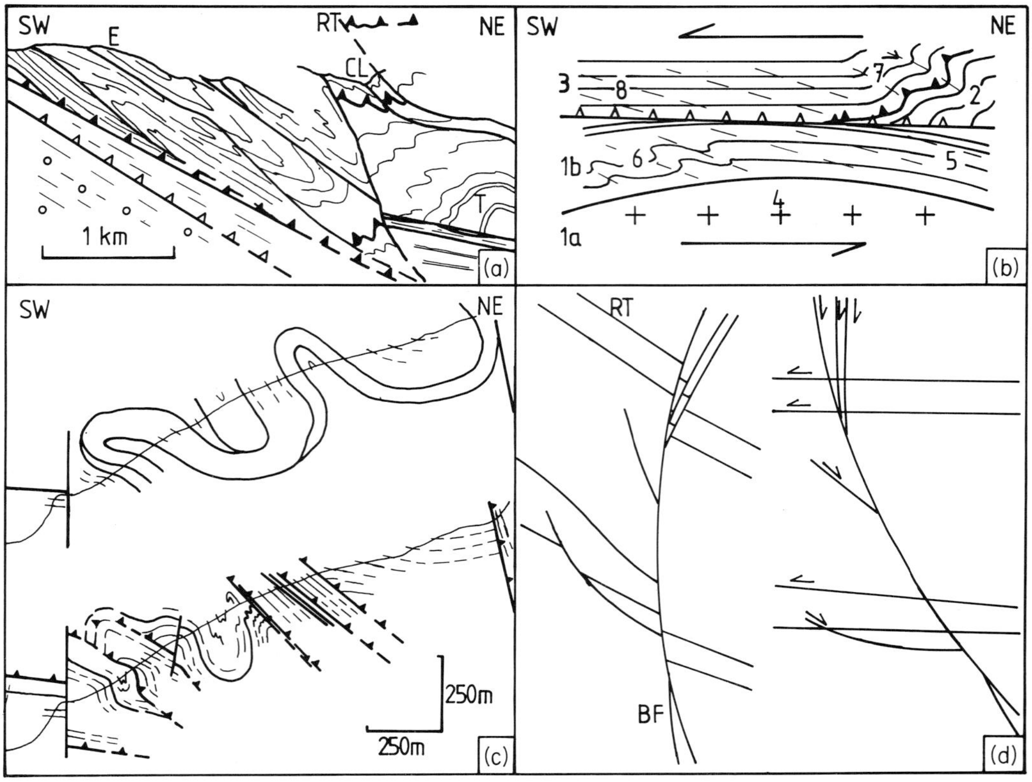

FIG. 10. Sketches illustrating interpretations of structure in the area in Fig. 9.
(a) Section from Enclaouse to Peyrassin (c.f. Fig. 1 of Sturani & Kerckhove 1963). Location letters as Fig. 9. All thick lines are thrusts. Solid teeth mark the base of Parpaillon flysch, on Briançonnais. Hollow teeth mark the base of the combined Parpaillon–Briançonnais mass. The Rouchouse thrust RT is extrapolated to show Parpaillon flysch in its hanging wall.
(b) Simplified sketch to show localization of folds associated with late cleavage. 1: External zone. la: Crystalline basement. 1b: Sedimentary cover. 2: Briançonnais zone. 3: Parpaillon flysch. (Thrust teeth as in (a).) 4: Incipient Argentera anticline, causing bed extension at 5 (Bersezio) but shortening, with folds at 6 (Roubines negres). In Parpaillon flysch, beds are tilted into the shortening field and folded ahead of steep ramping in Briançonnais lithologies at their NE margin (7) but less so further W (8).
(c) Comparison of interpretations of Briançonnais structure on the NE side of the Stura valley 1.5 km NNW of Bersezio. The upper section shows relationships implied by published maps, with continuous folds, major overturning and great variations in formation thicknesses. The lower section shows relationships as interpreted by N. Fry, V. M. Davies and M. A. Jones, with many thrusts causing variations of thickness and orientation and truncated ramp folds but little overturning.
(d) On the left is a down-plunge projection of faults between Becca della Sabiera and Pontebernardo, with Rouchouze Thrust (RT) and Bersezio Fault (BF) labelled. On the right is a restoration by untilting and removal of assumed dip-slip displacement. A clear distinction is apparent between originally sub-horizontal thrusts and later normal faults falling away from the Argentera Massif (to left). These do not provide a realistic continuity of surfaces for significant wrench displacements.

(4) intraformational thrusting as a potential explanation of thickness variations.

Two geometric features indicate that there has been more than one thrust system producing the present geometry. One is the variation in thickness of formations, suggesting a low angle of climb through bedding of intraformational thrusts. This contrasts with relatively steep climb through these variably thick formations of mappable thrusts with substantial stratigraphic separation. The other is the simple geometry and continuity of the mappable thrusts compared to the locally visible complexity of thrust geometries between them. In addition to these geometric features, changes in metamorphic state are visible across some mappable contacts, for example between Cretaceous rocks SW of (above) and NE of (beneath) the Upper Roburent Line (UR on Fig. 9) even though both are supposed to be part of the 'Rouchouze unit'. Davies (1983) reports north of Briançon a history in which early thrusts give rise to stacking and metamorphism and are cut by mappable later retrograde thrusts. The observations reported here on the nature of the frontal Briançonnais near Argentera accord with such a history.

Mappable thrusts, though visible for several kilometres of outcrop trace, may lose stratigraphic separation. The boundary supposed to separate Rouchouze from underlying units is important in this study. Traced southeastwards to older Mesozoic rocks, it loses stratigraphic separation and becomes untraceable within the Mont Oserot imbricate stack, east of Bersezio. The implications for method of study are the same as in the external zones. Attempts to trace continuity of the boundaries to named units of rock are ill-founded, because they distract from the information at locations other than at the allocated boundaries and they lead thereby to an underestimation of the shortening and displacement by thrusting.

Returning to the question of the continuity of Briançonnais and Parpaillon flysch units, the most important aspect of the new intepretation given here is that individual structures are continuous from within the Briançonnais zone to within the Parpaillon nappe. This is demonstrated particularly by the following:

(1) Mappable thrusts are continuous between the two units. One example is the base of the Rouchouze unit as displayed on the 1:50 000 geological map 'Larche'. It climbs in both walls from Trias through Jurassic and Cretaceous as it is traced northwestwards, and then continues over Becca della Sabiera. Here it becomes the NE-dipping contact between the 'Parpaillon nappe' and the 'Briançonnais zone', in fact carrying Cretaceous to Palaeogene (Briançonnais) over an overturned footwall syncline in the basal Parpaillon flysch E of Col de Larche (Fig. 9).

Another example is a thrust which has its character hidden by the interpretation of the 'Larche' map. This map joins as one structure the trace of a vertical fault at high angle to bedding at Grangie with the thrust carrying a hangingwall ramp of Briançonnais Trias over flysch which runs through Argentera village (see Fig. 9). The natural continuation of this thrust is shown on Sturani and Kerckhove's map (their Fig. 4) placing basal flysch over normal flysch near Encalouse (Fig. 9). This continuity from Briançonnais into flysch has been hidden by the imposition of a false structural continuity to the whole of the base of the Parpaillon helminthoid flysch mass.

(2) The gross synformal base to the Parpaillon flysch mass is a natural consequence of the thrust geometries. The thrusts of the type referred to as 'mappable', including both those mentioned above, have an angle of climb through Briançonnais formations mostly in a range from 10° to 30° (frequently 26°±1°). There are few thrust 'flats'. This angle decreases towards the top of the succession in the less competent Cretaceous–Palaeogene beds. Within the overlying Parpaillon flysch, ramp angles are very small and there is a marked increase in the proportion of flats to ramps. It is the thrust sheets bounded by these surfaces which are now stacked approximately parallel to the top surface of the NE-dipping external zone beneath. The result is a NE dip to the flysch part of the stack, but horizontal or SW dip to basal flysch on top of Briançonnais to the NE (Fig. 10a). This gross syncline is accentuated on the NE margin both by thrusts displacing the flysch base over to the SW and by the abnormally large ramp anticline in La Tinetta and Becca della Sabiera (Figs 9 & 10a). Thus, the gross synclinal form of the Parpaillon flysch mass is entirely explained by its structural continuity with the frontal Briançonnais zone.

(3) Two-phase fold geometries in the Parpaillon flysch between Argentera and Col de Larche correlate precisely with those in the higher stratigraphic levels of the frontal Briançonnais zone. The relationship between the two fold phases is most clearly displayed in the less competent Cretaceous sequence above the anticline of La Tinetta (T on Fig. 9). Here, both early and late phases are associated with axial planar cleavages. Traced higher up the succession, the beds become less competent, folds of both phases become tighter and the cleavages become more close in orientation to that of the bedding. However, the facing direction of the late folds, whether calculated from two limbs or measured as the facing direction of cleavage, remains 220°±10°. This cleavage is continuous with that in the Parpaillon flysch. The deformation of the Parpaillon flysch is so intense that in places it is not possible to

separate bedding and cleavage attitudes. Less deformed patches at Grangie and Col de Larche display late folds, facing 220°±10°. It is within these two patches that an earlier cleavage, and earlier folds to which that cleavage is axial planar, have been identified. Calculation of the original attitude of these folds gives facing towards approximately 270°. Similar calculations of early cleavage facing within the Briançonnais zone between Col de Larche and Bersezio give in all cases between 250° and 280°. On this evidence, it appears that the Parpaillon and Briançonnais elements have suffered the same two phases of folding and cleavage formation since their tectonic juxtaposition.

External zone contact effects

Two features of the top of the external zone are significant to the Parpaillon–Briançonnais relationship:

(1) The top of the sequence is locally thinned by the contact of Parpaillon and Briançonnais elements. The Sub-Briançonnais is cut down westwards by the Briançonnais floor SE of Bersezio, with some westward thinning of the Gres d'Annot. Sub-Briançonnais and Autapie materials exist only as a discontinuous thin 'smear' across the northern end of the Argentera Massif to Restefond, beneath the Parpaillon flysch. Briançonnais and Parpaillon elements play the same role, thinning underlying materials towards the crest of the incipient Argentera anticline.

(2) A strong cleavage exists beneath the contact, and shares orientation with the later of the two cleavages discerned in Parpaillon and frontal Briançonnais elements. West of Bersezio this cleavage is intense and is associated with apparent thinning of bedding. It faces 228°. Folds seen in the higher elements do not appear beneath the contact. However, their absence is entirely explicable in terms of angle of bedding to the plane of shear (Fig. 10b). The Parpaillon and Briançonnais beds are tilted by their underlying ramps towards the SW, into the direction of shear, and hence fall in the shortening field. The external zone beds are tilted away from the SW into the extension field. Furthermore, west of the crest of the Argentera Massif, the external succession thickens, with overfolds to the SW visible (Roubines Negres) in Jurassic slates, to which the cleavage is axial planar. This appears to confirm both the existence of the incipient Argentera anticline and the hypothesis that folds are localized according to angle of bedding to shearing.

Summary

From the above it is believed that the Parpaillon helminthoid flysch mass was emplaced on to a Briançonnais terrain at an early stage of Alpine tectonics, perhaps as the Autapie flysch arrived on the Sub-Briançonnais. At least the two cleavage-forming phases of deformation have occurred in which the distinction between Parpaillon and Briançonnais elements is only one of competence. The last of these phases corresponds to the horizontal shear which emplaced the Parpaillon–Briançonnais wedge on to the external zone. The floor of the overriding wedge is a hangingwall ramp through both Briançonnais and Parpaillon lithologies. In its footwall it leaves thin remnants of Sub-Briançonnais and Autapie flysch on an approximate footwall flat. This horizontal shear produced widespread cleavage, but folds are localized by bedding attitudes. Shear directions are not well constrained by cleavage or fold facing. However, the general picture is one of earlier deformation which may correspond to overshear to the W followed by the later 'emplacement' overshearing to the SW. Ramps in the western part of the Parpaillon nappe and the shape of the base of the Parpaillon nappe give an emplacement direction between 210° and 230° (Lawson 1987) consistent with reactivation of the earlier 230° transport direction. This work does not accord with the view that the Parpaillon flysch was emplaced, even in part, on to the external zones before the overthrusting by the Briançonnais zone, as required by Tricart (1980) and Merle & Brun (1984). It suggests that the Briançonnais zone accompanied the Parpaillon nappe in a single Miocene overthrusting of a Miocene erosion surface (as Kerckhove 1969). This author is not aware of any convincing argument why this should be emplacement by gravity rather than the straightforward thrusting from depth indicated by the westward climb on thrust ramps.

The Argentera uplift and backthrusting

The phrase 'incipient Argentera anticline' (above) represents an anticlinal feature with an amplitude of perhaps 100 m. The present structure has an amplitude exceeding 3 km. The 'incipient Argentera anticline' should be considered a minor feature which predisposes its area to become a pop-up during later Neogene thrusting, rather than implying that such thrusting was underway at the time of Parpaillon–Briançonnais emplacement. This is indicated by the way the Parpaillon–Briançonnais mass has since been (1) warped over the top of the massif, and (2) downfaulted by the faults falling away from it (Fig. 10d).

The present Argentera structure is poorly described. It is in essence a pop-up, with SW-overshearing structures on its SW flank and NE-overshearing structures on its NE flank. The latter include major thrusts rising from the pre-Triassic basement, including one placing external Mesozoic succession over Sub-Briançonnais at Demonte. In Vallone dell'Arma (north of Demonte), this places Gres d'Annot and Sub-Briançonnais over Briançonnais and then splays to cause repetitions of Briançonnais in Monte Omo. Such backthrusts from external basement may be more widespread and include those on the eastern margin of the Briançonnais zone. There are backfolds associated with these backthrusts, the largest of which overturns the upper half of the Mesozoic succession plus overlying Sub-Briançonnais and Briançonnais sheets around Monte Nebius. These structures demonstrate the lateness of the Argentera structure at depth, and the involvement of NE–SW thrusting in its uplift. Details of the structure are complex, for example including a detachment in Triassic Cargneule separating SW-oversheared cover from a NE-facing oversheared and pinched-in syncline in the Triassic unconformity over the top of the crystalline massif, 2 km south of Bersezio in the lower Ferriere valley (X on Fig. 9).

Lack of wrench faulting

Ricou (1980, 1981) and Ricou & Siddans (1986) consider the 'front' of the Briançonnais zone to be a transcurrent fault. There is also a suggestion that the Bersezio fault links to it. Such interpretations can be refuted on three grounds.

(1) From the Col de Larche area, it is clear that the external margin of Briançonnais lithologies consists of different sectors of different age and origin. A single explanation is simplistic, and in particular neglects the difference between contacts with Parpaillon flysch and those with underlying Sub-Briançonnais and external zones.
(2) Steep boundaries considered now to be candidates for transcurrent faults are steep due to the late tilt of all units to the E. In particular, most of Ricou's 'Briançonnais front' is nearly bed-parallel and shares its steepness with the bedding of the external zone. He records lineations indicative of sinistral shear with plunge to the south of about 45°. On removal of the external zone tilt, these faults become sub-horizontal and the shear sense and direction recorded by Ricou restore overthrusting to approximately NW. Ricou's data is therefore included here with other evidence for thrusting towards the NW quadrant. His localities include both structures within the Briançonnais zone and between Briançonnais and Parpaillon elements, indicating that these two elements have a long history in common, including thrusting to NW. If this is the same deformation as the northwestward phase recorded by Merle & Brun (1984) in the Parpaillon flysch, it would indicate that this was not the phase which first juxtaposed the Parpaillon and Briançonnais masses, as Merle and Brun suggest.
(3) In the Col de Larche area, candidates for transcurrent faults have been drawn on maps by stringing outcrop traces of fault surfaces which have different age and which cut one another at a high angle in three dimensions. For example, the otherwise superbly informative 1:250 000 geological sheet 'Gap' (BRGM) indicates (on its 'schema tectonique') dextral motion transferring from the late Bersezio fault to the thrust which carries the Rouchouze Briançonnais over the Parpaillon flysch. In three dimensions, the Bersezio fault drops this thrust down off the Argentera Massif anticline at almost right-angles (Figs 9 & 10d).

Far from providing evidence of transcurrent faulting, the frontal structures of the Briançonnais zone in this sector, including those described by Merle & Brun (1984) and Ricou & Siddans (1986), are unable to accommodate major transcurrent motion. The gross geometry is dominated by thrusts and by patterns of later normal faults.

Discussion of thrusting in the western Alps

(1) Constancy of displacement direction

Recent work has highlighted the magnitude and continuity of thrusts in the NW sector of the Alps (Butler *et al.* 1986; references therein). Butler (1984a) reports evidence for displacements towards 290° in Beaufortain, with reported accuracy of a few degrees. This is identical to the direction of lineations associated with thrusting near Vieux Chaillol, and indistinguishable from that reported as 'WNW' near Faucon–Turriers, more than 150 km south of Beaufort. Such a constancy of displacement direction is a profound constraint on models of Alpine tectonics.

Such constancy spans different times, styles and units. This author has recorded stretching lineations towards 290° in a bed-parallel mylonite zone in frontal Briançonnais 15 km NW of Briançon, above Monetier-les-Bains (Fig. 8). The overshear direction of sheathfolds N of Monte Rosa (Swiss Pennine Alps) is 290° (Malavieille *et al.* 1984, Lacassin & Mattauer 1985). The earlier overshear in the Vanoise shares this same direction (Platt this volume, Peel 1985).

This paper has concentrated on SW-directed thrusting. A direction of 230°, not quite so well constrained as the 290° direction, is indicated by ductile overshear in Monviso, the Embrunais–Ubaye main thrust phase and by later movements both on the Digne thrust and emplacing the Parpaillon nappe. Again, the constancy of direction spans times, styles and structural units.

(2) Tram-lines

The reason for the 290° direction in the Western Alps is not clear. However, in the SW sector there are regional features other than thrusts which have an alignment 050°–230°. A few of the larger lineaments are shown in Fig. 1. These are not faults at outcrop. They are, or were, the traces of monoclines at which upper crustal layers, whether sedimentary or tectonic, warp over what is taken to be a major line of weakness at depth. Vialon (1974) points out their parallelism to late Hercynian faults of the 'failles cevenoles' system. Lines A and B on Fig. 1 correspond approximately to faults mentioned by Vialon (1974, 1976), but the present author does not accept the precise position proposed by Vialon for line B which follows closely, and misleadingly in the view of this author, faults at the present erosion level within Mesozoic and later sediments. This author is also sceptical of the hypothesis that large transcurrent motions on these lines are responsible for Alpine structures (Vialon 1974, 1976, Boudon *et al.* 1976, Graham 1978). It is their role as weak lines, able to accommodate differential loading by vertical displacements, which is believed here to give rise to widespread monoclines on this trend.

Of these structures, the longest is a present-day monocline, considered by Vialon (1974) to be a continuation of the Nimes fault. This line A on Fig. 1 is responsible for the warp of Schistes Lustrés over the SE margin of the Ambin Massif, for the drop down of Montgenèvre ophiolites just south of Chaberton, and for part of the tilt in SE Champsaur. This latter section is important, as failure to take the regional monocline into account could lead to overestimation of the thickness of stacked thrust sheets in the southern part of the Pelvoux Massif. The ends of this line are of interest. In the SW there is an abrupt reversal near Gap, so that it continues as the monoclinal drop down to the NW from the Remollon–Barles–Clamensane high towards Devoluy. As pointed out earlier, the present geometry is complex, but in Oligocene times it was an essentially simple tilt to the NW.

The NE end of this monocline closely overlies the western margin of the Ivrea Body, as determined by seismic reflection (Berckhemer 1968) (see Fig. 1). This suggests that the line of deep-seated weakness is in 'the French crust' passing beneath the Ivrea Body, whereas the shallow units in which we see the monocline to the SW rise over the Ivrea Body crest. Such an hypothesis accords with the uplift of deep units along the crest of the Ivrea Body, and with the preponderence of backthrusts banking up over this uplift, and is supported by work in the Sesia and Ivrea zones.

The other lineaments in Fig. 1 are more speculative. Of many small structures running 050°–230° between the Ubaye and Provence, which might either constrain thrusting or be the products of thrusting (pers. comm. G. Apps, K. Lawson), the largest is one which acts repeatedly as a lateral boundary to thrust sheets in the northern half of the Castellane Arc (line B in Fig. 1). The other lines are interpreted by this author as monoclines which came into existence after a phase of initial thrusting but before further overthrusting to the SW, so leading to alignments of lateral and oblique ramps.

The importance of the lineaments in Fig. 1 is that they illustrate the existence of a regional 'grain' to deeper structure, running approximately 050°–230°, which is followed by the thrusting in this part of the Alps. It suggests that thrust transport direction is controlled by features internal to the region, rather than externally by plate motions. Such 'tram-lining' implies partitioning of relative plate motions into vector components, with 050°–230° being continuously available to accommodate one component. Whether the other component, acting simultaneously, is a 290° thrust system, or a steep wrench fault behind the 230° thrust wedge, is at present a matter of conjecture.

(3) Bending of the western Alpine arc

The constancy of displacement directions in the western Alps, at least in its main thrust systems, refutes a number of suggested explanations of its arcuate form. One hypothesis is that the front of an overthrusting wedge deforms and becomes bent by gradational lateral variation in displacement (whether accommodated by rotation, shear, or both). Such an hypothesis is not consistent with preservation, unrotated, in the most internal units, of both the 230° and 290° lineation directions corresponding to the external zone thrust directions. In addition, measurements by this author of along-strike stretching of only 2–3% near Bersezio, on what should be the outside of the arc, are too small for a bending model.

(4) Rotation of thrust sheets

The constancy of displacement directions, as measured now, indicates that there has been no discernible relative rotation across thrusts, when viewed on a regional scale, in the formation of the western Alps. The whole Alpine edifice may have rotated over some deeper, continent-wide detachment, but evidence within the Alps is against rotation being a significant factor in regional thrust systems.

This is a quite different question from that of a rotation within a thrust mass of the directions of extension and incremental shear. Such rotation is to be expected and is documented, for example, by Merle & Brun (1984) in this region and by Dietrich & Durney (1986) in the Helvetic nappes.

(5) Radial thrust systems

Malavieille *et al.* (1984) have compiled a regional pattern of stretching lineations induced by shear in the internal western Alps, radiating from 290° in the Pennine Alps, through 270° in northern Dora Maira Massif to about 250° in the S. Could this represent a set of true overthrusting directions, with divergence accommodated by lateral spreading of the overthrust sheet? This approach is dangerous. It has already been pointed out that internal elements often have more than one lineation, of different ages (see also Ellis *et al.* this volume). Subjectivity is introduced in making a choice which will fit a pattern, whether radial or not.

A lineation of 290° exists north of Monte Rosa and in Vanoise (Ellis *et al.*, Platt this volume). A different lineation of 270° exists in Vanoise, a direction seen also in Monviso and in the Dora Maira Massif. A 230° lineation also occurs in Monviso and Dora Maira. The gross radial pattern appears to result from overlapping of several regionally limited lineations, each of constant orientation. This would be in accord both with the generalization of constancy of direction of each system and the apparent lack of tangential extension. It is not the thrusting of a single Western Alpine system which creates a radial pattern, but the manner in which data from several non-radiating systems have been compiled.

The above statements are unfair to Malavieille *et al.* (1984) in one respect. These statements rely on critical areas where either structural or mineralogical relationships exist which separate the different strain increments in the deformation history of the rocks concerned. Where increments of deformation cannot be separated, the visible result of the superimposition of such increments is indeed a radial lineation pattern of 'the fabric'. The mistake is to consider the generalization of a regional finite pattern to be more fundamental than individual increments, and to favour the apparent simplicity of a single, finite fabric rather than the increased information content inherent in complex local relationships.

(6) Gravity nappes

There is no evidence within the rocks which outcrop in the SW Alps which is definitive of gravitationally driven nappe movement. Circumstantial evidence cited in favour of the gravity mechanism does exist, including: (1) the accumulation of Embrunais–Ubaye nappes in a regional 'low'; (2) the widespread olistostromes ('schistes a blocs') of overthrusting debris, demonstrating a footwall at sediment surface (Kerckhove 1969); (3) the overstepping emplacement sequence of Embrunais–Ubaye nappes.

None of these features is inconsistent with thrusting from depth in the hinterland. Despite being traditionally interpreted as the result of gravitational emplacement (see Debelmas & Kerckhove 1973), the lower Embrunais–Ubaye nappes present the firmest evidence against such a mechanism. Their emplacement direction is approximately 230°, the direction of ramping from depth of the early (first system) thrusts, and the direction which accords with control at depth within the Earth's crust ('tram-lines'). Their hinterland-derived structures are the piggy-back shortening, imbricating (second system) thrusts, with no sign of hinterland extension. The overstepping sequence of accretion is explicable on the basis that the Embrunais–Ubaye 'low' only developed during movement of the overthrust mass, simultaneous with the more local structures of the area. This is evident from the fact that it is these syn-emplacement local structures which comprise the regional Embrunais–Ubaye arc geometry, defining the core of the 'low'. The existence of a topographic low prior to over-thrusting is therefore not established and is also disfavoured by the Embrunais–Ubaye structural dome through which the lower nappes have cut.

A separate case could be made in favour of the gravitational hypothesis for pre-Oligocene emplacement of the helminthoid flysches (Autapie and Parpaillon) on to the Sub-Briançonnais and Briançonnais domains.

(7) Conservation of displacement in a thrust system

Earlier in this paper it was pointed out that thrust systems seldom, if ever, provide at outcrop full and unambiguous evidence of thrust linkage and of the amount of displacement on each thrust in the system. Butler (1984a, 1985) has made estimates of displacements by assuming constancy of displacement direction and lateral continuity, such that total displacements are conserved laterally through the system. This paper shows the strength of evidence favouring constancy of displacement direction on a regional scale. What of lateral conservation of displacement? Displacement may be lost in a variety of ways, including bending, rotation and radial thrusting, but all have been refuted in the western Alpine context. Problems will arise in measuring displacements and attempts to balance thrust systems where more than one system, of different direction, has migrated through the same rock mass. As far as each system is concerned, the geometric conditions necessary to justify Butler's approach are fulfilled in the western Alps. If Butler's method encounters difficulty, it is likely to result from limitations in the geological data, not from the method employed.

(8) Lateral transfer faults

Where a thrust system is developing and a propagating tip encounters an earlier steep zone of weakness aligned in the transport direction, it should be expected that this zone take up some relative displacement. Particularly at the sides of a thrust arc, such a boundary may persist, and a large displacement result. In the western Alps, such lateral transfer faults have been recognized in the 230° thrust system in the Pelvoux Massif (Gilchrist, in press). Their existence in the 290° direction is a matter of conjecture. Butler *et al.* (1986) and others have wanted a transfer of the displacement on this system in the external zone southwards of the Pelvoux Massif. Attempts to link this by a lateral transfer fault to transcurrent motion NE of the Argentera Massif (of Ricou & Siddans, 1986) fail because of the lack of such a structure cutting Embrunais–Ubaye. Butler *et al.* (1986) suggest that this displacement is transferred to the base of the Embrunais–Ubaye nappes, which is geometrically improbable as they were corrugated during, and indeed causing, their emplacement. Such a level is also hindward and higher than the early WNW thrusting in the Faucon–Turriers area.

There are three problems of lateral continuity of thrust displacements in the SW Alps.

(1) Where does the WNW displacement trace southwards of Faucon–Turriers? Are there WNW thrusting structures under the Valensole molasse or west of the Durance Line?
(2) To where does the displacement corresponding to Parpaillon emplacement, recognized in the S (Argentera), transfer N of the Embrunais? Is late southwestward displacement transferred on to thrusts within Briançonnais, external or intermediate levels in the structures around Argentiere and Vallouise (east of the Pelvoux Massif)?
(3) Linked to the last question, is the 290° thrusting in the Briançonnais Zone N of Briançon in direct continuity with 290° thrusting beneath? Or is it part of the general northwestward motion of Parpaillon and Briançonnais materials required to bring them to locations NE of where they are now emplaced?

(9) The arc of the western Alps

If rotation, bending and radial thrusting have not contributed to the arcuate form of the western Alps, how has it arisen? In essence, there are two arcuate features. One is the arcuate outcrop pattern at present erosion levels, the other is a fundamental deeper structure.

At outcrop, the external zone structures are not arcuate, nor do the most internal elements describe a clean arc. It is the outer part of the Briançonnais zone, just eastwards of its 'front', which describes the arc best. As already considered, the Briançonnais zone is a stack of thrust slices emplaced on to a substrate mainly of external zone, but with thin thrust slices of Sub-Briançonnais and flysch materials which contribute little to the regional geometry. The Briançonnais stack is underlain along most of its outcrop by either the 'Gres', 'flysch' or 'schistes a blocs' at the top of the external succession, or their place (in Embrunais–Ubaye) is taken in part by the Sub-Briançonnais and Autapie flysch units. When considered on a regional scale this constitutes a region-wide footwall flat, at or near surface. So, the arc shape is not a thrust lobe, delineated by a floorthrust rising from deeper in the external zones to break surface.

The essential feature of the external zone floor is that this footwall flat near its stratigraphic top has, with all its substrate, received an easterly tilt along an arcuate trace. This eastward downturn is responsible for the distinction between areas W, where most overthrust units have been eroded, and areas E, where the overthrust internal zones crop out as an arcuate belt. The best arc is just E of this outcrop boundary, where the Briançonnais slices are near vertical, so that the map pattern has not been perturbed by local topography and local vertical adjustments.

The other arcuate map feature is the western edge of the Po Plain, the internal limit to Alpine outcrops, also roughly followed by the crest of the sequence of internal massifs (Dora Maira, Gran Paradiso, Monte Rosa). These reflect the arcuate form of the underlying Ivrea Body (Fig. 1). The latter is nearly concentric with the eastward downturn at the edge of the external zones, and suggests that crustal loading by the arcuate crustal wedge, of which the Ivrea body is the front, is the cause of this flexure of the 'French' crust. Fundamentally, it is this major 'thrust' overlap between the two deep crustal wedges which is the arc-determining feature of the western Alps as a whole (see Karner & Watts 1983).

The gross geometry of the western Alps requires crustal elements (Helminthoid flysch, Briançonnais, Piémont), once lying S and E of an area now roughly represented by the Po Plain, to have moved northward and westward over this area, displacing the upper crust from that area (Valais, Sub-Briançonnais) to WNW and SW. Gross westward movement of the deeper Po crust has further shortened these elements and the external zones using previously established WNW, W and SW displacement directions. A variety of earlier arcuate features may have existed, but it is at a late stage that the western Alpine arc as we observe it today is imposed onto an earlier structural edifice.

References

Apps, G. 1987. *Evolution of the Gres d'Annot basin, SW Alps*. Unpublished PhD thesis, University of Liverpool.

Arnaud, H., Gidon, M. & Pairis, J.-L. 1977. Précisions sur la structure des Chaînes Subalpines meridionales dans la région de Faucon–Turriers–Clamensane (Alpes-de-Haute-Provence). *Géologie Alpine*, **53**, 5–34.

Berckhemer, H. 1968. Topographie des 'Ivrea–Körpers' abgeleitet aus seismischen und gravimetrischen Daten. *Schweizerische Mineralogische und Petrographische Mitteilungen*, **48**, 235–46.

Bodelle, J. 1971. *Les formations nummulitiques de l'arc de Castellane*. Thesis, Université de Nice.

Boudon, J., Gamond, J. F., Gratier, J. P., Robert, J. P., Depardon, J. P., Gay, M., Ruhland, M. & Vialon, P. 1976. L'arc alpin occidental: Réorientation de structures primitivement E–W par glissement et etirement dans un systeme de compression global N–S? *Eclogae geologicae helvetiae*, **69**, 509–19.

Butler, R. W. H. 1982. The terminology of structures in thrust belts. *Journal of Structural Geology*, **4**, 239–45.

—— 1983. Balanced cross-sections and their implications for the deep structure of the northwest Alps. *Journal of Structural Geology*, **5**, 125–37.

—— 1984a. *Evolution of thrust belts in the Alps (Savoy) and the Moine Thrust zone (NW Scotland)*. Unpublished PhD thesis, University College of Swansea.

—— 1984b. Balanced cross-sections and their implications for the deep structure of the NW Alps: reply. *Journal of Structural Geology*, **6**, 607–12.

—— 1985. The restoration of thrust systems and displacement continuity around the Mont Blanc Massif, NW external Alpine thrust belt. *Journal of Structural Geology*, **7**, 569–82.

——, Matthews, S. J. & Parrish, M. 1986. The NW external Alpine Thrust Belt and its implications for the geometry of the western Alpine Orogen. *In*: Coward, M. P. & Ries, A. C. (eds). *Collision Tectonics*. Special Publication of the Geological Society No. 19 pp. 245–60.

Cavelier, C. (coord.) *et al.* 1984. Paleogene. *In*: Debrand-Passard S. *et al.*, *Synthese géologique du Sud-Est de la France*. Memoire BRGM 125.

Chopin, C. 1984. Coesite and pure pyrope in high-grade bluechists of the western Alps: a first record and some consequences. *Contributions to Min-*

eralogy and Petrology, **86**, 107–18.

—— in press. Tectonic settings of regional metamorphism. Royal Society.

—— & Maluski, H. 1980. ^{40}Ar–^{39}Ar dating of high-pressure metamorphic micas from the Gran Paradiso area (western Alps): evidence against the blocking temperature concept. *Contributions to Mineralogy and Petrology*, **74**, 109–22.

Cooper, M. A. & Williams, G. D. (eds) *Inversion Tectonics*. Special Publication of the Geological Society.

Davies, V. M. 1983. *Alpine structure and metamorphism in a traverse from the Grandes Rousses Massif to the internal Briançonnais*. Unpublished PhD thesis, University of Liverpool.

Debelmas, J. & Kerckhove, C. 1973. Large gravity nappes in the French–Italian and French–Swiss Alps. *In*: De Jong, K. H. & Scholten, R. (eds). *Gravity and Tectonics*, pp. 189–200. Wiley, New York.

Dietrich, D. & Durney, D. W. 1986. Change of direction of overthrust shear in the Helvetic nappes of western Switzerland. *Journal of Structural Geology*, **8**, 389–98.

Dubois, P. & Curnelle, R. 1978. Resultats apportes par le forage Les Mees no. 1 sur le plateau de Valensole (Alpes-de-Haute-Provence). *Comptes-Rendus, Societé Géologique de France*, **4**, 181–4.

Ehtechamzadeh Afchar, M. & Gidon, M. 1974. Données nouvelles sur la structure de l'extrémité nord de la zone des chevauchements de Digne. *Géologie Alpine*, **50**, 57–69.

Evans, M. J. 1987. *Tertiary sedimentology and thrust tectonics in the SW Alpine foreland basin, Alpes-de-Haute-Provence, France*. Unpublished PhD thesis, University College of Swansea.

Fallot, P. & Faure-Muret, A. 1949. Sur un mode particulier de charriage. *Comptes-Rendus, Académie des Sciences, Paris*, **228**, 616–9.

Faure-Muret, A. 1955. Etudes géologiques sur le Massif de l'Argentera–Mercantour et ses enveloppes sedimentaires. *Memoire du service de la Carte Géologique de la France*.

Gidon, M. 1975. Sur l'allochtonie du 'Dome de Remollon' (Alpes françaises du Sud) et ses conséquences. *Comptes-Rendus, Académie des Sciences, Paris*, **280**, 2829–32.

—— & Pairis, J. -L. 1980–81. Nouvelles données sur la structure des ecailles de Soleil Boeuf (bordure sud du Massif du Pelvoux). *Bulletin du B.R.G.M. (deuxieme serie)*, **1**, 35–41.

——, ——, Arnaud, H., Aprahamian, J. & Uselle, J. -P. 1970. Les déformations tectoniques superposées du Devoluy meridional (Hautes-Alpes). *Géologie Alpine*, **46**, 87–110.

Gigot, P., Granjacquet, C. & Haccard, D. 1974. Evolution tectono–sedimentaire de la bordure septentrionale du bassin tertiaire de Digne depuis l'Eocène. *Bulletin, Societé Géoligique de France (7)*, **16**, 128–39.

Gilchrist, R. (in press.) *In*: Cooper, M. A. & Williams, G. D. (eds). *Inversion Tectonics*. Special Publication of the Geological Society.

Goguel, J. 1963. L'interprétation de l'arc des Alpes occidentales. *Bulletin, Societé Géologique de France*, **5**, 20–33.

Graham, R. H. 1978. Wrench faults, arcuate fold patterns and deformation in the southern French Alps. *Proceedings of the Geologists' Association*, **89**, 125–42.

—— 1981. Gravity sliding in the maritime Alps. *In*: McClay, K. R. & Price, N. J. (eds). *Thrust and Nappe Tectonics. Special Publication of the Geological Society of London*, **9**, 335–51.

Karner G. D. & Watts, A. B. 1983. Gravity anomalies and flexure of the lithosphere at mountain ranges. *Journal of Geophysical Research*, **88**, 10449–77.

Kerckhove, C. 1963. Schema structural de la nappe du Flysch a Helminthoides de l'Embrunais–Ubaye. *Travaux de Laboratoire de Géologie, Grenoble*, **39**, 7–24.

—— 1964. Mise en évidence d'une série à caractère d'olisthostrome au sommet des Gres d'Annot (Nummulitique autochtone), sur le pourtour de nappes de l'Ubaye (Alpes franco–italiennes: Basses-Alpes, Alpes-Maritimes, province de Cuneo). *Comptes-Rendus, Académie des Sciences, Paris*, **259**, 4742–45.

—— 1965. Structure du Massif du Pelat et des environs d'Allos. Problemes de paleogeographie subbriançonnaise au Sud de l'Ubaye (nappes de l'Ubaye, Basses-Alpes). *Travaux de Laboratoire de Géologie, Grenoble*, **41**, 33–55.

—— 1969. La 'zone du Flysch' dans les nappes de l'Embrunais–Ubaye (Alpes occidentales). *Géologie Alpine*, **45**, 5–204.

——, Debelmas, J. & Cochonat, P. 1978. Tectonique du soubassement parautochtone des nappes de Embrunais–Ubaye sur leur bordure occidentale, du Drac au Verdon. *Géologie Alpine*, **54**, 67–82.

Lacassin, R. & Mattauer, M. 1985. Kilometre-scale sheath fold at Mattmark and implications for transport direction in the Alps. *Nature*, **315**, 739–42.

Latreille, M. 1961. Les nappes de l'Embrunais entre Durance et haut Drac. These, Grenoble, 1958. *Memoire du service de la Carte géologique de la France*.

Lawson, K. D. 1987. *Thrust geometry and folding in the Alpine structural evolution of Haute Provence*. Unpublished PhD thesis, University College of Swansea.

Lemoine, M., Arnaud-Vanneau, A., Arnaud H., Bas, T., Bourbon, M., Dumont, T., Gidon, M., De Graciansky, P. C., Megard-Galli, J., Rudkiewicz, J. -L. & Tricart, P. 1986. The continental margin of the Mesozoic Tethys in the western Alps. *Marine Petroleum Geology*, **3**, 179–99.

Malavieille, J., Lacassin, R. & Mattauer, M. 1984. Signification tectonique de lineations d'allongement dans les Alpes Occidentales. *Bulletin, Societe Géologique de France*, **26**, 895–906.

Matthews, S. J. *Thrust sheet evolution in the Kinlochewe region and of the Moine thrust zone, NW Scotland and the Pelvoux Briançonnais,*

French Alps, Unpublished PhD thesis, University of Leeds.

MENARD, G. 1979. *Relations entre structures profondes et structures superficielles dans le sud-est de la France: essai d'utilisation de donées géophysiques*. These, Grenoble.

MERLE, O. & BRUN, J. P. 1984. The curved translation path of the Parpaillon Nappe (French Alps). *Journal of Structural Geology*, **6**, 711–9.

PAIRIS, J. -L. 1965. La demi-fenêtre d'Embrun (Hautes et Basses Alpes). *Travaux de Laboratoire de Géologie, Grenoble*, **41**, 97–132.

PEEL, F. J. 1985. *Tectonic Evolution of the Vanoise Massif, French Alps*. Unpublished PhD thesis, University of Oxford.

PLAN, J. 1964. Essai d'interprétation de la 'Fenêtre de Barcelonnette' (Basses-Alpes). *Travaux de Laboratoire de Géologie, Grenoble*, **40**, 7–35.

—— 1968. La 'fenetre' de Barcelonnette. *Bulletin du Service de la Carte Géologique de la France*, **280**, 437–530.

PLATT, J. P. 1984. Balanced cross-sections and their implications for the deep structure of the northwest Alps: discussion. *Journal of Structural Geology*, **6**, 603–6.

PLOTTO, P. 1977. *Structures et déformations des 'gres' du Champsaur au SE du Massif du Pelvoux*. Thesis, University of Grenoble.

RICOU, L. E. 1980. La zone subbriançonnaise des Alpes occidentales interpretée comme la trace d'un ample decrochement senestre sub-meridien. *Comptes-Rendus, Académie des Sciences, Paris*, **290**, 835–8.

—— 1981. Glissement senestre des nappes penniques le long de la bordure nord de l'Argentera: son role dans le jeu de l'arc alpin. *Comptes-Rendus, Académie des Sciences, Paris*, **292**, 1305–8.

—— & SIDDANS, A. W. B. 1986. Collision tectonics in the western Alps. *In*: COWARD, M. P. & RIES, A. C. (eds). *Collision Tectonics*, Special Publication of the Geological Society No. 19. pp. 229–44.

SCHNEEGANS, D. 1938. La géologie des nappes de l'Embrunais–Ubaye entre la Durance et l'Ubaye. *Memoire du service de la Carte Géologique de la France*.

SIDDANS, A. W. B. 1979. Arcuate fold and thrust patterns in the subalpine chains of SE France. *Journal of Structural Geology*, **1**, 117–26.

STURANI, C. & KERCKHOVE, C. 1963. Sur la terminaison sud-orientale de la nappe du Flysch a Helminthoides a proximite du Massif de l'Argentera (versant italien du col de Larche ou della Maddalena). *Travaux de Laboratoire de Géologie, Grenoble*, **39**, 213–29.

TANNER, G. 1985. Conference report: Tectonic studies group 14th annual meeting. *Journal of the Geological Society*, **142**, 573–7.

TINKLER, C. 1980. *The tectonic and metamorphic history of the Mon Viso ophiolite body, NW Italy (western Alps)*. Unpublished PhD thesis, University College of Swansea.

TRICART P. 1980. *Tectoniques superposeés dans les Alpes occidentales, au Sud du Pelvoux. Evolution structurale d'une chaine de collision*. Thesis, University of Strasbourg.

—— 1984. From passive margin to continental collision: a tectonic scenario for the western Alps. *American Journal of Science*, **284**, 97–120.

——, CARON, J. M. & VIALON, P. 1977. Relais de schistosité, structure en evential et discontinuités majeures sur la transversale du Pelvoux. *Bulletin, Societe Géologique de la France*, **19**, 873–81.

VANN, I. R., GRAHAM, R. H. & HAYWARD, A. B. 1986. The structure of mountain fronts. *Journal of Structural Geology*, **8**, 215–27.

VIALON, P. 1967. Etude géologique du Massif cristallin Dora Maira (Alpes cottiennes internes, Italie). *Géologie Alpine*, **43**, 245–88.

—— 1974. Les déformations synschisteuses superposées en Dauphine. Leur place dans la collision des elements du socle prealpin. *Schweizerische mineralogische und petrographische Mitteilungen*, **54**, 663–90.

—— 1976. L'arc alpin occidental: orientation des structures primitivement. *Eclogae geologicae helvetiae*, **69**, 509–19.

N. FRY, Department of Earth Sciences, University College, Swansea SA2 8PP, UK.

Tectonic Evolution of the Internal and Southern Zones of the Alps

Tertiary kinematics in the Lepontine dome

O. Merle, P. R. Cobbold & S. Schmid

SUMMARY: In the Pennine zone of the central (Lepontine) Alps, the mineral-zone boundaries of Tertiary metamorphism and the regional attitudes of foliation and thrust sheets define a dome. We investigate the Tertiary kinematic history of the dome using structural, metamorphic and geochronological data. Thus we distinguish two main ductile deformations, that occurred under differing metamorphic conditions. The first high-temperature deformation (HTD) developed under amphibolite facies conditions. It is responsible for (1) a radiating pattern of stretching lineation trends at the scale of the Lepontine area, and (2) structures indicating overthrusting towards the European foreland in the lineation direction (i.e. N and NW). The HTD event is attributable to NW–SE Tertiary convergence between Europe and the Adriatic sub-plate. The second retrograde deformation (RD) is responsible for differential uplift of the Pennine zone, with formation of an earlier subdome to the E (Ticino subdome, 25–20 Ma) and a later one to the W (Simplon subdome, 15–10 Ma).

In the Ticino subdome, kinematic indicators reveal a horizontal shearing top W. Some major structures such as the Maggia steep zone and the vertical duplication of the Antigorio basement nappe (Bosco–Gurin area) may be related to this westward phase. Dextral transcurrent shearing along the southern steep belt and backthrusting of the central Alps over the southern Alps along the N dipping Insubric mylonite belt are probably coeval with the retrograde deformation recorded in the Ticino subdome. In the Simplon subdome, kinematic indicators reveal a radial motion parallel to the lineation, which plunges towards the SSW in the Verampio window and towards the WSW at the Simplon Line.

Serial uplift of Ticino and Simplon subdomes is attributed to continuing dextral transpression at the boundary between Europe and the Adriatic sub-plate.

Finally, the back-folding event located at the northern limit of the Pennine zone postdates both HTD and RD deformational events described above. It is part of a later pop-up responsible for uplifting the external crystalline massifs.

Introduction

In this paper, we use structural, metamorphic and geochronological data to study the Tertiary kinematics of the Lepontine Alps. Many of the data we have taken from the abundant literature dealing with this area, but some structures (stretch lineation patterns and kinematic indicators) are presented here for the first time.

The Lepontine region appears as a single dome in terms of the mineral zone boundaries (isograds) of Tertiary metamorphism (Niggli 1970) and is usually referred to in this sense as the Lepontine dome (Fig. 1). The region is also a dome, albeit a composite one, on the basis of foliation and thrust sheet attitudes (Fig. 2). The Pennine zone within the dome shows ductile deformations of remarkable intensity and complexity, coeval with Tertiary metamorphism of amphibolite grade and the later cooling history. Thus the Lepontine dome provides an excellent and unusual opportunity for studying deep-seated processes of deformation (probably associated with plate convergence), followed by shallower processes (associated with uplift).

There are several reasons for studying the Lepontine dome. First, it occupies a critical position within the Pennine zone, which is often considered to represent an originally thinned continental margin (or margins) between the European plate proper to the N, and the southern Alps, belonging to an Adriatic sub-plate of African affinity, to the S (Fig. 1). In the Lepontine region, the current boundary between the Pennine zone and the southern Alps is the peri-Adriatic fault system, locally known as the Tonale, Insubric and Canavese Lines (see Schmid *et al.* this volume). Second, the eroded Lepontine dome provides a tectonic window through the Pennine zone. Third, the dome is also at a critical longitude, between the eastern Alps, roofed by Austro–Alpine thrust sheets, and the western Alps, which expose mostly Upper Pennine units (Monte-Rosa and Bernhard sheets, Fig. 1). Finally, in this general area, the peri-Adriatic fault system bends round from an EW attitude (Tonale line) to an almost N–S attitude (southern Canavese line). In such a context, it is perhaps not surprising that the internal structure of the Lepontine dome is complex. Nevertheless, we feel that an understanding of it should be helpful towards a

From: Coward, M. P., Dietrich, D. & Park, R. G. (eds), 1989, *Alpine Tectonics*, Geological Society Special Publication No. 45, pp. 113–134.

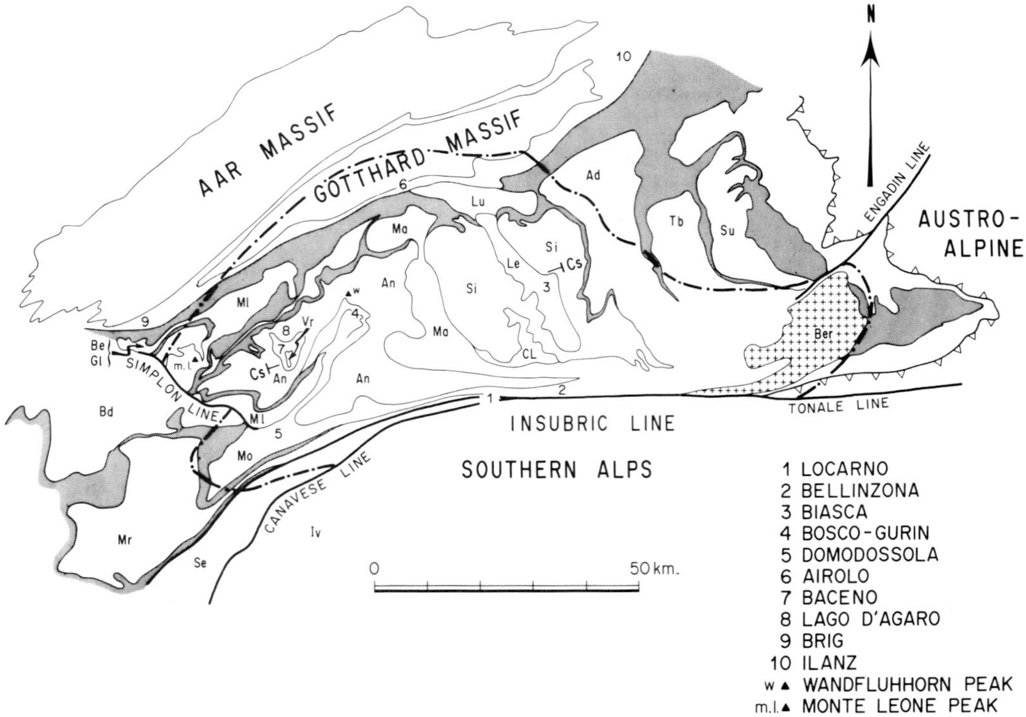

FIG. 1. Simplified geological map of the Lepontine dome. In the Pennine zone, cover rocks (stippled) are interleaved with slices of crystalline basement (Ad: Adula, An: Antigorio, Bd: Bernhard, Cl: Cima-lunga, Le: Leventina, Lu: Lucomagno, Ma: Maggia, Ml: Monte Leone, Mo: Moncucco, Mr: Monte Rosa, Se: Sesia, Si: Simano, Su: Suretta, Tb: Tambo,). Also shown are Verampio window (Vr), Berisal synform (Be), Glishorn antiform (Gl) and Bergell intrusion (Ber). Periadriatic faults (Canavese, Insubric and Tonale Lines) separate the Pennine zone from the southern Alps, which include the Ivrea zone (Iv). Thick dashed line encloses area of amphibolite facies metamorphism of Tertiary age. Cs indicates the line of the cross section shown in Fig. 9. For numbered localities and peaks see key at bottom right.

broader-scale understanding of Alpine structures and kinematics.

Since this contribution aims at regional kinematics, we primarily make use of the large scale patterns of foliations, lineations and shear indicators at amphibolite grade, or later greenschist grade. As a consequence, the reader familiar with the existing literature will find that we do little justice in this paper to the complexity of local deformation phases, especially the geometrical style, orientation data and interference patterns of folding on all scales (such as discussed in Huber *et al.* 1980). This shortcoming simply results from the fact that we are not yet able to fully integrate this kind of structural analysis into a regional kinematic history and that we are not aware of any previous attempts to do so.

We first describe the general structure of the Lepontine dome itself, including the thrust sheet sequence and the foliation pattern. Then we summarize what little is known of its pre-Tertiary history. In more detail, we discuss petrological and geochronological data on the Lepontine metamorphic event. This allows us to refer to metamorphic assemblages when describing linear and planar fabrics or kinematic indicators. Using the patterns of lineations and the associated senses of shear, we finally attempt a Tertiary kinematic interpretation of the Lepontine dome. The inferred motions are so thoroughly three-dimensional, that we are not yet able to use balanced cross-sections, or any other method, to calculate total amounts of displacement.

The structural dome

In the Lepontine region, a geological map of the Pennine zone (Fig. 1) looks much like a sliced onion, for several geometrical reasons.

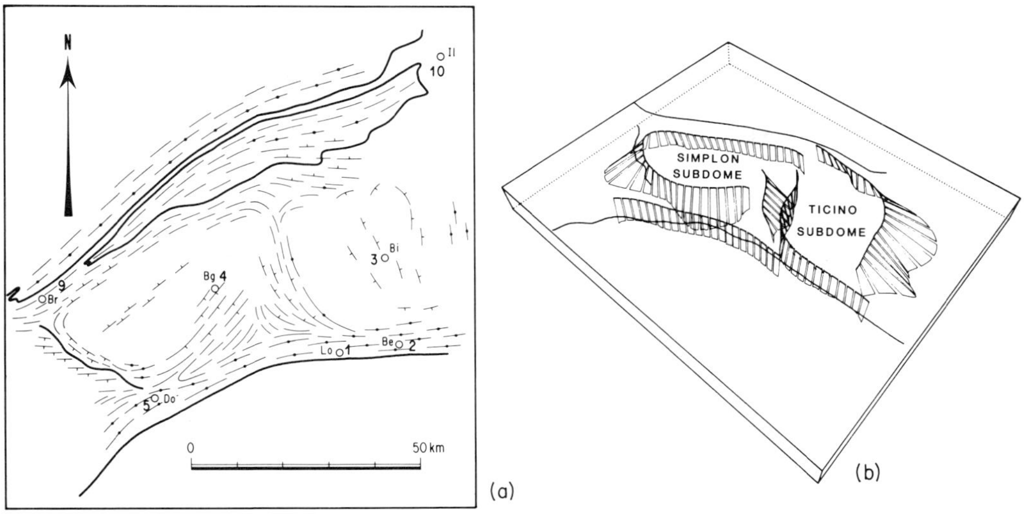

FIG. 2. Current attitude of main foliation in western and central parts of the Lepontine dome. Map at left shows smoothed strike and dip of foliation. Steep dips of 60° to 90° are shown by black dot centred on strikeline; moderate dips of 30° to 60°, by tick on downdip side of strikeline; flattish dips of O° to 30°, by lack of symbol. Numbered localities are as in Fig. 1: Biasca (Bi), Bellinzona (Be), Locarno (Lo), Bosco–Gurin (Bg), Domodossola (Do), Brig (Br) and Ilanz (Il). Data from Schmidt & Preiswek (1908), Knup (1958), Milnes (1974b), Steck (1984), Mancktelow (1985), supplemented by the authors' own measurements. Block diagram at right was drawn by computer using true perspective. Platelets are tangent to foliation as estimated at sea level, after regional averaging in areas 1 km×1 km square and interpolation using multiquadrics (Barbotin and Cobbold, work in progress).

First, in analogy with the shells of an onion, the region is composed of sheets of Hercynian basement, typically a few kilometres thick, interleaved with thinner sheets of Mesozoic sediments (Schmidt & Preiswerk 1908, Argand 1911, Staub 1924, Amstutz 1971, Milnes 1974a). Early authors described these structures as fold nappes, but Milnes (1974b) showed that large scale folding postdates the emplacement of thrust sheets.

Grossly parallel to the sheets themselves (and to the axial surfaces of major post-nappe isoclines) is a composite foliation at amphibolite grade. We have verified and, where necessary, supplemented the foliation attitudes measured by Schmidt & Preiswerk (1908), Knup (1958), Milnes (1974b), Steck (1984) and Mancktelow (1985). We have then smoothed the data (Fig. 2a) and constructed a simple block diagram (Fig. 2b), showing the central and western parts of the region. Thrust sheets and foliation attitudes form a regional dome.

To a first approximation, the dome has mirror symmetry about a central valley (the Val Leventina) running SSE from Airolo to Bellinzona (Fig. 1). At the bottom of this valley are exposed the lowermost granitic gneisses (Leventina, Lucomagno), considered infra-Pennine by some authors (e.g. Milnes 1974a). The overlying Simano sheet dips gently away to the E and W. Further E, the successively higher Adula, Tambo and Suretta sheets dip eastward by up to 25° on average. They are therefore predicted to occur at depths of up to 20 km or so beneath the overlying Austro–Alpine units which form the eastern boundary of the Lepontine dome, and are attributed to the Adriatic sub-plate. In the SW corner of the region, the uppermost Pennine sheets (Monte-Rosa and Bernhard units) are separated from lower ones by the Simplon Line, which is basically a SW dipping normal fault (Mancktelow 1985). This fault also appears to offset the metamorphic mineral zones (Fig. 1). In the N and S of the area are steep belts (Milnes 1974a), one bordering the External Crystalline Massifs (Aar and Gotthard), the other next to the Insubric Line. Both steep belts are generally considered to postdate the main deformational and metamorphic events (Milnes 1974a, Huber *et al.* 1980, Simpson 1982, Steck 1984, Mancel & Merle 1987). The southern steep belt is in fact the southern limb of an antiformal structure, most easily visible at the western and eastern sides of the dome, where the axis plunges gently to the W or E, respectively. In the W, this structure, known as the Vanzone antiform, folds the Monte-Rosa and Moncucco units. In the E, the units between the Bergell intrusion and the Austro–Alpine units are similarly folded. Notice that mineral zone boundaries have arcuate appearances in both these corners of the dome. Finally, to complete the description of the broadly symmetric features of the dome, we notice that the Simplon Line in the W and the Engadine Line in the E are symmetrically oriented and have similar positions at the two sides of the dome.

As well as these more symmetric features, the dome has some pronounced departures from mirror symmetry (Fig. 2b). Thus we distinguish the Ticino subdome in the E, separated from the Simplon subdome in the W by the N–S trending Maggia steep zone, termed the 'Querzone' by Preiswerk (1918) on account of its anomalous attitude. At its SE end, the Maggia steep zone curves into parallelism with the southern steep belt. Immediately SW of the Maggia steep zone, the foliation traces form a triangular transitional domain, between the Maggia steep zone itself, the southern steep belt and the NE striking Bosco–Gurin structure (Fig. 2a). Throughout much of the Simplon subdome, the foliation strikes NE whereas in the Ticino subdome, it strikes NNW.

Pre-Tertiary history

Relict high-pressure mineral assemblages (eclogites) are well known from the Sesia zone, but have also been described in the central Alps (Ernst 1976, Evans & Trommsdorf 1978, Heinrich 1982). Pressures and temperatures estimated from garnet and clinopyroxene and from the stability of jadeite, have a regional NS gradient, the pressure varying from 9–13 kb in the N to 20 kb in the S (Heinrich 1982, 1986). Although this metamorphic event is imperfectly dated, radiometric data suggest a Cretaceous age (Hunziker *et al.* this volume). The high-pressure metamorphism can be interpreted as resulting from crustal subduction before or during collision between two tectonic plates (Trümpy 1980, p. 80). For how long did this process continue? One possibility is that emplacement of basement slices continued without interruption from the late Cretaceous to the late Oligocene; but this is probably too simple a history. Thus, the P–T conditions of eclogite facies indicate a geothermal gradient lower than average, whereas the P–T conditions of Tertiary metamorphism indicate a geothermal gradient higher than average. A thickened crust with a low geothermal gradient is probably difficult to deform and so an alternative hypothesis is to consider a period between the Cretaceous and Tertiary events, during which there was little deformation. Tertiary deformation due to horizontal forces from the Adriatic sub-plate could perhaps have taken place as the crust became thin enough, by erosion after a late Cretaceous uplift (see Hurford & Hunziker 1985).

In any case, the kinematic record of this early stage of crustal collision is now completely overprinted by strong Tertiary deformations and metamorphism.

Tertiary metamorphism and cooling

Numerous petrological and metamorphic studies show that a high-temperature metamorphism took place during Tertiary times in the Pennine zone (Ernst 1973, Frey *et al.* 1974, 1980). Mineral assemblages are those of the amphibolite grade and mineral zones are nested in a concentric pattern on a map (Niggli 1970). Thus there is a systematic decrease of metamorphic grade from southern regions (Locarno and Bellinzona area) towards the N, E and W. Whereas there is no evidence against a possible sub-horizontal attitude of isotherms during the climax of metamorphism (Frey *et al.* 1980), isograd surfaces are now steeply dipping in the Simplon subdome, at the eastern margin of the Pennine zone and in the northern steep zone (Streickeisen & Wenk 1974, Fox 1975, Thompson 1976, Klaper & Bucher-Nurminen 1987). Thus, at first sight the large scale pattern of isograd surfaces looks like the one expected for a true thermal dome (see Turner 1968). However, large scale folding of originally sub-horizontal isotherms may have occurred after the climax of metamorphism (Milnes 1975, Frank 1983). The near coincidence of the structural and metamorphic domes is highly suggestive of such a folding.

The timing of metamorphic events in the Pennine zone is controversial and this in itself suggests that the notion of a simple thermal dome may be misleading. Geochronological data from muscovite–phengite (Rb–Sr) at the western and eastern periphery of the Lepontine dome (Hunziker 1970, Steinitz & Jäger 1981), outside the staurolite isograd, yield ages of 38–35 Ma. These ages have been interpreted as the peak of Lepontine metamorphism (Jäger 1973). Following this interpretation, the younger ages inside the straurolite isograd all have to be considered as cooling ages. This view assumes that there is a single peak in the time-temperature curve (referred to as the 'Lepontine event') and according to this concept the mineral isograds would approximate an isotherm at one and the same instant in time. The ages of 38–35 Ma are identical with those obtained for the greenschist metamorphic event in the western Alps (Bocquet *et al.* 1974, Frey *et al.* 1974, Chopin & Malusky 1980, Chopin & Monié 1984).

In contrast, ages obtained within the Lepontine dome using the U–Pb method on monazites or K–Ar on amphiboles are much more recent (around 25–20 Ma). Hence some workers have reappraised the chronology of the Alpine metamorphism, claiming that conditions sufficient to produce high grade assemblages were sustained until around 25–20 Ma (Hänny *et al.* 1975, Köppel & Grünenfelder 1978, Köppel *et al.* 1980, Deutsch & Steiger 1985, Monié 1985).

Furthermore, structural investigations (e.g. Thakur 1973, Huber *et al.* 1980, Klaper 1982) suggest that more than one episode of mineral growth occurred in the Lepontine area with respect to established deformation phases. Admittedly these deformation phases do not need to be contemporaneous all over the area considered. However, Rb–Sr ages on micas (Steiger & Bucher 1978, Steiger 1983) more directly suggest a second episode of mineral growth, particularly in the northern part of the Lepontine dome, S of the Gotthard Massif. This mineral growth does not necessarily indicate a second peak in temperature, but may indicate that mineral isograds did not form at the same time all over the Lepontine area.

Thus, the age of the so-called Lepontine metamorphic event remains unclear, possibly because the subject has been considered in terms that are too simple. We suggest that the notion of a single metamorphic event, synonymous with a single period of mineral growth and a single peak in the temperature–time curve throughout the entire Lepontine dome, should be abandoned in favour of a more complex and dynamic thermal history, which has yet to be worked out for particular domains within the Lepontine area. The complexity is due in part to very important post-38–35 Ma events, such as the intrusion of the Bergell batholith (30 Ma, Köppel & Grünenfelder 1975), an Oligocene phase possibly related to the intrusions along the Insubric Line (Laubscher 1983) and particularly the subsequent formation of the southern and northern steep belts (Hurford 1986, Schmid *et al.* 1987, Merle 1987).

Cooling of the Lepontine area has traditionally been attributed to 'uplift', without further explanations of its causes. The use of fission track data provided a major breakthrough in specifying the cooling history. For a region N of Locarno, Hurford (1986) demonstrated a period of very rapid cooling between 23 and 19 Ma, followed by slower cooling rates compatible with current uplift rates (Fig. 3, see also Werner *et al.* 1976, Steiner 1984). Since the formation of the southern steep belt appears to be intimately related to backthrusting along the Insubric Line, the period of rapid cooling is best attributed to uplift caused by backthrusting and backfolding during an Insubric phase (Argand 1916), of Oligocene–Miocene age and contemporaneous with northwards thrusting in the Helvetic nappes (Schmid *et al.* 1987). In this interpretation, the emplacement of the Lepontine area into shallow levels of the Earth's crust is the result neither of isostatic uplift nor of underplating and crustal extension (Platt 1986), but of a Neo-Alpine compressional phase.

The geochronological data also indicate that cooling occurred earlier in the E (Ticino subdome) and later in the W (Simplon subdome) (Jäger *et al.* 1967, Wagner *et al.* 1977). This has led to a model of differential uplift within the Pennine zone (Bradbury & Nolen-Hoeksema 1985). Westward propagation of uplift may be due to dextral transpression in front of a rigid indenter as, for example, the Adriatic subplate with the Ivrea zone at its NW corner (Schmid *et al.* this volume).

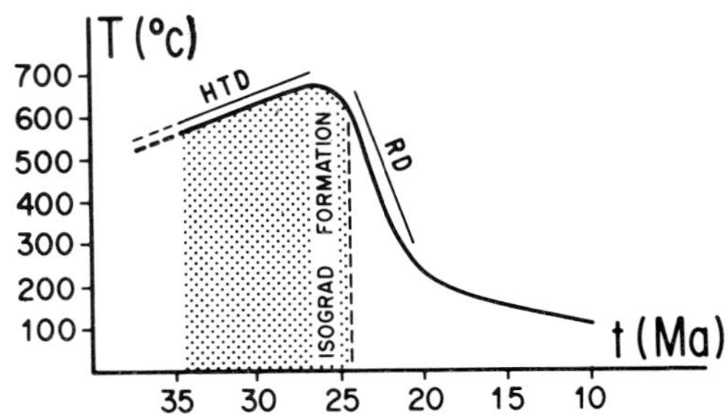

FIG. 3. History of cooling and deformation in the Lepontine dome. Curve gives temperature (T) as a function of time (t) for the last 35 Ma according to model B (II) of Hurford (1986), applicable to the Maggia valley only. There, high-temperature deformation (HTD) ended at about 25 Ma and was followed by retrograde deformation (RD). Within the Simplon subdome, rapid cooling started later than 25 Ma ago (see Schmid *et al.* this volume, Fig. 3).

High-temperature deformation (HTD)

In this section, we refer to deformation that occurred before the onset of rapid cooling. Provisionally, we consider the HTD throughout most of the Pennine zone to have occurred between 40 and 25 Ma. This age bracket is consistent with geochronological data discussed in the previous section and with the culmination of Meso-Alpine shortening in the late Eocene to early Oligocene (38–33 Ma) according to several authors (e.g. Ayrton & Ramsay 1974, Milnes 1978, Hsü 1979, Trümpy 1980, p. 90). Because of the long period of high temperature (40–25 Ma) we cannot further constrain the age of the HTD using metamorphic arguments. We cannot even exclude the possibility that some areas of the Lepontine dome inside the staurolite isograd reached amphibolite grade conditions previous to the event dated at around 38–35 Ma outside this isograd. Furthermore, although the onset of rapid cooling is well dated in the Maggia area at 25 Ma (Köppel & Grünenfelder 1975, Werner *et al.* 1976, Steiner 1984, Hurford 1986), some areas to the N (S of the Gotthard Massif) and W (Simplon domain) were at high temperatures considerably later than 25 Ma ago (Jäger *et al.* 1967, Purdy & Jäger 1976, Wagner *et al.* 1977). Thus we do not consider the HTD to be a well-timed tectonic event, synchronous over the entire area.

Stretch lineation

The main lineation observed in the Pennine zone is a preferred orientation of grain aggregates, individual grains, fold axes, rods and boudins. This structure seems to track the direction of maximum total stretch. Fold axes parallel to the stretch lineation occur at small to intermediate scales (Wenk 1955). Occasional eye structures are probably tranverse sections through sheath folds. We have also observed definite sheath folds in oblique sections (Cobbold 1980).

Optical observations in thin section show that the main foliation and lineation are defined by mineral assemblages of amphibolite grade, such as quartz+plagioclase(An>17)+biotite+staurolite+kyanite+garnet. Thin sections frequently show shear bands associated with the growth of metamorphic minerals. A study of inclusions within garnets from the Simplon area indicates that growth of garnet is contemporaneous with rising temperature and predates the peak of metamorphism in this particular area (Merle *et al.* 1986).

We have measured the attitude of HTD stretch lineations at 882 localities throughout the Lepontine dome (Fig. 4), thus confirming the unusual radiating pattern first described by Wenk (1955). If we compare the attitude of foliation and lineation (see Figs 1, 2 and 4), we see that in the central areas of flat-lying foliation (Ticino and Simplon subdomes), the lineations fan out in directions approximately perpendicular to the peri-Adriatic fault system. Even in the anomalous Maggia steep zone, this radial pattern of flat-lying lineations is not disturbed, the lineations following the strike of the steep zone.

As one approaches the northern steep belt near Airolo, both lineation and foliation acquire steep attitudes, suggesting that they have become folded together about a horizontal E–W axis during later events. Similar geometrical relationships occur as one approaches the southern steep belt, either in the Bellinzona area and further E, where the lineation is nearly vertical, or in the Domodossola area, where it now pitches at about 45°E. This suggests that foliation and lineation were folded together about horizontal axes during a later event in the southern steep belt as well. However the situation is not as simple as this, because for about 50 km between Domodossola and Bellinzona, passing through Locarno, the HTD lineation is flat-lying, often pitching gently E, but sometimes even gently W. If the foliation is restored to the horizontal by rotating about the E–W strike, the lineation remains E–W; whereas immediately to the N, in the flat-lying area, it currently strikes NW. We therefore conclude that the steep belt between Domodossola and Bellinzona cannot have undergone rigid rotation alone during later events. Either the HTD lineation formed in a flat-lying E–W attitude, or it became reorientated during later events as a result of superimposed ductile deformation (strain). This matter is further discussed in the following section.

Kinematic indicators

In our general experience, kinematic indicators are less clear and abundant in rocks of amphibolite grade than they are in rocks of greenschist grade. Nevertheless we have found high-temperature shear indicators in the Lepontine dome, both at outcrop scale (Fig. 5a) and in thin section (Figs 5b, c and d). In deformed granites, we mainly used the relative attitudes of foliation and shear bands with amphibolite

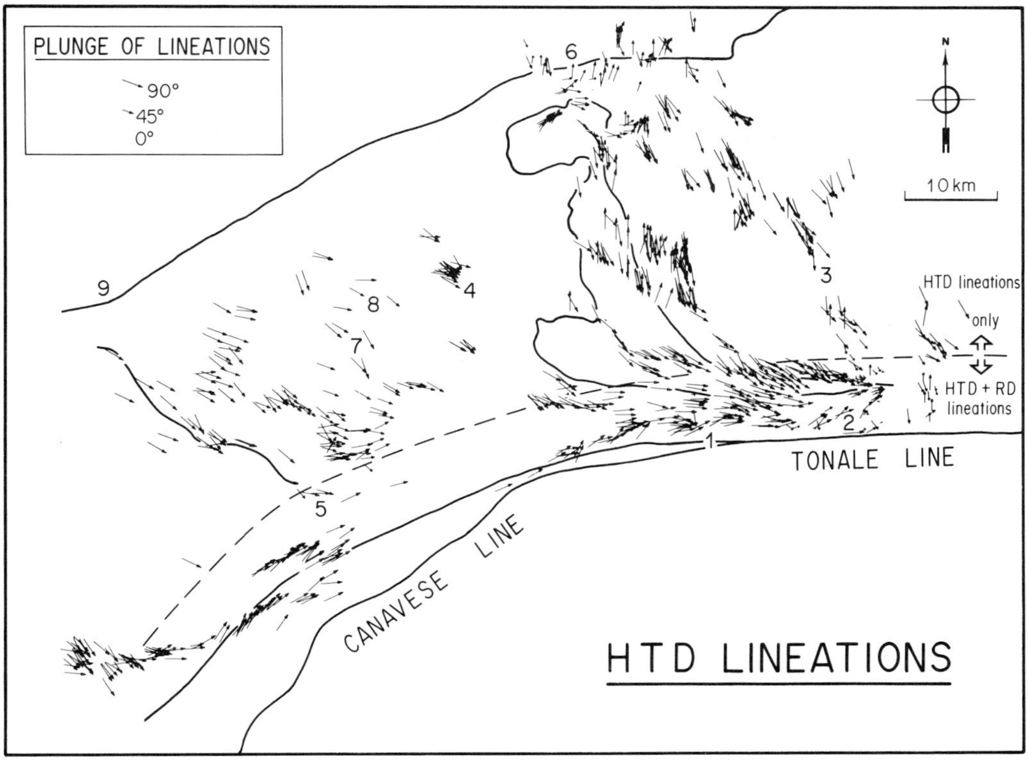

FIG. 4. Current attitude of high-temperature stretch lineations in central and western parts of Lepontine dome (new data only). Each arrow represents a single measurement. Length of arrow is inversely proportional to plunge in degrees (see key). Solid curves are a selection from the geological boundaries of Fig. 1. Numbered localities are as in Fig. 1. The age of the lineations in the southern steep belt (south of the solid broken line) cannot always be inferred with certainty: there are HTD lineations rotated by the backfolding as well as lineations formed during the RD event. For lineations within the Insubric mylonite belt see Schmid *et al.* (1987) (data from the Monte-Rosa area by E. Barbotin).

grade assemblages (C/S structures of Berthé *et al.* 1979), as well as the sigmoidal shapes of recrystallized tails around old grains (Simpson & Schmid 1983, Passchier & Simpson 1986). Such structures, together with asymmetric folds of the foliation, are especially clear in the Leventina orthogneiss (Fig. 5). In paragneisses we had greater difficulty in determining shear senses, but sometimes, like Mawer (1987), we were able to use shear bands, asymmetric folds, rolling structures or rotated boudins (Van Den Driessche & Brun 1987).

In the flat-lying areas of the Lepontine dome, HTD shear senses are consistently top to N to WNW in the lineation direction (Fig. 6). In most areas, there is little or no retrograde overprint, so that these shear directions appear to be original ones, operative during the HTD event. An exception is the Simplon area, where there is a retrograde overprint, discussed in a later section.

In the northern steep belt around Airolo, HTD shear indicators appear to have become folded into an upright position, together with foliation and lineation. Restored to a horizontal attitude, by rigid rotations about E–W axes, the shear indicators have top to N senses, as in the centre of the Ticino subdome. Further evidence for such rotations will be discussed later.

In the southern steep belt, HTD shear indicators have been found so far at two localities (Fig. 6). One is in the Monte-Rosa unit at Villadossola, near Domodossola (Schmid *et al.* 1987); the other is in the Monccucco unit at Ponte Brolla, near Locarno. Both yield dextral strike-slip shear senses in almost perfectly horizontal directions. Were these horizontal shear directions operative during the HTD event, or have the structures been rotated? If foliation is restored to the horizontal, by rigid rotation about strike, the shear senses become top to W, compatible in sense, but not in direction, with

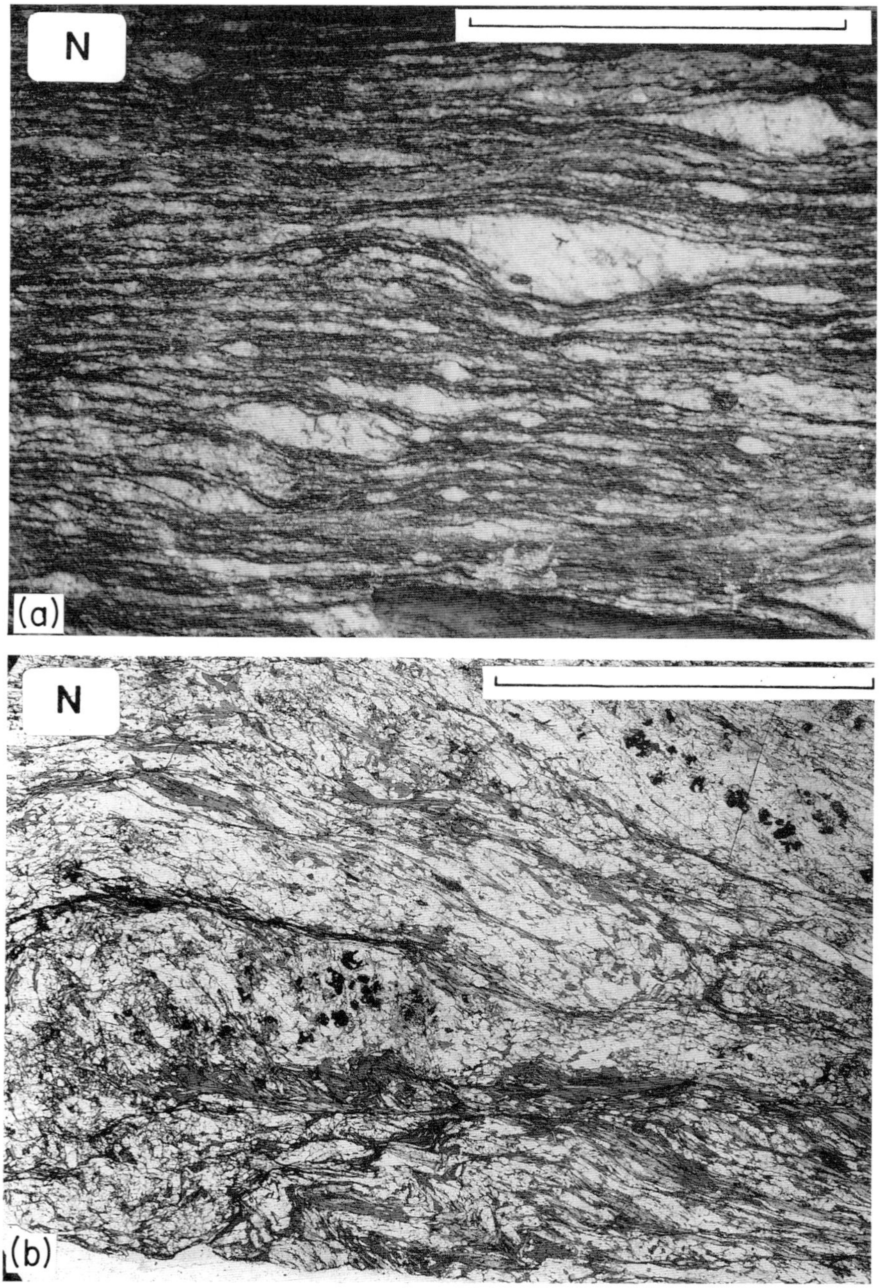

FIG. 5. Kinematic indicators for high-temperature deformation. (a) Shear bands and sigmoidal porphyroclasts in the flat-lying Leventina gneiss. Photograph of vertical outcrop (N at left), scale bar 5 cm long. (b) C–S structures and folds in the flat-lying Leventina gneiss (thin section, vertical, N at left, scale bar 2.5 cm long). (c) and (d) Detail from (b), showing shear bands of top to N sense, containing assemblages of two micas, feldspar and quartz. Dynamic recrystallization of potash feldspar (c) and dynamic recrystallization of quartz by grain boundary migration (d) indicate high temperatures during formation of shear bands. Scale bars 0.83 cm long (c) and 0.20 mm long (d).

the top to NW shears observed in the flat-lying areas immediately to the N. Thus a component of rigid rotation during later events is not impossible; but it does not appear to be sufficient in itself to account for the present attitudes, as discussed in detail in the previous section on HTD lineations. We conclude that there has been additional ductile strain in this part of the southern steep belt. At the Ponte-Brolla locality, shear indicators (shear bands and asymmetric folds and boudins) are coeval with the beginning of anatexis (Gapais, personal communication, 1987). The rocks are migmatitic and there is no evidence of a retrograde ductile overprint. We therefore conclude that dextral strike-slip at Ponte-Brolla occurred at high

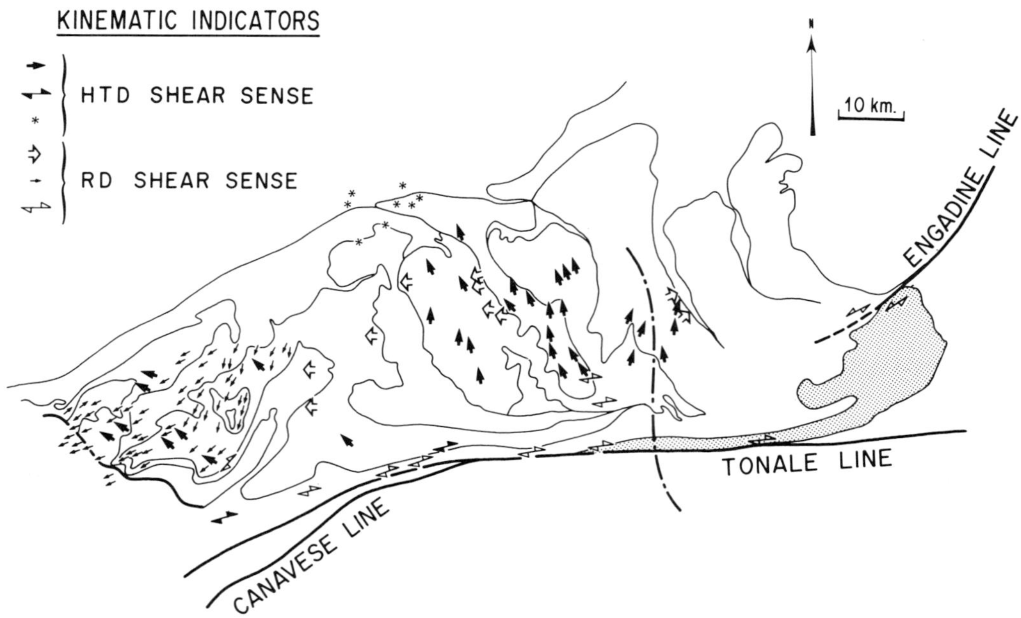

FIG. 6. Shear senses in the Lepontine dome. Each arrow represents a sense of shear, determined from a population of kinematic indicators at outcrop and in thin section, where such a population gave consistent results at a scale of about 100 m. In central areas of flat-lying foliation, horizontal shear components are shown by single arrows representing motion of roof (large black arrows for HTD, wide empty arrows for RD in Ticino subdome, small black arrows for RD in Simplon subdome). Note the reversed sense of shear in the Berisal and Glishorn backfolds. In the southern steep belt and along the Engadine Line, wrench components are shown by zigzag arrows (full black arrows for HTD, empty arrows for RD). In the northern steep belt, asterisks show localities where HTD shear directions are currently vertical (for interpretation, see Fig. 10). East of dashed line, shear-sense indicators have been observed recently at outcrop (by P. R. Cobbold, M. Ballevre, D. Gapais and E. Le Goff), but not yet in thin section, so that metamorphic conditions are not yet reliably known.

temperature. Unfortunately we do not know whether it occurred in the age bracket of 40–25 Ma, or at the beginning of the Insubric phase.

Retrograde deformations (RD)

In this section, we refer to deformations coeval with the cooling of the Lepontine area, i.e. from 23 Ma to 19 Ma in the Ticino subdome (Hurford 1986), and significantly later in the Simplon subdome (Bradbury & Nolen-Hoeksema 1985).

Geological mapping of stretch lineations (Fig. 7), shear indicators (Fig. 8a, b) and other structures reveal late ductile deformations (Steck 1984) superimposed under retrograde metamorphic conditions upon the HTD (Merle 1987, Merle *et al.* 1986). The closing stages of retrograde deformation (RD) demonstrably occurred under greenschist facies conditions in the Simplon area (Mancktelow 1985, Mancel & Merle 1987). Further E, the RD is only occasionally associated with syntectonic growth of chlorite. From studies of the RD, we infer different kinematic histories in the Simplon and Ticino subdomes. The southern and northern steep belts also have rather special histories.

Southern steep belt

The formation of the southern steep belt has been shown to be contemporaneous with mylonitization under greenschist facies conditions along the Insubric Line (Zingg & Schmid 1983, Schmid *et al.* 1987), during the Insubric phase in the sense of Argand (1916). Because the Insubric mylonites formed at the interface between the warm Lepontine domain and the earlier cooled southern Alps (>500°C vs. 150°C at 25 Ma, see Hurford 1986, fig. 7), this retrograde overprint is particularly obvious within the mylonite belt, which simply represents the southern margin of the steep belt. In the SW of the region, the central Alps were backthrust

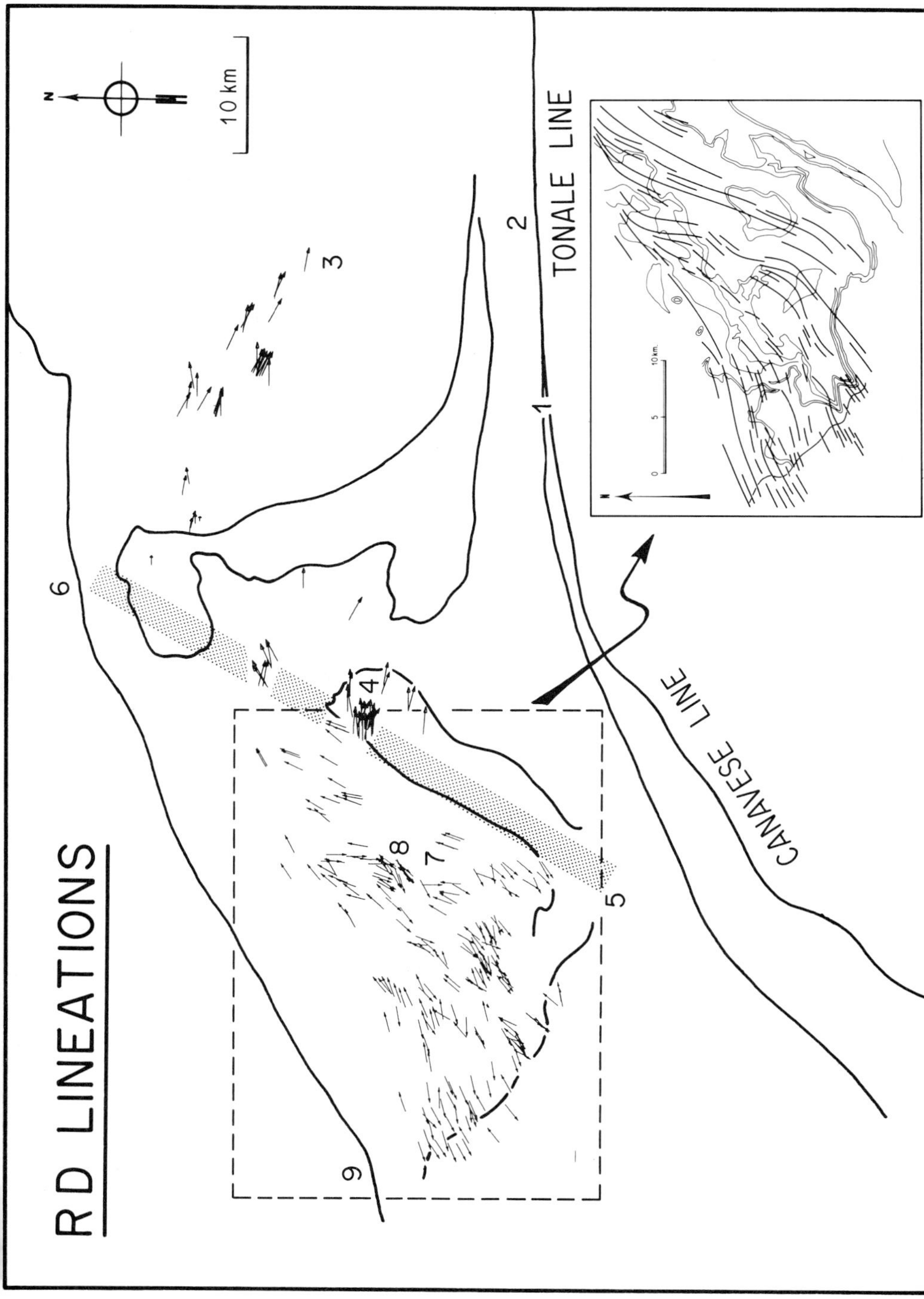

FIG. 7. Current attitude of retrograde stretch lineations in central and western parts of the Lepontine dome (new data, and old data from Merle 1987 and Merle & Le Gal 1988). Stippled zone separates lineations in the E (associated with formation of Ticino subdome, Maggia steep zone and Bosco–Gurin thrust) from later lineations in the W (associated with formation of Simplon subdome and normal faulting along Simplon Line). For legend, see Fig. 4. Inset at bottom right shows lineation trajectories (drawn by hand) in the Simplon domain. For senses of shear, see Fig. 6.

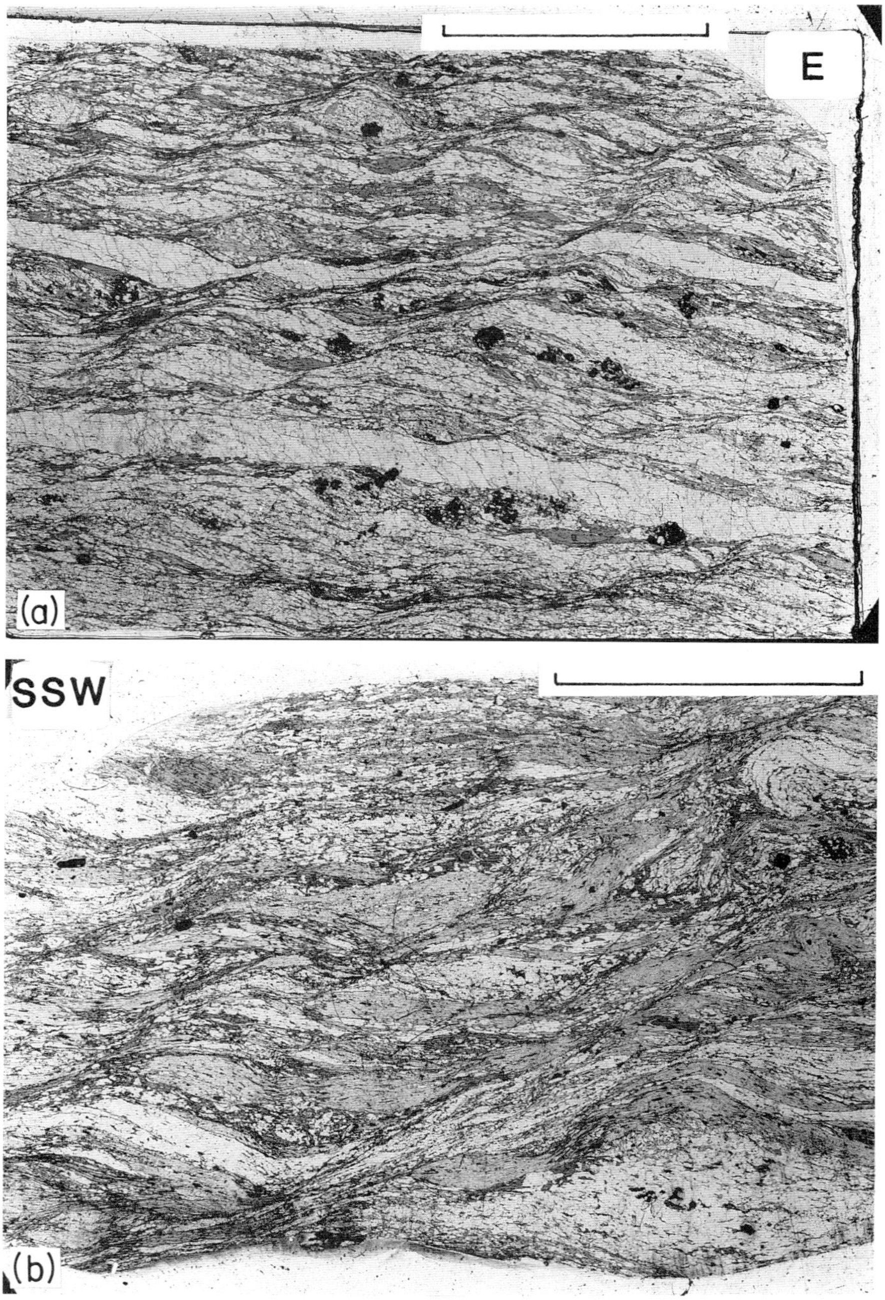

FIG. 8. Kinematic indicators for retrograde deformation. (a) Shear bands at contact between Simano nappe and Leventina gneiss (vertical thin section, E at right, scale bar 1 cm long, see Fig. 9 for locality). Sense of shear is top to W. (b) Shear bands and fold in Baceno schist, Verampio window (vertical thin section, SSW at left, scale bar 2 cm long). Sense of shear is top to SSW. Chlorite content about 20% (c) Shear bands within a quartzitic gneiss near base of Bosco-Gurin thrust (see Fig. 9 for locality). Microstructure of coarse-grained quartz outside a shear band (left) indicates migration recrystallization typical of high temperature deformation. Within same shear band (right), serrate grain boundaries and incipient recrystallization by mechanism of subgrain rotation indicate lower temperatures. Since shear bands represent late instabilities along a continuous strain path, we deduce that shearing occurred while rocks rapidly cooled (compare Fig. 3). Scale bars 0.83 mm long (left) and 0.25 mm long (right). (d) Flat-lying shear zone in Antigorio granite, Lago d'Agaro (locality 8, Fig. 1). Vertical outcrop, SSW at left, hammer gives scale. Shear sense is top to SSW.

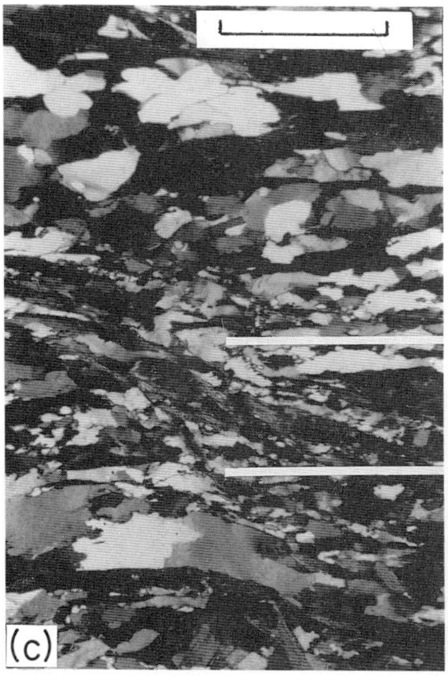

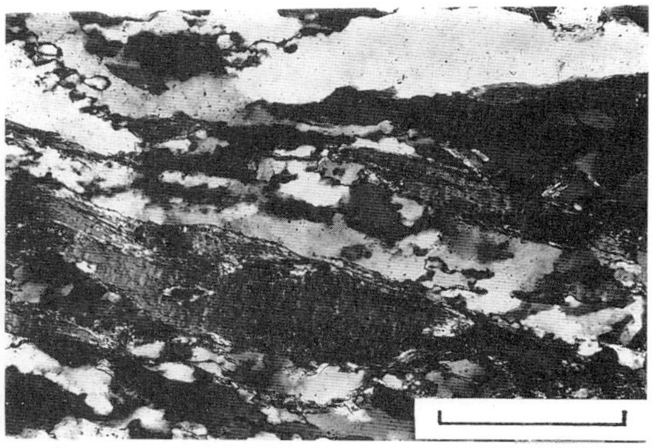

over the southern Alps along the N-dipping Insubric mylonite belt, as demonstrated by small scale shear criteria and steep stretch directions (Heitzmann 1987b, Schmid *et al.* 1987). Kinematic studies indicate that backthrusting was followed by dextral shearing in these mylonites. Although this sequence of events can be established locally, both events are best thought of in terms of a continuous process of dextral transpression on a larger scale (Schmid *et al.* this volume). Only during the very latest stages along the Insubric Line (cataclastic deformation at the Tonale and Centovalli lines, greenschist facies mylonitization of Lepontine marble

Laubscher 1971, Fumasoli 1974, Heitzmann 1987a, Schmid *et al.* 1987) did oblique convergence across the belt change into almost pure strike-slip.

Because of the existing N–S temperature gradient during the Insubric phase, it is dangerous to use ambient temperatures to decide on the relative ages of deformation. The retrograde nature of movements N of the mylonite belt is not always obvious, since higher temperatures prevailed for a longer period of time in the northern parts of the Lepontine area (Hurford 1986). In the Valle d'Ossola section, dextral strike-slip (Monte-Rosa nappe at Villadossola near Domodossola, Schmid *et al.* 1987) and backfolding (Vanzone antiform, Klein 1978, Milnes *et al.* 1981) can be attributed to this Insubric phase. Dextral shearing occurred under retrograde conditions all along the southern steep belt (Fig. 6 and Heitzmann 1987b), but also under high-temperature conditions (Villadossola and Ponte Brolla localities). The time relationships between dextral shearing and backfolding are not established yet and it is likely that both components of motion were synchronous in many parts of the southern steep belt, as postulated for the Insubric mylonites. In any case, the vertical uplift associated with backthrusting appears to have reached a maximum between Locarno and Bellinzona.

Ticino subdome

In the W, from Biasca to Bosco–Gurin, the Ticino subdome displays a set of RD stretch lineations of nearly EW trend, superimposed on the HTD stretch lineations (Fig. 7). This second deformation is scarce at the scale of the subdome, but where it occurs it is usually associated with flat-lying ductile mylonites. Thus, mylonites at least 10 m thick occur at the contact between the Leventina gneiss and the Simano sheet (Swiss National Grid 706.2–141.45) and between the Antigorio granite and Bosco–Gurin units (SNG 678.3–128.8). Shear senses determined at outcrop or from thin sections (Fig. 8a) are all top to W.

The retrograde character of this deformation is apparent in thin section. Chlorite has grown preferentially within shear bands that postdate the growth of amphibolite-grade porphyroblasts. However, we have never observed completely retrograde assemblages (as described later for the Verampio window or the Simplon line). Quartz microstructures within microscale shear bands often indicate progressive deformation at decreasing temperatures (Fig. 8c) and this can probably be correlated with the cooling recorded by geochronological data.

Westward shearing may be a key for understanding major structures such as the Maggia steep zone and at the duplication of the Antigorio granite in the Bosco–Gurin area. An E–W cross section through the Maggia steep zone (Fig. 9) shows a large fold overturned towards the W (see Merle & Le Gal 1988 for a fuller description). This structure appears to fold the earlier high-temperature foliation. Retrograde mylonite zones and later brittle faults all show top to W senses of shear or slip. We therefore interpret the overall structure as being modified, if not created, by top-W shearing during the RD event.

In the Bosco–Gurin area, the Antigorio granite is duplicated (Figs 1 and 9). There is a well developed RD stretch lineation striking E–W, parallel to the fold axis of the Wandfluhhorn fold (Hunziker 1966, Hall 1972, see Wandfluhhorn peak located on Fig. 1). In the core of the upper Antigorio sheet, strain is homogeneous and shear indicators are lacking, except at the base, where the granite, in contact with underlying paragneisses (Bosco–Gurin unit), becomes mylonitic in a thin zone dipping gently E. The sense of shear is top to W. Thus, the duplication of the Antigorio granite may be due to overthrusting towards the W (Merle & Le Gal 1988). The northern limit of the Bosco–Gurin thrust remains to be recognized in the core of the Antigorio granite itself, but the Wandfluhhorn fold may be a ductile termination of the Bosco–Gurin thrust. To the S, next to the southern steep belt, the Antigorio granite is refolded in a synformal structure (the Masera synform). On a map, the hinge of this fold curves in a manner consistent with a dextral sense of shear in the southern steep belt.

The westward deformation in the Ticino subdome is considered to be coeval with the formation of the southern steep belt. We believe that both events occurred at the western and southern margin of the uplifting Ticino subdome between 25 and 20 Ma, a period of rapid cooling in the Ticino subdome according to geochronological data (e.g. Werner *et al.* 1976, Wagner *et al.* 1977, Hurford 1986) (Fig. 3).

On the eastern side of the Ticino subdome, within the Adula unit, but still within the limit of Tertiary amphibolite-facies metamorphism, retrograde shear indicators with normal-fault sense (top to ENE) have recently been observed (Fig. 6). The metamorphic conditions of their formation and their age relationships with the shear indicators in the western part of the dome are not yet clearly determined.

Finally, in the steep zone that prolongs the

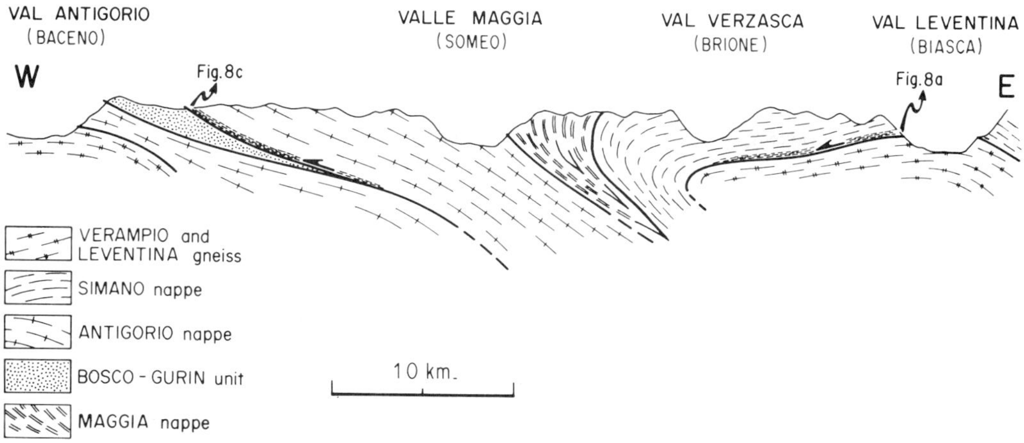

FIG. 9. Schematic geological cross-section running EW from Val Leventina to Val Antigorio across Maggia steep zone and Bosco-Gurin thrust (after Merle & Le Gal 1988). Section line is shown in Fig. 1. The Wandfluhhorn fold cannot be seen as its axis is parallel to the cross section. Arrows show shear senses in retrograde mylonites (see Figs 6, 8a, c).

Engadine Line to the SW (Fig. 1), as well as in the northernmost exposures of the Bergell intrusion, retrograde shear indicators reveal left-lateral wrench components along strike (Fig. 6). Once again, the metamorphic conditions of formation are not yet known, but the structures postdate the Bergell intrusion (30 Ma).

Simplon subdome

Throughout a well-defined triangular area between the Simplon Line and the northern lobe of the Maggia unit, later flat-lying foliations and lineations partially overprint older HTD structures (Steck 1984). The second deformation occurred under retrograde conditions, at temperatures lower than in the Ticino subdome (Merle 1987).

The stretching lineation in the field is mainly defined by an alignment of quartz aggregates, muscovite and chlorite. Coeval shear bands postdate the growth of garnet, staurolite or kyanite porphyroblasts; chlorite aggregates are usually located within microscopic shear bands. Thin sections of samples from the Verampio window and the Simplon line indicate P–T conditions of greenschist facies (Fig. 8b). Shear bands developed within the domain of stability of biotite but generally were contemporaneous with the breakdown of this mineral.

The RD lineations form a consistent, slightly radiating pattern (Fig. 7), restricted to an area of nearly flat-lying units, surrounded by zones of steeper attitude (including the Bosco–Gurin structure). Towards the W, the extent of this area of retrograde deformation beyond the Simplon Line is not yet known (Merle 1987, Mancel & Merle 1987, Steck 1987). The sense of shear, demonstrated by C/S structures, shear bands and asymmetric pressure shadows and porphyroclasts, is top down and SSW in the Verampio window, top down and WSW at the Simplon Line (Fig. 6) (Merle 1987, Mancel & Merle 1987). We correlate this deformation with the Simplon phase described by Mancktelow (1985, 1987) along the Simplon Line. At a few localities in the core and front of the Antigorio granite nappe, the shape of quartz aggregates indicates that the strain ellipsoid is oblate (Greco 1985, Daniel 1986). This strain is due to the retrograde deformation alone, high-temperature aggregates having been observed only at the base and rear of the granite sheet. We suggest that the oblate strain is due to centrifugal radial motion along the divergent RD trajectories of Fig. 7. The strain intensity appears to increase towards the Simplon Line, but is also very high in some ductile units (Baceno schist) exposed in the Verampio window (Merle 1987).

Folds with axes nearly perpendicular to the stretch direction and asymmetries consistent with the bulk sense of shearing are frequently observed in the Simplon subdome. Non-cylindrical folds can sometimes be seen, but the strain is not strong enough for complete reorientation of fold axes into the stretch direction.

The RD in the Simplon subdome must have occurred later than about 15 Ma, an age which represents the end of the amphibolite grade in this area (Jäger *et al.* 1967, Purdy & Jäger 1976, Wagner *et al.* 1977). This age is consistent with the time bracket of 19–6 Ma suggested by Mancktelow (1985) for the deformation recorded along the Simplon Line. On the basis of young ages (Frank & Stettler 1979) in fine fractions of mica, the deformation along the Simplon Line may indeed have lasted until 9 Ma. Thus, it is likely that the Simplon phase postdated the formation of the southern steep belt and the RD deformation in the Ticino subdome (20–25 Ma, see discussion in the previous section).

Northern steep belt

The northern steep belt is a region of backfolding located at the boundary between the Pennine zone and the external crystalline massifs (Huber *et al.* 1980). The backfolds have been described as late structures postdating the high-grade metamorphism in the Pennine zone (Higgins 1964, Chadwick 1968, Cobbold 1969, Thakur 1973, Huber *et al.* 1980, Simpson 1982).

The chronological relationships between high-temperature deformation (HTD) and backfolds are especially clear N of the Ticino subdome (Airolo area and Lucomagno section). Small scale HTD shear indicators (C/S structures in the Tremola series of the southern Gotthard Massif) show an apparent uplift of southern units with respect to northern units along the vertical stretch lineation (Fig. 10). Low-temperature flat-lying shear bands and shear zones (top to S) postdate the HTD indicators and are associated with the backfolding event (Fig. 10). They become prominent in the orthogneiss of the Gotthard Massif which seems to have been less affected by the HTD deformation (see Choukroune & Gapais 1983, Marquer & Gapais 1985). Thus we conclude that the HTD indicators were rotated into an upright position during backfolding. Backfolds postdate well-developed biotites which have been dated at 15 Ma (Steiger & Bucher 1978, Ramsay in Steck *et al.* 1979).

North of the Simplon subdome, the senses of shear associated with the retrograde stretching lineations (depicted in Fig. 6) are reversed between the Berisal synform and the Glishorn antiform (Fig. 1) (Mancel & Merle 1987, Fig. 8). This reversal is interpreted to be not original, but due to backfolding about axes near-parallel to the earlier retrograde lineations. Hence, a very young age of less than 10 Ma is inferred for backfolding within the northern steep belt.

Kinematic history

Using the structural, metamorphic and geochronological data presented or discussed above, we now interpret the kinematics of the Lepontine dome in three stages, which we consider in reverse order (younger to older):

(1) During the youngest stage (about 10 Ma to present), the Lepontine dome appears to have been rigid, except at its margins. In the northern steep belt between Brig and Ilanz, ductile deformation was expressed as backfolding and southwards overthrusting (Fig. 10). Ductile deformation also occurred in the external zones to the N and is responsible for the N6OE striking foliation of the crystalline massifs (Choukroune & Gapais 1983, Marquer & Gapais 1985). To the S of the Lepontine dome, dextral strike-slip motions occurred on the Insubric and Tonale faults. All these ductile and brittle motions we attribute to dextral transpression across the Insubric Line.

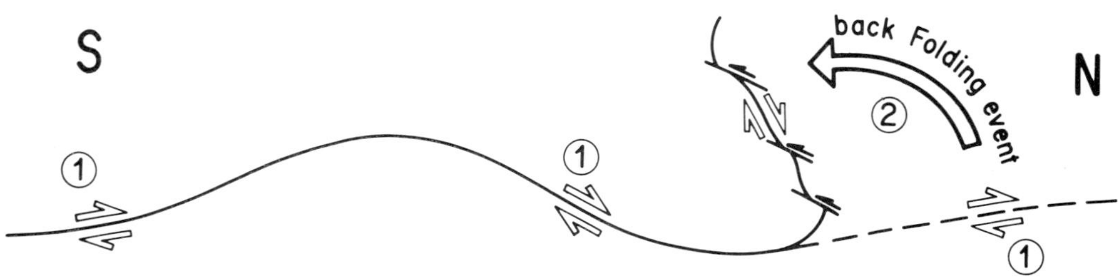

Fig. 10. Schematic N–S section through northern steep belt (Airolo and Lucomagno area, see Fig. 1), illustrating shear indicators (large hollow arrows) of earlier HTD event (1), flat-lying in the S, but steepened and overturned in the N, as a result of later back-folding event (2). Late retrograde shear bands (small arrows) are associated with backfolding.

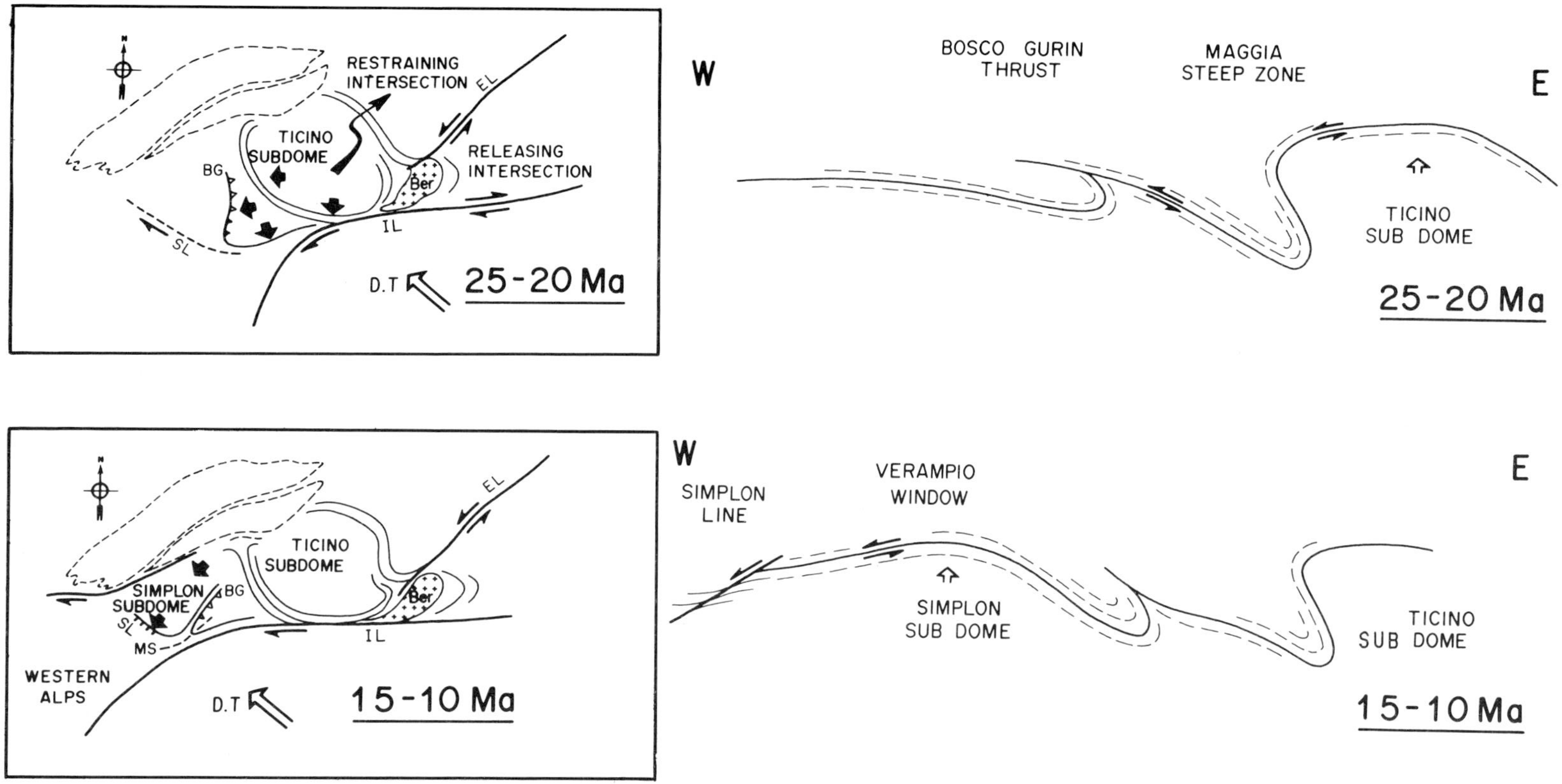

FIG. 11. Tertiary kinematics of the Lepontine dome during uplift (schematic interpretation). Two stages are shown at top (25–20 Ma) and bottom (15–10 Ma), in plan (left) and E–W section (right). In plan, approximately northwestward displacement (DT) of Adriatic subplate relative to Europe results in northwestward shortening within Lepontine area, dextral transpression across Insubric line (IL), and serial uplift of Ticino subdome (top), followed by Simplon subdome (bottom). Present day shape of external massifs merely serves as frame of reference. Full black arrows indicate horizontal components of shearing (roof motion), double arrows indicate wrench components. Ber is Bergell intrusion; BG, Bosco-Gurin thrust; SL, Simplon Line; EL, Engadine Line. The axial-plane trace (dashed line) of Masera synform (MS) curves in a manner consistent with dextral shearing parallel to Insubric Line. Cross sections show enveloping surfaces of foliation, main structures, retrograde shear indicators (black arrows) and uplift of subdomes (empty arrows).

(2) The Insubric phase (about 25–10 Ma) is best considered as twofold (Fig. 11). The kinematics are closely related in time to the uplift history of the Lepontine dome. If we attribute the back-thrusting along the southern steep belt to the converging Adriatic sub-plate, it seems sensible to expect some lateral escape of rocks within the Pennine zone at the same time, especially at releasing intersections of conjugate strike-slip faults such as the Engadine and Insubric Lines. The two regional subdomes of the Pennine zone (see the map of foliation attitude, Fig. 2a) could be due to these major retrograde deformations, coeval with uplift, whose importance seems to have been underestimated until now. An earlier dome (Ticino dome, 25–20 Ma) to the E and a later one (Simplon dome, 15–10 Ma) to the W of the Pennine zone might thus be attributed to continuing dextral transpression at the boundary between both tectonic plates (Fig. 11). The domes could have formed in an area of excess shortening, located in front of the indenting Adriatic sub-plate and at the restraining intersection of the Insubric and Engadine strike-slip fault zones, whereas horizontal extensions were concentrated at the releasing intersections to the E and W of the domes (Fig. 11). The NE–SW direction of extension in the Simplon subdome is kinematically compatible with NW–SE shortening and dextral shearing. The radial stretch trajectories accompanying the extension in the Simplon subdome are perhaps attributable to strong gravitational effects (i.e. a spreading process) contributing to unroofing of the subdome. Physical and numerical models of such gravitational processes characteristically show (a) divergence of stretch trajectories from topographic highs (b) oblate strain ellipsoids resulting from a combination of divergent horizontal shearing and vertical shortening and (c) strain intensities that increase downslope (see Hudleston 1983, Gilbert & Merle 1987).

(3) Before about 25 Ma, the Lepontine area underwent radial shearing (top to N or NW). Restoration of the younger backthrusting along the Insubric Line suggests that the major tectonic units were southerly dipping before backthrusting. If so, the radial motion implies an overthrusting to the N and NW at a level deep enough to sustain amphibolite facies metamorphism. Thus the high-temperature deformation is attributable to Tertiary convergence between Europe and the Adriatic sub-plate. The general direction of this convergence would appear to have been NW–SE, as during the later Insubric phase; but because motion was roughly radial, centrifugal motion of the hangingwall would have led to arc-parallel extension. Radial motions of this kind have been described in experiments scaled for gravity (Davy & Cobbold in press). In such experiments, the hangingwall deforms more readily than the footwall. If we apply these ideas to the Pennine zone, we infer arc-parallel extension at upper levels. Oblate mineral aggregates of amphibolite grade, frequently observed in flat-lying areas of the Ticino subdome, suggest that arc-parallel extension may indeed have operated at this stage, preventing strong crustal thickening and perhaps facilitating initial uplift of the Lepontine dome. The Insubric and Engadine Line may also have been active at this early stage, their releasing intersection facilitating emplacement of the Bergell intrusion some 30 Ma ago (see Trümpy 1980, p. 77).

To conclude, all three stages in the Tertiary kinematic history inferred for the Lepontine dome can be attributed to dextral transpression across the Insubric and Tonale Lines. Uplift of the dome was mainly due to backthrusting along the southern steep belt, but was aided by lateral escape of rocks and an associated component of horizontal regional extension within the Lepontine area.

Concluding remarks

From our structural study of the Lepontine area, we conclude that the central Alps during Tertiary times underwent a tectonic history rather different from those usually described for full frontal collision. Whereas the western Alps appear to have undergone frontal collision with the Adriatic subplate (Menard & Thouvenot 1984), the central Alps were subjected to dextral transpression, with relatively little crustal thickening. This study also suggests that post-collisional late Oligocene and early Miocene deformations have so far been underestimated in the Lepontine area.

A reconstruction of the main tectonic features in N–S section cannot show the effects of strike-slip motion and lateral escape, but it does indicate that backthrusting of the Pennine zone over the southern Alps is one of the main features of central Alpine history during the

Tertiary (see, e.g. Fig. 8 in Schmid *et al.* this volume, and fig. 3 in Milnes 1978). Taking into account the northwestward overthrusting in the external domain, the history of the Pennine zone at this time is that of an erosional pop-up, which is probably the best model for explaining the rapid cooling recorded at around 25–20 Ma and the emplacement of amphibolites into shallow levels of the Earth's crust.

During the closing stages of deformation, a similar but smaller pop-up uplifted the external massifs and formed the northern steep belt, with associated minor backthrusting in the northern Pennine zone.

ACKNOWLEDGEMENTS: This work was started by the Rennes group in 1979. At the time, Martin Huber and John Ramsay very kindly provided copies of maps, theses and other publications difficult to obtain outside Switzerland. The authors have benefited from useful discussions with colleagues at Zürich (especially Geoff Milnes and Neil Mancktelow), Basel (André Zingg), Bern (Johannes Hunziker) and Rennes (Pierre Choukroune, Michel Ballevre and Philippe Davy). Geoff Milnes and John Platt provided careful and constructive reviews.

References

AMSTUTZ, 1971. Carte tectonique de l'Ossola aux Grisons. *Archives des Sciences de Genève*, **24**, 5–9.

ARGAND, E. 1911. Les nappes de recouvrement des Alpes Pennines et leurs prolongements structuraux. *Beiträge zur Geologischen Karte der Schweiz*, **31**, 1–26.

—— 1916. Sur l'arc des Alpes Occidentales. *Eclogae Geologicae Helvetiae*, **14**, 145–204.

AYRTON, S. N. & RAMSAY, J. G. 1974. Tectonic and metamorphic events in the Alps. *Schweizerische Mineralogische und Petrographische Mitteilungen*, **54**, 609–39.

BERTHE, D., CHOUKROUNE, P. & JEGOUZO, P. 1979. Orthogneiss, mylonite and non coaxial deformation of granites: the example of the South Armorican Shear Zone. *Journal of Structural Geology*, **1**, 31–42.

BOCQUET, J., DELALAYE, M., HUNZIKER, J. C. & KRUMMENACHER, D. 1974. K–Ar and Rb–Sr dating of blue amphibole, micas and associated minerals from the western Alps. *Contributions to Mineralogy and Petrology*, **47**, 7–26.

BRADBURY, H. J. & NOLEN-HOEKSEMA, M. C. 1985. The Lepontine Alps as an evolving metamorphic core complex during A-Type subduction: evidence from heat flow, mineral cooling ages and tectonic model. *Tectonics*, **4**, 187–211.

CHADWICK, B. 1968. Deformation and metamorphism in the Lukmanier region, central Switzerland. *Geological Society of American Bulletin*, **79**, 1123–50.

CHOPIN, C. & MALUSKI, H. 1980. $^{39}Ar-^{40}Ar$ dating of high pressure metamorphic micas from the Gran Paradiso area (western Alps): evidence against the blocking temperature concept. *Contributions to Mineralogy and Petrology*, **74**, 109–22.

—— & MONIE, P. 1984. A unique magnesiochloritoid-bearing high-pressure assemblage from the Monte-Rosa, western Alps: petrologic and $^{40}Ar-^{39}Ar$ radiometric study. *Contributions to Mineralogy and Petrology*, **87**, 388–98.

CHOUKROUNE, P. & GAPAIS, D. 1983. Strain pattern in the Aar granite (central Alps): orthogneiss developed by bulk inhomogeneous flattening. *Journal of Structural Geology*, **5**, 1–10.

COBBOLD, P. R. 1969. *Geology of the Val Camadra – Greina area, Ticino, Switzerland.* Unpublished BSc Thesis, Imperial College, London.

—— 1980. Sheath folds and large strains in rocks. *In*: Large strain deformation: report on a tectonic studies group Meeting held at Imperial College on 14 November 1979. *Journal of Structural Geology*, **1**, 338–9.

DANIEL, T. 1986. *Analyse de la déformation des granites: utilisation du traitement de l'image.* Unpublished diploma thesis, Rennes University.

DAVY, P. & COBBOLD, P. R. (in press). Indentation tectonics in nature and experiment. 1. Experiments scaled for gravity. *Bulletin of the Geological Institutions of Uppsala*, N. S., **14**.

DEUTSCH, A. & STEIGER, R. H. 1985. Hornblende K–Ar ages and the climax of Tertiary metamorphism in the Lepontine Alps (south-central Switzerland): an old problem reassessed. *Earth and Planetary Science Letters*, **72**, 175–89.

ERNST, G. 1973. Interpretative synthesis of metamorphism in the Alps. *Geological Society of America Bulletin*, **84**, 2053–78.

ERNST, W. G. 1976. Mineralogic study of eclogitic rocks from Alpe Arami, Lepontine Alps, southern Switzerland. *Journal of Petrology*, **18**, 371–98.

EVANS, B. W. & TROMMSDORFF, V. 1978. Petrogenesis of garnet lherzolite, Lima di Gagnone, Lepontine Alps. *Earth and Planetary Science Letters*, **40**, 333–48.

FOX, J. S. 1975. Three dimensional isograds from the Lukmanier pass, Switzerland, and their tectonic significance. *Geological Magazine*, **112**, 547–64.

FRANK, E. 1983. Alpine metamorphism of calcareous rocks along a cross-section in the central Alps: occurrence and breakdown of muscovite, margarite and paragonite. *Schweizerische Mineralogische und Petrographische Mitteilungen*, **63**, 37–93.

FRANK, M. & STETTLER, A. 1979. K–Ar and $^{39}Ar/^{40}Ar$ systematics of white K-mica from an alpine metamorphic profile in the Swiss Alps. *Schweizerische Mineralogische und Petrographische Mitteilungen*, **59**, 375–94.

FREY, M., HUNZIKER, J. C., FRANK, W., BOCQUET, J., DAL PIAZ, G. V., JÄGER, E. and NIGGLI, E. 1974. Alpine metamorphism of the Alps. *Schweizerische Mineralogische und Petrographische Mitteilungen*, **54**, 247–90.

——, BUCHER, K., FRANK, E. & MULLIS, J. 1980. Alpine metamorphism along the geotraverse Basel–Chiasso: a review. *Eclogae Geologicae Helvetiae*, **73**, 527–47.

FUMASOLI, M. V. 1974. Geologie des Gebietes nördlich und südlich der Torio–Tonale–Linie im Westen von Gravedona. *Mitteilungen des geologischen Instituts der Eidgenössischen Technischen Hochschule und der Universität Zürich*, neue Folge **194**.

GILBERT, E. and MERLE, O. 1987. Extrusion and radial spreading beyond a closing channel. *Journal of Structural Geology*, **9**, 481–90.

GRECO, A. 1985. Analisi strutturale della parte frontale del ricoprimento pennidico dell' Antigorio in Val Formazza (Novara, Italia). *Schweizerische Mineralogische und Petrographische Mitteilungen*, **65**, 299–23.

HALL, W. D. H. 1972. *The structural geology and metamorphic history of the Lower Pennine nappes, Valle di Bosco, Ticino, Switzerland*. Unpublished PhD thesis, University of London.

HÄNNY, R., GRAUERT, B. & SOPTRATANOVA, G. 1975. Paleozoic migmatites affected by high-grade tertiary metamorphism in the central Alps (Valle Bodengo, Italy). *Contributions to Mineralogy and Petrology*, **51**, 173–96.

HEINRICH, C. A. 1982. Kyanite–eclogite to amphibolite facies evolution of hydrous mafic and pelitic-rocks, Adula nappe, central Alps. *Contributions to Mineralogy and Petrology*, **81**, 30–38.

—— 1986. Eclogite facies regional metamorphism of hydrous mafic rocks in the central Alpine Adula nappe. *Journal of Petrology*, **27**, 123–54.

HEITZMANN, P. 1987a. Calcite mylonites in the central Alpine 'root zone'. *Tectonophysics*, **135**, 207–15.

—— 1987b. Evidence of late Oligocene/early Miocene backthrusting in the central Alps. *Geodynamica Acta*, **1**, 183–92.

HIGGINS, A. K. 1964. *The structure and metamorphic geology of the area between Nufenenpass and Basodino, Ticino*. Unpublished PhD thesis, Imperial College, London.

HSÜ, K. J. 1979. Thin skinned plate tectonics during neo-alpine orogenesis. *American Journal of Sciences*, **279**, 353–66.

HUBER, M., RAMSAY, J. & SIMPSON, C. 1980. Deformation in the Maggia and Antigorio nappes, Lepontine Alps. *Eclogae Geologicae Helvetiae*, **73**, 2, 593–606.

HUDLESTON, P. J. 1983. Strain patterns in an Ice cap and implications for strain variations in shear zones. *Journal of Structural Geology*, **5**, 455–63.

HUNZIKER, J. C. 1966. Zur Geologie und Geochemie des Gebietes zwischen Valle Antigorio und Valle di Campo. *Schweizerische Mineralogische und Petrographische Mitteilungen*, **46**, 473–552.

—— 1970. Polymetamorphism in the Monte Rosa, western Alps. *Eclogae Geologicae Helvetiae*, **63**, 151–61.

HURFORD, A. & HUNZIKER, J. C. 1985. Alpine cooling history of the Monte Mucrone eclogites (Sesia–Lanzo zone): fission track evidence. *Schweizerische Mineralogische und Petrographische Mitteilungen*, **65**, 325–34.

HURFORD, J. 1986. Cooling and uplift patterns in the Lepontine Alps, south-central Switzerland, and an age of vertical movement on the Insubric fault line. *Contributions to Mineralogy and Petrology*, **92**, 413–27.

JÄGER, E. 1973. Die Alpine Orogenese im Lichte der Radiometrischen Alterbestimmung. *Eclogae Geologicae Helvetiae*, **66**, 11–21.

——, NIGGLI, E. & WENK, E. 1967. Rb/Sr Alterbestimmungen an Glimmern der Zentralpen. *Beiträge zur Geologischen Karte der Schweiz*, **134**.

KLAPER, E. M. 1982. Deformation und Metamorphose in der nördlichen Maggia-zone. *Schweizerische Mineralogische und Petrographische Mitteilungen*, **62**, 47–76.

—— & BUCHER-NURMINEN, K. 1987. Alpine metamorphism of pelitic schists in the Nufenen pass area, Lepontine Alps. *Journal of Metamorphic Geology*, **5**, 175–94.

KLEIN, J. A. 1978. Post-nappe folding southeast of the Mischabelrückfalte (Pennine Alps) and some aspects of the associated metamorphism. *Leidse Geologische Mededelingen*, **51**, 233–312.

KNUP, P. 1958. Geologie und Petrographie des Gebietes zwischen Centovalli, Valle Vigezzo und Onsernone. *Schweizerische Mineralogische und Petrographische Mitteilungen*, **38**, 217–232.

KÖPPEL, V. & GRÜNENFELDER, M. 1975. Concordant U–Pb ages of monazite and xenotime from the central Alps and the timing of high temperature Alpine metamorphism, a preliminary report. *Schweizerische Mineralogische und Petrographische Mitteilungen*, **55**, 129–32.

—— & —— 1978. The significance of monazite U–Pb ages: examples from the Lepontine area of the Swiss Alps. *In*: ZARTMAN, R. E. (ed). *Fourth International Conference, Geochronology, Cosmochronology, Isotope Geology*, 226–7.

——, GÜNTHERT, A. & GRÜNENFELDER, M. 1980. Patterns of U–Pb zircon and monazite ages in polymetamorphic units of the Swiss central Alps. *Schweizerische Mineralogische und Petrographische Mitteilungen*, **61**, 97–119.

LAUBSCHER, H. P. 1971. The large scale kinematics of the western Alps and the northern Apennines and its palinspastic implications. *American Journal of Science*, **271**, 193–226.

—— 1983. The late Alpine (periadriatic) intrusions and the Insubric Line. *Memorie della Società Geologica Italiana*, **26**, 21–30.

MANCEL, P. & MERLE, O. 1987. Kinematics of the northern part of the Simplon Line. *Tectonophysics*, **135**, 265–75.

MANCKTELOW, N. 1985. The Simplon Line: a major

displacement zone in the western Lepontine Alps. *Eclogae Geologicae Helvetiae*, **78**, 73–96. 73–96.

—— 1987. Quartz textures from the Simplon fault zone, southwest Switzerland and north Italy. *Tectonophysics*, **135**, 133–53.

MARQUER, D. & GAPAIS, D. 1985. Les massifs cristallins externes sur une transversale Guttanen–Val Bedretto (Alpes centrales): structure et histoire cinématique. *Comptes-Rendus de l'Académie des Sciences de Paris*, **301**, II, 543–6.

MAWER, C. K. 1987. Shear criteria in the Grenville Province, Ontario, Canada. *Journal of Structural Geology*, **9**, 531–9.

MENARD, G. & THOUVENOT, F. 1984. Ecaillage de la lithosphère européenne sous les Alpes occidentales: arguments gravimétriques et sismiques liés à l'anomalie d'Ivréa. *Bulletin de la Société Géologique de France*, **XXVI**, 875–84.

MERLE, O. 1987. Histoire de la déformation dans les Alpes lepontines occidentales (Alpes centrales). *Bulletin de la Société Geologique de France*, **III**, 183–90.

——, LE GAL, PH. & MANCEL, P. 1986. Déformation et métamorphisme dans la région du Simplon (Alpes centrales). *Eclogae Geologicae Helvetiae*, **79**, 705–18.

—— & —— 1988. Post-amphibolitic westward thrusting and fold vergence in the Ticino domain. *Eclogae Geologicae Helvetiae*, **81**, 215–36.

MILNES, A. G. 1974a. Structure of the Pennine zone (central Alps): a new working hypothesis. *Geological Society of America Bulletin*, **85**, 1727–32.

—— 1974b. Post-nappe folding in the western Alps. *Eclogae Geologicae Helvetiae*, **67**, 2, 333–48.

—— 1975. On steep isogradic surfaces in the Simplon area: a discussion. *Contributions to Mineralogy & Petrology*, **53**, 65–8.

—— 1978. Structural zones and continental collision, central Alps. *Tectonophysics*, **47**, 364–92.

——, GRELLER, M. & MÜLLER, R. 1981. Sequence and style of major post-nappe structures, Simplon-Pennine Alps. *Journal of Structural Geology*, **3**, 411–20.

MONIE, P. 1985. La méthode $^{39}Ar-^{40}Ar$ appliquée au métamorphisme alpin dans le massif du Mont Rose (Alpes occidentales). Chronologie détaillée depuis 110 Ma. *Eclogae Geologicae Helvetiae*, **78**, 487–516.

NIGGLI, E. 1970. Alpine Metamorphose und Alpine Gebirgsbildung. *Fortschritte der Mineralogie*, **47**, 16–26.

PASSCHIER, C. W. & SIMPSON, C. 1986. Porphyroclast systems as kinematic indicators. *Journal of Structural Geology*, **8**, 831–44.

PLATT, J. 1986. Dynamics of orogenic wedges and the uplift of high-pressure metamorphic rocks. *Geological Society of America Bulletin*, **97**, 1027–53.

PREISWERK. H. 1918. Geologische Beschreibung der Lepontinischen Alpen. *Beiträge zur geologischen Karte der Schweiz*, **26**.

PURDY, J. N. and JÄGER, E. 1976. K–Ar ages on rock forming minerals from the central Alps. *Memorie Istituto di Geologia e Mineralogia dell' Università di Padova*, **30**, 75–7.

SCHMID, S. M., ZINGG, A. & HANDY, M. 1987. The kinematics of movements along the Insubric Line and the emplacement of the Ivrea zone. *Tectonophysics*, **135**, 47–66.

SCHMIDT, C. & PREISWERK, H. 1908. Geologische Karte der Simplongruppe. *Spezialkarte n°48 mit Erläuterungen*.

SIMPSON, C. 1982. The structure of the northern lobe of the Maggia nappe, Ticino, Switzerland. *Eclogae Geologicae Helvetiae*, **75**, 495–516.

—— & SCHMID, S. M. 1983. An evaluation of criteria to deduce the sense of movement in sheared rocks. *Geological Society of America Bulletin*, **94**, 1281–8.

STAUB, R. 1924. Der Bau der Alpen: Versuch einer Synthese. *Beiträge zur geologischen Karte der Schweiz*, **NF 52**.

STECK, A. 1984. Structures et déformations tertiaires dans les Alpes centrales. *Eclogae Geologicae Helvetiae*, **77**, 55–100.

—— 1987. Le Massif du Simplon–réflexions sur la cinématique des nappes de gneiss. *Schweizerische Mineralogische und Petrographische Mitteilungen*, **67**, 27–45.

——, RAMSAY, J. G., MILNES, A. G. & BURRI, M. 1979. Compte-rendu de l'excursion de la Sociéte Géologique Suisse et la Société Suisse de Minéralogie et Pétrographie en Valais et en Italie Nord du 2 au 5 Octobre 1978. *Eclogae Geologicae Helvetiae*, **72**, 287–311.

STEIGER, R. H. 1983. Isotopic dating of metamorphic events in the central Alps. *Terra Cognita*, **3**, p. 140.

—— & BUCHER, I. 1978. Are Rb–Sr biotite ages in the central Alps necessarily cooling ages? *Fourth International Conference, Geochronology, Cosmochronology, Isotope Geology, U.S. Geological Survey, Open file report*, 414–15.

STEINER, H. 1984. Radiometrische Alterbestimmungen an Gesteinen der Maggia Decke (Penninikum der Zentralalpen). *Schweizerische Mineralogische und Petrographische Mitteilungen*, **64**, 227–59.

STEINITZ, G. & JÄGER, E. 1981. Rb–Sr and K–Ar studies on rocks from the Suretta nappe, eastern Switzerland. *Schweizerische Mineralogische und Petrographische Mitteilungen*, **61**, 121–31.

STRECKEISEN, A. & WENK, E. 1974. On steep isogradic surfaces in the Simplon area. *Contributions to Mineralogy and Petrology*, **47**, 81–95.

THAKUR, J. C. 1973. Events in Alpine deformation and metamorphism. *Geologische Rundschau*, **62**, 549–63.

THOMPSON, P. H. 1976. Isograd patterns and pressure-temperature distributions during regional metamorphism. *Contributions to Mineralogy and Petrology*, **57**, 277–95.

TRÜMPY, R. 1980. *An Outline of the Geology of Switzerland*. Wepf & Co, Basel.

TURNER F. J. 1968. *Metamorphic Petrology*. McGraw-Hill, New York.

VAN DEN DRIESSCHE, J. & BRUN, J. P. 1987. Rolling structures at large shear strain. *Journal of Struc-*

tural Geology, **9**, 691–704.

WAGNER, G. A., REIMER, G. M. & JÄGER, E. 1977. Cooling ages derived by apatite fission track, mica Rb–Sr and K–Ar dating: the uplift and cooling history of the central Alps. *Memorie Istituto di Geologia e Mineralogia dell' Università di Padova*, **30**.

WENK, E. 1955. Eine Strukturkarte der Tessinenalpen. *Schweizerische Mineralogische und Petrographische Mitteilungen*, **35**, 311–19.

WERNER, D., KÖPPEL, V., HÄNNY, R. & RYBACH, L. 1976. Cooling models for the Lepontine area (central Swiss Alps). *Schweizerische Mineralogische und Petrographische Mitteilungen,* **56**, 661–7.

ZINGG, A. & SCHMID, S. 1983. Mylonite zone along the Insubric Line west of the Lago Maggiore. *Terra Cognita*, **3**, NB 24.

O. MERLE & P. R. COBBOLD, Centre Armoricain d'Etude Structurale des Socles, CNRS, Univ. Rennes 1, 35042 Rennes Cedex, France.

S. SCHMID, Geologisches Institut, ETH-Zentrum, 8092 Zurich, Switzerland.

Thrusting and backthrusting in the Briançonnais domain of the western Alps

J. P. Platt, G. S. Lister, P. Cunningham, P. Weston, F. Peel, T. Baudin & H. Dondey

SUMMARY: The Briançonnais domain comprises the deformed remnants of a semi-independent microcontinental strip that lay close to the European margin of the Mesozoic Tethys. A Triassic carbonate platform of south-Alpine type was broken up by extension in Early to Middle Jurassic time, producing a syn-rift sequence characterized by small basins with marginal breccia wedges, internal unconformities, and local terrestrial sediments. The pre-Triassic basement was probably reduced to about half its original thickness during this extensional phase. Late Jurassic and Cretaceous subsidence was accompanied by pelagic sedimentation. Flysch sedimentation began shortly before the onset of compressional tectonics in late Palaeocene or Eocene time.

Underthrusting and accretion of the Briançonnais domain to the developing Alpine thrust wedge caused an early phase of thrusting, producing antiformal stacks of kilometre-scale thrust sheets in parts of the external Briançonnais. The internal margin of the domain was underplated at depths of up to 40 km, and was affected by high-P low-T metamorphism and regional ductile deformation. Stretching lineations and slip-vectors related to this stage trend NW in the northern French Alps, but swing to WSW in the S.

Backthrusting in Oligocene time was probably associated with the start of collision with the European margin proper. Families of back thrusts and associated recumbent folds cut, deform or reactivate older structures. The members of each family appear to have been activated in sequence towards the internal margin of the Alps, forming large-scale duplexes, the floor and roof thrusts of which are in part reactivated early thrusts.

The trend of stretching lineations and slip-vectors in the Briançonnais domain vary in both space (NW in the N, SW in the S), and time (later trends lie consistently anticlockwise of early ones). These trends and their changes are incompatible with calculated Africa/Europe plate motions for the Tertiary. Independent NW to WNW motion of Adria relative to Europe is probably required. Anticlockwise rotation and bending of the internal Alps during collision could explain the radial pattern of the early vectors. The radial pattern of the later backthrusting matches the radial pattern of coeval forethrusting in the external zones, however; this is therefore unlikely to be a result of later rotations, and may reflect the dominance of gravitational body forces during the later stages of mountain-building.

Introduction

The Briançonnais domain forms part of the Alpine Pennine zone, which appears to comprise the deformed and metamorphosed contents of a branch of the Mesozoic Tethys that separated Europe from Adria during Mesozoic time (Debelmas & Lemoine 1970, Lemoine *et al.* 1986). Fragments of both oceanic and continental basement and their sedimentary cover were assembled by subduction and accretion to the SE (Adrian) margin of the ocean during late Cretaceous to Eocene time, and were then emplaced as a major allochthon on to the imbricated European margin during the collisional phase in Oligocene–Pliocene time (Milnes 1978, Platt 1986). The Briançonnais domain itself is defined by its distinctive Mesozoic stratigraphy, and represents a palaeogeographic realm within the Pennine zone. Mesozoic sequences from this domain are now found in the pre-Alps, and together with several pre-Mesozoic basement massifs, form a continuous belt extending from the Simplon fault in the central Alps to Liguria (Fig. 1).

The Briançonnais domain probably formed a relatively high-standing system of horst blocks within the Mesozoic Tethys, close to and possibly attached to the European margin (Debelmas and Lemoine 1970, Trümpy 1980, Tricart 1984, Lemoine 1986), but separated from the latter by the Sub-Briançonnais basin in the western Alps and by the Valais trough in the central Alps. These basins were probably underlain by thin continental crust, and possibly in part by oceanic crust; and it is noteworthy that the floor of the Sub-Briançonnais basin is nowhere exposed. For this and other reasons

From: Coward, M. P., Dietrich, D. & Park, R. G. (eds), 1989, *Alpine Tectonics*, Geological Society Special Publication No. 45, pp. 135–152.

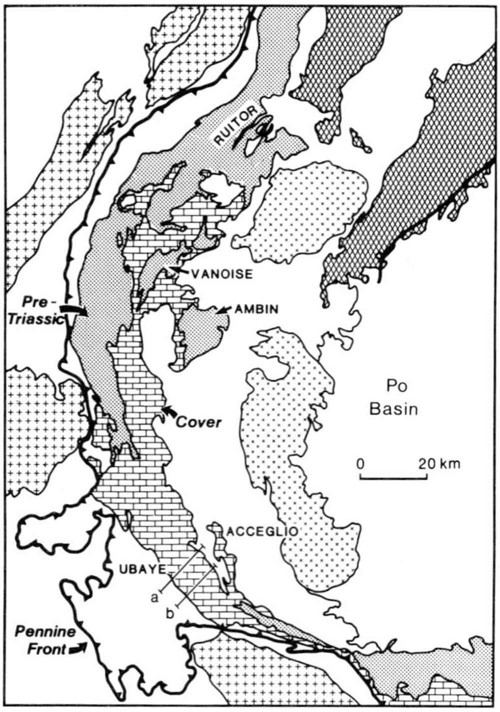

FIG. 1. Geological sketch of the western Alps to show location of the Briançonnais domain. Stipple, crystalline basement and pre-Triassic sedimentary rocks; brick pattern, Triassic–Eocene cover. Also ornamented are the External Crystalline Massifs (+ pattern), internal massifs (x pattern), and Austro–Alpine and southern Alpine domains (cross-hatched). a,b: section lines in Fig. 7.

discussed later, the Briançonnais domain is best regarded as having been a microcontinental strip within the Mesozoic Tethys, independent of either continental margin (Platt 1986).

The distinctive Mesozoic stratigraphy, and the abundance of easily recognizable lithostratigraphic markers, greatly facilitates detailed analysis of the complex structural history; and the continuity of the domain allows comparison of the history and kinematics of deformation around the Alpine arc.

Geological outline

Pre-Mesozoic basement

In the central Alps the Briançonnais domain consists largely of a slab up to 10 km thick of pre-Mesozoic rocks, comprising the Grand St Bernard nappe of classical Alpine geology. The frontal part of this body is composed of Carboniferous clastic sediments (the zone Houillère), which continues far down into the western Alps. The rest of the Grand St Bernard nappe, and the Ruitor, Vanoise, and Ambin massifs further south, consist largely of metasediments and gneisses affected by Hercynian and earlier metamorphic events (Desmons 1977). Thick sequences of coarse clastic sediments and volcanics of probable Permian age occur in the Vanoise and in Liguria.

The crystalline rocks vary widely in their grade of pre-Alpine metamorphism: lower greenschist facies in the northern Vanoise (Guillot 1987), upper greenschist facies in the southern Vanoise and Ambin Massifs (Desmons 1977), lower amphibolite facies in the Ruitor (Baudin 1987), and locally up to sillimanite grade in parts of the Grand St Bernard (Burri 1983). These different bodies are therefore presumably derived from different levels within the pre-Alpine continental crust. Their present disposition is partly a result of Permian tectonism – rifting or possibly late Hercynian thrusting (Peel 1985) – as Permian sediments rest unconformably on both Carboniferous sediments and older crystalline rocks. A more important control on basement geometry was Mesozoic extension, however, and Mesozoic sediments rest stratigraphically on all basement rock-types including those of the highest grade.

The overall assemblage of basement rock-types is similar to that of the Dora Maira Massif, and to parts of the southern Alpine and Austro–Alpine domains (Desmons 1986), and lacks obvious affinities to the External Crystalline Massifs. This may have some bearing on the provenance of the domain (see below).

Cover

The Mesozoic stratigraphy of the Briançonnais domain commences with a distinctive Lower Triassic shallow-marine orthoquartzite that usually lies in apparent stratigraphic conformity on Permian or Permo–Triassic clastics. It is followed by up to 800 m of Triassic platform carbonates with intercalations of evaporites (Megard-Galli & Baud 1977). There is evidence for intra-Triassic rifting: slumps, breccias, and internal unconformities are common (Baud & Mégard-Galli 1977, Jaillard *et al.* 1986) and Triassic carbonates locally rest unconformably on Permian, Carboniferous, or pre-Alpine crystalline basement.

Up to this point the sequence resembles those of the southern Alpine and Austro–Alpine domains, and is quite distinct from the Helvetic or Dauphinois Triassic. The Lower and Middle Jurassic are thin or absent over much of the Briançonnais, and parts of the area were emergent at times, as indicated for example by mid-Jurassic bauxites in the Vanoise (Goffé 1977). Emergence was associated with, and possibly caused by, substantial extensional tectonism: localised basins accumulated thick sequences of Liassic shaly and siliceous limestones (Jaillard *et al.* 1986, Lemoine *et al.* 1986), breccia lenses are common, and in many places the Upper Jurassic rests unconformably on older units right down to the basement (Fig. 2). The Upper Jurassic consists of a uniform clean light-grey marble (Malm), probably of pelagic origin, that locally passes up into mottled red and green nodular limestone that may reflect very low rates of deposition. This deposit almost certainly reflects both subsidence and isolation of the Briançonnais domain from sources of clastic detritus, as a result of the final break-up of the early Mesozoic carbonate platforms in the Tethys and the initiation of sea-floor spreading (Bernoulli & Jenkyns 1974, de Graciansky *et al.* 1979, Lemoine *et al.* 1986). Water depths cannot be determined, but the presence of chert-rich horizons led Bourbon (1980) to suggest that deposition was locally below the CCD.

The Lower Cretaceous is absent; and the Malm is followed by a thin sequence of Upper Cretaceous and Palaeocene marly limestones that include hard-ground horizons (Ellenberger & Raoult 1979), probably deposited in deep water. The youngest unit is a black shale with minor detrital intercalations (referred to as flysch), which is usually attributed to the Eocene.

Early Mesozoic extension, and the associated periods of emersion and erosion, caused significant variations in stratigraphic sequence over short distances (Fig. 2). These variations, which can be seen within individual thrust sheets, illustrate the danger inherent in the tradition of identifying and correlating Alpine nappes on the basis of their stratigraphy.

Pre-convergent tectonic history

The similarity of both the pre-Triassic basement and the Triassic cover to that of the Adrian margin suggests that the Briançonnais domain may have the same provenance as Adria, and that in pre-Jurassic time it lay S and W of its present position, close to Iberia. Rifting began in the Triassic, extension became substantial in the lower to Middle Jurassic time, and the domain was effectively separated from both continental margins as the ocean basin developed. The crust, initially of normal continental thickness, may have been reduced to 10–20 km by the end of the Middle Jurassic. During Cretaceous time, the domain formed a deep submarine swell within the Tethys, scoured by deep marine currents and receiving only pelagic sediment.

Convergent tectonic history

The precise time at which convergence started in the Briançonnais is difficult to establish. The youngest formation involved in the deformation is the flysch, which is commonly assigned to the middle Eocene (Lutetian). This is based on the occurrence of *Nummulites* in the Ubaye area (Gidon 1962), and by comparison with dated sequences in the Briançonnais nappes of the pre-Alps (e.g. Ellenberger 1958). The pre-Alpine nappes had a very different tectonic history, however: they were accreted at the front of the pre-collisional accretionary wedge, rather than being subducted and underplated, as was the rest of the Briançonnais domain. Sedimentation could have continued for some time in the pre-Alps in a setting analogous

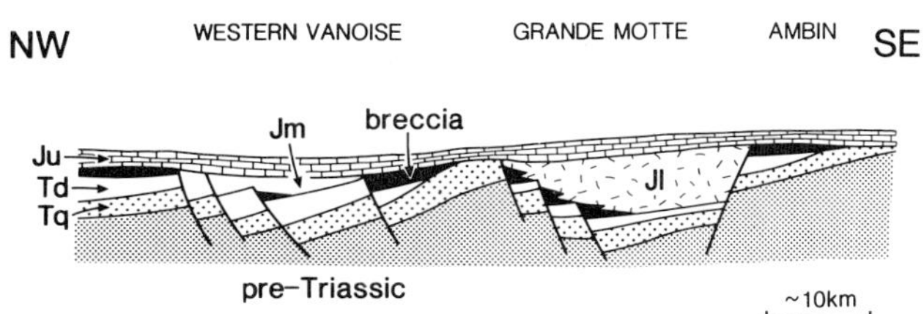

FIG. 2. Schematic section to illustrate primary stratigraphic relationships, Vanoise–Ambin Massifs (modified after Jaillard *et al.* 1986). Tq, Lower Triassic quartzite; Td, Middle to Upper Triassic carbonates; Jl, Lias; black, Lower to Middle Jurassic carbonate breccia; Jm, Middle Jurassic; Ju, Upper Jurassic.

to that of a present trench-slope basin (Karig 1974). With the exception of the Ubaye region, the youngest fossils in the main Briançonnais domain occur in the pelagic marls beneath the flysch, and constrain their age only to the period Palaeocene to mid-Eocene (Ellenberger & Raoult 1979). Radiometric dates (see below) suggest that parts of the domain had already experienced a prolonged metamorphic history by 40 Ma (late Eocene), so convergence is likely to have started some time before this, possibly as early as the Palaeocene, and certainly not later than the mid-Eocene.

Convergence started with the underthrusting of the domain beneath the Alpine accretionary wedge: the more internal parts were tectonically buried to depths of as much as 40 km. Some of the sedimentary cover was detached during this process and accreted at the wedge front (pre-Alps), and some of the basement may have been carried to greater depths and is no longer visible. Much of the thrusting seen within the Briançonnais is likely to have been caused by this process of subduction and accretion. High P/T-ratio metamorphism was caused by deep burial; further deformation accompanied uplift, and the metamorphic conditions evolved into the greenschist facies. Early in Oligocene time the domain was strongly affected by thrusting directed towards the interior of the arc (backthrusting), and this appears to be related to the onset of continental collision.

The main tectonic problems addressed by this paper are (1) the geometry and kinematics of the thrusting and backthrusting, and their relationship to metamorphic and tectonic events, (2) structural variations around the Alpine arc, and (3) the role of strike-slip or transpressive tectonism during convergence.

Metamorphism and dating

Much of the internal Briançonnais has been affected by blueschist-facies metamorphism; but a number of difficulties have arisen in interpreting this event. Critical mineral assemblages are largely confined to the crystalline basement massifs, where they partly overprint pre-Alpine parageneses. Glaucophane and ferroglaucophane are common in the Grand Saint Bernard, Ruitor, southern Vanoise and Ambin Massifs (Bocquet 1974), and the Arpont schist of the southern Vanoise also contains lawsonite pseudomorphs (Platt & Lister 1985a) and rare jadeitic pyroxene+quartz (Bocquet 1974, Saliot 1979). Desmons (1977) estimates that the high-pressure peak of metamorphism occurred at >10 kb, <350°C. By contrast, the cover sequences from the same regions do not show evidence for such high pressures of metamorphism. Crossite and lawsonite pseudomorphs occur in parts of the Vanoise (Ellenberger 1960), and in the western Vanoise Goffé (1977) has reported occurrences of magnesiocarpholite, for which he estimates metamorphic conditions of 7 kb, 360°C (Goffé 1982). Both pressure estimates are minima: whether there is really a difference in metamorphic pressure between basement and cover is unclear. What is clear, however, is that the immediately overlying Schistes Lustrés nappe in the Vanoise lacks any evidence for high-pressure metamorphism, so that there is an increase in the apparent pressure of metamorphism of several kilobars down through a structural pile only a few kilometres thick. This led Platt (1986) to suggest that some of the tectonic contacts in the area may be extensional structures formed during uplift.

Further S, in the Acceglio zone, Mesozoic rocks contain jadeite+quartz (Lefèvre 1974, Houfflain & Caby 1987), confirming that at least locally, Briançonnais cover rocks were carried to depths of 35–40 km. High pressure minerals have not been reported from elsewhere in the Briançonnais cover; and rare occurrences of lawsonite in the Zone Houillère (Bocquet 1974) and in Permian rocks in the Ubaye region (Cunningham 1987) provide the only evidence for high P/T ratio metamorphism in the external Briançonnais.

Excess argon in Briançonnais blue amphiboles has so far hindered radiometric determination of the age of the high-pressure metamorphism, but it cannot be younger than the Meso-alpine event at 35–40 Ma (Bocquet *et al.* 1974, Desmons 1977), and for stratigraphic reasons it is unlikely to be older than Palaeocene (but for an alternative opinion, see Desmons 1977).

The high-pressure event in the internal Briançonnais has been variably overprinted by greenschist-facies metamorphism, characterized in particular by largely static growth of albite porphyroblasts and green biotite (Platt & Lister 1985a). Rb/Sr and K/Ar dates on white micas suggest that this event occurred at 35–40 Ma (Bocquet *et al.* 1974), and correlates closely with the Meso-alpine event established in the central Alps (Frey *et al.* 1974). The external Briançonnais shows signs of widespread lower-greenschist-facies Alpine metamorphism, which has not been dated.

Structure

This section is based on detailed structural studies in the Ruitor, Vanoise, Ambin and Ubaye regions (Fig. 1). There are sufficient similarities in structural style and history from these areas to justify making some generalizations; there are also significant differences, which we will bring out in discussions of the individual areas.

Early thrusting

There is abundant evidence for an early phase of thrust tectonics throughout the Briançonnais domain. Cover sequences are duplicated across relatively discrete faults that are parallel to or at a low angle to the stratigraphy; basement and cover are commonly intercalated, and the original cover has been largely stripped off many of the basement massifs and replaced by cover derived from elsewhere. Significant stratigraphic differences from one thrust sheet to another suggest that many thrusts either reactivated or cut Mesozoic extensional faults. In the internal Briançonnais thrusting was associated with recumbent folding and substantial ductile deformation that produced a regional subhorizontal schistosity. Folds are commonly isoclinal and attenuated, and their axes are difficult to measure with any precision. Examples of such folds and thrusts are shown in Figs 3–5, which also illustrate a common problem with these structures: overprinting by later deformation has complicated the geometry and obscured detailed relationships.

The precise relationship between the early thrusts and folds is difficult to establish. Neither overprints the other; but there is no sign of a marked increase in ductile strain adjacent to thrusts or on the overturned limbs of folds, as is seen in structures such as the Morcles nappe where thrusting and recumbent folding are genetically linked (Ramsay 1981). It may be that the folds were initiated as relatively open structures during thrusting, and were subsequently deformed into their present state by an essentially later phase of ductile deformation.

The relationship between deformation and the metamorphism has a significant bearing on this question. It is striking that in the Ruitor, southern Vanoise, and Ambin Massifs the main phase of glaucophane growth was static, producing rosettes of prismatic crystals that helicitically enclose pre-Alpine fabrics. The glaucophane was subsequently mechanically rotated to define the first Alpine foliation and lineation, accompanied by dynamic recrystallization and further growth of blue amphibole (Platt & Lister 1985a). These relationships can perhaps be most easily explained in terms of the following history.

(1) Thrusting occurred under cold, relatively brittle, conditions during subduction and underplating.
(2) After underplating, glaucophane-schist facies metamorphism occurred under static conditions at high pressure, once the temperature had risen sufficiently (to about 300°C) to activate metamorphic reactions.
(3) Regional ductile deformation then commenced under metamorphic conditions, producing the regional schistosity and tightening pre-existing folds.

This suggested history implies that the thrusting (event 1) and the ductile deformation (event 3) were not genetically related. The ductile deformation, which postdates the high-pressure metamorphic peak, may have been related to the exhumation of the high-pressure rocks rather than to subduction (Platt 1986).

Thrusting in the more external parts of the Briançonnais domain is less obscured by the later deformation. In the western Vanoise, for example, a thrust with a fairly clearly defined footwall ramp is preserved in the overturned limb of a major backfold (Figs 5 and 6). In the Ubaye region, the cover is deformed into an antiformal stack of Dl thrust sheets, the highest one of which includes a well-developed hangingwall anticline (Fig. 7). The thrusts are associated with a slaty cleavage, for the most part parallel to bedding, and a WSW-trending stretching lineation. Small-scale imbrication within the thrust sheets is commonly associated with a crenulation cleavage.

Thrust surfaces are commonly decorated by layers of cargneule (locally up to 50 m thick), mylonite (rare), or fibrous calcite rock (up to 3 m thick) formed by repeated hydraulic fracturing during thrusting (Cunningham 1987). Resistance to motion on many thrusts must have been greatly reduced by the presence of evaporites (now represented by cargneule, Warrak 1974) or by high pore-fluid pressure.

Some relatively early subhorizontal tectonic contacts place younger rocks on older, and therefore appear to be extensional. Examples include the contact between inverted Liassic and basement in the southern Vanoise (Fig. 4c), the contact between Anisian and basement in the western Vanoise (Figs 5 and 6), and the contact between Schistes Lustrés (probably post-Triassic in age) and Palaeozoic basement in the Ruitor (Fig. 8). The first two examples

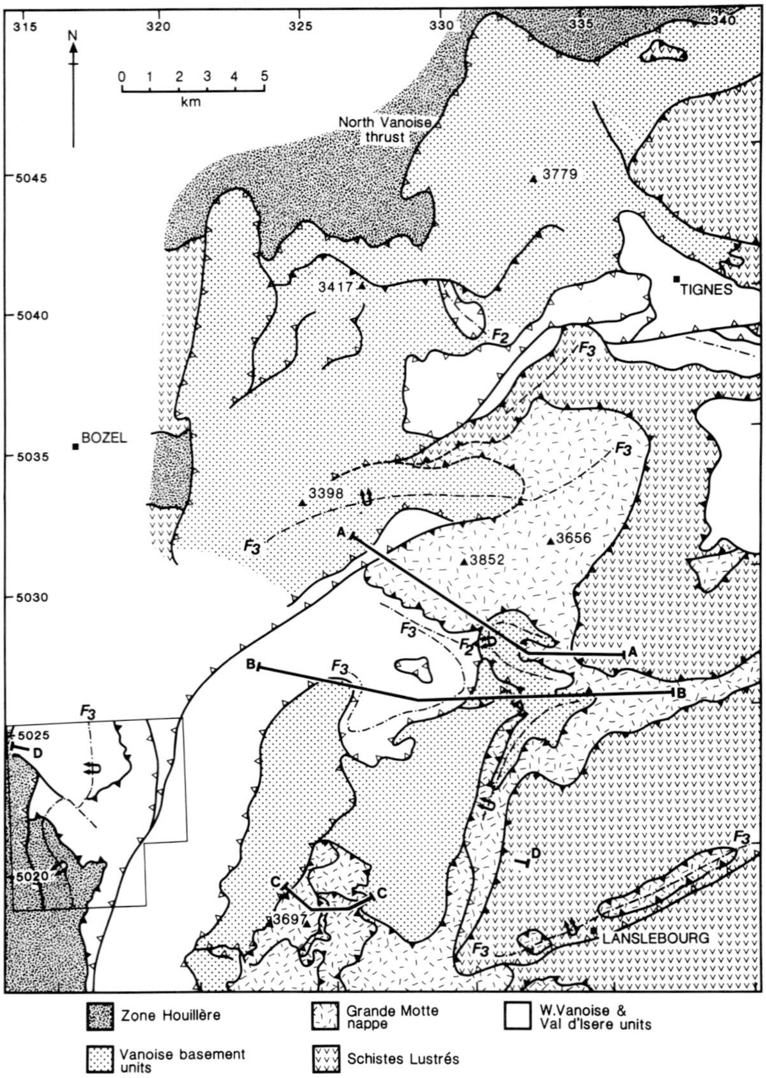

FIG. 3. Structural sketch map of the southern Vanoise Massif, showing the location of major early forethrusts (solid teeth), late backthrusts (open teeth), and axial traces of major folds. The north Vanoise thrust, which is a late forethrust, is also shown. All thrusts are shown with the teeth on the original upper plate, regardless of their present orientation. Zone Houillère and Vanoise basement units shown here include attached cover sequences, and the Schistes Lustrés includes Triassic evaporitic and carbonate units at its base.

are marked by thin zones of mylonite. In some of these cases stratigraphic section may have been removed by Mesozoic extension or erosion, and the contact subsequently reactivated as a thrust (Dondey 1986); others may represent local plucking or stripping of rocks from thrust footwalls (Platt & Leggett 1986); but some may be a result of extension during the post-subduction phase of uplift and exhumation (Platt 1986).

Kinematics of the early deformation

The earliest potential kinematic information from the internal Briançonnais is provided by lineated glaucophane. In the southern Vanoise, this trends consistently NW–SE (Fig. 9a) except where locally reoriented by later deformation. In the Ruitor it trends towards 295° (Fig. 9b), but has been everywhere affected by later strain, and in the Ambin Massif the lineation originally

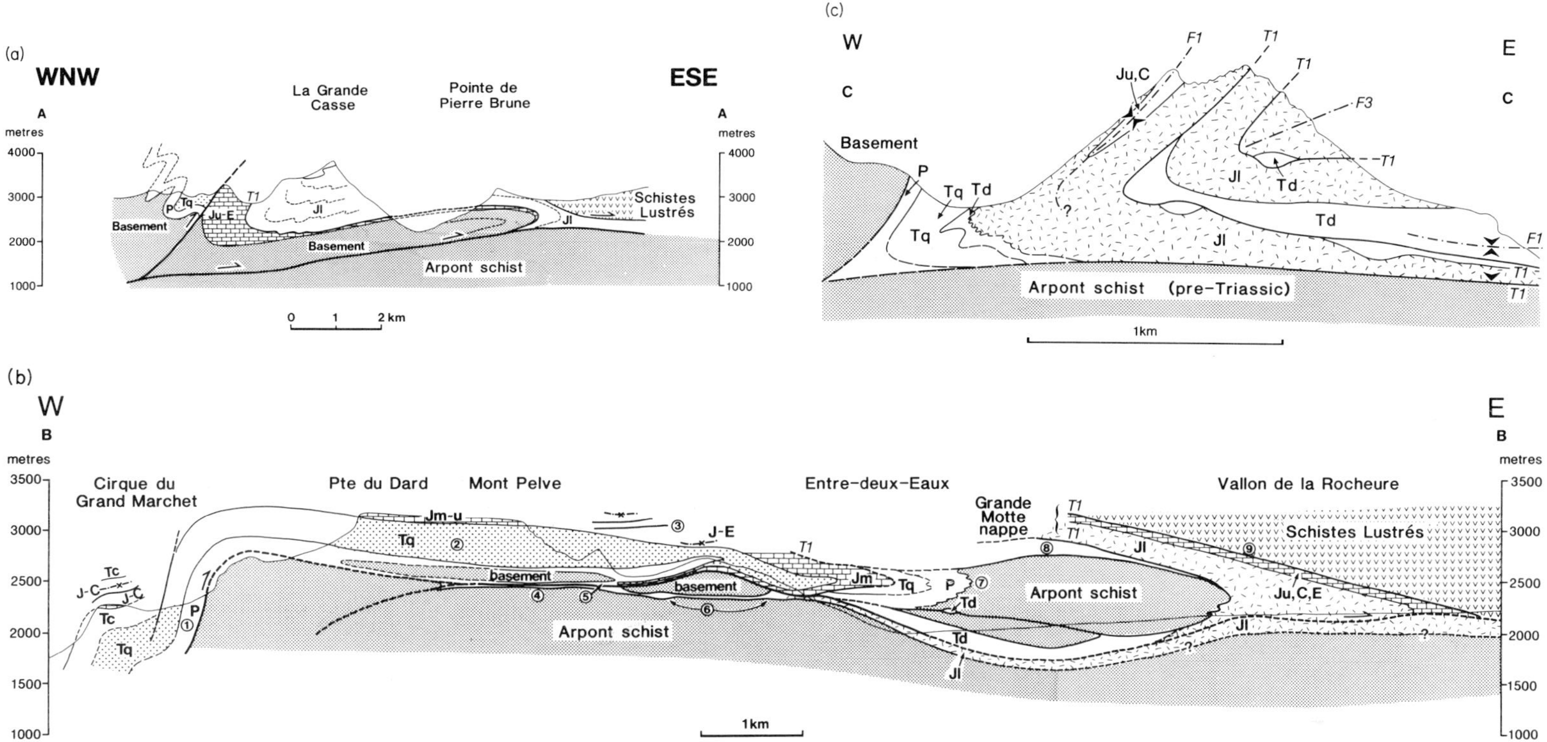

FIG. 4. Structural sections across the southern Vanoise Massif. For locations see Fig. 3. (a) Grande Casse (AA), (b) Mont Pelve (BB), (c) La Dent Parrachée (CC). (c) is modified after Platt and Lister (1985b). P, Permian; Tq, Lower Triassic quartzite; Td, Triassic carbonate rocks, including cargneule; Jl, Lias; Jm, Middle Jurassic; Ju, Upper Jurassic; C, Cretaceous to Palaeocene pelagic limestone; E, Palaeocene or Eocene flysch. *T1*, early thrusts; *F1*, early folds; *T3* backthrusts; *F3* backfolds. Notes on the Mont Pelve section (b): (1) D_3 shear-zone, Cirque du Grand Marchet. Slices of basement, P, Tc and probable Schistes Lustrés. Minimum displacement of 12 km on D_3 shear-zone to hanging wall cut-off of Schistes Lustrés in Vallon de la Rocheure. (2) 200 m scale tight folding in Tq, Mont Pelve. (3) Early? thrusts and isoclines in Middle Jurassic to Eocene, projected S from Pte de la Rechasse. Strong D_3 overprint. (4) Crête de la Frêt: 1 m Lias of Grande Motte nappe (inverted by D_1) lies on Arpont schist; overlain by 10 m Schistes Lustrés, then by the inverted limb of the Pelve structure. Cut-offs of Grande Motte nappe indicate minimum 8 km displacement on Pelve thrust. (5) Mylonitic inverted limb of Pelve 'fold-nappe' (D_3): from bottom up, Ju, Tc, Tq, P, basement. (6) Boudins of inverted Norian dolomite above Rhaetic and Lias (Grande Motte nappe). (7) Permian and Arpont schist in early thrust slice folded in axial zone of major D_3 fold. Projected in from Pointe de Pierre Brune. (8) Slices and isoclines of P, Tc, Ju, C, E at base of Grande Motte nappe on Pointe de Pierre Brune. Strong D_3 overprint. (9) Base of Schistes Lustrés nappe on Pointe de Lanserlia strongly reactivated by D_3.

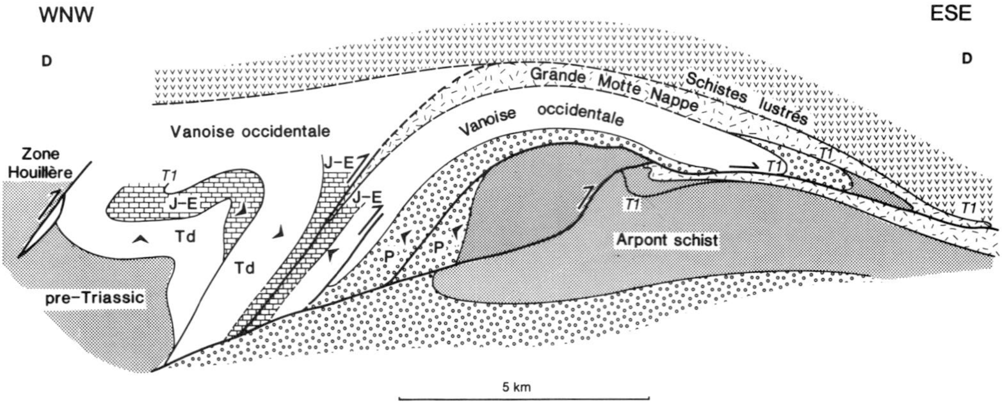

FIG. 5. Synoptic structural section across the western and southern Vanoise (DD). For location see Fig. 3. Abbreviations as in Fig. 4. Arrowheads indicate younging directions.

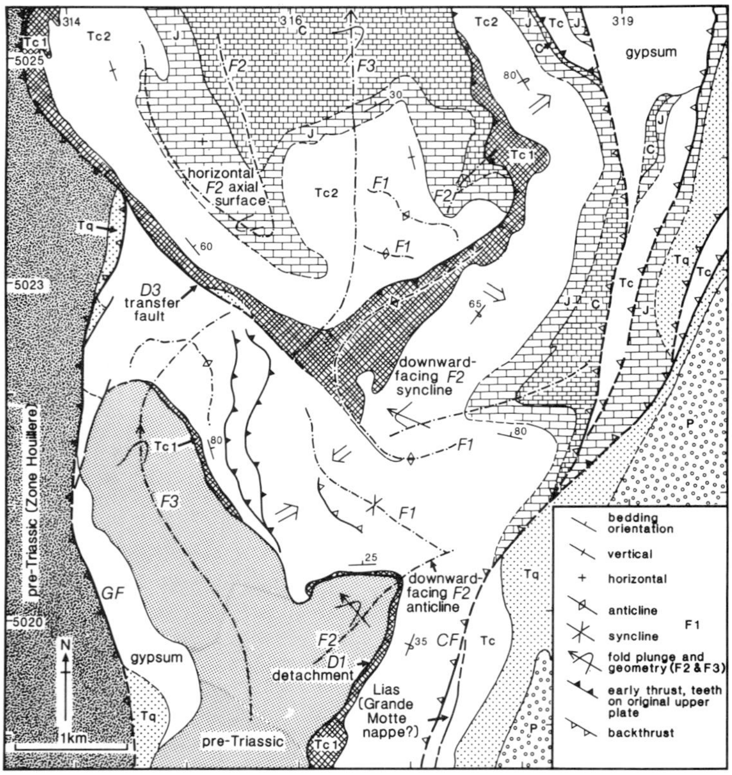

FIG. 6. Structural map of the western Vanoise, based on detailed mapping by Jaillard (1985), and structural data from Peel (1985). For location see Fig. 3. The structural interpretation presented here differs from that of Jaillard, and is not the only one possible. CF: Chaviere fault. GF: Gebroulaz fault.

trended NW, but has been widely scattered by later deformation (Fig. 9c). Further round the Alpine arc, in the Acceglio zone, Houfflain & Caby (1987) report predominantly WSW-trending glaucophane lineations. These lineations represent the finite extension direction during the early ductile deformation that accompanied or followed thrusting. We have found no clear-cut sense-of-shear indicators; and, as discussed above, the ductile deformation

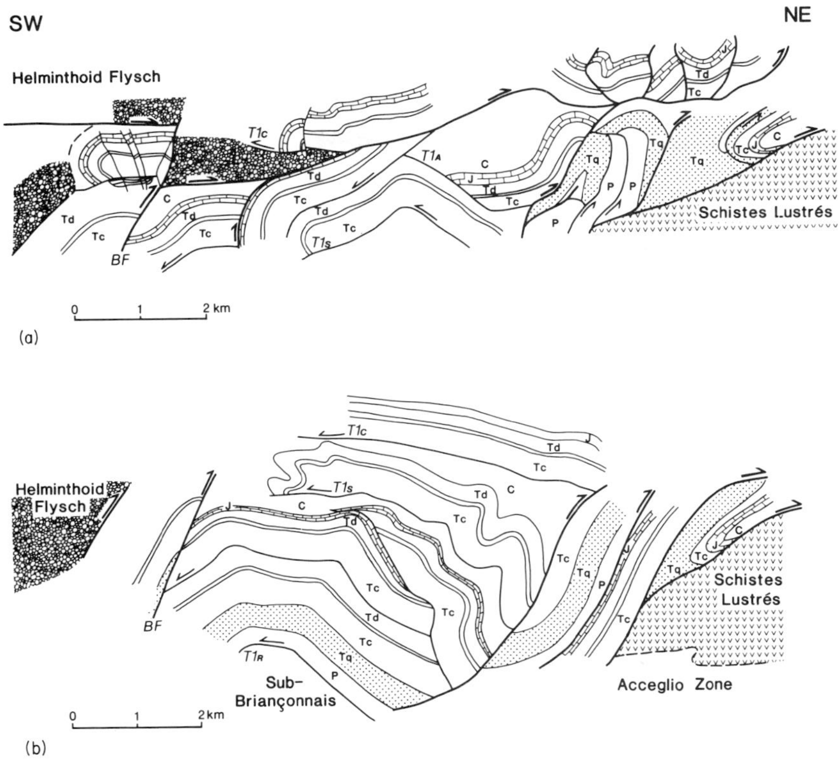

FIG. 7. Structural sections across the Briançonnais domain in the Ubaye region. See Fig. 1 for locations. Abbreviations as in Fig. 4, except: Tc, Anisian to early Ladinian limestone; Td, late Ladinian–Carnian dolostone. These two units are separated by a 30 m marker horizon of white limestone. Early thrusts: T_{IR}, Rouchouze thrust; T_{IA}, Aiguille thrust; T_{IS}, Sautron thrust; T_{IC}, Châtelet thrust. BF: Bersezio fault.

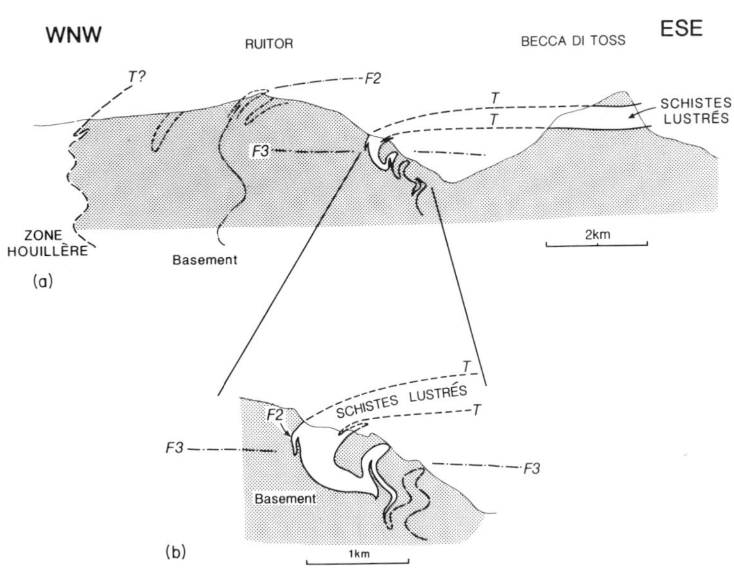

FIG. 8. Schematic section across the Ruitor Massif, after Baudin (1987). See text for discussion.

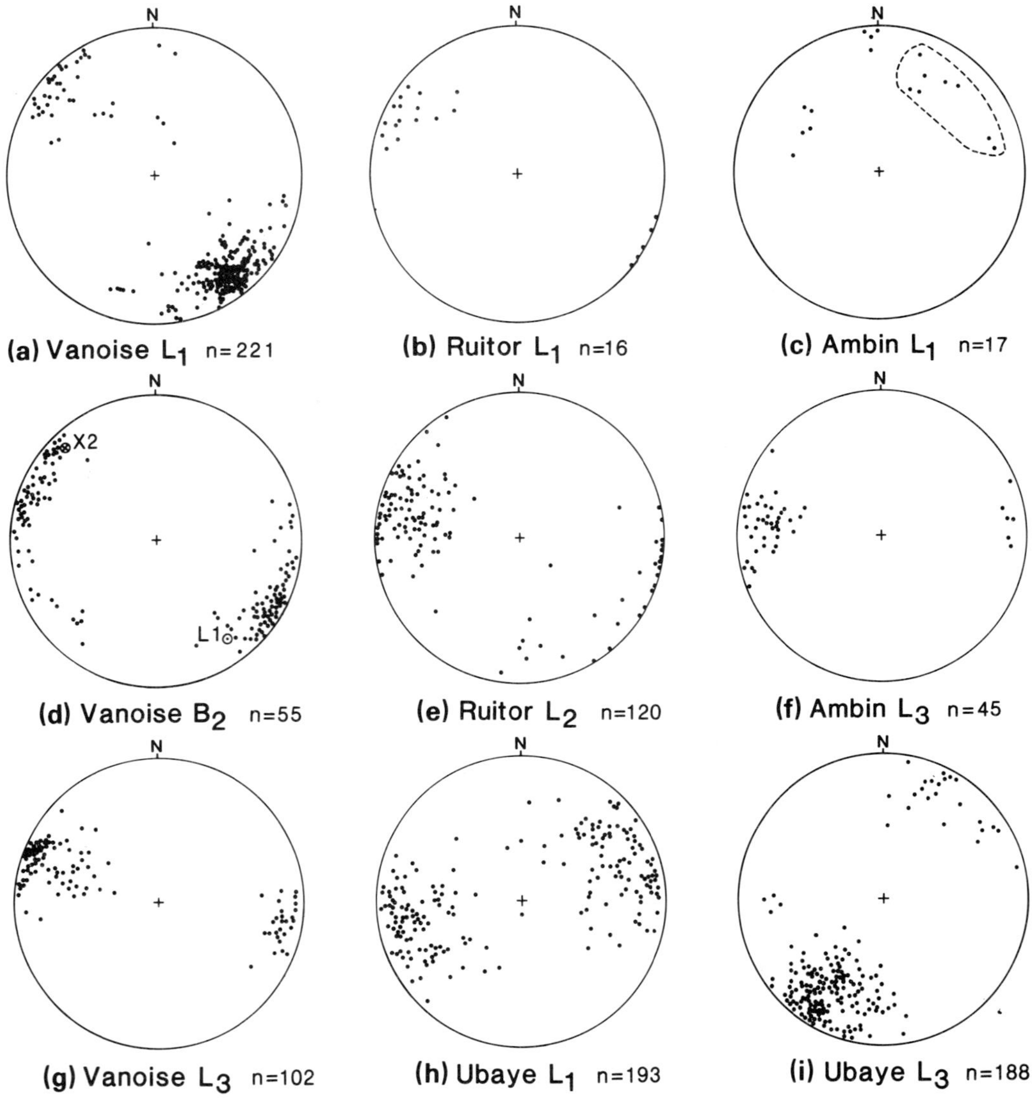

FIG. 9. Lineations and fold axes from the Briançonnais domain. (a) Early lineations (L_1), Vanoise; (b) L_1, Ruitor; (c) L_1, Ambin (the data surrounded by a dashed line have been reoriented by later deformation); (d) second phase fold axes (B_2), Vanoise, with mean L_1 and probable D_2 stretching direction (X_2) indicated; (e) L_2 stretching lineation, Ruitor; (f) L_3 stretching lineation, Ambin; (g) backthrust-related stretching lineation, Vanoise, (h) early thrust-related slip vectors, Ubaye; (i) backthrust-related slip-vectors, Ubaye.

was not necessarily associated with thrusting. It should therefore be used with great caution as a kinematic indicator.

The orientation of the glaucophane lineation in the Vanoise contrasts significantly with the dominant E–W stretching lineations characteristic of much of the adjacent Pennine zone (Malavieille *et al.* 1985, Choukroune *et al.* 1986), and we suspect the latter are younger.

In the external Briançonnais of the Ubaye region, stretching lineations are widespread on cleavage and crenulation-cleavage surfaces associated with the early thrusting and imbrication. These trend towards 260° (Fig. 9h); asymmetric porphyroclasts, mica fish, and shear-bands in thrust zones indicate WSW shear, and thrust geometry, fold vergence and cleavage vergence confirm this.

Post-thrust, pre-backthrust deformation

Parts of the internal Briançonnais have been affected by folds and ductile structures that formed under greenschist-facies conditions, and that fold and deform the early thrusts, folds, and high-pressure fabrics. These second-phase folds trend SW to NW (Fig. 9d), vary from upright to inclined N-asymmetric, and have S-dipping axial-plane crenulation cleavages. In the Ruitor the folds trend SW, are NW-asymmetric, and there is a strong stretching lineation trending to 285° (Fig. 9e) (Baudin 1987). Shear-bands and quartz *c*-axis fabrics indicate a westerly sense of shear. In the Vanoise these folds change from SW-trending with steep S-dipping axial planes, to NW-trending with subhorizontal axial planes in regions of higher strain (Fig. 9d). These features of the fold geometry led Platt & Lister (1985b) to propose a NW sense of shear.

D_2 folds range up to kilometre-scale folds such as those in Figs 4(c) and 5, one of which refolds an early thrust (the fold is S-vergent because it has been refolded by D_3). Their overall distribution is difficult to assess, because where the high-pressure metamorphic assemblages are not developed, they are difficult to distinguish from the early folds; and backthrust-related folds can also trend E–W in areas of high strain. They appear, however, to occur in E–W-trending zones, some of which coincide with changes in the stacking order of the early thrusts, and which may have been branch-lines or oblique ramps during the early deformation. Examples include the col de la Vanoise and Val d'Isère (Fig. 3). This suggests that D_2 may have transferred displacement on to and reactivated some of the early thrusts. The fact that the folds were initiated at a small anticlockwise angle to these E–W trending zones suggests that there may have been a component of dextral shear along them. A possibility exists that these zones may have served to transfer dextral transpressive displacement from the Insubric fault zone into the western Alps. In this respect they may be analogous to the D_2 structures described by Steck (1984) from the Simplon region and the central Alps, which he relates to dextral motion on the Insubric Line. Steck's structures are later (Oligocene–Miocene), however, and are associated with a SW-trending stretching lineation. Even in the Vanoise, we should not necessarily assume that all D_2 structures formed at the same time or for the same reasons. We have not found comparable structures in the Ubaye region, and they may be restricted to the more northerly parts of the domain.

In the Vanoise D_2 was followed by static growth of albite and green biotite porphyroblasts, which probably represents the 35–40 Ma meso-Alpine metamorphic event. In the Ruitor D_2 was synchronous with the greenschist-facies maximum.

Backthrusting

Backthrusts, backfolds, and related ductile strain dominate the present structure of much of the Briançonnais domain (Debelmas 1976, Tricart *et al.* 1977, Debelmas & Kerckhove 1980, Tricart 1984), and it is rather surprising that recent kinematic analyses of the western Alpine arc appear to underestimate the importance of this deformation (Malavielle *et al.* 1985, Choukroune *et al.* 1986). The Vanoise Massif, for example, is an antiformal stack of backthrust structures (Fig. 5). The Gebroulaz and Chavière faults in the western Vanoise are now very steep (Fig. 5) with minimum displacements of the order of 2 km (Peel 1985). The hangingwall of the Chavière fault forms a 2 km scale tight antiform overturned to the E that refolds an early thrust and a major D_2 fold. The Pelve thrust emplaces western Vanoise rocks eastwards over the Arpont schist and Grande Motte nappe of the southern Vanoise, and is folded right over the culmination (Fig. 4b). Cutoffs of the Schistes Lustrés and Grande Motte nappes indicate a minimum of 8 km, and possibly 12 km, of displacement. These early thrusts have been refolded around an E-closing recumbent fold-'nappe' in the hangingwall of the Pelve thrust (Fig. 4b); further south a kilometre-scale synform refolds early thrusts and D_2 folds in the foot wall (Fig. 4c) (Platt & Lister 1985b). The lower backthrusts are horizontal, so the overall geometry suggests a 'piggyback' style of thrust initiation and activity migrating towards the east. The Pelve thrust and lower structures roof on to and therefore reactivate the older thrust at the base of the Schistes Lustrés nappe, and these backthrusts therefore constitute a large-scale duplex (Fig. 5). The more external backthrusts may also roof on to the Schistes Lustrés thrust, as the latter only appears to be breached on the eastern side of the Vanoise (Fig. 3). A characteristic feature of these backthrusts is that they reactivate older thrust surfaces along significant parts of their trajectories, which makes them difficult to identify and follow.

The sense of shear is indicated by the large-scale structural geometry (Fig. 10), but small-scale structures provide more precise information. Stretching lineations trend 110° in the Vanoise (Fig. 9g); asymmetric porphy-

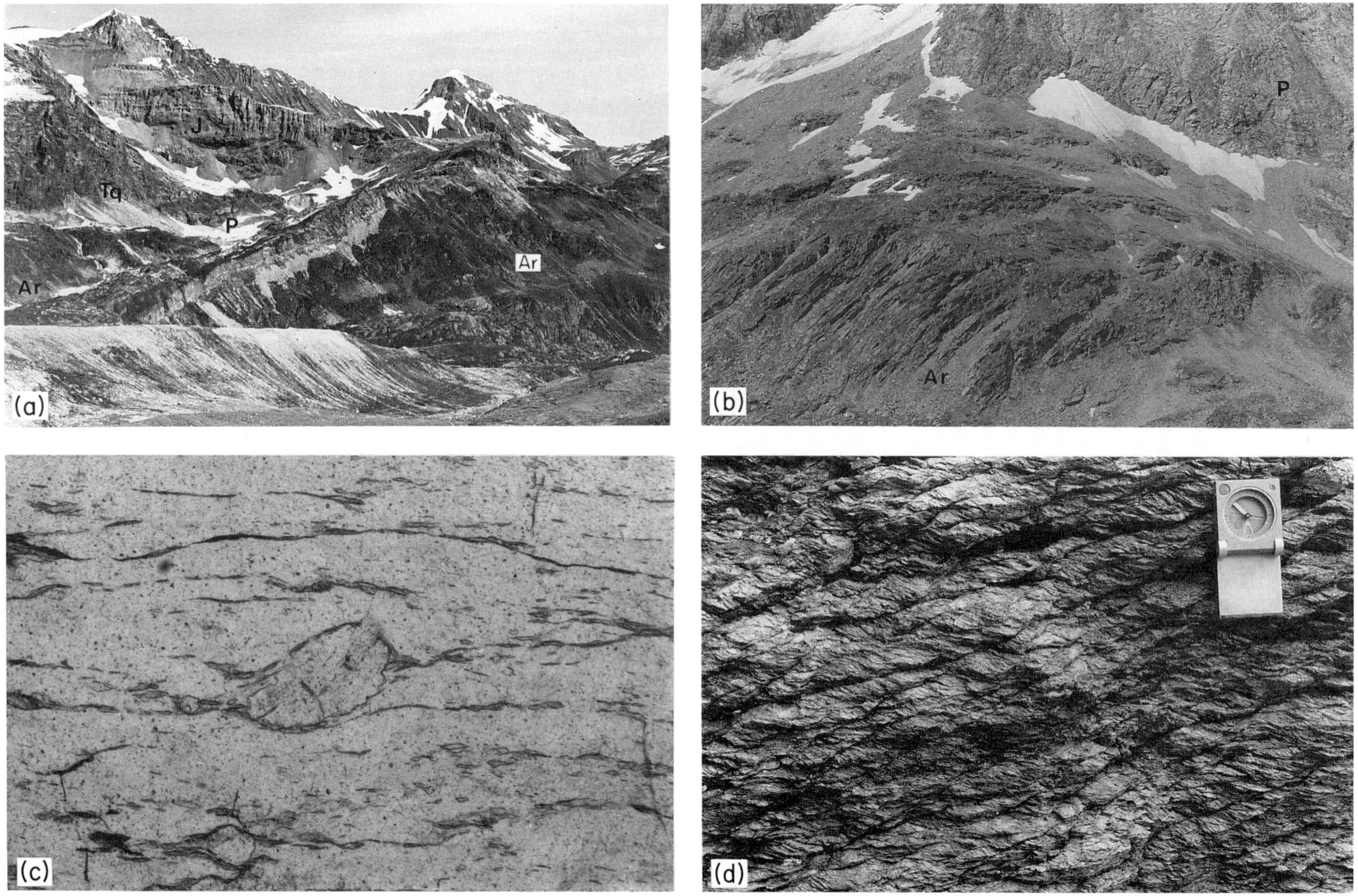

FIG. 10. (a) The Pelve backthrust and recumbent fold, looking NE from Crête de la Frêt towards Plateau de la Rechasse (location 5 in Fig. 4b). Basement (Ar) in the core of the fold is overlain by Permian phyllite (P), Triassic quartzite (Tq), and Jurassic limestone (J) of the western Vanoise unit. This unit has been emplaced eastwards over Arpont schist (at right) along the mylonitic inverted limb of the fold (dipping left). Liassic–Cretaceous limestone of the Grande Motte nappe on skyline. (b) Tête d'Aussois backthrust, looking west from Fond d'Aussois. This thrust emplaces Permian clastics (P) over Arpont schist (Ar), and is part of the same major movement zone as the Pelve thrust (see Fig. 5). S_2 foliation in the Arpont schist dips left, and has been transposed by D_3 mylonitization below the thrust. (c) Asymmetric porphyroclast from the Pelve thrust zone (location 5 in Fig. 4b). Inferred shear sense is top to right (ESE). (d) Late shear-bands (extensional crenulation cleavage). Plan de la Gorma shear zone. This is the lowest backthrust in Fig. 5. Inferred shear sense is top to left (ESE).

roclasts, shear-bands (Fig. 10), and quartz c-axis fabrics (Fig. 11) all indicate easterly shear. The foliation generally dips W or NW, but flattens in zones of high strain, and minor folds are reoriented into the extension direction (Platt & Lister 1985b), again indicating easterly shear. The fabrics are readily distinguished from early structures by their microstructural style: quartz and calcite show unannealed effects of crystal-plastic deformation and dynamic recrystallization, and albite porphyroblasts are rotated. In the Ambin Massif, there is a strong late deformation with a microstructural style very similar to that of the backthrusting deformation in the Vanoise, but as yet we have not been able to establish the sense of shear. Stretching lineations trend 090° (Fig. 9f).

A similar history is recorded in the Ubaye region. The early thrust stack has been strongly reimbricated by backthrusts across the entire width of the domain, and both the external contact (with the Helminthoid flysch nappes) and the internal contact (with the Schistes Lustrés) are now backthrusts. The more external structures are steeper, and the geometry suggests a northeastward migration of activity. Several major overturned backfolds are developed near the internal margin. Crenulation cleavages axial-planar to the folds, and associated with minor folds adjacent to the backthrusts elsewhere, carry a good stretching lineation trending 030° (Fig. 9i). Asymmetric pressure-shadows and shear-bands confirm the NE sense-of-shear suggested by the thrust and fold geometry. Both in the Vanoise and Ubaye region backthrusts appear to have been activated in sequence towards the internal margin of the domain; that is, a piggyback thrusting sequence with respect to the backthrust sense of transport. It is likely, however, that on a larger scale, successive families of such backthrusts were initiated in the reverse sequence, towards the external margin of the Alps, as suggested by Tricart *et al.* (1977) and Tricart (1984). Such a pattern has been predicted on experimental and theoretical grounds (Davis *et al.* 1983).

Ruitor Massif

The apparent exception to the importance of backthrusting is the Ruitor Massif. The structure is dominated by a major late recumbent W-closing fold (Fig. 8). This is associated with D_3 minor folds which resemble in style and associated microstructure the backfold-related structures of the Vanoise. The folds are W-asymmetric, however, and although clear sense-of-shear criteria are lacking, they are overprinted by a strong W-directed set of shear bands. This led Baudin (1987) to conclude that the structure as a whole was W-directed. The structure could be conjugate to or later than the backthrusts seen elsewhere.

It is noteworthy that in this interpretation, the geometry of the thrusts between basement and Schistes Lustrés in the Ruitor Massif (which are refolded by both D_2 and D_3) suggests an early phase of backthrusting, as suggested by Caby *et al.* (1978). The Schistes Lustrés is sandwiched between two bodies of basement (the so-called Avise syncline), and the structure opens out to the east (Fig. 8). The thrusts probably postdate the glaucophane fabric (D_1) but predate D_2.

An alternative explanation of the geometry is that the upper limb of the major D_3 fold is overturned, and that it is either a footwall synform beneath a backthrust (not identified), or the lower part of a E-vergent D_3 fold-pair. In this interpretation the Avise syncline has been inverted, and the original structure would have opened out to the W. A difficulty with this explanation is that the contact between the Ruitor basement and the Carboniferous sediments of the zone Houillère to the W, generally thought to be stratigraphic, would have had to be inverted prior to D_3. This is possible but has not been confirmed. In the absence of clear-cut sense-of-shear criteria the interpretation of this structure is best left open.

Tectonic interpretation

One of the most striking results of this study is that the Briançonnais domain has a remarkably consistent history around the whole of the western Alpine arc. Compressional deformation began at about the same time (late Palaeocene to early Eocene); it involved an early phase of thrusting; and was accompanied or followed in the internal Briançonnais by high P–T ratio metamorphism and polyphase ductile deformation. The whole domain was then affected by backthrusting and backfolding, probably in Oligocene time. There is at present no evidence for significant diachroneity in these events, even though they affect a belt that changes trend through about 80°. This in itself lends some support to the idea that the internal zones of the Alps were significantly less arcuate at the time these deformations occurred.

A second important result is that the lineations representing extension or slip directions change orientation in both time and space. The temporal variation is clearly seen in both the

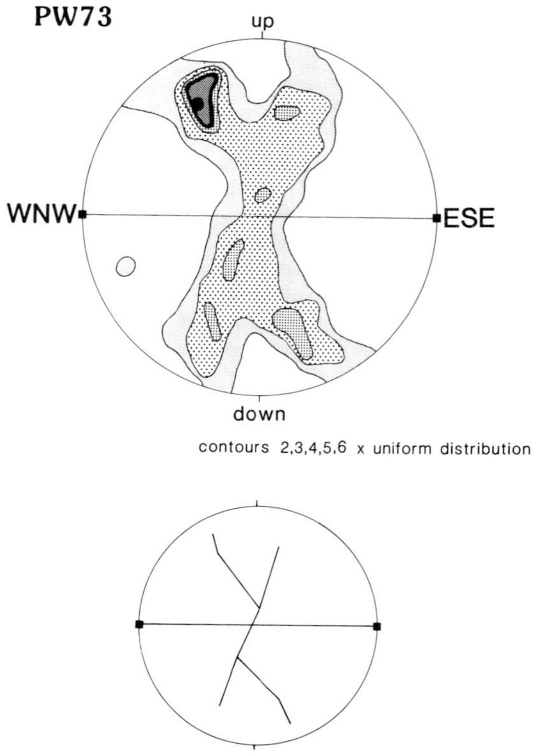

FIG. 11. Quarts *c*-axis fabric from the Pelve thrust zone (location 5 in Fig. 4b). Inferred sense of shear is top to right. Fabric skeleton shown below.

Ubaye and Vanoise regions, for example (Figs 6 and 10), where early (thrust-related?) and late (backthrust-related) vectors differ by up to 50°. Note that late vectors lie systematically anticlockwise to early ones. Both spatial and temporal variations are brought out on Fig. 12. Early and late lineations swing in sympathy with the change in trend of the Alps as a whole, so that they remain at a high angle to the regional trend. The backthrust-related lineations are nearly at right angles to the regional trend, and compare well with forethrust-related lineations from the external zones. The early lineations, however, lie clockwise to the trend normal.

A third important result of this work is a negative one; i.e. the lack of evidence for strike-slip faulting in any part of the domain we have studied. We have examined several major vertical fault zones, such as the Gebroulaz and Chavière faults (Fig. 6) and the Bersezio fault (Fig. 7), for evidence of strike-parallel movement, and they all appear to be over-steepened backthrusts. Strike-parallel or transpressive deformations are likely in those parts of curved orogens that lie close to the plate-convergence vector; and have figured largely in several recent hypotheses for the origin of the western Alps (e.g. Ricou & Siddans 1986). The lack of geological evidence for this mode of deformation provides an important constraint on these ideas.

Relation to plate motions

The most recent and most precise analysis of Africa/Europe plate motions during the Mesozoic–Tertiary period (Dewey *et al.* this volume) identify a period of sinistral motion, with Africa moving ESE relative to Europe, from 175–(roughly) 92 Ma, a period of ENE motion from 92–65 Ma, a period of slow and not very precisely determined motion from 65–51 Ma, renewed but slow NE convergence from 51–38 Ma, N–S convergence from 38–9 Ma, and NW convergence from 9 Ma to the present (Fig. 13). It appears impossible to reconcile these motions with our kinematic data from the Briançonnais domain, or with those from the Alps as a whole (Malavieille *et al.* 1985, Choukroune *et al.* 1986, Platt 1986). There

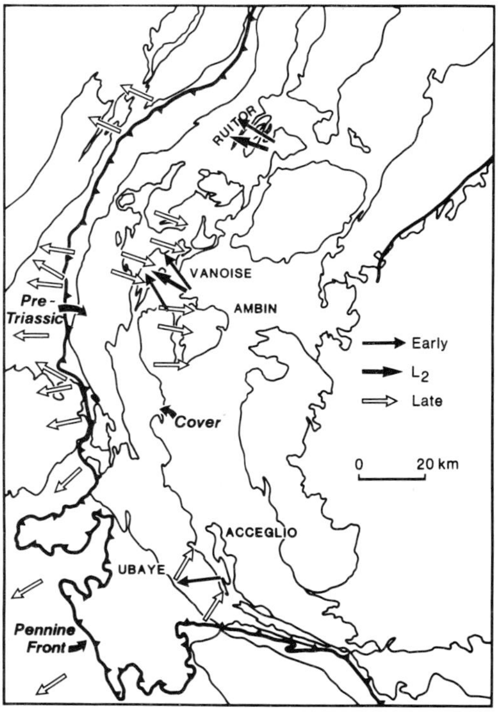

FIG. 12. Spatial and temporal variation in slip-vectors and stretching lineations in the Briançonnais domain. Early linear elements include glaucophane lineations and early thrust-related slip-vectors. Late elements include stretching lineations and slip-vectors related to backthrusts in the Briançonnais domain, and forethrusts in the external zones (after Platt 1986).

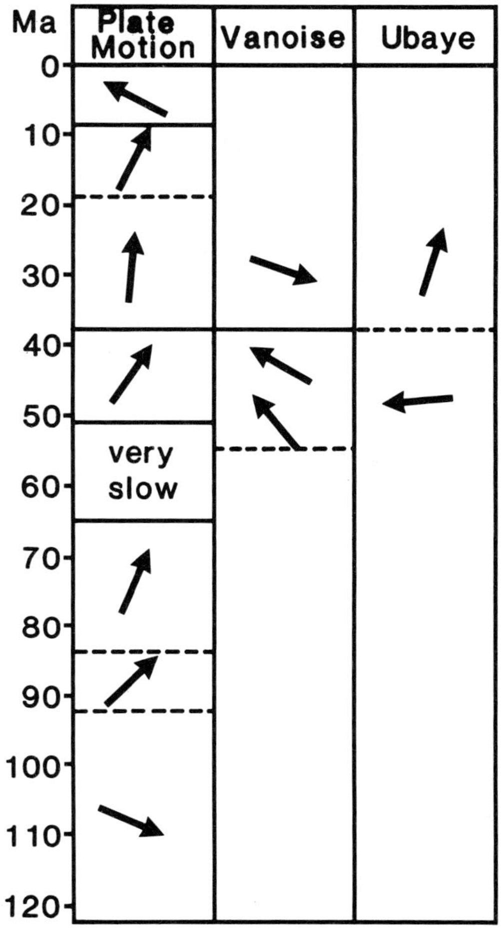

FIG. 13. Comparison of Africa–Europe relative plate motion vectors during Mesozoic–Tertiary time (after Dewey *et al.* 1987, fig. 2) with lineation trends from the Briançonnais domain.

are three possible reasons for this discrepancy:

(1) The linear elements we measure may not represent the direction of thrusting. We have already pointed out that the early glaucophane lineations in the internal Briançonnais probably represent the finite extension direction during uplift-related ductile deformation, rather than the direction of convergence during subduction. It is also possible that early linear elements may be reoriented by later deformation. If this is accompanied by ductile strain, it is usually detectable; but rigid body rotations are difficult to constrain. It is possible, for example, that the early SW-trending lineations from the Ubaye and Acceglio regions were originally NW-directed, and have been affected by wholesale rotation of this segment of the arc during collision. It is less likely, however, that the great bulk of W to NW directed early lineations from right around the arc (Malavielle *et al.* 1985, Choukroune 1986, Platt 1986) could be reconciled with a NE or N-directed plate convergence vector in this way.

(2) It may be that the linear elements do reveal the thrusting directions but that these do not coincide with the plate-convergence vector. The forces that cause thrusting are produced by a combination of the convergence between the underthrusting slab and the thrust wedge, and body forces produced by the geometry of the thrust wedge itself (Chapple 1978, Davis *et al.* 1983, Platt 1986). No rigorous dynamic analysis has yet been published of the situation where these are not parallel, as will generally be the case in an arcuate belt, but it is possible that

the local thrusting direction may differ significantly from the plate convergence vector. The divergence, however, should always be much less than 90°. The difficulty in the Alps is therefore to reconcile NW or WNW early lineations with a NE-directed Africa–Europe convergence vector.

(3) The third possibility is that Adria–Europe convergence was different to the Africa–Europe motion calculated from the Atlantic magnetic anomaly data. The idea that Adria moved as a promontory of Africa has been widely accepted since it was proposed by Channell *et al.* (1979) on the basis of palaeomagnetic data from Adria. The data are limited, however, and discrepancies between palaeomagnetic vectors from parts of stable Adria and Africa have been reported by Besse *et al.* (1984), Marton & Veljovic (1983), and Vandenberg (1983). Lowrie (1986) has re-examined the question, and concludes that Adria may have rotated by 15–30° in an anticlockwise sense during the Tertiary relative to Africa. Anderson (1987) has also concluded on the basis of seismological data that Adria is currently rotating about a pole in northern Italy relative to Africa.

A complete resolution of this problem is beyond the scope of this paper, but our evidence suggests that the possibility of independent motion of Adria should be taken seriously.

The following tentative suggestions may be made on the basis of the available information:

(1) Early (pre-collisional, pre-end Eocene) lineations trend generally NW or WNW in the Briançonnais and elsewhere in the internal Alps (Platt 1986), and this may approximately represent the Adria–Europe convergence direction at this time. The WSW-directed lineations in the Ubaye and Acceglio regions could be reconciled with this if a roughly 50° counterclockwise rigid-body rotation of the SW part of the Alpine arc occurred during later WNW-directed collision with the irregular European continental margin.

(2) Late (syn-collisional, Oligocene) back-thrusting lineations are roughly radial, and parallel to the forethrusting direction in immediately adjacent parts of the external Alps. This radial pattern of thrusting, which extends right on to the European foreland, is unlikely to be an effect of later rotation. Possible causes include rotation of Adria during convergence; independent motion of different blocks within the collision zone; and gravitational effects. The fact that the later thrusting directions are almost everywhere normal to strike suggests that locally-generated gravitational body forces dominated this period of thrusting.

References

ANDERSON, A. 1987. Is the Adriatic an African promontory? *Geology*, **15**, 212–5.

BAUDIN, T. 1987. *Étude Géologique du Massif du Ruitor (Alpes franco–italiennes): Evolution structurale d'un socle briançonnais.* Unpublished doctoral thesis, University of Grenoble.

BERNOULLI, D. & JENKYNS, H. C. 1974. Alpine, Mediterranean, and central Atlantic Mesozoic facies in relation to the early evolution of the Tethys. *In*: DOTT, R. H. & SHAVER, R. H. (eds). *Modern and Ancient Geosynclinal Sedimentation: Special Publication–Society of Economic Palaeontology and Mineralogy*, **19**, 129–60.

BESSE, J., POZZI, J. -P., MASCLE, G. & FEINBERG, H. 1984. Palaeomagnetic study of Sicily: consequences for the deformation of Italian and African margins over the last 100 million years. *Earth and Planetary Science Letters*, **67**, 377–90.

BOCQUET, J. 1974. *Études minéralogiques et pétrologiques sur les métamorphismes d'âge alpin dans les alpes françaises.* Thèse d'État. University of Grenoble.

——, DELALOYE, M., HUNZIKER, J. C. & KRUMMENACHER, D. 1974. K–Ar and Rb–Sr dating of blue amphiboles, micas, and associated minerals from the western Alps. *Contributions to Mineralogy and Petrology*, **47**, 7–26.

BURRI, M. 1983. Le front du Grand Saint Bernard du Val d'Hélens au Val d'Aoste. *Eclogae Geologicae Helvetiae*, **76**, 469–90.

CABY, R., KIENAST, J.-R. & SALIOT, P. 1978. Structure, metamorphisme et modéle d' évolution tectonique des Alpes Occidentales. *Revue de Géologie Dynamique et de Géographie Physique* **20**, 307–22.

CHANNELL, J. E. T., D'ARGENIO, B. & HORVATH, F. 1979. Adria, the African promontory, in Mesozoic Mediterranean palaegeography. *Earth Science Reviews*, **15**, 213–92.

CHOUKROUNE P., BALLÈVRE, M., COBBOLD, P., GAUTIER, Y., MERLE, O. & VUICHARD, J. P. 1986. Deformation and motion in the western Alpine arc. *Tectonics*, **5**, 215–26.

CUNNINGHAM, P. C. 1987. *The structural evolution of the Ubaye–Briançonnais belt in the French–Italian Alps.* Unpublished thesis, Oxford University.

DAVIS, D., SUPPE, J. & DAHLEN, F. A. 1983 Mechanics of fold-and-thrust belts and accretionary wedges. *Journal of Geophysical Research*, **88**, 1153–72.

DEBELMAS, J. 1976. Deux coupes transversales des Alpes franco–italiennes. *Schweizerische Mineralogische Petrographische Mitteilungen*, **56**, 561–5.

—— & LEMOINE, M. 1970. The western Alps: palaeogeography and structure. *Earth Science Reviews*, **6**, 221–56.

—— & KERCKHOVE, C. 1980. Les Alpes franco–italiennes. *Géologie Alpine*, **56**, 21–58.

DE GRACIANSKY, P. -C., BOURBON, M., CHARPAL, O. DE CHENET, P.-Y. & LEMOINE, M. 1979. Genèse et évolution comparées de deux marges continentales passives: marge Ibérique de l'ocean Atlantique et marge Européenne de la Tethys dans les Alpes Occidentales. *Bulletin de la Société Géologique de la France*, **21**, 663–74.

DESMONS, J. 1977. Mineralogical and petrological investigations of Alpine metamorphism in the internal French western Alps. *American Journal of Science*, **277**, 1045–66.

—— 1986. The Alpine metamorphisms and their environments in the western Alps: unsolved problems. *Schweizerische Mineralogische Petrographische Mitteilungen*, **66**, 29–40.

DONDEY, H. 1986. *Étude Geologique des Unités de Vanoise Méridionale de Pralognan à Aussois*. Unpublished doctoral thesis, University of Grenoble.

ELLENBERGER, F. 1958. *Étude géologique du pays de Vanoise*. Mémoires explic. de la service carte géologique de la France. 15. Bureau de Recherches Géologiques et Minières. Orléans, France.

—— 1960. Sur une paragénèse éphémère à lawsonite et glaucophane dans le métamorphisme alpin en Haute–Maurienne (Savoie) *Bulletin de la Société Géologique de la France*, **9**, 190–4.

—— & RAOULT, J. F. 1979. Les enseignements géologiques des rochers de la Loze, Pralognan (Massif de la Vanoise, Savoie). *Travaux Scientifiques du Parc Nationale de la Vanoise*, **10**, 37–69.

FREY, M., HUNZIKER, J. C., O'NEIL, J. R. & SCHWANDER, H. W. 1976. Equilibrium-disequilibrium relations in the Monte Rosa granite, western Alps: petrological, Rb–Sr and stable isotope data. *Contributions to Mineralogy & Petrology*, **55**, 147–79.

GIDON, M. 1962. La Zone Briançonnais en Haute Ubaye (Basses Alpes) et son prolongement au sudest. *Memoire Service Carte Géologique de la France*. Bureau de Recherches Géologiques et Minières, Orléans, France.

GOFFÉ, B. 1977. Succession de subfacies métamorphiques en Vanoise méridionale (Savoie). *Contributions to Mineralogy & Petrology*, **62**, 23–42.

—— 1982. Definition du faciès à Fe, Mg–carpholite — chloritoide, un marquer du métamorphisme de HP–BT dans les métasediments alumineux. *Memoires des Sciences de la Terre Université Marie Curie*, Paris.

GUILLOT, F. 1987. *Géologie de l'Antepermien de Vanoise septentrionale* (*Zone Briançonnaise interne, Alpes occidentales, Savoie, France*). Unpublished thesis, University of Lille Flandres Artois.

HOUFFLAIN, B. & CABY, R. 1987. Rétrocharriages précoces en climat schiste bleu à lawsonite-grenat: la 'Bande d'Acceglio–Longet' (Alpes Cottiennes). *Comptes Rendues de l'Académie de Sciences, Paris*, **304**, 199–204.

JAILLARD E. 1984. *Étude géologique des unités briançonnaises de Vanoise occidentale au Sudouest de Pralognan (Savoie)*. Unpublished thesis, University of Grenoble.

——, DONDEY, H. & DEBELMAS, J. 1986. Reconstructions paléogéographiques de la zone briançonnaise de Vanoise (Alpes occidentales): nouveaux arguments pour une origine intra-briançonnaise de l'unite de la Grande Motte. *Comptes Rendues de l'Académie de Sciences Paris*, **302**, 1091–4.

KARIG, D. E. 1974. Evolution of arc systems in the western Pacific. *Annual Reviews of Earth & Planetary Science*, **2**, 51–75.

LEMOINE, M., BAS, T., ARNAUD-VANNEAU, A., ARNAUD, H., DUMONT, T., GIDON, M., BOURBON, M., DE GRACIANSKY, P. -C., RUDKIEWICZ, J. -L., MEGARD-GALLI, J. & TRICART, P. 1986. The continental margin of the Mesozoic Tethys in the western Alps. *Marine and Petroleum Geology*, **3**, 179–99.

LEFÉVRE, R. 1974. Microanalyse par émission ionique secondaire d'un pyroxène jadéitique d'âge alpin de la Bande d'Acceglio (Alpes cottiennes italiennes). *Bulletin de la Société Géologique de la France*, **16**, 248–254.

MALAVIELLE, J., LACASSIN, R. & MATTAUER, M. 1985. Signification tectonique des linéation d'allongement dans les Alpes occidentales. *Bulletin de la Société Geologique de la France*, **26**, 895–906.

MARTON, E. & VELJOVIC, D. 1983. Palaeomagnetism of the Istria peninsula, Yugoslavia. *Tectonophysics*, **91**, 73–87.

MEGARD-GALLI, J. & BAUD, A. 1977. Le Trias moyen et supérieure des Alpes Nord-Occidentales: données nouvelles et corrélations stratigraphiques. *Bulletin du Bureau de Recherches Géologiques et Minières de la France*, **3**, 233–50.

MILNES, A. G. 1978. Structural zones and continental collision, Central Alps. *Tectonophysics*, **47**, 369–92.

PEEL, F. J. 1985. *Structural evolution of the Vanoise Massif (French Alps)*. Unpublished DPhil thesis, Oxford University.

PLATT, J. P. 1986. Dynamics of orogenic wedges and the uplift of high-pressure metamorphic rocks. *Bulletin of the Geological Society of America*, **97**, 1037–53.

—— & LISTER, G. S. 1985a. Structural history of high-pressure metamorphic rocks in the Vanoise Massif, French Alps, and their relation to Alpine tectonic events. *Journal of Structural Geology*, **7**, 19–36.

—— & —— 1985b. Structural evolution of a nappe complex, southern Vanoise Massif, French

Penninic Alps. *Journal of Structural Geology*, **7**, 145–160.

—— & LEGGETT, J. K. 1986. Stratal extension in thrust footwalls, Makran accretionary prism: implications for thrust tectonics. *Bulletin of the American Association of Petroleum Geologists*, **70**, 191–203.

RAMSAY, J. G. 1981. Tectonics of the Helvetic nappes. *In*: MCCLAY, K. R. & PRICE, N. J. (eds). *Thrust and Nappe Tectonics, Special Publication of the Geological Society of London*, No. **9**, 293–310.

RICOU, L. E. & SIDDANS, A. W. B. 1986. Collision tectonics in the western Alps. *In*: COWARD, M. P. & RIES, A. C. (eds). *Special Publication of the Geological Society of London* No. **19**, 229–53.

SALIOT, P. 1979. La jadeite dans les Alpes françaises. *Bulletin Mineralogique*, **102**, 391–401.

STECK, A. 1984. Structures de déformation tertiares dans le Alpes centrales (transversale Aar-Simplon-Ossole). *Eclogae Geologicae Helvetiae*, **77**, 55–101.

TRICART, P. 1984. From passive margin to continental collision: a tectonic scenario for the western Alps. *American Journal of Science*, **284**, 97–120.

——, CARON, J. M., GAY, M., & VIALON, P. 1977. Relais de schistosités, structures en éventail et discontinuitiés majeures sur la transversale du Pelvoux (Alpes occidentales). *Bulletin Société Géologique de la France*, **14**, 873–81.

TRÜMPY, R. 1980. *Geology of Switzerland*. Wepf & Co., Basel.

VAN DEN BERG, J. 1983. Reappraisal of palaeomagnetic data from Gargano (south Italy). *Tectonophysics*, **98**, 29–41.

WARRACK, M. 1974. The petrography and origin of dedolomitized, veined or brecciated carbonate rocks, the 'cornieules' in the Fréjus region, French Alps. *Journal of the Geological Society of London*, **130**, 229–47.

J. P. PLATT, P. C. CUNNINGHAM & P. WESTON, Department of Earth Sciences, Parks Road, Oxford OX1 3PR, England.

G. S. LISTER, Department of Earth Sciences, Monash University, Clayton, Victoria 3168, Australia.

F. PEEL, BP Ltd, Britannic House, Moore Lane, London EC2Y 9BU, England.

T. BAUDIN & H. DONDEY, Institut Dolomieu, Université de Grenoble, Rue Maurice-Gignoux, F-38031 Grenoble Cedex, France.

The role of the Periadriatic Line in the tectonic evolution of the Alps

S. M. Schmid, H. R. Aebli, F. Heller & A. Zingg

SUMMARY: The Periadriatic Line and related lineaments formed as a result of post-collisional deformations which severely modified the Alpine chain. This post-late Oligocene deformation is the result of dextral transpression between the Adriatic sub-plate and the European foreland. Indentation of the western edge of the southern Alps caused uplift, related to backthrusting and associated deformations of the Lepontine region combined with E-directed escape of the central Alps. In the eastern Alps the response to dextral transpression is mainly by lateral escape along conjugate strike slip zones with minor or no vertical movements. Older deformations along this essentially late Alpine lineament can still be inferred locally and include: extension and transfer faulting in the late Palaeozoic to early Mesozoic, Cretaceous deformations, and Tertiary phases of compression (Eocene) and possibly extension (Oligocene). The geometry of crustal thinning associated with the formation of the passive continental margin of the southern Alps (associated with initial uplift of the Ivrea zone) has a profound influence on strain localization and the kinematics of movements along and north of the present day Periadriatic Line.

The polymorphism of the Periadriatic Line

The Periadriatic Line as the northern boundary of the southern Alps

Traditionally, the Periadriatic Line (Fig. 1a) is considered to mark a first order tectonic boundary between the southern Alps with only minor S-vergent thrusting and folding and the main body of the Alps with its W- to N-verging nappe structure. The change in tectonic style is very impressive across the western sector of the Periadriatic Line (Canavese and Tonale Lines) where this lineament separates the penetratively deformed area of the Lepontine region from the southern Alps (including the Ivrea zone) which lack an Alpine penetrative deformation. However, this contrast in tectonic style mainly reflects the metamorphic conditions during deformation. South-vergent backfolding also affected the central Alps (Argand 1916) and the 'minor' S-vergent thrusting in the southern Alps has recently been found to be rather impressive (Laubscher 1985), causing about 40 km shortening.

Further to the E the boundary between Austro–Alpine nappes and southern Alps was rather artificially drawn along the Pusteria and Gailtal Lines and S of the Drauzug (Fig. 1a) rather than N of the Drauzug and along the Defereggen–Anteselva–Vals (DAV) Line (see Bögel 1975 for an extensive discussion of the literature). Finally, the Periadriatic Line runs into the central zone of the bivergent Karawanken (Bauer 1984, Bauer & Schermann 1984), interpreted in terms of a flower structure by Laubscher (1983b). Thus, it seems that the notion of a major subdivision into southern Alpine and Austro–Alpine domains, each characterized by their own tectonic history and later juxtaposed along a first order tectonic boundary is misleading. Over most of its distance the Periadriatic Line transsects the crust of the southern continental margin (Fig. 1a) and it is merely a matter of definition as to where exactly (N or S of the Drauzug) one chooses to draw this 'first order' tectonic subdivision.

The Periadriatic Line as the southern limit of Alpine metamorphism

Ahrendt (1980) puts emphasis on another aspect of the Periadriatic Line, namely on that of a lineament which marks the southern limit of Tertiary metamorphism. Such a definition is very useful in regions with a strong metamorphic overprint (northern part of the Canavese Line and the western portion of the Tonale Line, Fig. 1b). However, as pointed out by Ahrendt (1980), it is rather the DAV Line which marks the southern boundary of Alpine biotite ages in the region S of the Tauern window than the Pusteria Line further to the S. Problems with this definition arise at the southwestern end of the Periadriatic Line as well. There, Zingg *et al.* (1976) report low grade metamorphism and Cretaceous ages from the cover of the southern Alps (Canavese zone s. str., Fig. 2), S of the

From: Coward, M. P., Dietrich, D. & Park, R. G. (eds), 1989, *Alpine Tectonics*, Geological Society Special Publication No. 45, pp. 153–171.

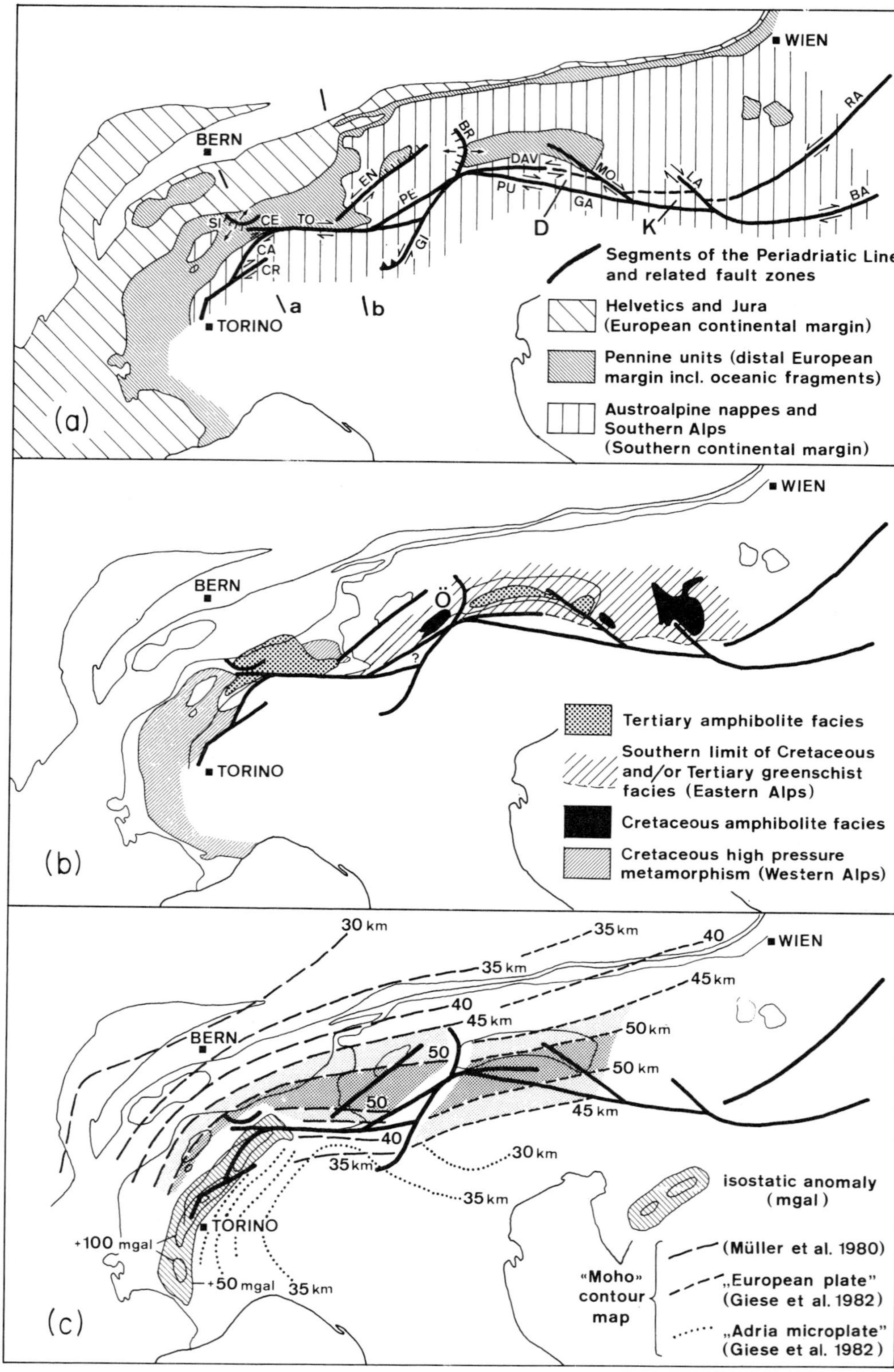
WIEN
BERN
TORINO
BR
DAV
MO
LA
RA
BA
EN
PE
PU
GA
SI
CE
TO
CA
CR
GI
D
K
a
b
(a)
Segments of the Periadriatic Line and related fault zones
Helvetics and Jura (European continental margin)
Pennine units (distal European margin incl. oceanic fragments)
Austroalpine nappes and Southern Alps (Southern continental margin)
WIEN
BERN
Ö
?
TORINO
(b)
Tertiary amphibolite facies
Southern limit of Cretaceous and/or Tertiary greenschist facies (Eastern Alps)
Cretaceous amphibolite facies
Cretaceous high pressure metamorphism (Western Alps)
30 km
35 km
40
WIEN
35 km
45 km
40
45 km
BERN
50 km
50
50 km
50
45 km
40
30 km
35 km
35 km
TORINO
+100 mgal
+50 mgal
35 km
(c)
isostatic anomaly (mgal)
(Müller et al. 1980)
„European plate" (Giese et al. 1982)
«Moho» contour map
„Adria microplate" (Giese et al. 1982)

Periadriatic Line. Two additional problems arise with this definition: (1) The impressive jump in radiometric ages ceases to exist east of the DAV Line; (2) the same lineament marks the southern and eastern boundary of Tertiary (Lepontine dome, Tauern window) as well as Cretaceous cooling ages (Sesia zone, eastern Alps). One feature emerging from inspection of Fig. 1(b) is highly conspicuous: Both the Lepontine dome and the Tauern window with their amphibolite grade Tertiary metamorphism are situated at the interference points of SW–NE and E–W-trending parts of the Periadriatic lineament. This geometry suggests a causal relationship between updoming and movements along the Periadriatic Line.

The Periadriatic Line as the site of a root zone

A third aspect of the Periadriatic Line is related to the confusing notion of a 'root zone'. The Periadriatic Line marks the southern boundary of this so-called root zone, which is better referred to as the 'southern steep belt' (Milnes 1974), characterized by steeply N- and NW-dipping foliation planes. The term 'root' is misleading because it comprises both the vision of an origin for far-travelled nappes and at the same time the subvertical orientation of the nappe contacts. This old-fashioned concept of a highly squeezed root which runs in a subvertical orientation through the entire crust (as suggested in most classical cross sections through the Alps) led several authors to consider the Periadriatic Line as a suture zone (e.g. Dewey & Bird 1970). Mesozoic sediments along the Periadriatic Line (e.g. the Canavese) have classically been considered to represent the root of the Austro–Alpine nappes (Argand 1910), although they simply represent the cover of the southern Alps (Ahrendt 1980). In fact, the southern steep belt is genetically linked to late Tertiary post-nappe emplacement folding and backthrusting (Milnes 1974).

The Periadriatic Line as a strike-slip zone

Laubscher (1971a, b) placed emphasis on yet another aspect of the Periadriatic Line by postulating some 300 km dextral strike slip at or in the vicinity of this line. This postulate results from large scale kinematic considerations concerning the polarity of thrusting in the Alps, Dinarides and Appennines. As the Alpine belt swings into the N–S oriented strike of the western Alps the westward motion of the Adriatic sub-plate was held responsible for EW directed shortening in the western Alps including obduction of the Ivrea geophysical body (Laubscher 1984).

The Periadriatic Line in the light of geophysical data

Figure 1(c) suggests that the western sector of the Periadriatic Line marks the western edge of the Ivrea gravity anomaly. Kissling (1980, 1984) showed that the northwestern boundary of the Ivrea geophysical body is identical with the projection of the Periadriatic Line down to a depth of at least 10 km. Due to the NW-dip of the Canavese Line (as shallow as 40°–60° across the Valle d'Ossola) the gravity anomaly extends below the Sesia zone. The rather sharp bend into the E–W strike of the Tonale Line near Locarno coincides with the northeastern termination of this gravity anomaly. This suggests a causal relationship between the geometry of the Ivrea gravity anomaly and the position of the Periadriatic Line.

The same bend from a NNE–SSW into an E–W strike is noticable in the contour map for the Moho under the western and central Alps. Figure 1(c) incorporates older published data (Müller *et al.* 1980, Giese *et al.* 1982), and according to a more recent re-evaluation of seismic data (Ansorge, pers. comm.) this bend appears even more discrete. Thus, both the Canavese and Tonale Lines are situated S and

FIG. 1. The Periadriatic Line in the context of Alpine geology.
(a) The segments of the Periadriatic Line and related late Alpine faults. The segments of the Periadriatic Line are from W to E: CR–Cremosina; CA–Canavese; CE–Centovalli; TO–Tonale; GI–Giudicarie; PU–Pustertal; DAV–Defereggen–Anteselva–Vals; GA–Gailtal. Related faults are from W to E: SI–Simplon; EN–Engadine; PE–Pejo; BR–Brenner; MO–Mölltal; LA–Lavanttal; BA–Balaton; RA–Raba (Balaton and Raba lines after Kazmer & Kovacs 1985). D–Drauzug; K–Karawanken.
(b) Extremely simplified metamorphic map of the Alps (after Niggli & Zwart 1973, Frank 1987). Ö: thermal dome in the southern Ötztal nappe.
(c) Contour maps of the crust–mantle boundary (after Müller *et al.* 1980 and Giese *et al.* 1982) and isostatic anomalies of the Ivrea geophysical body (after Veccia 1968).

SE of the Moho trough. This close relationship is completely lost further to the E. For a short distance the axis of the Moho depression changes into the southern Alps. This suggests that the character of the Periadriatic Line changes rather drastically towards the E.

Terminology

Although it became customary to use the terms 'Periadriatic' and 'Insubric' synonymously we use the term 'Periadriatic' to denote the entire system of fault zones as depicted in Fig. 1(a). This figure also indicates the names used for various segments of the Periadriatic fault system. It is one of the aims of this contribution to show that portions of the Periadriatic Line accommodated different displacements at different times. The northeastern portion of the Canavese Line and the western sector of the Tonale Line play a rather well defined role during the uplift of the Lepontine region associated with backthrusting (Schmid *et al.* 1987), and following Spitz (1919) we reserve the term 'Insubric' to this portion of the Periadriatic Line which was active during the late Oligocene–Miocene postcollisional 'Insubric phase' in the sense of Argand (1916).

Dextral transpression during the Insubric phase and associated uplift of the Lepontine region

The Insubric mylonite belt south of the Lepontine region

A greenschist-facies mylonite belt of approximately 1 km thickness defines the NE portion of the Canavese Line which curves from a NNE–SSW into an E–W strike over a short distance (Fig. 2). The mylonites dip with values of 40°–60° to the NW across the Val d'Ossola. The dip angle increases towards the E as the mylonite belt swings towards an E–W strike (values around 70° are typical for the region around Locarno). The same mylonite belt can be followed further to the E along the Tonale Line (Wiedenbeck 1986, Heitzman 1987a,b). There, the mylonitic foliation dips at 60°–70° to the N or NNW. However, a late brittle fault (the Tonale Fault, which will be discussed below) is in a subvertical orientation and runs exactly E–W.

West of Locarno the mylonite belt comprises three types of protoliths (Schmid *et al.* 1987), being from the SE towards the NW: Ivrea-derived mylonites (mainly former paragneisses), Canavese sediments (Permo–Mesozoic cover of the southern Alps), and mylonites derived from the Austro–Alpine Sesia zone (which here lacks evidence for a high pressure Eoalpine event). East of Locarno mylonitization affects the Tonale zone (a characteristic association of lithologies which are part of the pre-Triassic basement of the Austro–Alpine nappes, Cornelius & Cornelius-Furlani 1931 Lardelli 1981) and parts of the tail of the Oligocene Bergell intrusion (Jorio tonalite, Fumasoli 1974, Heitzmann 1987b, Vogler & Voll 1981). In this region cover and basement of the southern Alps are only moderately overprinted by mylonitization but affected by cataclastic movements (Fumasoli 1974).

Manifestations of brittle faulting and cataclasis are rarely found within the mylonites of the Canavese Line W of Locarno. The E–W oriented Centovalli Line, however, represents a rather broad zone of cataclasis and fault breccias affecting the gneisses of the southern steep belt which lack signs of earlier mylonitization. Thus, it appears that the site of earlier ductile movements is spacially separated from the site of later brittle movements. This separation no longer exists E of Locarno. A system of dextral Riedel shears affects the southern steep belt of the central Alps (Fig. 2) including the mylonites along the Tonale Line (Fumasoli 1974, Heitzmann 1987a,b). A sharp brittle fault (Tonale Fault) forms the southern boundary of (i) the Riedel shears, and, (ii) the mylonites. This distinction between Tonale mylonite belt (N-dipping) and Tonale Fault (subvertical brittle fault) will turn out to be useful for later discussions of the kinematics of movements and the eastwards extension of the Periadriatic Line.

The kinematics of movement along the Canavese Line have recently been analysed (Schmid *et al.* 1987): mylonitized Canavese sediments with variably oriented stretching lineations separate an earlier formed, more northerly located mylonite belt which indicates backthrusting of the Sesia zone to the SE from a later, more southerly located mylonite belt formed during dextral shearing. The change in orientation of the stretching lineations from down-dip plunging to subhorizontal orientations is rather abrupt. However, overprinting relationships suggest that at least parts of the southern mylonite belt were previously affected by backthrusting as well.

The change in the orientation of stretching lineations is much more gradual E of Locarno. In the area of Passo S. Jorio (Fig. 2) the stretch-

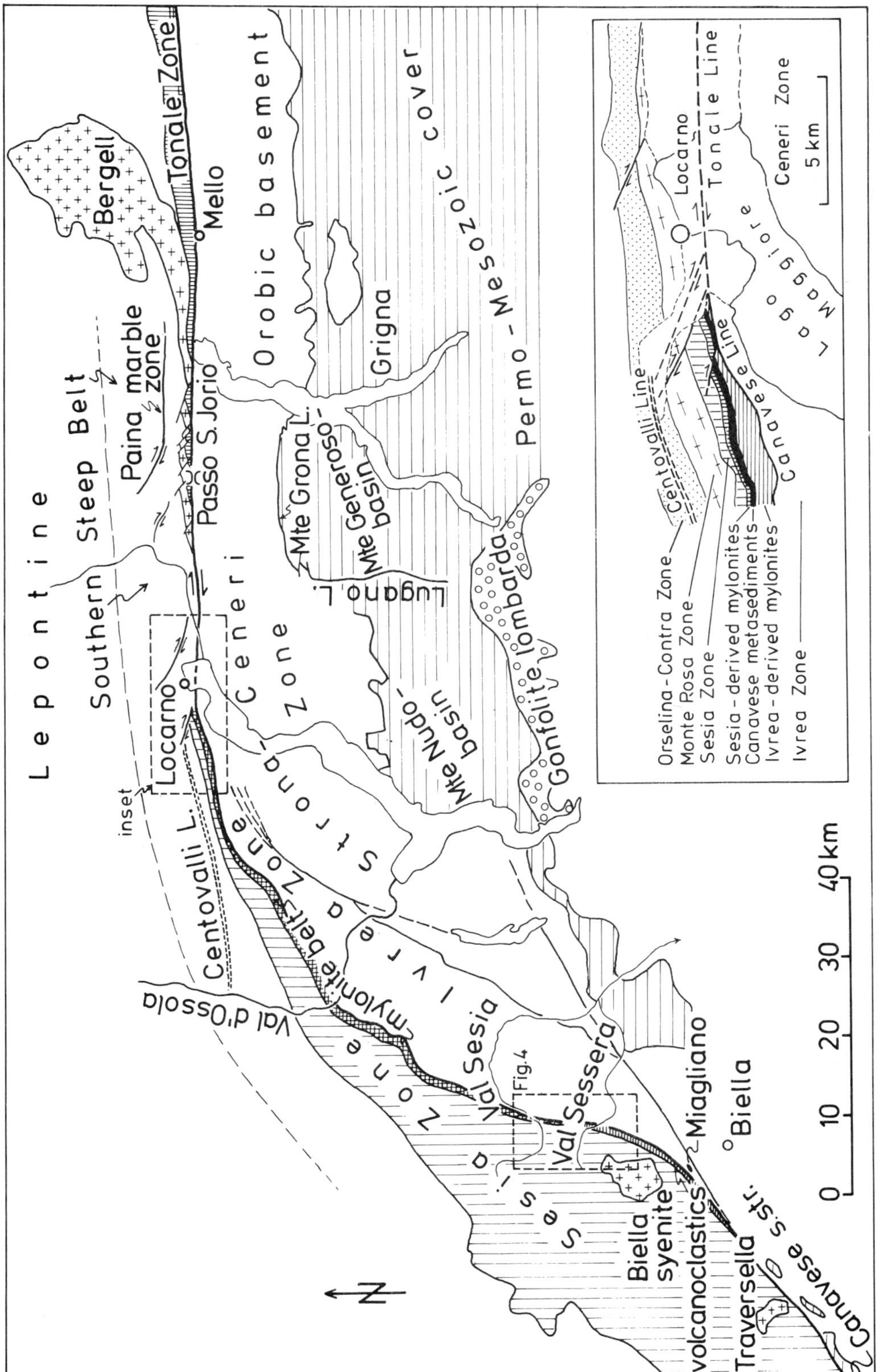

FIG. 2. Schematic map of the Canavese Line and its transition into the western Tonale Line indicating the location of geological and geographical names used in the text.

ing lineations are in a down-dip orientation (backthrusting) within the strongly deformed tonalites of the Bergell intrusion. Towards the S and within the mylonitized Tonale zone they gradually change from the same down-dip orientation to a pitch to the W (dextral combined with backthrusting movement) and finally into a subhorizontal orientation (dextral strike-slip, Vogler & Voll 1981, Heitzmann 1987a,b). This continuous change in the orientation of the lineation goes hand in hand with decreasing temperatures (from N to S) during the mylonitization event. Still further to the E and S of the main body of the Bergell intrusion (near Mello, Fig. 2) only subhorizontal stretching lineations indicating dextral shear are observed within the mylonitized Tonale zone (Wiedenbeck 1986).

Stretching lineations formed under conditions of simple shearing with a constant direction of shearing asymptotically approach the displacement vector. The above mentioned changes in orientation of the stretching lineations along and across strike suggest that the displacement vector changed in time and space. This leads to considerable difficulties in inferring the exact movement path because, strictly, the lineations merely record the finite stretch accumulated over some unknown duration in time at a given locality. Although Schmid *et al.* (1987) recognized two mylonite zones with distinct orientations of stretching lineation and sense of shear we interpret this configuration to indicate one continuous transpressive dextral strike-slip motion rather than two strictly separate events. Within the mylonite belt a volume of rock will first be affected by backthrusting to the SE and perpendicular to the NE–SW strike of the Canavese Line and it will then be either completely overprinted by dextral strike-slip movements (horizontal lineations in the southern mylonite belt), or it will be passively carried to the E due to later strain concentration within the southern mylonite belt (preserved steep lineations in the northern mylonite belt). The very gradual change in the orientation of the lineations within the Insubric mylonites as observed E of Locarno supports such a view. Additional arguments for a continuous movement during the Insubric phase will be given later when synchronous deformations outside the mylonite belt will be discussed.

Only during the final stages of deformation did dextral strike-slip outlast transpression. The brittle deformation within the Centovalli fault zone finds its eastward continuation in the Tonale Fault with its associated Riedel shears. This same event also caused strike slip deformation under greenschist facies conditions within more internal parts of the southern steep belt (Paina marble zone, Heitzmann 1987).

Formation of the southern steep belt and verticalization of the northwestern rim of the Ivrea zone

The contemporaneity of backfolding within the Monte Rosa and Sesia units with backthrusting within the Insubric mylonite belt was postulated in a section along the Val d'Ossola by Schmid *et al.* (1987). The rapidly and continuously decreasing amount of retrograde overprint towards the NW and away from the Insubric Line reflects a horizontal temperature gradient during backfolding and backthrusting. This gradient results from the rapid juxtaposition of a hot hangingwall (central Alps) over a relatively cooler footwall (southern Alps).

Verticalization is not restricted to the southern steep belt. Recent work in the Ivrea zone and along its contact to the Strona-Ceneri zone by Brodie & Rutter (1987) and by Handy (1986, 1987) indicates an Alpine age for late folding and tilting within the Ivrea zone around a NE–SW oriented axis. However, the sedimentary cover of the southern Alps, further to the S and E, was not affected by this verticalization. Thus, a hinge zone somewhere within the Strona–Ceneri zone or at the contact to the sedimentary cover has to be postulated.

Contemporaneity of backfolding and backthrusting N of the Insubric Line was also inferred for the region E of Locarno (Heitzmann 1987b). It is very tempting to also relate uplift of the Orobic basement S of the Tonale Line (De Sitter & De Sitter-Koomans 1949) to the same event. According to Bertotti (1987) this uplift can be interpreted to have tilted the northward continuation of Mesozoic normal faults, such as the Lugano Line, situated between the M. Nudo and Generoso basins. The continuation of the Lugano Line (the M. Grona Line) appears as a sinistral strike-slip zone in present day map view (see Figs 2 and 7 for localities). The age of thrusting in the Grigna mountains (Laubscher 1985) is still uncertain since Brack (1981) reports a pre-Adamello (Cretaceous?) phase of deformation further to the E.

Both the southern steep belt and the Ivrea zone were also affected by dextral shearing but a clear sequence of backthrusting followed by dextral shearing cannot be established yet. Some of this dextral shearing within the southern steep belt occurred under amphibolite facies conditions (Steck 1984, Merle *et al.* this volume) whereas E of Locarno dextral shearing occurred under greenschist facies conditions (Heitzmann

1987). A large portion of the very complex deformation N of the southern steep belt can also be attributed to the Insubric phase (Merle *et al.* this volume).

The cooling history of the Lepontine region

Radiometric data from the Lepontine area are so numerous by now that the application of the cooling age concept allows for a rather precise evaluation of temperature vs time curves as depicted in Fig. 3. With K/Ar and Rb/Sr dating on biotite and white micas combined with fission track data on apatite and zircon, Hurford (1986) was able to constrain the cooling curves for 19 samples along a N–S profile through the Lepontine area N of Locarno to such detail that two periods of uplift can be distinguished. Within the southern steep belt rapid cooling (65°C/Ma corresponding to an uplift rate of about 2.2 mm/a) from 23 to 19 Ma was followed by significantly slower cooling (14°C/Ma corresponding to an uplift rate of around 0.4 mm/a) compatible with present day uplift rates. The same pattern is recorded for the flat-lying central and northern parts of the Lepontine except that the onset of rapid cooling becomes younger. Since the southern limit of this area of rapid cooling coincides with the Insubric Line, a correlation of the period of rapid cooling with uplift due to backthrusting along the steeply inclined Insubric Line is obvious. Because the onset of rapid cooling is expected to postdate the beginning of rapid uplift (Werner 1980), backthrusting along the Insubric Line may have

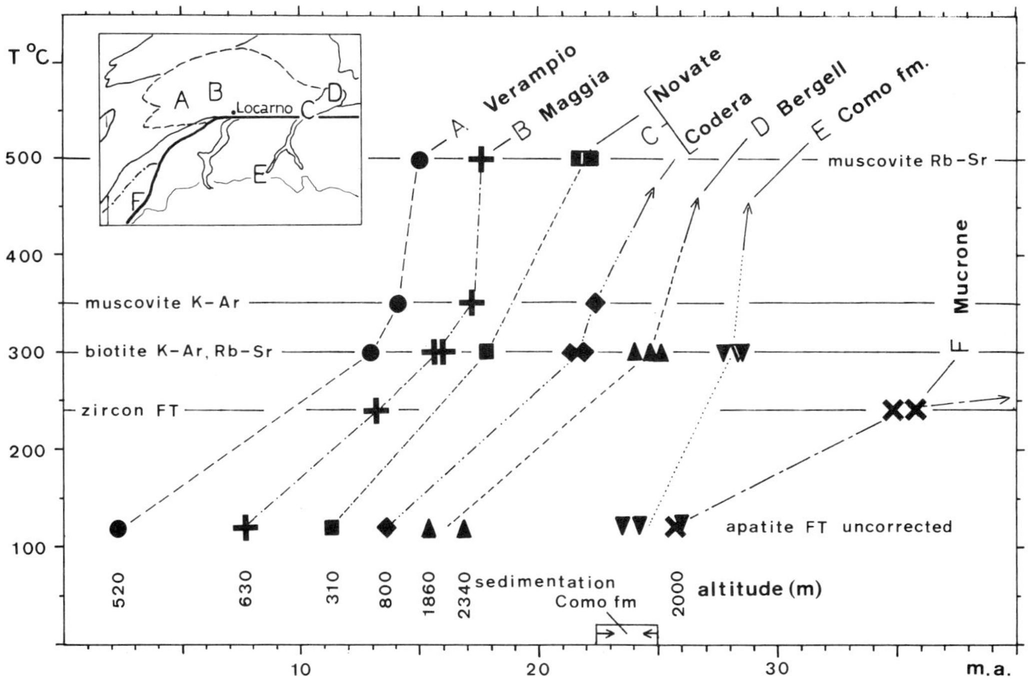

FIG. 3. Cooling history derived from published radiometric age determinations.
Tertiary amphibolite facies terrains: A and B Oligocene intrusions: C and D Bergell boulders from the Como formation: E. Cretaceous high pressure terrains of the Sesia zone: F.
A: Verampio (KAW 201), reference (2), (3), (4).
B: Maggia (KAW 1887), ref. (7).
C: Novate (KAW 132), ref. (2), (4).
Codera (KAW 553), ref. (3), (4).
D: Bergell (Z 6, BM 2, BM 3), ref. (1), (2), (5).
E: Como fm. (KAW 1003, 1004, 1005), ref. (4), (5).
F: Mucrone (KAW 987, 988, 989), ref. (6).
References: (1) Armstrong *et al.* (1966), (2) Jäger *et al.* (1967), (3) Purdy & Jäger (1976), (4) Wagner *et al.* (1977), (5) Wagner *et al.* (1979), (6) Hurford & Hunziker (1985), (7) Hurford (1986).

started somewhat earlier than 23 Ma ago. Based on an estimated amount of 9 km rapid vertical uplift during 4 Ma (Hurford 1986) a minimum shear strain rate of 8 x 10^{-14} sec^{-1} may be estimated during mylonitization. Later cooling and uplift affects both central Alps (14°C/Ma) and Ivrea zone (7°/Ma, Hurford 1986) and may partly be attributed to isostatic uplift.

The N–S age trend for the onset of rapid cooling (19 Ma vs 23 Ma for the southern steep belt and the northern and central Lepontine areas respectively) supports the idea of a substantial horizontal temperature gradient during backthrusting. This allows mylonitization under greenschist-facies conditions and along the Insubric Line to be contemporaneous with more distributed deformation under amphibolite-facies conditions further to the N.

The age differences along an E–W traverse N of the Insubric Line are much more pronounced. Cooling ages for most radiometric systems systematically decrease from the Bergell area in the E to the Verampio–Simplon area in the W (Fig. 3). From these data, combined with the pattern of mineral isograds which excludes simple tilting around a N–S axis it is clear that the Lepontine region cannot have been uplifted as a rigid block without contemporaneous internal ductile deformation (Merle *et al.* this volume).

This cooling pattern can be interpreted in the light of the earlier discussed concept of continuous dextral transpression with dominant backthrusting NW of the Canavese Line followed by dominant strike slip along and N of the Tonale Line. Thereby the Bergell region would have suffered its very rapid uplift at an early stage while positioned somewhere around Locarno whereas the Verampio area was uplifted while the Bergell area escaped sideways by dextral strike-slip. In fact, the displacement of the roof of the Bergell intrusion in relation to the 'Gonfolite lombarda' containing Bergell boulders (Gunzenhauser 1985) suggests dextral strike slip by some 60 km (Fumasoli 1974, Heitzmann 1987b) and is compatible with such an idea. Fission track data on these boulders (Fig. 3) suggest very rapid cooling and erosion (Wagner *et al.* 1977). The deposition of the Bergell boulders (latest Oligocene) predates most of the subsequent strike-slip movement. Later uplift of the area W of the Bergell intrusion was responsible for a considerable tilt of this intrusion around a N–S oriented axis resulting in a well documented contact aureole E of the Bergell batholith, while no such aureole can be observed in the western parts of the same intrusion (Trommsdorff & Nievergelt 1983). Based on hornblende geobarometry, Reusser (1987) determined a pressure difference of 2.5 kbar between eastern and western Bergell at the time of intrusion. This would correspond to a later differential uplift of the western parts of this intrusion in the order of 9 km.

Earlier deformations along the Canavese Line

Two mylonite belts along the southwestern portion of the Canavese Line

SW of the Valle Sesia the character of the Canavese Line and the adjacent zones changes significantly. The region considered is situated at the transition between the Upper Cretaceous high P–low T metamorphism and the Tertiary greenschist-facies metamorphism in the Sesia zone. The Ivrea zone consists of Palaeozoic granulite-facies rocks with a border of (Permian?) metatonalites and metagranites towards the Canavese Line (see Figs 4 and 5). Both Ivrea and Sesia zone show Oligocene magmatism. The Sesia zone was intruded by the Biella syenite. Along the Canavese Line remnants of related andesites and tuffites are found in a steeply dipping position. Boulders of Upper Cretaceous high-pressure rocks, found within the tuffite, indicate that this volcano–sedimentary series represents the cover of the Sesia zone (Bianchi & Dal Piaz 1963). Its Oligocene age has been confirmed by K–Ar age dating (Scheuring *et al.* 1974). This cover is now at the same topographic level as the intrusives. In the Ivrea zone an Oligocene tonalite was found near Miagliano (Carraro & Ferrara 1968) and mafic dykes of inferred Tertiary age are observed in the granitoid rim of the Ivrea zone in the Valle Sessera region.

In the region between the Valle Sessera and the Valle Sesia (Figs 2 and 4) the Canavese Line swings into a SSW–NNE strike. At the same time the inclination increases to about 70° to the WNW. Further to the SSW, a steep ESE dip is observed, possibly as a consequence of interference with the younger Cremosina Line (Fig. 1). Two mylonite belts can be distinguished. The earlier mylonitization affects the tonalitic rim of the Ivrea tectonic unit. This earlier mylonite belt (belt 1) also contains tectonically emplaced slivers of Canavese sediments. Based on sense of shear criteria this mylonite belt 1 indicates a relative uplift of the Ivrea zone combined with a dextral component of shear (Fig. 4). Mafic dykes of inferred Oligocene age

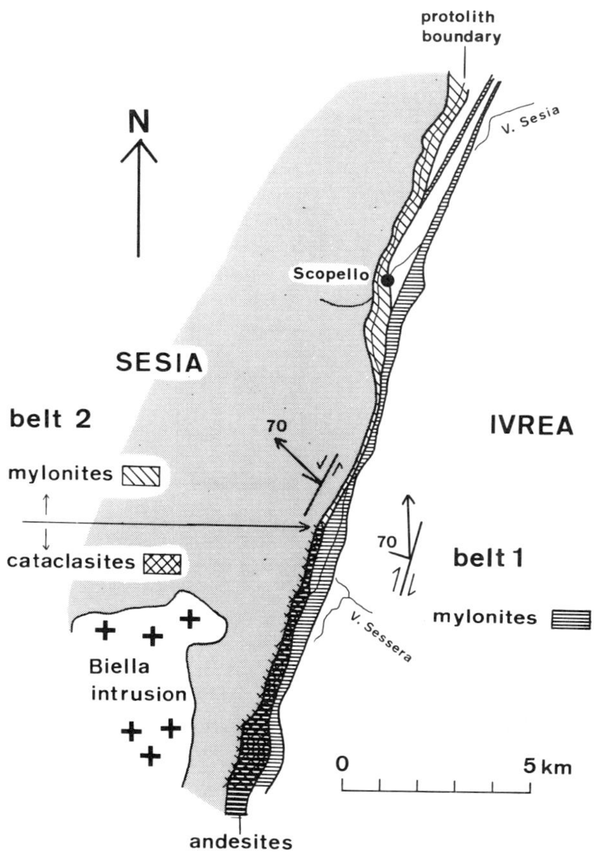

FIG. 4. Sketch map of the southwestern part of the Canavese line

intrude the western margin of the Ivrea zone and mylonite belt 1 and they are clearly seen to postdate mylonitization.

Granitic rocks (Sesia zone) and the above mentioned Oligocene andesite bearing formation are affected by later mylonitization and/or cataclasis within mylonite belt 2. It is only this post-Oligocene mylonite belt 2 which represents the continuation of post-late Oligocene backthrusting movements along the Insubric Line, as discussed in the previous chapter. Sense of shear criteria within mylonite belt 2 indicate a relative uplift of the Sesia zone in respect to the Ivrea zone, combined with a minor sinistral strike-slip component. South of the Sesia valley the tectonites of the post-late Oligocene mylonite belt 2 pass through the transition from crystal plasticity into cataclasis. This indicates that shallower levels of the post-Oligocene Periadriatic Line are exposed in this region and that the amount of post-Oligocene backthrusting and uplift is minor. This view is supported by the cooling path deduced by Hurford & Hunziker (1985).

Palaeomagnetic evidence for a large scale rotation of Ivrea and Sesia units near the Canavese Line

The remanent magnetism of 26 specimens taken from 4 mafic dykes intruding the mylonitized rim of the Ivrea zone (belt 1) was measured. The results of these investigations are plotted in Fig. 6 together with other data taken from the literature. A 60° rotation of the palaeopoles around a horizontal axis with a strike parallel to the Canavese Line (N20E) in an anti-clockwise sense viewed to the north brings these poles into a position near to that of stable Europe in the Oligocene. A similar rotation has to be applied to the data obtained from the Oligocene intrusion of Traversella, the mafic dykes found in the internal part of the Sesia zone and the andesites of the Sesia cover, whereas the mafic

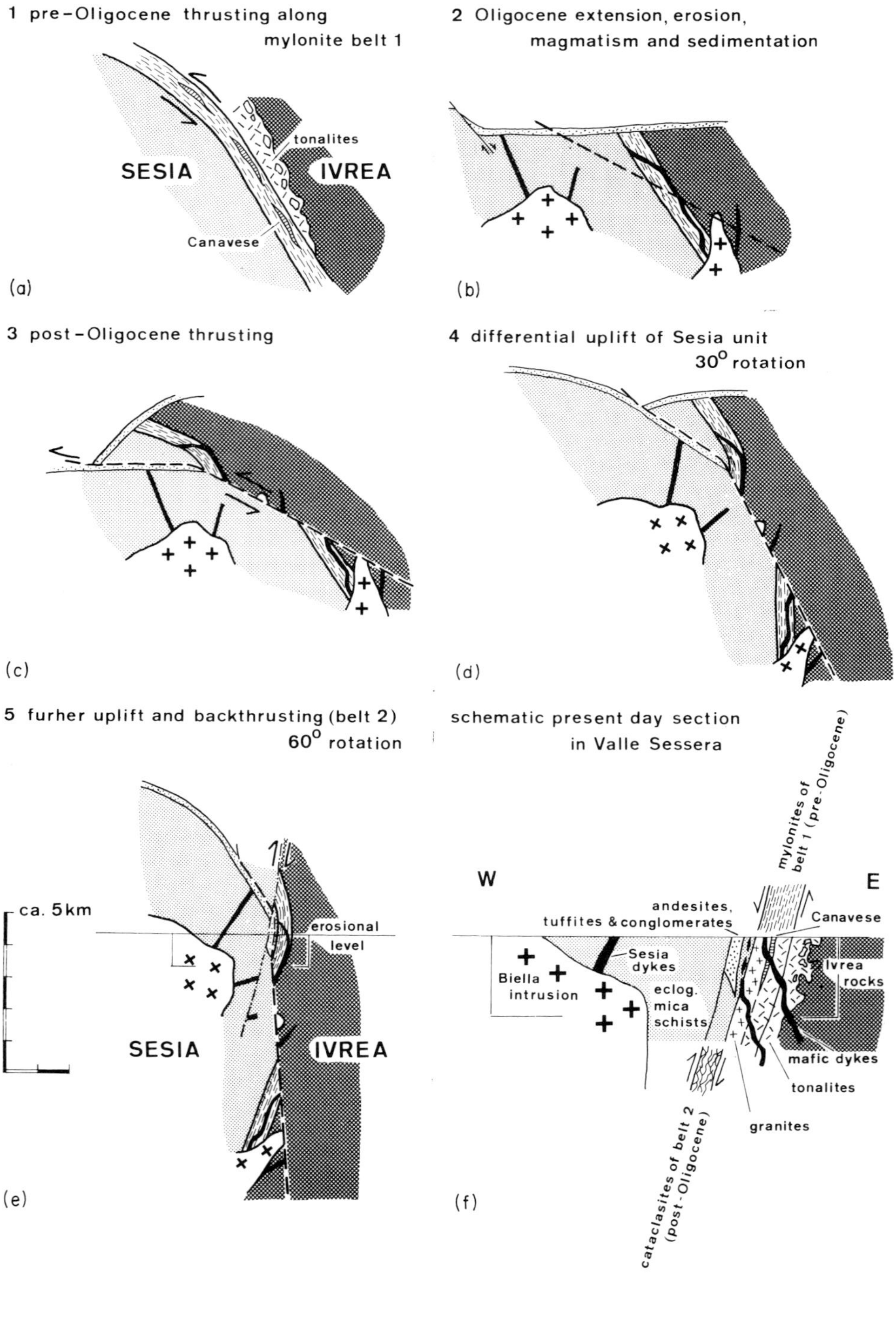

FIG. 5. A possible kinematic reconstruction (a–e) for the profile across the Canavese Line in the Valle Sessera (f). For further explanations see text.

dykes of more external parts of the Sesia zone require a smaller amount of rotation (Lauza, 1977). After this backrotation the palaeopoles cluster with other measurements from the Piemonte and the central Alps.

These findings very strongly support the idea of post-Oligocene verticalization of the Ivrea zone during the Insubric backfolding event, so far based on rather circumstantial evidence (Handy 1987, Schmid *et al.* 1987). Since the mafic dykes intrude mylonite belt 1, a backrotation has to be applied to this mylonite belt as well in order to restore the pre-Oligocene situation depicted in Fig. 5.

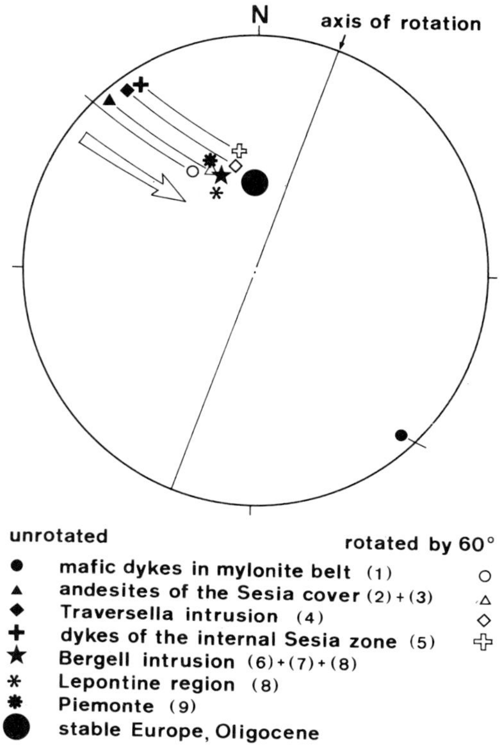

FIG. 6. Summary of a palaeomagnetic data suggesting a 60° rotation around a horizontal pole positioned N20°E (lower hemisphere). References: (1) new data, (2) Lanza (1979), (3) Heller & Schmid (1974), (4) Lanza (1984), (5) Lanza (1977), (6) Heller (1971), (7) Heller (1973), (8) Heller (1980), (9) Van den berg (1979).

A kinematic model for the southwestern part of the Canavese Line

Figure 5 (f) shows the present-day configuration. Although the kinematic evolution is not completely constrained, the following episodes can be reconstructed combining palaeomagnetic and structural data: pre-Oligocene dextral strike-slip combined with thrusting of Ivrea rocks in a northwesterly direction lead to the formation of mylonite belt 1 (Fig. 5a). This mylonite belt contains slivers of Canavese sediments and mainly affects the (Permian?) tonalitic rim of the Ivrea zone. Both Canavese sediments and tonalites are interpreted to represent shallow crustal levels of the southern Alps, strongly tectonized and overridden by the rest of the Ivrea zone during thrusting and dextral strike slip (Figs 4, 5a). Both subduction and exhumation of the high pressure Sesia rocks to the W of mylonite belt 1 occurred in Cretaceous times (Hurford & Hunziker, 1985, report mica ages suggesting 300°C at 60 Ma ago). Mylonite belt 1 cannot be held responsible for subduction (lack of high pressure overprint) nor exhumation (opposite sense of shear in Fig. 5a) of the Sesia high pressure rocks. Thus, movements along mylonite belt 1 are interpreted to postdate exhumation. Tentatively we relate these movements to the Mesoalpine collision event of the western Alps, but the regional significance of these movements remains enigmatic.

Oligocene magmatism is very widespread all along the Periadriatic Line and Laubscher (1983b) relates these intrusions to a scenario of extension affecting a large region all around the Alps. Although direct evidence for such an extensional event has not yet been found along this lineament, this scenario could well explain rapid erosion and redeposition (Fig. 5b) of huge Sesia boulders well exposed NW of Biella near Sordevolo (Bianchi & Dal Piaz 1963).

In order to explain the sandwiching of Oligocene andesites between the erosional surface of the Sesia zone and mylonite belt 1 a second post-Oligocene thrusting event is required (Fig. 5c). Subsequently, the Insubric phase produces large scale rotation of both the internal Sesia zone and the western rim of the Ivrea zone (Fig. 5d, e). The component of backthrusting must have been minor. This explains the above mentioned transition into cataclastic deformation along mylonite belt 2.

Although the local geology is not yet fully understood the following conclusions can be drawn from the observations in the Sessera area:

(1) Pre-Oligocene deformations are documented in mylonite belt 1.
(2) There is a lateral transition from crystal plasticity into cataclastic deformation along belt 2 in post-Oligocene times, i.e. during the Insubric phase.

(3) Palaeomagnetic data indicate post-Oligocene rotation and verticalization of the Ivrea zone and the internal part of the Sesia zone.

In the next section a possible model will be discussed which helps to explain this rapid change in amount and direction of shearing along the Insubric Line.

A model of indentation and sideways escape of the Lepontine region during the Insubric phase

In an attempt to understand the complexity of late to post-Oligocene movements along the Insubric Line and the associated deformation within central and southern Alps the role of the Ivrea zone (and its extension to depth, the Ivrea geophysical anomaly) as a rigid buttress with restricted internal deformation has to be emphasized. Schmid *et al.* (1987) interpret the vertical uplift of this slice of deep continental crust and upper mantle to result essentially from crustal thinning culminating in the early Jurassic (possibly starting in the late Palaeozoic, Handy 1986 1987, Brodie & Rutter 1987) combined with Cretaceous subduction. Radiometric data from the Ivrea zone (McDowell & Schmid 1968, Hunziker 1974) indicate temperatures at around 300°C 180 Ma ago. Hence mantle material whose rheology is dominated by olivine, and in addition, deeper crustal levels dominated by feldspar rheology, were present in shallower crustal levels and before the onset of orogenesis. This had severe consequences for later compressional movements in that the rheology of the Ivrea rocks must have been dominated by the flow properties of olivine and feldspar. Both these minerals are significantly harder to deform at shallow depth than quartz, the mineral which will determine the rheology of upper crustal levels under normal circumstances (see Ranalli & Murphy 1987 for a compilation of depth vs strength curves).

The mylonites of the Insubric Line closely follow the pre-existing northwestern margin of the Ivrea zone. Their localization is the product of a pre-existing configuration created by Mesozoic rifting (Fig. 7), i.e. a NNE–SSW oriented axis of spreading offset by E–W running transfer faults (Weissert & Bernoulli 1985) modified by Cretaceous subduction and underplating of the Sesia zone (Schmid *et al.* 1987). We interpret the relatively sharp bend of the Canavese Line into the E–W strike as it merges with the Tonale Line to be predetermined by a former transfer fault. No deep crustal or upper mantle rocks are preserved along the continuation of this passive continental margin into the Lower Austro–Alpine units of eastern Switzerland. The exceptionally good preservation of a passive continental margin in the southern Alps, including lower crust and upper mantle is attributed to another anomaly, namely the subduction and exhumation of continental crust in the form of the Sesia zone. The quartz-bearing Sesia zone acted as a buffer inhibiting the deformation and subduction of the deeper parts of this continental margin at its western edge during collision.

Dextral transpression during the Insubric phase led to highly variable amounts of uplift related to a very complex displacement path along the curved Insubric Line (Fig. 7) and in front of a rigid indenter in the form of the Ivrea zone which is bounded by a transfer fault in the north. The lateral escape of the central Alps occurred almost entirely in an eastern direction because, (i) the southwestern Canavese Line (mylonite belt 2) acommodates only minor amounts of backthrusting and sinistral strike-slip, (ii) a substantial component of dextral strike-slip is recorded along the Canavese Line northeast of Valle Sesia, i.e. in a region where this line strikes NE–SW and perpendicular to the inferred direction of indentation (Fig. 7), and (iii) further to the E and S of the Bergell intrusion the component of backthrusting rapidly decreases in favour of dextral strike-slip which clearly outlasts backthrusting (Heitzmann 1987b).

The complex cooling history of the Lepontine region with the earlier mentioned migration of late Tertiary cooling and uplift from E to W (Fig. 3) can now be readily explained by this model which suggests a point of indentation somewhere at the northwestern edge of the Ivrea body. Parts of the central Alps which were previously uplifted and backthrusted at this northwestern edge of the Ivrea zone would successively move to the E. Extension across the Simplon Line (Mancktelow 1985) postdates most of the backthrusting and backfolding. These movements along the Simplon Line indicate a substantial tangential stretch in front of the indenter during the closing stages of dextral transpression. For a more detailed discussion of the internal deformation within the Lepontine dome the reader is referred to Merle *et al.* (this volume).

Viewed in cross section (Fig. 8) the modification of the Insubric movements to the previously existing Alpine edifice are impressive.

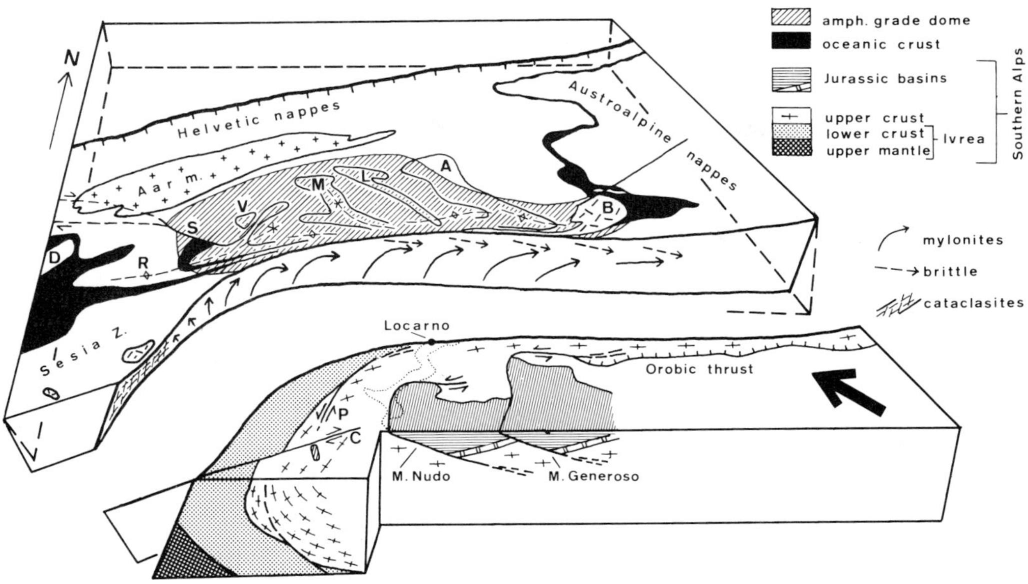

FIG. 7. Three-dimensional sketch illustrating the model of asymmetric escape of the Lepontine region as a result of indentation of the passive continental margin in the Southern Alps (Ivrea geophysical body in particular). Arrows indicate the displacemant path of the central Alps in relation to the southern Alps as inferred from the analysis if stretching lineations (Insubric mylonites) and slickensides (brittle overprint of the Insubric line by the Tonale fault).

Backthrusting and backfolding in the south are largely contemporaneous with the age of major nappe transport and folding in the Helvetic nappes as given by Trümpy (1969) and Pfiffner (1986). Pfiffner (1985) pointed out that some 15–30 km displacement are still required as far back as point X in Fig. 8(b) along the Garvera thrust, which accomodates part of the 50 km displacement along the Glarus thrust. Two solutions are possible for the continuation of this thrust: it either runs down into the Moho as a ductile shear zone or it has to be kinematically connected with backthrusting along the Insubric Line. We prefer the second option which would imply that the central Alps form an asymmetric wedge or orogenic lid (Laubscher 1983a) with complete decoupling between lower and upper crust. The asymmetry of the shape of this wedge, i.e. low angle thrusts in the Helvetics and steeply inclined backthrusting along the Insubric Line is also reflected in a similar but less pronounced asymmetry of the northern and southern slope of the Moho (Figs 1c, 8b). It thus appears that the present day topography of the Moho is closely related to and caused by post-collisional shortening during the Insubric phase. The question as to whether post-collisional subduction during the Insubric phase and below the central Alps is needed (as depicted in the model of Laubscher 1984) or not cannot be decided before a serious attempt in volume balancing is made under consideration of the amount of syntectonic erosion.

Whereas Laubscher envisaged a similar geometry during the Eocene collision (fig. 16b in Laubscher 1983a) we prefer a model of S-sloping subduction during the Eo- and Mesoalpine stages. We propose that the Insubric backthrust merely displaces the suture between Austro–Alpine and Pennine units and that the same suture extends further to the S and under the southern Alps. This reasoning can be supported by the following arguments:

(1) The density inversion at the base of the Ivrea geophysical body (Fig. 8a) is readily explained by Eoalpine subduction of an originally much larger volume of continental crust as compared to the present-day Sesia zone (Schmid *et al.* 1987).

(2) The palaeomagnetic results presented earlier give firm support for late or post-Oligocene verticalization of the Ivrea zone. On the other hand most of the cover of the southern Alps further to the S and E remained in a subhorizontal attitude. This

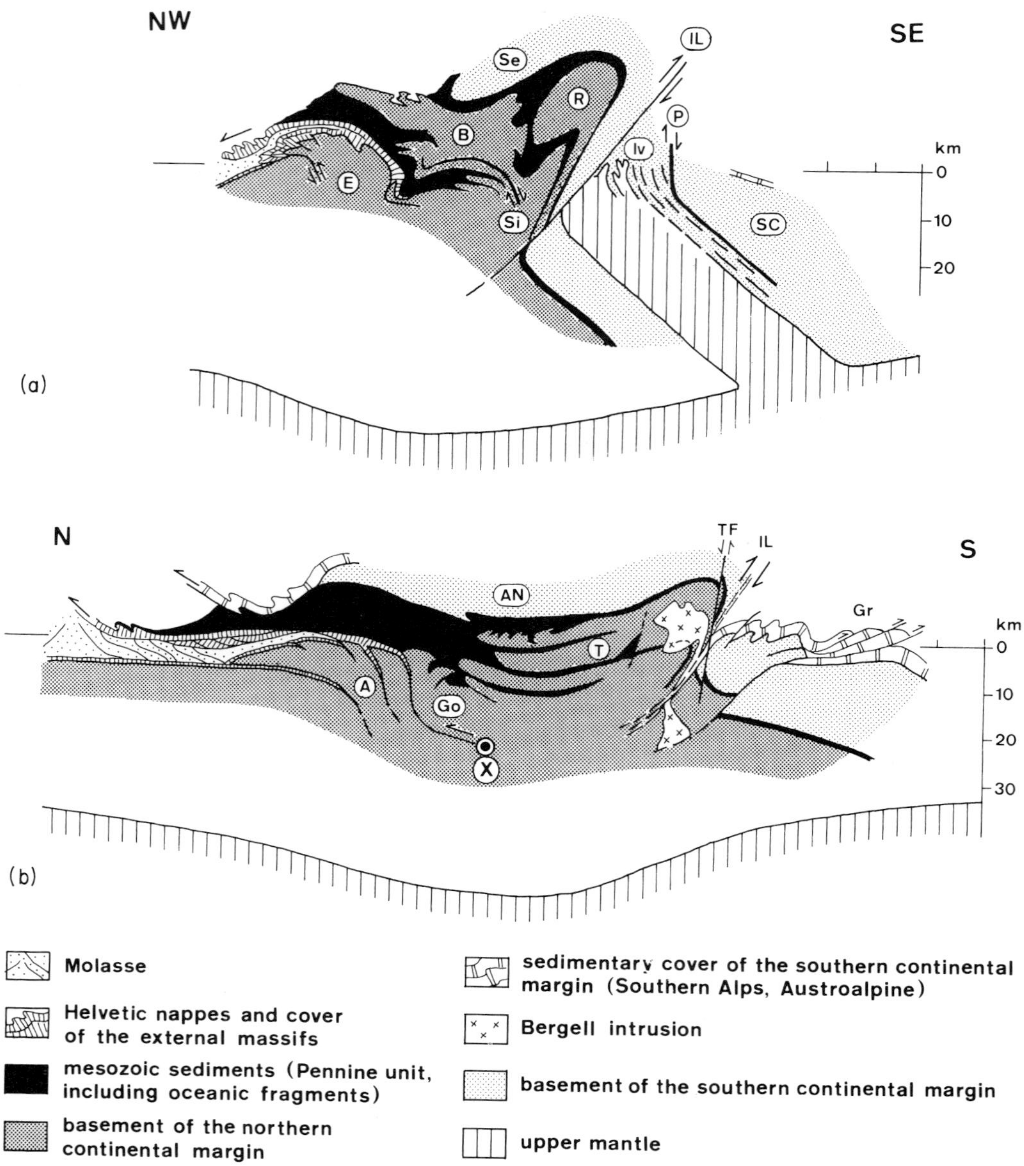

FIG. 8. Two scaled profile sketches through the Alps (profile trace indicated in Fig. 1a). (a) Profile through the Simplon area (after Kissling 1980, Schmid *et al.* 1987, Escher *et al.* 1988). E–external massifs; B–Bernhard nappe; R–M Rosa nappe; Se–Sesia–Dent Blanche nappe; Iv–Ivrea zone; SC–Strona Ceneri zone; Si–Simplon Line; IL–Insubric Line; P–Pogallo Line. (b) Profile through eastern Switzerland (after Trümpy 1985 and Pfiffner 1985). A–Aar Massif; Go–Gotthard 'Massif'; T–Tambo nappe; AN–austro–Alpine nappes; Gr–Grigna mountains; TF–Tonale fault (brittle); IL–Insubric Line (mylonites); X–see text.

situation calls for a hinge zone somewhere SE of the Periadriatic Line (Fig. 7). Thus, it is tempting to bring the suture back into a moderately S-dipping orientation as shown in Fig. 8(b).

(3) Müller *et al.* (1980) provide evidence for a deep seated lithospheric root S of the Periadriatic Line, and Giese *et al.* (1982) interpret their seismic section through a region further to the E in terms of crustal doubling.

(4) In the absence of substantial vertical displacements across the Periadriatic lineament further to the E (Gailtal Line) a major subdivision into Austro–Alpine nappes and southern Alps becomes rather dubious, at least in profile view (we will discuss possible strike-slip components later).

The Periadriatic Line and related lineaments in the eastern Alps

The following discussion is largely based on reviewing the literature. Considerable uncertainties about the kinematics of movements still exist because of the paucity of modern microstructural work eludicating the sense of shear. The increasing importance of Cretaceous deformations and metamorphism in the eastern Alps is another complicating factor.

The exact nature of interference between Giudicarie and Tonale Lines is still unknown. A simple offset of the Tonale and Pustertal Lines by the Giudicarie Line can be ruled out because of the non-existence of an extension of the Giudicarie Line into the central Alps. It is possible that the eastern parts of the Tonale Line only take up the cataclastic components observed along the Tonale Fault and that the mylonite belt, which takes up most of the 60 km dextral shear continues along the Pejo Line where Andreatta (1948) described greenschist-facies mylonites.

The Giudicarie Line runs immediately to the W and parallel to a palaeogeographic boundary between the Trento swell and the Lombardian basin (Castellarin 1972). This boundary persisted from the early Jurassic into the Eocene and forms the eastern edge of deposition of upper Cretaceous flysch deposits (Bernoulli *et al.* 1981). Furthermore, Brack (1981) showed that N–S compression within the southern Alps west of the Giudicarie Line predates the Eocene Adamello intrusion. Further to the E it appears that this pre-Eocene (Cretaceous?) deformation is transformed into sinistral transpression along the Giudicarie Line. All this suggests that this line may have acted as a transpressive sinistral strike-slip zone in Cretaceous times, although it certainly was also active after the Oligocene (Castellarin & Sartori 1979).

Cretaceous movements along or in the vicinity of the northern segment of the Giudicarie Line are indicated by east directed thrusting of the Oetztal nappe (Schmid & Haas 1987), contemporaneous with Cretaceous amphibolite grade metamorphism (Thöni 1981, 1986). This thermal dome within the southern Oetztal nappe (Fig. 1b) must have rapidly cooled during Cretaceous times (Thöni 1986). Pre-Alpine radiometric ages (Del Moro *et al.* 1982) are found in a region S of the Oetztal nappe but still N of the traditional location of the northern branch of the Guidicarie Line. All this suggests a scenario of backthrusting and westward escape of the Oetztal nappe in Late Cretaceous times, associated with movements along or in the vicinity of the Periadriatic Line (i.e. the northern Giudicarie Line).

Renewed compression during the Insubric phase produced tangential stretching along the Brenner fault (Behrmann 1987, Selverstone 1988) combined with updoming of the Tauern window (Fig. 1a). The localization of tangential stretching and updoming in front of an indenter (southern Alps E of the Giudicarie Line) is very similar to what is observed at the western edge of the Lepontine dome. There, tangential stretching is recorded along the Simplon Line (Mancktelow 1985, Merle *et al.* this volume) and in front of the indenting Ivrea zone (Fig. 1a).

Based on a modern microstructural and textural analysis of the greenschist-facies mylonites along the DAV line, Kleinschrodt (1987a, b) demonstrated sinistral strike-slip movements combined with a minor component of uplift north of the DAV line, responsible for an age jump in terms of biotite ages (Borsi *et al.* 1978). These biotite cooling ages and the deformation of dykes associated with Oligocene intrusives in that region date these movements to be contemporaneous with the Insubric phase. Additionally, Stöckhert (1984) provides evidence that the same lineament was also active during the Cretaceous. Dextral movements along the Pusteria and Gailtal Lines are indicated by a series of large scale Riedel shears: the Mölltal and Lavanttal Lines (Fig. 1a).

Thus, we are faced with evidence for both dextral and sinistral strike-slip movements forming at a very acute angle during post-Oligocene times. Palaeogeographic evidence can substantiate such strike-slip movements. Bechstädt (1978) calls for a very substantial amount of strike-slip between the northern calcareous Alps and the Drauzug based on palaeogeographic arguments. Kleinschrodt (1987a) proposed the DAV Line as a possible site for such sinistral movements. On the other hand there are considerable facies differences across the Gailtal Line between Drauzug and the southern Alps

(Prey, in Oberhauser 1980 and Bechstädt 1978) calling for compression and/or dextral strike-slip.

In a recent attempt to reconstruct the palaeogeography of a very large region including the Transdanubian central ranges (Bakony, Hungary) Kazmer & Kovacs (1985) postulate lateral escape of Drauzug and Bakony unit to the E between the DAV and Raba Lines operating sinistrally and the Gailtal and Balaton Lines operating dextrally (Fig. 1a). Although these authors date this lateral escape of some 450 km to have taken place from middle Eocene to late Oligocene times, the closing stages of this escape may still coincide in time with the Insubric phase of the western Alps. The fact that the Mölltal and Lavanttal Lines dextrally transect a large area to the N of the Gailtal Line and the younger age for transpression in the Karawanks (late Miocene, Laubscher 1983b) may indicate that dextral transpression outlasted this continental escape.

We regard the total amount of dextral strike-slip movement N of the Adriatic sub-plate (300 km) to be rather well constrained by the arguments of Laubscher (1971a). However, recent structural work in the Austro–Alpine nappes (Ratschbacher 1986, Schmid & Haas 1987) calls for a regime of dextral transpression already during Cretaceous orogeny of the eastern Alps. It thus appears that a large portion of the 300 km total strike-slip predates the post-late Oligocene movements along the Periadriatic Line.

In summary, it appears that the response of the eastern Alps to post-Oligocene dextral transpression is of a fundamentally different nature as compared to the western Alps: lateral escape between conjugate strike-slip zones rather than substantial uplift associated with foreland thrusting (Helvetic nappes) and backthrusting (Insubric Line). As shown in Fig. 1c the Moho trough no longer follows the direction of the Periadriatic Line E of the Tonale Line and its topography seems to be unrelated to and unaffected by Giudicarie, Pusteria and Gailtal Lines.

Acknowledgements: This contribution would not have been possible without a lively exchange of ideas with many of our colleagues, particularly G. Bertotti, P. Cobbold, M. Handy, P. Heitzmann, J. Hunziker, R. Kleinschrodt, H. Laubscher, N. Mancktelow, O. Merle, A. G. Milnes, L. Ratschbacher, and V. Trommsdorff. U. Gerber and A. Uhr kindly helped to prepare the figures. H. Stünitz, D. Dietrich, M. Johnson and an anonymous reviewer substantially helped to improve the manuscript. We also acknowledge support by the Schweizerische Nationalfonds (project No. 2. 851–0.80 as well as project No. 4. 903–0.85.20 which is part of the National Research Programme NFP 20).

References

Ahrendt, H. 1980. Die Bedeutung der Insubrischen Linie für den tektonischen Bau der Alpen. *Neues Jahrbuch Geologische Paläontologische Abhandlungen*, **160**, 336–62.

Andreatta, C. 1948. La 'Linea di Peio' nel massiccio dell'Ortler e le sue miloniti. *Acta Geologica Alpina*, **1**, 1–63.

Argand, E. 1910. Sur la racine de la nappe rhetique. *Beiträge Geologische Karte Schweiz NF*, **24**, 17–19.

—— 1916. Sur l'arc des Alpes Occidentales. *Eclogae geologicae Helvetiae*, **14**, 145–91.

Armstrong, R. L., Jäger, E. & Eberhardt, P. 1966. A comparison of K–Ar and Rb–Sr ages on Alpine biotites. *Earth and Planetary Science Letters*, **1**, 13–19.

Bauer, F. K. 1984. Zur Geologie der westlichen Karawanken und zum Verlauf des Periadriatischen Lineamentes. *Jahrbuch Geologische Bundesanstalt*, **127**, 289–97.

—— & Schermann, O. 1984. Das Periadriatische Lineament in den Karawanken. *Jahrbuch Geologische Bundesanstalt*, **127**, 299–305.

Bechstädt, Th. 1978. Faziesanalyse permischer und triadischer Sedimente des Drauzuges als Hinweis auf eine grossräumige Lateralverschiebung innerhalb des Ostalpins. *Jahrbuch Geologische Bundesanstalt*, **121**, 1–122.

Behrmann, J. H. 1987. Large scale subhorizontal extension in a convergent orogen: The Sterzing–Steinach mylonite zone in the eastern Alps. *Terra Cognita*, **7**, 89.

Bernoulli, D., Bichsel, M., Bolli, H. M., Häring, M. O., Hochuli, P. A. & Kleboth, P. 1981. The Missaglia Megabed, a catastrophic deposit in the Upper Cretaceous Bergamo Flysch, northern Italy. *Eclogae geologicae Helvetiae*, **74**, 421–42.

Bertotti, G. 1987. *Relationships between sediments and crystalline basement in the area between Lago di Como and Lago di Lugano*. Abstract 5. réunion du Groupe Tectonique Suisse, Fribourg.

Bianchi, A. & Dal Piaz, G. 1963. Gli inclusi di 'micaschisti eclogitici' della zona Sesia nella formazione porfiritica permiana della zona del Canavese fra Biella ed Oropa; charatteristiche ed eta dei fenomeni metamorfici. *Giornale Geologico Serie 2a*, **31**, 39–76.

Bögel, H. 1975. Zur Literatur über die 'Periadriatische Naht'. *Verhandlungen Geologische Bundesanstalt 1975*, **2–3**, 163–99.

BORSI, S., DEL MORO, A., SASSI, F. P., ZANFERRARI, A. & ZIRPOLI, G. 1978. New geopetrologic and radiometric data on the Alpine history of the Austride continental margin south of the Tauern window. *Memorie Istituto Geologia Mineralogia Universita Padova*, **32**, 1–17.

BRACK, P. 1981. Structures on the southwestern border of the Adamello intrusion (Alpi Bresciane, Italy). *Schweizerische Mineralogische Petrographische Mitteilungen*, **61**, 37–50.

BRODIE, K. H. & RUTTER, E. H. 1987. Deep crustal extensional faulting in the Ivrea zone of northern Italy. *Tectonophysics* **140**, 193–212.

CARRARO, F. & FERRARA, G. 1968. Alpine 'Tonalite' at Magliano, Biella (zona Diorito–Kinzingitica). A preliminary note. *Schweizerische Mineralogische Petrographische Mitteilungen*, **48**, 75–80.

CASTELLARIN, A. 1972. Evoluzione paleotettonica sinsedimentaria del limite tra 'piattaforma veneta' e 'bacino lombardo' a nord di Riva del Garda. *Giornale Geologico, Annale Museo Geologico Bologna 2a*, **38**, 11–171.

—— & SARTORI, R. 1979. Struttura e significato della Linea delle Giudicarie Sud. *Rendiconti Societa Geologica Italiana*, **2**, 29–32.

CORNELIUS, H. P. & CORNELIUS-FURLANI, M. 1931. Die Insubrische Linie vom Tessin bis zum Tonalepass. *Denkschrift Akademie der Wissenschaften Mathematische Klasse*, **102**, 208–301.

DEL MORO, A., SASSI, F. P. & ZIRPOLI, G. 1982. New radiometric data on the Alpine thermal history in the Ötztal–Merano area (eastern Alps). *Memorie Scienze Geologiche Padova*, **35**, 319–25.

DE SITTER, L. U. & DE SITTER-KOOMANS, C. M. 1949. The Geology of the Bergamasc Alps, Lombardia, Italy. *Leidse Geologische Mededelingen*, **148**, 1–257.

DEWEY, J. F. & BIRD, J. N. 1970. Mountain belts and the new global tectonics. *Journal of Geophysical Research*, **75**, 2625–47.

ESCHER, A., MASSON, H. & STECK, A. 1988. Coupes géologiques des Alpes occidentales suisses. *Mémoires de géologie Lausanne*, 1988, 2.

FRANK, W. 1987. Evolution of the Austro–Alpine elements in the Cretaceous. *In*: FLÜGEL, H. W. & FAUPL, P. (eds). *Geodynamics of the eastern Alps*, 379–406. Franz Deuticke, Vienna.

FUMASOLI, M. W. 1974. Geologie des Gebietes nördlich und südlich der Jorio–Tonale-Linie im Westen von Gravedona. *Mitteilungen Geologisches Institut ETH NF 194*.

GIESE, P. REUTTER, K. -J., JACOBSHAGEN, V. & NICOLICH, R. 1982. Explosion seismic crustal studies in the Alpine Mediterranean region and their implications to tectonic processes. *In*: BERCKHEMER, H. & HSÜ, K. J. (eds). *Alpine-Mediterranean Geodynamics: Geodynamical Series American Geophysical Union*, **7**, 39–73.

GUNZENHAUSER, B. A. 1985. Zur Sedimentologie und Paläogeographie der oligo-miocänen Gonfolite Lombarda zwischen Lago Maggiore und der Brianza (Südtessin, Lombardei). *Beiträge Geologische Karte Schweiz NF*, **159**, 1–114.

HANDY, M. 1986. *The Structure and rheological evolution of the Pogallo Fault zone, a deep crustal dislocation in the Southern Alps of northwestern Italy (Prov. Novara)*. Unpublished PhD thesis, University of Basel.

—— 1987. The structure, age and kinematics of the Pogallo Fault zone: southern Alps, northwestern Italy. *Eclogae geologicae Helvetiae*, **80**, 593–632.

HEITZMANN, P. 1987a. Calcite mylonites in the central Alpine 'root zone'. *Tectonophysics*, **135**, 207–15.

—— 1987b. Evidence of late Oligocene/early Miocene backthrusting in the central Alpine 'root zone'. *Geodynamica Acta*, **1**, 183–92.

HELLER, F. 1971. Remanent magnetization of the Bergell granite. *Zeitschrift Geophysik*, **37**, 557–71.

—— 1973. Magnetic anisotropy of granitic rocks of the Bergell massif (Switzerland). *Earth and Planetary Science Letters*, **20**, 180–88.

—— 1980. Paleomagnetic evidence for late Alpine rotation of the Lepontine area. *Eclogae geologicae Helvetiae*, **73**, 607–18.

—— & SCHMID, R. 1974. Paläomagnetische Untersuchungen in der Zone Ivrea–Verbano (Prov. Novara, Norditalien): Vorläufige Ergebnisse. *Schweizerische Mineralogische Petrographische Mitteilungen*, **54**, 229–42.

HUNZIKER, J. C. 1974. Rb–Sr and K–Ar age determination and the Alpine tectonic history of the western Alps. *Memorie Istituto Geologia Mineralogia Universita Padova*, **31**, 1–54.

HURFORD, A. J. 1986. Cooling and uplift patterns in the Lepontine Alps (south-central Switzerland) and an age of vertical movement on the Insubric fault line. *Contributions to Mineralogy and Petrology*, **92**, 413–27.

HURFORD, J. & HUNZIKER, J. C. 1985. Alpine cooling history of the Monte Mucrone eclogites (Sesia–Lanzo zone): fission track evidence. *Schweizerische Mineralogische Petrographische Mitteilungen*, **65**, 325–34.

JÄGER, E., NIGGLI, E. & WENK, E. 1967. Rb–Sr Altersbestimmungen an Glimmern der Zentralalpen. *Beiträge Geologische Karte Schweiz NF*, **134**, 1–67.

KAZMER, M. & KOVACS S. 1985. Permian–Paleogene paleogeography along the eastern part of the Insubric–Periadriatic lineament system: evidence for continental escape of the Bakony–Drauzug unit. *Acta Geologica Hungarica*, **28**, 71–84.

KISSLING, E. 1980. *Krustenaufbau und Isostasie in der Schweiz*. Unpublished PhD thesis, ETH, Zürich.

—— 1984. Three-dimensional gravity model of the northern Ivrea–Verbano zone. *Contributions Geology of Switzerland, Geophysics*, **21**, 53–61.

KLEINSCHRODT, R. 1987a. Quarzgefügeanalyse im Altkristallin südlich des westlichen Tauernfensters (Südtirol/Italien). *Erlanger geologische Abhandlungen*, **114**, 1–82.

—— 1987b. Evolution and kinematic interpretation of alpine fabrics in the Altkristallin south of the western Tauern Window (south Tyrol/Italy). *Terra Cognita*, **7**, 80.

LANZA, R. 1977. Paleomagnetic data from the andesite and lamprophyric dykes of the Sesia–Lanzo zone (western Alps). *Schweizerische Mineralogische Petrographische Mitteilungen*, **57**, 281–90.

—— 1979. Paleomagnetic data on the andesitic cover of the Sesia–Lanzo zone (western Alps). *Geologische Rundschau*, **68**, 83–92.

—— 1984. Paleomagnetism in the Traversella Massif. *Bollettino di Geofisica Teorica ed Applicàta*, **26**, 115–24.

LARDELLI, T. 1981. *Die Tonalelinie im unteren Veltlin*. Juris Druck und Verlag, Zürich.

LAUBSCHER, H. P. 1971a. Das Alpen–Dinariden-Problem und die Palinspastik dėr südlichen Tethys. *Geologische Rundschau*, **60**, 813–33.

—— 1971b. The large-scale kinematics of the western Alps and the northern Apennines and its palinspastic implications. *American Journal of Science*, **271**, 193–226.

—— 1983a. Detachment, shear and compression in the central Alps. *Geological Society of America Memoir*, **158**, 191–211.

—— 1983b. The late Alpine (Periadriatic) intrusions and the Insubric Line. *Memorie Societa Geologica Italiana*, **26**, 21–30.

—— 1984. The tectonic problem of the Ivrea Body and the Insubric Line. *Annales Geophysicae*, **2**, 169–70.

—— 1985. Large-scale, thin-skinned thrusting in the southern Alps: Kinematic models. *Geological Society of America Bulletin*, **96**, 710–18.

MANCKTELOW, N. 1985. The Simplon Line: a major displacement zone in the western Lepontine Alps. *Eclogae geologicae Helvetiae*, **78**, 73–96.

MCDOWELL, F. W. & SCHMID, R. 1968. Potassium–Argon ages from the Valle d'Ossola section of the Ivrea–Verbano zone (northern Italy). *Schweizerische Mineralogische Petrographische Mitteilungen*, **48**, 205–10.

MILNES, A. G. 1974. Structure of the Pennine zone (central Alps): A new working hypothesis. *Geological Society of America Bulletin*, **85**, 1725–32.

MÜLLER, S., ANSORGE, J., EGLOFF, R. & KISSLING E. 1980. A crustal cross section along the Swiss Geotraverse from Rhinegraben to the Po plain. *Eclogae geologicae Helvetiae*, **73**, 463–83.

NIGGLI, E. & ZWART, H. J. 1973. *Metamorphic Map of the Alps, scale 1:1 000 000, sheet 17 of the Metamorphic Map of Europe*. Leiden / UNESCO Paris.

OBERHAUSER, B. 1980. *Der geologische Aufbau Österreichs*. Springer-Verlag, Wien.

PFIFFNER, O. A. 1985. Displacements along thrust faults. *Eclogae geologicae Helvetiae*, **78**, 313–33.

—— 1986. Evolution of the north Alpine foreland basin in the central Alps. *Special Publications of the International Association of Sedimentologists*, **8**, 219–28.

PURDY, J. W. & JÄGER, E. 1976. K–Ar ages on rock forming minerals from the central Alps. *Memorie Istituto Geologia Mineralogia Universita Padova*, **30**, 1–31.

RANALLI, G. & MURPHY, D. C. 1987. Rheological stratification of the lithosphere. *Tectonophysics*, **132**, 281–95.

RATSCHBACHER, L. 1986. Kinematics of Austro–Alpine cover nappes: changing translation path due to transpression. *Tectonophysics*, **125**, 335–56.

REUSSER, E. 1987. *Phasenbeziehungen im Tonalit der Bergeller Intrusion*. Unpublished PhD thesis, ETH, Zürich.

SCHEURING, B., AHRENDT, H., HUNZIKER, J. C. & ZINGG, A. 1974. Paleobotanical and geochronological evidence for the alpine age of the metamorphism in the Sesia zone. *Geologische Rundschau*, **63**, 305–26.

SCHMID, S. M. & HAAS, R. 1987. The transition from near surface thrusting to intra-basement décollement along the Schlinig thrust. *Terra Cognita*, **7**, 68.

——, ZINGG A. & HANDY, M. 1987. The kinematics of movements along the Insubric Line and the emplacement of the Ivrea Body. *Tectonophysics*, **135**, 47–66.

SELVERSTONE, J. 1988. Evidence for east–west crustal extension in the eastern Alps: implications for the unroofing history. *Tectonics*, **7**, 87–106.

STECK, A. 1984. Structures de déformation tertiaires dans les Alpes centrales. *Eclogae geologicae Helvetiae*, **77**, 55–100.

SPITZ, A. 1919. Fragmente zur Tektonik der Westalpen und des Engadins. *Verhandlungen Geologische Reichsanstalt Wien*, **1919**, 104–22.

STÖCKHERT, B. 1984. K–Ar determinations on muscovites and phengites from deformed pegmatites and the minimum age of the Old Alpine deformation in the Austridic basement to the south of the Tauern window (Ahrn valley, southern Tyrol, eastern Alps). *Neues Jahrbuch Mineralogische Abhandlungen*, **150**, 103–20.

THÖNI, M. 1981. Degree and evolution of the Alpine metamorphism in the Austro–Alpine unit W of the Hohe Tauern in the light of K/Ar and Rb/Sr age determinations on micas. *Jahrbuch Geologische Bundesanstalt*, **124/1**, 111–74.

—— 1986. The Rb–Sr thin slab isochron method – an unreliable geochronologic method for dating geologic events in polymetamorphic terrains? *Memorie Scienze Geologiche Padova*, **38**, 283–352.

TROMMSDORFF, V. & NIEVERGELT, P. 1983. The Bregaglia (Bergell) Iorio intrusive and its field relationships. *Memorie Societa Geologica Italiana*, **26**, 55–68.

TRÜMPY, R. 1969. Die helvetischen Decken der Ostschweiz: Versuch einer palinspastischen Korrelation und Ansätze zu einer kinematischen Analyse. *Eclogae geologicae Helvetiae*, **62**, 105–42.

—— 1985. Die Plattentektonik und die Entstehung der Alpen. *Neujahresblatt Naturforschende Gesellschaft Zürich*, **187**, Orell Füssli, Zürich, 1–47.

VAN DEN BERG, J. 1979. Paleomagnetism and the changing configuration of the western Mediterranean area in the Mesozoic and early Cenozoic eras. *Geologica Ultraiectina*, **20**, 1–179.

VECCIA, O. 1968. La zone Cuneo–Ivrea–Locarno, element fondamental des Alpes. Géophysique et Géologie. *Schweizerische Mineralogische Petrographische Mitteilungen*, **48**, 215–26.

VOGLER, W. S. & VOLL, G. 1981. Deformation and metamorphism at the south-margin of the Alps, east of Bellinzona, Switzerland. *Geologische Rundschau*, **70**, 1232–62.

WAGNER, G. A., MILLER, D. S. & JÄGER, E. 1979. Fission track ages on apatite of Bergell rocks from central Alps and Bergell boulders in Oligocene sediments. *Earth and Planetary Science Letters*, **45**, 355–60.

——, REIMER, G. M. & JÄGER, E. 1977. Cooling ages derived by apatite fission track, mica Rb–Sr and K–Ar dating: the uplift and cooling history of the central Alps. *Memorie Istituto Geologia Mineralogia Universita Padova*, **30**, 1–27.

WEISSERT, H. J. & BERNOULLI D. 1985. A transform margin in the Mesozoic Tethys: evidence from the Swiss Alps. *Geologische Rundschau*, **74**, 665–79.

WERNER, D. 1980. Probleme der Geothermik im Bereich der Schweizer Zentralalpen. *Eclogae geologicae Helvetiae*, **72**, 263–70.

WIEDENBECK, M. 1986. Structural and isotopic age profile across the Insubric Line, Mello, Valtellina, N Italy. *Schweizerische Mineralogische Petrographische Mitteilungen*, **66**, 211–27.

ZINGG, A., HUNZIKER, J., FREY, M. & AHRENDT, H. 1976. Age and degree of metamorphism of the Canavese zone and of the sedimentary cover of the Sesia zone. *Schweizerische Mineralogische Petrographische Mitteilungen*, **56**, 361–75.

S. M. SCHMID & H. R. AEBLI, Geologisches Institut, ETH-Zentrum, 8092 Zürich, Switzerland.

F. HELLER, Institut für geophysik, ETH-Hönggerberg, 8093 Zürich, Switzerland.

A. ZINGG, Geologisch–Paläontologisches Institut, Bernoullistrasse 32, 4056 Basel, Switzerland.

Structural and metamorphic constraints on the tectonic evolution of the upper Pennine Alps

A. C. Ellis, A. C. Barnicoat & N. Fry

SUMMARY: A series of thrust and shear zone structures in the upper Pennine Alps dissect the area into a number of horsts and duplexes. The diversity of the mineral stretching lineation directions when combined with shear-sense indicators reveals substantial variation in the movement direction both between and within different units, and in some cases at different times within the same unit. The present-day geometry, illustrated in a new block diagram, is controlled largely by backthrusts. These backthrusts have overprinted, and in some cases reactivated, earlier, more ductile forethrusts and cut out all but one of the regional megascopic folds. While this backthrusting renders structural analysis of the early history of the area intractable, study of the petrogenesis of early, high-pressure assemblages yields some data. Deformation during subduction, leading to the subcretion of rock packets onto the over-riding plate is recorded by a 'hairpin' in the P–T paths of several units, and by a few relic structures within ophiolite units. The forms of P–T paths for units which have undergone high pressure metamorphism indicates that they underwent similar histories. Geochronological data for this area are as yet of insufficient resolution to decide if the tectonometamorphic events were continuous or episodic.

The upper Pennine Alps have long been considered one of the classic examples of the internal (high-grade) parts of an orogenic belt. Recent advances in understanding of both structural geology and metamorphic petrology have enabled the development of new ideas on the evolution of this part of the Alps. In this paper, inferences based on the metamorphic petrology, which largely reflect early events, have been combined with a structural analysis to give an integrated model for the region centered on Monte Rosa, as shown in Fig. 1.

Geological setting

Various aspects of the geology of the region considered here have been previously described by Bearth in a long series of meticulous papers (e.g. 1952, 1953, 1957, 1964, 1967, 1973, 1976, Bearth & Schwander, 1981) and also by Dal Piaz (1965, 1974), Dal Piaz, Lombardo & Gosso (1983), Dal Piaz & Lombardo (1986), Frey *et al.* (1976), Dal Piaz & Ernst (1978), Klein (1978), Milnes *et al.* (1981), Müller (1983), Chopin & Monié (1984), Monié (1985), Barnicoat & Fry (1986) and Lacassin (1987) together with other references cited in the text. A brief description of the structural sequence from bottom to top is given here. The area comprises a tectonic edifice of schists, gneisses and granites representing 'European' basement, with some sedimentary cover. This basement-cover complex is overlain by an ophiolitic sequence. Above this occurs a varied suite of metasediments. This whole assemblage of thrust slices is in turn overlain by more basement rocks (Sesia–Dent Blanche), representing 'Apulian' basement, which are largely outside the scope of this paper. Both the 'European' and 'Apulian' basement are dominantly Hercynian in age, their cover largely Triassic and Jurassic. The ophiolitic material is considered Jurassic in age. Much of the material was metamorphosed under high-pressure conditions during the Cretaceous, while the Apulian plate was still between 900 and 500 km from its current location (Fig. 2). In many areas the Eoalpine high-pressure event has been partially or completely overprinted by later, lower pressure assemblages.

Regional structure – the traditional view

Since the work of Argand (1911), the internal Alps have been subdivided into a series of what he called nappes, considered to be recumbent antiforms overlying zones of intense shear strain across which they were effectively detached from underlying units. The essential point about a nappe was that it was the fundamental unit which, although it could be internally subdivided, behaved structurally as a more or less coherent body of rock during orogenesis. Argand argued that the Monte Rosa unit

From: Coward, M. P., Dietrich, D. & Park, R. G. (eds), 1989, *Alpine Tectonics*, Geological Society Special Publication No. 45, pp. 173–188.

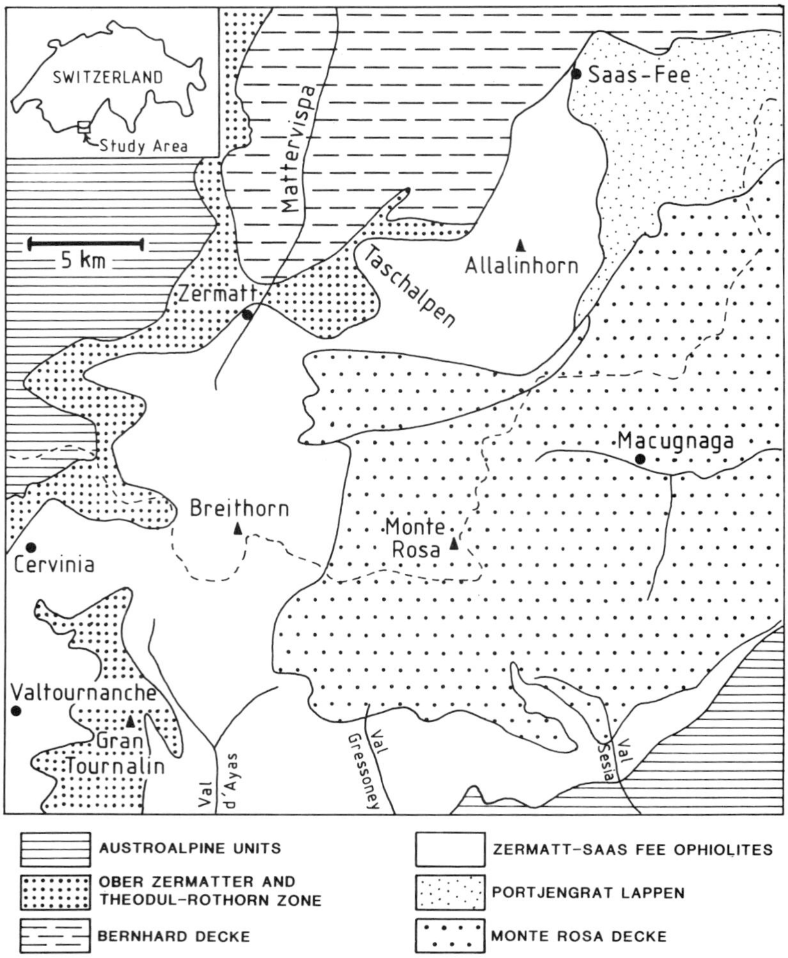

FIG. 1. Map of the study area, with traditionally recognized units. Modified from Bearth (1964).

showed true nappe morphology, possessing both right way up and inverted limbs, the Mesozoic calcschists and greenstones overlying it being considered an autochthonous or parautochthonous cover sequence.

Since that early description, work by many people has shown the structural picture to be rather different. Staub (1942) and Güller (1947) erected four subdivisions separated by thrust surfaces within the Mesozoic rocks around Zermatt and structurally separated the main body of the ophiolites from the overlying metasediments. Bearth (1967) called the ophiolites of this region the Zermatt–Saas Fee ophiolites. Dal Piaz (1965) and Dal Piaz & Ernst (1978) extended the concept of an ophiolite nappe onto the southern side of the Swiss–Italian border, correlating the Zona Corno del Camoscio–Breithorn there with the Zermatt–Saas ophiolites and arguing that the same divisions of the sedimentary cover could be recognized as in the Zermatt area. Thus the Pillonet–Becca di Nana and the Gressoney–Gran Tournalin units were correlated with the Obere Zermatter Schuppenzone and the Theodul–Rothorn zones of Bearth (1964) respectively. Wilson (1978) and Mazurek (1985) described the internal thrusts within the Mesozoic metasedimentary cover in more detail, highlighting the importance of early thrusts refolded during the later backfolding phase. Dal Piaz (1965) had earlier pointed out that all of these Mesozoic cover rocks were allochthonous with respect to the Monte Rosa basement. Dal Piaz & Ernst (1978), and Gosso *et al.* (1979) described the relationships between

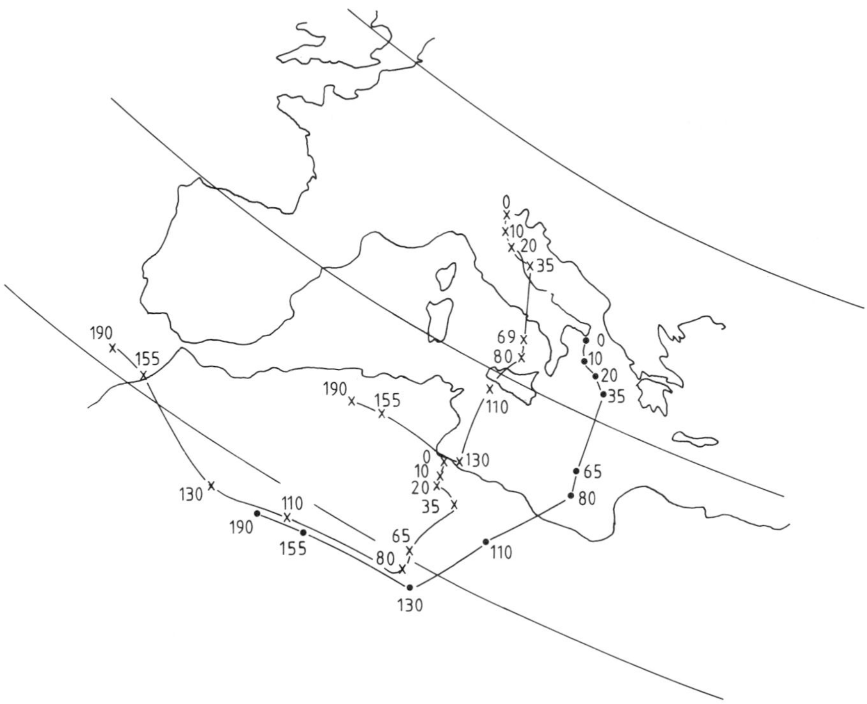

FIG. 2. Map showing the motion of parts of Apulia and Africa relative to a fixed Europe. Lines of latitude, drawn at 190 Ma, are at 10° intervals. Figures on movement paths are dates in Ma. Note that Apulia underwent a clockwise rotation of about 30° between 130 and 80 Ma. Prior and subsequent to this Apulia has moved with Africa. After Dercourt *et al.* (1986).

the basement gneisses of the Monte Rosa, the ophiolitic units and the Mesozoic metasedimentary cover and demonstrated that these are separated by late thrust contacts. Dal Piaz (1965) also noted the presence of thrust contacts internal to the ophiolites. Bearth (1952) described zones of intense schistosity associated with slices of Mesozoic rock types, the Furgg and Stelli zones, cutting across Argand's Monte Rosa nappe. These were further investigated by Wetzel (1972), but the importance and significance of these zones of intense shear remained enigmatic.

Sartori (1987) has remapped and reinterpreted the geology of the area around Zermatt and dismisses late thrusting as being relatively unimportant, contrary to the findings of Bearth (1953, 1964), Güller (1947), Staub (1942), Dal Piaz (1965), Kienast (1973) and this work.

Milnes *et al.* (1981) produced a model for the evolution of this part of the upper Pennine Alps, in which folds dominated the geometry of units and their structural interrelationships. At several localities throughout the present study area megascopic folds may be seen, and their presence can be inferred on the basis of detailed mapping at many others. However, all of these folds, with the exception of the Vanzone antiform and its parasitic folds, are bounded by discrete shear zone and thrust surfaces.

Early evolution, deduced from metamorphic history

Several recent studies have used metamorphic petrology to construct P-T paths for various portions of the upper Pennine and Austro–Alpine Alps. Of particular relevance here are the works of Frey *et al.* (1976), Chopin & Monie (1984), Monie (1985) and Dal Piaz & Lombardo (1986) on the rocks of the Monte Rosa area, Meyer (1983), Barnicoat & Fry (1986) and Fry & Barnicoat (1987) on the ophiolites and Koons (1982), Lardeaux *et al.* (1982), Ungaretti *et al.* (1983), Rubie (1984) and Oberhänsli *et al.* (1985) on rocks from the Sesia zone. The peak metamorphic conditions for all of these units are compatible with their generation within a thermally mature subduction zone (see Fry & Barnicoat 1987 for a discussion of this). Analyses of the available plate movement and palaeomagnetic data (Savostin *et al.* 1986, Westphal *et al.* 1986, Dercourt *et al.* 1986) show that the movement of Apulia and hence also the rate of subduction, were slow during the Cretaceous: this is compatible with the comparatively high temperatures recorded by the high-pressure metamorphism. Their analyses of the closure of Tethys do not allow a lengthy period of subduction prior to 130 Ma and it may be that the thermal maturity inferred from the metamorphic data was apparent rather than real. The evidence that subcretion or underplating of successively lower units took place while subduction continued is consistent with the very limited geochronological database for peak metamorphic assemblages (summarized in Fry & Barnicoat 1987), as is the lack of any detectable heating post-peak pressure (Rubie 1984). The arrival in the subduction zone of the low density European basement, now seen in the Monte Rosa area, for example, appears to have caused subduction to cease. Structures relating to the juxtaposition of ophiolitic units can be seen (Fry & Barnicoat 1987); the age of the overall infrastructure in the area is more equivocal (see below). Because of the complex and differing structural histories of the various units, differences in later stages of the estimated P-T paths are to be expected. Additionally, even the relative chronology attributable to later parageneses is poorly defined, so structural data have been used to constrain this part of the geological history.

Later evolution, deduced from structural analysis

Regional geometry

In order to study the regional structure of the area, several traverses were made. These involved mapping strips of ground, usually at a scale of 1:25 000 (ACE). Figure 4 presents the results of this survey. The cross sections drawn on the basis of this mapping (Fig. 3) were used to construct a block diagram (Fig. 5). This shows how the individual units making up the structural pile are slabs and lenses of rock, bounded by approximately planar thrust surfaces. In this treatment of the internal Alps, units are named according to the principal thrust which carries them, following the suggestion of Elliot & Johnson (1980). For example, the Mischabel thrust carries the Mischabel unit and the Portjengrat thrust carries the Portjengrat unit.

Types of principal thrust and tectonic units

As defined in this paper, a principal thrust must fulfill one of two criteria. It is either out-of-sequence to structures in its footwall (i.e. oversteps its footwall; Butler 1987) and persists for more than a few kilometres on a map, or it is a thrust surface across which an abrupt change in penetrative linear fabric occurs. The tectonic units defined on the basis of such a framework of principal thrusts are shown in Figs 3 and 5. Principal thrusts between units may separate units of similar or differing lithology. A given principal thrust may separate similar lithologies at one point on its surface and completely different lithologies elsewhere. Thus a unit may change in lithological content and metamorphic grade along its outcrop, and correlations based on such features may be erroneous. Comparing Figs 1 and 4 it is clear that the thrusts separating units correspond, at least locally, to previously defined unit boundaries. Table 1 summarizes the relationships between our structurally defined units and the units used by previous workers.

Transport directions

The relationship between mineral stretching lineations and transport direction has been extensively debated. In general the stretching lineation will develop parallel to the overshearing direction at the time of its development.

FIG. 3. (a) Cross-section A–A[1] (Fig. 4) of the upper part of the Macugnaga unit and the lower part of the Monte Rosa unit, separated by the Monte Rosa thrust. Note the NE-facing folds in the Macugnaga unit. (b) Cross-section B–B[1] (Fig. 4) showing, from top to bottom, the Bettaforca, Breithorn and Monte Rosa units, together with a small part of the Macugnaga unit. All are separated by thrusts, indicated with heavy lines. (c) Cross-section C–C[1] (Fig. 4) through parts of the Theodul–Mischabel, Gran Tournalin, Ayas and Breithorn units and their respective principal thrusts. The section indicates how late thrusts are important internally, as well as at the boundaries of backthrust units. It also indicates clearly the presence of lateral ramps used to constrain the thrust movement direction in the Gran Tournalin unit.

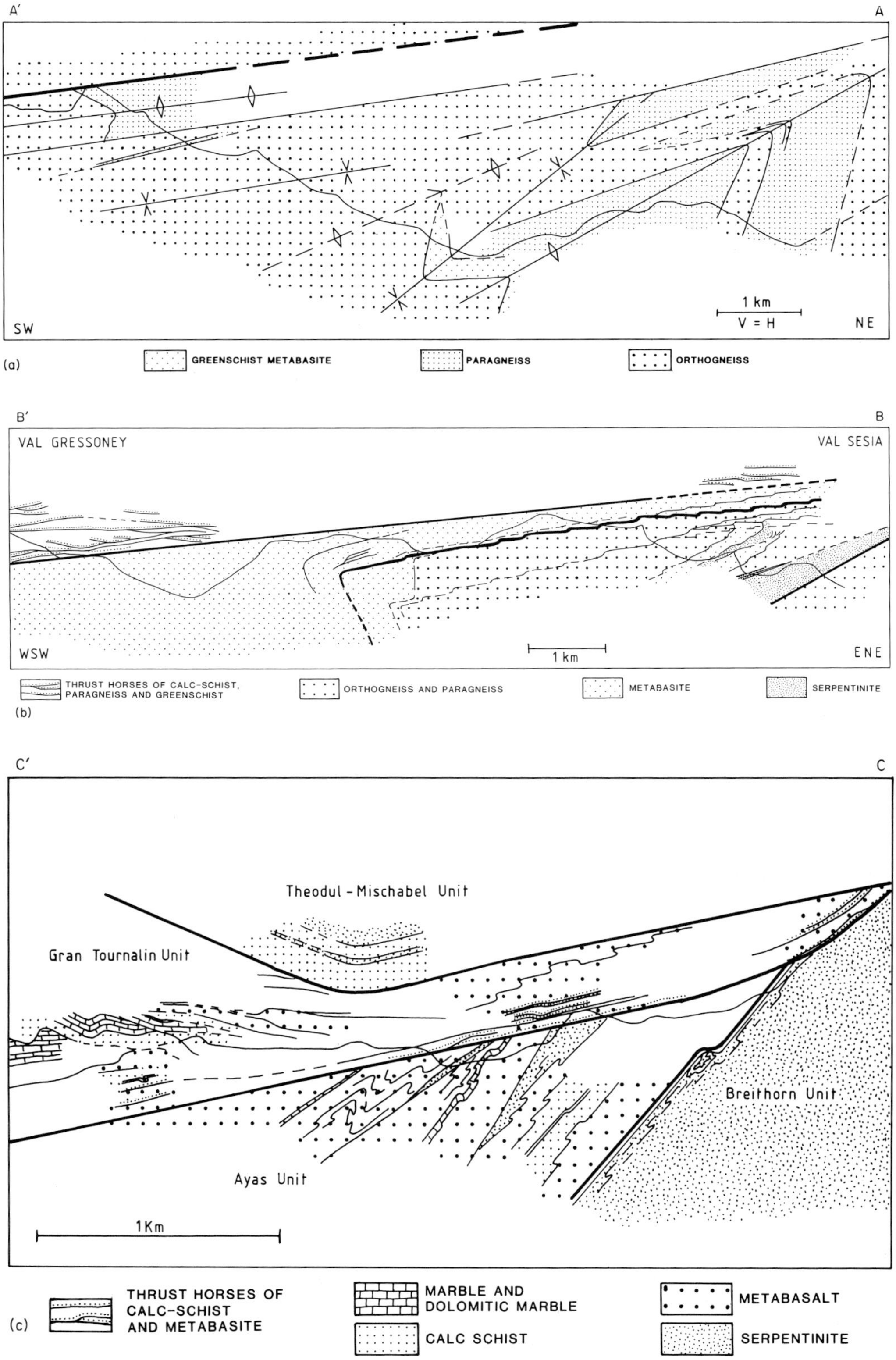
A′
A
SW
NE
1 km
V = H
(a)
GREENSCHIST METABASITE
PARAGNEISS
ORTHOGNEISS
B′
B
VAL GRESSONEY
VAL SESIA
WSW
ENE
1 km
THRUST HORSES OF CALC-SCHIST, PARAGNEISS AND GREENSCHIST
ORTHOGNEISS AND PARAGNEISS
METABASITE
SERPENTINITE
(b)
C′
C
Theodul - Mischabel Unit
Gran Tournalin Unit
Breithorn Unit
Ayas Unit
1Km
THRUST HORSES OF CALC-SCHIST AND METABASITE
MARBLE AND DOLOMITIC MARBLE
CALC SCHIST
METABASALT
SERPENTINITE
(c)

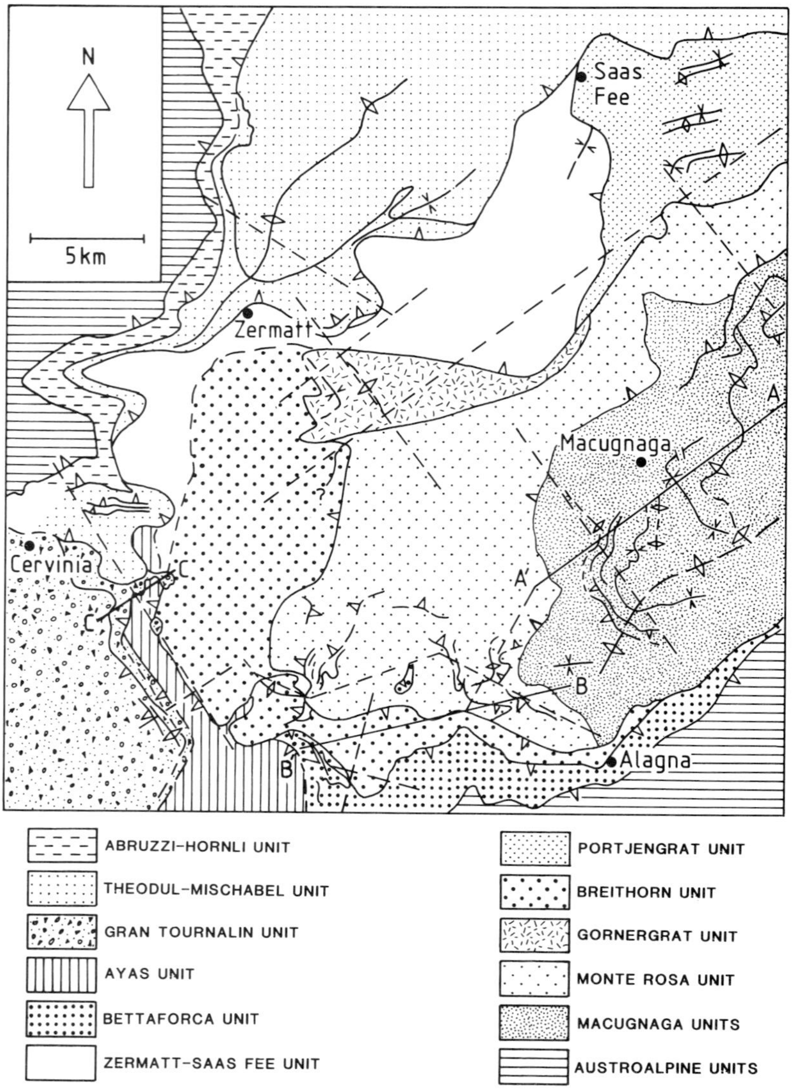

FIG. 4. Regional structural map of the Zermatt–Saas Fee–Macugnaga–St Giacomo–Cervinia area. Data from this study: Bearth (1953, 1957, 1964), Klein (1978) and Martin (1982). A–A[1], B–B[1] and C–C[1] mark the lines of section shown in Fig. 3.

Baird & Dewey (1986) and Lacassin (1984) have used stretching lineations from the area discussed here to indicate transport directions. Lacassin (1983, 1984) and Lacassin & Mattauer (1985) confined their detailed discussion of kinematics to the Portjengrat, Gornergrat and Monte Rosa units, whilst Baird & Dewey (1986) concentrated on the area around Val d'Ayas, Valtournanche and Zermatt. In this paper, new lineation data are presented and these are discussed together with the data from previous works.

Five distinct sets of lineations can be distinguished on the basis of structural association, orientation and metamorphic grade. Lineations of different sets overprint one another and the relative ages of lineations can be determined, as described below. Lineations are discussed from the base to top of the pile and Fig. 6 shows a map of the units, on which are plotted the azimuth readings of mineral elongation lineation data collected during this work. Overlain on this are the trends of the different lineation sets described in the following paragraphs. The

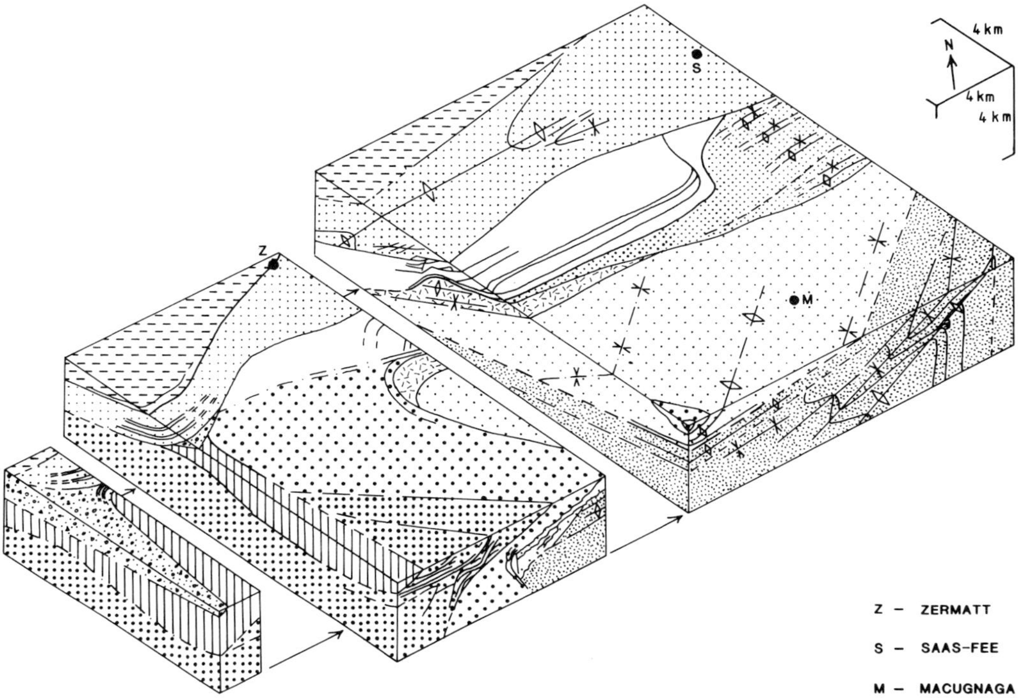

FIG. 5. Three-dimensional block diagram from the regional map and cross-sections. This is drawn from approximately the same viewpoint as Argand's (1911) block diagram. It shows the principal thrust-bounded units which make up the tectonic edifice. Place names are projected on to the surface of the block.

TABLE 1. *Correlation between units used in this work with those of previous workers.*

Abruzzi–Hörnli unit	Obere Zermatter schuppen zone (Bearth 1962)
Theodul–Mischabel unit	Includes Theodul–Rothorn zone of Bearth (1962), Mischabel unit of Bearth (1964), Milnes *et al.* (1981) & Müller (1983).
Gran Tournalin unit	Part of Gran Tournalin unit of Dal Piaz (1965). Does not equate with Theodul–Rothorn unit of Bearth (1962).
Ayas unit	Part of Corno de Camoscio–Breithorn unit of Dal Piaz (1965).
Bettaforca unit	Part of Gran Tournalin unit of Dal Piaz (1965). Does not equate with Theodul–Rothorn unit of Bearth (1962).
Zermatt–Saas unit	Zermatt–Saas Fee ophiolites *sensu stricto* of Bearth (1967).
Portjengrat unit	Part of Mischabel nappe, Bearth (1952), Bearth (1964); Portjengrat zone of Dal Piaz & Ernst (1978); Portjengrat unit of Bearth (1964). Portjengrat complex of Steck *et al.* (1979); Portjengrat unit of Martin (1982).
Breithorn unit	Part of Corno de Camoscio–Breithorn unit of Dal Piaz (1965).
Gornergrat unit	Gornergrat complex of Steck *et al.* (1979); Gornergrat zone of Lacassin (1984).
Monte Rosa unit	Monte Rosa unit of Bearth (1952), Milnes (1974) and Martin (1982).
Macugnaga unit	Macugnaga unit of Bearth (1952), Milnes (1974) and Martin (1982).

trends have been computed using a moving point average technique, which has the effect of both averaging data within a given square kilometre and smoothing this with respect to the immediately adjacent four square kilometres. The arrowheads on these trends indicate the associated sense of overshear.

Mineral elongation lineations studied by Lacassin (1984) were shown by him to fall into two groups, those which record penetrative

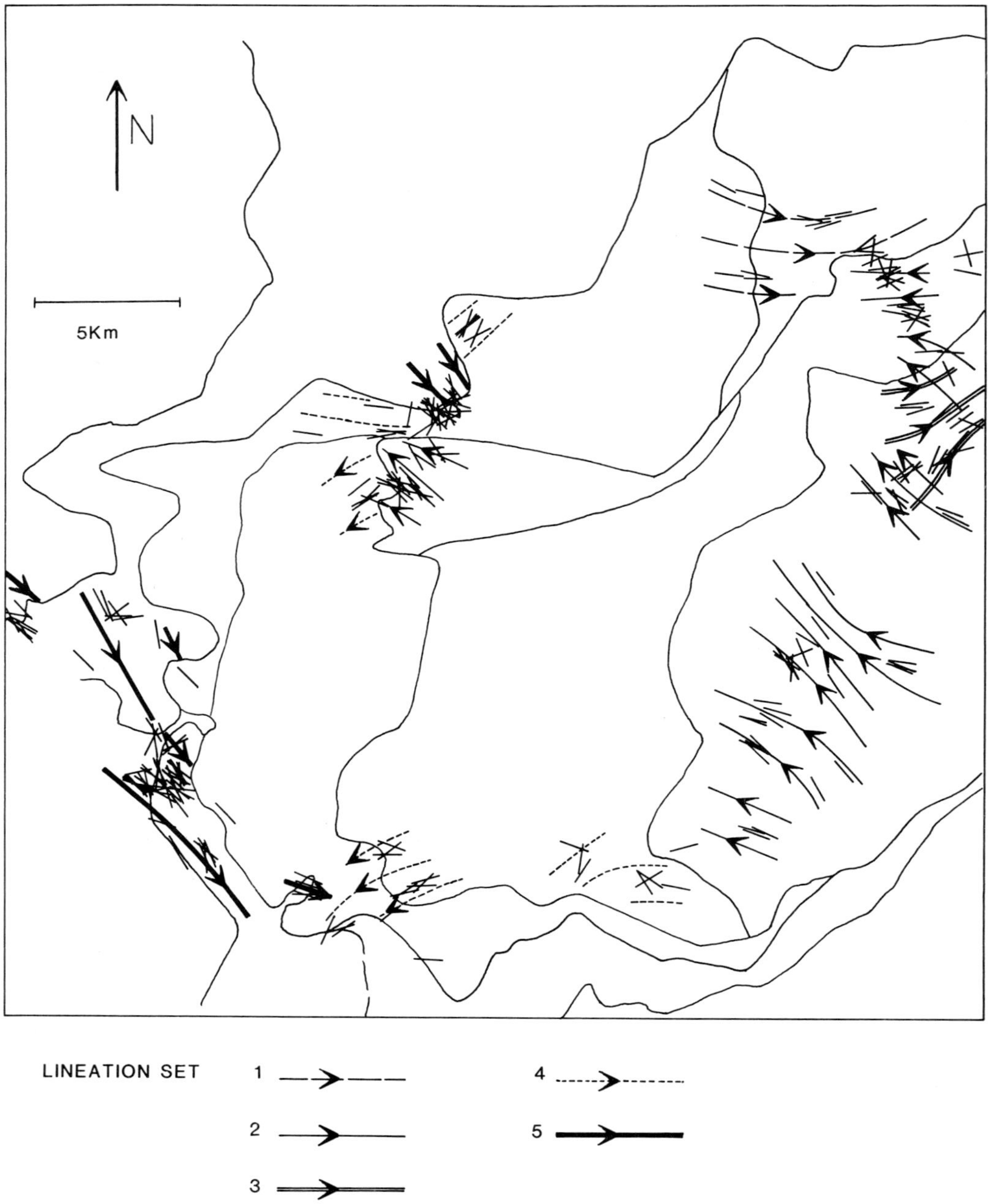

FIG. 6. Mineral stretching lineation azimuths. A smoothing/averaging routine was applied to each of the lineation generations discussed in the text. The smoothed/averaged trends have been superimposed on the data set with arrows indicating the sense of overshear.

overshear towards the ESE or ENE, and those which record overshear towards the NW. Of these two lineations, the former are preserved within the Portjengrat and the base of the Zermatt–Saas unit, the latter preserved in the Macugnaga, Monte Rosa and Gornergrat units.

Lacassin (1984) based his interpretation of different senses of overshear on the asymmetries of recrystallized pressure shadows developed around feldspar porphyroclasts at greenschist facies. In this work the trends of these linear fabrics have been mapped over a larger area than that covered by Lacassin. The penetrative linear fabric within the Portjengrat unit is shown to be older than that of the Macugnaga, Monte Rosa and Gornergrat units by type-three fold interference patterns developed in association with the linear fabrics.

A third set of linear fabrics is developed in the lowest unit studied, the Macugnaga unit. These demonstrably overprint the earlier set found in this unit and are usually confined to discrete shear zone surfaces at various scales. Locally however, adjacent to important shear zone surfaces, complete reorientation of the early linear fabric and associated fold hinges into the overprinting orientation is observed. Both reorientated and new lineations are shown on Fig. 6 as double barred arrows. We assign a north-easterly overshear to them as C/S mylonites developed within the same shear zones indicate transport to the NE (Fig. 7). Regionally, Steck (1980, 1987) described a late NE–SW linear fabric, but ascribed a SW overshear to it. His late linear fabric (l'etirement longitudinal) is a very widespread fabric affecting both the internal and external Alps, different to that described here, which is much more limited. It appears that they were generated by different tectonic mechanisms. The C/S fabric described here, is defined by white mica and quartz ribbons, this shear fabric being readily distinguishable in the field from the earlier fabric described above.

The fourth set of lineations includes the majority of lineations measured within the Zermatt–Saas and Breithorn units, plus lineations measured in the S of the Monte Rosa unit and the Ayas unit. All these units appear at some stage to have undergone high pressure metamorphism (Bearth 1967, Barnicoat & Fry 1986, Dal Piaz *et al.* 1986) though all the lineations measured are defined by minerals which have grown at greenschist facies. All the lineations in this set are approximately SW- or W-plunging and the set may be of the same age as the linear fabric formed of re-orientated glaucophanes described by Chadwick (1974) from Täschalp. Snowball garnets from Täschalp have been examined in an attempt to determine the shear kinematics (Fig. 7b), but no consistent sense of overshear has been found and it appears that the fabrics observed may have developed by pure shear. However, quartz c-axis fabrics and snowball garnets consistently indicate a SW overshear on lineations at the base of the Breithorn unit.

The final set of lineations is developed penetratively throughout the uppermost five units and is distinguished on Fig. 6 with heavy bars

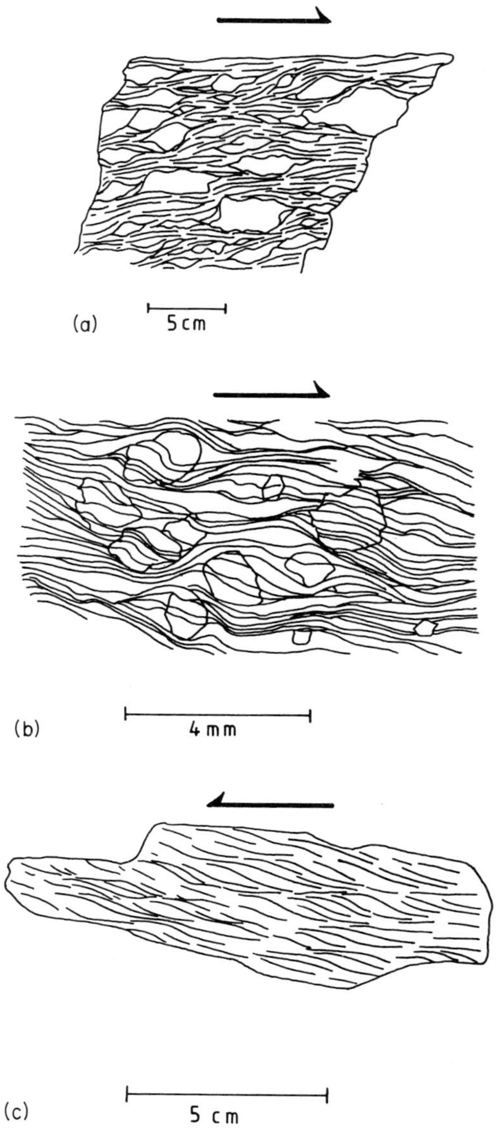

FIG. 7. Examples of sense of shear indicators. (a) asymmetric recrystallized pressure shadows around feldspar porphyroclasts (from Lacassin, 1984); (b) garnet porphyroblasts with inclusion trails in actinolitic metabasite from the Zermatt–Saas unit; (c) c/s fabric defined by quartz ribbons and white mica flakes.

and arrowheads. Lineations in the four uppermost units, the Ayas, Gran Tournalin, Theodul–Mischabel and Abruzzi–Hörnli units, plunge uniformly towards the NW. There are insufficient data to show the mean plunge direction in the next unit down, the Bettaforca unit, but this is nonetheless grouped with the overlying four units on the basis of its internal structural similarity (see later). Lineations are developed at low greenschist facies in mylonitic epidote actinolite metabasalts of the Ayas units. They are parallel to lineations in thrust mylonites cutting the epidote actinolite mylonitic fabric. It would appear therefore that the lineations record stages in a progressive deformation at low greenschist facies.

Lineations of this set have been observed in thrust zone mylonites on lateral ramp surfaces beneath the Gran Tournalin, the lateral ramp orientation confirming the direction of thrusting indicated by the linear fabric. As this fabric must have developed during movement on these thrust surfaces, the most unambiguous guide as to their sense of overthrusting is the well-known Mischabel backfold (Milnes 1974, Milnes *et al.* 1981, Müller 1983), which is a hangingwall antiform to the Theodul–Mischabel thrust, as mapped by Bearth (1952, see also Barnicoat in press, Fig. 1b). This megafold implies overthrusting towards the SE, which is confirmed by the sense of asymmetry of C/S bands in thrust zone mylonites and asymmetric dolomite fish developed within and adjacent to thrust zones, as well as extensional crenulations developed within thrust horses.

Movement on principal thrust surfaces

The previous section described penetrative linear fabrics developed within the body of rock units. Clearly, such a penetrative lineation within a unit does not necessarily indicate the movement direction on the principal thrust of that unit. The principal thrust may be of different age to the penetrative lineation and may have resulted from different movements. In order to show that movements on a principal thrust are parallel to the linear fabric within the body of a unit, lineations within principal thrust mylonites and the orientations of any lateral ramps on these thrusts must also be examined (Elliot & Johnson 1980). Problems may still arise if principal thrusts have been used more than once during orogenesis, especially if earlier increments of movement were in a different direction to later ones. In such cases, only the linear fabric associated with the latest movements may be preserved on the thrust surface.

In practice, it is often difficult to measure fabric orientations directly on principal thrusts. A best approximation of such movement directions may be obtained from the orientation of the latest penetrative fabric developed within a unit, especially if this can be shown to increase in intensity towards the base of the unit. Figure 8 is a map of principal thrust surfaces with overthrust movement directions assigned to these. Where the direction of overthrusting is known with confidence from direct evidence, a solid arrow symbol is shown. Where the direction of overthrusting is known with less confidence, or is implied by reference to the last linear fabric preserved within the body of the overlying unit, the overthrust direction is indicated by an open arrow. If there is evidence for two movement directions at different times on a single principal thrust then each is shown individually.

Internal deformation of units

As implied previously the internal character of units is very variable. In the following section, some of the important characteristics, as highlighted in the cross sections (Fig. 4) and the block diagram (Fig. 5) will be described. The lowest two units, the Macugnaga and Monte Rosa units, are separated by a thrust (Figs 3, 4 and 5). This thrust is believed to have significant displacement, as it separates rocks in the Upper Val d'Ayas and Upper Val Sesia of the Monte Rosa unit which record high pressure metamorphism (Gosso *et al.* 1979, Dal Piaz & Lombardo 1986, Chopin & Monie 1984) from rocks in the Macugnaga unit below, which do not appear to have undergone this Eoalpine event. Kilometre-scale, close to tight folds have been recorded during this work in the Macugnaga, Monte Rosa and Portjengrat units, with fold hinges parallel to the early mineral stretching lineation. It can be seen, from both Figs 5 and 6 that there is a relatively rapid change in orientation of both early linear fabrics and fold hinges across the Portjengrat thrust. Following Laccasin (1984) and the observations presented above, the Portjengrat (principal) thrust is considered to underlie what has traditionally been called the Furgg zone (Bearth 1952, Wetzel 1972).

Within the dominantly ophiolitic units, folding at many scales can be directly observed (e.g. Chadwick 1974, Martin 1982, Gosso *et al.* 1979). Whereas folds in the units described above are often of somewhat simple geometry, due to their relative rheological simplicity, folds in the lithologically more varied ophiolites are of a more complex nature. Outcropping above Saas Fee, the large-scale Mittaghorn synform (Bearth & Schwander 1981, Milnes *et al.* 1981, Trümpy 1980) may be directly observed. This structure is underlain by imbricated Mesozoic and older metasediments and Hercynian augengneisses and feldspathic schists. It is from within this imbricate zone that Lacassin (1983)

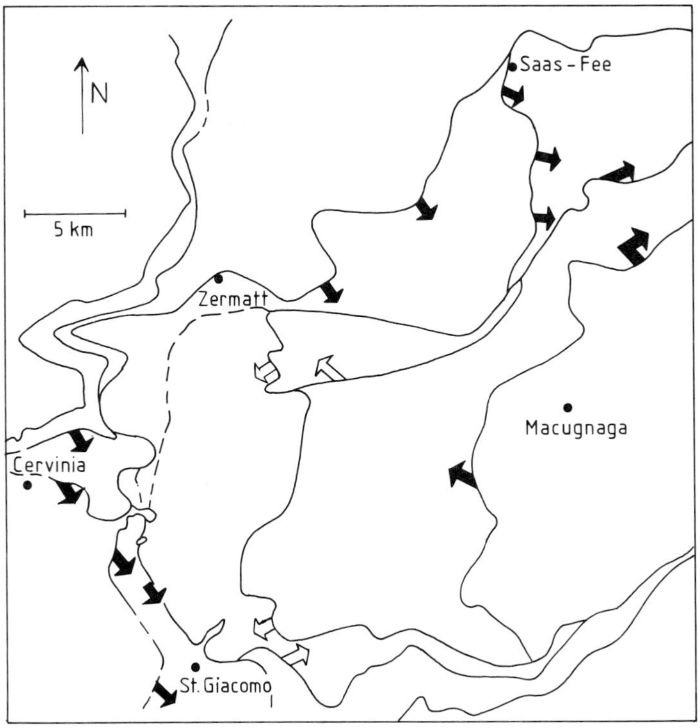

FIG. 8. Map of principal thrusts with overthrust directions marked.

and Lacassin & Mattauer (1985) described kilometre-scale sheath folds, suggesting high values of simple shear parallel to the stretching lineation. The Mittaghorn synform affects only the Zermatt–Saas ophiolites and dies out in the underlying imbricates (which overlie the Zermatt–Saas principal thrust). It appears from this geometry that the synform developed prior to, or during, development of the imbricate zone and subsequent principal thrust.

The ophiolites of the Breithorn unit are folded, with the underlying Monte Rosa unit, by the Vanzone antiform of Argand (1911) and Bearth (1957). This fold is a very late structure and as seen on Figs 4 and 5, it folds the Breithorn thrust. It is the only fold within the study area which affects a unit boundary as defined here. Geometrically it may be regarded as a late backfold.

To the NE of Zermatt, the Mischabel backfold deforms the boundary between Gran St Bernhard gneisses and the Permo–Carboniferous and Mesozoic sedimentary cover sequence, all of which are contained within the Theodul–Mischabel unit. The thrust to which this fold is a hangingwall antiform is a flat-lying planar structure which cuts out the Zermatt–Saas ophiolites above Saas Fee.

As described in an earlier section, the Theodul–Mischabel thrust is one of a family of principal thrusts with SE directed transport. Internally these units are best described as imbricate stacks of thrusts (Fig. 3b, c), although the sense of thrust propagation with regard to the backthrust direction (piggy back or break-back, Butler 1982) of these internal thrusts cannot usually be determined. What is obvious is that all thrusts, within this system of SE directed thrusts, cut through an intensely deformed and highly folded sequence of rocks including metabasites and metasediments. Similarity of thrusting direction and the earlier mineral stretching lineation suggests that the internal strain of the units and its subsequent localisation onto discrete thrust surfaces may be linked in a progressive deformation sequence. In the cliffs to the NE of the Gran Tournalin for example, an isoclinal fold with a wavelength of several hundred metres is exposed. Thrusts internal to the unit within which this occurs, the Theodul–Mischabel unit, are not folded, so this fold must have developed during earlier stages of the deformation.

Tectonic evolution

In order to examine the tectonic evolution of the area, data obtained from the high pressure metamorphic assemblages preserved in some units may be used to infer the early part of the structural evolution (the nature and duration of the high pressure subduction event). The later evolution may be constrained by considering the later structures described above.

The P–T path leading to peak high-pressure conditions is (poorly) constrained by pre-Alpine relics in basement rock units, inclusions within garnets in ophiolitic metabasalts of the Zermatt–Saas unit and variably preserved early assemblages in the Allalin gabbro. The comparability of the P–T paths implied by these parageneses and the array of P–T data for all W Alpine high-pressure assemblages (Fig. 9) suggests that metamorphism occurred in a regionally constant environment. Fry & Barnicoat (1987) suggested that this was a thermally mature and stable subduction zone. The nature of the P–T paths, with a 'hairpin' at peak conditions (Fig. 9), together with the nature and distribution of kyanite, chloritoid and lawsonite-bearing veins and fractures, suggests that subduction of individual rock packets was terminated by significant tectonic events. Fry & Barnicoat (1987) suggested that these events were the underplating (subcretion) of successive rock packets. This involved the transfer of thrust displacement onto progressively lower surfaces as more material is accreted onto the overlying plate. This process, also discussed by Rubie (1984) and Platt (1987), abruptly terminated the up-pressure, up-temperature portion of the P–T path travelled by each rock packet, but at different times. The limited geochronological data available for the high-pressure metamorphism (discussed briefly in Barnicoat & Fry 1986) support this model. Cooling occurring directly after the maximum pressure due to the restriction of heat transfer from below, should be recorded in the P–T paths of the rocks if subduction continues for *c.* 20 Ma or more (Rubie 1984). The estimated P–T paths for all units plotted on Fig. 9 imply, (or in the case of the Zermatt–Saas ophiolite unit, document) a decrease in temperature immediately postdating the maximum pressure. The general form of the post-peak paths for rocks from Monte Rosa, the Zermatt–Saas unit and the Sesia zone are all very similar, suggesting a common history for these units subsequent to accretion. The lack of notable post-peak heating in the rocks of the Monte Rosa unit implies that either subduction continued after its subcretion or that it underwent a very rapid uplift towards the surface and therefore cooled. The absence of jadeite from the granitic rocks of Monte Rosa (in contrast to those of the Sesia zone) intimates either that the P–T estimates plotted on Fig. 9 are in error (too cold?) or that some warming of the subducted/underplated material occurred. Structures reflecting juxtaposition of units during subcretion can be identified locally within the Zermatt–Saas ophiolite unit (Fry & Barnicoat 1987), though they have suffered reactivation.

Locally, structures associated with the high pressure metamorphism can be identified. As deformation progressed, it generally became more localised so that synchronous with much of the low pressure metamorphism, the area became dissected by a series for thrusts and shear zones.

It will be obvious from the discussion earlier that it is not possible to draw balanced or geometrically restorable cross sections across the present study area, even for the structures developed in the later part of this evolution. There are several reasons for this:

(1) Movements on different principal thrusts have occurred in different directions.
(2) There is no internal stratigraphy which may be correlated unambiguously from unit to unit, due largely to the high internal strain of units prior to thrusting.
(3) Many of the principal thrusts are out of sequence to one another.
(4) There is little structural record of the high pressure part of the evolution, and those structures which do exist are often too complex and dismembered to unravel.

Despite these difficulties, some conclusions concerning the regional structural evolution may be drawn from the information available.

Baird & Dewey (1986) have argued that the stretching lineation reflecting the direction of overshear should be related to the movement vector of Africa against Europe as deduced from mid-Atlantic ocean-spreading magnetic anomaly patterns. The movement vectors of Africa and Apulia relative to a fixed European reference are shown in Fig. 2. The paths plotted suggest that S–N lineations should overprint SW–NE lineations if the stretching lineations have developed parallel to the movement vector from Upper Cretaceous to Recent times. N–S lineations are not generally found in the western Alps, and many Alpine geologists argue that the N–S shortening predicted for the later stages of the Alpine orogeny must have been partitioned into shortening across several micro-

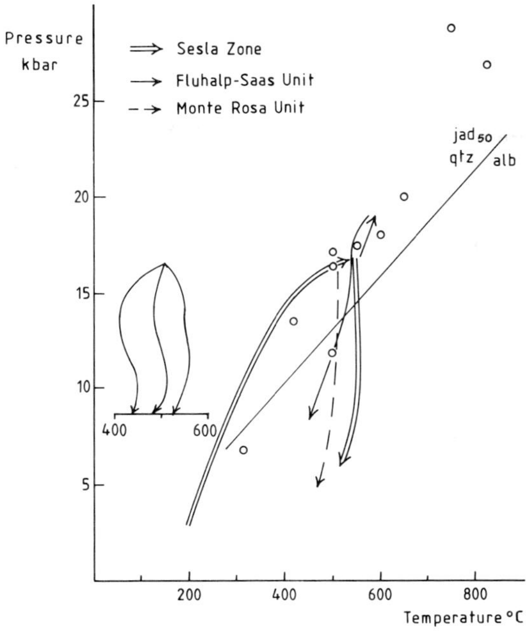

FIG. 9. P-T diagram showing estimated peak conditions from throughout the western Alps (from the summary of Fry & Barnicoat, 1987). Also plotted are P-T paths for the Sesia zone (after Ungaretti *et al.* 1983), the Zermatt–Saas ophiolite unit (after Fry & Barnicoat, 1987) and the Monte Rosa unit (after Monie, 1985). The inset shows three P–T paths calculated by Rubie (1984) for his subduction/accretion model. The different lines represent paths for different descending slab temperatures. Open circles are other P–T estimates from the Alps as collated in Fry & Barnicoat (1987).

plate margins. In this work, it is demonstrated that NE-directed lineations overprint NW-directed lineations. This clockwise rotation in overshear direction is in direct contradiction to the anticlockwise rotation implied by the plate movement analysis, suggesting that there is no direct correlation between transport directions observed in the field and convergence vectors of the major plates in the Alpine region.

Discussion

Within the upper Pennine Alps, the earliest, deep-level tectonics are characterised by ductile structure. While frequently mineralogically overprinted, this stage of the evolution has left a clear imprint on the recorded P–T paths. In areas with well-preserved high-pressure parageneses, eclogitic and glaucophane-bearing assemblages frequently carry intense fabrics which have only been locally examined. The bulk of observed fabrics comprise mineralogies characterised in the field as greenschist. Detailed examination on one set of assemblages, associated with a suite of albite veins, has shown that they formed under conditions estimated to be 8±2 kbar and 425±50°C (Barnicoat in press). It is thus possible that the tectonic evolution outlined above occurred while pressures and temperatures were continuously changing from those of eclogite formation to those of greenschist development, during Lepontine times. Alternatively, there may have been a significant time period between Cretaceous subduction and underplating-related processes and renewed activity in the Eocene. Such a break could have occurred between blueschist and greenschist conditions or within the period characterised by greenschist (s.l.) assemblages. Almost all of the dated greenschist minerals from the Monte Rosa, Gornergrat and Portjengrat units collected by Hunziker (1969, 1970, 1974), Hunziker & Bearth (1979) and Monie (1985) are Tertiary, but without knowing the range of P–T conditions and structural setting represented by their specimens, the continuity or otherwise of the tectonometamorphic evolution still remains debatable.

ACKNOWLEDGEMENTS: We are grateful to NERC for a studentship (ACE) and research grant GR3/5107 (ACB). We have profited from comments on the manuscript by Dr W. R. Fitches and discussions with J. C. Hunziker and R. Lacassin.

References

ARGAND, E. 1911. Les nappes de recouvrements des Alpes Pennines et leurs prolongement structureaux. *Beitrage zur Geologischer Karte Schweiz*, N.F. **31**.

BAIRD, A. W. & DEWEY, J. F. 1986. Structural evolution in thrust belts and relative plate motions: the upper Pennine Piémont zone of the internal Alps, southwest Switzerland and northwest Italy. *Tectonics*, **5**, 375–87.

BARNICOAT, A. C. In press. The mechanism of veining and retrograde alteration of alpine eclogites. *Journal of Metamorphic Geology*.

—— & FRY, N. 1986. High-pressure metamorphism of the Zermatt–Saas Fee ophiolite zone, Switzerland. *Journal of the Geological Society, London*, **143**, 607–18.

BEARTH, P. 1952. Geologie und Petrographie des Monte Rosa. *Beitrage zur Geologischer Karte Schweiz*, N.F. **96**.

—— 1953. *Geologisches Atlas der Schweiz 1:25 000, Blatt Zermatt (29) mit Erläutungen*. Schweizerische Geologische Kommission.

—— 1957. *Geologisches Atlas der Schweiz 1:25 000, Blatt Saas und Monte Moro (30 & 31) mit Erläutungen*. Schweizerische Geologische Kommission.

—— 1964. *Geologisches Atlas der Schweiz 1:25 000, Blatt Randa (43) mit Erläutungen*. Schweizerische Geologische Kommission.

—— 1967. Die Ophiolithe der Zone von Zermatt–Saas Fee. *Beitrage zur Geologischer Karte Schweiz*, N.F. **132**.

—— 1973. Gesteins- und Mineralparagenesen aus den Ophiolithen von Zermatt. *Schweizer Mineralogische und Petrographischer Mitteilungen*, **44**, 15–26.

—— 1976. Zur Gliederung der Bündnerschiefer in der Region von Zermatt. *Eclogae Geologicae Helvetiae*, **69**, 149–61.

—— & SCHWANDER, H. 1981. The post-Triassic sediments of the ophiolite zone of Zermatt–Saas Fee and the associated manganese mineralisations. *Eclogae Geologicae Helvetiae*, **74**, 189–205.

BUTLER, R. W. H. 1982. The terminology of structures in thrust belts. *Journal of Structural Geology*, **4**, 239–45.

—— 1987. Thrust sequences. *Journal of the Geological Society, London*, **144**, 619–34.

CHADWICK, B. 1974. Glaucophane fabric in the cover of the Monte Rosa nappe, Zermatt–Saas Fee, southwest Switzerland. *Geological Society of America Bulletin*, **85**, 907–10.

CHOPIN, C. & MONIE, P. 1984. A unique magnesiochloritoid-bearing, high pressure assemblage from Monte Rosa, western Alps: petrologic and $^{40}Ar-^{39}Ar$ radiometric study. *Contributions to Mineralogy and Petrology*, **87**, 388–98.

DAL PIAZ, G. V. 1965. La formazione mesozoica dei calcischisti con pietre verdi fra la Valsesia et la Valtournanche ed isuoi rapporti struturalli con il ricoprimento Monte Rosa e con la zona Sesia–Lanzo. *Bollettino della Societa Geologica Italiano*, **84**, 67–104.

—— 1974. Le metamorphisme de haute pression et basses temperature dans l'évolution structurale du bassin ophiolitique alpino–appeninique (lere partie: considerations paleogeographiques). *Bollettino della Societa Geologica Italiano*, **93**, 437–68.

—— & ERNST, W. G. 1978. Areal geology and petrology of eclogites and associated metabasites of the Piémont ophiolite nappe, Breuil–St Jacques area, Italian western Alps. *Tectonophysics*, **51**, 99–126.

——, LOMBARDO, B. & GOSSO, G. 1983. Metamorphic evolution of the Mt Emilius klippe, Dent Blanche nappe, western Alps. *American Journal of Science*, **273A**, 438–58.

—— & —— 1986. Early alpine eclogite metamorphism in the Penninic Monte Rosa–Gran Paradiso basement nappes of the northwestern Alps. *Geological Society of America, Memoir*, **164**, 249–65.

DERCOURT, J., ZONENSHAIN, L. P., RICOU, L. -E., KAZMIN, V. G., LE PICHON, X., KNIPPER, A. L., GRANDJACQUET, C., SBORTSHIKOV, I. M., GEYSSANT, J., LEPVRIER, C., PECHERSKY, D. H., BOULIN, J., SIBUET, J. -C., SAVOSTIN, L. A., SOROKHTIN, O., WESTPHAL, M., BAZHENOV, M. L., LAUER, J. P. & BIJU-DUVAL, B. 1986. Geological evolution of the Tethys belt from the Atlantic ocean to the Pamirs since the Lias. *Tectonophysics*, **123**, 241–315.

ELLIOT, D. & JOHNSON, M. R. W. 1980. Structural evolution in the northern part of the Moine thrust belt, NW Scotland. *Transactions of the Royal Society of Edinburgh: Earth Sciences*, **71**, 69–96.

FREY, M., HUNZIKER, J. C., O'NEIL, J. R. & SCHWANDER, H. W. 1976. Equilibrium–disequilibrium relations in the Monte Rosa granite, western Alps: petrological, Rb–Sr and stable isotope data. *Contributions to Mineralogy and Petrology*, **55**, 147–79.

FRY, N. & BARNICOAT, A. C. 1987. The tectonic implications of high pressure metamorphism in the western Alps. *Journal of the Geological Society, London*, in press.

GOSSO, G., DAL PIAZ, G. V., PIOVANO, V. & POLINOR, R. 1979. High pressure emplacement of early-alpine nappes, post nappe deformations and structural levels (internal northwestern Alps). *Memorie degli Istituti di Geologia e Mineralogia dell'Universita di Padova*, **32**, 5–15.

GULLER, A. 1947. Zur Geologie der Südlichen Mischabel- und der Monte-Rosa-Gruppe. Mit Einschluss des Zmutt–Tales westlich Zermatt. *Eclogae Geologicae Helvetiae*, **40**, 39–161.

HUNZIKER, J. C. 1969. Rb–Sr-Altebestimmungen aus den Walliser Alpen. Hellglimmer und Gesamtgesteinsalterswerte. *Eclogae Geologicae*

Helvetiae, **62**, 527–42.
—— 1970. Polymetamorphism in the Monte Rosa, western Alps. *Eclogae Geologicae Helvetiae*, **63**, 151–61.
—— 1974. Rb–Sr and K–Ar age determination and the tectonic history of the western Alps. *Memorie degli Istituti di Geologia e Mineralogica dell'Universita di Padova*, **32**, 54pp.
—— & Bearth, P. 1969. Rb–Sr-Altebestimmungen aus den Walliser Alpen. Biotitalterswerte und ihre Bedeutung für die Abkühlungsgesichte der alpinen Metamorphose. *Eclogae Geologicae Helvetiae*, **62**, 205–22.
Klein, J. A. 1978. Post-nappe folding southeast of the Mischabelrückfalte (Pennine Alps) and some aspects of the associated metamorphism. *Leidse Geologische Mededilingen*, **51**, 233–312.
Koons, P. O. 1982. *An investigation of experimental and natural high pressure assemblages from the Sesia zone, western Alps, Italy*. Doctor of Natural Sciences dissertation, E.T.H., Zurich.
Lacassin, R. 1983. Mise en évidence d'une zone de cisaillement ductile majeure dans la couverture de la nappe du Mont-Rose: étude microstructurale. *Compte-Rendus de l'Academie des Sciences, Paris*, **297**, 613–18.
—— 1984. *Etude des mechanism de déformation dans le versant nord de la nappe du Mont Rose (Alpes Suisses) et relation avec les grandes chevauchements. Approaches metholodique de la déformation des roches gneissiques*. These, 3eme cycle, Montpelier.
—— 1987. Kinematics of ductile shearing from outcrop to crustal scale in the Monte Rosa nappe, western Alps. *Tectonics*, **6**, 69–88.
—— & Mattauer, M. 1985. Kilometre scale sheath fold at Mattmark and implications for transport directions in the Alps. *Nature*, **316**, 739–42.
Lardeaux, J. -M., Gosso, G., Kienast, J. -R. & Lombardo, B. 1982. Relations entre le metamorphisme et la déformation dans la zone de Sesia–Lanzo (Alpes occidentales) et le probleme de l'eclogitisation de la croute continentale. *Bulletin de la Societe Geologique de France*, **24**, 793–800.
Martin, B. A. 1982. *Structural and metamorphic studies on the ophiolitic envelope of the Monte Rosa nappe, Pennine Alps*. Ph.D. thesis, University College Swansea; University of Wales.
Mazurek, M. 1985. Structural evolution and metamorphism of the Dent Blanche nappe and Combin zone west of Zermatt (Switzerland). *Eclogae Geologicae Helvetiae*, **79**, 41–56.
Meyer, J. 1983. The development of the high-pressure metamorphism in the Allalin metagabbro (Switzerland). *Terra Cognita*, **3**, p. 187.
Milnes, A. G. 1974. Structure of the Pennine zone (central Alps) a new working hypothesis. *Geological society of America Bulletin*, **85**, 1727–32.
——, Greller, M. & Müller, R. 1981. Sequence and style of major post-nappe structures, Simplon-Pennine Alps. *Journal of Structural Geology*, **3**, 411–20.
Monie, P. 1985. La methode $^{39}Ar-^{40}Ar$ appliquée au metamorphisme alpin dans le massif du Mont-Rose (Alpes occidentales). Chronologie detailleé depuis 110 Ma. *Eclogae Geologicae Helvetiae*, **78**, 487–516.
Müller, R. 1983. Die Struktur der Mischabelfalte (Penninischer Alpen). *Eclogae Geologicae Helvetiae*, **76**, 391–416.
Oberhänsli, R., Hunziker, J. C., Martinotti, G. & Stern, W. B. 1985. Geochemistry, geochronology and petrology of Monte Mucrone: an example of Eoalpine eclogitization of Permian granitoids in the Sesia–Lanzo zone, western Alps, Italy. *Chemical Geology* (Isotope Geosciences section), **52**, 165–84.
Platt, J. P. 1987. The uplift of high pressure–low temperature metamorphic rocks. *Philosophical Transactions of the Royal Society of London*, **A321**, 87–103.
Rubie, D. C. 1984. A thermal-tectonic model for high pressure metamorphism and deformation in the Sesia zone, western Alps. *Journal of Geology*, **92**, 21–36.
Sartori, M. 1987. Structure de la zone du Combin entre les Diablons et Zermatt (Valais). *Eclogae Geologicae Helvetiae*, **80**, 789–814.
Savostin, L. A., Sibuet, J. -C., Zonenshain, L. P., Le Pichon, X. & Roulet, M. -J. 1986. Kinematic evolution of the Tethys belt from the Atlantic Ocean to the Pamirs since the Triassic. *Tectonophysics*, **123**, 1–35.
Searle, M. P. 1985. Sequence of thrusting and origin of culminations in the northern and central Oman mountains. *Journal of Structural Geology*, **7**, 129–43.
Staub, R. 1942. Gedanken zum Bau der Westalpen zwischen Bernina und Mitelmeer. *Vjschr. natf ges. Zürich*. **82**.
Steck, A. 1980. Deux directions principales de flux synmetamorphiques dans les Alpes centrales. *Bulletin de la Societe Vaudoise des Science naturelles, Lausanne*, **75**, 141–9.
—— 1984. Structures et déformation tertiaires dans les Alpes centrales. *Eclogae Geologicae Helvetiae*, **77**, 55–100.
—— 1987. Le massif du Simplon – reflexions sur la cinematique des nappes de gneiss. *Schweizerische Mineralogische und Petrographishe Mitteilungen*, **67**, 27–45.
Trümpy, R. 1980. *Geology of Switzerland. Part A*. Wepf, Basel, 104 pp.
Ungaretti, L., Lombardo, B., Domeneghetti, C. & Rossi, G. 1983. Crystal-chemical evolution of amphiboles from eclogitised rocks of the Sesia–Lanzo zone, Italian western Alps. *Bulletin Mineralogie*, **106**, 645–72.
Westphal, M., Bazhenov, M. K. Lauer, J. P., Pechersky, D. M. & Sibuet, J. -C. 1986. Paleomagnetic implications on the evolution of the Tethys belt from the Atlantic Ocean to the Pamirs since the Triassic. *Tectonophysics*, **123**, 37–82.
Wetzel, R. 1972. Zur Petrographie der Furgg Zone (Monte Rosa–Decke). *Schweizerische Miner-*

alogische und Petrographische Mitteilungen, **52**, 161–236.

Wilson, C. J. L. 1978. Deformation in the Theodul–Rothorn zone (Zermatt, Switzerland). *Eclogae Geologicae Helvetiae*, **71**, 517–49.

A. C. Ellis & A. C. Barnicoat, Department of Geology, University College of Wales, Aberystwyth SY23 3DB, Wales, UK.

N. Fry, Department of Geology, University College, Swansea SA2 8PP, Wales, UK.

Structural and metamorphic history of a segment of the Sesia–Lanzo zone, and its bearing on the kinematics of Alpine deformation in the western Alps

J. Ridley

SUMMARY: The Sesia–Lanzo zone of the western Alps is a zone of heterogeneous crystalline basement that was reworked and remobilised during the Alpine orogeny, partly under eclogite-facies metamorphic conditions. Reworking took place in three separable phases of deformation, each under different metamorphic conditions and each localised predominantly within a particular zone in the crust. The structures and fabrics produced by the phase of deformation post-dating the early eclogite-facies event, and associated with glaucophane–albite parageneses, suggest a strike-slip tectonic setting. Within the Sesia–Lanzo zone, up to a few tens of kilometres of N–S strike-slip movement is inferred for the late Cretaceous–early Tertiary. Movement within a strike-slip system may have been responsible for much of the uplift of the eclogitic rocks from 60 km depth.

The Sesia–Lanzo zone is regarded as a slice of continental crust that underwent subduction in the Alpine orogeny (Compagnoni *et al.* 1977, Dal Piaz *et al.* 1978). Parts, but not all, of the zone show evidence of deep burial (>50 km depth) at an early stage of the Alpine orogeny. Geochronological studies give dates for the three successive Alpine metamorphic episodes; an eclogite event (P > 15 kb, T > 550°C, Koons 1986) at 130–100 Ma, lower pressure metamorphism (in the stability field of albite) at 80–60 Ma, and greenschist-facies metamorphism at around 40 Ma (Oberhänsli *et al.* 1985, Hurford & Hunziker 1985, Stöckhert *et al.* 1986).

This paper is mainly concerned with the kinematic interpretation of the deformation associated with the second metamorphic phase. The interpretation is based on examination of differences in the metamorphic history across the zone, combined with structural observations, such as the orientation of planar and linear fabrics. Particular reference is made to indicators of differences in metamorphic pressure at stages in the history, as these give information about relative vertical movements within the zone. The study was undertaken to try to understand how deeply-buried crustal rocks return to the Earth's surface and also to give information about the geometry of the early (Eoalpine) stage of Alpine tectonism.

Geological setting

The Sesia–Lanzo zone consists of remobilized pre-Alpine crystalline basement composed predominantly of metapelites, metabasites and metagranitoids. The granitic rocks have been dated as Permian (Oberhänsli *et al.* 1985). Relict mineralogies in the other rock types indicate pre-Alpine upper-amphibolite to granulite-facies metamorphism, similar to that seen in the neighbouring Ivrea zone (e.g. Dal Piaz *et al.* 1983), hence assumed to have been formed at around 400 Ma (Pidgeon *et al.* 1968)

The zone has been split into three major units on the basis of contrasting tectonothermal histories (Fig. 1, after Dal Piaz *et al.* 1971). To the E, the eclogitic micaschist unit (EMS) shows evidence of all three Alpine metamorphic stages outlined above, with the early, eclogitic assemblages generally well preserved. To the W, adjacent to the Piémont zone, is the Gneiss Minuti (GM). This also shows a polyphase Alpine history, but the rocks show predominantly greenschist-facies assemblages – although relicts of earlier high-pressure assemblages are found locally. Passchier *et al.* (1981) and Williams & Compagnoni (1983) show that pressures in the GM during the early stages of Alpine metamorphism were significantly lower than in the EMS and that the western portion of the GM may never have experienced high pressures. Structurally above these two units, in places covering their mutual contact, is the seconda zona Dioritica–kinzigitica (IIDK). This shows only restricted Alpine deformation and metamorphic recrystallization and pre-Alpine amphibolite-facies assemblages are often preserved. High-pressure assemblages have been considered to be absent from this unit except within mylonitic zones along its margins (Gosso *et al.* 1979, Lardeaux *et al.* 1982). Gosso

From: Coward, M. P., Dietrich, D. & Park, R. G. (eds), 1989, *Alpine Tectonics*, Geological Society Special Publication No. 45, pp. 189–201.

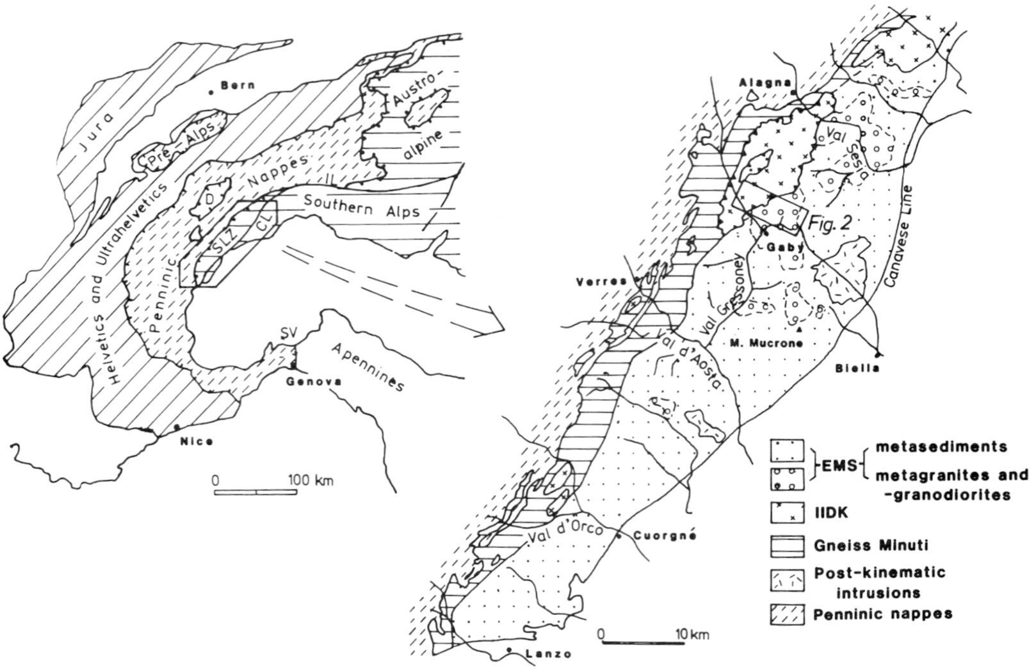

FIG. 1. Major divisions of the western Alps and the Sesia–Lanzo zone. Data for the Sesia–Lanzo zone comes from Dal Piaz *et al.* (1971) and Compagnoni *et al.* (1977). CL, Canavese line; D, Dent Blanche nappe; IL, Insubric line; SLZ, Sesia–Lanzo zone; SV, Sestri–Voltri line.

(1977) maps and describes a further unit with well-developed glaucophane-bearing assemblages forming a 'klippe' above the IIDK, which he ascribes to the EMS. In addition to these differences in history in an E–W section through the zone, there is, in general, decreasing evidence of high-pressure metamorphism northwards (Compagnoni *et al.* 1977, Reinhardt 1966, Isler & Zingg 1974).

Critical to the understanding of the Sesia–Lanzo zone, is the nature of the contacts between these units. The contacts are moderately inclined to sub-vertical. They are frequently marked by mylonitic zones, generally 1–10 m (Williams & Compagnoni, 1983), but locally up to 100 m thick (Dal Piaz *et al.* 1971). Each unit contains similar rock types, and relict pre-Alpine assemblages are everywhere similar. The differences in Alpine tectonic and metamorphic histories of the units has generally been taken to imply that the units were significant distances apart during the early stages of the Alpine orogeny and were brought together into the present disposition, by large movements along the contacts after the main phase of high-pressure metamorphism and associated deformation (Gosso *et al.* 1979). This idea is contested here on consideration of the metamorphic and structural features in the IIDK–EMS contact zone.

The IIDK–EMS contact

Both the mylonitic rocks of the contact between the IIDK and EMS, and the features seen in a 7 km traverse across the contact N of Gaby, Val Gressoney, are described here (Figs 2 and 3). The contact trends NE to ENE, and is steeply NW-dipping. Reconnaisance work suggests that the essential characteristics described here are also present in sections through the contact to the N and S.

The IIDK outcrops to the NW of the contact and consists largely of biotite-garnet-bearing metapelitic gneisses with migmatitic and pegmatitic segregations (kinzigites), together with minor amphibolites and marble layers and pods. The penetrative fabric in these rocks is defined microscopically by biotite and sillimanite. By analogy with dated rocks from the neighbouring Ivrea zone, it is therefore assumed to be from a pre-Alpine orogenic event. In general this fabric is oblique to the contact.

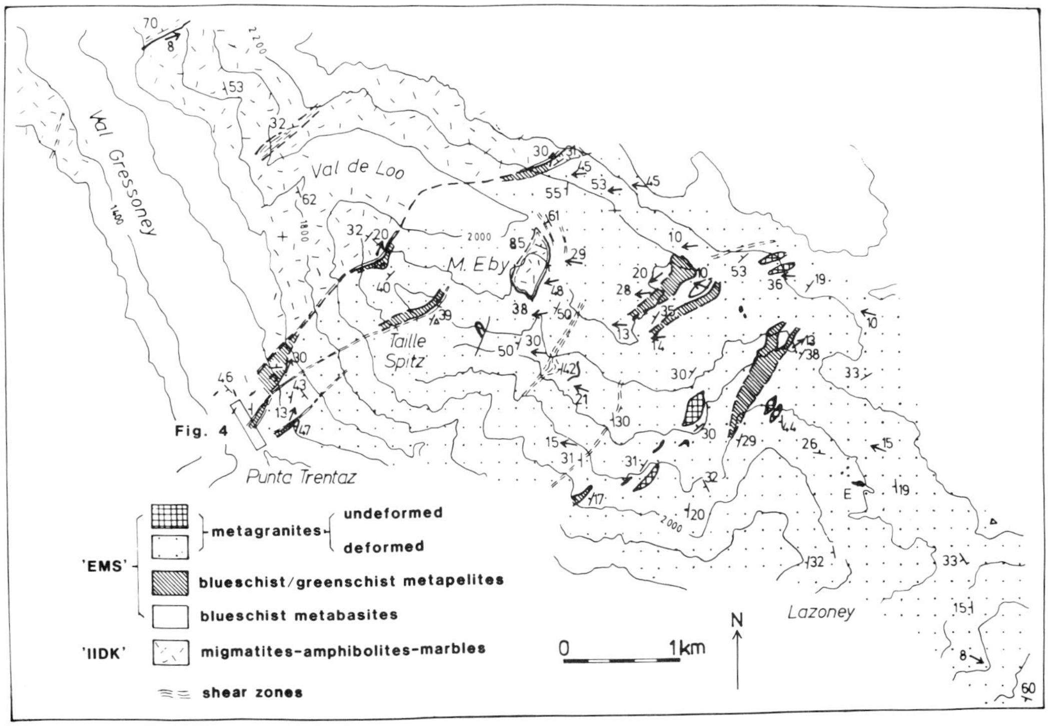

FIG. 2. Map of a segment of the EMS–IIDK contact north of Gaby, Val Gressoney. For locality see Fig. 1. Contours are at 200 m intervals. Dip and strike symbols refer to the dominant planar fabric at each locality, arrows the most prominent lineation. Lithological variations within the IIDK are not distinguished.

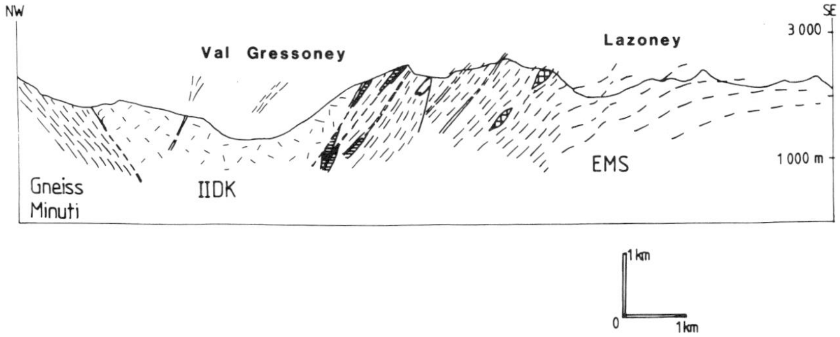

FIG. 3. Cross section through the area of Fig. 2, extended northwestwards to show the position and orientation of the contact between the IIDK and Gneiss Minuti. Shear zones in the IIDK above the land surface have been projected on to the plane of the section. Symbols are as in Fig. 2, except that within the EMS the trend of the dominant planar fabric is shown and its intensity schematically by the density of symbols.

Throughout the IIDK there are isolated discrete shear zones up to 10 m wide which deform this fabric. These shears are NE-trending and steeply dipping, i.e. are sub-parallel to the contact. They show subhorizontal stretching lineations.

The EMS to the SE of the contact is composed largely of metagranitic rocks in the area investigated, but includes pods and lenses of metabasites and irregular, often cuspate bodies of metapelites (Fig. 2). For 5 km horizontal distance from the contact, the dominant foliation in the EMS is sub-parallel to the contact, although variable in intensity and orientation (Fig. 2). Within this area there are lenses up to 100 m across of undeformed granite and one 500 m wide lens of migmatitic metapelites with textures and assemblages similar to those seen in the IIDK. Stretching lineations, mainly from quartz aggregates plunge at a shallow to moderate angle to between WSW and WNW (Fig. 2).

At greater than 5 km distance from the contact, the foliation in the EMS is generally weaker and flatter lying and no longer sub-parallel to the contact. In this area the lineation plunges at a shallow angle WNW or ESE. This is the orientation of the dominant lineation over large areas of the EMS to the SE (Hy 1984).

Mylonites in the contact zone

Figure 4 shows a detailed sketch map of mylonites exposed in the Torrente Lys 300 m upstream from Punta Trentaz. The mylonites closely follow a lithological contact between pelitic gneisses and metagranitoids.

North of the contact zone the metapelites show relict high temperature assemblages and migmatitic segregations, and are therefore asigned to the IIDK. The metagranitoids to the S show a strong penetrative foliation subparallel to the lithological contact. These granitic rocks include lenses and pods, some in the form of rootless folds, of metabasites with glaucophane-bearing assemblages.

Mylonites are developed in the metapelitic

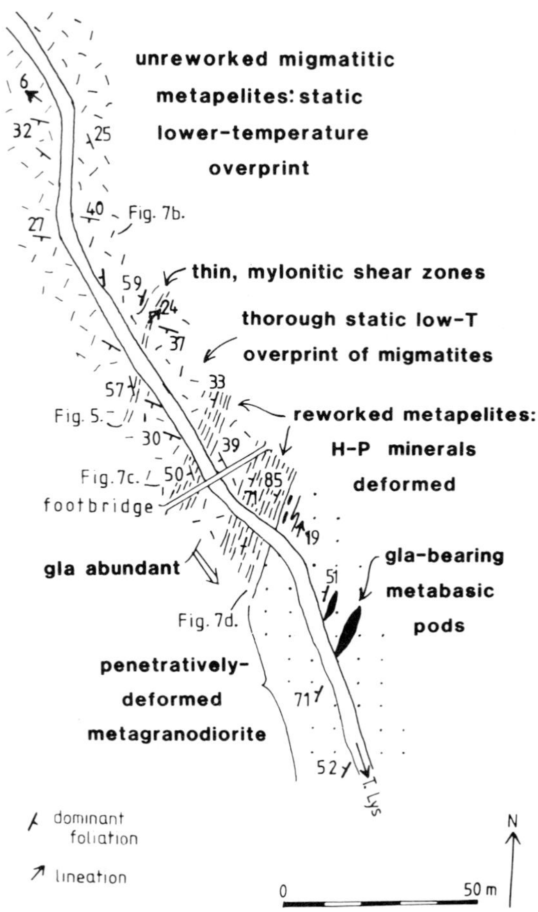

FIG. 4. Large-scale sketch map of the contact between the IIDK and EMS exposed in the bed of the Torrente Lys 300 m above Punta Trentaz. For locality see Fig. 2. The contact separates penetratively-deformed metagranitic rocks of the EMS from essentially unreworked migmatitic metapelites of the IIDK. The IIDK metapelites are cut by shear zones and partially reworked over a 50 m wide zone at the contact.

gneisses within 50 m of the contact. The mylonitisation is heterogeneous, but has generally resulted in transposition and partial overprinting of the migmatitic-gneiss fabric. In general, the degree of overprinting increases towards the contact. Fifty metres from it there are discrete 2–50 cm wide shear zones cutting rocks with a preserved gneissic fabric (Fig. 5). Within 10 m of the contact, overprinting is essentially penetrative, and the newly formed foliation is itself folded. Despite the intense deformation, the migmatitic leucosome/melanosome segregations remain discernible throughout.

The mylonitic foliation is sub-parallel to, or within 30° of the trend of the lithological contact and of the penetrative foliation within the adjacent metagranitoids. Lineations in the mylonites plunge at a low angle NNE (Fig. 4). These are mineral lineations (glaucophane or quartz aggregates), and are parallel to the axes of locally-developed, weakly-asymmetric folds.

The sense of displacement for small shears in this contact zone can readily be determined by examining the deflection and displacement of the earlier fabric. The implied shear is sinistral with minor vertical movement (Fig. 5). S-C relationships, and the obliquity of elongate quartz grains are consistent with this inference.

Other sections through the contact do not show identical sequences. Five-hundred metres NE of the Torrents Lys section, on the N side of Val de Loo, a traverse from SE to NW passes from metagranitoids through a 20–50 m thick layer of phengite-quartz-garnet metapelites with a well-developed foliation parallel to the contact. These are interpreted as reworked pelitic gneisses, and are hence asigned to the EMS. They are separated from unreworked and unrecrystallized migmatitic gneisses by a 1–2 m layer of foliated metabasite (the 'eclogitic mylonite' of Gosso 1979), rich in glaucophane or barrositic amphibole. This metabasite marks the contact. Although it is thoroughly reworked with a strong tectonite fabric, there is nowhere the extreme grain-size reduction seen in the Torrente Lys section, hence the intensity of deformation along the contact is probably much lower.

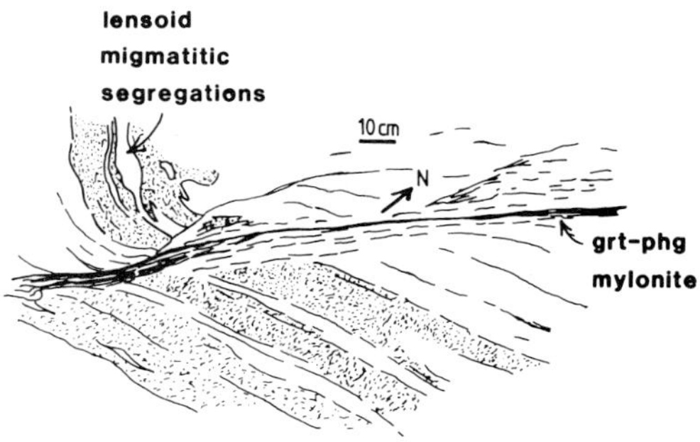

FIG. 5. Sketch from field photograph of a minor shear zone in the IIDK at Punta Trentaz, 50 m from the contact with the EMS showing transposition of the migmatitic fabric and sinistral offset (for locality, see Fig. 4). The outcrop is horizontal.

Metamorphism

Within the Sesia zone as a whole, there is a time sequence of at least four metamorphic imprints (e.g. Lardeaux *et al.* 1982): a pre-Alpine upper-amphibolite–granulite facies event, an early-Alpine eclogite-facies event, an event in which glaucophane + albite was a stable assemblage and a greenschist-facies imprint.

Through most of the EMS, there is a regional development of eclogitic assemblages – including jadeite (X_{jad}^{pyx} >90%) + quartz and sodic-pyroxene + garnet + quartz parageneses. Neither of these assemblages have been found in the area shown in Fig. 2. The assemblage in metabasites, both within this part of the EMS and in the mylonitic rocks along the contact is typically barroisite/glaucophane–epidote–garnet–sphene.

The microfabric of these rocks shows a mosaic of elongate grains of epidote and sodic/subsodic amphibole. Augen of partially recrystallised brown (high-T, hence pre-Alpine) hornblende are common (Fig. 7a). The presence of these relicts is taken to imply that the rocks could not have recrystallized as eclogites prior to the formation of the presently observed amphibole-bearing assemblages.

In EMS metagranites, the syn-deformational assemblage is generally quartz–albite–phengite–clinozoisite–garnet–glaucophane/actinolite–sphene.

The assemblage is the same in deformed and undeformed rocks, but the textures differ in such a manner as to imply that this assemblage was stable during deformation. In undeformed rocks, the igneous phases are pseudomorphed: original K-feldspar is replaced by 'checker-board' albite, plagioclase by clinozoisite + phengite + albite, and biotite by phengite + garnet. In the deformed rocks original K-feldspar grains can be recognized in elongate clots of strained grains of albite of the same overall dimensions. White mica and epidote are aligned parallel to the penetrative foliation.

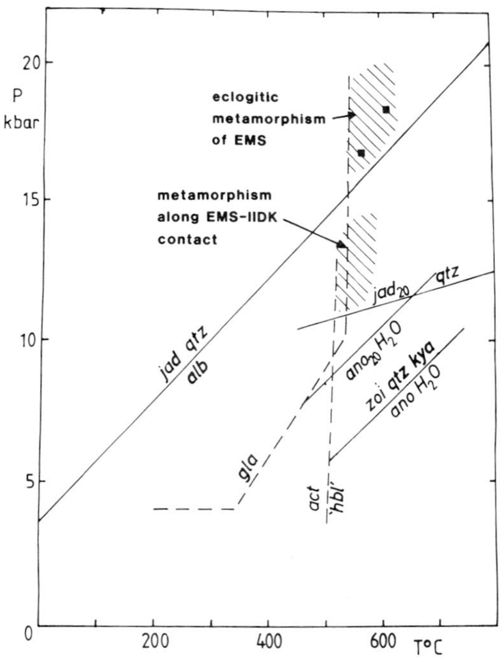

FIG. 6. P–T conditions for the eclogite-facies metamorphism of the M. Mucrone area (Hy 1984, Koons 1986), and suggested conditions for the glaucophanitic metamorphism of the IIDK–EMS contact. In the latter area: Ca-bearing plagioclase is unstable relative to zoisite-bearing assemblages, albite remains stable, glaucophane is a stable phase in suitable rock compositions. The presence of hornblende-barroisite rather than actinolite in metabasites implies temperatures equivalent to the epidote–amphibolite facies. Curves: maximum stability of glaucophane (Maresch 1977), albite–jadeite–quartz (Holland 1980), anorthite and ano_{20} stability in the presence of a hydrous fluid (Goldsmith 1982), actinolite–hornblende (Spear 1981), and albite–$jadeite_{20}$ + quartz (Holland 1983). The squares give determinations of Koons (1986) for eclogitic assemblages from the eastern margin of the Sesia–Lanzo zone.

Metapelites enclosed within the EMS granites have similar assemblages as the metagranitoids, but are phengite-rich, and have less albite and clinozoisite. In some metapelites relict high-T garnets with Alpine overgrowths can be recognized. These assemblages suggest that the dominant metamorphic imprint corresponds with the glaucophanitic phase of Gosso (1977), ie. in terms of the metamorphic sequence for the Sesia–Lanzo zone as a whole, is post-eclogitic. There are no assemblages that tightly constrain the P–T conditions for this phase of metamorphism, nor are there any reliable geothermometers/barometers that may be used to estimate these conditions. The available constraints suggest T = 500–600°C at 9–14 kb (Fig. 6). There is some evidence of increasing pressures of metamorphism towards the SE. Locality 'E' (Fig. 2) is an omphacite–garnet–phengite metabasite without quartz. Metagranites in the same area show albite–epidote–actinolite/hornblende aggregates, interpreted to be pseudomorphs after omphacite.

In the mylonitic EMS–IIDK contact zone, in shear zones within the IIDK, and in the unreworked IIDK, incomplete transitions are often seen from pre-Alpine textures and mineralogies to high-pressure assemblages. The existence of partial overprinting implies that many assemblages seen are not equilibrium assemblages (cf. Koons *et al.* 1987). Interpretation of the metamorphic conditions in these rocks is therefore potentially uncertain. Interpretation is also complicated by evidence of two contrasting partial overprints, the glaucophanitic phase and a greenschist event.

In general, rocks within shear zones show a greater degree of mineralogical transformation than undeformed rocks, and undeformed rocks close to shear zones show more transformation than those at a greater distance. Table 1 summarizes the mineralogies seen in metapelites and metabasites in the three structural settings.

In metapelites at a distance from any shear zone, typical reaction textures are: biotite is rimmed by <50 μm coronas of white mica and the grain centre may be partially overprinted topotactically by chlorite (Fig. 7b), plagioclase and sillimanite are replaced by sericite–albite–clinozoisite, and garnets remain unaffected or are partially replaced by chlorite.

Close to shear zones (Fig. 7c), original biotite–sillimanite–plagioclase aggregates are replaced by fine-grained (10–20 μm) white mica. There is new garnet growth, both as rims around pre-Alpine garnets and as new small euhedral grains. Glaucophane is present but rare, seen normally along quartz – mica interfaces.

In shear zones there is greater textural re-equilibration (Fig. 7d). The fine-grained mica aggregates recrystallized to give a decussate texture. Glaucophane is more common and is seen as grains aligned parallel to the foliation.

In metabasites the assemblage away from shear zones is hornblende–clinozoisite–garnet, with plagioclase normally partially saussaritized. Close to shear zones, clinozoisite replaces plagioclase and there are glaucophane overgrowths around the original brown hornblende. Although the textures and mineralogies vary, they are all consistent with mylonitization and

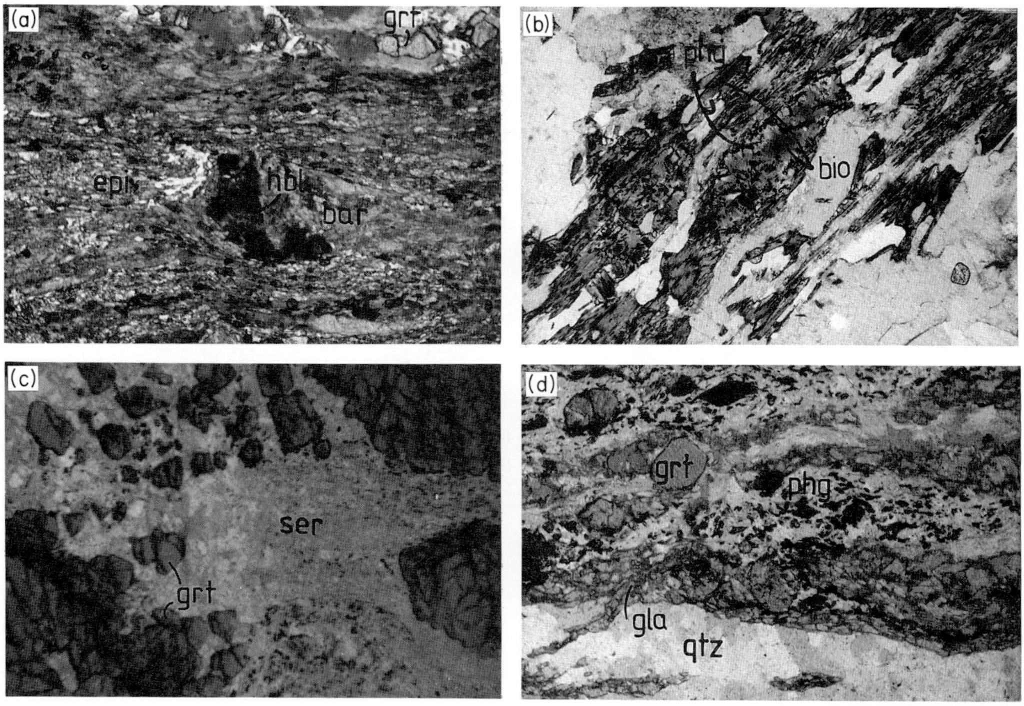

FIG. 7. Photomicrographs of textures from rocks along or close to the IIDK–EMS contact. Long dimension of each plate is 3 mm. For localities of (b–d), see Fig. 4.
(a) Barroisite–clinozoisite–garnet metabasite from a shear zone within the EMS 100 m north of Punta Trentaz. A large hornblende porphyroclast with fine inclusions of a high-relief mineral (probably rutile) is set in a foliated matrix of epidote and barroisite, and shows recrystallisation to barroisite along its margins. A garnet porphyroblast, upper right, is altered to chlorite.
(b) A partially overprinted migmatitic metapelite showing brown biotite with phengite coronas and topotactic replacement inside the coronitic rims to chlorite.
(c) Completely overprinted migmatitic metapelite showing a fine-grained (10–20 μm) sericite aggregate, presumed to be a replacement of biotite + sillimanite + plagioclase, and two phases of garnet growth, the latter as small grains and rims on earlier porphyroblasts delineated by inclusion trails, e.g. $x-x^1$ in bottom left of view. The sericite shows no preferred orientation or fabric.
(d) Metapelite deformed under glaucophanitic metamorphic conditions. The glaucophane is as polygonal grains of similar grain size. The quartz/glaucophane + phengite + garnet + clinozoisite banding visible is presumed to be an inherited migmatitic segregation banding.

the formation of shear zones being related to the glaucophanitic phase of deformation seen in the neighbouring EMS.

Inferences about the metamorphic history

Two points may be made about the metamorphic history of the various units which are important in any tectonic reconstruction.

(1) The whole of the IIDK went through high pressure (glaucophanitic, but not necessarily eclogitic) metamorphic conditions, although most of the rocks show incomplete or no recrystallization under these conditions. As there are isolated shear zones throughout the IIDK with high-pressure mineralogies, it is clear that the rocks between these shear zones also went through the same conditions. The lack of recrystallization is presumably related to kinetic barriers to reaction (cf. Koons *et al.* 1987).

(2) The rocks in the EMS within 2–3 km of the IIDK contact, and also the IIDK, were metamorphosed at significantly lower pressures (at a similar temperature) than those of the rest of the EMS. In contrast to

TABLE 1. *Mineral assemblages and typical replacement reactions in the IIDK*

	Structural sites		
Underformed	Adjacent to shear zones	Shear zones	
Metapelites			
Assemblages	bio-plg-qtz-grt-sil	qtz-clz-grt-phg (-gla)	qtz-clz-grt-phg-gla
Pseudomorphing and replacement reactions	sil→kya bio→phg+chl plg→clz+ser+alb grt→chl	grt→chl	grt→chl
Metabasites			
Assemblages	hbl (brown) -plg-grt	hbl-gla-clz-grt	grt-clz-qtz, bar-clz-grt, gla-clz-grt
Pseudomorphing and replacement reactions	grt→chl plg→ser+clz+alb	hbl→gla gla→chl grt→chl	grt→chl

the response of the IIDK to the glaucophanitic phase of metamorphism, for which there is widespread partial or sporadic recrystallisation, no sample has been found in this area showing evidence of recrystallization to eclogitic assemblages. Albite, for instance, is never replaced by jadeite–quartz though this replacement invariably takes place at Monte Mucrone 10 km SE, even in undeformed rocks (Compagnoni & Maffeo 1973). It is therefore considered unlikely that the rocks passed through eclogite conditions and failed to recrystallize. The differences in reaction histories between the EMS close to the IIDK contact and that further SE, are taken to indicate that the former never reached extreme high pressures.

Recrystallization under greenschist-facies conditions

Late recrystallization under greenschist-facies conditions becomes more widespread and important towards the western margin of the Sesia–Lanzo zone (Compagnoni *et al.* 1977). In the area studied here the greenschist overprint is static and sporadic except along narrow NE-trending subvertical shear zones seen in the westernmost 1–2 km of the EMS. More widespread deformation under greenschist-facies conditions is seen immediately west of the area shown in Fig. 2. The whole of the Gneiss Minuti was intensely deformed under greenschist/epidote-amphibolite facies conditions (Compagnoni 1977).

Discussion

Significance of the IIDK – EMS contact

It has been argued that there was no significant difference between the metamorphic P–T histories of the IIDK and of the EMS immediately SE of their contact. It is also noted that the mylonitic deformation along the contact took place under the same P–T conditions as the penetrative deformation in this part of the EMS.

The following features of the contact are noted: there are no 'exotic' rocks within the contact mylonites, that is, rocks not otherwise

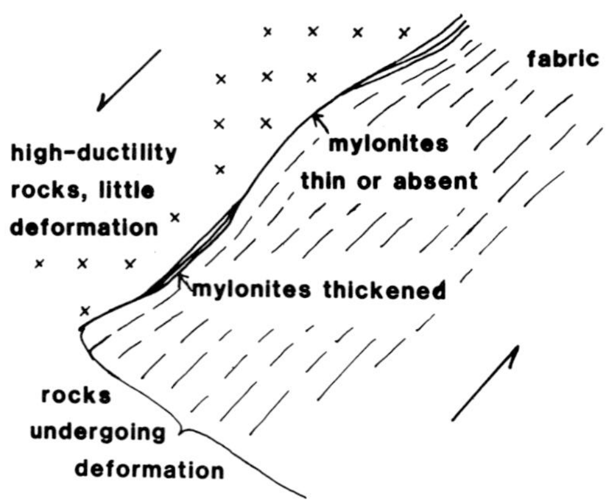

FIG. 8. Variations in mylonite thickness along a slightly irregular contact between deforming and effectively non-deforming rocks. Around any 'protrusion' of the undeformed rocks there will be more intense strain in the immediately adjacent deformed rocks.

observed in the area of study. The contact is not exactly planar (Fig. 2). There are changes in thickness and intensity of mylonitisation along the contact: as exposed on the N side of Val de Loo, for instance, the zone is marked only by a 2 m wide layer of deformed metabasites. The lineation orientation in the mylonites is oblique by about 40° to that in the neighbouring EMS metagranites.

From these features, it is suggested that the contact is a deformed pre-Alpine lithological contact between a granitic pluton and metapelitic gneisses and that the presence of mylonites along the margins of the IIDK is a result of strain compatibility effects where a rock body undergoing deformation (the EMS metagranites) has a slightly irregular contact with rocks, which under the prevailing metamorphic conditions were much more rigid (the IIDK migmatitic gneisses) – see Fig. 8. Experimental investigations of strain adjacent to rigid bodies in a deforming matrix (Celma 1982, Selkman 1983) show rotation of principal strain directions. This is suggested as an explanation for the differences in lineation orientation between the contact-zone mylonites and the regional trend in the neighbouring EMS metagranites.

Significance of the glaucophanitic phase of metamorphism and kinematics of the associated deformation

The dominant Alpine metamorphic event in the area of Fig. 2 is a moderate to high-pressure event, with P–T conditions intermediate between the eclogite and epidote-amphibolite facies. To the E and W of this area, this phase of metamorphism is less important and the associated assemblages are only locally or partially developed. To the E, this phase of metamorphism is seen as a sporadic overprint of earlier eclogite-facies assemblages, to the W, the later, upper-greenschist-facies event is dominant with relicts of earlier glaucophanitic and, locally, eclogitic assemblages (see Compagnoni & Williams 1983). The belt within the Sesia–Lanzo zone over which the glaucophanitic event is dominant is structurally well defined. It is the belt within which foliation is sub-parallel to the EMS–IIDK contact, i.e. dipping 30–50° WNW, including discrete shear zones within the IIDK.

The differences in metamorphic histories of the rocks considered here and those of the rest of the EMS imply that during the early stages of Alpine orogenesis the two were at different depths, the former never having reached pressures high enough to give the breakdown of albite. The exact magnitude of the pressure difference is uncertain; the available constraints suggest 2 to 8 kbars (Fig. 6).

The two areas have been mapped as part of a single continuous unit by previous workers in the area; there are no lithological breaks (Compagnoni *et al.* 1977), nor are there any large-scale post-metamorphic structures along which vertical movements could have taken place so as to juxtapose to the same present structural level, rocks that were earlier at different levels in the crust.

Two alternative explanations for the different metamorphic histories are possible:

(1) The glaucophanitic metamorphism close to the IIDK contact and the eclogitic metamorphism in the rest of the EMS were essentially synchronous and the whole Sesia–Lanzo zone was tilted after high-pressure metamorphism, such that different original levels are exposed in a traverse across the zone. This is a possible explanation if the pressure differences across the EMS are close to the minimum of the likely range, but would be inconsistent with the patterns proposed for the eclogite event by Koons (1986), of increasing pressure northwards along the eastern margin of the zone.

(2) The glaucophanitic metamorphism is later than the eclogitic event and relative vertical movements took place during the glaucophanitic phase of metamorphism. This is the preferred history. In this model, the 3 km wide zone of NW-dipping foliation along the NW margin of the EMS would be a major zone of movement with a N-down component of motion. Stretching lineations within this zone plunge at a low to moderate angle to the W, implying predominantly strike-slip movement. If downthrow is to the N, the lineation pattern indicates sinistral strike-slip movement with a subordinate normal component. Sinistral movement is consistent with the minor structures within the mylonitic contact zone (see above).

The metamorphic pressure differences of 2 to 8 kbars suggest vertical movements of 7 to 25 km. If we assume that all this relative vertical movement is taken up along this zone, the average lineation plunge of about 30° (Fig. 2), would indicate strike-slip displacements of 15–50 km.

Can the Gneiss Minuti be similarly interpreted?

To the W of the IIDK is the Gneiss Minuti. Reconnaisance work by the author and examin-

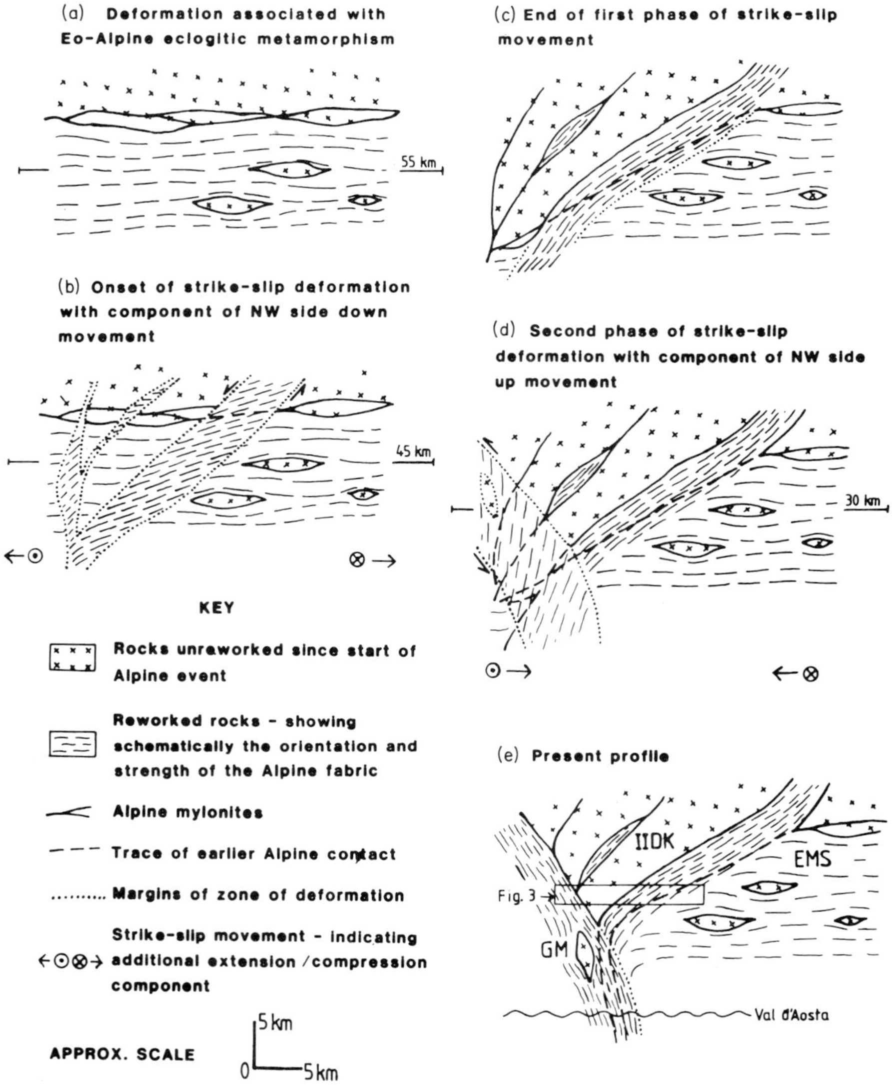

FIG. 9. Proposed Alpine tectonic history for the central Sesia–Lanzo zone.
(a) First major deformation (130–100 Ma?) producing a relatively flat-lying fabric. This deformation was synchronous with eclogite-facies metamorphism in the EMS. The deformation is indicated as being concentrated in the lower levels of the crustal pile because only rocks showing eclogite facies metamorphism have given the earliest Alpine dates. The deformation was not completely penetrative – undeformed lenses preserve pre-Alpine fabrics, e.g. Monte Mucrone.
(b & c) Beginning and end of the first phase of strike-slip movement (80–60 Ma). The deformation in this phase is along NW-dipping zones and affects in part rocks that were deformed in the earlier phase, and in part rocks that were not. Metamorphic pressures in the EMS were reduced as a result of this deformation. In this model the 'klippe' of EMS at Punta Plaida (Gosso 1977) is viewed as a body of rock within the IIDK which underwent penetrative deformation at this time.
(d & e) Beginning and end of the deformation phase associated with greenschist–amphibolite facies metamorphism (40 Ma). Deformation in this phase is concentrated in the Gneiss Minuti. In a similar fashion as the earlier phases, deformation is not completely penetrative. The Sparone 'klippe' (Minnigh 1977) may be a large 'augen' left undeformed during this phase.

ation of the results of Passchier *et al.* (1981) and Williams & Compagnoni (1983), suggest that there may be similarities between the Gneiss Minuti and the glaucophanitic zone of deformation outlined in this paper.

The Gneiss Minuti shows a strong syndeformational metamorphic imprint under upper-greenschist to lower-amphibolite facies conditions. Metamorphic recrystallization under similar conditions is recorded in neighbouring rocks, but as a static overprint, or syntectonic only along narrow shear zones.

There are SE-dipping mylonitic zones along the GM-IIDK contact, with assemblages showing that these mylonites formed under the same, greenschist-facies metamorphic conditions as deformation in the GM. The mylonites are 100 m thick at Monte Nery (Dal Piaz *et al.* 1971), but die out to almost nothing 1 km along strike. These mylonites could therefore also be the result of strain compatibility effects along the edge of a deforming body of rock, in this case the Gneiss Minuti deforming and the IIDK effectively rigid.

Conclusions

Proposed tectonic history for the Sesia–Lanzo zone

Figure 9 summarizes a possible history from 130–30 Ma for the Sesia–Lanzo zone which takes into account the localization and nature of the deformation at the various stages in the history. The distinction in each diagram is between reworked and unreworked rocks at that point in time and not necessarily between the units as mapped at the present day.

This paper is not specifically concerned with deformation associated with the eclogite-facies metamorphism. Studies in the eastern portion of the EMS (Hy 1984) have shown a polyphase fold history, but generally a relatively flat-lying foliation and a NW-trending lineation. Such a lineation orientation has been correlated with the regional plate convergence direction by Choukroune *et al.* (1986) and Baird & Dewey (1986).

It has been argued above that deformation during the glaucophanitic phase of metamorphism was concentrated within a NW-dipping zone of deformation along the present EMS–IIDK contact and was related to sinistral strike-slip motion with a subordinate NW-down normal motion. Almost all 80–60 Ma dates for this phase of metamorphism have been obtained from the eastern edge of the Sesia–Lanzo zone, where the phase is seen as a static overprint of eclogites, hence it is uncertain whether this is the age of the deformation along the IIDK contact. The age must, however, be intermediate between the eclogitic and greenschist phases of metamorphism.

During the greenschist-amphibolite phase of metamorphism (at 40 Ma) deformation was concentrated within the Gneiss Minuti, forming a steeply SE-dipping foliation. South of the area studied here, the Gneiss Minuti truncates the glaucophanitic zone and hence cuts out the IIDK, though areas of unreworked 'IIDK' metapelites are seen within the Gneiss Minuti further S (Minnigh 1977). Passchier *et al.* (1981) report NE-trending lineations in the Gneiss Minuti, but it is not clear if these are stretching lineations (H. Stünitz, pers. comm.). Nevertheless, a component of strike-slip motion for this deformation fits the fabrics observed.

The final geometry (Fig. 9e) is constructed so as to reproduce sections across the Sesia–Lanzo zone through the upper Gressoney valley (cf. Gosso 1977, Passchier *et al.* 1981) and along the Val d'Aosta.

The idea of strike-slip motion in the western Alps is not new (see reviews by Ricou 1984, Ricou & Siddans 1986). Nicolas (1974), from microstructural observations, suggested sinistral strike-slip movement at the western margin of the Lanzo Massif along the Viu–Locana zone, the southward strike continuation of the Gneiss Minuti. The model presented here is consistent with the main inference of these works, that deformation in the internal western Alps was dominated by N–S directed sinistral strike-slip motion within the period 90–40 Ma. The fault lines are arcuate, following the present overall structure of the western Alps and are hence NE-trending at the latitude of the area considered in this paper. This arcuation is believed to result from later Alpine phases of deformation (Ricou & Siddans 1986). Total strike-slip displacements are unknown, but the results of this paper suggest a few tens of kilometres within the Sesia–Lanzo zone itself.

Mechanisms for uplift of deeply-buried rocks

Ricou & Siddans (1986) point out that the best preserved high pressure assemblages in the Alps appear to come from the regions of strike-slip deformation. The results of the present work suggest that a possible reason for this correlation is related to the uplift mechanisms of deeply-buried rocks.

Progressive thrusting can not by itself give rise to unroofing of deeply-buried units; there must be in addition deep erosion over large areas (Platt 1986). Strike-slip faults are rarely single fault strands, but characteristically splay upwards to give 'flower' structures (Emmons 1969, Wilcox *et al*. 1973). Any irregularity along the fault line will give rise to compressional or extensional strains across it and hence, respectively, uplift or downthrow of blocks within the fault zone (e.g. Woodcock & Fischer 1986). The work of Ricou (1984) shows that there is a series of sub-parallel strike-slip fault zones in the western Alps. The rocks between these zones can be regarded as fault-bounded blocks and during regional strike-slip deformation may potentially undergo relative uplift or downthrow. In both stages of strike-slip history illustrated in Fig. 9, the eclogitic rocks are represented as having been uplifted relative to the block on the other side of the active fault line. The important difference between this as a mechanism for exhumation of eclogites compared to progressive thrusting, is that deep erosion need only be localized, concentrated on the uplifted fault block, possibly the area of present exposure of eclogites, rather than over the whole orogenic belt.

ACKNOWLEDGEMENTS: The fieldwork for this paper was carried out during the tenure of a Royal Society European Programme Research Fellowship. I thank G. Gosso, J. -M. Lardeaux and D. C. Rubie for initial guidance in the Sesia–Lanzo zone, and S. Schmid and H. Stünitz for comments on earlier versions of the manuscript.

References

BAIRD, A. W. & DEWEY, J. F. 1986. Structural evolution in thrust belts and relative plate motion: the upper Pennine Piémont zone of the internal Alps, southwest Switzerland and northwest Italy. *Tectonics*, **5**, 375–87.

CELMA, A. G. 1982. Domainal and fabric heterogeneities in the Cap de Creus quartz mylonites. *Journal of Structural Geology*, **4**, 443–55.

CHOUKROUNE, P., BALLEVRE, M., COBBOLD, P., GAUTIER, Y., MERLE, O. & VUICHARD, J. -P. 1986. Deformation and motion in the western Alpine Arc. *Tectonics*, **5**, 215–26.

COMPAGNONI, R. 1977. The Sesia–Lanzo zone: high pressure–low temperature metamorphism in the Austro–Alpine continental margin. *Rendiconti Societa Italiana di Mineralogia e Petrologia*, **33**, 335–74.

—— & MAFFEO, B. 1973. Jadeite-bearing metagranites l.s. and related rocks in the Mount Mucrone area (Sesia–Lanzo zone, western Italian Alps). *Schweizerische Mineralogische Petrographische Mitteilungen*, **53**, 355–78.

——, DAL PIAZ, G. V., HUNZIKER, J. C., GOSSO, G., LOMBARDO, B. & WILLIAMS, P. F. 1977. The Sesia–Lanzo zone, a slice of continental crust with Alpine high pressure–low temperature assemblages in the western Italian Alps. *Rendiconti Societa Italiana di Mineralogia e Petrologia*, **33**, 281–334.

DAL PIAZ, G. V., GOSSO, G. & MARTINOTTI, G. 1971. La II zona dioritico–kinzigitica tra la Valsesia e la Valle d'Ayas (Alpi Occidentali). *Memorie della Societa Geologica Italaliana*, **10**, 257–76.

——, HUNZIKER, J. C. & STERN, W. B. 1978. The Sesia–Lanzo zone, a slice of subducted continental crust. *U.S. Geological Survey Open File Report*, **78.707**, 83–6.

——, LOMBARDO, B. & GOSSO, G. 1983. Metamorphic evolution of the Mt Emilius Nappe, Dent Blanche Nappe, western Alps. *American Journal of Science*, **283A**, 435–58.

EMMONS, R. C. 1969. Strike-slip patterns in sand models. *Tectonophysics*, **7**, 71–87.

GOLDSMITH, J. R. 1982. Plagioclase stability at elevated temperatures and water pressures. *American Mineralogist*, **67**, 653–75.

GOSSO, G. 1977. Metamorphic evolution and fold history in the eclogitic micaschists of the upper Gressoney valley (Sesia–Lanzo zone, western Alps). *Rendiconti Societa Italiana di Mineralogia e Petrologia*, **33**, 389–407.

——, DAL PIAZ, G. V., PIAVANO, V. & POLMO, R. 1979. High pressure emplacement of early-Alpine nappes, post-nappe deformations and structural levels. *Memorie degli Instituti di Geologica Mineralogia dell Universita di Padova*, **32**, 5–15.

HOLLAND, T. J. B. 1980. The reaction albite = jadeite + quartz determined experimentally in the range 600–1200°C. *American Mineralogist*, **65**, 129–34.

—— 1983. The experimental determination of activities in disordered and short-range ordered jadeitic pyroxenes. *Contributions to Mineralogy and Petrology*, **82**, 214–20.

HURFORD, A. J. & HUNZIKER, J. C. 1985. Alpine cooling history of the Monte Mucrone eclogites (Sesia–Lanzo zone): fission track evidence. *Schweizerische Mineralogische Petrographische Mitteilungen*, **65**, 325–34.

HY, C. 1984. *Métamorphisme polyphasé et évolution tectonique dans la croûte continentale éclogitisée: les séries granulitiques et pélitiques du Monte Mucrone (Zone Sesia–Lanzo, Alpes Italiennes)*. Thèse 3ème cycle, University of Paris, 198 pp.

ISLER, A. & ZINGG, A. 1974. Geologie der Sesia-Zone zwischen Rimella und der Valle Anzasca (Nord Italien). *Schweizerische Mineralogische Petrolgraphische Mitteilungen*, **54**, 81–96.

KOONS, P. O. 1986. Relative geobarometry from high-pressure rocks of quartzo-feldspathic compositions from the Sesia zone: western Alps.

Contributions to Mineralogy and Petrology, **93**, 322–44.

——, RUBIE, D. C. & FRUEH-GREEN, G. 1987. The effects of disequilibrium and deformation on the mineralogical evolution of quartz diorite during metamorphism in the eclogite facies. *Journal of Petrology*, **28**, 679–700.

LARDEAUX, J. M., GOSSO, G., KIENAST, J. R., & LOMBARDO, B. 1982. Relations entre le métamorphisme et la déformation dans la zone de Sesia–Lanzo (Alpes occidentales), et le problème de l'éclogitisation de la croûte continentale. *Bulletin Societé géologique de France*, **24**, 793–800.

MARESCH, M. V. 1977. Experimental studies of glaucophane: an analysis of present knowledge. *Tectonophysics*, **43**, 109–23.

MINNIGH, L. D. 1977. A new klippe of 'II Zona Dioritica-Kinzigitica' in the Sesia–Lanzo zone. *Rendiconti Societa Italiana di Mineralogia e Petrologia*, **33**, 409–12.

NICOLAS, A. 1974. Mise en place des péridotites de Lanzo (Alpes Piémontaises). *Schwiezerische Mineralogische Petrographische Mitteilungen*, **54**, 449–70.

OBERHÄNSLI, R., HUNZIKER, J. C., MARTINOTTI, G. & STERN, W. G. 1985. Geochemistry, geochronology and petrology of Monte Mucrone: an example of eo-alpine eclogitisation of Permian granitoids in the Sesia–Lanzo zone, western Alps, Italy. *Chemical Geology*, **52**, 165–84.

PASSCHIER, C. W., URAI, J. L. VAN LOON, J. & WILLIAMS, P. F. 1981. Structural geology of the central Sesia–Lanzo zone. *Geologie en Mijnbouw*, **60**, 497–507.

PIDGEON, R. T., KÖPPEL, V. & GRÜNENFELDER, M. 1968. Isotopic U–Pb ages from the Ceneri zone, southern Alps. *Schweizerische Mineralogische Petrographische Mitteilungen*, **48**, 213–14.

PLATT, J. P. 1986. Dynamics of orogenic wedges and uplift of high pressure metamorphic rocks. *Geological Society of America Bulletin*, **97**, 1037–53.

REINHARDT, B. 1966. Geologie und Petrographie der Monte Rosa-Zone, der Sesia-Zone, und des Canavese im Gebiet zwischen Valle d'Ossola und Valle Loana (Prov. Novara, Italien). *Schweizerische Mineralogische Petrographische Mitteilungen*, **46**, 553–678.

RICOU, L. E. 1984. Les Alpes Occidentales: chaîne de décrochement. *Bulletin Societé géologique de France*, **26**, 861–74.

—— & SIDDANS, A. W. B. 1986. Collision tectonics in the western Alps. *In*: COWARD, M. P. & RIES, A. C. (eds). *Collision Tectonics. Special Publication of the Geological Society of London* No. 19, 229–44.

SELKMAN, S. O. 1983. Stress and displacement distributions around pyrite grains. *Journal of Structural Geology*, **5**, 47–52.

SPEAR, F. 1981. Amphibole–plagioclase equilibria: an empirical model for the relation albite + tremolite = edenite + 4 quartz. *Contributions to Mineralogy and Petrology*, **77**, 355–64.

STÖCKHERT, B., JÄGER, E. & VOLL, G. 1986. K–Ar age determinations on phengites from the internal part of the Sesia zone, western Alps, Italy. *Contributions to Mineralogy and Petrology*, **92**, 456–70.

WILCOX, R. E., HARDING, T. P. & SEELY, D. R. 1973. Basic wrench tectonics. *Bulletin of the American Association of Petroleum Geologists*, **57**, 74–96.

WILLIAMS, P. F. & COMPAGNONI, R. 1983. Deformation and metamorphism in the Bard area of the Sesia–Lanzo zone, western Alps, during subduction and uplift. *Journal of Metamorphic Geology*, **1**, 117–40.

WOODCOCK, N. H. & FISCHER, M. 1986. Strike-slip duplexes. *Journal of Structural Geology*, **8**, 725–35.

J. RIDLEY, Departement für Erdwissenschaften, ETH-Zentrum, CH-8092, Zürich, Switzerland. *Present address*: Department of Geology, University of Zimbabwe, PO Box MP 167, Mount Pleasant, Harave, Zimbabwe.

On the age of deep crustal extensional faulting in the Ivrea zone, northern Italy

K. H. Brodie, D. Rex & E. H. Rutter

SUMMARY: High temperature fault zones in the Valle d'Ossola region of the Ivrea zone in northern Italy appear to accommodate extension of the deep crustal section. $^{40}Ar-^{39}Ar$ age spectra indicate that this extensional event occurred prior to 280 Ma, and initiated a period of crustal thinning and cooling. Interpretation of the closure temperature in this system is dependent on grain size and cooling rate.

The Ivrea zone in northern Italy exposes a sequence of amphibolite to granulite-facies metasedimentary, metabasic and ultrabasic rocks, thought to represent an upended section through the lower continental crust as it was at the end of the Hercynian orogeny. Recent work (Brodie & Rutter 1987) has demonstrated the existence of high temperature fault zones in the deeper part of the sequence that appear to accommodate extension of the deep crustal section. The aim of this contribution is to summarize the evidence for deep crustal extension in this region, and to present and discuss new evidence concerning the age of the extensional episode.

Regional setting

The Ivrea zone consists of a NE–SW–trending, steeply dipping sequence of metasedimentary and meta-igneous rocks (Fig. 1) which have been deformed and metamorphosed during the Hercynian and Alpine orogenies, and possibly during pre-Hercynian times (Schmid 1967, review by Zingg 1983). Metamorphic grade increases from amphibolite facies in the SE to granulite facies in the NW where the zone is truncated by the Insubric Fault, which juxtaposes the Ivrea rocks against Alpine metamorphic rocks of the Canavese zone. In the SE the zone is separated by the Pogallo Fault from the Strona–Ceneri zone which contains greenschist to amphibolite-facies ortho- and paraschists and gneisses (e.g. Boriani & Giobbi 1981).

Within the Ivrea zone the existence of mylonitic fault zones was recognized by Schmid (1967). Brodie (1981) showed that the higher temperature fault zones formed subsequent to the peak of the high temperature regional deformation and metamorphism, and while the rocks were still at high, or locally even rising, temperatures (in some cases greater than 650°C, by an amount dependent on water activity). This is evidenced by the late-tectonic growth of orthopyroxene in the Anzola metabasic shear zone (Brodie & Rutter 1985). Some of the shear zones may have formed at slightly lower temperatures, subject to the requirements that the temperature was at least high enough to favour the development of hot-work microstructures in metabasic rocks (at least 500°C) and that water in sufficient quantities to cause hydration reactions remained excluded.

In a more recent study (Brodie & Rutter 1987) we have mapped the distribution and geometry of the high temperature shear zones in the structurally lower part of the Ivrea zone. We have argued that, after removal of the effects of later, lower temperature deformation which was responsible for large scale flexural folding of the sequence and tilting of the layering to a vertical attitude, the high temperature shear zones represent low-angle extensional faults which divide the lower Ivrea zone into three tectonic blocks. Metamorphic grade increases with depth across the fault boundaries.

Evidence for the extensional nature of the faulting comes from the consistent way in which the faults transect the regional lithological banding with the sense of shear such as to excise part of the stratigraphic sequence. Movement sense on the faults was inferred largely from the way in which the banding of the host rock is deflected into the mylonite zones and from the geometry of shear bands within them. The pattern of local extension of the rocks was interpreted as part of a crust-wide pattern of extension on the basis that:

(1) The movement picture inferred for the structurally overlying Pogallo fault (Handy 1987), which may have been contemporaneously active, is almost identical.
(2) The fault displacements emplaced cooler rocks on to hotter ones.
(3) Such a displacement pattern best explains

From: Coward, M. P., Dietrich, D. & Park, R. G. (eds), 1989, *Alpine Tectonics*, Geological Society Special Publication No. 45, pp. 203–210.

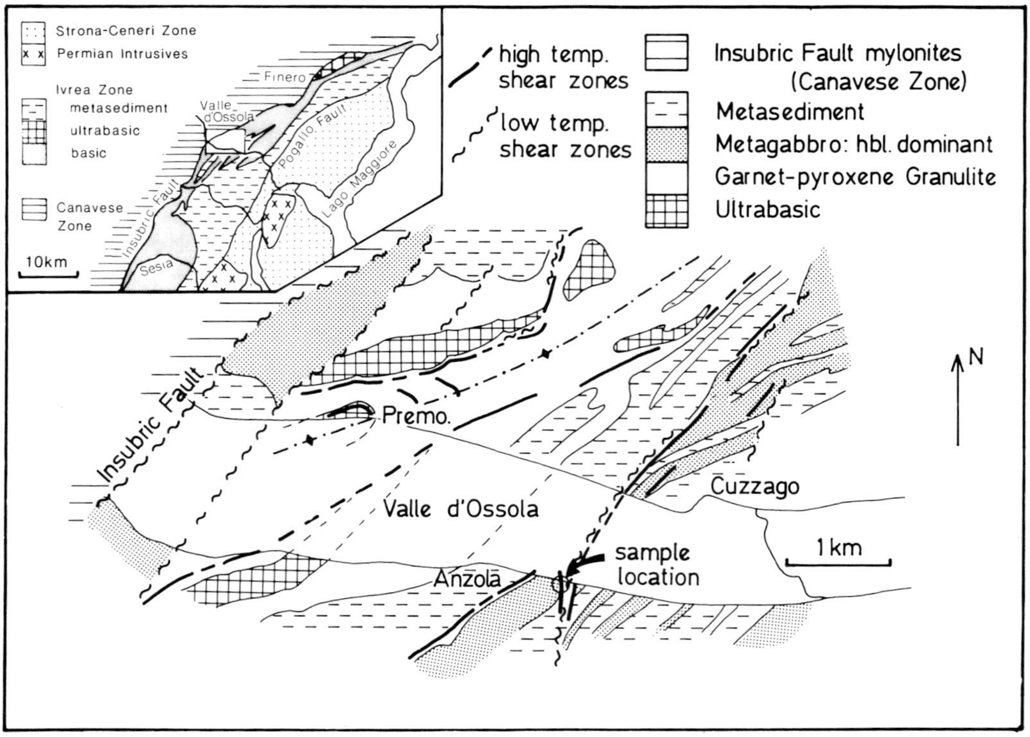

FIG. 1. Simplified geology of the Valle d'Ossola region of the Ivrea zone showing the high temperature shear zones, and the location of the samples dated. Premo = Premosello (after Brodie & Rutter 1987).

the microstructural and mineralogical evidence for heating of at least one of the fault zones (the Anzola shear zone) at the end of the slip period.

(4) Provided the age of the faulting is late Palaeozoic or early Mesozoic, the regional tectonic framework for much of that period appears to have been dominated by crustal distension (e.g. Winterer & Bosellini 1981).

Age of regional metamorphism

Some uncertainty exists over the age of the regional granulite-facies metamorphism within the Ivrea zone (for detailed discussion see Zingg 1983). Whole-rock Rb–Sr determinations on para-gneisses (Hunziker & Zingg 1980) have been interpreted in terms of the peak of the metamorphism being Ordovician in age (478±20 Ma), with the Ivrea zone subsequently undergoing very slow cooling until Hercynian times (temperatures of ~ 700°C persisting over 150 Ma). It has been suggested that this metamorphic peak was associated with intrusion of basic magma into the lower crust and partial melting of the para-gneisses (Schmid & Wood 1976). More recent Sm–Nd data (Voshage *et al.* 1987) suggest that a crustal formation event occurred at approximately 600 Ma, and these authors also present evidence for younger (~290 Ma) intrusive activity. Pin & Vielzeuf (1983) consider that the granulite-facies metamorphism occurred at the end of the Hercynian orogeny (300±20 Ma) rather than the rocks having been at granulite facies temperatures since the Ordovician. In general, the radiometric dating evidence points to a multiple intrusion history for the basic rocks (Voshage *et al.* 1987).

Younger dates have been obtained on individual minerals and are thought to date the cooling history of the region (Fig. 2). These mineral cooling ages suggest that relatively rapid cooling of the rocks occurred between 270 and 200 Ma. This has been interpreted in terms of crustal extension which would have initiated such a period of cooling (Zingg 1983, Hodges & Fountain 1984). Evidence from the sedimentary cover of the southern Alps region is also consistent with Permo–Carboniferous extensional or transtensional faulting and basin formation

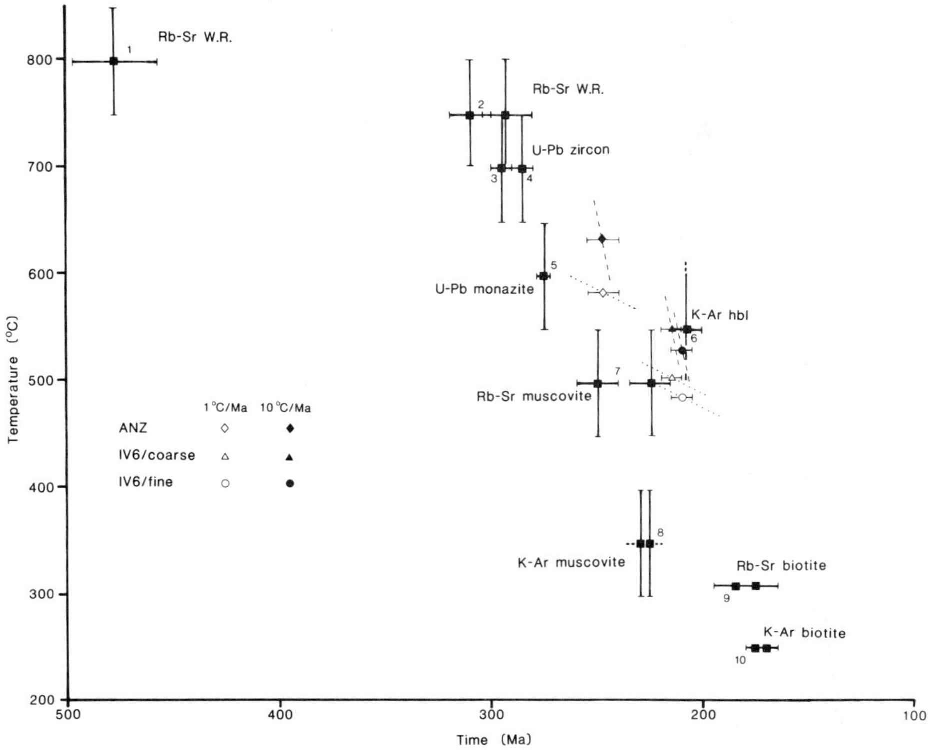

FIG. 2. Temperature–time curve based on published radiometric dates showing the cooling history of the Ivrea zone (partly after Zingg 1983). The data for samples ANZ and IV6 are included, using the closure temperature for Ar in hornblende estimated at cooling rates of 1° and 10° C/Ma for the grain size of the hornblendes dated (see Fig. 4). The dotted and dashed lines through these points represent cooling rates of 1° and 10° C/Ma respectively. Additional data points are identified by numbers as follows – 1: Hunziker & Zingg 1980; 2: Graeser & Hunziker 1968, Voshage *et al.* 1987; 3 & 5: Koppel, 1974; 4: Pin 1986; 6 & 10: McDowell & Schmid 1968; 7 & 8: Hunziker 1974; 9: Jäger *et al.* 1967, Graeser & Hunziker 1968. The error bars shown are as given in the cited references. However, the uncertainty in temperatures associated with the Rb–Sr data points 1 and 2 is probably much greater than shown (Zingg 1988, pers. comm.).

and associated volcanic activity (Winterer & Bosellini 1981, Laubscher & Bernoulli 1982). Alternatively, the high temperature extensional faults could relate to the period of crustal thinning associated with the formation of the Tethyan Ocean during late Triassic to Lower Jurassic time (230–180 Ma). Recent work on the Pogallo Fault (Handy 1987), which transects at a small angle the south-eastern boundary of the Ivrea zone, has shown it to be a mid-crustal extensional fault, with an almost identical movement picture to the high-temperature extensional faults within the Ivrea zone. Schmid *et al.* (1987) and Handy (1987), have argued that the Pogallo fault is related to the later period of crustal extension and, given the similarity of movement pictures, it is tempting to suggest that the Ivrea zone faults are a deeper crustal expression of this same period of extension. In an attempt to resolve this conflict, but bearing in mind the problems associated with dating of fault rocks, one of the metabasic shear zones was dated using the $^{40}Ar/^{39}Ar$ age spectrum technique.

Samples dated

In order to try to place constraints on the age of the high temperature faulting, amphibole grains from two samples were analysed, one from a metabasic shear zone and the other from the adjacent metagabbro host rock, both from Anzola quarry in the Valle d'Ossola.

Sample ANZ represents the relatively undeformed metagabbro from Anzola. It is a medium grained, equigranular rock with a granoblastic texture and is composed of calcic amphibole and plagioclase, with minor Fe–Ti oxides and clinopyroxene. It possesses a weak foliation defined by grain-shape. The grain size of the amphibole is ~ 2 mm, and it lies compositionally between a magnesio- and an edenitic-hornblende ($Mg/Mg+Fe^{2+}$ ~ 0.68).

Sample IV6 is from a high temperature metabasic shear zone, also from Anzola Quarry. It is a fine-grained mylonite with porphyroclasts of amphibole (ranging in size from 1 mm down to ~ 400 μm) in a matrix of amphibole, plagioclase and minor Fe–Ti oxides. Minor alteration along grain boundaries is locally observed. The grain size of the amphibole in the matrix is in the order of 150 μm. The amphibole is also of edenitic-hornblende composition. Two separate grain-size fractions were analysed from this sample, a coarse fraction of grains between 250 and 300 μm, and a fine fraction of grains less than 140 μm. The grain size of the coarser fraction corresponds to the porphyroclasts, whilst the fine fraction relates to the matrix amphibole. The deformation producing this shear zone is inferred to have occurred at temperatures of at least 650°C (Brodie & Rutter 1985), but evidence is seen within parts of this sample for the effects of cataclastic deformation superimposed upon intracrystalline-plastic deformation microstructures. Relatively high temperatures outlasted the high temperature deformation causing recovery and annealing of the microstructure. The superimposed cataclastic effects (cleavage cracking present in all phases) may have been imposed during Alpine orogenesis and has implications for the interpretation of the radiometric age, according to the temperature at which cracking occurred.

Experimental technique

Full experimental details of the irradiation and argon analysis used at Leeds are to be found in Parsons *et al.* (1988). The hornblende samples were purified using a Franz magnetic separator and the final 100% purity of the c. 75 mg samples irradiated were checked under a binocular microscope. International standard minerals, HB3GR hornblende, LP–6 biotite and Nancy Fe biotite, together with a Leeds laboratory standard hornblende FY12a, were used as 'J' monitors for the irradiation (Roddick 1983). Errors quoted on the original steps are analytical errors while the errors quoted on the total fusion ages include the 'J' value errors (Table 1).

Results and discussion

All three dated samples produce U-shaped spectra (Fig. 3), indicating the presence of small amounts of excess radiogenic argon in the low and high temperature steps. The undeformed metagabbro has a minimum age of 247±7.1 Ma with 53% of the gas. There is no reason to

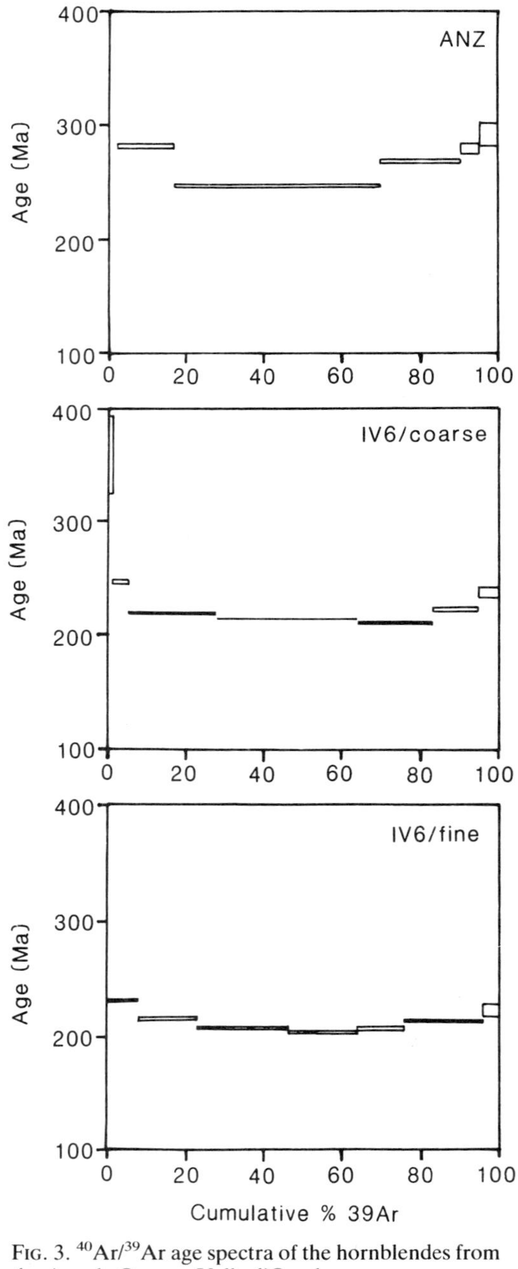

FIG. 3. $^{40}Ar/^{39}Ar$ age spectra of the hornblendes from the Anzola Quarry, Valle d'Ossola.

TABLE. 1. $^{40}Ar/^{39}Ar$ *results for hornblende from Anzola Quarry.*

Sample ANZ Hornblende

Temp (°C)	39K	37Ca	38Cl	Ca/K	*40/39K	%Atm 40	App.Age (Ma)	Err	%39
810	0.036	0.753	0.101	41.2536	425.628	28.1	2319.74	90.05	0.5
875	0.022	0.545	0.035	49.0492	120.830	27.1	1002.48	31.99	0.3
930	0.094	1.481	0.041	31.2378	45.766	23.8	447.41	16.28	1.2
980	1.129	19.404	0.230	34.2105	27.451	11.8	281.43	1.88	14.9
1040	3.981	71.767	0.639	35.8782	23.890	7.7	247.32	0.61	52.7
1090	1.575	27.201	0.290	34.3706	26.034	13.3	267.93	1.18	20.8
1155	0.357	6.436	0.066	35.8860	27.344	43.9	280.42	5.02	4.7
1350	0.364	4.903	0.061	26.8279	28.686	65.2	293.12	9.46	4.8

Integrated values
*40/39K 27.750
Age 284.27 Ma. Errors: Analytic 1.80 and J 5.56 Ma.
Wt %K 0.231 *40 27.61 x 10-7 cc/gm

Sample IV6/Coarse hornblende

Temp (°C)	39K	37Ca	38Cl	Ca/K	*40/39K	%Atm 40	App.Age (Ma)	Err	%39
785	0.109	0.677	0.054	12.4122	119.874	73.6	1000.12	76.03	0.6
890	0.091	0.855	0.029	18.7252	35.584	58.1	358.61	34.24	0.5
950	0.756	6.198	0.148	16.3138	23.690	20.0	246.50	2.53	4.4
990	3.863	31.406	0.652	16.1796	20.817	7.2	218.33	0.61	22.4
1020	6.285	51.120	1.022	16.1856	20.326	4.3	213.48	0.49	26.5
1070	3.318	26.333	0.547	15.7921	19.909	5.0	209.34	0.75	19.3
1170	2.006	15.823	0.353	15.6954	21.071	8.5	220.85	1.20	11.6
1310	0.798	5.626	0.118	14.0212	22.592	37.6	235.78	4.21	4.6

Integrated values
*40/39K 21.403
Age 224.11 Ma. Errors: Analytic 0.91 and J 4.31 Ma.
Wt %K 0.506 *40 46.80 x 10-7 cc/gm

Sample IV6/Fine hornblende

Temp (°C)	39K	37Ca	38Cl	Ca/K	*40/39K	%Atm 40	App.Age (Ma)	Err	%39
830	0.178	1.667	0.077	18.6514	97.248	66.7	849.05	34.08	0.9
940	1.441	8.679	0.485	11.9896	22.049	17.8	230.46	1.43	7.3
970	3.010	16.635	0.967	10.9961	20.467	9.2	214.88	0.58	15.2
1000	4.630	25.536	1.451	10.9760	19.676	6.2	207.03	0.44	23.3
1020	3.510	19.472	1.088	11.0412	19.348	4.6	203.76	0.95	17.7
1060	2.377	13.259	0.747	11.0988	19.611	6.3	206.38	1.46	12.0
1175	4.061	23.173	1.272	11.3564	20.316	6.1	213.38	0.73	20.5
1300	0.630	3.596	0.182	11.3634	21.300	37.0	223.10	3.45	3.2

Integrated values
*40/39K 20.781
Age 217.98 Ma. Errors: Analytic 0.62 and J 4.15 Ma.
Wt %K 0.714 *40 64.28 x 10-7 cc/gm

believe that the spectrum would have come below 250 Ma if the temperature step had been further divided. The very shallow U-shape spectrum for the shear zone samples point to ages of 215±5.2 Ma (90% gas) for the coarse grained amphibole fraction and 210±5 Ma (92% gas) for the finer grained amphibole fraction.

In order to relate these results to the regional temperature-time path, it is necessary to consider the closure temperature of hornblende to argon diffusion. The closure temperature is dependent on both the effective grain size of the amphibole and the rate of cooling of the rock (Dodson 1973, 1979, Harrison 1981).

Closure temperature is defined as the temperature through which the rock passed at the determined date (Dodson 1973), and is given by

$$T_c = \frac{E/R}{\ln((ART_c^2\, D_o/a^2)/(E\, dT/dt))}$$

in which A is a geometric factor (=55 for spherical grains), E is the activation enthalpy for argon diffusion in the mineral. D_o is the Ar diffusivity extrapolated to infinite temperature on an Arrhenius diagram, a is the 'effective' grain radius, R is the gas constant, T_c is the closure temperature and dT/dt is the cooling rate. In Fig. 4 this equation has been evaluated for T_c of amphibole in order to show the effect of cooling rate and effective grain-size on closure temperature. Diffusion data for argon in hornblende are as given by Harrison (1981).

The cooling history for the rocks of the Ivrea zone can be estimated by plotting closure temperature versus age for the various mineral and whole-rock ages available (e.g. Zingg 1983). Our new hornblende ages are shown on this plot in Fig. 2, using observed grain-sizes of the rocks in question. The effect of variations in assumed cooling rate between 1° and 10°C/Ma are also shown. The ages obtained are clearly broadly consistent with pre-existing age determinations, the variations in grain-size between samples and the apparent cooling rate when the cooling rate is about 4°C/Ma. Note that part of the scatter of data shown on this figure must arise from the fact that the assembled data shown are not taken from the same depth level within the crustal section.

In the above discussion we have used the term 'effective' grain-size. The effective grain-size controlling diffusive loss of argon during cooling of a mineral grain need not be the true grain-size. Consider, for example, a large hornblende crystal in a rock in which the pore fluid pressure is low, hence the effective pressure is high. The existence of high-permeability fluid pathways along cleavage cracks in the grain will be suppressed, and the effective grain-size will equal the true grain-size. On the other hand, if the pore-fluid pressure approaches the total pressure, high permeability pathways may exist within a grain containing cleavage cracks (Brace *et al.* 1968), thus the effective grain-size may be much smaller than the true grain-size. This will cause anomalously young ages to be obtained.

The above argument based on the effects of

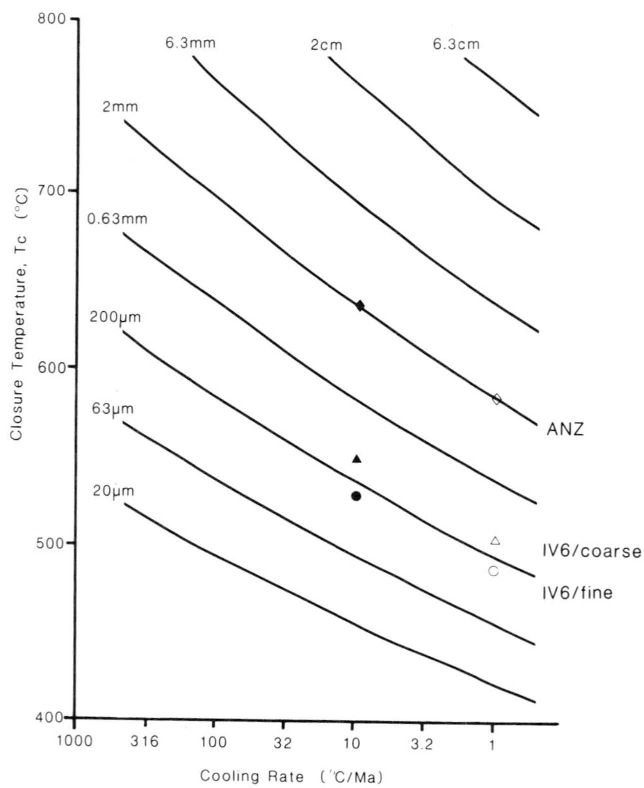

FIG. 4. Graph of closure temperature against cooling rate, contoured for grain size, for Ar diffusion in hornblende, using the diffusion data given by Harrison (1981). The equation for T_c is given in the text. The approximate location on this graph of the samples dated in this paper are indicated showing the values of T_c used in Fig. 2.

pore fluid pressure variations applies equally to grains whether or not they have been subjected to deformation effects after crystallization. Provided the temperature has not dropped too low relative to the expected range of closure temperatures, deformation effects may reduce the effective grain-size of minerals. This may be accomplished by the effects of cataclasis, even under elevated effective pressures. High temperature plastic deformation may result in enhanced argon diffusivity as a result of increased dislocation and point defect densities, and as a result of their motion. It is likely that one or other of these effects has had some influence on the dates obtained from the sheared metagabbro sample reported here. Of these however, high temperature plastic deformation is unlikely to have been a problem in this case because it occurred at a temperature higher than the expected range of closure temperatures. Onstott & Peacock (1987) have shown that the presence of fine-grained phyllosilicate inclusions within hornblende, probably produced as a result of alteration during fluid migration through the hornblende, may similarly reduce the 'effective' grain size, depending on whether the alteration occurred at temperatures above or below the closure temperature. Such intergrowths may be too fine grained to be observed optically or by scanning electron microscopy. Amphiboles from Ivrea zone samples similar to those dated here do not show such alteration intergrowths when examined by transmission electron microscopy, which is consistent with the observation that such alteration is more common in Fe-rich amphibole relative to Mg-rich amphibole (Onstott & Peacock 1987).

The ages obtained for the sheared sample and its immediate host rock are less than that of the undeformed metagabbro host rock, which in turn yields an age that must be less than that of the time of metamorphic crystallization. Extrapolation of the cooling rate backwards to the higher temperatures required for crystallization of the metamorphic mineral assemblage (about 650°C) suggests that it could have formed not less than 300 Ma ago. The whole-rock Rb–Sr age of 478 Ma, interpreted by Hunziker & Zingg (1980) for the peak of metamorphism, provides a maximum boundary for the age of the high temperature fault rocks, but the probable association of the faulting with the cooling history means that the younger age for the onset of faulting is more likely. This inference is also supported by the fact that high temperature shear zones also occur in the phlogopite–peridotite of Finero (Brodie 1980). These cut rocks which had previously suffered granulite-facies metamorphism and which gave a Rb–Sr whole-rock date of *c.* 290 Ma (Graeser & Hunziker 1968). A similar age is attributed to the phlogopite metasomatism (Voshage *et al.* 1987).

The extensional faulting appears to have caused cooling corresponding to about 400°C over 100 Ma. Brodie & Rutter (1987) estimated that the Anzola shear zone, which is about 10 m wide, accommodated about 5 km of displacement (shear strain = 500) and caused about 2 km of crustal thinning. A minimum estimate for the shear strain-rate is therefore (500/10) Ma^{-1} $= 1.7\times10^{-13}\ \text{sec}^{-1}$. It may easily have been up to 100 times greater if crustal extension were episodic or concentrated into the early part of the cooling history.

Conclusions

High temperature shear zones which accommodate extension of the layered rocks of the structurally lower part of the Ivrea zone were initiated prior to 280 Ma, probably around 300 Ma, and began a period of crustal thinning and cooling of more than 100 Ma duration. The considerably younger dates (215 and 210 Ma) obtained from sheared metabasic rocks in which the grain-size has been reduced are interpreted in terms of the effects of grain-size reduction on the closure temperature for argon in hornblende, and help further to chart the cooling history of the rocks. The dates obtained in this study are consistent with previously published radiometric ages which document the early Mesozoic cooling of the region.

ACKNOWLEDGEMENTS: Thanks go to Mr T. Oddy and Mr D. Pickles for their painstaking mineral separation. Mr P. Guise for invaluable work with the argon data handling, and Mr T. Walters of AWRE Aldermaston for the reactor facilities. Leeds Isotope Laboratory acknowledges NERC Grant No. GR3/5932. K.H.B and E.H.R acknowledge NERC Grant No. GR3/5636.

References

Boriani, A. & Giobbi, E. O. 1981. Composition, level of intrusion and age of the 'Serie dei Laghi' orthogneisses (northern Italy–Ticino, Switzerland). *Societa Italia Mineralogia e Petrologia*, **38**, 191–205.

Brace, W. F., Walsh, J. B. & Frangos, W. T. 1968. Permeability of granite under high pressure. *Journal of Geophysical Research*, **73**, 2225–36.

Brodie, K. H. 1980. Variations in mineral chemistry across a phlogopite peridotite shear zone. *Journal of Structural Geology*, **2**, 265–72.

—— 1981. Variation in amphibole and plagioclase composition with deformation. *Tectonophysics*, **78**, 385–402.

—— & Rutter, E. H. 1985. On the relationship between deformation and metamorphism with special reference to the behaviour of basic rocks. *In*: Thompson, A. B. & Rubie, D. C. (eds). *Metamorphic Reactions: Kinetics, Textures and Deformation*. Advances in Physical Geochemistry 4, Springer, Berlin, 138–79.

—— & —— 1987. Deep crustal extensional faulting in the Ivrea Zone of Northern Italy. *Tectonophysics*, **140**, 193–212.

Dodson, M. H. 1973. Closure temperature in cooling geochronological and petrological systems. *Contributions to Mineralogy and Petrology*, **43**, 55–70.

—— 1979. Theory of cooling ages. *In*: Jäger, E. & Hunziker, J. C. (eds). *Lectures in Isotope Geology*. Springer, 194–202.

Graeser, S. & Hunziker, J. C. 1968. Rb–Sr- und Pb-Isotopenbestimmugen an Gesteinen und Mineralien der Ivrea–zone. *Schweizerische Mineralogische und Petrographische Mitteilungen*. **48**, 189–204.

Harrison, T. M. 1981. Diffusion of ^{40}Ar in hornblende. *Contributions to Mineralogy and Petrology*, **78**, 324–31.

Handy, M. R. 1987. The structure, age and kinematics of the Pogallo Fault Zone; southern Alps, northwestern Italy. *Eclogae geologicae Helveticae*, **80**, 593–632.

Hodges, K. V. & Fountain, D. M. 1984. Pogallo Line, south Alps, northern Italy: an intermediate crustal level, low angle normal fault? *Geology*, **12**, 151–5.

Hunziker, J. C. 1974. Rb–Sr and K–Ar age determination and the Alpine tectonic history of the western Alps. *Memorie degli Istituti di Geologiae Mineralogia dell'Universita di Padova*, **31**.

—— & Zingg, A. 1980. Lower Palaeozoic amphibolite to granulite facies metamorphism in the Ivrea zone (southern Alps, N Italy). *Schweizerische Mineralogische und Petrographische Mitteilungen*, **60**, 181–213.

Jäger, E., Niggli, E. & Wenk, E. 1967. Rb–Sr-Altersbestimmungen an Glimmern der Zentralpen. *Beitrage zur geologischen Karte der Schweiz. N. F.*, **134**, 67 pp.

Koppel, V. 1974. Isotopic U–Pb ages of monazites and zircons from the crust-mantle transition and adjacent units of the Ivrea and Ceneri zones (southern Alps, Italy). *Contributions to Mineralogy and Petrology*, **43**, 55–70.

Laubscher, H. P. & Bernoulli, D. 1982. History and deformation of the Alps. *In*: K. J. Hsü (ed). *Mountain Building Processes*, 169–80. Academic Press, London.

McDowell, F. W. & Schmid, R. 1968. Potassium Argon ages from the Valle d'Ossola section of the Ivrea–Verbano zone (N Italy). *Schweizerische Mineralogische und Petrographische Mitteilungen*, **48**, 205–10.

Onstott, T. C. & Peacock, M. W. 1987. Argon retentivity of hornblendes: a field experiment in a slowly cooled metamorphic terrane. *Geochimica el Cosmochimica Acta*, **51**, 2891–903.

Parsons, I., Rex, D. C., Guise, P. & Halliday, A. N. 1988. Argon loss by alkali feldspars. *Geochimica et Cosmochimica Acta* **52**, 1097–112.

Pin. C. 1986. Datation U–Pb sur zircons a 285 Ma du complexe gabbro–dioritique du Val Sesia–Val Mastallone et age tardihercynien du metamorphisme granulitique de la zone Ivrea–Verbano (Italie). *Comptes-Rendus Academic de Science de Paris*, **303**, 827–30.

—— & Vielzeuf, D. 1983. Granulites and related rocks in Variscan Median Europe: a dualistic interpretation. *Tectonophysics*, **93**, 47–74.

Roddick, J. C. 1983. High precision intercalibration of ^{40}Ar–^{39}Ar standards. *Geochimica et Cosmochimica Acta*, **47**, 887–98.

Schmid, R. 1967. Zur Petrographie und Struktur der Zone Ivrea–Verbano zwischen Valle d'Ossola und Val Grande (Prov. Novara, Italien). *Schweizerische Mineralogische und Petrographische Mitteilungen*, **47**, 935–1117.

—— & Wood, B. J. 1976. Phase relationships in granulitic metapelites from the Ivrea–Verbano zone (northern Italy). *Contributions to Mineralogy and Petrology*, **54**, 255–79.

Schmid, S. M., Zingg, A. & Handy, M. R. 1987. The kinematics of movements along the Insubric line and the emplacement of the Ivrea zone. *Journal of Structural Geology*, **135**, 47–66.

Voshage, H., Hunziker, J. C., Hofmann, A. W. & Zingg, A. 1987. A Nd and Sr isotopic study of the Ivrea zone, southern Alps, N Italy. *Contributions to Mineralogy and Petrology*, **97**, 31–42.

Winterer, E. L. & Bosellini, A. 1981. Subsidence and sedimentation on Jurassic passive continental margin, southern Alps, Italy. *Bulletin of the American Association of Petroleum Geologists*, **65**, 394–421.

Zingg, A. 1983. The Ivrea and Strona–Ceneri zones (southern Alps, Ticino and N Italy)–a review. *Schweizerische Mineralogische und Petrographische Mitteilungen*, **63**, 361–92.

K. H. Brodie & E. H. Rutter, Department of Geology, Imperial College, London SW7 2BP, UK.

D. Rex, Department of Earth Sciences, University of Leeds, LS2 9JT, UK.

South-Alpine thrusting and trans-Alpine convergence

D. Roeder

SUMMARY: The south front of the Alps in northern Italy contains a Neogene S-vergent fold-thrust belt. It affects the crust of the Adria plate, a series of shelf sediments, and S Alpine foredeep fill. The belt shows that after collision, the Alps assumed the bivergent wedge shape of an intracratonic compressional belt, notwithstanding its long prior history of N-vergent subduction and thrust imbrication.

Parts of the S Alpine fold-thrust belt are buried beneath Apennine foredeep fill. Thrust transport is about 60 km to the SE; it has produced detached foothills folds, a triangle zone of stacked blind and emergent thrusts, and a basement front with out-of-sequence thrusting above a line of duplex anticlines. The structure of the foreland can be described by the curvature of an elastic line-load deflection. Details not predicted by the load model include Mesozoic rifting, foreland upthrusts, and perhaps complex map patterns of thrust fronts.

A match between trans-Tethys convergence and Alpine kinematics suggests either that a 150 km wide oceanic gap north of the Tauern was closed in late Miocene. Or it suggests an origin of the Tauern terranes on the S margin of a sub-Apennine crustal slab. The first choice is consistent with traditional Alpine crustal data but inconsistent with Mesozoic palaeogeography. The second choice is consistent with newer, EGT-generated, crustal data from the Ligurian–Apennine area. It admits 26% compression or 80 km bulk strain within the upper crustal slab while plate convergence generates 380 km of thrust overlap.

A conceptual model of Neogene Alpine convergence, consistent with the second choice, assumes increased traction where the basal detachment of the upper crustal slab passes upward into the brittle shell. Part of the viscous overlap at plate convergence rates is translated upward into compression within a back-to-back pair of thickening and over-critical Coulomb wedges.

South-vergent thrusting in the Italian Alps has been described on maps (Castellarin 1981 and many others), in classical cross sections (Gortani & Desio 1925, De Sitter & De Sitter-Koomans 1949, Pieri & Groppi 1981), on reflection-seismic profiles (Chierici *et al.* 1979, Pieri in Bally 1983), and in tectonic analyses (Gaetani & Jadoul 1979, Castellarin 1983, Doglioni 1985, Laubscher 1985, Cassano *et al.* 1986). South-vergent thrusting is causing a major seismic hazard in part of the Italian Alps (Gebrande *et al.* 1978, Slejko *et al.* 1987). South-vergent thrusts involving Mesozoic strata have been encountered in two wells (Belluno and Bernadia) drilled by petroleum industry.

However, there are few published attempts to quantify S Alpine thrusting. One reason is that the central Alpine architecture of north-vergent crystalline nappes is perceived to accommodate only a thick-skinned belt of minor thrusts (Errico *et al.* 1980, Trumpy 1980, plate 1, profiles 1 and 2). A more severe reason is that a large part of the S Alpine thrust terrane is buried under post-kinematic basin fill.

The present dynamic analysis of this belt is based on published data and on current analytical techniques. Available data have been condensed into structure cross sections which describe a narrow thrust belt in the footwall of a basement thrust, as well as its setting, stratigraphy, and structural style. In subsequent steps, I test the interpretation by simple modelling techniques. It is tested further by matching it against specifically Alpine data sets, including the Alpine crustal model (Mueller *et al.* 1980), newer crustal data from the south part of the European Geotraverse (EGT), (Galson & Mueller 1985, Giese 1985), and unpublished results from the Nato Advanced Research Workshop in Gradisca d'Isonzo (Giese & Nicolich 1987). The interpretation finally is matched against local adaptations of the plate convergence model (Pitman & Talwani 1972 and others).

Setting

Figure 1 is a tectonic map of the southern Alps showing thrust fronts, basement highs, the extent of Plio–Pleistocene cover, and the location of cross sections.

Castellarin & Vai (1982) and Fig. 1 show the south Alpine thrust belt extending along the Alpine–Dinaride arc. The belt forms two smaller arcs that are concave to the S and are

From: Coward, M. P., Dietrich, D. & Park, R. G. (eds), 1989, *Alpine Tectonics*, Geological Society Special Publication No. 45, pp. 211–227.

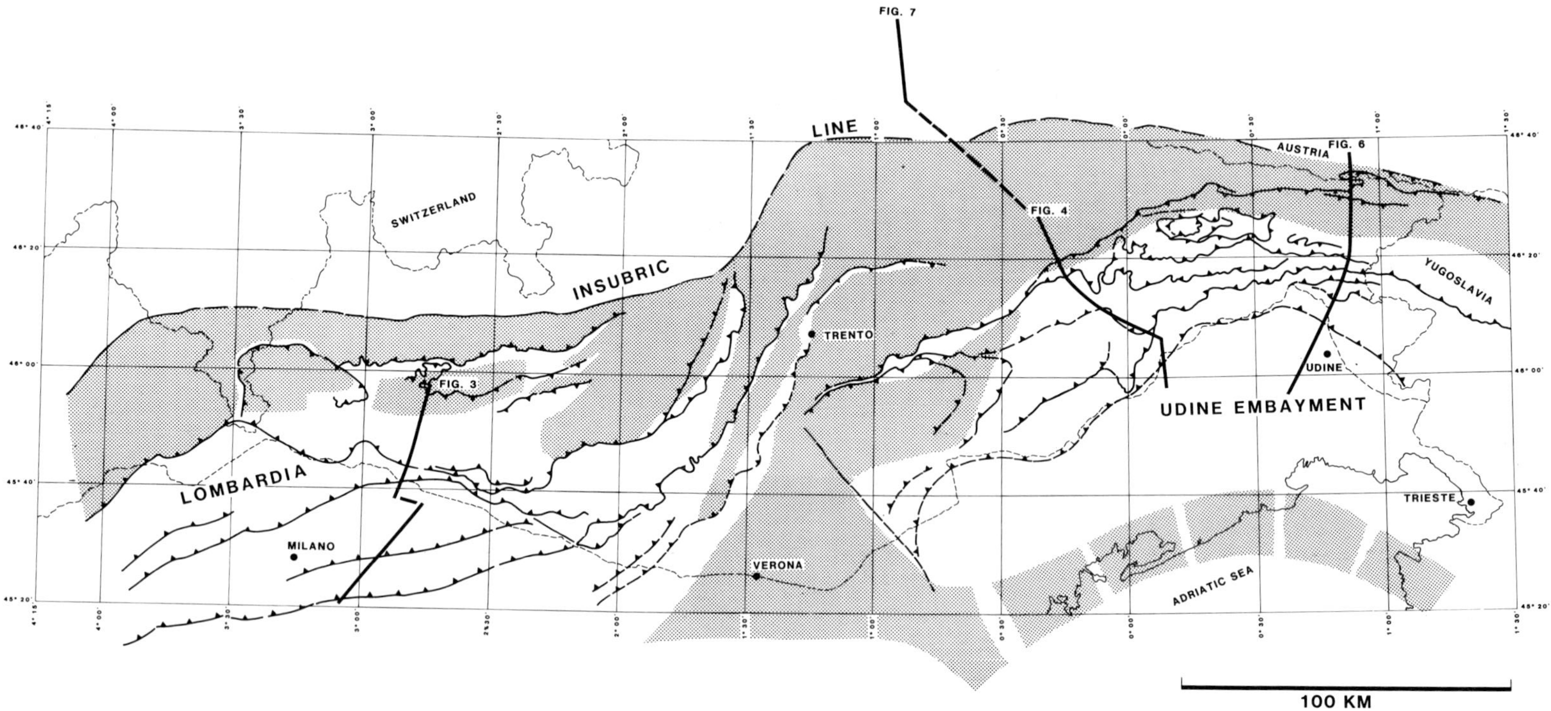

FIG. 1. Tectonic map of the S Alpine fold-thrust belt, after Castellarin (1981) and many others, showing location of structure cross sections, major thrusts with barbs on hangingwall, the Insubric Line and other minor faults. Dot pattern signifies high, but not necessarily outcropping, basement uplifted on thrusts, on foreland blocks, and in a flexural forebulge.

offset or relayed at the Verona–Trento high. The western of these arcs is covered by the location name of Lombardia. The eastern arc conforms to the Udine embayment. The arc pattern is evident in thrust fronts curving northeasterly from the Po Plain into the Alpine area N of Milano. Thrust fronts also curve easterly and southeasterly around the Udine embayment. The front of the belt shows a variety of styles. Emergent thrust fronts and/or blind thrusts predominate in the Udine embayment. A plainward-dipping series above an upper detachment or underthrust prevails in Lombardia.

Near the internal edge of the south-Alpine belt, there is an en-echelon group of basement thrusts with discordant geometry suggesting either a strike-slip component or out-of-sequence kinematics. The eastern segment of this group of thrusts is the Valsugana line, recently discussed by Doglioni (1985). A western segment is inferred from basement uplifts named Orobic, Sub-orobic, and Valtrompia. The connection between the Valsugana segment and the western, Lombardian, segment is interrupted at, or obscured by, the tilted and faulted W flank of the Verona–Trento high. At depth and toward the N, perhaps all S Alpine and NW Dinaride basement fronts merge into a major transcrustal thrust. This has been assumed without evidence in the cross sections of the present paper. The assumption is a simple requirement of the thrust model used. The requirement follows from a reasonable (assumed) range of ramp angles, from the finite vertical offset between foreland and basement fronts, and from a reasonable (assumed) ratio between dip overlap and strike extent of the S Alpine thrusts.

For convenience, the N edge of the S Alpine belt is defined at the Insubric line. It is a vertical fault parallel to the gross Alpine grain with Neogene right-lateral offset. The cross sections imply that the seismically active Insubric Line is detached on S Alpine thrusts at 5–10 km of depth.

The foreland or Po plain shows regional S dip S of the buried Foothills front (Pieri & Groppi 1981, Pieri in Bally 1983). High basement is suggested by Mesozoics of the Verona–Trento high and by well control (Pieri & Groppi 1981) and seismic data (Cassano *et al.* 1986, Slejko *et al.* 1987) S of the Udine embayment.

Stratigraphy

Good reviews of S Alpine stratigraphy include De Sitter & De Sitter-Koomans (1949), Trumpy (1980), Pieri & Mattavelli (1986), Squyres (1975), and Winterer & Bosellini (1981). As summarized in Fig. 2, the S Alpine thrust belt involves, (a) Hercynian and older continental crust at a reduced thickness, (b) a repeatedly rifted, 5–7 km thick passive-margin series with more than 50% carbonate of Permian to Cretaceous age, and (c) a variety of thin, locally derived, non-orogenic, fining-upward flysches of Cretaceous to Eocene age. It also includes (d) a poorly known and differentiated, up to 4 km thick clastic body of Oligocene age referred to as S Alpine Molasse (Trumpy 1980) or Gonfolite (Pieri & Groppi 1981). Finally, it includes (e) a 3–5 km thick Mio-Pliocene clastic foredeep wedge.

Figure 2 and the cross sections contain known stratigraphic thicknesses and a lithostratigraphy in letter symbols summarized as follows: PC, P are pre-Triassic crystalline and sedimentary rocks. T_1 to T_7 include Triassic units, with the Ladinian Schlern dolomite labelled T_3, and with the Norian Principale dolomite in black and labelled T_6. Jurassic marine rift fill and pelagic sediments are labelled JL, JD, JM. Cretaceous units CI and CS include pelagic Maiolica, Scaglia, and Biancone, formations. One of the pre-orogenic or retrograde flysch units is labelled TE. The Gonfolite, not persistently distinguished, is labelled TO. The Alpine foredeep fill is labelled TM. East of the Verona–Trentino platform, the foredeep fill continues, with unchanged geometry, into the Plio–Pleistocene, TP. West of this platform, an unconformity of early Pliocene age marks a major change in foredeep geometry. There, the Plio–Pleistocene thickens southward and forms the Apennine foredeep fill. The present morphological Alpine front is the zero edge of Plio–Pleistocene sedimentation.

Structural style

Over its average width of 80 km, the S Alpine thrust belt is articulated into a foothills belt with blind thrusts soling above the Mesozoic carbonates, a frontal imbrication of three or four thrust sheets detached near the base of the Mesozoics, and a basement-involving thrust containing a line of duplex structures in its footwall.

Figures 3 to 7 show hand-drawn (not computer-drawn) structure cross sections which illustrate most, but not all, elements of S Alpine structural style. The sections are not balanced because on their internal edges, no control of displacement and deformation is available. Construction techniques include maintaining

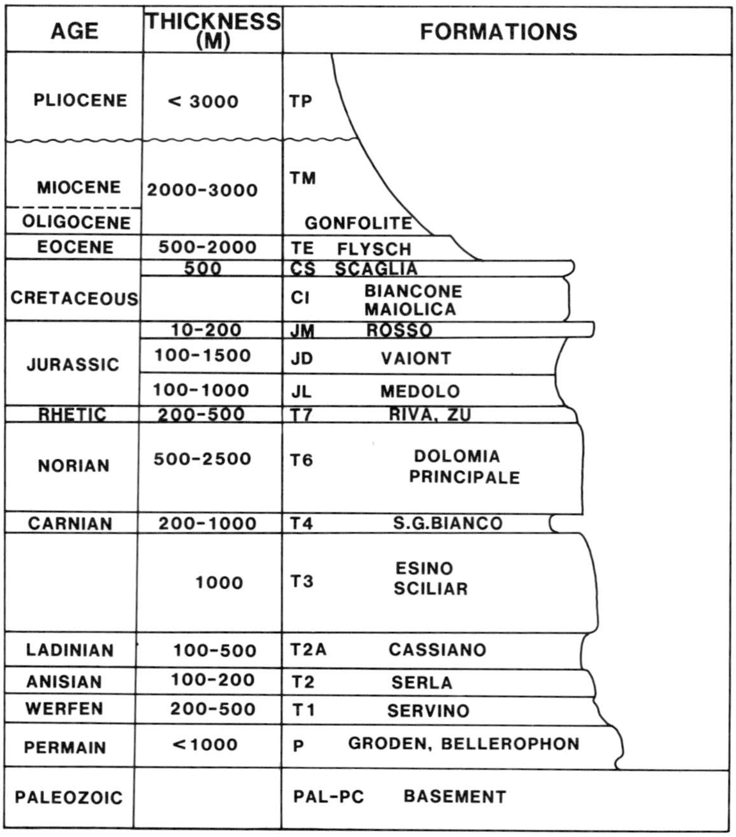

FIG. 2. Stratigraphic column of the southern Alps, showing names, thickness ranges, and graphic appearance in the cross sections of the present paper. The stratigraphy is simplified and ignores strong regional variations.

line length, interpolating stratigraphic thickness, honouring known dip attitudes, and estimating ramp cutoff angles. For only part of the cross sections, restorations to depositional states are included in the present paper.

The supracrustal bulk strain depends in part on the depth extent of supracrustal rocks. The depth to the base of deformation and to the basement top has been assumed by elastic-flexure modelling of the foreland crust (Karner & Watts 1983). The method used is the iterative shape fitting based on available data (Turcotte & Schubert 1982, Roeder 1980, 1987). A series of more closely spaced cross sections would suggest that the depth to the base of deformation may change abruptly along strike of the elastic flexure. I have assumed that such changes are local and reflect Mesozoic rifting, not variations in lithospheric rigidity.

Foothills

Figure 3 is a cross section redrawn after Fig. 5 in Errico *et al.* (1980). It crosses the Lombardian Alps W of the Brembo valley and the northern Po Plain E of Milano. The section is based on the geological quadrangle map of Bergamo (Carta Geologica d'Italia 1:100 000) and on data in De Sitter & De Sitter-Koomans (1949). The subsurface structure is derived from published Malossa field data (Chierici *et al.* 1979, Errico *et al.* 1980, Pieri & Groppi 1981), from published regional basin data (Cassano *et al.* 1986), and from the flexural-load model used in Roeder (1985) and in the present paper.

The section shows a foothills belt external to a major underthrust or delta structure (Jones 1982). The amount of thrusting is unknown, but in the foothills, the shortening below the upper detachment is reasonably well controlled.

The foothills belt, with three faulted anticlines named Pandino, Malossa, and Brignano, is unconformably covered by Pliocene. A generally S-dipping, 3 km-thick Eocene to Miocene series has an upper detachment at its base and is draped over the foothills anticlines. The upper detachment is not known in outcrop; it may be bed-parallel and may carry younger over older within the poorly exposed Lombardian frontal hills. A sparsely documented S-dipping homocline north of Brignano implies a delta structure, unknown in any degree of detail, with room for a stack of perhaps three imbrications. As shown, the total amount of blind thrusting on foothills thrusts is 10.5 km.

Toward the E, the change in foothills structure from blind thrusting with upper detachment to blind thrusting without upper detachment appears to coincide with the Verona–Trento high. It also coincides with the easterly thinning of the pre-kinematic Gonfolite series.

Two cross sections in the Udine embayment follow the courses of the rivers Piave (Figs 4, 5) and Tagliamento with Fella (Fig. 6). They are based on the geological quadrangles of Cortina, Pieve di Cadore, Belluno, Udine, and Pontebba, on published cross sections (Roeder & Bogel 1978, Bally *et al.* 1985), on data by Bosellini & Sarti (1978), and on subsurface data by Pieri & Groppi (1981), Cassano *et al.* (1986), and Carulli *et al.* (1980).

Both of these sections show foothills structures detached at a shallow level within the Jurassic. In Fig. 6, this follows from the small displacement on thrusts underlying the wells Terenzano and Lavariano, from thrust repetitions of Eocene, Cretaceous, and Jurassic in the Bernadia well (Martinis in Bogel & Schmidt 1972), and from the resulting tapered thrust wedge. In Fig. 4, the shallow depth of the detachment is less certain. Of the two subthrust imbrications shown, the more southerly is adapted from Cassano *et al.* (1986). The more northerly is a northeasterly extension of a footwall imbrication poorly controlled in the area SW of the cross section. Possibly, there is only one footwall imbrication present. It may sole out in the Raibl evaporite (T_4) or in Jurassic strata. A shallow sole-out of the twin subthrust imbrications appears more conformable with the foothills detachment in Fig. 6, but it causes problems of balancing and thrust kinematics, to be discussed below.

The W flank of the Verona–Trento high along the Adige valley and Lago di Garda (Fig. 1) serves as inclined (at 15° or more) foreland to Alpine thrusts in Lombardia (Castellarin 1981). The steep dip of this flexure is considered excessive for a flexural bulge (Turcotte & Schubert 1982). Rather, it suggests an intracrustal detachment at shallow depth or a steeply dipping crustal fault along the E flank of the Verona–Trento high.

Frontal imbrication

The frontal imbrication is a vaguely defined zone in the middle part of the thrust belt. Surface geology suggests that style elements are consistent with Dahlstromian thrust models and their successors (Boyer & Elliott 1982, Suppe 1983). There are four groups of style elements:

(1) Stacks of N-dipping thrust sheets are shown on the cross section of Fig. 3 N of the Brignano well and on the cross section of Fig. 6 N of the Bernadia well.

(2) Indicators of dip compression within thrust sheets include kink folds and backthrusts. As shown by the Bernadia well, folds may affect emplaced stacks of small thrust slices. Backthrusts commonly show repeated section, but S-dipping faults with missing section may also be backthrusts as suggested by Laubscher (1985). By location, faults cutting up-section northward suggest an origin at back limbs of fault-bend folds.

(3) S-dipping homoclines of 5–15 km dip extent are interpreted as front limbs of fault-bend folds (Suppe 1983), as shown in Figs 3 and 6. Major thrusts with low dips are assumed to underlie these homoclines in ramp position with respect to the hangingwall. Backthrusts (Laubscher 1985) apparently originate with preference at or near ramp tops. In general, however, the ramp interpretation depends on acceptance of, rather than proves, the thrust scenario proposed here.

(4) Anticlinally warped roof thrusts above duplex stacks of imbrication (Dahlstrom 1970, Mitra 1986) are not common. The cross section in Fig. 4 shows an anticlinally warped duplex structure underlying the foreland of the Valsugana line. Its roof thrust underlies the entire width of the Alps S of the Valsugana line. This interpretation is based on the well Belluno no. 1 drilled in 1960 in the Piave valley near Soverzine. The well encountered thickened Dolomia Principale overlying an imbrication of Jurassic. The nearest surface trace of a thrust fault is the Piancavallo–Barcis thrust

22 km to the SE. This thrust, however, soles out upward in the Cretaceous of the Bosco del Cansiglio; it is probably a local backlimb branch of the roof thrust. A more likely outcrop trace of the Belluno roof thrust is the mountain front thrust juxtaposing lower Cretaceous and undivided Mio–Pliocene, 30 km SE of the Belluno well.

The subthrust imbrication encountered in the Belluno well is located 3.2 km above its regional elevation. This suggests one or two major footwall thrust units below the roof thrust, likely involving the entire series above the base of deformation. Size and shape of the footwall imbrication are unknown; they depend in part on the ramp-flat geometry of the Belluno roof thrust.

Problems of balancing and thrust kinematics

Figure 5 combines three partial restorations of cross sections 4 and 6. Restorations are made to help understand further the poorly known subsurface structure of the thrust belt.

All cross sections in this paper represent a pre-seismic stage of understanding. Therefore, they are speculative, non-unique, and unbalanced accounts of the subsurface geology. However, in the cross section of Fig. 4, an obvious mismatch of thrust geometry suggests an anomalous thrust sequence of two independently progressive steps. As shown in Fig. 5, this kinematic sequence may merit further study.

The shallow-soled twin subthrust imbrications cannot be bedlength-restored by restoring the deep-soled footwall imbrication N of the Belluno well. However, the twin imbrications could have resulted from toe pressure during emplacement of the crystalline frontal thrust, in particular from toe pressure at the ramp through the lower and middle Triassic. This implies that the twin imbrications are older than the deep footwall imbrication. Its sole fault ramps upsection from the Permian and dissects the detachments of the twin imbrications.

This kinematic sequence would imply that ahead of the crystalline front, toe addition takes place in steps and retrogresses into deeper layers, rather than progresses into shallower layers.

Thrust geometry at the crystalline front

Out-of-sequence thrusting (Boyer & Elliott 1982) is suggested by cross-cutting relationships in some areas along the crystalline front, but has not yet been studied systematically. If substantiated, this kinematic behaviour will add insight into post-collisional Alpine geodynamics.

The southern edge of the crystalline belt is a major thrust with local complications. It is not continually exposed, and its interpreted map trace is not continuous. At depth the crystalline frontal thrust is likely to be continuous. It ramps up from near the Moho into the sediment stack. Its displacement is assumed to be equal to the bulk strain in its sedimentary foreland. Major segments of the crystalline frontal thrust carry the names Sub-Orobic, Valtrompia, Sub-Brenta, and Valsugana lineaments.

In the cross sections of Figs 4 and 6, the crystalline front contains a major thrust ramping up through an unknown thickness of basement, continuing in a ramp-flat ascent through the sediment column, and arriving at the surface as the main mountain-front thrust. In the cross section of Fig. 4, a steeper backlimb thrust with smaller displacement carries the regional name, Valsugana line. It dissects S-dipping strata containing a bed-parallel imbrication of Norian Dolomia Principale. In the cross section of Fig. 6, the backlimb thrust is mapped as the Valresia thrust. It dissects the main ramp at nearly 90° to the bedding and juxtaposes ramp strata against frontal imbricates. For the cross section of Fig. 6, a restoration to flat crystalline frontal thrust is shown in Fig. 5.

In the cross section of Fig. 6, the structures dissected by the Valresia thrust are consistent with a compressional ramped roof thrust showing imbrication, detached folding, and backthrusting. An inset below the cross section shows the possible shape and size of the root part of the backthrust dissected by the Valresia backlimb thrust. This fragment of Triassic rocks may occupy part of the space available for the footwall imbrication shown. In the cross section of Fig. 4, the earlier structures are distinctly thin-skinned and may be part of the system of detachment and ramp thrusting affecting the sediment skin covering the Dolomites crustal high (Heissel 1982, Doglioni 1985).

Critical taper at the crystalline front

Out-of-sequence thrusting on a major crustal

thrust has been ascribed to a mechanical differentiation into a rising complex affected by erosion, compression, and possibly strike-slip, and into a foreland complex with subsidence and transport (Roeder 1988). A mechanical linkage between these complexes is offered by the theory of critically tapered wedges (Chapple 1978, Davis *et al.* 1983, Dahlen *et al.* 1984). Applying a critical-taper interpretation to the southern Alps constrains speculation about the unknown subsurface structure of the crystalline front, but the lack of data allows only a conceptual approach.

The geometry of the crystalline front suggests that thrusting originated with a break across the upper crust at a shear angle of perhaps 25°, possibly detaching near the base of the brittly reacting upper crust. Foreland subsidence and a ramp top near the basement top convert the crustal reverse faulting into bed-parallel décollment, frontal growth, and blind thrusting. The finite shear angle in the supracrustal segment is assumed to be in the order to 10° or less. The crystalline front assumes the homoclinal shape of a fault-bend fold (Suppe 1983).

Stresses acting on the crustal segment of the crystalline front can be understood as independent of topography. The supracrustal segment however, absorbs the mountain-front topography and conforms to the critically-tapered wedge. This wedge is thickening downward by footwall imbrications below the main S Alpine thrust. Nearly every factor tends to make the supracrustal segment an undercritical wedge, as listed: increasing basal slope, foreland sedimentation, toe addition, erosion increasing toward the internal side of the wedge, decrease in pore-fluid pressure within and at the base of the wedge.

Two factors contribute to an out-of-sequence thrust within the allochthonous crystalline complex:

(1) the main thrust sheet is located within an undercritical wedge under compression and thickening by folding, imbrication, and back thrusting, and
(2) transport incorporates the crystalline front into the subsiding supracrustal wedge while intracrustal reverse faulting keeps supplying and uplifting crustal material.

The late thrusts dissect earlier compressive-wedge structures. Because they are branch thrusts, they can be discontinuous in map view but do reflect a continuous geodynamic situation. They may not have much or any strike-slip.

This scenario of roof-thrust transport at the sediment base suggests that footwall imbrications in the foreland of the crystalline front do not involve basement. A possible exception, however, is the inferred detachment below the Verona–Trento high, as discussed.

Strike-slip tectonics

Right-lateral displacement of 50–80 km has been suggested on the Insubric Line with a peak during the late Oligocene and with historic seismicity (Trümpy 1980, Schmid *et al.* 1987, Slejko *et al.* 1987). Strike-slip and its timing are consistent with the thrust model, and they imply that the Insubric Line is cut off by the crustal thrust at 5–15 km of depth. The offset of the wrench by the thrust implies decoupling of oblique convergence into the vectors of strike-slip and dip slip. This is analogous to the decoupling of oblique subduction (Fitch 1970, Seeber & Armbruster 1979).

Integrating crustal and supracrustal data

The S Alpine thrust interpretation must be consistent with crustal data which are unusually abundant in the Alps (Mueller *et al.* 1980, Giese *et al.* 1982, Galson & Mueller 1985 and many others). Most of the available crustal data describe the depth to Moho and allow the mapping of crustal slabs, their thrust overlap, and their vertical deflection. The analytical tools to use for integrating these data with supracrustal thrust models are elastic load modelling, plate tectonic convergence paths, and crustal balancing in the sense of Butler *et al.* (1986) and Roeder (1988). The areal extent required for this task comprises the entire Alps. Accordingly, the present paper includes a supracrustal cross section of the Alpine N side (Fig. 7) and a trans-Alpine crustal section (Figs 8, 9, 11).

Measuring the load deflection

The foredeeps on both Alpine flanks are wedge-shaped in cross section and suggest an origin by flexure under the load of Alpine overthrusting and subduction, not only of the northern foreland (Roeder 1980, Karner & Watts 1983, Royden & Karner 1984), but also of the south-Alpine foreland (Roeder 1985, Cassano *et al.* 1986). In detail, the tectonic history of the foredeeps is diverse and complicates the study of their long-term elastic properties. However, the method of geometric fitting of line loads (Roeder 1980, 1988, Turcotte & Schubert 1982) is empirical, based on structural data, and bypasses physical properties. Its interpretive value is limited, but it has been used to develop additional constraints for Alpine tectonics.

Elastic flexure of northern foreland

Figure 7 shows a cross section of the northern Alps near the meridian of Munich and 25 km E of the meridian of the section in Fig. 4. This N Alpine section is based on Muller (1978), Muller-Wolfskeil (1981), Bachmann *et al.* (1982), Frisch (1984), and field work by Frisch, Bogel, Lammerer and Roeder in 1975 to 1987.

The Molasse base, of late Eocene to Oligocene age, outlines a downwarp controlled by a line of zero deflection along the Danube and by several wells in the Alpine foothills belt (Muller 1978, Bachmann *et al.* 1982). If this downwarp has an elastic component, its data points should reflect the sum total of its history of loading and unloading. Hence it can be modelled by the equation of line-load deflection (Turcotte & Schubert 1982). Therefore, the curvature of this downwarp can be matched by assumed combinations of three variables: (a) the location and (b) the amount of the maximum downwarp and (c) the lithospheric rigidity and/or the elastic wavelength. Two extreme combinations have geological significance:

(1) A maximum deflection assumed beneath the Tauern would be consistent with the base-molasse curvature, if the deflection were 10.2 km deep and the rigidity of the European lithosphere beneath the Bavarian foreland were $D=3.3 \times 10^{23}$ Nm.
(2) Alternatively, a maximum deflection near the orographic median line of the Alps S of Bressanone would match the Molasse base if it were 16.7 km deep and if the foreland lithosphere had a rigidity of $D=6.9 \times 10^{23}$ Nm.

The second model is preferred; it conforms better to the curvature of the foreland Moho (Giese *et al.* 1982) than the first model. The crustal slab outlined by the Moho and by the Molasse base dips more steeply than suggested by either load model. The excess area enclosed by the Moho data strongly supports the main thesis of the present paper. It suggests a major crustal overlap within the S Alpine and Austro–Alpine crust. This overlap is in addition to the overlap caused by Alpine subduction itself.

Two stages of flexing in the southern foreland

Po Valley subsurface data (Pieri & Groppi 1981, Pieri in Bally 1983) show that the W part of the S Alpine foreland has had a dip reversal in early Pliocene. This reversal can be ascribed to a change in the flexural dynamics and can be used to approximate the flexural parameters.

Pre-Pliocene Alpine thrusting was associated with northward flexure of unknown wavelength and/or rigidity, unknown location, and unknown amount of maximum depression. Pliocene to Recent Apennine thrusting is reversing the northward tilt. The Apennine foredeep fill thickens southward and shows a northern pinch-out line along the W part of the Alpine southern front. A distance of 150 km between the zero edge and the most likely location of the Apennine load centre suggests a rigidity of $D=9.34 \times 10^{22}$ Nm. This is considerably less than the rigidity of the N Alpine foreland. However, curvature based on this rigidity is consistent with seismic data in the Apennine foothills (Bally *et al.* 1985).

The pre-Pliocene flexure beneath the Alpine load has been modelled by assuming the same rigidity/wavelength as obtained for the younger stage. The Alpine-age line of zero deflection is unmapped beneath the Plio–Pleistocene wedge. A reasonable fit of structure data has been obtained by locating the Alpine maximum deflection near the Insubric Line.

The amount of maximum deflection needed to obtain a reasonable fit of structure data varies along strike between 9.5 and 13 km. This is substantially less than the 16.7 km obtained for the cross section in Fig. 7. The difference reasonably expresses the cumulative effect of flexure since late Eocene on the Alpine N side, and only since early Miocene on the Alpine S side.

East of the Verona–Vicenza high, the early Pliocene reversal of load flexure does not affect the Alpine foothills because of the easterly increasing distance to the apennine load. A subsurface high inland of the Adriatic coast near Venice (Fig. 1) is documented by some wells and appears to represent the peripheral bulge of Neogene-age Alpine loading. Its location relative to the Friuli–Veneto thrust belt suggests that the foreland rigidity/flexural wavelength is unchanged from the western (Lombardia) area.

Line loads, point loads and ring loads

Some discussion is needed to define the limits to the method of geometric fitting, to see whether or not limits may be set by the geometry of forelands and loading orogens.

The estimate of the rigidity and/or flexural wavelength for the Adria plate under Alpine load has been derived by comparing Alpine and Apennine geodynamic settings, both of considerably different geometry. Simplified, the Alpine load is ring-shaped, and the load of the Ligurian Apennine is point-shaped. In the west part of the Udine embayment, the Alpine body can be approximated as line-shaped. The elastic rigidity does not vary with the geometry of the load (Heiskanen & Vening Meinesz 1958, Le Pichon *et al.* 1973). However, the load itself, and thus the amount of maximum deflection, varies with the geometry of the loading body (G. Karner, pers. comm. 1985). By the method used in the present study, the amount of maximum deflection is not predicted from the load, but results from fitting a rigidity value to foreland data. Therefore, the comparison of Alpine and Apennine settings is appropriate if it is limited to geometric fitting.

Crustal thickness

Figure 8 shows a composite crustal cross section of the eastern Alps, the western Po Plain, and the Ligurian area. There is an offset in the cross section separating two segments. The Alpine segment of the cross section extends along a line near Munich, Innsbruck, the Tauern window, and the Belluno well. It combines Moho data as compiled by Mueller *et al.* (1980) and Giese *et al.* (1982) with surface-derived information discussed in the foregoing. The segment covering the Po Plain and the Ligurian area is oriented N–S and goes through Milano; it is a redrawn and geologically interpreted version of Giese's (1985) fig. 4. The composite section contains graphic statements about crustal thicknesses, as well as support for a major crustal overlap within the upper, southern plate of the Alpine collision. Crustal units identified geologically and in refraction-seismic surveys are reviewed separately for each segment. Subsequently, contacts between crustal units in both segments are interpreted geologically.

Alpine segment

The northern foreland crust (label A) shows congruence between Moho and foredeep base. A constant crustal thickness of 28–29 km can be extrapolated downdip nearly to the Alpine Moho root. In an older series of data (Mueller *et al.* 1980, Giese *et al.* 1982), a strongly N-dipping Moho segment or line of interpolation limited the Moho root to the S. This segment is carried as a dashed line (Figs 8 and 11). Present data (fig. 9 of Giese 1985) support a 65 km deep Moho root 20–50 km farther S.

The core complex (label B) is represented by the Tuxer and Zillertaler central gneiss bodies as flexural folds within an allochthonous crustal flake. The pre-Alpine location of the core complex is not evident from crustal data. Paleogeography, however, suggests that the core complex came from the S edge of the European foreland plate (Frisch 1977 and many others).

The Austro–Alpine crust (label D) is assumed to have been attached in pre-Oligocene time to the Adriatic or S Alpine foreland crust (label C), with separation at the S Alpine thrust belt and the Insubric Line. The crust of the Adria plate thins northward from about 22 km below Milano (Giese 1985) to 12 km at the Tauern window (Roeder & Bogel 1978) and to a zero edge somewhere S of the Vorderriss well in the northern Alps (Bachmann *et al.* 1982).

Southern segment

Refraction-seismic surveys in the western Po Plain and in the Ligurian Apennines show two stacked slabs of crust, each identified by crustal velocities and an M-type velocity jump (Giese 1985). The upper slab, labelled 'crust 1' in Fig. 8 and in Giese's (1985) fig. 4, is the foreland basement of the southern Alps and is identified as crust of the Adria plate. A southerly dip parallel to the base of the Apennine foredeep is suggested. A northerly dip beneath the Alps is not evident, perhaps because it is compensated by northward crustal thinning. The lower slab, labelled 'crust 2', ties westward into European crust, and therefore also northward into the N Alpine foreland. The Alpine Moho root, mapped to attain 65 km (Giese 1985, fig. 9) is developed in crust 2. In Fig. 8, the Moho root appears as a combination of an elastic line-load depression and a northward thickening from 18 to 30 km.

A third and highest Moho segment or crustal base appears at $h = -20$ km in the Ligurian sea. Following Giese (1985), I assume a transcrustal thrust at the Apennine front between this segment and crust 1. The amount of northward transport on this thrust depends on the S dip of the crust 1 slab, and therefore on the location of the subthrust trailing edge of crust 2.

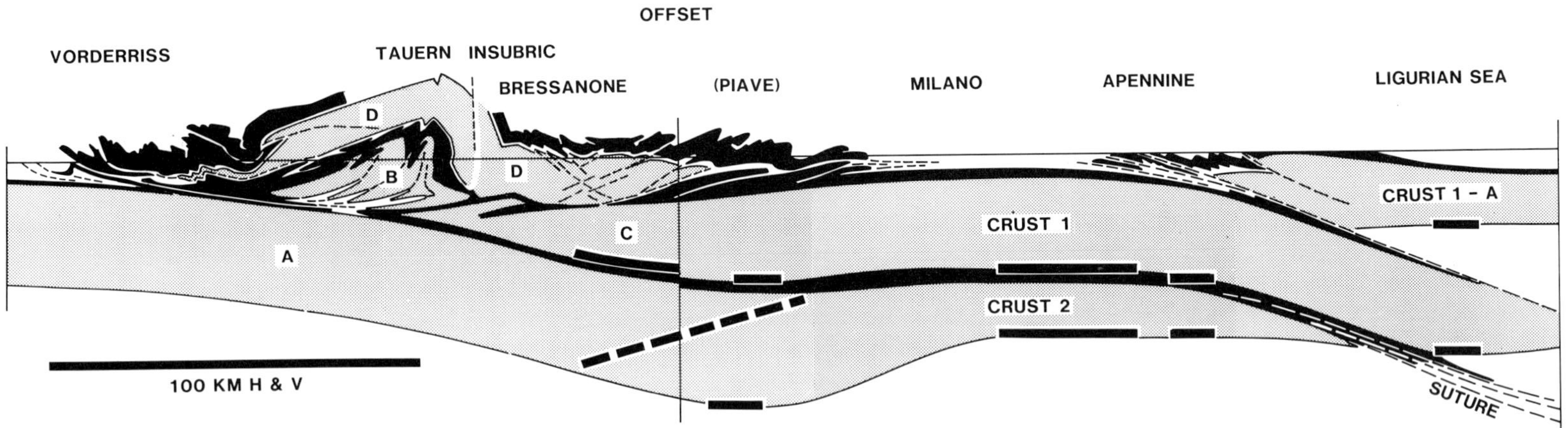

FIG. 8. Crustal cross section of eastern Alps along traces of sections in Figs 7 and 4, based on references made in the text. Vertical line is an E–W offset of 250 km. Shaded: crust. Black: Mesozoic sediments. Present topography is not shown. Letter symbols denote crustal domains discussed in the text. M: Moho.

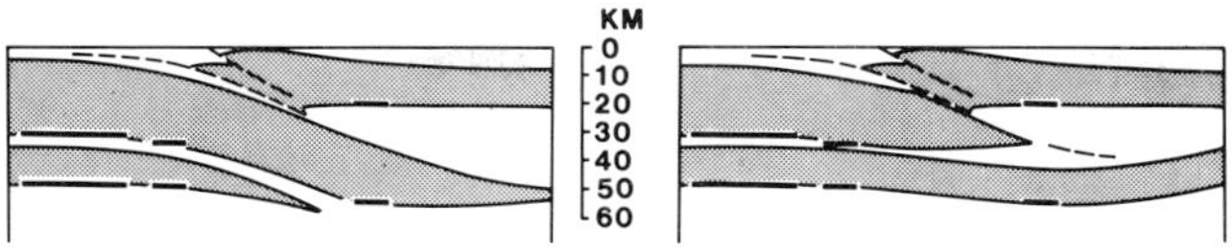

FIG. 9. Two possible interpretations of crustal data in the Apennine–Ligurian area, near the S edge of section in Fig. 8. Left: short-slab interpretation as carried in Fig. 8. Right: long-slab interpretation possibly preferred in the N Tyrrhenian area.

Crustal contacts

Possible geological types of crustal contacts include basement tops, ramping thrust faults, intracrustal or subcrustal décollements, and undisturbed crustal bases. A coherent tectonic model is needed to sort out these types. The model is, at present, confined to its N–S cross section. Certainties and uncertainties about its E–W extent are well illustrated in figs 2, 7, and 9 of Giese (1985).

The base of units labelled C and D is a thrust fault of Cretaceous to late Eocene age, and it is referred to as the Austro–Alpine sole thrust. It juxtaposes terranes of strongly contrasting geological history, and it is part of the Alpine collision suture. This implies that initially it dissected a continuous panel of crust, but its initial footwall and its whereabouts are not known. Fragments of it, carried along as horses, are mapped as lower and middle Austro–Alpine units, not shown in Figs 8 and 11.

The ramp-flat pattern of the Austro–Alpine sole thrust starts as a décollement in the mantle just at or below the Moho of crust 1 (for a different view, see Butler *et al.* 1986). The mantle material near this décollement is assumed to contain metamorphosed and tectonized oceanic rocks of the Penninic domain. Along the cross section of Fig. 8, the near-Moho detachment is 150 km or more than 200 km long, depending on the choice of correlation between Moho segments defining crusts 1 and 2. Two choices are illustrated in Fig. 9. A short trailing edge of crust 2 would accommodate a S-dipping crust 1 and 80 km or more of northward thrust transport on the Apennine or Tuscan slab. A southward extension of crust 2 beyond Fig. 8 would leave room for only 40 km of Apennine crustal transport.

The Moho segments outlining the Alpine Moho root and the shallow sub-Apennine Moho are interpreted as undisturbed crustal bases.

The tectonic model in Fig. 8 suggests that complete crustal overlap is required at the S Alpine crystalline front to fill the space above the Alpine Moho root. It also suggests that, as a S–N component, the Tauern gneiss cores travelled as a subthrust horse complex for a minimum of 300 km.

Charting trans-Alpine plate convergence

The last test for the S Alpine thrust interpretation is Alpine plate tectonics, comparing bulk strain or cumulative thrust transport with dated amounts of plate convergence, using the convergence since collision at 38 Ma.

Figure 10 shows four plate paths describing the motion of the site of Verona, or a point in Africa, respectively, relative to Europe considered fixed in the diagram. The motion paths are to scale and in map view. Poles of rotation to converge Europe and Africa (EU–AF) are published by Dewey *et al.* (1973), by Biju-Duval *et al.* (1977), and by Dewey *et al.* (this volume). I have converted these data into distance travelled along small circles passing through Verona, using the law of cosines (Morgan 1969). Numbered dots are dated positions on the plate convergence paths.

The strain path labelled 'Roeder 1984 Trans-Alpine' is based on a plate model of the western Mediterranean (Roeder 1984). In this model, the trans-Alpine path is a finite-difference composite of Europe–Africa convergence, right-lateral strike-slip on the Saharan Atlas, and divergence across the Strait of Tunis. The trans-Alpine convergence is measured between Europe and the Adria terrane, an area which has been considered as an independently moving plate (Biju-Duval *et al.* 1977, Gorler & Giese 1978) or as an African promontory (Argand 1924, Horvath & Channell 1977). Poor data control suggests that the Adria unit was successively rift field, promontory and plate. The strain path by Roeder (1984) tries to account for this succession.

With some differences, the two first-quoted plate paths and the finite-difference solution

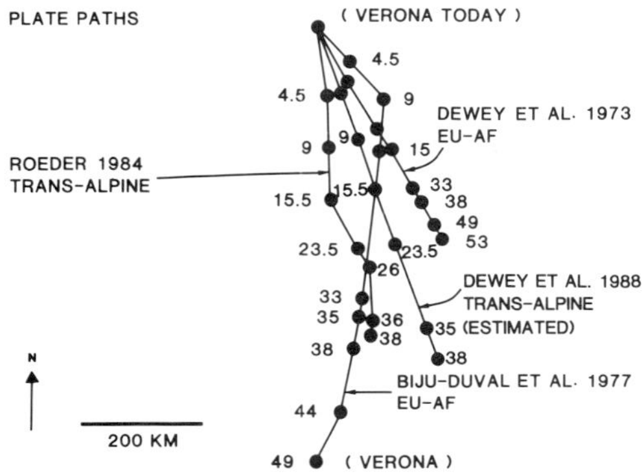

FIG. 10. Map chart showing dated plate paths of trans-Alpine or Tethyan (EU–AF) convergence projected into small circles passing through Verona, near location of cross section Fig. 8. Black dots are positions dated in million years.

(Roeder 1984) suggest a cumulative and reasonably steady trans-Alpine and N-directed convergence of 500, 500 and 340 km at the meridian of Verona. This is equivalent to 1.3–0.9 cm per year. A fourth plate path can be estimated from Dewey *et al.* (this volume).

Crustal restoration through Neogene

Figure 11 shows a series of restorations, from the present back in time to late collisional, of the crustal section shown in Fig. 8. The restorations follow in direction, timing, and extent, the trans-Alpine plate path shown in Fig. 10.

A scale in time and distance extends the present-state section southward. Three earlier post-collisional stages are dated at 4.5, 15.5 and 25.7 Ma. The amount of palinspastic restoration at each stage is taken from the time-distance scale. The restorations are approximately area-balanced but details are limited by the graphic resolution of the restoration.

The Alpine post-collisional restoration reconfirms the consensus (Tollmann 1970, Laubscher 1983, Butler *et al.* 1986, and many others) that plate convergence greatly exceeds the traditionally assumed bulk strain, as well as the cross-sectional length of crust documented in older refraction-seismic surveys. The restoration is consistent with the model of stacked crustal slabs beneath the Po Plain (Giese 1985) and with the assumption of major regional sub-Moho detachment (present paper). However, the documentation of this model is confined to the western Alps; the model is applied without data support to the eastern Alps.

The earliest restoration describes the Alpine configuration 12 Ma after collision defined as thrust juxtaposition between the Austro–Alpine crust and the rocks in the Tauern window. The time–distance scale of plate convergence is reasonably consistent with the short-slab assumption made for crust 2 in Fig. 9, if one assumes that the Tauern gneiss and its sediment cover represent the outer reaches of the Jurassic extensional or transtensional S margin of Europe. However, the slab of crust 2 must extend farther to the S, if thrusting in the tectonic units above the Tauern gneiss had ceased already in the Palaeocene (Ratschbacher *et al.* 1987, Selverstone 1988).

The crustal fragments of the Tauern Massif are shown at a depth range of 16–22 km which is suggested by their meso-Alpine Barrovian metamorphism (Raith *et al.* 1978). Ductile deformation in the fragments coeval with brittle uplift and right-lateral strike-slip on the Insubric Line and within the Austro–Alpine unit (Selverstone 1988, Lammerer 1987) are consistent with oblique subduction of crustal Europe beneath the Adria terrane and with the unit stacking as illustrated.

The second reconstruction shows the beginning of central Alpine buckling and compressive uplift accompanied by continued strike-slip at the Insubric lineament. This tectonism is dated by the cooling of the Lepontine area W of the cross section (Schmid *et al.* 1987). The third reconstruction shows the south-Alpine thrust belt about halfway toward its present state of strain. Enough foredeep sediments have accumulated on both Alpine flanks to indicate a shallow line-loaded Moho root beneath the Alpine core, a feature that resulted from the stacking of continental crust. The fourth reconstruction is a partial repetition, in silhouette form, of the crustal cross section of Fig. 8.

Speculations about post-collisional Alpine thrusting

Available crustal data strongly support the interpretation of the southern Alps as a detached fold-thrust belt. The Moho geometry can be used to estimate the size of the unknown buried part of the thrust wedge. Available paths of trans-Alpine plate convergence, however vaguely constrained, are consistent with the newest set of regional Moho data. The remaining task is to appraise the post-collisional strain

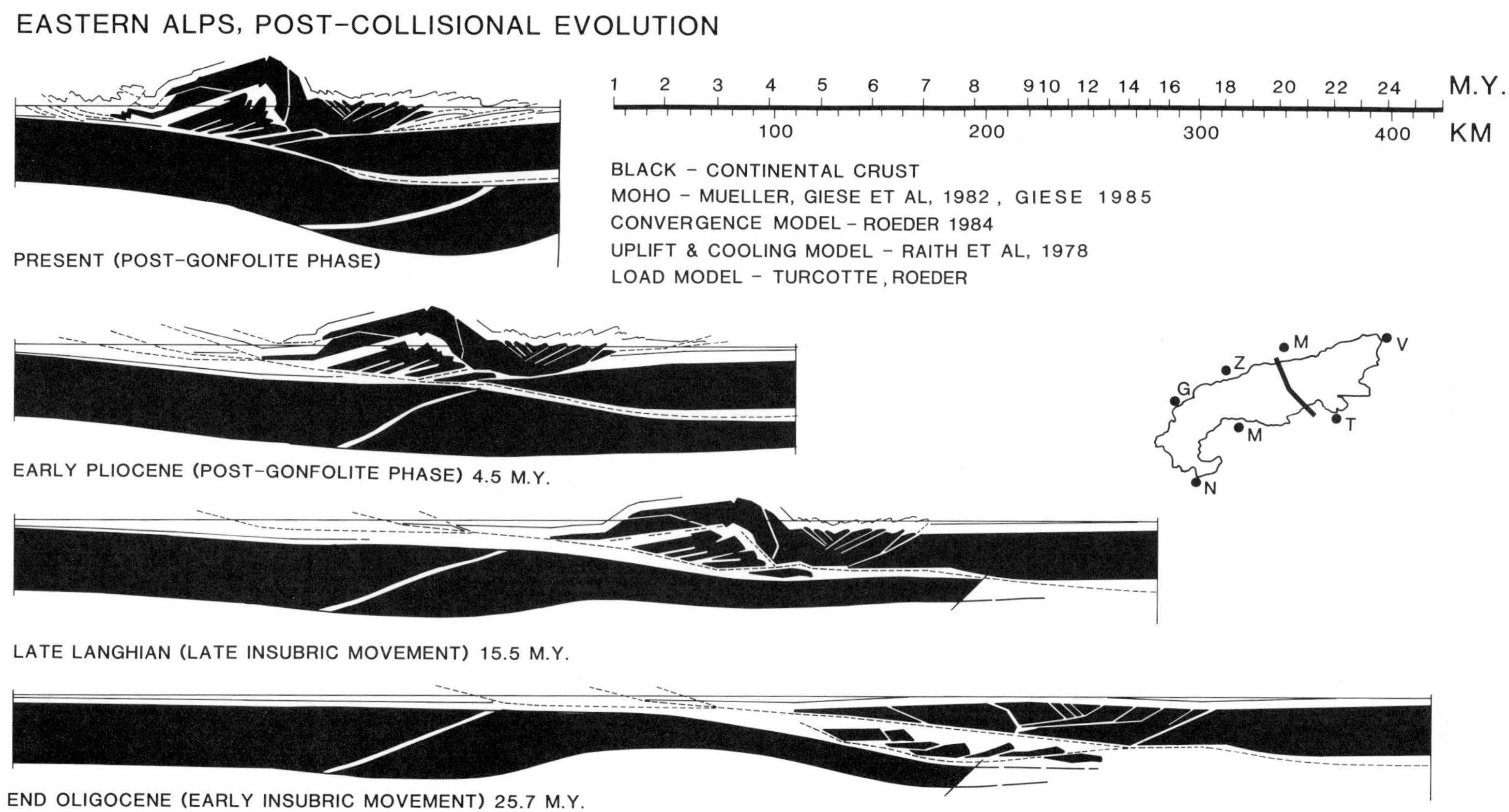

FIG. 11. East Alpine crustal cross section as in Fig. 8 restored along the dated plate convergence path of Roeder 1984 (see Fig. 9). Crust is shown as black silhouettes. Dashed lines are major faults. Letters on location map are cities, counter-clockwise: Vienna, Munich, Zurich, Geneva, Nice, Milano, Trieste.

within the hangingwall of the Alpine crustal overlap.

As shown graphically in Fig. 8, a Moho root of 35 km is to be filled with a thrust repetition within crust that is 20 km or less in thickness. At a finite shear angle of 15°, this repetition would require 75 km of thrust overlap to fill the Moho root. This figure is similar to an estimated 60 km of fold-thrust shortening in the sedimentary part of the S-Alpine belt. Averaged over a 100 km wide S-Alpine belt, the thrusting would amount to 38% strain. Central Alpine unroofing by compressive buckling and Neogene-age imbrication in the calc-Alpine belt also gives an estimate of 38% strain. If the calc-Alpine imbrication is pre-collisional in age (Ratschbacher *et al.* 1987 and many others), the total post-collisional compressive strain in the Alpine upper plate is only 26%, at a bulk shortening of 80 km. This compression developed while the crustal slab travelled 380 km northward over European crust.

Cooling models (Raith 1978, Selverstone 1988) and large-scale kinematics (Figs 8, 11) suggest that the detached base of this compressional tectonics coincides with the ductile-brittle transition at 10–15 km of depth. Following Platt (1987), I suggest that the compression within the rising hanging wall is caused by the changeover from a viscous flow law at depth to a plastic or Coulomb flow law and increased traction above 10 or 15 km and within the Alpine body proper. Above the ductile-brittle transition, traction at the interface between crust 1 and crust 2 begins to be governed by the equilibrium of critical taper, in the form of a bilaterally symmetric, back-to-back, wedge pair. Two groups of observations suggest that the Alpine double wedge tends to be overcritical and responds by lowering its taper: (a) thrust belts on all Alpine flanks grow outward by blind thrusting and possibly involve out-of-sequence thrusting (discussed in the present paper), and (b) deep-reaching normal faulting juxtaposes unrelated thrust sequences and possibly indicates gravitational collapse of excess topography (Platt 1987, Selverstone 1988).

Of possible factors that could maintain the Alpine double wedge in overcritical state, one is crustal in nature, and one is dynamic: (a) Outward thrusting transports Alpine material over increasingly thick, and probably increasingly rigid, crust. This lowers the slope of the wedge base and may produce an overcritical topographic slope. (b) Viscous detachment of crust 1 over crust 2 may generate a topographic surplus faster than erosion and coulomb-type pure shear and wedge growth can absorb and disperse it. The adequate mechanism for topographic dispersal may indeed be gravitational collapse.

Deep-reaching normal faults in the Tauern strike across the Alpine grain and are interpreted (Selverstone 1988) as part of the model of lithospheric extension (Wernicke 1985). This introduces a three-dimensional aspect into post-collisional Alpine convergence which goes beyond the immediate scope of the present paper.

ACKNOWLEDGEMENTS: The present paper updates an earlier synthesis (Roeder 1985). It has been presented in a less advanced form before the EUG (Roeder 1987) and at the Nato Advanced Research Workshop in Gradisca d'Isonzo (Giese & Nicolich 1987). It is published by permission of the Anschutz Corporation and Trend International Exploration, Ltd.

I owe much understanding to Drs Anelli, Bally, Casero, Crane, De Genevraye, Didomenico, Fuller, Gauriat, Groppi, Laubscher, Lewis, Martinelli, Mattavelli, Perrier, Pieri, Sarti, and Zappaterra. Harold Fyffe has field-reviewed the Bergamo part of my work. Bernd Lammerer has shown me his work in the Tauern window.

Colleagues at the Gradisca d'Isonzo workshop (1987), in particular Drs Buness, Della Vedova, Gebrande, Giese, Hirn, Klingele, Morelli, Nadir, and Nicolich, have shared with me their unpublished reprocessing of refraction-seismic and potential-field data.

The paper has been improved by the advice of Michael Coward and two anonymous reviewers, as well as by the editorial help of Annamarie Argandona.

References

ARGAND, E. 1924. La tectonique de l'Asie, *13th International Geological Congress Comptes-Rendus*, **5**, 171–372.

BACHMANN, G., DOHR, G. & MULLER, M. 1982. Exploration in a classical thrust belt and its foreland: Bavarian Alps, Germany. *Bulletin of the American Association of Petroleum Geology*, **66 (12)**, 2529–42.

BALLY, A. W. 1983. Seismic expression of structural styles, a picture and work atlas. *American Association of Petroleum Geology Studies in Geology Series* No. 15, 3 volumes.

——, CATALANO, R. & OLDOW, J. 1985. *Elementi di Tettonica Regionale. Evoluzione dei Bacini sedimentari e delle Catene montuose*. Pitagora Editrice Bologna, 1–276.

BIJU-DUVAL, B., DERCOURT, J. & LEPICHON, X. 1977. From the Tethys ocean to the Mediterranean seas: a plate tectonic model of the evolution of the western Alpine system. *In*: *Structural History of the Mediterranean Basins*, 143–64. Symposium Split.

BOGEL, H. & SCHMIDT, K. 1976. *Kleine Geologie der Ostalpen*. Ott Verlag Thun, 1–231.

BOSELLINI, A. & SARTI, M. 1978. Geologia del gruppo M. Cuar M. Covria (Prealpi Carniche): *Giornale di Geologia, Annali del museo geologia di Bologna, 2A*, **43**, I, 47–73.

BOYER, S. E. & ELLIOTT, E. 1982. Thrust systems. *Bulletin of the American Association of Petrology and Geology*, **66 (9)**, 1196–230.

BUTLER, R. W. H., MATTHEWS, S. J. & PARISH, M. 1986. The NW external Alpine thrust belt and its implications for the geometry of the western Alpine orogen. *In*: *Collision Tectonics*, COWARD M. P. & RIES, A. C. (eds). Special Publication of the Geological Society No. 19. 245–260.

CARULLI, G. B., CAROBENE, L., CAVALLIN, A., MARTINIS, B. & ONOFRI, R. 1980. Evoluzione strutturale Plioquaternario del Friuli e della Venezia Giulia. *In*: *Autori Vari*, Contrib. Realizz. Carta Neotettonica d'Italia. CNRPF Geodinamica pubbl. 356, Giannini, Napoli, 498–545.

CASSANO, E., ANELLI, L., FICHERA, R. & CAPPELLI, V. 1986. Pianura Padana Interpretazione integrata di dati Geofisici e Geologici. *73 Congresso SGI, AGIP*, 1–28.

CASTELLARIN, A. 1981. *Carta tettonica delle Alpin Meridionali 1:200 000*. CNR progetto finalizzato Geodinamica, **441**, 1–220.

—— 1983. Schema delle deformazioni tettoniche sudalpine, *in Atti del convegno internazionale 'I problemi del mare adriatico'. Bollettino Oceanologia Teorica et Applicata 2*, **2**, 105–14.

—— & VAI, G. B. 1981. Importance of Hercynian tectonics within the framework of the southern Alps. *Journal of Structural Geology*, **3**, **4**, 477–86.

CHAPPLE, W. M. 1978. Mechanics of thin-skinned fold-and-thrust belts. *Bulletin of the Geological Society of America*, **89**, 1189–98.

CHIERICI, G. L., BILGERI, D., VAHI, G. C., TORRICELLI, L. & PULGA, M. 1979. Exploration and Production in the Very Deep Malossa Field. *10th Work Petrological Congress Bucarest*, **2**, 1–14.

DAHLEN, F. A. & SUPPE, J. 1984. Mechanics of fold-and-thrust belts and accretionary wedges: cohesive Coulomb theory. *Journal of Geophysical Research*, **89**, 10087–101.

DAHLSTROM, C. D. A. 1970. Structural geology in the eastern margin of the Canadian Rocky Mountains, *Canadian Bulletin of Petroleum Geology*, **18**, 332–402.

DAVIS, D., SUPPE, J. & DAHLEN, F. A. 1983. Mechanics of fold and thrust belts and accretionary wedges. *Journal of Geophysical Research*, **88**, 1153–72.

DE SITTER, L. V. & DE SITTER-KOOMANS, C. M. 1949. The geology of the Bergamasc Alps, Lombardia, Italy. *Leidse Geol. Med.*, **14**, 1–257.

DEWEY, J. F., PITTMAN III, W. C., RYAN, W. B. F. & BONNIN, J. 1973. Plate tectonics and the evolution of the Alpine system. *Bulletin of the Geological Society of America* **84**, 3137–80.

DOBEN, K. 1981. 8. Alpenraum. *In*: *Geologische Karte von Bayern 1:500 000*, HAUNSCHILD, H. & JERZ H. (eds). Bayerisches Geologishes Landesamt, I–168.

DOGLIONI, C. 1985. The overthrusts in the Dolomites: ramp-flat systems. *Eclogae Geologicae Helvetiae*, **78**, (2), 335–50.

ERRICO, G., GROPPI, G., SAVELLI, S. & VAGHI, G. C. 1980. Malossa field: a deep discovery in the Po valley, Italy. *In*: *Giant Oil and Gas Fields of the Decade: 1968–1978*. Memoir of the American Association of Petroleum Geology, **30**, 525–38. 38.

FITCH, T. J. 1970. Earthquake mechanisms in the Himalayan, Burmese, and Andaman regions and continental tectonics in central Asia. *Journal of Geophysical Research*, **75 (14)**, 2699–709.

FRISCH, W. 1984. Sedimentological response to late Mesozoic subduction in the penninic windows of the eastern Alps. *Geologische Rundschau*, **73(1)**, 33–45.

—— 1977. Der alpidische Internbau der Venedigerdecke im westlichen Tauernfenster. *Neues Jahrb. Geol. Palaeontol. Monatsh.*, H., **11**, 675–96.

GAETANI, M. & JADOUL, F. 1979. The structure of the Bergamasc Alps. *Rend. Acad. naz. Lincei*, 8, 66, **5**, 411–16.

GALSON, D. A. & MULLER, ST (eds). 1985. *Proceedings of the second workshop on the European Geotraverse (EGT) Project, the southern Segment*. European Science Foundation, Strasbourg. pp. 1–367.

GEBRANDE, H., HAGE, H., MILLER, H., MULLER, G. & SCHMEDES, E. 1978. Aftershock investigations and fault-plane solutions of the Friuli earthquakes 1976. *In*: *Alps Apennines Hellenides, IUCG Scientific Report No. 38*, 173–80.

GIESE, P. 1985. The structure of the upper lithosphere between the Ligurian Sea and the southern Alps. Part B: The consolidated crust and the uppermost mantle. *In*: *Second workshop European Geotraverse*, European Science Foundation, 143–54.

——, REUTTER, K. J., JACOBSHAGEN, V. & NICOLICH, R. 1982. Explosion seismic crustal studies in the Alpine Mediterranean region and their implications to tectonic processes. *In*: BERCKHEMER, H. & HSÜ, K. (eds). *Alpine–Mediterranean Geodynamics*, Geodynamics Series, **7**, 39–74.

—— & NICOLICH, R. 1987. Report on Nato Advanced Research Workshop 'Improvement of Joint Interpretation of Geophysical and geological Data', Gradisca d'Isonzo Oct 1–8.

GORLER, K. & GIESE, P. 1978. Aspects of the evolution of the Calabrian Arc. *In*: CLOSS, ROEDER & SCHMIDT (eds). *Alps Apennines Hellenides*, IUCG Scientific Report, **38**, 374–388.

Gortani, M. & Desio, A. 1925. Sheet Pontebba of Carta Geologica delle Tre Venezie 1:100000, Rome.

Harding, T. & Lowell, J. D. 1979. Structural styles, their plate tectonic habitats, and hydrocarbon traps in petroleum provinces. *Bulletin of the American Association of Petroleum Geology*, **63**, 1016–58.

Heiskanen, W. A. & Vening Meinesz, F. A. 1958. *The Earth and its Gravity Field*, McGraw-Hill, New York, pp. 1–470.

Heissel, W. 1982. Südtiroler Dolomiten. *Sammlung geologischer Fuhrer*, **71**, 1–172. Borntraeger Verlag, Berlin.

Hellinger, S. J. & Sclater, J. G., 1983. Some comments on two-layer extensional models for the evolution of sedimentary basins. *Journal of Geophysical Research*, **88, B10**, 8251–69.

Horvath, F. & Channell, J. E. T. 1977. Further evidence relevant to the African/Adriatic promontory as a paleogeographic premise for Alpine orogency. *In*: Biju-Duval, B. (ed.). *Structural History of the Mediterranean Basins*, pp. 133–42. Technip.

Jones, P. B. 1982. Oil and gas beneath east-dipping underthrust faults in the Alberta foothills, Canada. *In*: Powers, R. B. (ed). *Geologic Studies of the Cordilleran Thrustbelt*, 61–74. RMAG, Denver.

Karner, G. D. & Watts, A. B. 1983. Gravity anomalies and flexure of the lithosphere at mountain ranges. *Journal Geophysical Research*, **88, B12**, 10449–77.

Lammerer, B. 1987. Thrust-regime and trans-pression regime tectonics in the Tauern Window (eastern Alps). *Geologische Rundschau*, (in press).

Laubscher, H. P. 1983. Detachment, shear, and compression in the central Alps. *In*: Hatcher, R. D. (ed). *Contributions to the tectonics and geophysics of mountain chains, Geological Society of America Memoir*, **158**, 191–211.

—— 1985. Large-scale, thin-skinned thrusting in the southern Alps; kinematic models. *Bulletin of the Geological Society of America* **96, 6**, 710–18.

Le Pichon, X., Francheteau, J. & Bonnin, J. 1973. *Plate Tectonics*. Developments in Geotectonics, **6**, 1–300. Elsevier, Amsterdam.

McKenzie, D. 1978, Some remarks on the development of sedimentary basins. *Earth and Planetary Science Letters*, **40**, 25–32.

Mitra, S. 1986. Duplex structures and imbricate thrust systems: Geometry structural position, and hydrocarbon potential. *American Association of Petroleum Geology Bulletin*, **70(9)**, 1087–112.

Morgan, J. W. 1968. Rises, trenches, great faults, and crustal blocks. *Journal of Geophysical Research*, **73, 6**, 1959–82.

Mueller, St. Ansorge, J., Egloff, R. & Kissling, E. 1980. A crustal cross section along the Swiss Geotraverse from the Rhine-graben to the Po Plain. *Eclogae Geologicae Helvetiae*, **73, 2**, 463–85.

Muller, M. 1978. Miesbach 1 an Staffelsee 1 – two basement tests below the folded molasse. *In*: Closs, Roeder, D. & Schmidt, (eds), *Alps Apennines Hellenides*. IUCG Scientific Report No. 38, 64–8.

Muller-Wolfskeil, P. 1981. Tektonische Analyse des Wamberger Sattels im Bereich Vorderriss. *Geological Bavarica*, **81**, 235–50.

Pieri, M. & Groppi, G. 1981. Subsurface geological structure of the Po Plain, Italy, *CNR Publicazione No.* **414** *del Progetto Finalizzato Geodinamico*, 1–11.

—— & Mattavelli, L. 1986. Geologic framework of Italian petroleum resources. *Bulletin of the American Association of Petroleum Geology*, **70 (2)**, 103–30.

Pitman, III, W. C. & Talwani, M. 1972. Sea-floor spreading in the north Atlantic. *Geological Society of America Bulletin*, **83, 3**, 619–46.

Platt, J. P. 1987. The uplift of high-pressure low-temperature metamorphic rocks. *Philosophical Transactions of the Royal Society of London, Series* **A. 321**, 87–103.

Raith, M., Raase, P., Kreuzer, H. & Muller, P. 1978. The age of the alpidic metamorphism in the western Tauern Window, Austrian Alps, according to radiometric dating. *In*: Closs, Roeder, D. & Schmidt, (eds), *Alps, Appennines Hellenides*, IUCG Scientific Report No. **38**, 140–8.

Ratschbacher, L., Frisch, W. & Ring, U. 1987. Stacking, dispersion, and extension as a consequence of Alpine transpression in the eastern Alps. *Terra Cognita*, **7 (2–3)**, p. 118.

Roeder, D. H. 1980. Geodynamics of the Alpine–Mediterranean system — a synthesis. *Eclogae Geologicae Helvetiae*, **73(2)**, 353–77.

—— & Bogel, H. 1978. Geodynamic interpretation of the Alps: *In*: Closs, Roeder & Schmidt (eds). *Alps, Apennines, Hellenides,* 191–212. IUCG, Scientific Report, **38**.

—— D. 1984. Tectonic evolution of the Apennines (abs). *In: Fossil fuels of Europe Conference and Exhibition. July 15–18, Geneva, Switzerland.* American Association of Petrology and Geology Bulletin, **68, 6**, 798.

—— 1985. *Geodynamics of southern Alps*, seminar paper given at University of Milano, November 28, 1985. The Anschutz Corporation, Exploration Research Division, Denver, 1–22.

—— 1987. South-Alpine Neogene fold-thrust belt (abstract). *Terra Cognita*, **7 (2–3)**, SO1. 10, 87.

—— 1988. Andean-age structure of eastern Cordillera (Province of La Paz, Bolivia). *Tectonics*, **7 (1)**, 23–39.

Royden, L. & Karner, G. D. 1984. Flexure of lithosphere beneath Apennine and Carpathian foredeep basins; evidence for an insufficient topographic load. *Bulletin of the American Association of Petroleum Geology*, **68 (6)**, 704–12.

Schmid, S. M., Zingg, A. & Handy, M. 1987. The kinematics of movement along the Insubric line and the emplacement of the Ivrea zone. *Tectonophysics*, **135**, 47–66.

Seeber, L. & Armbruster, J. 1979. Seismicity of the Hazara Arc in northern Pakistan: decollement *vs*. basement faulting. *In*: AbulFarah & De Jong, (eds). *Geodynamics of Pakistan*, 131–47. Geological Survey of Pakistan, Quetta.

Selverstone, J. 1988. Evidence for east-west crustal extension in the eastern Alps: implications for the unroofing history of the Tauern window, *Tectonics*, **7** (**1**), 87–105.

Slejko, D., Carulli, G. B., Carraro, F., Castaldini, D., Cavallin, A., Doglioni, C., Iliceto, V., Nicolich, R., Rebez, A., Semenza, E., Zanferrari, A. & Zanolla, C. 1987. Modello sismotettonico dell'Italia nord-orientale. *Consiglio Naz. delle Ricerche, gruppo difesa terremoti, rendus* **1**, 1–82.

Squyres, C. H. (ed.) 1975. Geology of Italy, volume I, *Earth Science Society of Libyan Arabian Republic, Tripoli*, 1–402.

Suppe, J. 1983. Geometry and kinematics of fault-bend folding. *American Journal of Science*, **283**, 684–721.

Tollmann, A. 1970. Fur und wider die Allochthonie der Kalkalpen sowie ein neuer Beweis fur ihren Fernschub. *Verh. Geol. Bundesanst.* (**2**), 324–45.

Trümpy, R. (ed.) 1980. *Geology of Switzerland, Part A: an outline of the geology of Switzerland*, Wepf & Co., Basel. 1–104.

Turcotte, D. L. & Schubert, G. 1982. *Geodynamics. Application of continuum physics to geological problems*, 1–450.

Werner, D., Koppel, V., Hanny, R. & Rybach L. 1977. Cooling models for the Lepontine area (central Swiss Alps). *Schweizerische Mineralogische Petrographische Mitteilungen* **56, 3**, 661–8.

Wernicke, B. 1985. Uniform-sense normal simple shear of the continental lithosphere: *Canadian Journal of Earth Sciences*, **22**, 108–25.

Winterer, E. L. & Bosellini, A. 1981. Subsidence and sedimentation on Jurassic passive continental margin, southern Alps, Italy. *Bulletin of the American Association of Petroleum Geology*, **65**, 394–421.

D. Roeder, The Anschutz Corporation, 2400 Anaconda Tower, Denver, CO 80202, USA.

The tectonics of the southern Alps and the Austro–Alpine nappes: a comparison

H. P. Laubscher

SUMMARY: The southern Alps as well as the Austro–Alpine nappes originate from the southern margin of Tethys with close analogies both in basement petrography and stratigraphic sequence. Although they played vastly different roles in Alpine orogeny they developed similar tectonic features. In both domains there is a ramp-flat geometry with associated folds, analogous to those found in thrust belts worldwide, and the basement consists of thrust plates not more than 5 km thick with a conspicuous frontal ramp fold, complicated by imbrications. Regional décollement levels are at the base of the middle Triassic, the Carnian, and the Rhaetian. Both the southern Alps generally and at least the western end of the Upper Austro–Alpine Lechtal nappe are affected by transverse segmentation which in the southern Alps is demonstrably due to reactivation of Jurassic normal fault zones. Tectonic wedging into the regional décollement layers was a frequent phenomenon. In contrast to the southern Alps, the Silvretta–Lechtal nappe forms part of the Alpine nappe system with total translations of many hundreds of kilometres. It now overlies nappes deriving from different palaeogeographic domains (Lower Austro–Alpine, south Pennine, mid-Pennine, and north Pennine nappes in the Rätikon). Reactivation of the equivalent of the original Lechtal sole thrust, although at a somewhat deeper level within the south Pennine Arosa zone, led to the introduction of exotic Arosa masses into the Carnian interval of the Lechtal nappe, and reactivation of the transverse faults and their propagation to deeper levels resulted in non-cylindrical kinematics of Penninic nappes along the Rhine line.

Introduction

For some years now I have been attempting to establish the kinematic tectonics of the southern Alps beginning in the Orobic Alps (Laubscher 1985, Keller *et al.* 1987). In map view, the southern Alps appear strongly segmented by pre-existing transverse fault zones, mostly of Jurassic age. The kinematic style differs from segment to segment, but essentially it is of a typical overthrust belt with ramp-flat geometry and associated folds. A dominant feature is the basement ramp fold – the Orobic anticline being the clearest example.

During the process of analysing and synthesizing the south Alpine tectonics it was unavoidable to be constantly reminded of the western part of the Austro–Alpine nappes in the Rätikon (Fig. 1). In the first place, the stratigraphic sequence of the Austro–Alpine Lechtal nappe (Fig. 2) is similar and consequently gave rise to décollement on equivalent layers. Secondly, there is something resembling the basement ramp fold called the 'Arlberg line' complicated by imbrications and oblique faults, and thirdly, there is a pronounced transverse segmentation. There is, of course, one striking difference. The Austro–Alpine nappes have been transported over their northern foreland by probably hundreds of kilometres, in the later phases riding piggyback as part of an 'orogenic lid' (Laubscher 1983a) on the lower nappes which originated in entirely different facies domains. In contrast, the south Alpine overthrusts, though by no means negligible in throw, overrode their southern foreland by merely tens of kilometres individually, and, perhaps, 100 km cumulatively, the hangingwall and footwall masses consisting of rocks of similar facies.

In this article I want to reassess the tectonics of the Rätikon in the light of what has become known about the tectonics of overthrust belts elsewhere and particularly of my work in the southern Alps. In so doing, I am aware of the excellent work done in this classical if exceedingly complex area by generations of geologists (e.g. Trümpy 1916, Arni 1926, Stahel 1926, Leutenegger 1928, Verdam 1928, Ampferer 1934, Blumenthal 1936, Heissel *et al.* 1965, Oberhauser 1968, Kobel 1969). The complexities are at the root of the many persisting differences of opinion and controversies, and I have no illusion that these will be solved by this article. However, I am convinced that some new perspectives will be useful.

From: Coward, M. P., Dietrich, D. & Park, R. G. (eds), 1989, *Alpine Tectonics*, Geological Society Special Publication No. 45, pp. 229–241.

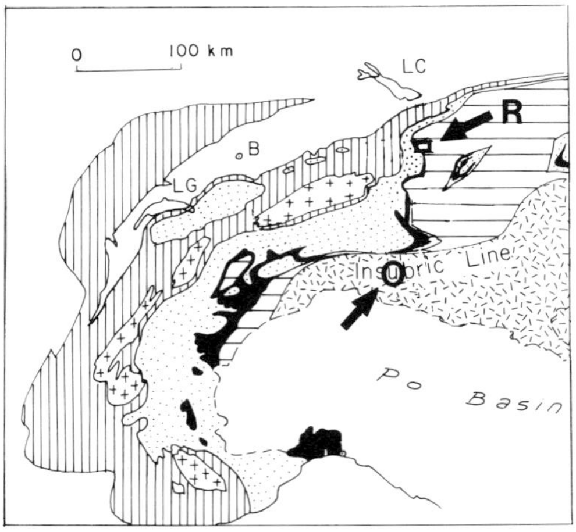

FIG. 1. Tectonic map of the Alps. Vertically ruled: Helvetic nappes and Jura. Crosses: external massifs. Dots: Penninic nappes. Black: Ophiolite nappe. Horizontally ruled: Austro–Alpine nappes. Random dashes: southern Alps. B = Bern; LG = Lake Geneva; LC = Lake Constance; O: Orobic Alps; R: location of the Rätikon (see Fig. 2).

An overview of the tectonics of the southern Alps

Palaeogeographically, both the southern Alps and the Austro–Alpine nappes are a part of the southern margin of Tethys (e.g. Laubscher 1970). During the collisional phases of the Africa–Europa Plate interactions they followed vastly different paths, which are only imperfectly understood (Laubscher & Bernoulli 1982). They are now separated by the Insubric fault zone or Periadriatic Line, the detailed interpretation of which is still controversial although its essential role as an important dextrally-transpressive boundary (Laubscher 1970, 1984) has been documented by now in many places (e.g. Fumasoli 1974, Schönlaub 1980, Schmid *et al.* 1987) and is widely accepted.

The common palaeogeographic origin of the Austro–Alpine nappes and the southern Alps is inferred from great similarities in both stratigraphy and basement petrography. Arguably this similarity in the physical characteristics of the rock sequence resulted, under similar P–T conditions, in similarities of structural style. As to the southern Alps, they are a srtiking example of ramp-flat frontal thrust belt tectonics (Laubscher 1985). This is expressed with exemplary clarity in the Grigna area E of Lake Como, which consequently was used by Laubscher (1985) as the type area from which, by lateral extrapolation, the kinematics of other and more difficult segments was inferred. In the Grigna segment, a basement plate not more than 5 km thick ramps on to a first flat at the base of the middle Triassic which is a regional décollement horizon. A second ramp connects this flat across the thick middle Triassic limestones, with another regional flat in the Carnian, and there certainly were higher ramps and flats now removed by erosion. This simple picture is complicated by imbricate masses, particularly at the sole of the basement thrust (Keller *et al.* 1987).

Laubscher (1985) modelled the Grigna segment kinematically and projected its kinematics laterally into other segments across transverse fault zones, developed by reactivation of mostly Jurassic normal faults (e.g. Bernoulli 1964, Castellarin 1982). He inferred that several of the faults generally interpreted as normal faults had to be the base of the roof of tectonic wedges (Price 1986), commonly either in the Middle Triassic or Carnian décollement horizons. Preliminary inspection suggests the presence of transverse wedges, e.g. in the Carnian W of the Val Brembana transverse zone (Gaetani & Jadoul 1979, Laubscher 1985). It is these wedge structures that Alpine tectonicians apparently (according to a number of discussions) find difficult to accept, although they have been described long ago from the northern front of the Alps (Habicht 1945) and the Jura mountains (Schardt 1894, compare Laubscher 1977), and recently from the south Alpine molasse (Bernoulli *et al.* 1987). They have been found to be common in other fold- and thrust-belts (e.g. Banks & Warburton 1986). In my experience they are a common mode of failure, on a variety of scales, in sedimentary sequences with pronounced contrasts in competence.

An overview of the tectonics of the Rätikon

Figure 1 shows the location of the Rätikon at the northwestern corner of the Austro–Alpine nappes. It borders that obvious though ill-defined western boundary of the eastern Alps—the 'Rhine line'. Figure 2 is a tectonic sketch map of the area, based on the maps by Heissel *et al.* (1965) and Allemann *et al.* (1985). The main elements are the Upper Austro–Alpine Silvretta basement nappe and its sedimentary décollement part, the Lechtal nappe. The frontal masses of the Silvretta nappe are complexly folded and imbricated and separated from the Lechtal nappe by an imbricated ramp

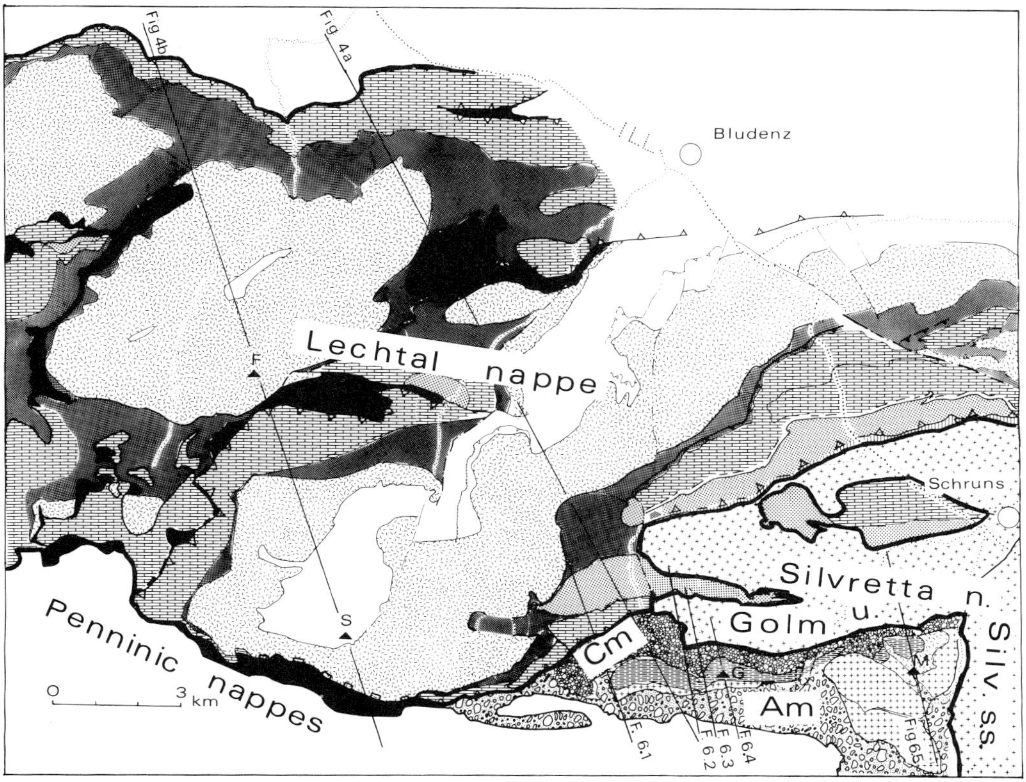

FIG. 2. Tectonic sketch map with an overview of the tectonic units of the Rätikon. F = Fundelkopf; S = Schesaplana; G = Geissspitze; M = Mittagspitze, Cm = 'Carnian mélange'; Am = Arosa mélange. Stratigraphic ornamentations are those of Fig. 3. Post-Hauptdolomit signature shown only in Mittagspitz–Geissspitz zone. Also shown are the traces of the cross-sections of Figs 4 and 6. After Heissel *et al.* (1965) and Allemann *et al.* (1985), simplified.

fold (Figs 3 and 4). Underlying the Silvretta and Lechtal nappes are mélange zones, 'Schuppenzonen' in German, which are characterized by elements of diverse origin, of all sizes, sometimes preserving a certain order of thrusting, mostly however associated in a practically chaotic fashion. The higher of the two mélanges, here termed the 'Carnian mélange', contains many large and mappable imbrications of Lechtal nappe elements that grade into smaller blocks floating in a matrix of the highly mobile Carnian interval. The Carnian mélange is here treated as a tectonic unit as I have become convinced that attempts at ordered separation of the elements has led to confusion rather than clarification. As explained below, reasons may be set forth for a common origin of this mélange. Below the Carnian mélange there is the Arosa mélange, characterized by south Penninic and lower Austro–Alpine elements, with an occasional addition of elements from higher Austro–Alpine units. Again, the size of the elements varies, some being of dimensions approaching small nappes and deserving a special name. The northern parts of the Arosa mélange have been introduced into the Lechtal nappe by late thrusts issuing from the interior of this zone when the nappe already overlay the Arosa zone. The final residence of these Arosa elements within the Lechtal nappe is very puzzling in places and has been the subject of controversy. The Arosa mélange overlies middle to north Penninic elements dealt with only marginally in this paper.

In addition to this essentially cross-sectional characterization of the tectonics of the Rätikon there is an obvious array of non-cylindrical, transverse to oblique, map-view elements. The Silvretta nappe changes character, and disintegrates to the W, across two N–S trending fault zones. The Lechtal nappe contains, besides N–S to NNE–SSW trending lineaments that

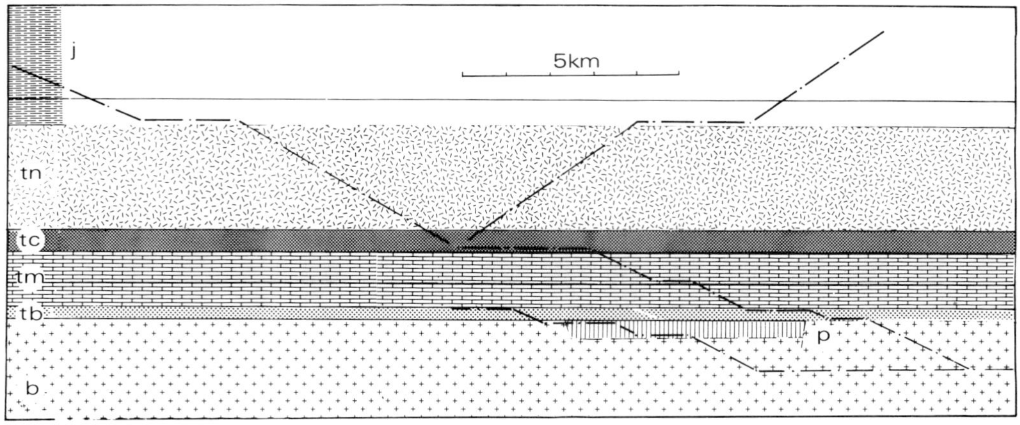

FIG. 3. Tectono-stratigraphic units of the Rätikon, simplified. Maximum original thicknesses after Pliöchinger (1980). Dash-dot lines are the main ramp-flat systems. b = basement; p = Palaeozoic, tb = basal Triassic; tm = middle Triassic; tc = Carnian; tn = Norian Hauptdolomit; j = Rhaetian and Lias.

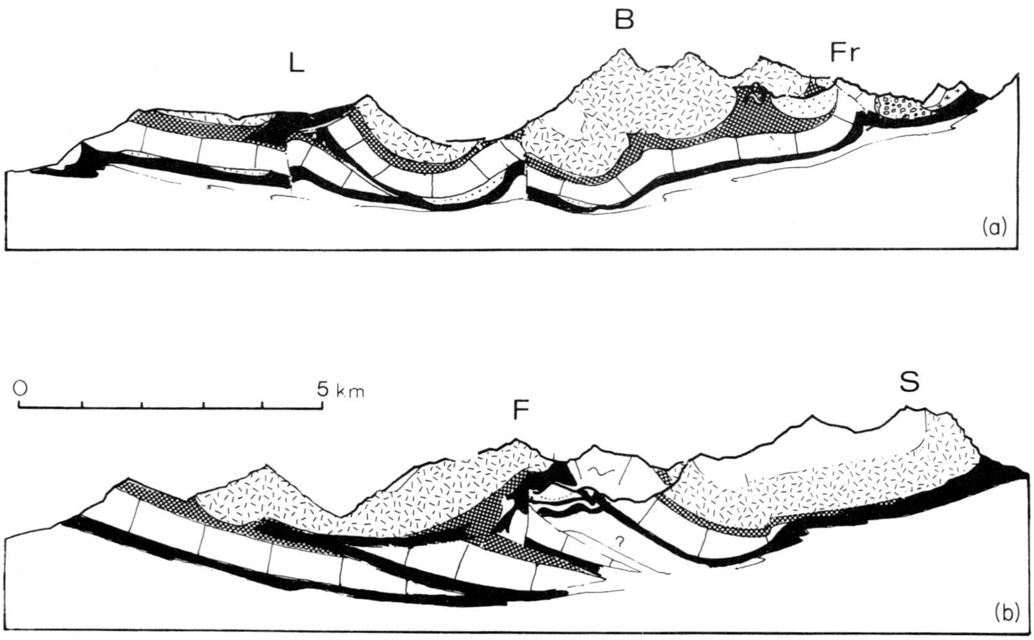

FIG. 4. Two cross-sections through the Lechtal nappe, after Kobel (1969, modified). Ornament similar to Fig. 3. Black: Arosa zone. L = Loischkopf; B = Branderspitze; F = Frescalot or Freschluakopf; F = Fundelkopf; S = Schesaplana. For location see Fig. 2.

exhibit a rather obvious sinistral element a series of NE–SW trending high zones separating large synclinal blocks – 'Schollen' or 'floes' – dominated by the Norian Hauptdolomit. These transverse to obliquely trending zones are often interpreted as indications of a special phase of W to NW-directed transport. In addition, there are NW-trending dextral faults which, together with the complementary sinistral faults define a wedge-shaped domain with its apex S of Bludenz (Fig. 2). These transverse elements appear to have been active during several phases but their present appearance is due to very late, post-nappe movements.

The latest of the movements affecting the area is the updoming of the 'Prättigau half-window', which affects all the nappes and seems to be associated with the development of the late Alpine crustal brachyanticlines which are responsible for the more important windows in the Alpine nappe edifice (Laubscher 1983b, 1988). The following discussion of the various domains begins with the sedimentary décollement part and then takes up the basement nappe, the mélange substratum and finally the map-view kinematics.

Characteristics of the Lechtal nappe

The main décollement of the Lechtal nappe usually took place at the base of the Anisian where the Reichenhall rauhwacke of Austrian or Punt-la-Drossa beds of Swiss authors shows vestiges of evaporites. However, in some places décollement occurred within or at the base of the lower Triassic or even some late Palaeozoic sediments. The most important higher décollement zone in the Carnian on top of which the dominant competent interval, the Hauptdolomit, is often deformed disharmonically. It is subdivided into large synclinal blocks or 'floes' which seem to float with predominantly tectonic contact on the Carnian. This latter contains incompetent shales and evaporites, and these have been accumulated tectonically into large irregular masses which are often associated with exotic terrains introduced by thrusting. Of these exotic terrains the most puzzling are those of the Arosa mélange, but they ought to be viewed together with those consisting of imbrications of Lechtal Triassic and of tectonic accumulations of lower Triassic SE of the Schesaplana block. Most of these introductions of exotic terrains are concentrated along the southeastern borders of the Hauptdolomit blocks, whereas in the N or NW the passage from the Middle Triassic across the Carnian into the Hauptdolomit appears to be uninterrupted even if tectonically disturbed.

The southeastern border of the Schesaplana block is detached from its middle Triassic underpinnings and turned up in the fashion of a delta structure. Space is filled with large accumulations of gypsum and two striking bodies of lower Triassic. These are much thicker than figures reported from stratigraphically undisturbed sections (Fig. 3) would suggest, due to duplex formation or some similar process, tied to décollement at both the bottom and the top of a formation. Duplex formation of the basal Triassic and the introduction of the thickened masses into various peculiar tectonic positions is modelled schematically in Fig. 5. Figure 5(f) in particular, shows the introduction of a tectonic wedge into the Carnian interval. The relative sense of shear at the base of the upturned Hauptdolomit is that of a S-directed backthrust. At the southeastern border of the Schesaplana block this shear is transmitted to the base of the units covering the masses of lower Triassic – a process that seems to be quite common elsewhere (compare Laubscher 1985). The N–S striking segments of this border indicate a sinistrally transpressive scenario.

Figure 4 also shows the occurrence of the other exotic terrains in the Carnian interval. The map picture may be interpreted in various ways, as a straightforward cylindrical projection of the outcrops onto the profiles is not warranted. In my interpretation I have chosen the subsidiary 'floes' or imbrications of the Garvion, the Augstenberg, and the Heupiel which abut against the Hauptdolomit of the Fundelkopf 'floe', separated from it only by various amounts of Carnian (plus or minus Arosa exotics), to be the imbricated infill of the Fundelkopf delta structure. In so doing I have kept in mind the indications that the Naafkopf unit of the mid-Penninic Falknis nappe is involved in the late motions of the Gorvion imbrication (see Heissel *et al.* 1965) although without attempting to project them specifically onto the profile (Fig. 4b). There the situation is simplified as an exact cylindrical projection is not feasible. One thing is certain: the pronounced SE limb of the Fundelkopf syncline coincides with the occurrence of a pile of imbrications. The picture is reminiscent of the tectonic wedges reported from the frontal Canadian Rockies, e.g. by Price (1986). Such delta structures (or triangle structures, or wedge-into-split-apart structures or tectonic wedges) have been postulated by Laubscher (1985) in the southern Alps, and there particularly in the basal Triassic and Carnian intervals. These late imbrications that originated in décollement surfaces of the Arosa zone (or, more rarely, even in the underlying mid- to north Penninic units), were ramped over the middle Triassic onto the Carnian décollement flat and in places wedged into the Carnian interval. Where the covering Triassic of the Lechtal nappe was stripped off by the back-thrusting of the Hauptdolomit, the exotics remained intermingled with the Carnian, and particularly with amassments of Carnian evaporites to form what may be termed an 'Arosa-Carnian melange'. The following types of Arosa exotics may be observed: Simple sole association (e.g. at the northeastern base of the Schesaplana

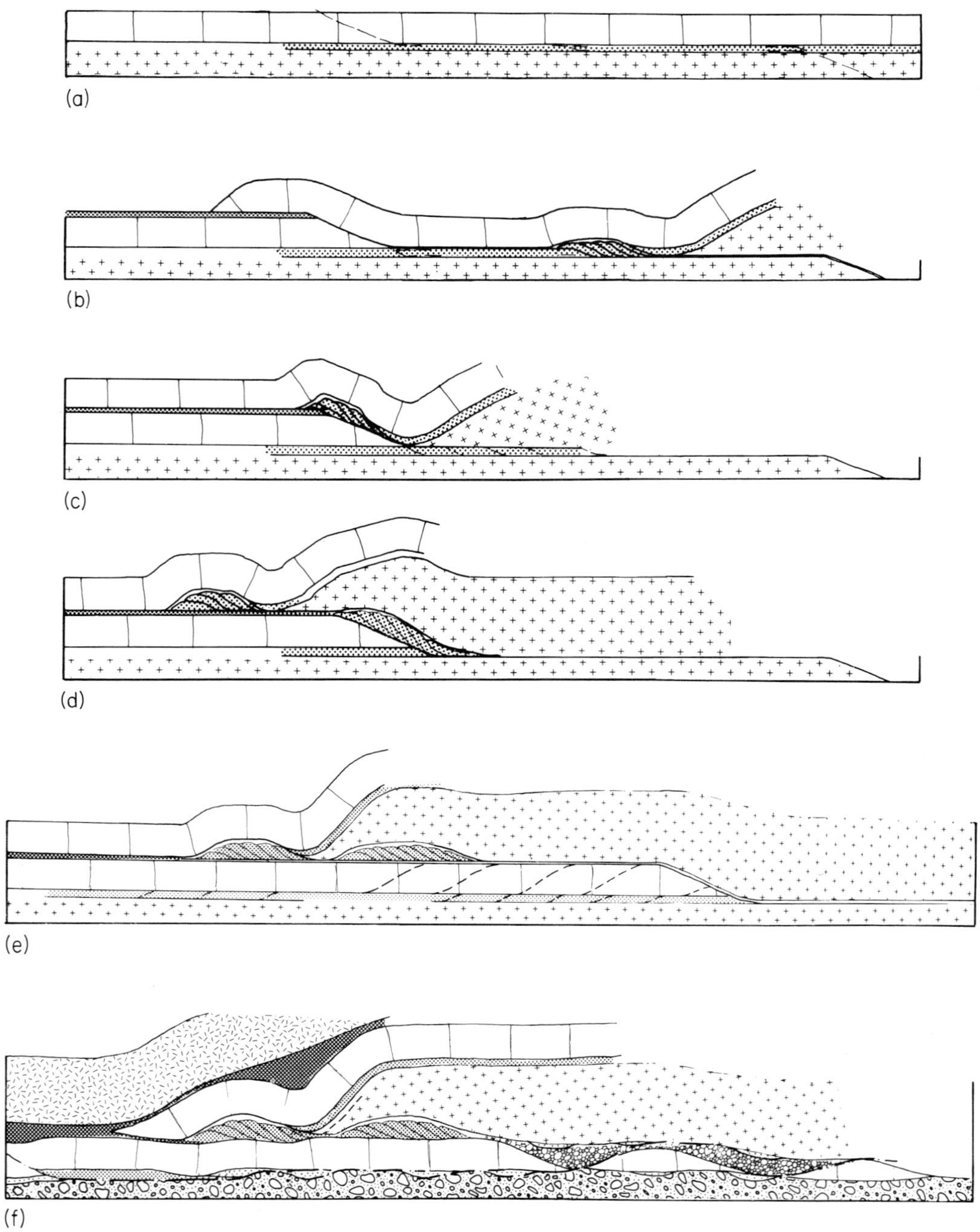

FIG. 5. (a) to (d) Kinematic sequence showing the thickening of basal Triassic by repeated 'duplex' formation during ramp-flat thrusting. Ornament is similar to Fig. 3. (e) shows potential down-stepping branches of the main thrust, with independent systems in the middle and the basal Triassic (extensional ramp-flat systems or extensional 'duplexes'). (f) shows a kinematically advanced stage of these processes. In addition, a tectonic wedge (delta structure) into the Carnian interval, where the stratigraphic sequence is split apart, is illustrated. At the base of the overriding basement a 'Carnian mélange' develops and through extension enters into contact with the Arosa mélange. Beds younger than Carnian have been stripped off the wedge by backthrusting. This may be the situation at the front of the Silvretta nappe.

block), the association with imbrications at the base of a Hauptdolomit delta structure (e.g. at the sole of the northern part of the Gorvion imbrication), and the 'Arosa-Carnian mélange' (farther north along the southwestern to northwestern border of the Fundelkopf block). The latter occurrences may even be projected eastward across the Fundelkopf block into the conspicuous accumulation of mélange in the Loischkopf (Fig. 4a).

All these occurrences are complicated by the oblique tectonics of the welts separating the Hauptdolomit blocks which in addition to elements of thrusting and folding also contain steep cross-cutting faults such as that of the Amatschon Joch, see below.

The complex distribution of Arosa exotics within the Lechtal nappe was attributed by Ampferer (1934) to 'Reliefüberschiebung', i.e. late thrusting over a previously irregularly eroded surface. This notion, however, was rejected by most of the other workers in the area. Another favourite process advocated, e.g. by Kobel (1969) was that of diapiric intrusions. However, Arosa material hardly lends itself to diapiric intrusion, it must have been transported and distributed as a part of the surrounding tectonic mass movements.

The transition of the Lechtal to the Silvretta nappe

The southern border of the Lechtal nappe changes character from W to E. Whereas S of the Schesaplana the middle Triassic is rudimentary and involved in imbrication with the Arosa zone, it reappears as a continuous band further E with increasing stratigraphic completeness and importance. It becomes thickened by folding and thrusting (Fig. 4a; compare Kobel 1969, figs 3 and 15, Blumenthal 1936). Along the Zaluanda fault it is apparently displaced sinistrally by several kilometres to reappear N of the Rellstal, where it is underlain by a thick accumulation of basal Triassic and smithereens of late Palaeozoic imbricated within the basement. The latter is badly tectonized ('Phyllitgneis'). Thus the apparent transition from the sedimentary décollement part across a ramp fold into the basement part is not smooth at all, and some further complications have to be considered in addition to those already mentioned. In particular, there are the outcrops of Triassic appearing in windows beneath the basement (compare Fig. 6, profile 2; Burger in Trümpy 1980, fig. 36). Formerly, these outcrops of Triassic had often been interpreted as synclines on top of the basement (e.g. Leutenegger 1928), but this interpretation was hardly tenable in view of the detailed mapping by Heissel *et al.* (1965). These windows imply that Lechtal elements have been overrun by Silvretta basement. Kobel (1969) interprets them as the continuation of the southern border of the Schesaplana farther W, and as a general notion this is probably the simplest solution. Logical continuation of this postulate leads to the conclusion that, in this case, the basement to middle Triassic masses E of the Zaluanda fault were thrust on to the décollement flat of the Carnian, very possibly forming a delta structure. The Zaluanda fault thus terminates in the N at the thrust front of middle Triassic into the Carnian.

While this appears plausible as a general notion and is well-nigh inescapable if one wants to avoid much more complicated schemes, it cannot be the whole story. The outcrops in the windows are three-dimensionally discontinuous and imbricated. In particular the basal Triassic seems to be thickened by repetitions similar to those of the Silvretta front. A model has been devised in Fig. 5 that demonstrates a possible sequence of duplex formation that would lead to such accumulations of basal Triassic in various places. In Fig. 5(e) the Silvretta nappe is wedged into the Carnian interval as conjectured above, and the middle Triassic (in Fig. 5f already resting on Arosa melange) is subjected to stretching and discontinuous block formation by repeated downstepping of the thrust – a sort of 'extensional duplexing'. Thereby a melange already produced in the overridden Carnian interval is further mixed with the dissected and dismembered middle and lower Triassic, resulting in what here is termed loosely the 'Carnian melange' on top of the Arosa mélange.

This idea has been pursued to the situation farther S, the belt of Lechtal Triassic N of the Mittagsspitze–Geisspitze zone. Here the Carnian melange would be expected to reappear from below the Silvretta basement. Usually, the eastern part of this belt is associated with the Mittagsspitze unit, and Kobel (1969), e.g. takes this as an indication of the upper Austro–Alpine origin of the Mittagsspitze zone. However, the contacts are tectonic, and the remainder of the Mittagsspitze unit shows an underdeveloped middle Triassic as well as Liassic(?) breccias, both typical of lower Austro–Alpine units, e.g. the Gotschnagrat–Casanna unit farther S, near Klosters (Weissert 1975). To the W, the inferred belt of Carnian melange continues into the various domains of

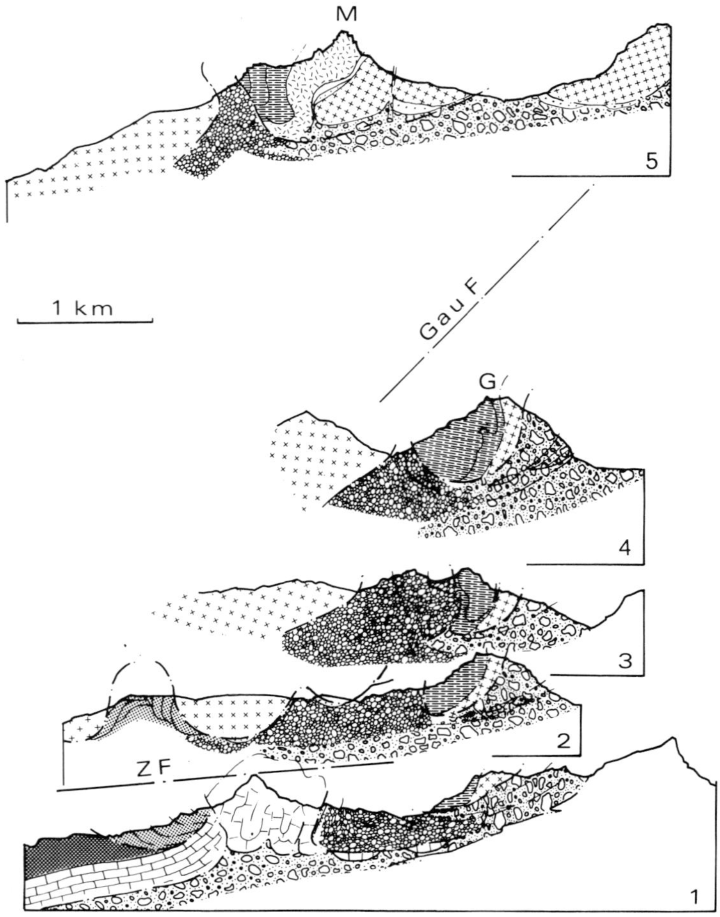

FIG. 6. Profiles through the Mittagspitze–Geissspitze zone, after Blumenthal (1936), simplified and modified. M = Mittagspitze; G = Geissspitze; Gau. F = Gauertal fault; ZF = Zaluanda fault. Location of the profiles in Fig. 2.

imbrications discussed extensively in the literature (e.g. Leutenegger 1928, Blumenthal 1936, Kobel 1969) which fit into the disintegration of the southern border of the Lechtal nappe depicted schematically in Fig. 5(e, f).

Trails of isolated elements of Mesozoic and Palaeozoic, some of them definitely of Lechtal facies, continue to the S and reappear at the base of the Silvretta nappe in the Lower Engadin window (e.g. Oberhauser 1980) and in the Madrisa zone a few kilometres S of Fig. 2. Here an inverted sequence of late Palaeozoic to middle Triassic in Lechtal facies is found (Trümpy & Haccard 1969, Haldimann 1975). Décollement, stripping of beds younger than middle Triassic and inversion of the sequence all demand a special kinematics which is illustrated in Fig. 7. The initial phase of this development is similar to a situation in the Bergamasc Alps at Passo San Simone (Keller *et al.* 1987).

The Silvretta nappe

The Silvretta nappe in Fig. 2 is subdivided by the Gampadels fault into the Silvretta nappe s.s. and the Golm unit, the latter pierced by the windows discussed above and bounded in the W by the Zaluanda fault. These two faults testify to a stepwise disintegration of the Silvretta nappe westwards, the regional implications of which will be discussed below. The tectonic and kinematic situation depicted so far calls for the Lechtal, Golm, and Silvretta s.s. units to belong to the same nappe. They are separated by a series of fascinating, somewhat puzzling, but by no means unintelligible tectonic complications.

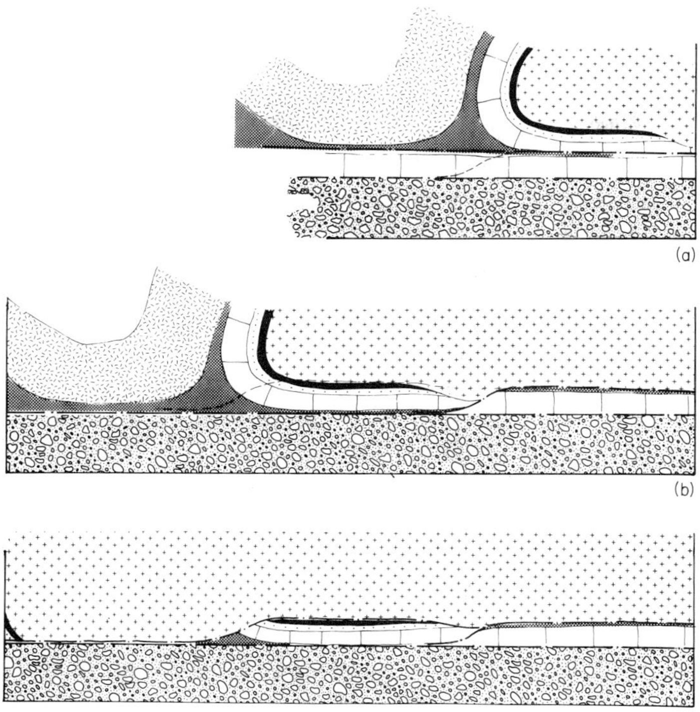

FIG. 7. Development of the inverted sequence in the Madrisa zone (Haldimann 1975). Black: Palaeozoic.

It would appear highly artificial to introduce here a boundary between an upper Austro–Alpine Lechtal and a middle Austro–Alpine Silvretta nappe (discussion in Oberhauser 1980; compare Tollmann 1978).

Map view kinematics

In Fig. 8 the oblique shear zones and faults are superposed on the tectonic sketch map of Fig. 2. It shows several remarkable associations. The most obvious one is formed by complementary sinistral and dextral faults in the Schesaplana block, labelled S in Fig. 8. They define a wedge with an apex SE of Bludenz that was differentially pushed to the NNW. In part this wedge structure is younger than the internal thrusting and imbrication of the Lechtal nappe. However, at least the Zaluanda fault (ZF) and the Saula shear zone (SZ in Fig. 8) which fit into the pattern are as old or older than these internal tectonics where they functioned as lateral rails. These faults apparently had a long history only the later parts of which are immediately obvious.

If we consider these oblique faults together with the anticlinal axes and thrusts they appear to bound laterally, their importance is further enhanced. The Zaluanda fault not only bounds the Silvretta basement during its differential northwards thrusting, it also forms the western limit of the anticlinal wrinkles that give rise to the windows in the Golm unit (Fig. 6, profile 2). For example the southern of these wrinkles is juxtaposed to the synclinal mass of lower Triassic to the W of Lüner Alp. The southern boundary of the Golm unit is the N limb of another fold (although superimposed on the N limb of the later Prättigau half-window, it predates it and is a structure in its own right). In the core of this fold the Zaluanda fault first gets lost in the chaos of the Carnian mélange but possibly reappears as the western boundary of the Geissspitze unit (G: Geissspitze in Fig. 8) and the concomitant larger elements of the Arosa zone S of it (not represented in the tectonic map; see Heissel *et al.* 1965). The straightforward interpretation of this situation would call for a differential sinistral northwards translation of the Geissspitze and associated units into the Golm unit, much in the fashion of

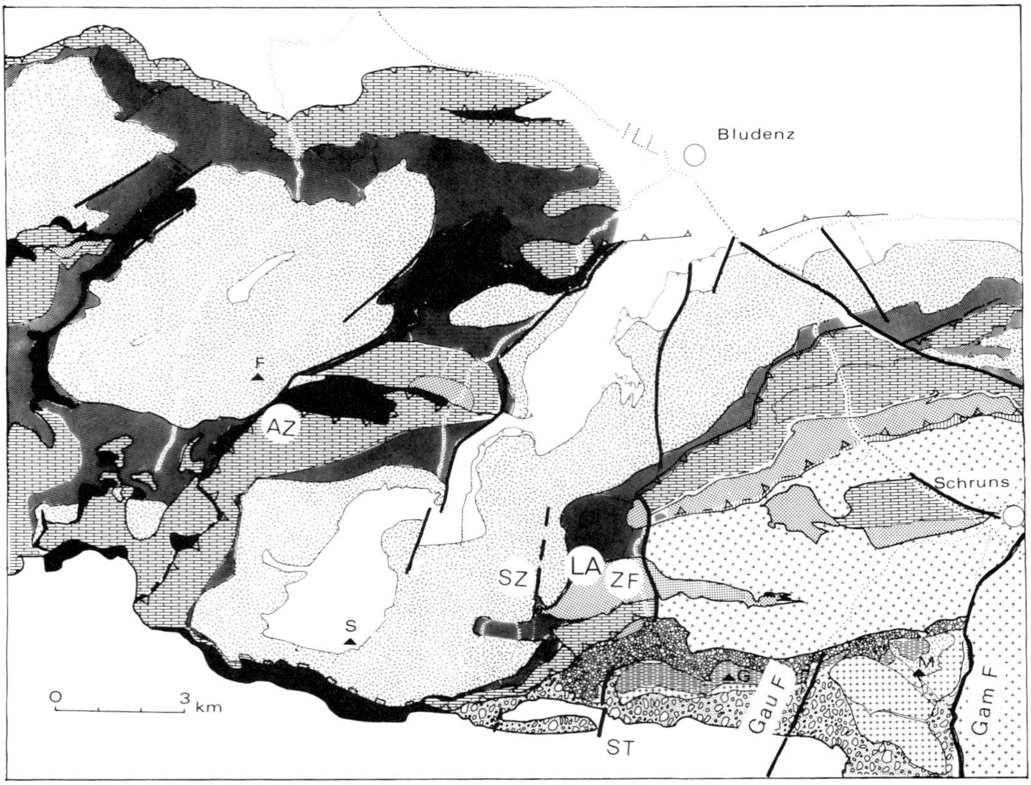

FIG. 8. The main oblique faults and shear zones in the Rätikon, superimposed on Fig. 2. Only those names used in the text are shown. AZ = Amatschonjoch zone; SZ = Saula zone; ZF = Zaluanda fault; ST = Schweizer Tor; GauF = Gauertal fault; GamF = Gampadelsertal fault; LA = Lüner Alp.

a delta underthrust and quite comparable to the Golm emplacement into the Carnian of the Lechtal nappe farther N. It is even possible that such processes affected subunits, e.g. that the Geissspitze basement was thus emplaced in the Rhaetian as the contact is tectonic and most of the Triassic sequence is missing. I think it probable that the Mittagsspitze domain is separated from the Geissspitze by another cross-structure, the Gauertal fault (Gau F in Fig. 8), as its architecture is quite different; however, the critical area is not exposed. Adding together all these differential northwards movements on the E side of the Zaluanda fault it may be concluded that this was one of the more important, regional, western boundary elements of the Silvretta–Lechtal nappe. Although probably initiated during the Cretaceous development of the Austro–Alpine system, it probably continued into the Tertiary. This may be gathered from the fact that the western end of the Sulzfluh nappe proper in the Kirchlispitzen appears to be somehow integrated into the differential sinistral motion, although it seems to be limited by a shear zone W of the Zaluanda fault. A closer look at the internal tectonics of the Sulzfluh nappe E of the cross-faults of the Schweizer Tor (ST in Fig. 8), however, reveals that part of the Sulzfluh nappe has been differentially piled up on the E flank of an apparent continuation of the Zaluanda fault, on top of rare outcrops of Arosa zone (see Heissel *et al.* 1965). Carrying this argument somewhat farther to the E, it may be conjectured that the Gauertal fault between the Geissspitze and Mittagspitze units continues to the S and, after cutting off the serpentinites, flysch units and granites of the Bilkengrat, heads directly into the sinistral fault W of the Sulzfluh. The Gauertal fault too would have had an important early history and a weaker reactivation during the emplacement of the Sulzfluh nappe. The nearly parallel Gampadels fault that borders the Silvretta nappe proper, belongs to the same system, being possibly its most important element.

Proceeding now to the W of the Zaluanda fault, we perceive that the oblique elements within the Lechtal nappe have similar characteristics; they appear to belong to the same system. For instance, they limit laterally approximately E–W striking features, with a differential sinistral northwards motion of the E flank. They are transpressive to a certain extent, and the compressive N to NW-directed component has been interpreted as due to a separate phase of thrusting in that direction (e.g. Trümpy 1916). However, the kinematic relationships in the area do not bear out this conjecture.

A brief digression into regional tectonics

(Consult the Tectonic Map of Switzerland and the Geologic Map of Switzerland, 1:500 000.) The western boundary of the Austro–Alpine nappes in the Rätikon is not merely an erosional feature; it is a consequence of the internal tectonics and particularly of the system of cross faults and the stepwise disintegration of the Austro–Alpine nappes. It is a regional feature that may be followed southward in the direction of the Rhine line to the area between Davos and Lenzerheide-Tiefencastel where it interferes with the Ortler–Albula zone. An important point here is that this 'Rhine line' was reactivated in the Tertiary during the emplacement of the Sulzfluh nappe, as already mentioned. Something analogous happened along its southwards continuation between Lenzerheide and Tiefencastel, E of which the Falknis and Sulzfluh units are replaced by the Schams units. These units are stratigraphically quite similar but tectonically very difficult to tie to the Falknis and Sulzfluh nappes. Cylindrical projections (e.g. Streiff 1962) fail in the area of Tiefencastel ('The Schams dilemma, one of the major headaches of Alpine geology,' Trümpy, 1980, p. 70). Realization that the Rhine line was tectonically active during the Mesoalpine nappe emplacement would appear to go a long way towards a solution of this kinematic puzzle.

Conclusions

Palinspastically, both the southern Alps and the Austro–Alpine nappes belong to the southern margin of Tethys. This is inferred from the close analogies in basement petrography as well as stratigraphic sequence. Similarity in the physical properties of the rocks at similar low P, low T conditions arguably resulted in similar tectonic features during compressive phases of Africa–Europa Plate interaction, although the two domains played vastly different roles in the construction of the Alpine edifice. Similar features of both domains are particularly a ramp-flat tectonics with associated folds, analogous to those found in thrust belts worldwide. In both domains the basement thrust mass is a plate not more than 5 km thick with a conspicuous frontal basement ramp fold, complicated by imbrications. The first regional décollement level is at the base of the middle Triassic although lower Triassic and even late Palaeozoic are dragged along occasionally. The second regional décollement layer is in the Carnian, and the third is in the Rhaetian. Both the southern Alps generally and at least the western end of the Upper Austro–Alpine Lechtal nappe are affected by transverse segmentation; in the Southern Alps the transverse fault zones are demonstrably due to reactivation of Jurassic normal fault zones. A feature believed important in both domains, though still controversial, is tectonic wedging particularly into the regional décollement layers. The main difference between the domains is that the Lechtal nappe forms part of the Alpine nappe system with total translations of many hundreds of kilometres, whereas the south-Alpine thrust masses overrode their immediate foreland by no more than tens of kilometres. One of the consequences is that the Silvretta–Lechtal nappe came to overlie nappes of very different rocks deriving from different palaeogeographic domains (Lower Austro–Alpine, south-Pennine, central-Pennine, and north-Pennine nappes in the Rätikon). During later phases, transverse faults and décollement-thrust systems were reactivated in various degrees and at various levels within the nappe system. In particular, reactivation of the equivalent of the Lechtal sole thrust, although at a somewhat deeper level within the south-Pennine Arosa zone, led to the introduction of exotic Arosa masses into the Carnian interval of the Lechtal nappe. Reactivation of the mostly sinistral transverse faults and their propagation to deeper levels in the Midpenninic nappes resulted in dilaceration of the latter, with non-cylindrical kinematics along the Rhine line – the conspicuous western boundary of the eastern Alps. This boundary is obviously tectonic rather than merely erosional, and apparently was active during the various phases of Alpine collisional tectonics.

ACKNOWLEDGEMENT: The author is indebted to Dorothee Dietrich for many suggestions which helped improve the manuscript.

References

Allemann, B., Schwizer, B. & Martin, B. 1985. *Geologische Karte des Fürstentums Liechtenstein, 1:25 000. Government of Liechtenstein.*

Ampferer, O. 1934. Geologische Bauformen und Baufragen des Rätikon-Gebirges. *Jahrbuch Geologische Bundesansstalt*, **84**, 13–40.

Arni, P. 1926. *Geologische Forschungen im mittleren Raetikon.* Dissertation, University of Zürich. *Buch- und Kunstdruckerei Vogt-Schild Solothurn*, 85 pp.

Banks, C. J. & Warburton, J. 1986. 'Passive-roof' duplex geometry in the frontal structures of the Kirthar and Sulaiman mountain belts, Pakistan. *Journal of Structural Geology*, **8**, 229–37.

Bernoulli, D. 1964. *Zur Geologie des Monte Generoso (Lombardische Alpen). Beiträge zur geologischen Karte der Schweiz, NF 118*, 135 pp.

——, Bertotti, G., Gelati, R., Napolitano, A., Valdisturlo, A. & Zingg, A. 1987. The relations of the Gonfolite Lombarda (south-alpine molasse) to the Mesozoic sequence of the southern Alps. *Nationales Forschungsprogramm 20, Bulletin*, **4**, 13–14.

Blumenthal, M. M. 1936. *Der Sedimentzug der Tschaggunser Mittagspitze und die Gliederung des oberostalpinen Deckenkerns im Vorarlberger Rätikon.* Jahresbericht der Natforschende Gesellschaft Graubünden, LXXIV, N.F. 1935/36.

Castellarin, A. 1982. Lineamenti ancestrali sudalpini. *In*: Castellarin, A. & Vai, G. B. (eds). Guida alla geologia del sudalpino centro-orientale. *Societa geologica italiana, Guide geologiche regionali, 41–53.*

Fumasoli, M. W. 1974. Geologie des Gebietes nördlich und südlich der Jorio–Tonale Linie im Westen von Gravedona. *Mitteilungen Geologisches Institut ETH & Universität Zürich, N.F. 194.*

Gaetani, M. & Jadoul, F. 1979. The structure of the Bergamasc Alps. Rendiconti. *Accademia Nazionale dei Lincei VIII*, **66**, 411–16.

Habicht, K. 1945. Geologische Untersuchungen im südlichen sanktgallisch–appenzellischen Molassegebiet. *Beiträge zur geologischen Karte der Schweiz*, **NF 83**, 1–166.

Haldimann, P. A. 1975. Arosa- und Madrisa-zone im Gebiet zwischen Klosters und dem Gafiental GR. *Eclogae geologicae Helvetiae*, **68**, 301–10.

Heissel, W., Oberhauser, R., Reithofer, O. & Schmidegg, O. 1965. *Geologische Karte des Rätikon, Voralberg, 1:25 000. Geologische Bundesanstalt.*

Keller, P., Schönborn, G., Schumacher, M. & Laubscher, H. 1987. The Suborobic imbrications in the Bergamasc Alps. *Terra Cognita*, **7**, 60.

Kobel, M. 1969. Erläuterungen zur Tektonik des Rätikon mit spezieller Berücksichtigung des Kalkalpenanteils. *Jahrbuch des Vorarlberger Landesmuseumsvereins 1968/69*, 245–60.

Laubscher, H. P. 1970. Das Alpen–Dinaridenproblem un die Palinspastik der südlichen Tethys. *Geologische Rundschau*, **60**, 813–33.

—— 1977. An intriguing example of a folded thrust in the Jura. *Eclogae Geologicae Helvetiae*, **70**, 97–104.

—— 1983a. Detachment, shear and compression in the central Alps. *Memoirs of the Geological Society of America*, **158**, 191–211.

—— 1983b. The late Alpine Periadriatic intrusions and the Insubric Line. *Memorie della Societa Geologica Italiana*, **26**, 21–30.

—— 1984. The tectonic problem of the Ivrea body and the Insubric line. *Annales Geophysicae*, **2**, 169–70.

—— 1985. Large scale, thin-skinned thrusting in the southern Alps. Kinematic models. *Bulletin of the Geological Society of America*, **96**, 710–18.

—— 1988. Décollement in the Alpine system: an overview. *Geologische Rundschau*, in press.

Leutenegger, W. O. 1928. *Geologische Untersuchungen im mittleren nordöstlichen Rätikon.* Unpublished dissertation, University of Zürich.

Oberhauser, R. 1968. *Zur Geologie der West–Ostalpen–Grenzzone im Vorarlberg und im Prätigau. Verhandlungen der geologischen Bundesanstalt. Sonderblatt A.* Wien

—— (ed.) 1980. *Der geologische Aufbau Österreichs.* Geologische Bundesanstalt, Springer Verlag, Wien, New York.

—— 1980. Das Unterengadiner Fenster. *In*: Oberhauser, R. (ed). *Der Geologische Aufbau Österreichs*, 291–99. Springer, Wien.

Plöchinger, B. 1980. Die nördlichen Kalkalpen. *In*: Oberhauser, R. (ed.) *Der geologische Aufbau Österreichs*, 218–64. Geologische Bundesanstalt, Springer Verlag, Wien.

Price, R. A. 1986. The southeastern Canadian Cordillera. Thrust faulting, tectonic wedging, and delamination of the lithosphere. *Journal of Structural Geology*, **8**, 239–54.

Schardt, H. 1894. Excursions géologiques dans le Jura Méridional. *In*: Payot, F. (ed.). *Jura et Alpes de la Suisse, livret-guide géologique*, 1–10. Lausanne.

Schmid, S. M., Zingg, A. & Handy, M. 1987. The kinematics of movements along the Insubric Line and the emplacement of the Ivrea zone. *Tectonophysics*, **135**, 47–66.

Schönlaub, H. P. 1980. Die Gailtallinie. *In*: Oberhauser, R. (ed.). *Der Geologische Aufbau Österreichs*, 422–5. Geologische Bundesanstalt, Springer Verlag, Wien.

Stahel, A. H. 1926. *Geologische Untersuchungen im nordöstlichen Rätikon.* Unpublished dissertation, University of Zürich.

Streief, V. 1962. Zur östlichen Beheimatung der Klippendecken. *Eclogae Geologicae Helvetiae*, **55**, 77–132.

Tollmann, A. 1978. Plattentektonische Fragen in den Ostalpen und der plattentektonische Mechanismus des mediterranen Orogens. *Mitteilungen der Österreichischen Geologischen Gesellschaft*, **69**, 291–351.

Trümpy, D. 1916. Geologische Untersuchungen im

westlichen Rätikon *Beiträge zur geologischen Karte der Schweiz N.F.*, **46/2**, 43–163.

TRÜMPY, R. 1980. An outline of the Geology of Switzerland. *In*: *Geology of Switzerland, a guide-book*, Schweizerische Geologische Kommission. Wepf & Co., Basel, 102 pp.

—— and HACCARD, D. 1969. Réunion extraordinaire de la Société géologique de France. Les Grisons, du 14 au 21 septembre 1969. *Compte-rendu sommaire des séances de la Société géologique de France*, **9**, 329.

VERDAM, J. 1928. *Geologische Forschungen im nördlichen Rätikon*. Unpublished dissertation, University of Zürich.

WEISSERT, H. 1975. Zur Geologie der Casanna bei Klosters. *Eclogae Geologicae Helvetiae*, **68**, 222–9.

H. P. LAUBSCHER, Geologisch-paläontologisches Institut, Bernoullistr. 32, CH-4056 Basel, Switzerland.

West-directed décollement of Austro–Alpine cover nappes in the eastern Alps: geometrical and rheological considerations

L. Ratschbacher & F. Neubauer

SUMMARY: The Austro–Alpine unit of the eastern Alps constitutes a thrust belt characterized by multiple deformations under changing external stress regimes during the Alpine orogeny. Problems arise in recognizing direction and geometry of initial detachment tectonics and flow behaviour of mylonites accompanying detachment thrusts.

In the first part of this article, we analyse the actual distribution of Austro–Alpine cover sediments as acquired during the Alpine orogeny and studied detachment thrusting on three sections through thrust systems occupying different vertical positions in the Austro–Alpine stack. In the second part, we study ductile rock flow in mylonites along individual thrusts in the basement-cover décollement zone.

The ramp-flat geometry, the dip of frontal ramps, and the imbrication of youngest sediments at the leading edge of thrust systems show that detachment is directed towards the W or NW and is, thus, in agreement with displacement directions derived from the analysis of microstructures in ductile mylonites of the Austro–Alpine unit. This direction deviates from the current N-ward hypothesis of Alpine décollement.

Cover imbrication occurred along brittle thrust faults, whereas broad belts of mylonites accompany the basement-cover detachment thrusts. Mylonite generation may have started under differential stress conditions as high as 100 MPa. Stress relaxation was achieved in monomineralic rocks mainly by geometric softening, producing highly ordered phyllosilicate, quartz and calcite textures, and, in polymineralic rocks, by pressure solution and grain boundary sliding mainly along quartz–phyllosilicate and phyllosilicate–phyllosilicate contacts. Imposed strain rates range from $10^{-11}sec^{-1}$ to $10^{-15}sec^{-1}$, and ductile flow lasted up to 8 Ma.

Thrust tectonics has been recognized for more than eighty years in the eastern Alps. Imbrication of the Austro–Alpine unit ('Austro–Alpine'), the uppermost in the Alps, has since been explained by thin-skinned tectonic concepts (e.g. Tollmann 1959, Clar 1965). The tectonic subdivision into Upper, Middle and Lower Austro–Alpine subunits reflects the main décollements in this piece of continental crust. In the eastern Alps, an imbrication by stacking of internally rigid slices and a northward motion of nappes has been incorporated into this concept. Recently, however, multidirectional displacement paths and high internal strains have been reported for thrust sheets in the eastern Alps (e.g. Fritz & Kralik 1987, Krohe 1987, Neubauer 1987, Ratschbacher 1984, 1986, 1987, Slapansky & Frank 1987).

This paper deals with direction and geometry of the Cretaceous (Eoalpine) décollement of Austro–Alpine cover nappes and discusses strain and flow regime in the décollement zone between the Upper Austro–Alpine cover and Middle Austro–Alpine basement. To describe direction and geometry of the detachment tectonics we first considered the actual distribution of the Permo–Mesozoic cover rocks as acquired during the Alpine orogeny of the eastern Alps. This distribution points to a top to (N)W imbrication of the Austro–Alpine. Then, we selected three areas E of the Tauern Window to discuss detachment thrusting of Austro–Alpine cover nappes in detail: we found (north)westward imbrication and flat-ramp geometry. Finally, we describe ductile rock flow in mylonites accompanying thrusts of the basement-cover décollement and derive some qualitative to semiquantitative information about the rock rheology involved in the detachment of Austro–Alpine cover nappes.

Synopsis of the tectonogenesis of the eastern Alps

After a stretching-rifting phase during the Triassic and Jurassic, compressional tectonics started in the eastern Alps in the early Cretaceous (±130 Ma, Frank *et al.* 1987). It is correlated with subduction of the Penninic oceanic basin under the Austro–Alpine continental crust. The latter forms a rim along the northern part of the Adriatic microplate which is the upper plate during Alpine orogeny (e.g. Hawkesworth *et al.* 1975). Subduction of the

From: COWARD, M. P., DIETRICH, D. & PARK, R. G. (eds), 1989, *Alpine Tectonics*, Geological Society Special Publication No. 45, pp. 243–262.

Penninic crust and overriding of the Austro–Alpine units were accompanied by a counterclockwise rotation of the Adriatic plate and by collision with crustal slivers such as the Briançonnais (e.g. Dercourt *et al.* 1986). Continental collision occurred during the Tertiary with the underthrusting of the Helvetic shelf zone of the European Plate. This succession of events resulted in a nappe pile having the Austro–Alpine at the top, the Penninic (including the continental slivers) in the middle, and the parautochthonous Helvetic at the bottom (e.g. Frisch 1979).

Reconstructions of the movement pattern of the Adriatic plate (e.g. Biju-Duval *et al.* 1977, Dercourt *et al.* 1986) indicate large dextral strike-slip components along its northern margin between 130 Ma and 80 Ma due to the rotation. This stimulated work on a structural response to the rotation, which should have resulted in dextral transpression with NW directed thrusting at the northern edge of the Adriatic plate. Displacement analyses based on microstructures in ductile mylonites in thrust and shear zones confirm transpressional NW-directed motions in the Austro–Alpine during early stages of the Alpine orogeny (Ratschbacher *et al.* 1987, Schmid & Haas 1987). Thereafter, however, the motion direction changed in time to N–S and frontal and subsequently to E–W and lateral, relative to the European margin.

General geological setting

The Upper Middle and Lower Austro–Alpine subunits comprise each a medium-grade crystalline and a low-grade Palaeozoic basement with a Permo–Mesozoic sedimentary cover. Figure 1 (a, b) summarizes the main structural units referred to in this article.

The Upper Austro–Alpine comprises the unmetamorphosed to low-grade Permo–Mesozoic cover, the low-grade Palaeozoic basement and some slices of medium-grade basement at its very base. The Northern Calcareous Alps occupy the highest tectonic position forming a wide belt of Permo–Mesozoic rocks N of the Graywacke-zone. The latter is considered to be the sedimentary base of parts of the Northern Calcareous Alps. Other low-grade basement rocks are accumulated in the Gurktal thrust system and the thrust system of the so-called Palaeozoic of Graz.

The Middle Austro–Alpine encompasses medium-grade crystalline basement and relict Permo–Mesozoic cover.

The Lower Austro–Alpine is composed of sedimentary rocks and crystalline basement, and originally formed the margin of the Austro–Alpine to the Penninic zone.

During the Alpine orogeny, the cover was detached in most places and formed imbricates including cover as well as low- and medium-grade basement.

Along the crest of the eastern Alps, tectonic windows expose Penninic units illustrating the allochthonous nature of the Austro–Alpine. In the S the Austro–Alpine is bounded by the Periadriatic Lineament, which is an important fault system, and forms the border to the southern Alps.

Detachment geometry and motion direction from thrust geometries

Distribution of Permo–Mesozoic cover imbricates in relation to Alpine deformation zones

Figure 1(b) depicts the distribution of medium- and low-grade basement and Permo–Mesozoic cover together with the major Eolapine (Cretaceous) thrust zones E of the Tauern Window. The Permo–Mesozoic sediments, especially those sandwiched between different basement imbricates, are accumulated along the western and northwestern edge of the Alpine thrust systems. They are missing along the eastern and southern edges; there the thrusts separate only basement sheets. Ramps generally dip toward the SE.

We explain this distribution by thrusts which climb from SE toward NW out of the basement or out of the basement-cover detachment into the Permo–Mesozoic cover, and by imbrication in a piggy-back thrust sequence with the youngest rocks at the leading edge of the thrust systems. From both we infer a top to NW motion of Austro–Alpine nappes. The kinematic picture deduced from this overall distribution is consistent with that gained from the orientation of stretching lineations and the sense of vorticity of bulk flow derived from displacement analyses in ductile mylonites of some Alpine detachment zones. These are the Gurktal thrust system (Nowak 1986, Neubauer 1987, Ratschbacher 1987), the graywacke zone (Ratschbacher 1986, Neubauer *et al.* 1987), the Palaeozoic of Graz thrust-system (Fritz & Kralik 1987) and detachments in the eastern part of the eastern Alps (Ratschbacher 1987). Furthermore, it is consistent with the sense of displacement in basement shear zones like the Schlinig thrust (Schmid &

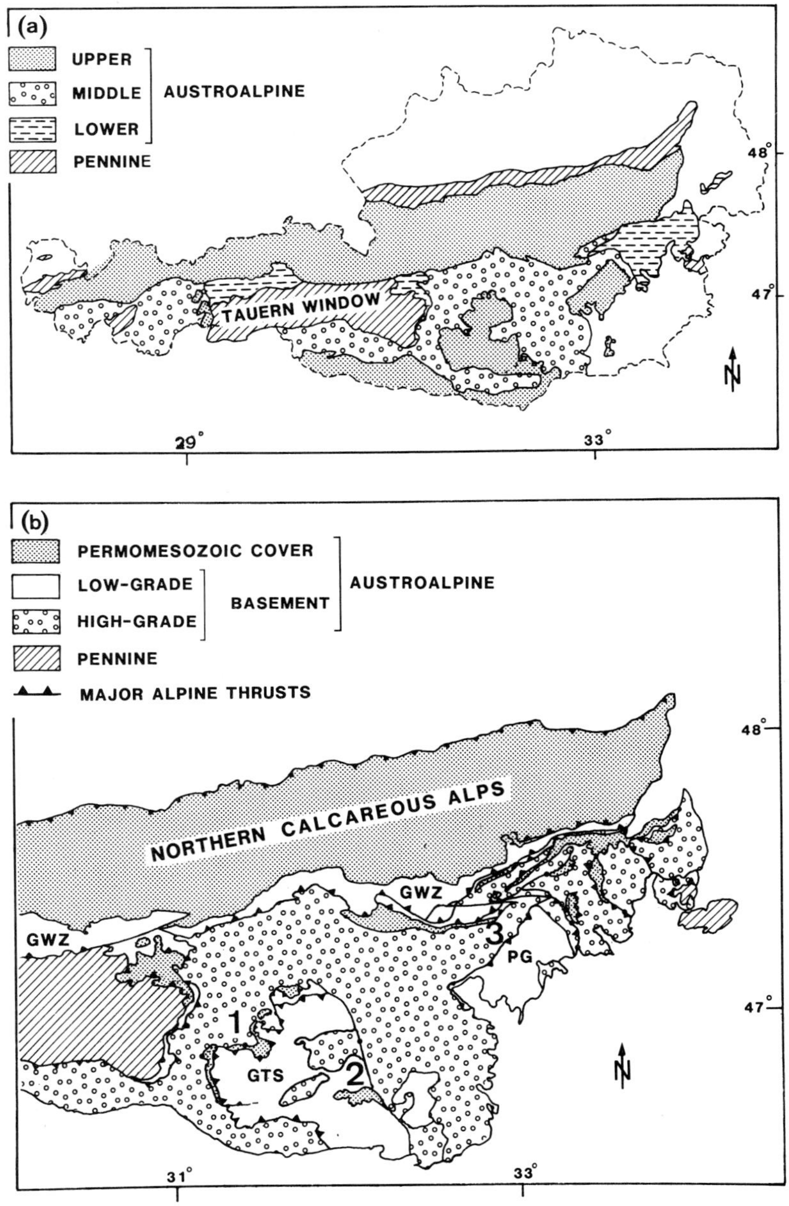

FIG. 1. (a) Main structural units of the eastern Alps in Austria. (b) Schematic structural map of the Austrian eastern Alps E of the Tauern Window with the distribution of Alpine cover and basement units and traces of major Alpine thrusts; ticks on the hangingwall. GWZ: Graywacke zone, PG: Palaeozoic of Graz, GTS: Gurktal thrust system. Locations of study areas, 1: Krappfeld section, 2: Ackerl section, 3: Graywacke zone.

Haas 1987) and shear zones between the Schladming and Wölzer basement imbricates (Rockenschaub 1986). It is inconsistent with the kinematics of the Plattengneis shear zone in the basement of the Koralpe region (Frank *et al.* 1983, Krohe 1987), which shows a top to N motion. We interpret the latter to represent a shear zone active during a later phase of deformation similar to the deformation zones which overprint the décollement-related shears (e.g. Nowak 1986, Neubauer 1987, Ratschbacher & Oertel 1987).

Figure 1(b) indicates that the Permo–Mesozoic sediments, outcropping in the area of the Austro–Alpine basement, taking erosion into account, insufficiently represent all its original cover. We suggest that those Permo–Mesozoic sediments originally covering the various basement imbricates are now mainly accumulated in the Northern Calcareous Alps.

Thrust geometries in Upper Austro–Alpine cover nappes

We selected three areas E of the Tauern Window of different vertical positions in the Upper Austro–Alpine stack, to describe thrust geometries in cover detachments. A detailed analysis is hindered and balancing of sections is impossible at the present stage of investigation due to effects of subsequent deformations. These involve folding, fold imbrication, low-angle normal and strike-slip faulting. However, from kinematic analyses in ductile mylonites accompanying detachment thrusts in or near the selected areas, we have control on extension directions and in some cases on displacement directions. Using this information and that offered by maps, we constructed profiles parallel to the displacement directions avoiding macrostructures of subsequent deformations. The studied areas are indicated in Fig. 1(b) and are:

(1) The 'Krappfeld Mesozoic' in the southeastern part of the Gurktal thrust system (Upper Austro–Alpine) (Wolter *et al.* 1982, Appold & Pesch 1984). In this area, which exposes a tectonically high level of detachment, we will concentrate on geometry and motion direction. Thrusting was not recognized until recently in this area (Wolter *et al.* 1982).

(2) The 'Ackerl section' along the northwestern edge of the Gurktal thrust system (Upper Austro–Alpine) (Neubauer 1980, 1987). This area shows imbrication of medium and low-grade basement nappes and Permo–Mesozoic slices. There, we will discuss detachment geometry.

(3) The southern edge of the Graywacke zone in eastern Austria (Ratschbacher 1984, Neubauer *et al.* 1987). It encompasses the detachment between the Upper and Middle Austro–Alpine. This area spans the zone of basement-involved detachment, and shows strong overprinting due to subsequent tectonic events. We will concentrate on geometry, and, later on, we will present an analysis of ductile strain and rock rheology involved in this detachment.

Detachments in the southeastern part of the Gurktal thrust system (Krappfeld Mesozoic)

The Krappfeld area depicts thrust geometries characteristic of the high level detachments in the Austro–Alpine thrust belt (Fig. 2). Ductile deformation is negligible in this section; it consists of a spaced cleavage in the Permian and Carnian slates. These slates acted as detachment horizons (flats) whereas rigid Permoscythian sandstones and middle to late Triassic carbonates form ramps along which thrusts climb upsection toward the WNW. The leading edge of the thrust system shows the stratigraphically youngest imbricates.

We infer WNW-directed thrusting. Some of the thrusts are clearly reactivated by normal faults which portray the widespread easterly-directed extensional event in the eastern Alps, which started in the uppermost Cretaceous (Neugebauer & Hammerschmid 1986, Nowak 1986, Krohe 1987, Ratschbacher *et al.* 1987).

Detachments in the northwestern part of the Gurktal thrust system (Ackerl section)

In this section, thrust imbrication incorporates an inverted slice of medium-grade basement and accompanying Permo–Mesozoic strata, two slices of low-grade basement of different sedimentary facies, and a late Palaeozoic cover slice (Fig. 3). This imbricated stack overrides crystalline basement as well as transgressive, E-wedging Permo–Mesozoic strata. In the latter, the stratigraphic range is always higher in the W than in the E. Deformed pebbles in the autochthonous cover indicate 30 to 50% thinning of strata subparallel to bedding and corresponding lengthening parallel to the WNW–ESE stretching lineation; strain is plane (Ratschbacher 1986). Strain intensity is highest in the low-grade basement slice at the base of the imbricated stack and diminishes toward the top of the stack as revealed by the intensity of Alpine foliation. No systematic strain study has been carried out until now.

We infer generally WNW-directed imbrication from thrust geometry (ramp-flat), occurrence of Permo–Mesozoic rocks outcropping exclusively along the northwestern rim of the thrust system, ESE-dipping ramps in the Permo–Mesozoic rocks, and basement imbrication. As in the Krappfeld Mesozoic, the area was affected by easterly-directed extension during a subsequent deformation event especially along its western tip.

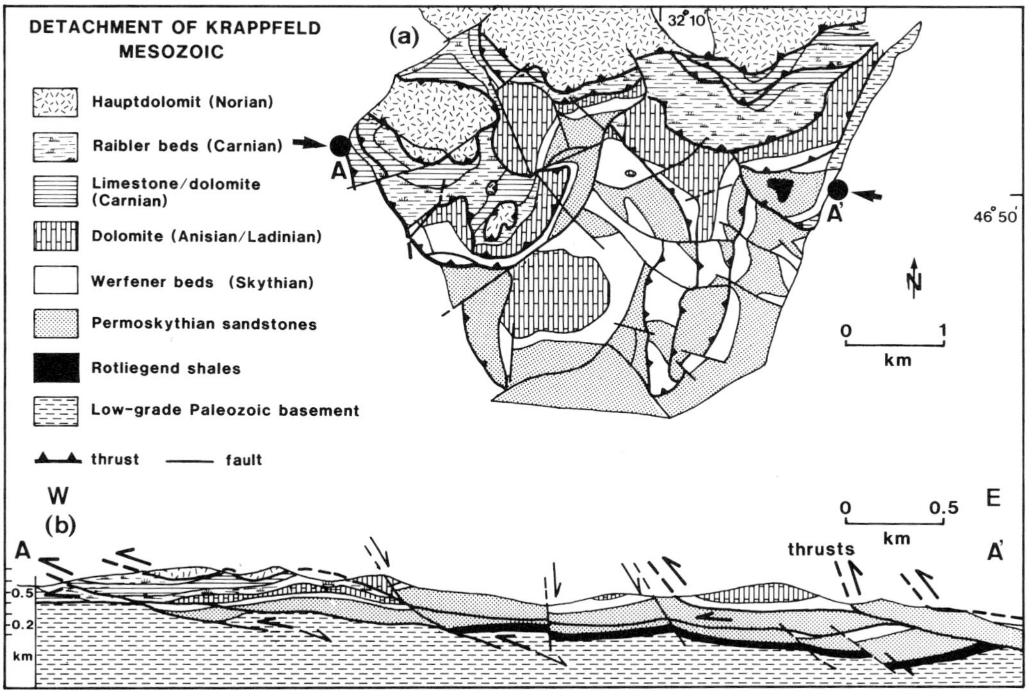

FIG. 2. (a) Simplified structural map (area 1 on Fig. 1b) of the Krappfeld Mesozoic in the eastern part of the Gurktal thrust system (modified after Appold & Pesch 1984). (b) Cross section (A–A′ in (a)) parallel to the thrust-transport direction. WNW directed thrust imbrication is overprinted by ESE directed normal faulting.

Detachment of the Graywacke zone

The Graywacke zone (Upper Austro–Alpine) consists of the lower Veitsch nappe and the upper Noric nappe (Fig. 4). The Veitsch nappe overlies Permo–Mesozoic strata which, in turn, rest transgressively on crystalline basement rocks of the Middle Austro–Alpine unit. The Permo–Mesozoic cover and the crystalline rocks may be imbricated too.

Figure 4 shows a simplified structural map along the northern edge of the Middle Austro–Alpine basement: the zone including the Permo–Mesozoic cover, the Veitsch nappe and the basal section of the Noric nappe generally dips northward and, with the thickness variation in its single thrust sheet, reflects an anastomosing thrust-surface configuration. The Veitsch nappe forms the strongly attenuated core of the Graywacke-zone décollement. Ductile deformation caused up to 60% thinning of stratigraphic units parallel to the bedding-parallel foliation and variable lengthening parallel to the ESE–WNW-trending stretching lineation. Rotational criteria indicate bulk rock flow toward the WNW (Ratschbacher 1984). Despite this ductile overprint, the detailed map of Fig. 5(a) and the cross-section of Fig. 5(b), which is parallel to the ESE–WNW-trending main-phase thrust-displacement direction, reveals thrust geometries of thin-skinned areas. Thrusts climb up-section to the WNW on frontal shear zones and show duplex formation by WNW directed imbrication. We consider the small cover-sediment bodies resting on medium-grade basement, where thrusts seemingly cut down-section, as basement troughs. Effects of later deformations include refolding of the thrusts and, although fold axes generally parallel the thrust-transport direction of the detachment in this area, it is difficult to avoid completely their effects in section construction.

Age of detachment deformation

Age constraints on time of thrusting and related deformation in the studied detachments are supplied by sedimentary deposits involved and by radiometric dating of the metamorphism accompanying deformation. In the Krappfeld area, Wolter *et al.* (1982) and Appold & Pesch (1984) showed that Gosau beds (Santonian to

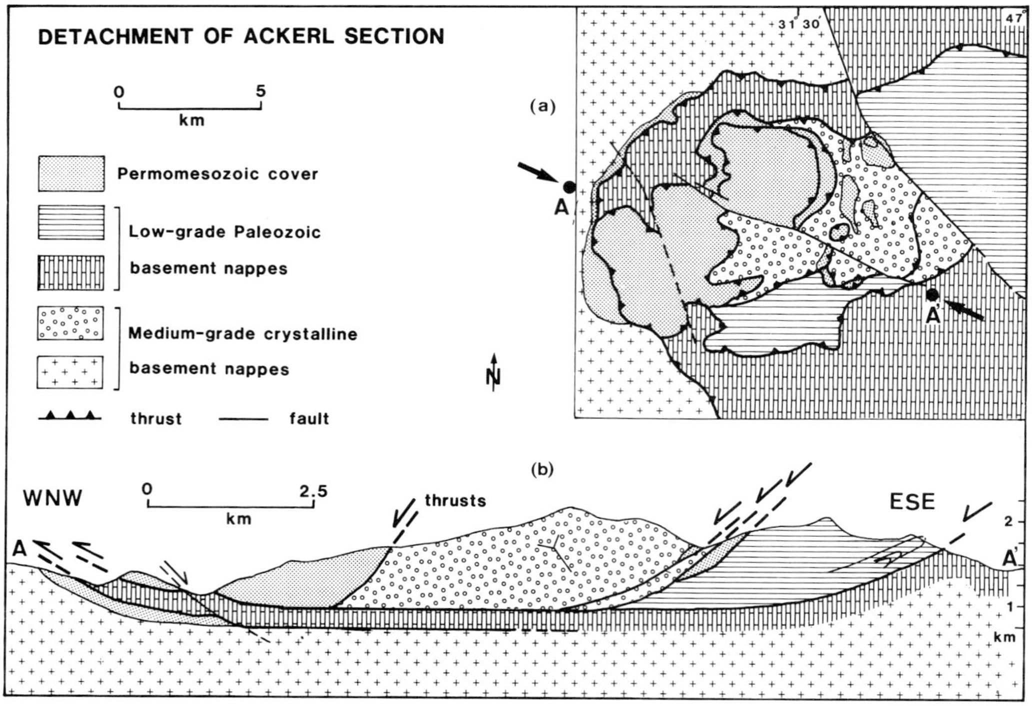

FIG. 3. (a) Simplified structural map (area 2 on Fig. 1b) of the northwestern edge of the Gurktal thrust system (Ackerl section; modified after Neubauer 1980). (b) Cross section (A–A′ in (a)) parallel to the thrust-transport direction. Imbricated stack contains an inverted slice (arrow is up) of medium-grade basement, two slices of low-grade basement of different sedimentary facies and slices of cover sediments in different tectonic positions.

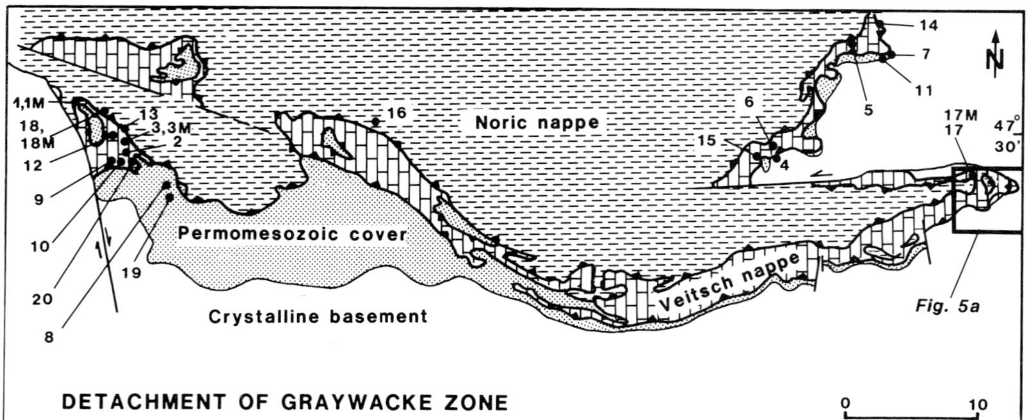

FIG. 4. Simplified structural map along the northern edge of the Middle Austro–Alpine basement (modified after Flügel & Neubauer 1984). Veitsch and Noric nappes: Upper Austro–Alpine. Crystalline basement and cover: Middle Austro–Alpine. Numbers: samples discussed in text. The décollement zone above the Middle Austro–Alpine basement is characterized by an anastomosing thrust-surface configuration.

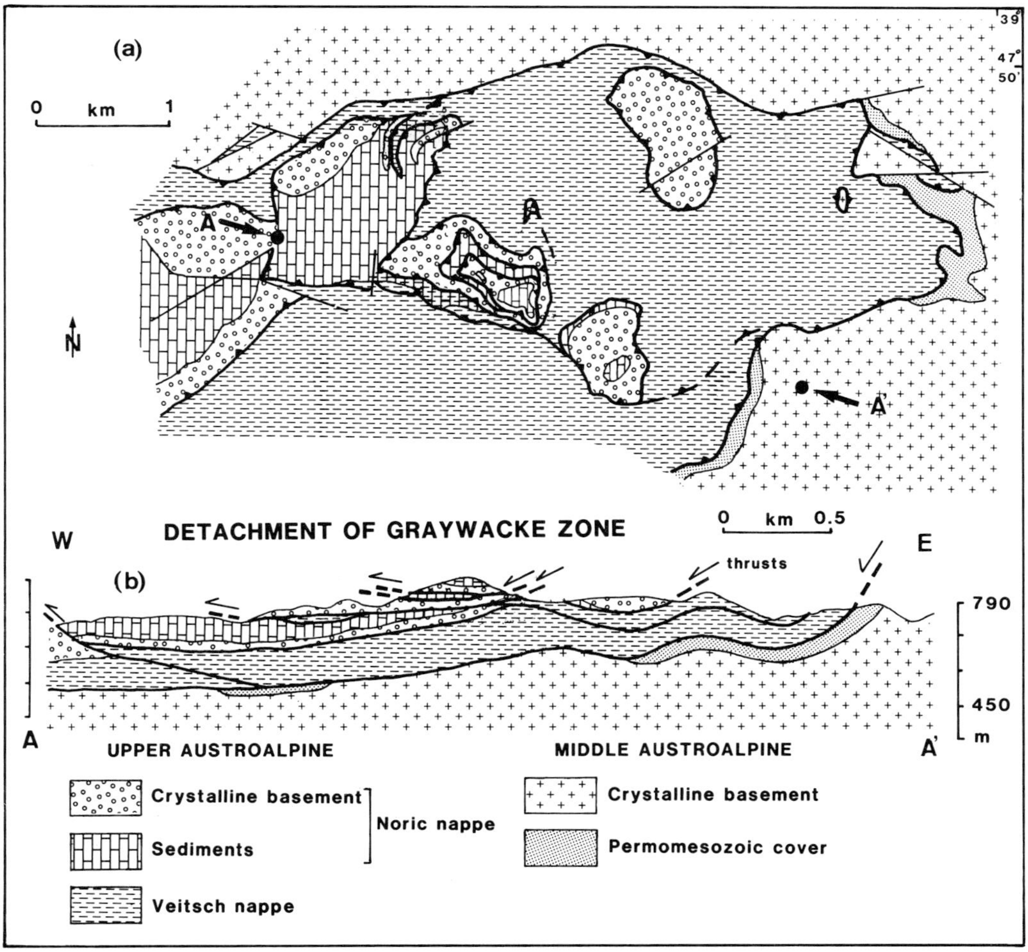

FIG. 5. (a) Structural map (modified after Neubauer *et al.* 1987), which corresponds with area 3 on Fig. 1 (b) and with rectangle on Fig. 4, and (b) cross-section parallel to the thrust-transport direction of the Graywacke zone E of Bruck/Mur. Note WNW-directed thrust imbrication.

Maastrichtian) unconformably overlie thrusts. Next to the Ackerl section, syntectonic micas were dated by Frimmel (unpublished data) and yielded Upper Cretaceous cooling ages. Ratschbacher & Klima (1985) reported similar cooling ages of syntectonic micas from the base of the Graywacke zone. Neubauer *et al.* (1987) found the lower intercept in a zircon discordia diagramm at 139±6 Ma and interpreted this age as the peak of Alpine metamorphism. Detachment belongs, therefore, to the Cretaceous (Eoalpine) orogeny.

Rheological considerations and detachment model in basement involved décollement: the Graywacke zone

In the eastern Alps, the cover sediments and the low-grade basement rocks of the Upper Austro–Alpine face the crystalline basement units of the Middle Austro–Alpine. The zone of thrusts and mylonites between these two units thus represents a major décollement and a rheologically weak zone in the Austro–Alpine. We selected the Graywacke zone to derive some qualitative and semiquantitative information about the rheological conditions involved in this basement-cover décollement, because structures, strain and metamorphism have been studied recently and simple lithologies are involved. Based on rheological properties of the main lithologies and the structures of the detachment zone, we finally refine a kinematic model proposed earlier for the cover-basement detachment in the Austro–Alpine (Ratschbacher 1986).

We start with a summary of structures, strain and metamorphism. A textural and microstructural approach is used for deformation-regime and mechanism analyses of the main lithologies of the detachment: quartz mylonites including pebble bearing rocks (>80% quartz), phyllosilicate-rich mylonites (≪80% quartz) and calcite mylonites. This forms the base for a semiquantitative estimate of stress, strain rate and duration of flow. The strength of these mylonites, which show varying strain and strain rates, governs, together with the geometry of thrusts, the final décollement model.

An idealized section through the Graywacke-zone detachment shows conglomeratic rocks at the base of the Permo–Mesozoic cover sediments. These are overlain by a quartzphyllitic and a competent quartzitic member; dolomites and marbles are mostly sheared off and preserved only locally. The Veitsch nappe has metapsammites and phyllites at the base, a massive marble member in the middle and phyllites and pebble–bearing quartzitic rocks at the top. Relicts of crystalline basement and quartz-phyllitic rocks form the basal part of the Noric nappe. Due to the ductile deformation and the outcrop conditions it is generally difficult to trace the exact location of ramps and flats of thrusts with respect to the main lithologies, but flats of thrusts generally run along phyllosilicate-rich mylonites in the quartz-phyllitic member and at the edges of the quartzitic member of the Permo–Mesozoic cover, in the basal quartzphyllites, at the rims of the marble member and at the top of the uppermost, competent quartz-rich unit of the Veitsch nappe. Ramps are generally located in quartz and calcite mylonites of the competent units of the Permo–Mesozoic cover and the Veitsch nappe; these are the basal coglomerates and the quartzitic member and the marble and pebble-bearing quartzitic member. Sample localities are given in Fig. 4.

Structures

The main structures which outline the deformation in the décollement zone are a regional foliation and a stretching lineation trending approximately E–W. Foliation is subparallel to the thrust surfaces and to bedding. In monomineralic rocks the foliation is marked by recrystallized primary minerals, deformed relicts, and grain-size variations. In polymineralic rocks, metamorphic minerals (in general phyllosilicates), metamorphic banding and residue seams occur in addition. Anastomosing foliation patterns are abundant on the micro- and mesoscale (Fig. 6a, b). A distinction between S- and C-surfaces (Berthé *et al.* 1979) is possible in weakly deformed, coarse-grained rocks. Fine grained, incompetent rocks usually show only one set of foliation surfaces, which we interpret as C-planes. In quartz and calcite-rich rocks, the grains are recrystallized, mostly with a subrectangular shape-fabric parallel to the macroscopic foliation.

We wish to stress the anastomosing configuration of planar structural elements on all scales (thrusts, foliations) and the lozenge-shape fabric of the domains they enclose.

Strain

Two different methodical approaches were employed to record strain. First, pebbles were used to calculate strain in pebble-bearing quartz-rich rocks. Second, the theory of March (1932) was applied to a series of quantitatively determined preferred orientation patterns of phyllosilicate grains in phyllosilicate-rich mylonites (Ratschbacher & Oertel 1987).

Pebble strains lie near plane strain or in the constrictional field of strain geometry; March strains are in the flattening field and near plane strain. The host rocks of the pebbles are characteristically competent and pebble strains are taken to give the finite strain in those rocks. Strain intensity interpreted from the preferred orientation of phyllosilicate grains may yield an indication of the amount of strain in the less competent rocks. March strain intensity always exceeds pebble strain intensity (Ratschbacher & Oertel 1987). Results from both approaches outline the Veitsch nappe as an area of high average strain (Fig. 7). Strain decreases with distance from thrusts and groups of nearly equal strain data are separated by intensity jumps (Fig. 8).

We interpret the observed variation in strain as due to heterogeneous deformation. It is represented by the anastomosing foliation pattern observed from macro- (thrust sheet) to micro-scale (foliation). The high strains in the Veitsch nappe outline it as the strongly attenuated core of the Graywacke-zone décollement.

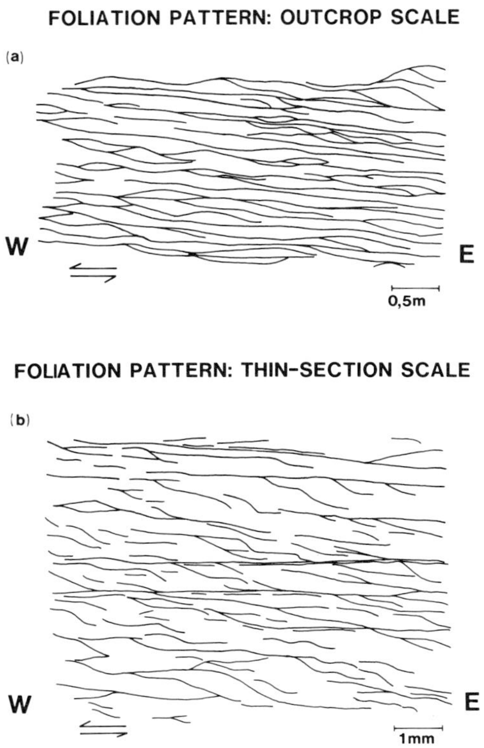

FIG. 6. Anastomosing foliation pattern on (a) the meso- and (b) the micro-scale: line drawings from outcrops and thin-section photographs.

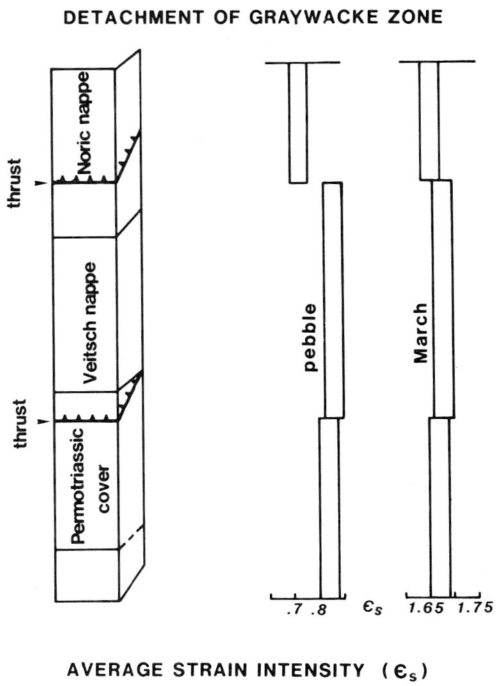

FIG. 7. Average pebble and March strain intensity of main tectonostratigraphic units of the Graywacke-zone décollement. Veitsch nappe outlined as area of high strain. Strain-intensity factor: $\epsilon_s = 1/\sqrt{3}[(\bar{\epsilon}_1 - \bar{\epsilon}_2)^2 + (\bar{\epsilon}_2 - \bar{\epsilon}_3)^2 + (\bar{\epsilon}_3 - \bar{\epsilon}_1)^2]^{1/2}$

Pressure–temperature conditions and crystallization history

Metamorphism reached temperatures and pressures of about 450°C and 4 kbar in the Permo–Mesozoic cover at the base of the Graywacke zone, and values of ≥300°C and 2–3 kbar at the base of the Northern Calcareous Alps in the studied area (Nievoll 1983, Ratschbacher & Klima 1985).

Minerals that recrystallized or grew during the Alpine metamorphism, such as muscovite, biotite, chlorite, actinolite, epidote, quartz and calcite, are preferentially aligned parallel to the Alpine foliation. In strain shadows, minerals grew as fibres parallel to foliation and stretching lineation. Minerals paralleling the structural elements (foliation, lineation) were bent and kinked during later deformations.

Low-grade metamorphic conditions and growth of main Alpine mineral assemblages are linked to the cover-detachment deformation (Ratschbacher & Klima 1985, Ratschbacher & Oertel 1987).

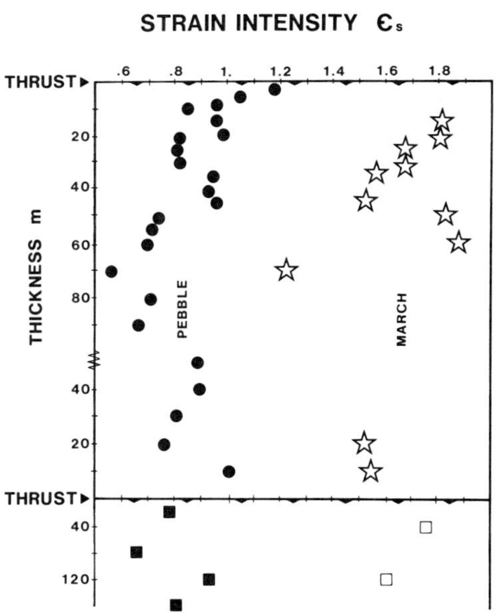

FIG. 8. Pebble and March strain intensity versus distance from thrust in a cross-section across the Graywacke-zone décollement. ϵ_s see Fig. 7.

Deformation regime

Many of the rocks of the detachment zone do not show any mesoscopic criteria of rotational versus irrotational deformation on the meso-scale. This may be due to their monomineralic nature (carbonates, quartzites, phyllites) and/or the high strain they suffered. In addition to our survey on the meso-scale, which indicates with the presence of S–C structures in the coarse grained rocks the existence of a component of rotational deformation in the overall flow, we performed a textural study in quartzites and phyllites.

Thirteen quartz c-axis fabrics of monomineralic quartz pebbles in meta-conglomerates of all units were measured in a cross-section across the Graywacke-zone décollement (Fig. 9a–c). They all show monoclinic symmetry. In Fig. 10, strain intensity, as measured from the shape fabric of deformed pebbles, is plotted against the degree of fabric asymmetry of the quartz type I cross-girdle textures (Lister 1977) measured in pebbles of the same samples. Asymmetry is characterized by the obliquity of the central girdle segment with respect to the YZ plane of the finite-strain reference frame (see Behrmann & Platt 1982). In general, fabric asymmetry decreases with increasing strain intensity, but however, there is some scatter of data points. Assuming deformation by simple shear alone and that the c-axis distribution is controlled kinematically (e.g. fig. 14c of Platt & Behrmann 1986, Lister *et al.* 1978), the degree of asymmetry (the angle α in Fig. 10) should become less pronounced with increasing strain. That is, for ideal simple shear and variable strain intensity, the data points should lie on the 'simple shear curve' of Fig. 10.

We interpret the deviation from this line as indication for a partly coaxial deformation. Accordingly, the fabric asymmetries of c-axis textures of samples in the décollement zone suggest a strain history combined with more or less rotation, and thus range from pure to simple shear. The coaxial strain component seems to have increased during the progressive strain history: less strained samples plot nearer to the simple shear line than higher strained ones.

This is corroborated by phyllosilicate textures measured across the décollement zone: half of them are of orthorhombic symmetry typical for homogeneous, coaxial strains (Fig. 9m, n). Non-coaxial deformation often caused typical deformation structures such as S–C surfaces and shear bands (e.g. White *et al.* 1980), which are reflected in phyllosilicate pole figures in an asymmetric girdle spacing and an asymmetric pole maximum position with respect to the foliation pole measured along the stretching lineation. Phyllosilicate grains, preferentially disposed along S- and C-surfaces, produce single distributions, which combine to form the asymmetrical total distribution (O'Brien *et al.* 1987, Fig. 9o). We therefore interpret asymmetric textures as indicating dominant non-coaxial deformation.

We interpret the deviation from a history of progressive simple shear with increasing strain intensity in terms of flow partitioning. At the onset of detachment tectonics, the overall deformation fairly accurately approached flow by simple shear. At later stages, a component of coaxial extensional strain along the stretching lineation may have compounded the flow to deviate from simple shear. This explains the prolate strain geometry recorded by the pebble shapes.

Deformation mechanisms

Quartz mylonites and monomineralic quartz pebbles

Monomineralic quartz pebbles (>94% quartz, Fig. 11a, b) and quartz mylonites (>80% quartz, Fig. 11c) show highly ordered c-axis

FIG. 9. Quartz, calcite, and phyllosilicate textures measured across the Graywacke-zone décollement. See Table 1 for microstructural parameters and Fig. 4 for sample sites. Equal area projections, lower hemisphere; s_1: foliation; l_1: stretching lineation. The projection plane is the X–Z plane of finite strain (X>Y>Z, principal strain axes) and the orientation is the same for all pole figures except for the phyllosilicate ones which are projected on the X–Y plane. Sense of rotation indicated where deduceable. Contour intervals: 1.5% per 1% area for quartz, 1% per 1% area for calcite; phyllosilicate orientation in multiples of uniform distribution: 0.5, 1, 5.5, 14.5, and 23.5 times uniform for (m), 0.5, 1, 5, 12, and 36 times uniform for (n), 1, 2, 4, and 6 times uniform for (o). (a)–(c) Quartz c-axis textures of monomineralic pebbles (1, 2, 3: Veitsch nappe); (d)–(f) Quartz c-axis textures of quartz-rich mylonites (4: Permo–Mesozoic cover; 5, 6: Veitsch nappe); (g)–(i) Quartz c-axis textures in phyllosilicate-rich mylonites (7: Permo–Mesozoic cover; 1M, 2M: Veitsch nappe). (j)–(l) Calcite twin-lamellae and c-axis textures of marbles and calcite mylonites (8: Permo–Mesozoic cover; 9, 10: Veitsch nappe); (m)–(o) Basal-plane pole figures (11: Permo–Mesozoic cover; 12, 13: Veitsch nappe).

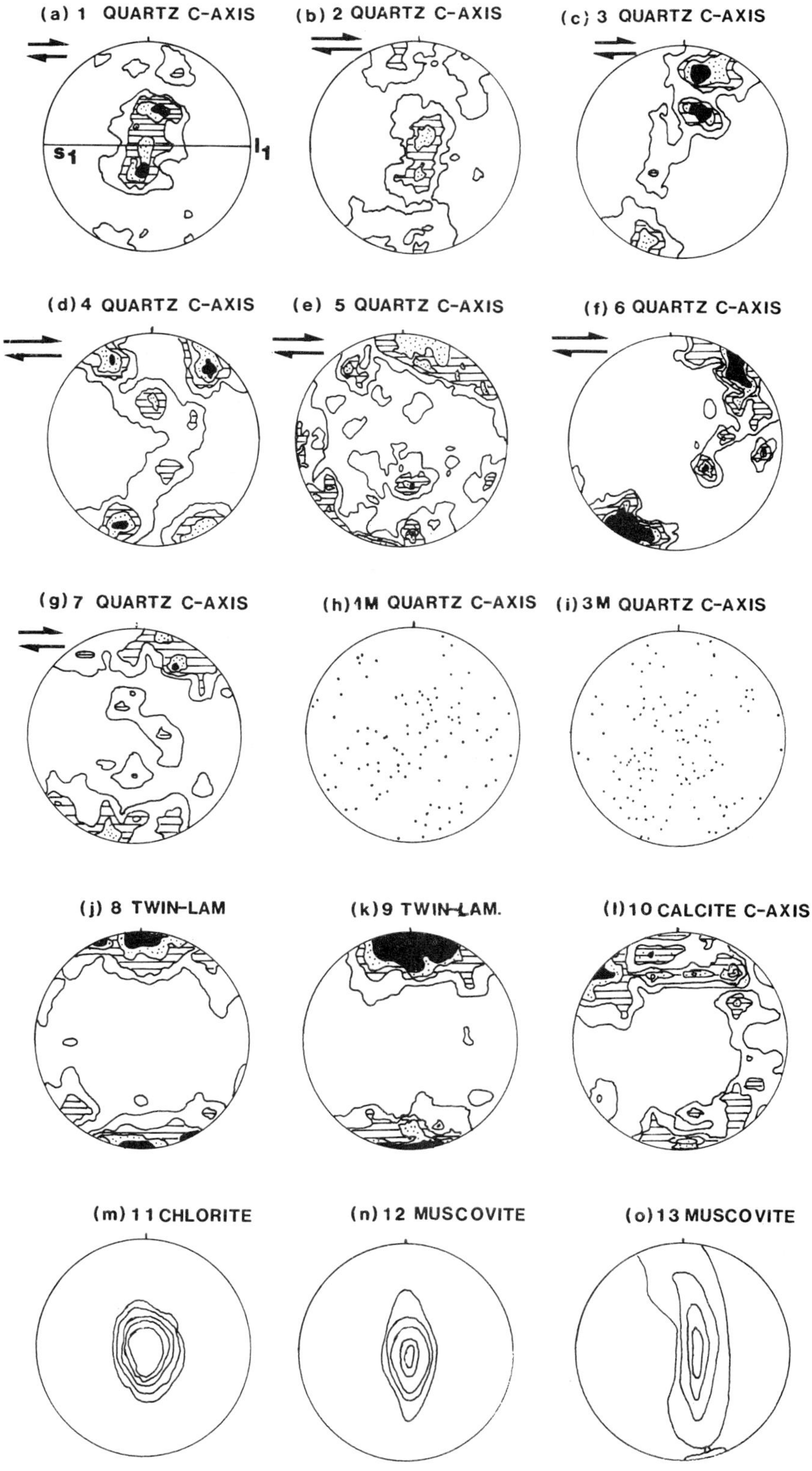
(a) 1 QUARTZ C-AXIS
S1
l1
(b) 2 QUARTZ C-AXIS
(c) 3 QUARTZ C-AXIS
(d) 4 QUARTZ C-AXIS
(e) 5 QUARTZ C-AXIS
(f) 6 QUARTZ C-AXIS
(g) 7 QUARTZ C-AXIS
(h) 1M QUARTZ C-AXIS
(i) 3M QUARTZ C-AXIS
(j) 8 TWIN-LAM
(k) 9 TWIN-LAM.
(l) 10 CALCITE C-AXIS
(m) 11 CHLORITE
(n) 12 MUSCOVITE
(o) 13 MUSCOVITE

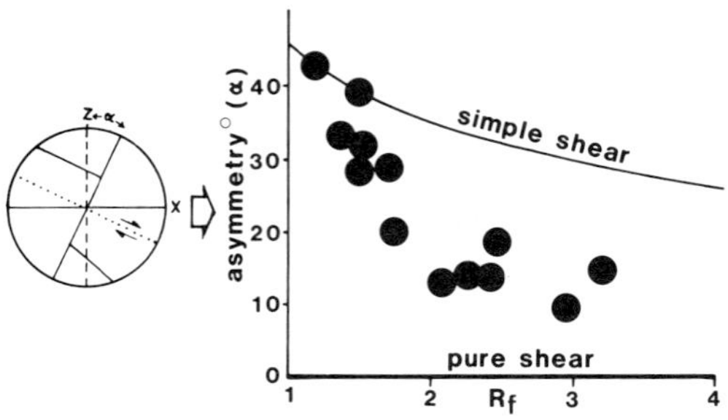

FIG. 10. Degree of fabric asymmetry of quartz c-axis textures (expressed by the obliquity of the central girdle segment of type I cross girdles to the Y–Z plane of the finite strain reference-frame, X>Y>Z, principal axes) versus pebble-strain intensity (expressed as strain ellipse axial ratio). Simple and pure shear lines indicate angles between the X–Y plane and the principal shear plane during progressive simple shear and pure shear deformations.

textures (Fig. 9a–f). In the quartz mylonites (Fig. 11c), quartz has the smallest grain size of all the rocks studied (Table 1). Grain boundaries are sometimes pinned by second phase minerals (opaques, phyllosilicates, Fig. 11d). Quartz-grain axial ratios (long/short) are smallest in the quartz mylonites. In some highly strained quartzites, layers with varying axial ratios occur showing nearly equal grain shapes in extremely fine-grained layers. In the pebbles, grain size is independent of strain intensity (Table 1). Undulose extinction, subgrains, and serrated grain boundaries in quartz are ubiquitous in quartz-rich rocks (Fig. 11a–c).

We consider crystal-plastic deformation mechanisms to be dominant in the quartz-rich rocks. Recrystallization, crystallographic preferred orientations, subgrains and serrated grain boundaries underpin this interpretation. Steady-state recrystallized grain size indicates in the pebbles the independence of grain size on strain. A contribution of grain boundary sliding can be indirectly inferred from the combination of fine grain size and low grain axial ratios in the thin, fine-grained layers of some quartz mylonites.

Phyllosilicate-rich mylonites

In contrast to the quartz mylonites quartz c-axis fabrics (Fig. 9g–i) are poorly developed in phyllosilicate-rich mylonites (≪80% quartz, Fig. 11e). This is most clearly shown by fabrics of samples from the same outcrops (Fig. 9a, b versus 9h, i): the first from pebbles, the latter from phyllosilicate-rich bands in the matrix. Quartz-grain axial ratios are highest in the phyllosilicate-rich rocks (Table 1), however, they may vary from layer to layer and show nearly equal shapes in extremely fine-grained layers. Strain shadows behind rigid inclusions and truncated quartz and/or feldspar grain-shapes are common (Fig. 11e). Phyllosilicate grains show bent or kinked ends, but slip along (001) planes and along contacts with other minerals is common as well.

In contrast to the quartz mylonites, pressure solution and, to an unknown degree, grain-boundary sliding were dominant in the phyllosilicate-rich mylonites. Truncated minerals with residue seams connected with high quartz-grain axial ratios and poorly developed crystallographic preferred orientations of quartz are the evidence for solution. The evidence for sliding is given by the slipped phyllosilicates, the small quartz axial ratios and the lack of preferred orientation of quartz in many of the phyllosilicate-rich mylonites. Although rigid body rotation of platy grains in the phyllosilicate-rich rocks is inferred as the major orientation process, sliding between and within the grains contributed as well.

Calcite mylonites

All specimens of calcareous rocks, except the coarsest marbles and the finest mylonites, have a preferred shape orientation of calcite grains parallel to the mesoscopic foliation (Fig. 11e, f, Table 1). Grain axial ratios (long/short) in the marbles strongly vary; in the mylonites they are dependent on grain size: up to 2.65 in relict large grains and <1.46 in the matrix (Fig. 11f, g, Table 1). The calcite mylonites thus depict a fabric of alternating layers varying in grain size and grain axial ratios (Fig. 11g). In the fine-grained mylonites, opaques and/or phyllosilicate-grains are commonly attached to the calcite grain-boundaries. Twinning is ubiquitous in the marbles and the in the relict large grains of the mylonites, but nearly absent in the fine grains (Fig. 11h).

Crystallographic preferred orientations of calcite c-axes and twin-lamellae, determined by the U-stage, are developed in marbles as well as in calcite mylonites (Fig. 9j–l). The very fine mylonites could not be measured because of limitations in optical-resolution. Twin-poles and c-axes form maxima surrounding the maximum shortening axis. The textures are thus indicative of twinning as a main deformation mechanism (Casey *et al.* 1978).

TABLE 1.

Spec. no.	Locality	ϵ_s	$\sqrt{}$	d	R	σ	$\dot{\epsilon}_i$	t_{dd}
(a) Metaconglomerates (quartz)								
1	Veitsch nappe, pebble	1.12	−0.55	209	1.8	11	5.27	0.67
2	Veitsch nappe, pebble	0.96	−0.26	135	2.4	18	17.3	0.17
3	Veitsch nappe, pebble	0.8	−0.68	229	1.6	10	4.36	0.58
16	Noric nappe, pebble	0.55	0.06	188	1.6	12	7.09	0.25
17	Veitsch nappe, pebble	0.89	0.14	204	2.4	11	5.63	0.5
18	Veitsch nappe, vein			179	2.4	13	8.12	
(b) Quartz-mylonites (quartz)								
4	Permo–Mesozoic cover			57	1.5	46	181	
5	Veitsch nappe			47	1.3	57	305	
6	Veitsch nappe			30		94	1030	
(c) Quartz-phyllosilicate mylonites								
1M	Veitsch nappe			68	2.7			
3M	Veitsch nappe			95	1.6			
17M	Veitsch nappe	ˣ1.75	0.5	57	1.8			
18M	Veitsch nappe			74	2.3			
7	Permo–Mesozoic cover	ˣ1.76	0.55	35	1.4	79	676	
(d) Marbles (calcite)								
19	Permo–Mesozoic cover			329	1.8	15	42.7°	
8	Permo–Mesozoic cover			207	2.0	16	46.6°	
9	Veitsch nappe	0.66	−0.27	99	2.5	45	0.27÷	7.8
(e) Calcite-mylonites (calcite)								
20	Veitsch nappe			50	1.79	98	101÷	
10	Veitsch nappe	1.54	0.31	23	2.65*	200	8.55■	0.57
					1.46⁺			
15	Permo–Mesozoic cover			16	≈1.35			

+ small grains
* large grains
x March strains

° regime 1; ÷ regime 2; ■ regime 2 — Carrara marble (Schmid *et al.* 1980)

Specimen number, lithotectonic unit, strain, microstructural parameters, stress, strain rate, and total duration for ductile deformation associated with detachment tectonics in the eastern Alps.

ϵ_s: strain-intensity factor = $1/\sqrt{3}[(\bar{\epsilon}_1-\bar{\epsilon}_2)^2(\bar{\epsilon}_2-\bar{\epsilon}_3)^2(\bar{\epsilon}_3-\bar{\epsilon}_1)^2]^{1/2}$
$\sqrt{}$: Lode's parameter = $(2\bar{\epsilon}_2-\bar{\epsilon}_1-\bar{\epsilon}_3)/(\bar{\epsilon}_1-\bar{\epsilon}_3)$
d: dynamically recrystallized grain size, in μm [monomineralic rocks: linear intercept method (Underwood 1970); polymineralic rocks: arithmetic mean of long and short grain diameters]
R: grain axial ratio (long/short) in the X–Z plane of total strain (X>Y>Z, principal axes)
σ: interpreted magnitude of palaeostress in MPA
$\dot{\epsilon}_i$: strain rate (x 10^{-14}sec^{-1})
t_d: total duration of ductile deformation (x 10^6yr); $t_{dd} = \epsilon_s/\dot{\epsilon}_i$

The transition from marble to calcite mylonite, we think, reflects a progressive deformation characterized by twinning and syntectonic recrystallization. Thus, the large and intermediate-sized grains show a preferred shape orientation, twinning, polygonized substructure and recrystallization. The decrease of grain axial ratios and grain size in the fine-grained mylonitic layers suggest a transition to grain-boundary dominated processes (diffusion, sliding) in some of these layers. There twinning is rare and the grain size is small either due to dynamic recrystallization or grain-growth inhibition by second-phase particles.

Stress and stress relaxation mechanism

As an outcome of a regional study of syntectonic

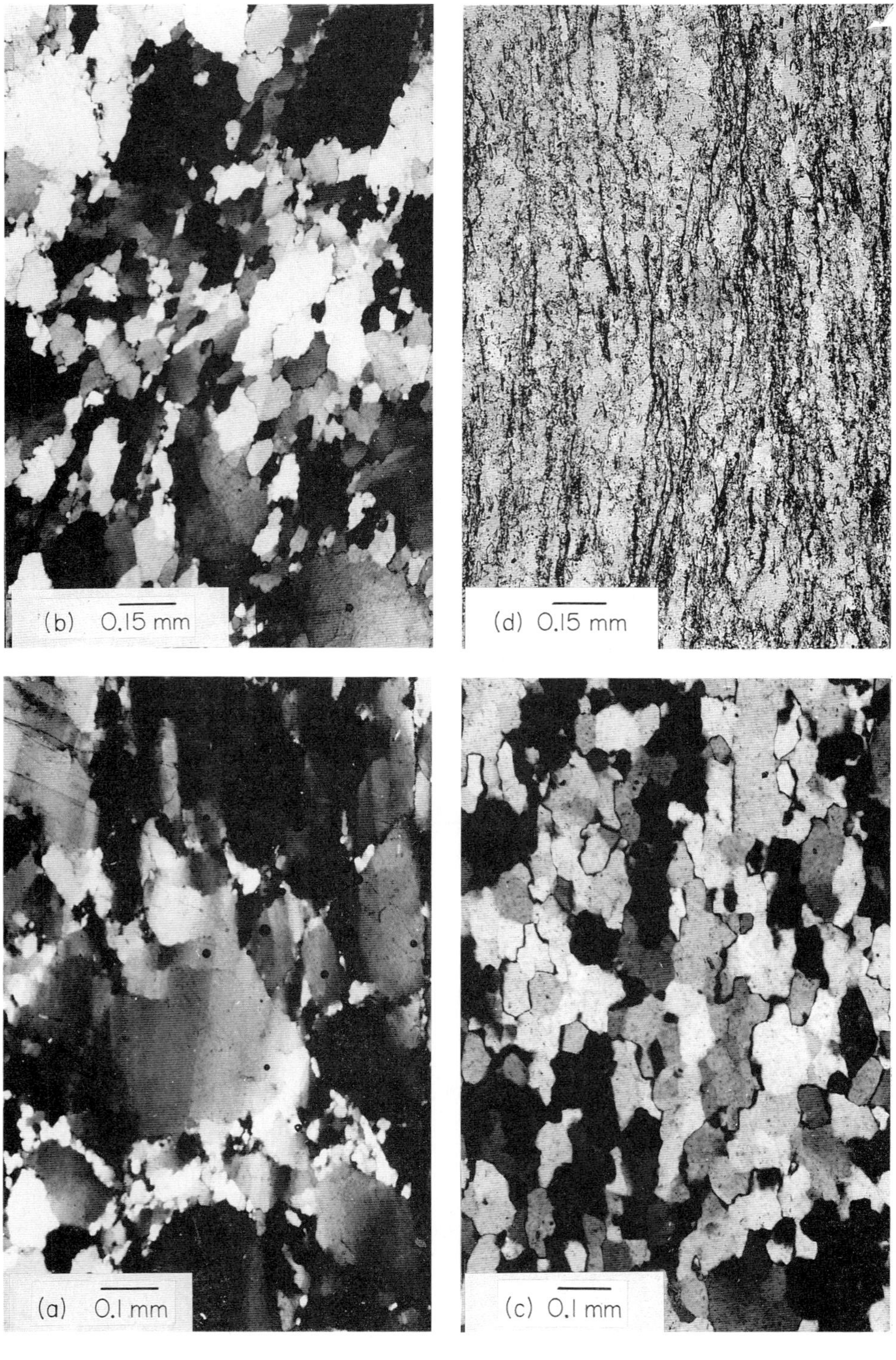

FIG. 11. Deformation microstructures in rocks of the Graywacke-zone décollement; sections are parallel to the X−Z plane of the finite strain. Sample sites see Fig. 4. (a) Strongly deformed relict quartz grains with recrystallization along grain margins and subgrain walls; quartz pebble, Veitsch nappe (14). (b) Dynamically

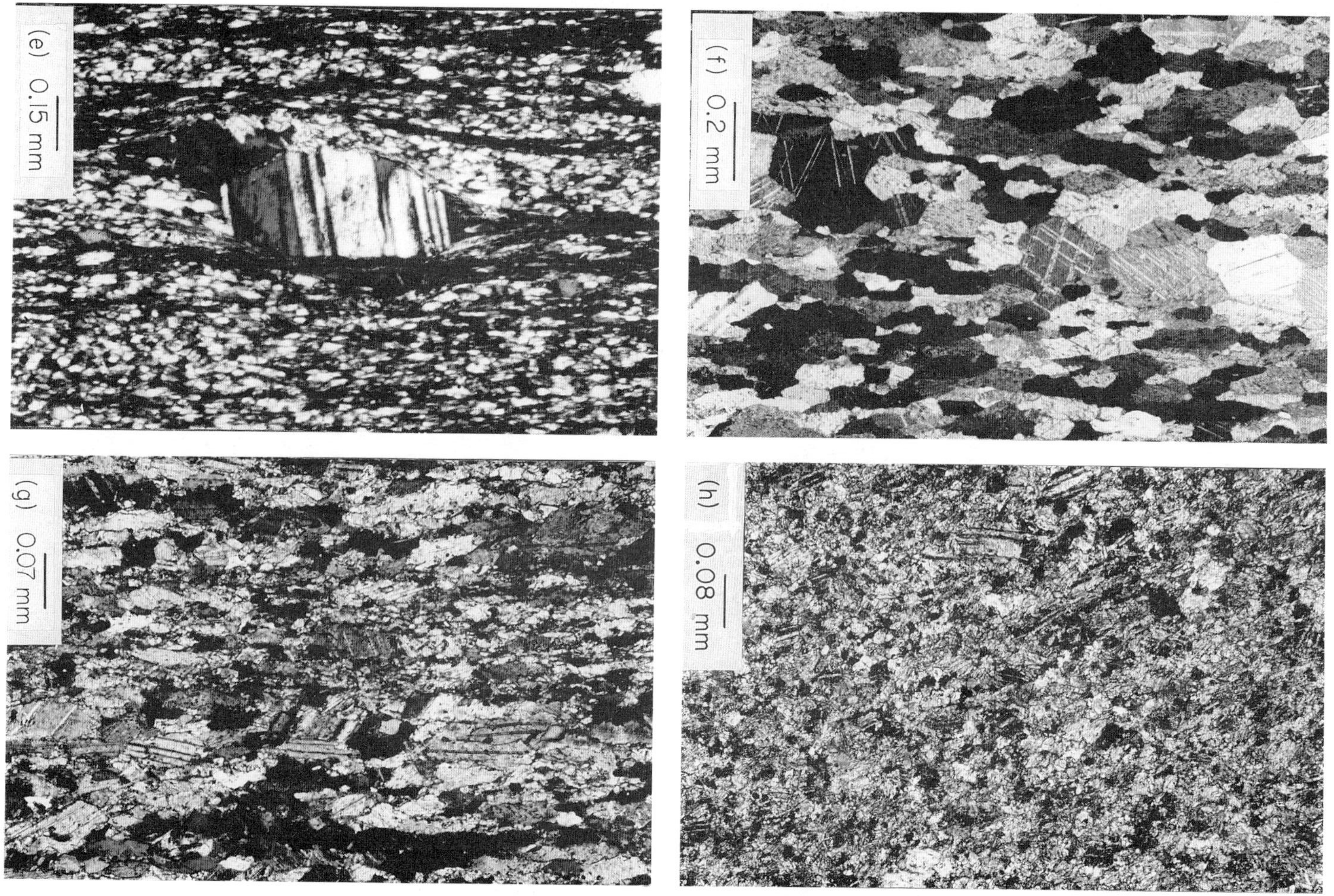

(e) Pressure solution and inhomogeneous neomineralization in strain shadows, quartz-phyllosilicate mylonite; Permo–Mesozoic cover (7). (f) Relict and flattened, recrystallized calcite grains and twinning, deformed marble, Veitsch nappe (9). (g) Alternation of highly flattened, twinned grains with small, nearly equidimensional, recrystallized, rarely twinned grains, calcite mylonite, Veitsch nappe (10). (h) Small, heavily twinned grains surrounded by very small (<20 μm) untwinned grains with equidimensional grain shape, calcite mylonite, Permo–Mesozoic cover (15).

microstructures, flow stress estimates were made across the Graywacke-zone décollement. We refer to the review literature existing in this field for the approach used (e.g. Ord & Christie 1984, Schmid 1982).

For the quartz-rich rocks we used the flow stress–grain size relation of Christie *et al.* (1980) for the dynamically recrystallized grains. Stresses were determined from the nearly monomineralic quartz pebbles (Fig. 11b) and from quartz-mylonites in the country rock (Fig. 11c). They range between 10 and 94 MPa (Table 1). The calculations are based on the following assumption (see also Ratschbacher 1984 for quartz): hot-working conditions (dynamic recrystallization), lack of static annealing, a heterogeneous stress history and monomineralic rocks.

Stress in calcite tectonites (Fig. 11f–h) of the Veitsch nappe was inferred from the experiments on Carrara marble of Schmid *et al.* 1980 (see also Schmid 1982, fig. 9, table 1). Stress magnitudes range between 15 and 45 MPa in the marbles and are thus similar to those measured in the quartz pebbles and go up to 100 MPa in the calcite mylonites. In the latter, stress as derived from grain size may be too high due to both pinning of grain boundaries and a switch to grain-boundary dominated deformation mechanisms. Samples in which diffusion and sliding were suspected to contribute to deformation were excluded from stress calculations.

Stress could have been close to 100 MPa in the quartz and calcite mylonites in the basal parts of the décollement and is up to an order of magnitude higher than in the obviously more flow-resistant wall rocks, the metaconglomerates and the coarse-grained marbles. We interpret the stress level recorded in the metaconglomerates and marbles to reflect the regional value and our peak values to be close to early stress concentrations, which later gave rise to mylonite zones. To relax these stress concentrations, specific softening mechanisms must have been operating in the rocks studied.

(1) The development of textures indicates that in many of the rocks of the décollement, the lattice of the main minerals was oriented for easy slip. Thus, the flow stress for intercrystalline slip was reduced on one or several slip systems (e.g. (001) in phyllosilicates, e-twins in calcite, basal plane in quartz). Reorientation of grains was thereby assisted by dynamic recrystallization and grain-boundary migration. Most of the nearly monomineralic mylonites in the décollement seem to portray this strain-softened state (e.g. Figs 9d–f, 11c, f).

(2) The dynamic recrystallization in monomineralic rocks points to a grain-size reduction of the starting material. In those rocks, where second phase particles are present in significant number, the main-phase grain boundaries are pinned during the grain-growth phase of recrystallization. This also led to a small grain size. The grain-size distribution in those rocks thus became bimodal: fully recrystallized domains in strongly strained rocks oppose domains where recrystallization is incomplete and a less strained, coarser-grained fabric is partly preserved. This may have led to a situation where different deformation mechanisms occurred simultaneously in different fabric domains: e.g. dislocation creep versus pressure solution and/or grain boundary sliding. We consider the microfabric of some of the calcite mylonites studied to reflect this situation (e.g. Fig. 11g, h).

(3) In the case of the polymineralic rocks, and in particular the phyllosilicate-rich mylonites, high preferred orientations of phyllosilicate grains triggered slip between grains. The lack of crystallographic orientation of quartz and the common two-phase grain contacts (quartz-phyllosilicate, quartz-opaques) suggest that sliding between distinct phases was active in these rocks. Finally, this process may have resulted in discontinuities (thrusts) along which the imposed strain rates are accommodated by cataclastic processes.

The microstructural parameters of the rocks studied document the following possible stress history: The décollement was affected by stresses up to 100 MPa until softening mechanisms led to high strains in the mylonitic zones. We consider both the weakening introduced by the preferred mineral orientations and the activation of grain-size sensitive deformation mechanisms as the main causes for stress relaxation in the detachment zone. In the case of the Graywacke-zone detachment, the former process may have given rise to many of the quartz and calcite mylonites along the thrust ramps of the Permo–Mesozoic cover and the Veitsch nappe; the latter to many of the calcite and, particularly, the phyllosilicate-rich mylonites along the flats of the detachment thrusts.

Deformation rate and duration of ductile deformation

Using the power flow-law of Koch *et al.* (1980) for dominant dislocation creep, we calculated deformation rates for the quartz mylonites. The temperature estimate for the Permo–Mesozoic cover, 450°C, was used in the calculations. Strain rates vary between 10^{-11} and 10^{-14}sec^{-1}.

According to the microstructural information reported above (Table 1, Fig. 11f–h), the flow laws of regimes 1, 2, and 3 of Carrara marble of Schmid *et al.* (1980) were used to constrain the deformation rates in the marbles, the calcite mylonites with the bimodal grain-size distribution, and the fine-grained marbles, respectively. Strain rates vary between 10^{-12} and 10^{-15}sec^{-1}.

Taking into account the experimental uncertainties in setting up the stress and strain-rate equations, the above numbers have to at best be regarded as semiquantitative. It is evident, however, that the strain rate varies across the décollement zone on a meso- to micro-scale.

By dividing the finite strain data (expressed by the strain intensity factor ϵ_s, which uses natural strains, see Table 1) through the strain rate, the duration of flow in each specimen of the décollement was calculated (Table 1). The estimates spread between 0.2 and 8 Ma, but encompass mostly the range between 0.5 and 1 Ma.

Model for basement-involved décollement (Graywacke zone)

A model for the basement-involved décollement of the Upper Austro–Alpine cover nappes in the eastern Alps has to explain, besides the ramp-flat thrust geometry (see also Ratschbacher 1986): (1) a high degree of ductile deformation in the zones between and along individual thrusts; (2) the lozenge-shape of many thrust sheets and the anastomosing configuration of foliations; and (3) the flow partitioning.

The kinematic framework of the décollement deformation is defined by the foliation which represents S-surfaces and the stretching lineation which gives the direction of maximum extension. The zones dominated by S-surfaces are cut by the zones of relatively intense strain containing C-surfaces, which extend the S-zones parallel to the stretching lineation (Fig. 12a).

In the décollement model, the S-surfaces correspond to the foliation planes of the ductile shear-zone model (Fig. 12b). On a larger scale,

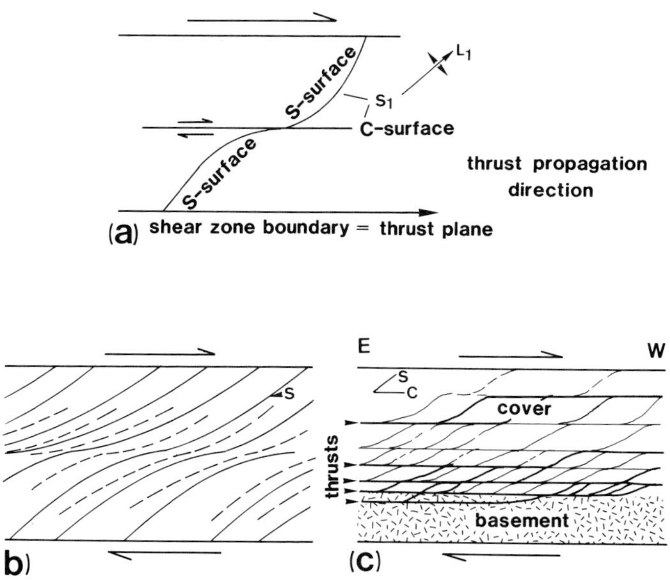

FIG. 12. Schematic décollement model of Austro–Alpine cover nappes in the eastern Alps (after Ramsay & Graham 1970, Ratschbacher 1986). (a) Kinematic framework; S-(foliation) and C-(shear) surfaces (Berthé *et al.* 1979); S_1, l_1: Alpine foliation and stretching lineation. (b) Ductile shear-zone model. (c) Extended shear-zone model including ramp-flat and S-C geometries. See text for discussion.

they represent zones of shearing along which, when eventually cohesion ceases to exist, the thrusts climb through the stratigraphic sequence (e.g. Eisenstadt & De Paor 1987). The cleavage formed by C-surfaces, penetratively developed in incompetent strata of the décollement, plays the same role on a smaller scale as the flats of the thrusts. A counterpart of the anastomosing foliation pattern on the microscale, the combination of S- and C-surfaces, is the lozenge-shaped geometry of some nappe bodies. This geometry was enhanced by flattening and rotation during the superposition of the coaxial component of the strain history (Fig. 12c).

Discussion

Based on a kinematic analysis in mylonites along major detachments in the Austro–Alpine unit, we suggested a Cretaceous top to (N)W directed décollement in the eastern Alps (Ratschbacher 1984, 1986, 1987, Neubauer 1987, Schmid & Haas 1987). The analysis of thrust geometries of three detachments having different vertical positions in the Austro–Alpine stack confirms this motion direction. Kinematic models for the eastern Alps, which incorporate solely N-directed displacement of thrusts, are, thus, incorrect and have to be replaced by multidirectional models (e.g. Ratschbacher 1986, Ratschbacher *et al.* 1987).

The type of shear zone proposed to fit the kinematic and structural observations in the Graywacke-zone detachment is the result of contrasts in strength of individual sediment layers. Zones of pre-existing inhomogeneities (e.g. layering) are activated during the displacement of a nappe pile by simple shear flow, and may become layers with varying strain rates. Zones of more intensive deformation are domains dominated by C-planes that show higher strain rates than zones in which the deformation proceeds more slowly and in which domains are dominated by S-planes. Examples of variable strain rates have been given by comparing deformation rates in rocks of different competence of the Graywacke-zone décollement. A comparison of the results of the pebble-based and March strain-analysis techniques yields a qualitative example for the variation of deformation intensity between S- and C-zones in the nappes: strain is always lower in the S-zones than in the C-zones. Strain based on the measurements of the pebbles may give an estimate of the amount of strain in the S-zones. March strains may yield an estimate of the strain in the less competent rocks, in which the bedding-parallel foliation is taken to represent C-zones.

Conclusions

(1) The imbrication of Permo–Mesozoic cover sediments along the northwestern edges of major Alpine thrust systems in the eastern Alps indicates a top to (N)W detachment of Austro–Alpine nappes.
(2) Thrust geometries of detachments involving cover as well as basement depict (N)W-directed imbrication by ramp-flat configurations, a climbing of thrusts toward the (N)W through the stratigraphic sequence, and an imbrication of youngest rocks at the (N)W edges of the thrust systems.
(3) Ductile mylonites accompanying thrusts along the basement-cover décollement depict dominant non-coaxial deformation, flow partitioning between coaxial and non-coaxial deformation in discrete deformation zones and stress relaxation. In monomineralic rocks, relaxation is achieved mainly by geometric softening and, in polymineralic rocks, mainly by diffusional mass transfer and grain boundary sliding.
(4) A model for the basement-cover detachment has to be based on thrust geometry as well as on strain and strain-rate variations in the décollement.

ACKNOWLEDGEMENTS: W. Frisch, H. Fritz, U. Ring, and A. Wetzel critically read the manuscript and Susanne Borchert additionally corrected the English text. Dorothee Dietrich as corresponding editor and the reviewers revealed many inconsistencies. All contributions are gratefully acknowledged.

References

APPOLD, T. & PESCH, P. 1984. Die Tektonik der postvariskischen Transgressionsserie im Krappfeld (Kärnten/Österreich). *Carinthia II*, **174/94**, 319–37.

BEHRMANN, J. H. & PLATT, J. 1982. Sense of nappe emplacement from quartz c-axis fabrics: an example from the Betic Cordilleras (Spain). *Earth and Planetary Science Letters*, **59**, 208–15.

Berthé, D., Choukroune, P. & Jegouzo, P. 1979. Orthogneiss, mylonite and non-coaxial deformation of granites: the example of the south Armorican shear zone. *Journal of Structural Geology*, **1**, 31–42.

Biju-Duval, B., Dercourt, J. & Le Pichon, X. 1977. From the Tethys ocean to the Mediterranean seas: a plate tectonic model of the evolution of the western Alpine system. *In*: Biju-Duval, B. & Montadert, L. (eds). *Structural History of the Mediterranean Basins*, pp. 143–64. Technip Paris.

Casey, M., Rutter, E. H., Schmid, S. M., Siddans, A. W. B. & Whalley, J. S. 1978. Texture development in experimentally deformed calcite rocks. *In: Proceedings of the 5th International Conference on Textures of Materials, Aachen*, 231–40, Springer, Berlin.

Christie, J. M., Ord, A. & Koch, P. S. 1980. Relationship between recrystallised grain-size and flow-stress in experimental deformed quartzites. *EOS*, **61**, 377.

Clar, E. 1965. Zum Bewegungsbild des Gebirgsbaues der Ostalpen. *Verhandlungen der geologischen Bundesanstalt Wien, Sonderheft G*, 11–35.

Dercourt, J., Zonenshain, L. P., Ricou, L. -E. *et al.* 1986. Geological evolution of the Tethys belt from the Atlantic to the Pamirs since the Lias. *Tectonophysics*, **123**, 241–315.

Eisenstadt, G. & De Paor, D. G. 1987. Alternative model of thrust-fault propagation. *Geology*, **15**, 630–33.

Flügel, H. W. & Neubauer, F. 1984. *Geologische Karte der Steiermark*, Geologische Bundesanstalt Wien.

Frank, W., Esterlus, M., Frey, I., Jung, G., Krohe, A. & Weber, J. 1983. Die Entwicklungsgeschichte von Stub- und Koralpenkristallin und die Beziehung zum Grazer Paläozoikum. *Jahresbericht 1982 Österreichischer Hochschulschwerpunkt S15*, **4**, 263–93.

——, Kralik, M., Scharbert, S. & Thönt, M. (1987). Geochronological data from the eastern Alps. *In*: Flügel, H. W. & Faupl, P. (eds). *Geodynamics of the eastern Alps*, 272–81, Deuticke, Vienna.

Frisch, W. 1979. Tectonic progradation and plate tectonic evolution of the Alps. *Tectonophysics*, **60**, 121–34.

Fritz, H. & Kralik, M. 1987. Kinematics and geochronology of early Cretaceous thrusting in the Paleozoic of Graz (eastern Alps). *Terra Cognita*, **7**, 97.

Hawkesworth, C. J., Waters, D. J. & Bickle, M. J. 1975. Plate tectonics in the eastern Alps. *Earth and Planetary Science Letters*, **24**, 405–13.

Koch, P. S., Christie, J. M. & George, R. P. 1980. Flow laws of wet quartzite in the alpha field. *EOS*, **61**, 376.

Krohe, A. 1987. Kinematics of Cretaceous nappe tectonics in the Austro–Alpine basement of the Koralpe region (eastern Austria). *Tectonophysics*, **136**, 171–96.

Lister, G. S. 1977. Discussion: crossed-girdle c-axis fabrics in quartzites plastically deformed by plane strain and progressive simple shear. *Tectonophysics*, **39**, 51–4.

——, Paterson, M. S. & Hobbs, B. E. 1978. The simulation of fabric development in plastic deformation and its applications to quartzite: the model. *Tectonophysics*, **45**, 107–58.

March, A. 1932. Mathematische Theorie der Regelung nach der Korngestalt bei affiner Deformation. *Zeitschrift für Kristallographie*, **81**, 285–97.

Neubauer, F. 1980. Zur tektonischen Stellung des Ackerlkristallins (Nordrand der Gurktaler Decke). *Mitteilungen der österreichischen Geologischen Gesellschaft*, **73**, 39–53.

—— 1987. The Gurktal thrust system within the Austro–Alpine region–some structural and geometrical aspects. *In*: Flügel, H. W. & Faupl, P. (eds). *Geodynamics of the Eastern Alps*, 226–36, Deuticke, Vienna.

——, Frisch, W. & Hansen, B. T. 1987. Time relations between Eoalpine metamorphism and thrusting: evidence from the crystalline basement of the eastern Greywacke zone. *In*: Flügel, H. W. & Faupl, P. (eds). *Geodynamics of the Eastern Alps*, 263–71, Deuticke, Vienna.

Neugebauer, J. & Hammerschmidt, K. 1986. Altalpidische Krustenausdünnung im Ostalpin – Neuinterpretation des Metamorphose- und Deformationsablaufes an der Grenze Mittel/Oberostalpin in der Saualpe (Ostalpen). *Symposium Strukturgeologie, Kristallingeologie, Tektonik, Tübingen*, Attempto, 39–41.

Nievoll, J. 1983. *Stratigraphische und strukturgeologische Untersuchungen in der Grauwackenzone bei Veitsch*. Unpublished PhD thesis, University of Graz.

Nowak, H. W. 1986. *Kristallisations- und Deformationsgeschichte am S-Rand der Gurktaler Decke (Ostalpen/Österreich)*. Unpublished PhD thesis, University of Vienna.

O'Brien, D. K., Ratschbacher, L., Wenk, H.-R. & You, Z. 1987. Preferred orientation of phyllosilicates in phyllonites and ultramylonites. *Journal of Structural Geology*, **9**, 719–30.

Ord, A. & Christie, J. M. 1984. Flow stress from microstructures in mylonitic quartzites of the Moine thrust zone, Assynt area, Scotland. *Journal of Structural Geology*, **6**, 639–54.

Platt, J. P. & Behrmann, J. H. 1986. Structures and fabrics in a crustal-scale shear zone, Betic Cordillera, SE Spain. *Journal of Structural Geology*, **8**, 15–33.

Ramsay, J. G. & Graham, R. H. 1970. Strain variation in shear belts. *Canadian Journal of Earth Sciences*, **7**, 786–813.

Ratschbacher, L. 1984. Kinematische Analyse der Deckenbewegungen und assoziierter Deformation im Deckgebirge der östlichen Ostalpen (Obersteiermark – Österreich). *Jahresbericht 1983 österreichischer Hochschulschwerpunkt S15*, **5**, 67–200.

—— 1986. Kinematics of Austro–Alpine cover nappes: changing translation path due to transpression. *Tectonophysics*, **125**, 335–56.

—— 1987. Strain, rotation, and translation of

Austro–Alpine nappes. *In*: Flügel, H. W. & Faupl, P. (eds). *Geodynamics of the Eastern Alps*, 235–43, Deuticke, Vienna.

—— & Klima, K. 1985. Übersicht über Geologie und Mineralgehalt in einem Querprofil vom Altkristallin zur Kalkalpenbasis (Triebener Tauernpaβ – Flitzenschlucht, Paltental, Steiermark, Österreich). *Jahrbuch der geologischen Bundesanstalt Wien*, **128**, 151–73.

—— & Oertel, G. 1987. Superposed deformations in the eastern Alps: strain analysis and microfabrics. *Journal of Structural Geology*, **9**, 263–76.

——, Frisch, W. & Ring, U. 1987. Stacking, dispersion and extension as a consequence of Alpine transpression in the eastern Alps. *Terra Cognita*, **7**, 118.

Rockenschaub, M. 1986. *Geologische und geochronologische Untersuchungen im Grenzbereich Wölzer Kristallin – Schladminger Kristallin in den nördlichen und östlichen Schladminger Tauern (Steiermark)*. Unpublished PhD thesis, University of Vienna.

Schmid, S. M. 1982. Microfabric studies as indicators of deformation mechanisms and flow laws operative in mountain building. *In*: Hsü, K. J. (ed). *Mountain Building Processes*, 95–110, Academic Press, London.

—— & Haas, R. 1987. The transition from near surface thrusting to intra-basement décollement along the Schlinig thrust. *Terra Cognita*, **7**, 68.

——, Paterson, M. S. & Boland, J. N. 1980. High temperature flow and dynamic recrystallization of Carrara marble. *Tectonophysics*, **65**, 245–80.

Slapansky, P. & Frank, W. 1987. Structural evolution and geochronology of the northern margin of the Austro–Alpine in the northwestern Schladming crystalline (NE Radstädter Tauern). *In*: Flügel, H. W. & Faupl, P. (eds). *Geodynamics of the eastern Alps*, 244–62, Deuticke, Vienna.

Tollmann, A. 1959. Der Deckenbau der Ostalpen auf Grund der Neuuntersuchung des zentralalpinen Mesozoikums. *Mitteilungen der Gesellschaft österreichischer Geologie und Bergbaustudenten*, **10**, 1–62.

Underwood, E. E. 1970. *Quantitative Stereology*. Addison-Wesley, Reading.

White, S. H., Burrows, S. E., Carreras, J., Shaw, N. D. & Humphreys, F. J. 1980. On mylonites in ductile shear zones. *Journal of Structural Geology*, **2**, 175–87.

Wolter, L., Thiedig, F., Pesch, P. Halamic, J. & Appold, Th. 1982. Geologie und Tektonik des Krappfeld-Mesozoikums (Ebersteiner Trias) in Kärnten/Österreich. *Mitteilungen des Geologisch-Paläontologischen Institutes der Universität Hamburg*, **53**, 207–48.

L. Ratschbacher, Institute of Geology, University, D-7400 Tübingen, FRG.
F. Neubauer, Institute of Geology, University, A-8010 Graz, Austria.

Models of the Development of the Alpine Chain

Kinematics of the western Mediterranean

J. F. Dewey, M. L. Helman, E. Turco, D. H. W. Hutton & S. D. Knott

SUMMARY: The kinematic understanding of the relationship between relative plate motion and the structure of orogenic belts depends upon a knowledge of relative plate motion across the plate boundary system, the relative motion of small blocks and flakes within the system, an evaluation of orogenic body forces, and an understanding of the thermomechanical evolution of the upper part of the orogenic lithosphere in determining strength and detachment levels. We have built a preliminary model for the Cenozoic kinematic evolution of the western Mediterranean oceanic basins and their peripheral orogens that integrates (1) the motion of Africa relative to Europe based upon a new study of Atlantic fracture zones using SEASAT data and the Lamont–Doherty magnetic anomaly database, (2) a new interpretation of the rotation of Corsica/Sardinia and the opening of the Balearic and Tyrrhenian oceanic basins, (3) sedimentary facies sequences in the Apennines, Calabria, and Sicily, and (4) Apennine/Calabrian structure and structural sequence.

Compressive motion between Africa and Europe was recognized as the causal mechanism for the Alpine orogeny early in this century (Termier 1903, Argand 1916, Heim 1922, Staub 1924). It was not, however, until 50–70 years later that the kinematics and timing of relative motion between these continental masses could be determined with any accuracy (Smith 1971, Dewey *et al.* 1973). This new insight was a direct consequence of the geometric constraints imposed by moving torsionally rigid plates on a spherical surface (Morgan 1968, Le Pichon *et al.* 1973) according to the developing plate tectonic paradigm.

Since 1971, many attempts have been made to directly relate geologic observations to plate motion. The underpinnings of such analyses are rather simple to understand. Plate boundaries are regarded as crustal scale shear zones whose geometry involves extension parallel with the shear direction determined by stretching lineations (Ramsay & Graham 1970). If these are found to have a consistent orientation on a regional scale, they may be considered an indicator of transport direction to be equated with the direction of plate motion (Shackleton & Ries 1984).

In western Europe, recent studies have related stretching lineations in the Alps (Baird & Dewey 1986, Choukroune *et al.* 1986) and various macro- and micro-stress indicators (tectonic stylolites, tensional joints, slickensides on fault planes or bed-to-bed surfaces, and orientation of fold axes) in the Alpine foreland (Letouzey 1986, Bergerat 1987), to plate motions. We note, however, that the conclusions of such studies are strongly dependent upon the model of relative plate motion between Africa and Europe to which they are tied. Baird & Dewey (1986) based their conclusions on the relative motion data of Livermore & Smith (1985) whereas the motions described by Savostin *et al.* (1986) are preferred by Choukroune *et al.* (1986), Letouzey (1986), and Bergerat (1987).

We address two broad problems. First, we deal with the relative motion path between the African and European Plates and second, the extent to which we may '. . . relate the complex response of the continental crust to the generally simpler relative motions among the larger plates. . .' (Dewey 1982). To this end, we will first present our framework for the relative motion of Africa with respect to Europe based upon a new study of Atlantic fracture zones and magnetic anomalies. Next, we will present a preliminary model for the late Cenozoic kinematic evolution of a portion of the western Mediterranean region. We will use this model to illustrate how geological structures, as well as the kinematics of microplates and small ocean basins, are constrained by, but not necessarily directly related to, the motions of large plates.

Africa-Europe kinematics

The first quantitative kinematic studies were made using only a few selected magnetic anomalies (Pitman & Talwani 1972). This was a consequence of a limited data base in which only a few anomalies were easily recognizable or because the anomalies were thought to correlate with important tectonic events. Subsequent kinematic studies have used these same anomalies (e.g. Dewey *et al.* 1973, Biju-Duval *et al.* 1977, Livermore & Smith 1985, Savostin *et al.*

From: Coward, M. P., Dietrich, D. & Park, R. G. (eds), 1989, *Alpine Tectonics*, Geological Society Special Publication No. 45, pp. 265–283.

1986). The result is a series of 'time-lapse' frames showing the relative positions of continental masses at 10–15 Ma intervals. However, the time interval may be shortened to 2–3 Ma (the average length of magnetic reversals in late Cretaceous and Tertiary time) by using all identified magnetic anomalies. This is possible because of the vastly increased data base that is currently available.

We have undertaken a new study of the relative motion of Africa with respect to Europe. In this study, we have re-identified all magnetic anomalies in the late Cretaceous–Tertiary sequence (Anomalies 1–34) in the north and central Atlantic Ocean using the Lamont-Doherty data bank. We have also remapped fracture zones in these oceans using bathymetric maps (Laughton *et al.* 1975, Laughton & Monahan 1978, Searle *et al.* 1982), SEASAT altimetry data, and SEASAT-derived gravity images (Haxby 1985). Lastly, we have determined rotation parameters for individual anomalies in each ocean and have computed the finite difference solution between them to determine the relative motion of Africa with respect to Europe. For the opening of the central Atlantic prior to Anomaly 34, we use the parameters of Pindell *et al.* (1988).

The palaeopositions of plates in convergent zones cannot be directly reconstructed. This is because the magnetic anomalies and fracture zones that give the timing and direction of motion between the plates have been destroyed, usually by subduction. In such situations reconstructions are made, generally, by completing the finite difference circuit between adjacent sets of plates separated by spreading oceans. This finite difference gives the relative motion between the two plates of primary interest. The methodology used in this approach is fully described in Dewey *et al.* (1973) and Ladd (1976).

A modified version of the African motion in the European frame of reference, from 175 Ma to the present, is shown in Fig. 1(a). To facilitate comparison, the position of Africa is shown only at time intervals equivalent to those used in other models (Anomalies 5, 6, 13, 21, 24, 30, 33, 34, M-O, M-4, M-10, M-16, M-21, M-25, and the Blake Spur Anomaly (BSA)). A complete kinematic model using all anomalies will be presented elsewhere. In Fig. 1(b) the motion is shown by locating the position of 3 points on the present day African coast at these past times.

Many older models (Smith 1971, Dewey *et al.* 1973, Biju-Duval *et al.* 1977), and to a lesser extent newer ones (Savostin *et al.* 1986), have been marked by drastic changes in the direction of Africa's motion with respect to Europe. In each successive model, however, there has been a tendency towards a smoother, less eccentric path. Our model shows, with one exception, that the overall motion is relatively smooth and without sudden or drastic changes in direction (Fig. 1).

Several phases of motion can be distinguished in our representations (Figs. 2 and 3; Table 1). The first phase is quite obvious. In the pre-drift (>175 Ma) until Anomaly M-O (118 Ma) interval, Africa's motion relative to Europe was one of sinistral strike-slip. This motion corresponds to the separation of Africa and North America and the opening of the central Atlantic Ocean. At this time separation had not begun in either the north or south Atlantic although rifting had (Pindell *et al.* 1988).

Between Anomaly M-O (118 Ma) and Anomaly 34 (84 Ma), the motion began to change to northeasterly directed compression. Dating the timing of this change is difficult because of the absence of large scale magnetic reversals during this interval (Cretaceous Quiet Zone). In the north Atlantic, the separation of Iberia from North America was initiated by seafloor spreading north of the Azores–Gibraltar fracture zone. This marked the start of differential motion between the north and central Atlantic. Anomaly M-O (118 Ma) has been identified off the Iberian coast by Group Galice (1979) and has been correlated with the J-Anomaly Ridge mapped off the Newfoundland coast. However, Anomaly 34 (84 Ma) is the first clearly-identifiable anomaly

FIG. 1. (a) Position of Africa with respect to Europe from 175 Ma to the present. Africa's position is shown at Closure (175 Ma), Blake Spur Anomaly (165 Ma), Anomaly M-25 (156.3 Ma), Anomaly M-21 (149.2 Ma), Anomaly M-16 (141.2 Ma), Anomaly M-10 (129.8 Ma), Anomaly M-4 (125.4 Ma), Anomaly M-0 (118 Ma), Anomaly 34 (84 Ma), Anomaly 33 (74.Y Ma), Anomaly 30 (66.7 Ma), Anomaly 24 (55.7 Ma), Anomaly 21 (48.8 Ma), Anomaly 13 (35.5 Ma), Anomaly 6 (19.4 Ma), Anomaly 5 (8.9 Ma), and Present.
(b) Flow lines depicting the path of Africa's motion for three points on the present day African coastline. The positions are shown for the same times as in A. The results shown have not been corrected or smoothed.

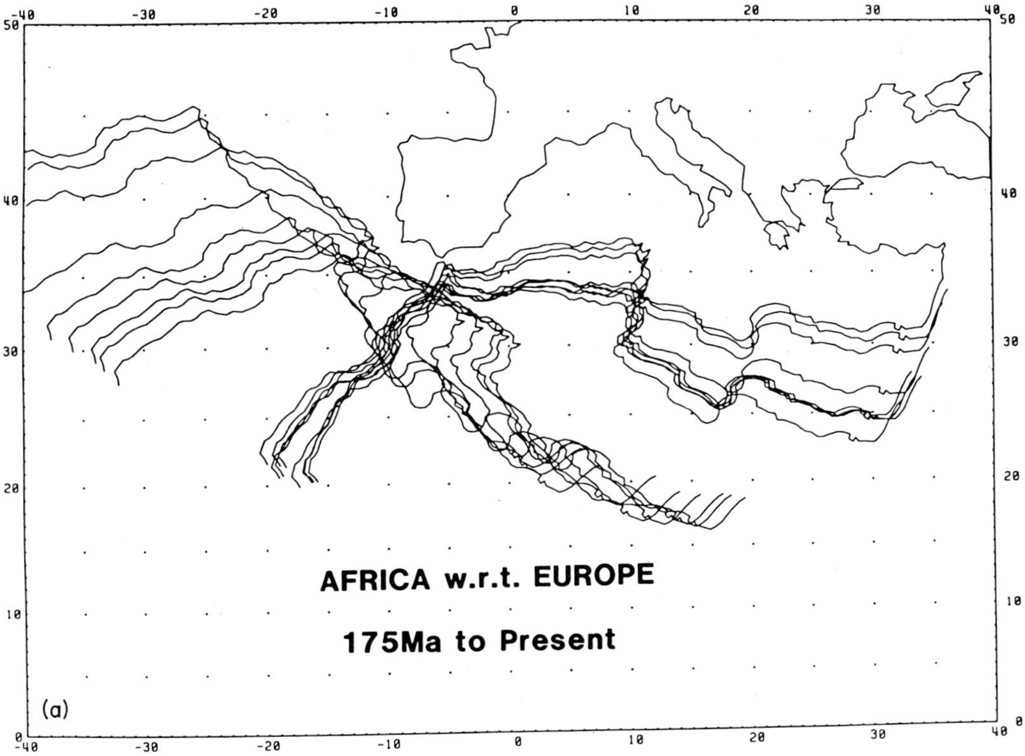
AFRICA w.r.t. EUROPE
175Ma to Present
(a)

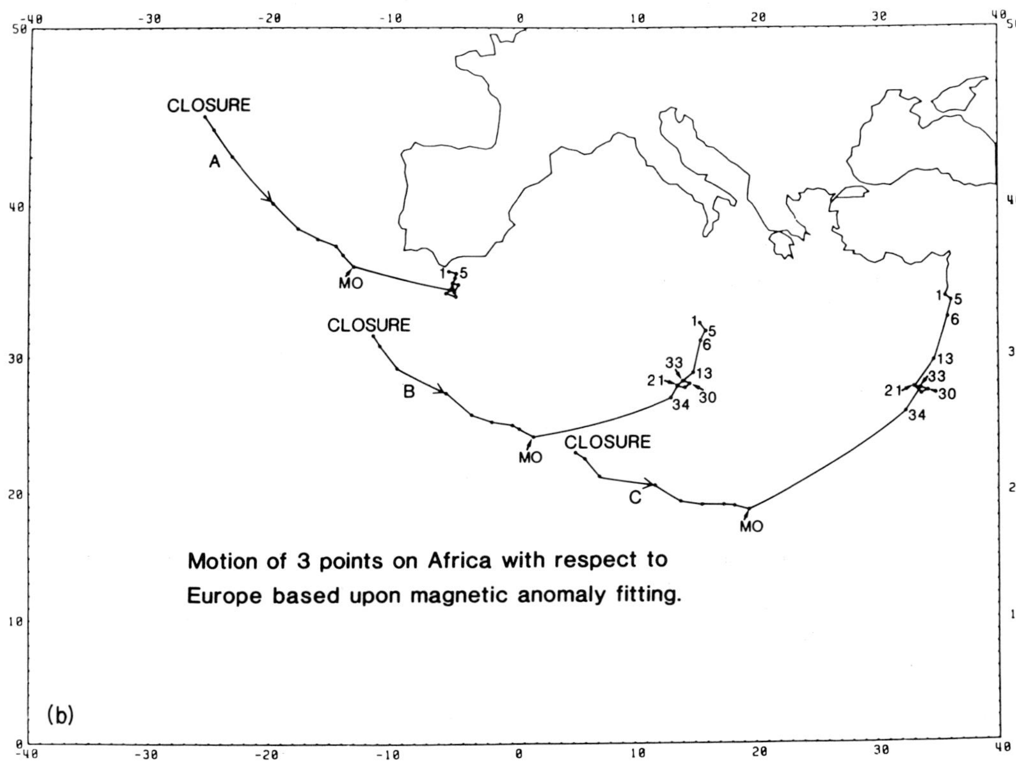
CLOSURE
A
MO
CLOSURE
B
MO
CLOSURE
C
MO
Motion of 3 points on Africa with respect to
Europe based upon magnetic anomaly fitting.
(b)

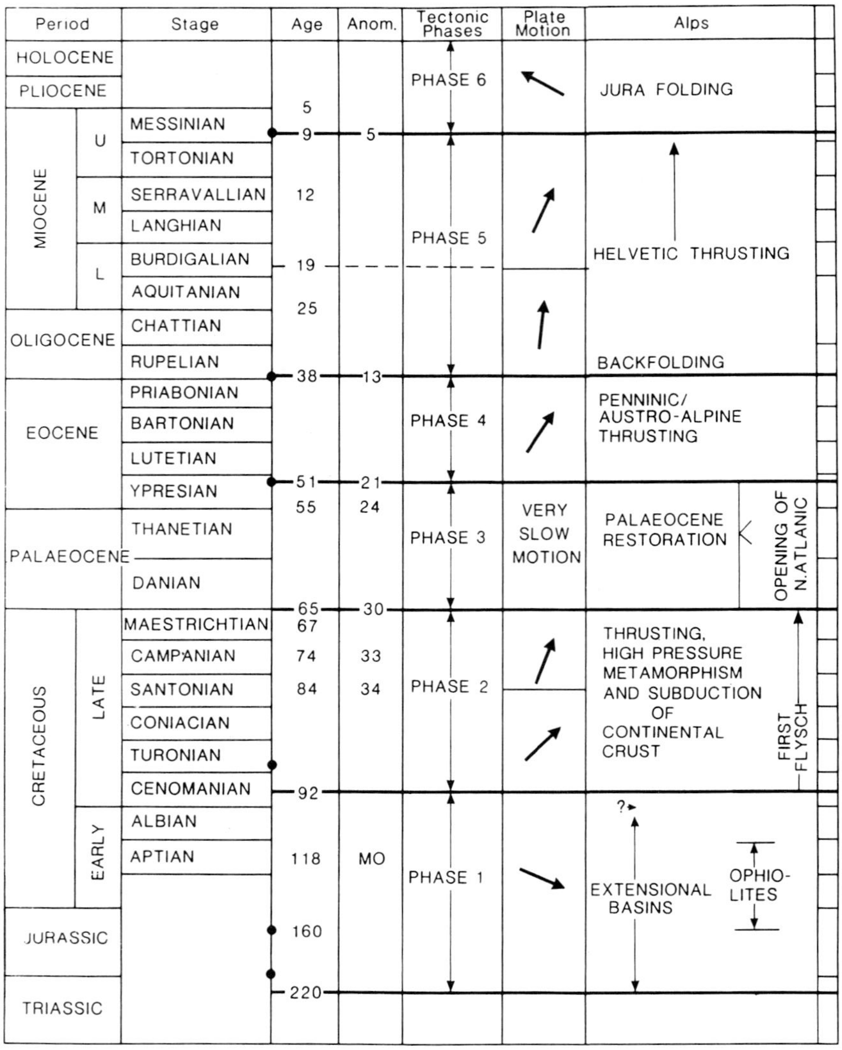

FIG. 2. Correlation chart showing the general direction of Africa's motion with respect to Europe and correlations with various tectonic events in the Alps, Mediterranean Sea, and the Atlantic Ocean.

seen on both sides of the Atlantic between the Azores–Gibraltar and Charlie–Gibbs fracture zones.

We suggest that 92 Ma (Cenomanian–Turonian boundary) is a time appropriate for this change. This is based on the occurrence of high pressure matamorphism in the Alps at this time (Bocquet *et al.* 1974; Ernst & Dal Piaz 1978). We note, however, that eclogite-facies metamorphism in the Sesia–Lanzo zone, dated at 100–110 Ma (Compagnoni *et al.* 1975), may be indicative of an earlier change of motion.

Shortly after Anomaly 30 (66.7 Ma), in the earliest Palaeocene, the motion changed again. The convergence of Africa and Europe slowed dramatically and the motion between the two continents became somewhat erratic. This correlates closely with a change in the orientation of the Charlie–Gibbs fracture zone, Thulean volcanism in the NE Atlantic, as well as renewed subsidence and sedimentation in parts of the North Sea (Ziegler 1975).

The apparent complexity of the Africa–Europe relative motion path at this time may be, partly, a consequence of the errors involved in combining north and central Atlantic rotation parameters. There are, as might be expected, many potential sources of error and uncertainty

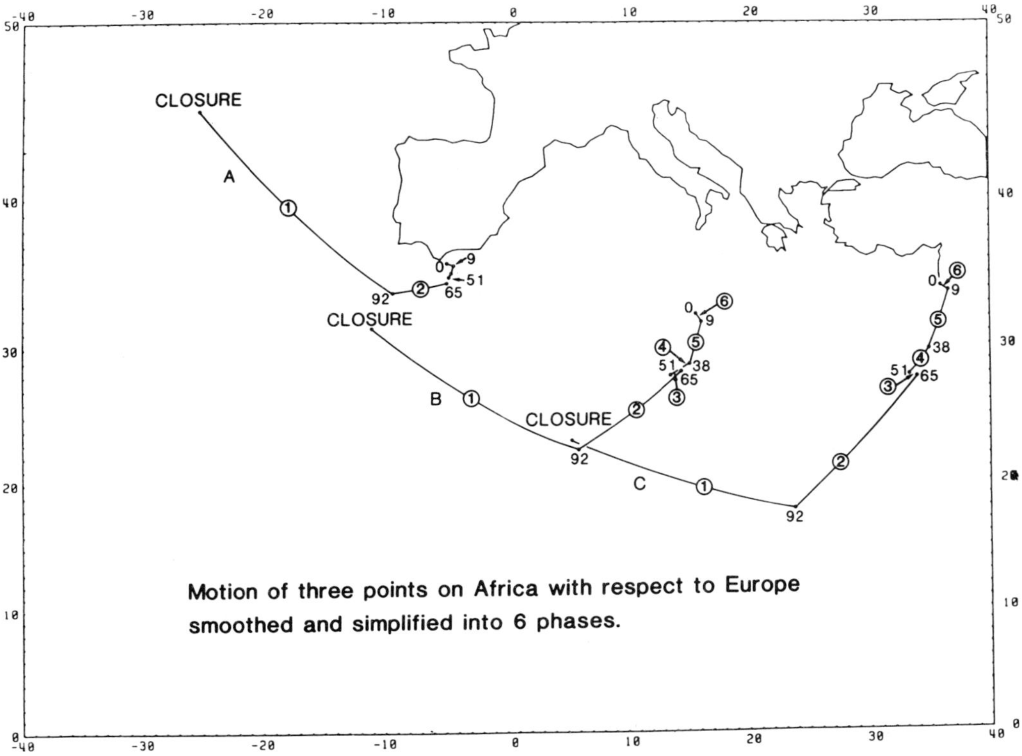

FIG. 3. Flow lines depicting the path of Africa's motion for the same three points as in Fig. 1 (a) after the motion path has been smoothed and corrected to correlate with the events and phases of motion shown in Fig. 2.

TABLE 1. *Rotation parameters used in construction of kinematic model. Rotation parameters were determined by finite difference solution of central and north Atlantic rotation parameters.*

Anomaly	Latitude	Longitude	Rotation
5	0.55	−15.78	− 0.94
6	24.04	−17.32	− 2.33
13	29.26	−20.67	− 6.65
21	35.08	−15.54	−10.39
24	33.04	−14.59	−10.77
30	31.52	−10.73	−11.29
33	33.51	−10.42	−11.61
34	34.82	−11.66	−14.64
92 Ma	40.50	−12.00	−33.50
M-O	43.90	− 9.08	−40.51
FIT	52.88	0.16	−59.86

in using finite rotations as a means of plate reconstruction. Although a discussion of such problems is beyond the scope of this paper, we recognize that '...uncertainties in reconstructions obtained by combining reconstructions of separate pairs of plates are often much larger than the misfits of data from any individual ocean...' (Stock & Molnar 1983). However, finite difference solutions computed using the recently published poles of Srivastava & Tapscott (1986) and Klitgord & Schouten (1986) give similar solutions to ours for this time interval. Similar changes in the Africa–Europe motion path are also suggested by Livermore & Smith (1985, see their figs. 5.12 and 5.13).

Following Anomaly 24 (55.7 Ma, early Eocene), the African motion 'resumes' its northward course, the azimuth of which lies, with only minor variation, between N and NNE. This continues until Anomaly 5 (8.9 Ma, late Miocene) at which time the motion becomes northwesterly (NNW to NW).

The apparent resumption of northward African motion after Anomaly 24 (55.7 Ma) correlates roughly with the start of seafloor spreading in the Norwegian–Greenland Sea concomitant with slowing and cessation of spreading in the Labrador Sea. Small changes in the African motion direction, such as at Anomalies 13 (35.5 Ma) and 6 (19.4 Ma), are correlative with geologic or tectonic events (Fig. 2).

Using the correlations shown in Fig. 2, the relative motion paths shown in Fig. 1 have been smoothed and adjusted and the resulting paths shown in Fig. 3. This iterative process, reconciling raw kinematic data with geological evidence, is an integral part of the model building process. An additional point to note is how the path of relative motion will differ at various points. A comparison of the eastern- and westernmost paths in Fig. 3 illustrates this quite well. Variation in motion vector from location to location may have important implications in kinematic and geological modelling.

Regional geology

The western Mediterranean Sea is composed of small, young basins that have formed from earliest Miocene time until the present. Some of the basins have formed on stretched or thin continental crust whereas others are floored by oceanic crust.

Surrounding the western Mediterranean is a nearly continuous system of orogenic belts that are only slightly older than the basins they surround. The structural polarity of these mountain belts is outward, away from the Mediterranean. The only exceptions are structures on islands within the Mediterranean and the double-sided orogen of the Pyrenees.

In a general sense, the age of basins and mountains is greatest to the N and W and decreases to the S and E. This pattern of ages is clearly related to the evolution of the region (see below). The major tectonic units and their ages are shown in Fig. 4. The westernmost part of the Mediterranean, the Alboran Sea, and the surrounding orogen, the Betic–Rif Mountains, are not included in this model.

Kinematic model of western Mediterranean evolution

The model presented for the Tertiary evolution of part of the western Mediterranean region is framed within constraints imposed by the motion of Africa and Europe. The model is presented as a series of reconstructions, at selected time intervals, from 38 Ma to the present. In each reconstruction, the direction of Africa's motion, as determined by finite difference solution (see above), is shown. Similarly, vector triangles show the direction and proportional amount of motion of different continental blocks and microplates; absolute magnitudes of motion are not shown. In all the reconstructions, Europe is fixed and other microplates and continental blocks move in the European frame of reference.

38 Ma Latest Eocene–Priabonian (Fig. 5a)

The origin of the present day western Mediterranean Sea postdates the cessation of several important orogenic events – Penninic thrusting in the Alps (Trümpy 1980), the Pyrenean orogeny (Fischer 1984), and collision in Corsica (Harris 1985). These events ended in latest Eocene or earliest Oligocene time. The relationships of various units at this time form the starting point for the reconstructions.

We have restored Corsica, Sardinia, and the Balearic Islands to their appropriate positions against the coasts of France and Spain. The Kabylies and Calabria sit outboard of these blocks. We prefer not to show these units because, with the data presently available, their exact location cannot be well-constrained. South and east of these blocks, on the northwest margin of a tract of Neotethyan (Jurassic) oceanic crust, was an accretionary wedge that was active prior to this time but with opposite vergence to the Alps (Knott 1987, Dietrich 1988). This oceanic fragment, roughly 300 km wide, separated the southeastern margin of the European plate from the North African rifted margin.

The African margin at this time consisted of a series of carbonate platforms and basins. Our paleogeographic reconstruction is considerably simpler than those previously published for the southern Apennines. D'Argenio *et al.* (1973) and Scandone *et al.* (1974) depicted three carbonate platforms (Panormide, Apulia, and an intermediate platform) separated by two basins (Lagonegro and Molise). The clear continuity of the Lagonegro into the Molise Basin to the north leads us to believe that the tectonic windows (Campagna, Monte Alpi), cited as evidence of this platform, are blocks of the Apulia platform pushed up along late Cenozoic strike-slip faults.

Moreover, we believe that a clear continuity exists between these basins, the Tellian (North Africa) and Imerese (Sicily) basins to the S, and the north Apennines basin. Therefore, our reconstruction shows only two main carbonate

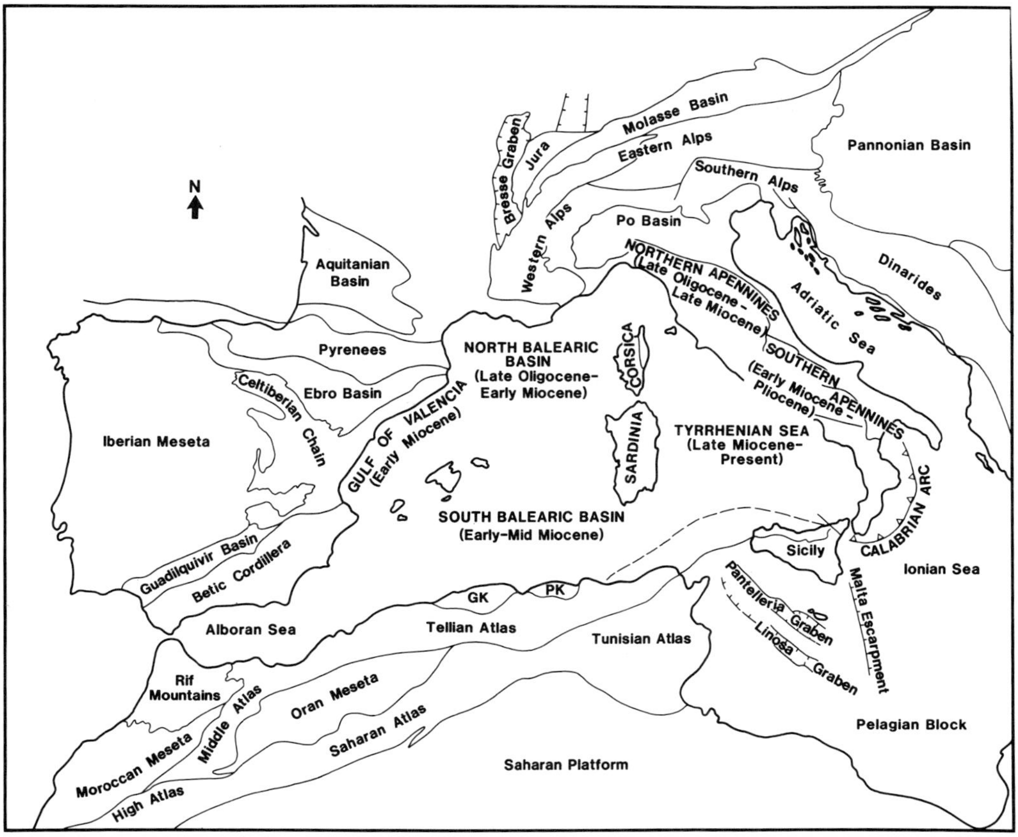

FIG. 4. Outline map of the western Mediterranean Sea and surrounding region showing the major tectonic units in the western Mediterranean and surrounding region. Basins and orogens of particular importance are labelled in upper case letters and their ages given. Not all units shown are referred to in the text.

platforms (Panormide and Apulia) separated by this basin system. We recognize nevertheless that the connection of the Panormide Platform in the southern Apennine domain with the Panormide Platform in Sicily is not clear as these carbonates do not outcrop in the southern part of Calabria (Scandone *et al.* 1974).

Eastward of the Imerese basin were the Trapanese seamounts and, further to the E, the Iblei block. The Iblean block underwent platformal-type sedimentation during the Triassic and Miocene but received basinal sediments in the intervening time. To the N and E, the Lagonegro–Molise basin passed into the Apulia Platform. The latter extended northwards to the northern Apennines interrupted only by the Gargano basin which received sediment until at least Eocene time, after which it was inverted (Martinis & Pavani 1967). The inversion of the Gargano basin, which accompanied fragmentation of the Apulian Platform, is attributed to continental collision in the Alps (Trümpy 1980).

From this time, the model shows the progressive closure of these older basins and deformation of the carbonate platforms concomitant with rifting of fragments of continent from the European mainland and opening of new basins as these continental fragments undergo translation to the SE.

23 Ma Earliest Miocene–Aquitanian (Fig. 5b)

By this time, extensional tectonics were evident throughout the region. In the Algero–Provencal or north Balearic basin Sardinia had begun to to drift from the French mainland, although Corsica had not. The pole about which Sardinia rotated was close to the northern tip of Corsica.

Various lines of evidence (field work, palaeomagnetics, magnetic anomalies, bathymetry,

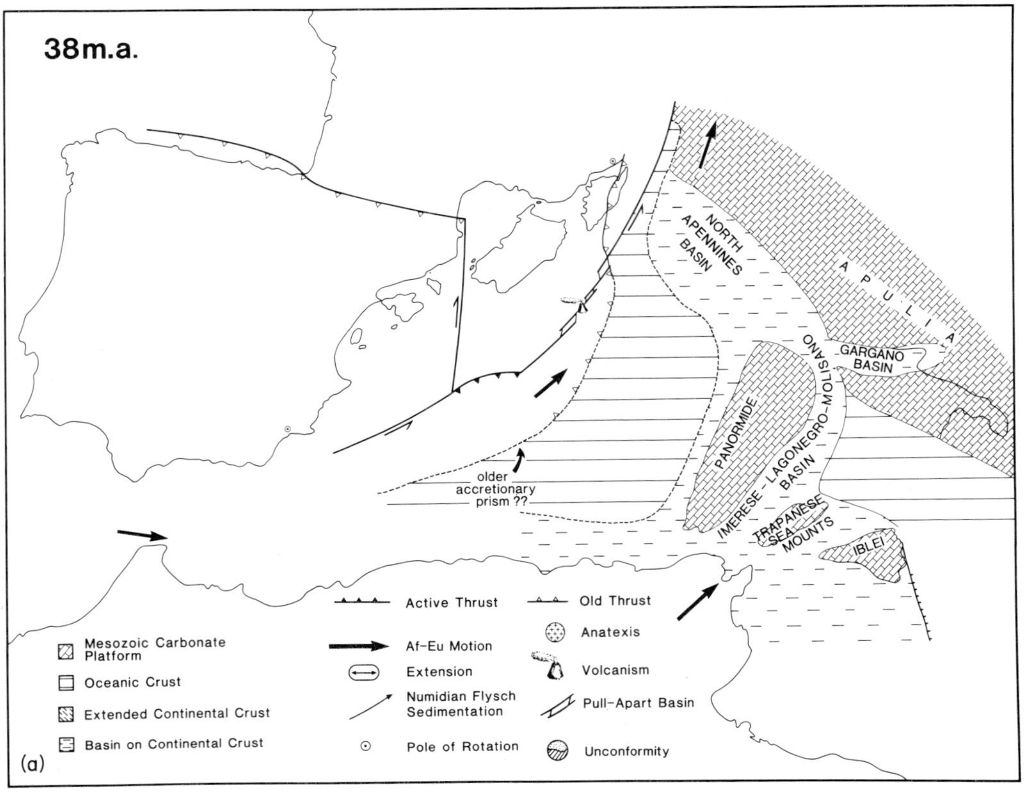
38m.a.
NORTH APENNINES BASIN
APULIA
GARGANO BASIN
PANORMIDE
IMERESE-LAGONEGRO-MOLISANO BASIN
TRAPANESE SEA MOUNTS
IBLEI
older accretionary prism ??
Active Thrust
Old Thrust
Mesozoic Carbonate Platform
Af-Eu Motion
Anatexis
Oceanic Crust
Extension
Volcanism
Extended Continental Crust
Numidian Flysch Sedimentation
Pull-Apart Basin
Basin on Continental Crust
Pole of Rotation
Unconformity
(a)

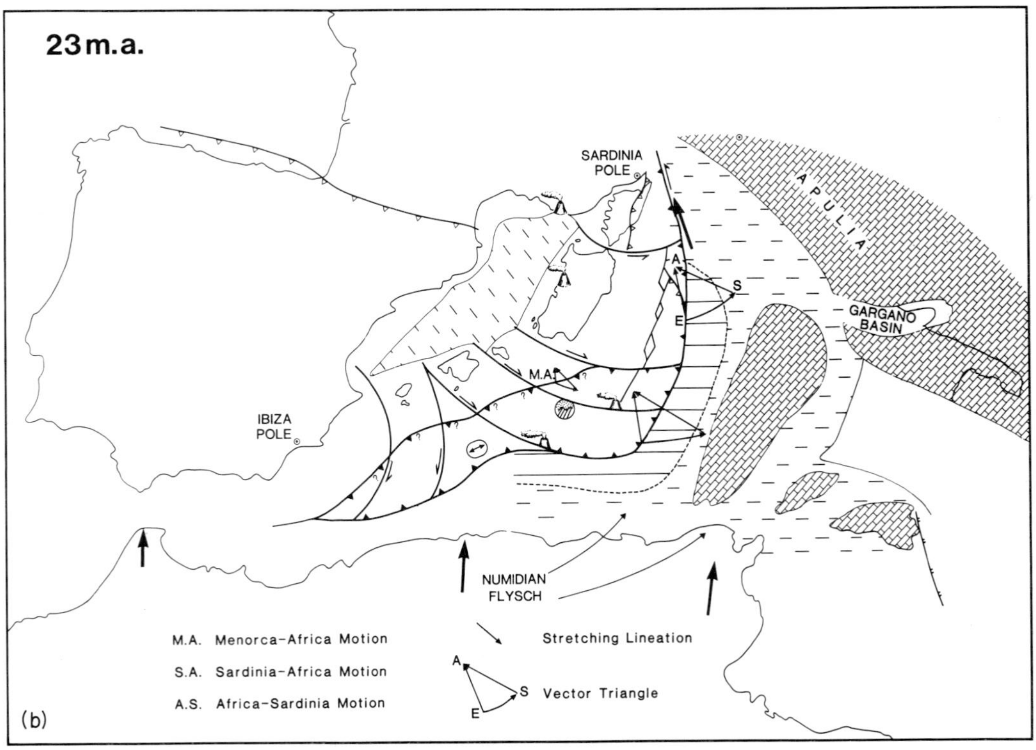
23m.a.
SARDINIA POLE
APULIA
GARGANO BASIN
A
S
E
M.A.
IBIZA POLE
NUMIDIAN FLYSCH
M.A. Menorca-Africa Motion
S.A. Sardinia-Africa Motion
A.S. Africa-Sardinia Motion
Stretching Lineation
Vector Triangle
(b)

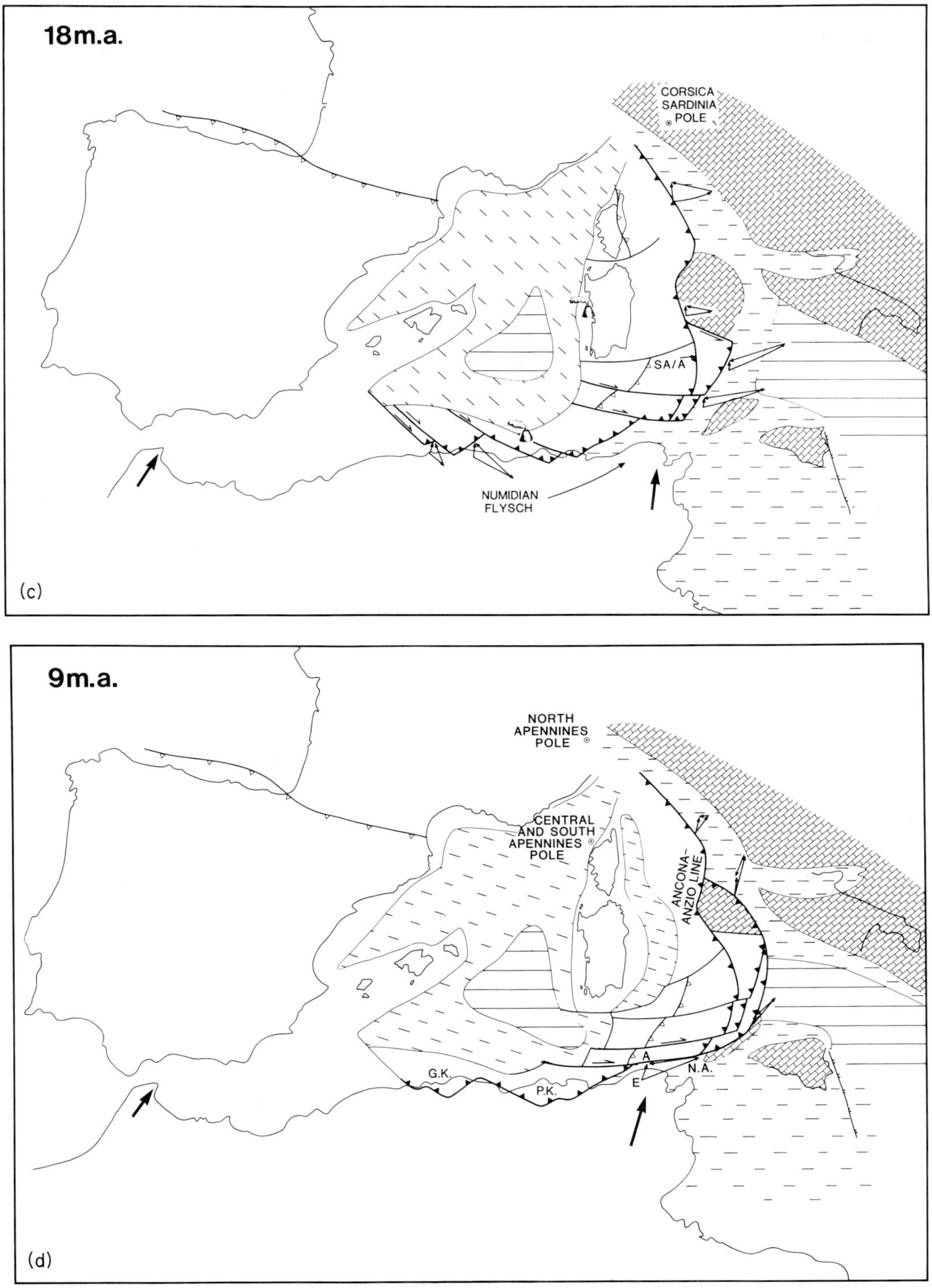

FIG. 5. Kinematic reconstructions depicting the evolution of part of the western Mediterranean region. The frame of the reconstruction shows the events occurring and the relations of tectonic elements at (a) 38 Ma latest Eocene–Priabonian, (b) 23 Ma earliest Miocene–Aquitanian, (c) 18 Ma early Miocene–Burdigalian, (d) 9 Ma late Miocene–Tortonian.

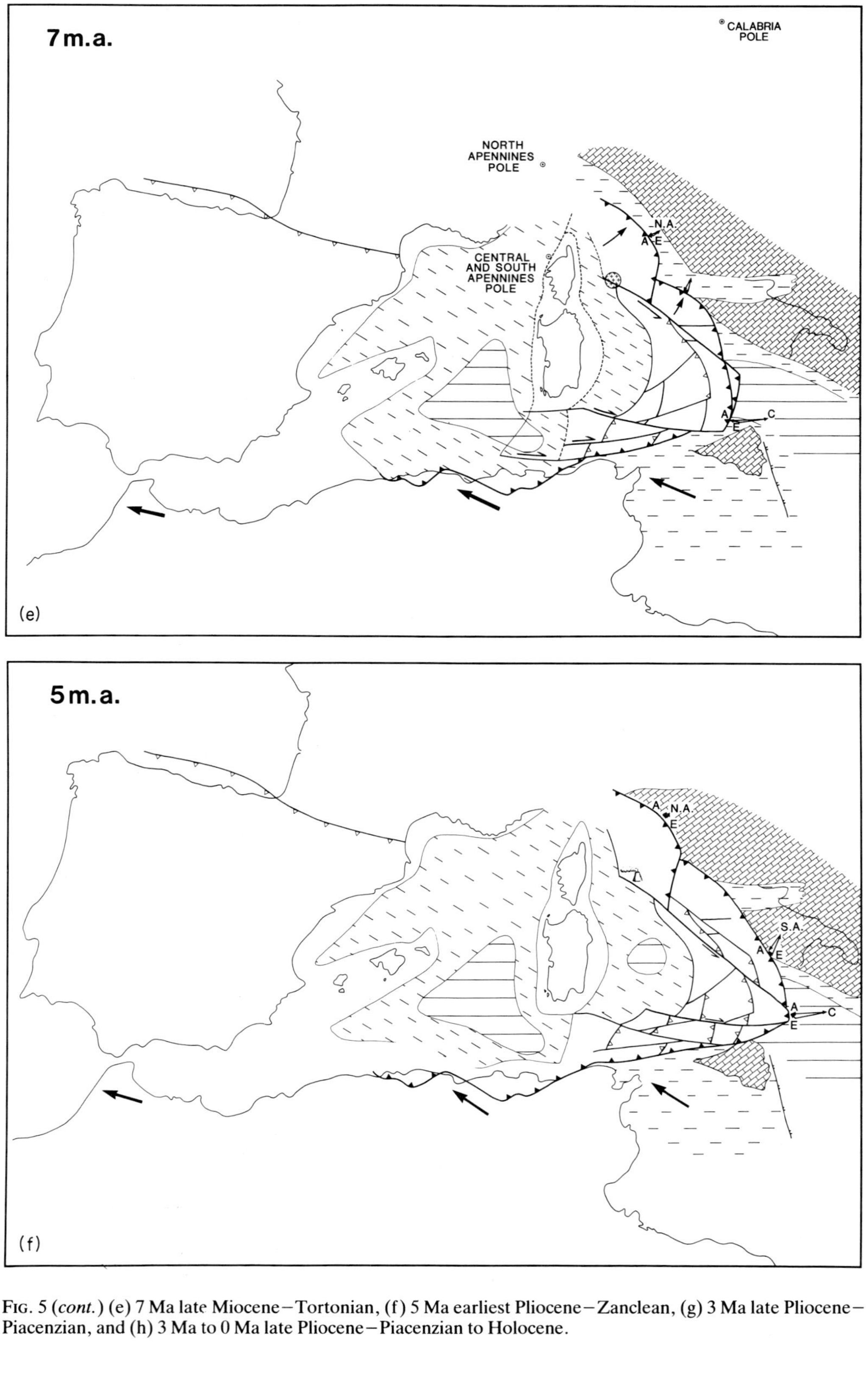

FIG. 5 (*cont.*) (e) 7 Ma late Miocene–Tortonian, (f) 5 Ma earliest Pliocene–Zanclean, (g) 3 Ma late Pliocene–Piacenzian, and (h) 3 Ma to 0 Ma late Pliocene–Piacenzian to Holocene.

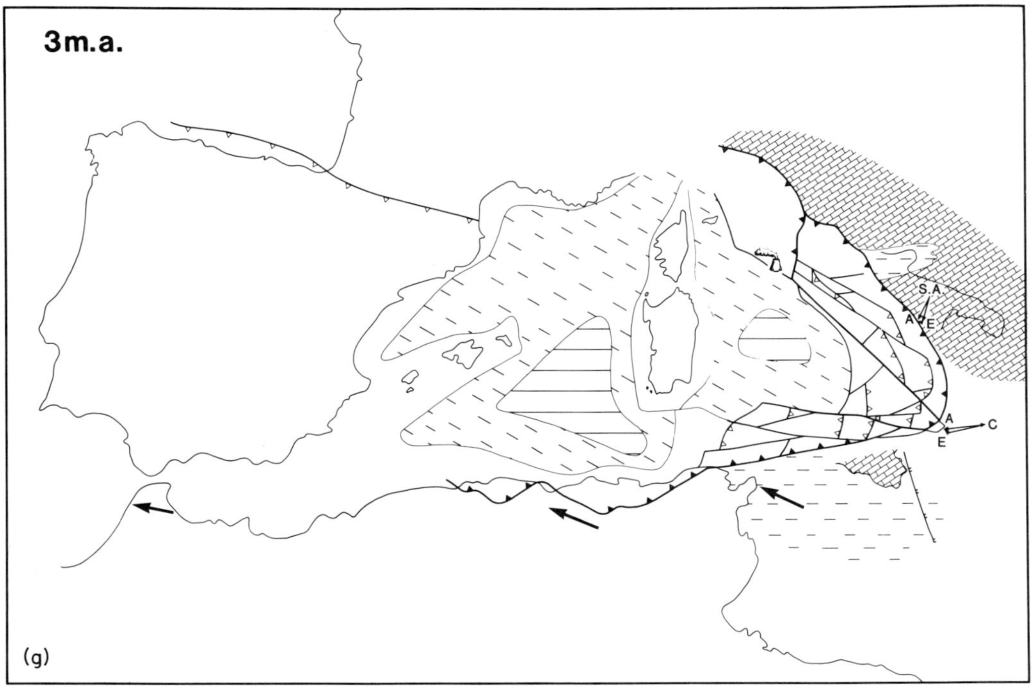

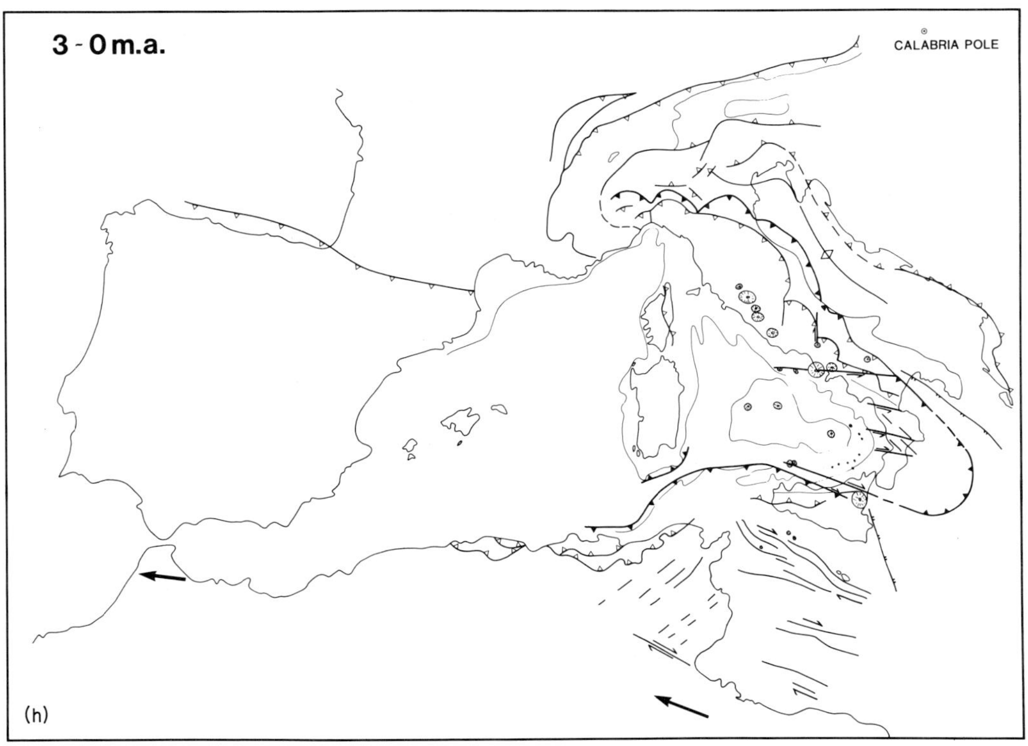

Legend (also given on a and b): Diagonal bricks = Mesozoic carbonate platform. Solid horizontal lines = oceanic crust. Dashed diagonal lines = newly extended continental crust. Dashed horizontal lines = basin on continental crust. Line with filled barbs = active thrust (barbs on overthrust side). Line with unfilled barbs = old thrust. Heavy arrows = direction of Africa's motion relative to Europe. Plus marks = anatexis. Double arrow within circles = extension. Volcanoes = volcanism. Long thin arrows = direction of Numidian Flysch sedimentation. Parallelogram = pull-apart basin. Circle with dot in centre = pole of rotation (unlabelled if site of future pole). Dots above diagonal lines (within circles) = unconformity. Short straight arrow = stretching lineation. Triangles = vector triangles showing relative motion of various blocks with respect to Europe and Africa. Explanation of each reconstruction is given in text.

reflection seismic) have been put forth by different authors to support the concept of differential rotation between Corsica and Sardinia. Some of these data are open to different interpretation (compare Bayer *et al.* 1973 with Burrus 1984) and none leads to unambiguous conclusions. The preponderance of evidence, however, clearly suggests that, at some time in their history, Corsica and Sardinia moved independently.

In the Gulf of Valencia, Mallorca and Menorca had begun to rotate away from Iberia about the Sardinia pole of rotation. However, we have found it necessary to use an independent rotation pole to restore Ibiza and Frontera to what we believe is their original position. Parallel with the length of the Kabylie–Calabria thrust belt, extension was also occurring.

Compressional tectonics went hand in hand with this extension. In the northern Apennines, the start of deformation has been dated at 27 Ma (Carmignani *et al.* 1978, Kligfield *et al.* 1986). Much of the older oceanic crust has been subducted or obducted. Subduction of oceanic lithosphere lead to the development of an accretionary wedge with transport direction to the present day NE (Knott 1987). At the same time, extension occurred in the internal part of the Kabylie–Calabria belt (Knott 1988, J. P. Platt pers. comm. 1988) and sinistral strike-slip motion occurred between Corsica and the northern Apennines (Waters 1986). Unconformities of this age have been recognized in south Calabria (Capo d'Orlando Formation), the Kabylies, and Sicily covering faults thought to be related to Alpine thrusting with European vergence (Bonardi *et al.* 1981).

Subduction was accompanied by abundant calc-alkaline volcanism in western Sardinia and southeastern France (Savelli *et al.* 1979), which shed volcaniclastic detritus into the Sicilian and Apennine sectors of the trench (Tufitidi Tusa of Ogniben 1967, 1969). Similar detritus deposited in the Kabylie trench (Durand-Delga 1981) was probably derived from transform-related volcanism. Deposition of quartzo-arenitic Numidian Flysch in the Immerese–Lagonegro–Molise Basin system and on the subsiding Panormide Platform (Bifurto Formation of Selli 1962) and its margins (Ogniben 1969, Giunta 1985) seems to have been related to the formation of a peripheral bulge on the North African margin.

18 Ma Early Miocene–Burdigalian (Fig. 5c)

By this time Corsica and Sardinia had completed their rotation about a combined pole. The completion of this rotation is dated by Montigny *et al.* (1981) at 19 Ma ±1 Ma using palaeomagnetic data. Similar dates have been obtained by Ferrara *et al.* (1972) on volcanic ash from DSDP Site 123 in the Gulf of Valencia dated at 22–19 Ma using the K–Ar method. This ash, derived from a 129.5+ m thick sequence is considered to be the slope accumulation of a volcano that was located in the Gulf of Valencia (Nesteroff *et al.* 1972). Only the southernmost part of the Algero–Provençal Basin (also known as the South Balearic Basin) remained to be formed as the Kabylies and Calabria migrated southeastward.

Deformation had now begun throughout most of the region. The final consumption of Neotethys led rapidly to a collision between the Liguride accretionary wedge and the Panormide Platform. What had been the SE margin of Neotethys was deformed in a greenschist facies ductile shear zone (see below). The southern Apennines thrust front propagated rapidly into the Lagonegro basin whereas collision was less advanced in the northern part of the Panormide Platform (Perrone 1982). The northern and southern portions of this platform became separated by an oblique ramp while at the same time the Anzio–Ancona Line developed as a frontal ramp.

Along the North African margin, emplacement of the Kabylies and deformation in Tunisia had just begun (Boullin 1984). Calc-alkaline volcanism continued in Sardinia but also occurred along the African margin from Dellys to Tunis (Bellon 1980). At the same time, deposition of Numidian flysch continued in the Imerese–Lagonegro–Molise basin system (Catalano & D'Argenio 1980).

9 Ma Late Miocene–Tortonian (Fig. 5d)

Throughout the Middle Miocene, compression continued throughout the Apennines. The motion of the north Apennines is described by a pole of rotation that is independent of that for the central and southern Apennines. This is indicated by the change of the Anzio–Ancona Line from a frontal to an oblique ramp (La Vecchia 1985).

The Apenninic thrusting was accompanied by the formation of a continuous foredeep in front of the chain. This foredeep and its piggy-back basins migrated toward the Apulian Platform as the thrust belt advanced (Pescatore 1978). Numidian flysch was resedimented from the piggy-back basins into the foredeep (Arenaric di Stigliano of Selli 1962).

Subduction of Ionian Sea oceanic crust resulted in formation of a passive roof duplex at the mountain front. Varicoloured clays were backthrust over the Lagonegro basin, Panormide Platform, and the Liguride and Calbride complexes, and Numidian flysch was carried on the varicoloured clay backthrust nappe.

In western Sicily, right lateral strike-slip faulting was a dominant mode of deformation as an accretionary prism was emplaced above the Trapanese SeaMounts. In North Africa, major deformation occurred as the Kabylies collided with and were emplaced upon the continental margin.

East of Corsica and Sardinia, stretching had begun in the region that would become part of the Tyrrhenian Sea (Hsü 1977, Scandone 1979). This extension was a harbinger of a major change in the Africa–Europe pole of rotation (see above.)

7 Ma Late Miocene–Tortonian (Fig. 5e)

A major change in boundary conditions occurred at Anomaly 5 (8.9 Ma) when Africa's motion became directed to the NW (Figs 1 and 3). This enhanced extension in the Tyrrhenian as well as in basins south of Sicily (Pantellaria, Linosa). At this time, the motion of Calabria and northeast Sicily became southeasterly-directed as these blocks began to rotate about an independent pole of rotation (Calabria Pole).

In the northern Apennines, deformation had largely ceased by this time with remaining deformation due essentially to bending (Eldredge *et al.* 1985). Crustal thickening led to anatexis and emplacement of the Elba Granite. In the southern Apennines, Sunda-style tectonics (Beck 1986) resulted in strike-slip faults throughout Calabria and into Sicily. The Calabrian Arc began to deform by heterogeneous simple shear. Along the North African margin and in Tunisia, older thrusts were reactivated along the new transform direction defined by the new Africa–Europe pole of rotation.

5 Ma earliest Pliocene–Zanclean (Fig. 5f)

This frame of our reconstruction depicts the situation immediately after the conclusion of the most dramatic event in Mediterranean evolution, the Messinian Salinity Crisis. During this event, the Mediterranean desiccated completely resulting in deposition of evaporites throughout the Mediterranean and in the Apennine fore-deep (Hsü *et al.* 1972). Blanketing the evaporites, after the refilling of the basin from the Atlantic, is a *Globigerina* ooze (Trubi Formation).

Extension in the Tyrrhenian continued throughout this event with the formation of oceanic crust. The locus of extension migrated to the SE, possibly a result of Calabria advancing through a break in the Apulian Platform into the Ionian Sea (Malinverno & Ryan 1986). In the southern Apennines and Calabria, compression continued accompanied by motion on strike-slip faults. In the northern Apennines, volcanism began in the Tuscany region as a result of anatexis (Barberi *et al.* 1974).

3 Ma–Present, Late Pliocene–Piacenzian to

Through the early Pliocene, the Tyrrhenian Basin continued to form as the locus of extension moved steadily to the SE as shown by 3.5 Ma old basalts in the Vavilov basin (Leg 107 Shipboard Party, 1986). Intramontane grabens developed in the Calabrian Arc and southern Apennines by back-arc extension, gravity collapse of overthickened crust, and strike-slip (pull-apart and compatibility basin) mechanisms.

In northeastern Sicily, right-lateral motion began on an older lineament, the Taormina Line, while compression continued in the southern Apennines and the Sicilian foredeep. Volcanism in the northern Apennines is well developed by this time.

3 Ma–Present Late Pliocene–Piacenzian to Holocene (Fig. 5h)

The locus of Tyrrhenian extension continued to shift to the SE as shown by oceanic basalts less than 2 Ma old in the Marsili Basin (Leg 107 Shipboard Party 1986). The segmentation and southeastward motion of the Calabrian Arc is aided by a series of strike-slip faults that display both left-lateral (southern Apennines, northern Calabria) and right-lateral (southern Calabria, Sicily) offsets. Block rotations, counterclockwise and clockwise respectively, occur along these faults as do pull-apart basins. In front of the Calabrian Arc a large accretionary wedge formed (Kastens 1981, Rossi & Sartori 1981). Behind the arc, in the Aeolian Islands, calc-alkaline and acidic volcanism occurred.

In the southern Apennines, as in the Calabrian Arc, deformation is controlled by strike-slip tectonics. In the southern part of the southern Appennines, these faults are oriented roughly ENE–WSW and to the north N–S. No compressional structures are found in Pliocene deposits around Mount Vulture (Mostardini & Merlini 1986). Thus the southern Apennines appears to be a transition region between the

Calabrian Arc and the slightly older northern Apennines arcuate belt (see 7 Ma above). Alkaline volcanism occurs in the southern Apennines (Roman Province, Mount Vesuvius, Mount Vulture) and in Sicily (Mount Etna).

Structural evolution of the Cetraro region

We have tested some of the structural predictions of our model on the southern Apennine thrust sheets in Calabria. In the Cetraro area (Fig. 6), Triassic to Miocene carbonates of the Panormide Platform are imbricated together and overthrust by sheets of Alpine ophiolite and basement (Dietrich 1976). We have made a detailed structural analysis of the well-exposed lower carbonate units along 36 km of coastline here and extended this inland over an area of 560 km^2 as a reconnaissance survey to check the regional thrust transport vectors. The work will be presented in full elsewhere.

In the coastal sections, the lower, mid-Triassic to lower Miocene, San Donato unit, composed of crystalline limestones, dolomites, carneugles, and pelites is overthrust by the younger upper Triassic to Miocene Verbicaro unit, which contains unmetamorphosed limestones, limestone breccias, and dolomites. Deformation in the San Donato unit is intense and dominated by thrusting. Although we have used a polyphase deformation methodology, we recognize that many of the deformations are related and progressive.

D_1 is a pervasive flat-lying cleavage that has transposed bedding. No folds of this age have been found but the facing/vergence contrasts in D_2 indicate that the earlier deformation entirely inverted the San Donato sequence. D_2 generated another penetrative, gently inclined, schistosity, which is mainly of submylonitic or mylonitic character.

Transport lineations, both here and inland, are constant in orientation with an average 068° trend. Major and minor, recumbent F_2 folds have NW–SE trends in low strain zones and passively rotated NE–SW trends in high strain zones. The sense of D_2 overshear, from the vergence of unrotated F_2 major folds, asymmetric boundins, and pressure shadows is consistently towards the NE. A 15–25 m thick mylonite zone within the San Donato unit, mapped over 30 km, has duplicated stratigraphy, truncated the short limbs of major F_2 folds, and created lateral cut-offs in the footwall. The D_3 and D_4 deformations produced transport-parallel mesoscopic folds with SE and NW vergence respectively, which located over lateral ramps in the thrust system.

Deformation during D_1–D_4 was in the lower greenschist facies (biotite, chlorite). D_5 structures are late, brittle, non-metamorphic thrusts that clearly displace but, in some cases, reactivate the D_2 mylonites. The geometries of ramps and flats, antiformal stacks, small- and large-scale duplexes, blind thrusts, and slickensides indicate that overthrusting was towards the ESE (106°). At Cetraro, this thrust system links into the contact between the San Donato unit and the overlying Verbicaro unit (Verbicaro thrust) and presumably is responsible for the emplacement of the upper unit.

The latest structures are a set of NW–SE-striking (154° average) steeply-inclined, ductile extensional shears and extensional faults. Although both NE and SW dipping conjugates are found the latter are more common.

We note the striking similarity between the structure and stratigraphy that we have described here with that detailed for the Apuane Alps by Carmignani *et al.* (1978). In the latter area, Triassic–Miocene carbonates are found in two nappes: a lower metamorphic Apuane nappe (equivalent to our San Donato unit) and an upper non-metamorphic Tuscan nappe (our Verbicaro unit). In the Apuane area, the metamorphic mylonites have NE (average 069°) transport lineations, NE overshear and the latest deformation is a set of NW–SE extensional faults relating to the Tyrrhenian Sea opening. The deformation sequence is dated as Upper Oligocene to Upper Miocene (27 Ma–10 Ma) (Carmignani *et al.* 1978, Kligfield *et al.* 1986). The dating of the deformation in Calabria is less well known but, from the age of the stratigraphy involved in the thrusting and the predictions of the model, it is within the same time range or only slightly younger.

In relating our deformation geometry and chronology at Cetraro to the larger-scale tectonic model, we have assumed that thrusting in the carbonates initiated at the earliest when the rotating southern extension of the Alpine system (*sensu lato*), the Apennine thrust system, first intercepted the western edge of the Panormide Platform and that the D_6 extensional faults are related to Tyrrhenian extension. Given these assumptions, the structures at Cetraro fit well into the overall model (Fig. 6).

D_1–D_4 overthrusting would have begun at about 20 Ma with an ESE transport lineation roughly normal to the thrust front. A little later (about 18 Ma) the northern kink in this system,

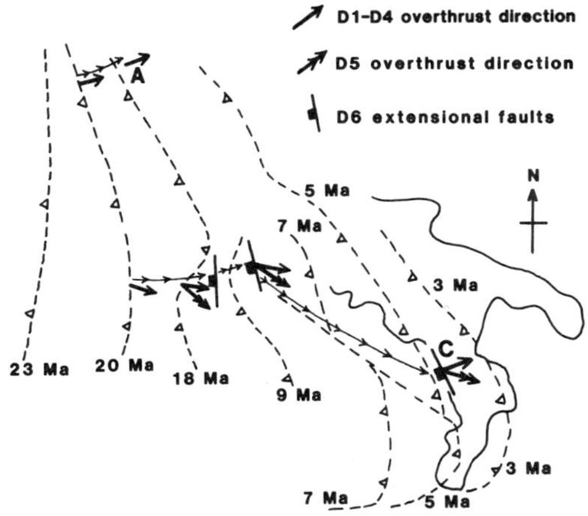

FIG. 6. Evolutionary diagram demonstrating how the structures observed at C (Cetraro area, Calabria) can be related to the sequential development of the Apennine thrust system (rotating southern extension of the Alpine system (*sensu lato*)) (D_1-D_4, D_5) and extension in the Tyrrhenian Sea (D_6). Once formed the structures are passively carried by the gross rotation of the thrust system and later strike-slip displacement. A = Apuane Alps (data from Carmignani *et al*. 1978). Dashed barbed lines = sequential position of the major thrust system. Thin, solid, arrowed lines = locii of movement for Cetraro and Apuane structural packages.

which may relate to the development of the Anzio–Ancona Line, generated SE overthrusting (D_5) about 40° oblique to D_1-D_3 transport. Following this, the structures were cut by extensional faults approximately parallel with the developing Tyrrhenian Sea margins (about 9–10 Ma). During this period, we envisage that each structural package, once formed, is rotated counterclockwise within the thrust system until, at about 7 Ma, the entire structural package is carried SE, mainly by strike slip, into eventual emplacement in Calabria (Fig. 6).

In the Apuane area, we envisage that thrusting in the northern part of the Apennine thrust system initiated in a more northeasterly direction, and somewhat earlier, than in Cetraro. Emplacement of the thrust sheets with a similar transport lineation to that seen further south could have been accomplished by a slower rotation closer to the northern Apennine pole.

Discussion

The western Mediterranean basins evolved between the converging European and African plates. Previous kinematic models have attempted to relate these large plate motions directly to the tectonics of the region. However, Africa's motion with respect to Europe is not always the factor that determines the geological evolution of this region. Rather, as shown by vector triangles in Figs 5(a–g), the motion of individual blocks with respect to each other often outweighs the importance of the Africa–Europe motion in creating the geology we observe today.

During the interval 23 Ma–18 Ma, the direction of Africa–Europe motion is between N and NNE. However, the Africa-Europe side of all vector triangles (Fig. 5c) indicates the least amount of motion. In the east, along the Apennines trend, ESE to ENE directions dominate whereas in the south, along the African margin, the direction of greatest motion is SE to SSE. These directions are in accord with orientations in which we have mapped in Calabria (see above) and which others have mapped in the northern Apennines (Carmignani *et al*. 1978) Tunisia (Carr & Miller 1979), and Algeria (Raymond 1977). Moreover, this time interval is highlighted by the formation of the Gulf of Valencia and the major part of the Algero-Provencal or north Balearic basin. The axes of these basins is generally perpendicular to the vectors Menorca/Africa and Sardinia/Africa respectively (Fig. 5b and c).

During other time intervals, the antithesis of this situation obtains. From Anomaly 5 (8.9 Ma) to the present the motion of Africa with respect to Europe seems to directly control the deformation observed in the western Mediterranean. Compressive features in the front of the Calabrian Arc are oriented perpendicular to the direction of Africa's motion. The axes of the Tyrrhenian Sea, the Linosa graben, and the Pantellaria graben are all parallel with this motion.

The lack of a direct relationship between the motion of Africa and deformation in the western Mediterranean region points out one of the dangers inherent in plate tectonic analyses, namely that knowing the plate kinematics of a region is not a simple panacea that obviously or directly explains all observed geology. However, the kinematics provide a starting point for prediction and for testing and refining the model. The Tyrrhenian extension serves to illustrate this point.

Extension in the Tyrrhenian region began prior to the change in Africa's motion at Anomaly 5 time (Fig. 5d), not after as might be expected. Analysis of our more detailed kinematic data shows why this occurred. Between Anomaly 5AC (13.7 Ma) and Anomaly 5A (11.6 Ma), the pole of rotation began to shift southeastward. This initiated the start of motion toward the NW. Thus the path of Africa's motion is somewhat smoother than we have depicted it and, more importantly, knowledge of the kinematics enables us to predict changes in the geology and to refine the model.

Despite the simple and smooth motion of Africa with respect to Europe (Figs 1a, 3), the geological sequences encountered in the region are complicated. The complexity results from the interaction of various blocks within the collisional realm. The kinematics of the major plates merely provide the framework within which these more complex interactions take place. Nevertheless, coherent, testable models can, as we have shown, be constructed that explain the features and relationships observed. The tasks that remain, with respect to our model, are to expand it areally, by including the Alboran Sea and Betic-Rif Cordillera, and temporally, by shortening the time intervals between some of our reconstructions.

ACKNOWLEDGEMENTS: MLH acknowledges partial financial support from ORS Award 85268 of The Committee of Vice-Chancellors and Principals of the Universities of the United Kingdom. MLH also thanks Walter Pitman, Steve Cande, and Jim Pindell for support, encouragement, and assistance while working on part of this project at Lamont-Doherty Geological Observatory. ET acknowledges the Consiglio Nazionale delle Ricerche (CNR) for financing a sabbatical visit to Oxford University to work on this project. DHWH would like to thank the University of Calabria for funding to do the fieldwork in Italy. SDK acknowledges a Research Studentship Award provided by the Natural Environment Research Council (UK). SDK also thanks Eugenio Turco for his hospitality during time spent in Calabria.

All of the authors greatly appreciate Claire Carlton's drafting skill and patience while making the many revisions on Fig. 5 required during the preparation of this paper.

References

ARGAND, E. 1916. Sur l'arc des Alpes occidentales: *Eclogae Geologicae Helvetiae*, **14**, 145–91.

BAIRD, A. W. & DEWEY, J. F. 1986. Structural evolution in thrust belts and relative plate motion: The Upper Pennine Piemont Zone of the Internal Alps, Southwest Switzerland and Northwest Italy. *Tectonics*, **5**, 375–87.

BARBERI, F., INNOCENTI, F., MARINELLI, G. & MAZZUOLI, R. 1974. Vulcanismo e tettonica a placche: esempi nell' area mediterranea. *Memorie della Societa Geologica Italiana*, **13**, 327–58.

BAYER, R., LEMOUEL, J. L. & LE PICHON, X. 1973. Magnetic anomaly pattern in the western Mediterranean. *Earth and Planetary Science Letters*, **19**, 68–76.

BECK, M. E. 1986. Model for late Mesozoic-early Tertiary tectonics of coastal California and western Mexico and speculations on the origin of the San Andreas Fault. *Tectonics*, **5**, 49–64.

BELLON, H. 1980. Chronologie radiometrique (K–Ar) des manifestations magmatique autour de la Mediterranee occidentale entre 33 et 1 Ma. *In*: WEZEL, F. C. (ed). *Sedimentary Basins of Mediterranean Margins*, C. N. R. Italian Project of Oceanography, Tecnoprint, Bologna, 341–60.

BERGERAT, F. 1987. Stress fields in the European platform at the time of Africa-Eurasia collision. *Tectonics*, **6**, 99–132.

BIJU-DUVAL, B., DERCOURT. J. LEPICHON, X. 1977. From the Tethys Ocean to the Mediterranean Seas: a plate tectonic model of the evolution of the western Alpine system. *In*: BIJU-DUVAL, B. & MONTADERT, L. (eds). *Structural History of the Mediterranean Basins*. Proceedings of an International Symposium, Editions Technip, Paris, 143–64.

BOCQUET, J., DELALOYA, M., HUNZICKER, J. C. & KRUMMENACHER, D. 1974. K–Ar and Rb–Sr dating of blue amphiboles, micas, and associated minerals from the western Alps. *Contributions to Mineralogy and Petrology*, **47**, 7–26.

BONARDI, G., GIUNTA, G., PERRONE, V., RUSSO, M., ZUPPETTA, A. & CIAMPO, G. 1981. Osservazioni sull'evoluzione dell'arco calabro–peloritano nel Miocene inferiore: la formazione di Stilo-Capo d'Orlando. *Bollettino della Societa Geologica Italiana*, **99**, 365–93.

BOULLIN, J. P. 1984. Nouvelle interpretation de la liason Apennin–Maghrebides en Calabre: consequences sur la paleogeographie tethysienne entre Gibraltar et les Alpes. *Revue de Geologie Dynamique et de Geographie Physique*, **25**, 321–38.

BURRUS, J. 1984. Contribution to a geodynamic synthesis of the Provencal Basin (northwestern Mediterranean). *Marine Geology*, **55**, 247–69.

CARMIGNANI, L., GIGLIA, G. & KLIGFIELD, R. 1978. Structural evolution of the Apuane Alps: an example of continental margin deformation in the northern Apennines, Italy. *Journal of Geology*, **86**, 487–504.

CARR, M. D. & MILLER, E. L. 1979. Overthrust emplacement of the Numidian Flysch Complex in the westernmost Modgod Mountains, Tunisia. *Geological Society of America Bulletin*, **90**, 886–923.

CATALANO, R. & D'ARGENIO, B. 1978. An essay of palinspastic restoration across western Sicily. *Geologica Romana*, **17**, 145–59.

CHOUKROUNE, P., BALLEVRE, M., COBBOLD, P., GAUTIER, Y., MERLE, O. & VUICHER, J. P. 1986. Deformation and motion in the western Alpine Arc. *Tectonics*, **5**, 215–26.

COMPAGNONI, R., DAL PIZA, G. V., HUNZIKER, J. C., GOSSO, G., LOMPARDO, B. T. & WILLIAMS, P. F. 1975. The Sesia-Lanzo Zone, a slice of continental crust with Alpine high-pressure, low-temperature assemblages in the western Italian Alps. *Rendiconti della Societa Italiana di Mineralogia e Petrologia*, **33**, 281–334.

D'ARGENIO, B., PESCATORE, T. & SCANDONE, P. 1973. Schema geologico dell'Appennino meridionale (Campania e Lucania): Convegno 'Moderne vedute sulla geologia dell'Apennino'. *Atti Accademia Lincei*, **183**, 49–72.

DEWEY, J. F. 1982. Plate tectonic evolution of the British Isles. *Journal of the Geological Society of London*, **139**, 371–412.

—— PITMAN, W. C. III, RYAN, W. B. F. & BONNIN, J. 1973. Plate tectonics and the evolution of the Alpine systems. *Geological Society of America Bulletin*, **84**, 3,137–80.

DIETRICH, D. 1976. La geologia della catena costiera calabra tra Cetraro e Guardia Piemontese. *Memorie della Societa Geologica Italiana*, **95**, 1–25.

—— 1988. Sense of overthrust shear in the Alpine nappes of Calabria, south Italy. *Journal of Structural Geology*, **10**, 573–81.

DURAND-DELGA, M. 1981. Overtures oceanique de Mediterranee occidentale et dislocation des chaines alpines. *In*: WEZEL, F. C., (ed). *Sedimentary Basins of Mediterranean Margins*. C. N. R. Italian Project of Oceanography, Tecnoprint, Bologna, 417–31.

ELDREDGE, S., BACHTADSE, V. & VAN DER VOO, R. 1985. Paleomagnetism and the orocline hypothesis. *Tectonophysics*, **119**, 153–79.

ERNST, G. W. & DAL PIAZ. G. V. 1978. Mineral paragenesis of eclogitic rocks and related mafic schists of the Piemont ophiolite nappe, Breuil–St Jacques area, Italian western Alps. *American Mineralogist*, **63**, 621–40.

FERRARA, G., BIGGAZI, G., BONADONNE, F. P. & GIULIANI, O. 1972. Radiometric dating of the Valencia volcanic rocks. *In*: RYAN, W. B. F., HSÜ, K. J. *et al.* (eds). *Initial Reports of the Deep Sea Drilling Project*, **13**, 773.

FISCHER, M. W. 1984. Thrust tectonics in the north Pyrenees. *Journal of Structural Geology*, **6**, 721–6.

GIUNTA, G. 1985. Problematiche e ipotesi sul Bacino Numidico nelle Maghrebidi siciliane. *Bollettino della Societa Geologica Italiana*, **104**, 239–56.

GROUP GALICE 1979. The continental margin off Galicia and Portugal: acoustical stratigraphy, dredge stratigraphy, and structural evolution. *In*: RYAN, W. B. F., SIBUET, J. C. *et al.* (eds). *Initial Reports of the Deep Sea Drilling Project*, **47**, 633–62.

HARRIS, B. L. 1985. Direction changes in thrusting the Schistes Lustrés in Alpine Corsica. *Tectonophysics*, **120**, 37–56.

HAXBY, W. F. 1985. The gravity field of the world's oceans. *U. S. Office of Naval Research*.

HEIM, A. 1922. Geologie de Schweiz II(2). *Die Schweizen Alpen II* Leipzig, Tauchnitz, 477–1118.

HSÜ, K. J. 1977. Tectonic evolution of the Mediterranean Sea basins. *In*: NAIRN, A. E. M., KANES, W. H. & STEHLI, F. G. (eds). *The Ocean Basins and Margins*, **4A**, *The eastern Mediterranean*, Plenum Press, New York, 29–75.

——, CITA, M. B. & RYAN, W. B. F. 1972. The origin of the Mediterranean evaporites. *In*: RYAN, W. B. F., HSÜ, K. J. *et al.* (eds). *Initial Reports of the Deep Sea Drilling Project*, **13**, 1,203–31.

KASTENS, K. A. 1981. *Structural causes and sedimentological effects of 'Cobblestone Topography' in the eastern Mediterranean*. Ph. D. Thesis, University of California, San Diego, 207 pp. (unpublished.)

KLIGFIELD, R., HUNZIKER, J., DALLMEYER, R. D. & SCHAMEL, S. 1986. Dating of deformation phases using K–Ar and $^{40}Ar/^{39}Ar$ techniques: results from the northern Apenines. *Journal of Structural Geology*, **8**, 781–98.

KLITGORD, K. D. & SCHOUTEN, H. 1986. Kinematics of the central Atlantic. *In*: VOGT, P. R. & TUCHOLKE, B. E. (eds). *The Geology of North America*, M, *The western north Atlantic Region*, DNAG–GSA Series, 351–78.

KNOTT, S. D. 1987. The Liguride Complex of southern Italy–a Cretaceous to Paleogene accretionary wedge. *Tectonophysics*, **142**, 217–26.

—— 1988. Structure, sedimentology, and petrology of an ophiolitic flysch terrain, Calabria, south Italy. *D. Phil. thesis*, *Oxford University*, 240 pp. (Unpublished).

LADD, J. W. 1976. Relative motion of South America with respect to north America and Caribbean tectonics. *Geological Society of America Bulletin*, **87**, 969–79.

LAUGHTON, A. S., ROBERTS, D. G. & GRAVES, R. 1975. Bathymetry of the northeast Atlantic. Mid-

Atlantic ridge to southwest Europe. *Deep-Sea Research*, **22**, 791–810.

—— & MONAHAN, D. 1978. *General Bathymetric Chart of the Oceans (GEBCO)*, Sheet 5.04, Scale 1:10 000 000, Canadian Hydrographic Service, Ottawa, Canada.

LA VECCHIA, G. 1985. Il sovrascorrimento dei Monti Sibillini analisi cinematica e strutturale. *Bollettino della Societa Geologica Italiana*, **104**, 161–94.

LEG 107 SHIPBOARD PARTY 1986. A microcosm of ocean basin evolution in the Mediterranean. *Nature*, **321**, 383–4.

LE PICHON, X., FRANCHETEAU, J. & BONNIN, J. 1973. *Plate Tectonics*. Elsevier, New York. 300 pp.

LETOUZEY, J. 1986. Cenozoic paleo-stress pattern in the Alpine foreland and structural interpretation in a platform basin. *Tectonophysics*, **132**, 215–31.

LIVERMORE, R. A. & SMITH, A. G. 1985. Some boundary conditions for the evolution of the Mediterranean. *In*: STANLEY, D. G. & WEZEL, F. C. (eds). *Geological Evolution of the Mediterranean Sea*, Springer-Verlag, New York, 89–105.

MALINVERNO, A. & RYAN, W. B. F. 1986. Extension in the Tyrrhenian Sea and shortening in the Apennines as a result of arc migration driven by sinking of the lithosphere. *Tectonics*, **5**, 227–45.

MARTINIS, B. & PAVAN, G. 1967. Note illustrative della Carta Geologica d'Italia alla scala 1:100 000, Foglio 157, Mont Sant' Angelo. *Servizio Geologico d'Italia, Roma*, 56 pp.

MONTIGNY, R., EDEL, J. B. & THUIZAT, R. 1981. Oligo–Miocene rotation of Sardinia: K–Ar ages and paleomagnetic data of Tertiary volcanics. *Earth and Planetary Science Letters*, **54**, 261–71.

MORGAN, J. W. 1968. Rises, trenches, great faults and crustal blocks. *Journal of Geophysical Research*, **73**, 959–82.

MOSTARDINI, F. & MERLINI, S. 1986. Appennino centro-meridionale: sezione geologiche e proposta di modello strutturale. *AGIP, 73rd Congresso della Societa Geologica Italiana, Roma*, 59 pp.

NESTEROFF, W. D., WEZEL, F. C. & PAUTOT, G. 1972. Summary of lithostratigraphic findings and problems. *In*: RYAN, W. B. F., HSÜ K. J. *et al.* (eds). *Initial Reports of the Deep Sea Drilling Project*, Vol. 13, 1021–1040.

OGNIBEN, L. 1964. Arenarie tipo Taveyannaz in Sicilia. *Geologica Romana*, **3**, 125–70.

—— 1969. Schema introduttivo alla geologia del confine calabro-lucano. *Memorie della Societa Geologica Italiana*, **8**, 453–763.

PERRONE, V. 1982. Considerazione sulla trasgressione del Miocene inferiore sulla Piattaforma Carbonatica campano-lucana. *Rendiconti della Societa Geologica Italiana*, **4**, 369–72.

PESCATORE, T. 1978. Evoluzione tettonica del Bacino Irpino (Italia meridionale) durante il Miocene. *Bollettino della Societa Geologica Italiana*, **97**, 783–805.

PINDELL, J. L., CANDE, S. C., PITMAN, W. C., III, ROWLEY, D. B., DEWEY, J. F., LABREQUE, J. & HAXBY, W. 1989. A plate-kinematic framework for models of Caribbean evolution. *Tectonophysics*, in press.

PITMAN, W. C., III & TALWANI, M. 1972. Seafloor spreading in the North Atlantic. *Geological Society of America Bulletin*, **83**, 619–43.

RAMSAY, J. G. & GRAHAM, R. H. 1970. Strain variation in shear belts. *Canadian Journal of Earth Sciences*, **7**, 786–813.

RAYMOND, D. 1977. Structure et evolution alpine d'un segment interne de l'oregene maghrebin: le Nord-Quest de la Grande Kabylie (Algerie). *Bulletin de la Societe Geologique de France*, **19**, 797–804.

ROSSI, S. & SARTORI, R. 1981. A seismic reflection study of the external Calabrian Arc in the northern Ionian Sea (Eastern Mediterranean). *Marine Geophysical Research*, **4**, 403–26.

SAVELLI, C., BECCALUVA, L., BERIU, M., MACCIOTTA, G. & MACCIONI, L. 1979. K–Ar chronology of the Tertiary 'calcalic' volcanism of Sardinia (Italy). *Journal of Volcanology and Geothermal Research*, **5**, 257–69.

SAVOSTIN, L. A., SIBUET, J. -C., ZONENSHAIN, L. P., LE PICHON, X. & ROULET, M.-J. 1986. Kinematic evolution of the Tethys belt from the Atlantic Ocean to the Pamirs since the Triassic. *Tectonophysics*, **123**, 1–35.

SCANDONE, P. 1979. Origin of the Tyrrhenian Sea and the Calabrian Arc. *Bollettino della Societa Geologica Italiana*, **98**, 27–34.

——, GIUNTA, G. & LIGUORI, V. 1974. The connection between the Apulia and Sahara continental margins in the southern Apennines and in Sicily. *24 Congres-Assemblee Plenaire C. I. E. S. M. Committee Geologique et Geophysique Marines, Monte Carlo, 1974*.

SEARLE, R. C., MONAHAN, D. & JOHNSON, G. L. 1975. General Bathymetric Chart of the Oceans (GEBCO), Sheet 5.08, Scale 1:10 000 000, *Canadian Hydrographic Service, Ottawa, Canada*.

SELLI, R. 1962. II Paleogene nel quadro della geologia dell'Italia centro-meridionale. *Memorie della Societa Geologica Italiana*, **3**, 737–89.

SHACKLETON, R. M. & RIES, A. C. 1984. The relation between regionally consistent stretching lineations and plate motions. *Journal of Structural Geology*, **6**, 111–17.

SMITH, A. G. 1971. Alpine deformation and the oceanic areas of the Tethys, Mediterranean, and Atlantic. *Geological Society of America Bulletin*, **82**, 2039–70.

SRIVASTAVA, S. P. & TAPSCOTT, C. R. 1986. Kinematics of the North Atlantic *In*: VOGT, P. R. & TUCHOLKE, B. E. (eds). *The Geology of North America*, M, *The western North Atlantic Region*, DNAG-GSA Series, 379–404.

STAUB, R. 1924. Der Bau des Alpen. *Beitr. Geologie Karte Schweiz*, **52**, 272.

STOCK, J. M. & MOLNAR, P. 1983. Some geometrical aspects of uncertainties in combined plate reconstructions. *Geology*, **11**, 697–701.

TERMIER, P. 1903. Les nappes des Alpes orientales et la synthese des Alpes. *Bulletin de la Societe Geologique de France*, **3**, 711–65.

TRUMPY, R. 1980. *Geology of Switzerland*. Wepf and Co., Basel, 104 pp.

WATERS, C. 1986. Evidence for Eocene strike-slip reactivation of Eoalpine compressive structures, Corsica. *Abstracts, Tectonic Studies Group, Annual Meeting, Hull, UK* 65.

ZIEGLER, W. H. 1975. Outline of the geological history of the North Sea. *In*: WOODLAND, A. W. (ed). *Petroleum and the Continental Shelf of North-West Europe*, **1**, *Geology*, John Wiley and Sons, New York, 165–87.

J. F. DEWEY, M. L. HELMAN & S. D. KNOTT, University of Oxford, Department of Earth Sciences, Parks Road, Oxford OX1 3PR, UK.

E. TURCO, Universita degli Studi di Calabria, Dipartmento di Scienze della Terra, 87030 Castiglione, Cosentino Stazione, Cosenza, Italy.

D. H. W. HUTTON, University of Durham, Department of Geological Sciences, Science Laboratories, South Road, Durham DH1 3LE, UK.

Delamination and asymmetric lithospheric thickening in the development of the Tyrrhenian Rift

J. E. T. Channell & J. C. Mareschal

SUMMARY: Late tectonic extensional basins are superimposed on the continental collision suture at certain points along the Alpine belt. The onset of extension in these basins (Alboran, Tyrrhenian, Pannonian and Aegean) postdates the continental collision by only a few million years. In the case of the Tyrrhenian region, the rifting of the collision suture is accompanied by contemporaneous shortening in the Sicilian/southern Apennine thrust belt. We use finite element models to test whether the rapid evolution of the collision suture into a zone of rifting, and the development of closely juxtaposed regions of compression and extension, can be explained by a mechanism involving collision-induced delamination of continental mantle lithosphere. With delamination underway, the models predict the development of a region of contraction flanked by regions of extension, but no appreciable migration of the delamination front. Investigation of various initial lithospheric geometries shows that a uniformly thickened lithosphere will not readily delaminate, and that delamination is triggered if crustal and lithospheric thickening are offset. Asymmetric, or offset, lithospheric thickening produces an asymmetry in underlying mantle flow which gives rise to extension which is largely restricted to one side of the region of compression. The juxtaposition of extension and shortening in the Tyrrhenian/Calabrian region, as well as the greater degree of extension in the Tyrrhenian relative to the Ionian basin, may be a result of asymmetric lithospheric thickening in the initial collision suture.

Late tectonic Mediterranean 'episutural' rift basins, such as the Tyrrhenian, Pannonian and Aegean (Fig. 1), are superimposed on the continental collision which closed the Mesozoic Tethys; and the beginning of rifting postdates the onset of collision by a few million years (Horvath & Berckhemer 1982). The rifting is accompanied by further contraction in closely adjacent thrust belts. The rapid evolution of the collision suture into regions of extension and adjacent contraction has not been adequately explained. This paper investigates the possibility that 'subduction' of continental margin mantle lithosphere (delamination) after the onset of the continental collision may account for the development of regions of extension and adjacent compression within the collision suture (see Reutter *et al.* 1980, Channell 1986).

Although horizontally layered mantle in continental interiors may be stable over geological time, the inherent instability of mantle lithosphere overlying a less dense asthenosphere may be triggered by faulting or plume erosion (Bird 1979), or by the process of lithospheric thickening in a continental collision zone (Houseman *et al.* 1981, Fleitout & Froidevaux 1982). Bird (1979) estimated the critical viscosity needed to initiate delamination, and proposed continental delamination as an explanation for the intra-plate volcanism and uplift of the Colorado Plateau. Bird & Baumgardner (1981) have shown that mantle delamination events may propagate at rates up to several cm/yr.

The evolution of the western Mediterranean

The western Mediterranean basins (Ligurian, Balearic, Tyrrhenian and Alboran Seas) are late Cenozoic extensional features partly floored by a thin ocean-type crust (see Morelli 1975). The large Bouguer anomaly in the Tyrrhenian Sea (Finetti & Morelli 1973) indicates a thin crust, and the seismicity suggests a northwestward dipping Benioff zone (Gasparini *et al.* 1982) (Fig. 2). Several contrasting models have been proposed for the evolution of the Tyrrhenian and other western Mediterranean basins (e.g. Van Bemmelen 1972, Boccaletti *et al.* 1974, Biju-Duval *et al.* 1976, Scandone 1979, Channell *et al.* 1979, Rehault *et al.* 1985, Wezel 1985, Dercourt *et al.* 1986). It is now generally accepted that the late Oligocene to Miocene development of the Ligurian Sea was the result of western Pacific-type back-arc basin extension during the northwestward-dipping subduction of Mesozoic Tethys beneath the Sardinia/Calabria island arc (Fig. 3). Ligurian extension is contemporaneous with calc-alkaline volcanism on Sardinia and with the rotation of

From: Coward, M. P., Dietrich, D. & Park, R. G. (eds), 1989, *Alpine Tectonics*, Geological Society Special Publication No. 45, pp. 285–302.

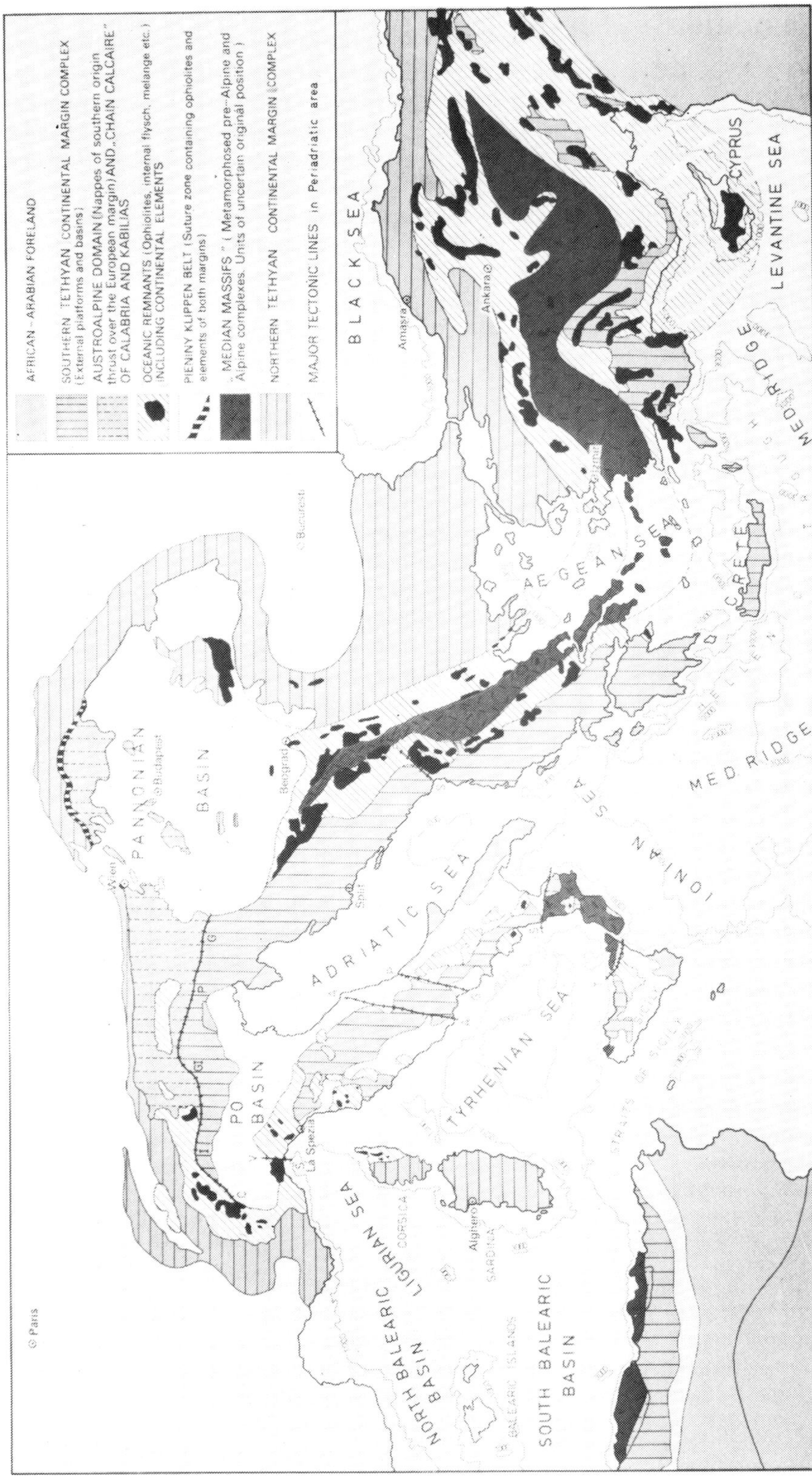

FIG. 1. Structural sketch of the Mediterranean area (after Channell *et al*. 1979).

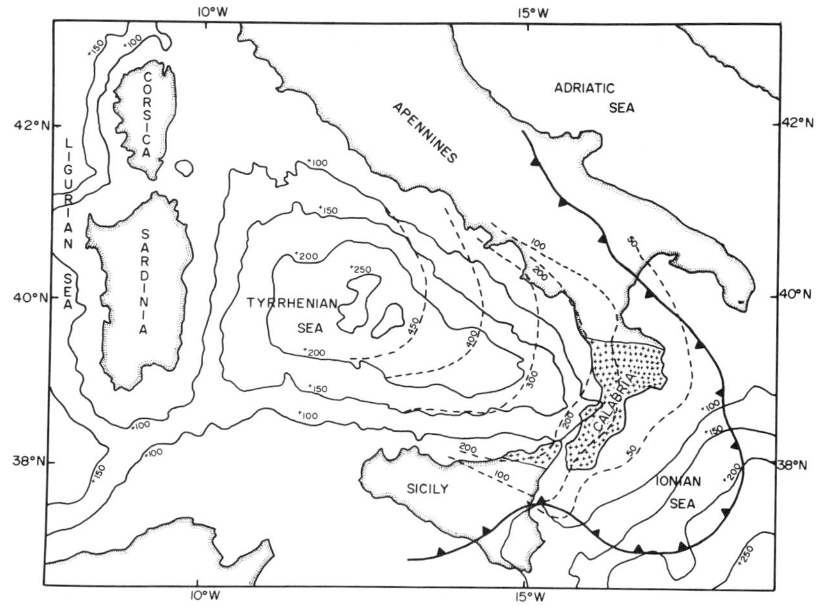

FIG. 2. Sketch map of the Tyrrhenian Sea area indicating the Calabrian massif emplaced on to the deformed Sicilian/Apenninic continental margin. Continuous contour lines represent the Bouguer anomalies in mgals (mainly after Finetti & Morelli 1973). The dashed contours represent the depth in kilometres to the Benioff zone, interpreted from the seismicity (mainly after Gasparini *et al.* 1982).

Sardinia away from the coast of Provence (Fig. 4). It is also generally accepted that the early-middle Miocene collision of this island-arc with the 'African' Mesozoic continental margin of Adria is responsible for most of the Apenninic deformation in the Southern Apennine/Sicilian thrust belt.

There is no clear consensus on the evolution of the western Mediterranean beyond this stage. The onset of rifting which initiated the development of the Tyrrhenian Sea is thought to be mid-Tortonian in age on the basis of the oldest marine sediments overlying a terrigenous pre-rift sequence in the western part of the Tyrrhenian Sea (Leg 107 Scientific Drilling Party 1986). The onset of rifting clearly postdates the Aquitanian to Langhian onset of collision (Fig. 4) of the Sardinia/Calabria island arc with the continental margin, but only by a few million years. Both extension in the Tyrrhenian and contraction in the surrounding thrust belt appear to have been more or less continuous from late Miocene to Recent.

Some of the traditional explanations for the rapid evolution of the Tyrrhenian Sea from a collisional suture zone to a region of extension invoked mantle diapirism, vertical motions and *in situ* oceanization (Van Bemmelen 1972, Selli 1985, Wezel 1985). These scenarios are now unfashionable and are not widely accepted tectonic processes. Other explanations consider the Tyrrhenian Sea as a western Pacific type back-arc basin, and require that the southern Apennine and Sicilian continental margins were separated by a narrow Mesozoic ocean (Mesogea) of which the Ionian Sea is a remnant. In this family of models (Biju-Duval *et al.* 1976, Moussat *et al.* 1985, Rehault *et al.* 1985, Dercourt *et al.* 1986, Malinverno & Ryan 1986, Rehault *et al.* 1987), the initial middle Miocene collision of the Calabria/Sardinia arc with the continental margin is an oblique convergence as the narrow Mesozoic ocean separating the Sicilian and the southern Apennine continental margins is consumed. Subduction of Ionian seafloor continues post-collision, hence Tyrrhenian extension is contemporaneous with subduction of ocean floor, and a W Pacific-type back-arc basin mechanism (see Toksoz & Hsui 1978) can be invoked. Malinverno & Ryan (1986) have proposed that the Sardinia/Calabria arc collided with a re-entrant in the passive margin. The 'rollback' (*sensu* Dewey 1980) of the Tethyan ocean floor trapped in the re-entrant, while the ends of the arc were pinned by the initial collision, is invoked as a means of generating the post-collisional extension in the arc. The model predicts that the onset of

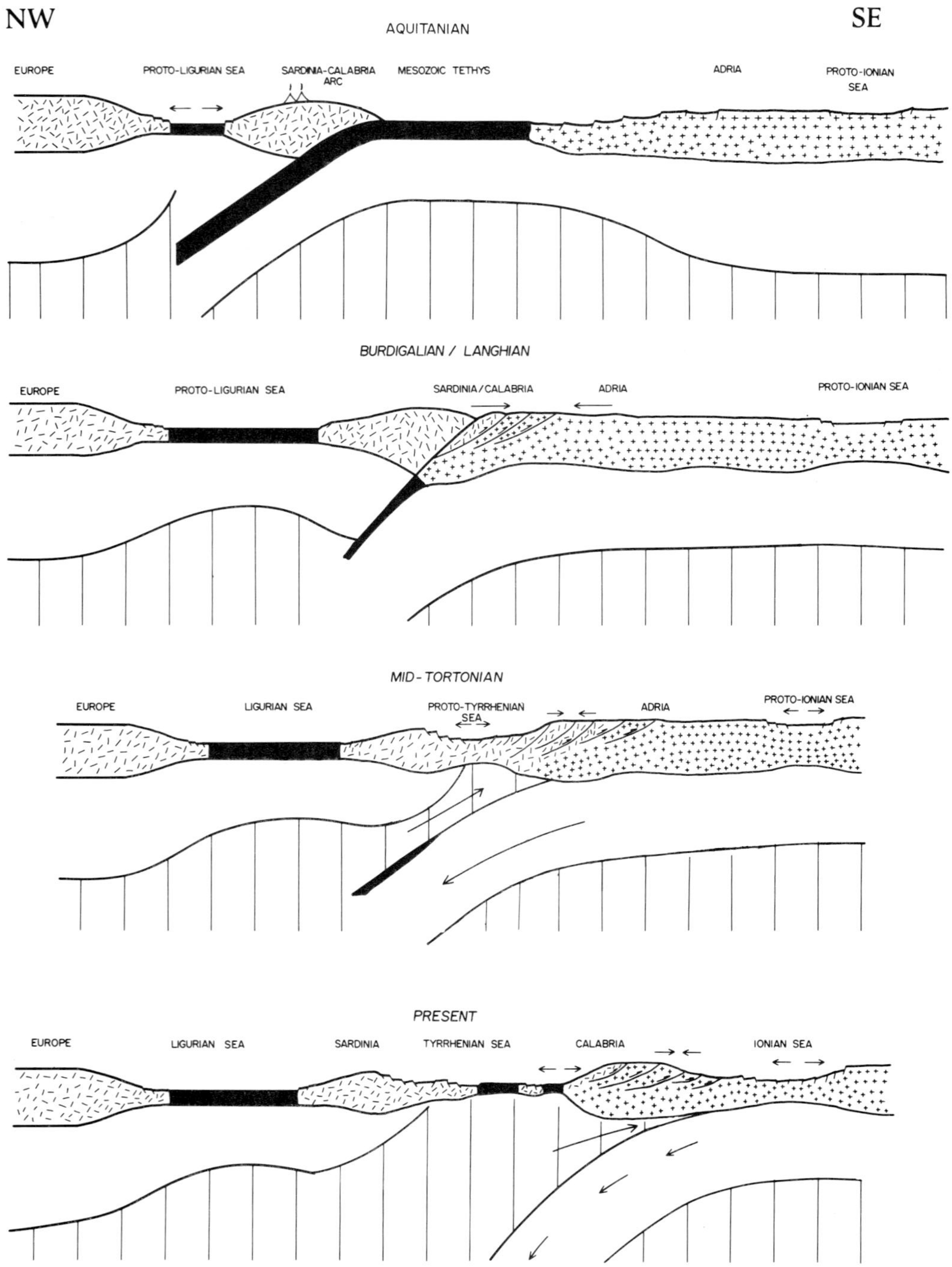

FIG. 3. Schematic cross-section from Provence (France) to the Ionian Sea passing through Sardinia and Calabria, indicating the timing of principal tectonic events and the hypothesized delamination of the continental margin mantle lithosphere.

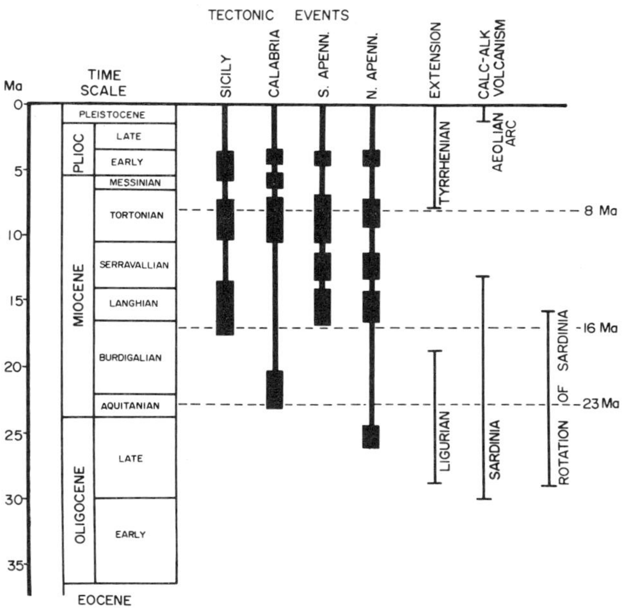

FIG. 4. Timing of tectonic events in the western Mediterranean (modified after Malinverno & Ryan 1986). The time-scale is from Berggren *et al.* (1985). Compressional events after Catalano & D'Argenio (1978), Bonardi *et al.* (1980), Pescatore (1977) and Kligfield *et al.* (1986). Extensional events after Burrus (1984) and Leg 107 Scientific Drilling Party (1986). Volcanic and rotational events after Savelli *et al.* (1979), Barberi *et al.* (1974) and Bellon *et al.* (1977).

collision of the arc with the continental margin was slightly *younger* in central Calabria than in the neighbouring Sicilian and Apenninic parts of the thrust belt, but there is no clear evidence for this. In the southern part of the Calabria–Peloritani arc, the initial collision appears to predate the deposition of the Stilo–Capo d'Orlando Formation of late Aquitanian–Langhian age (Bonardi *et al.* 1980). In western Sicily and in the southern Apennines (N of Calabria), the internal units of the continental margin were deformed from late Burdigalian–Langhian time (Cocco & Pescatore 1975, Pescatore 1977, Catalano & D'Argenio 1978). Therefore the onset of collision in the Calabrian–Peloritani area may be slightly *older* than that in western Sicily and the Southern Apennines (Fig. 4).

The family of models which involves the existence of the 'Mesogea' ocean is inconsistent with the advocated continuity of the Mesozoic continental margin from Sicily, beneath the Calabrian nappes, into the southern Apennines (D'Argenio 1976, Amodio-Morelli *et al.* 1976, Channell *et al.* 1979, Scandone 1979). The observed facies similarities and the analagous timing of development of the two continental margins can be explained either by continuity of the margin or by postulating that they represent a conjugate pair separated by an ocean basin. However, in view of the African origin of the Oligocene Numidian Flysch (Wezel 1970), its occurrence in stratigraphic continuity with the underlying Mesozoic–early Tertiary sediments both on the Mesozoic continental margin of the southern Apennines (Cocco *et al.* 1974) and of Sicily (Catalano & D'Argenio 1978) is a strong argument for the continuity of the flysch basin, and hence of the continental margin. An intervening ocean basin would have inhibited the transport of the 'African' flysch on to the southern Apennine continental margin. In addition, the palaeomagnetic data from the thrust belt in Sicily (Channell *et al.* 1980, 1986) indicate that palaeogeographic units which now trend approximately E–W were oriented almost N–S prior to deformation. The pattern of clockwise rotation of thrust sheets in Sicily and anticlockwise rotations in the southern Apennines (Catalano *et al.* 1976, Channell *et al.* 1980, 1986) can be incorporated in palinspastic reconstructions of the continental margins. These reconstructions show that the arcuate shape of the deformed margin is largely tectonic in origin, and not due to the original shape of the continental margin. Therefore, there is no compelling argument for the presence of a re-entrant or an intervening ocean basin in this part of the Mesozoic continental margin of Adria.

The finite element models

During an episode of continental convergence, the crust and mantle lithosphere are thickened in the collision zone. Horizontal density variations induce stresses that affect the subsequent evolution of the collision belt. The effect of crustal thickening is to induce tensional stresses (Bott 1971, Artyushkov 1973, Fleitout & Froidevaux 1982). In general, these stresses are counteracted by the effect of thickened mantle lithosphere. The mantle lithosphere is colder and denser than the asthenosphere and will be pulled down, thus generating compressional stresses in the collision zone. The sinking of the mantle lithosphere can perpetuate the collision process, and the presence of a thick lithospheric root maintains the convergent belt in a state of compression. Fleitout & Froidevaux (1982) have shown that the density moment:

$$M = \int_0^L g\rho\,(z)\mathrm{d}z$$

determines the stress regime. The moment of the deeper density heterogeneities is larger, and therefore deeper heterogeneities have a more important stress effect. In general, mantle lithospheric thickening has more influence than crustal thickening and will maintain the convergent belt in a state of compression even after convergence has ceased. The removal of the mantle lithosphere, at the end of the convergent episode, would transform the collision zone into a zone of extension. Rapid lithospheric thinning cannot occur by thermal conduction (Mareschal 1983, Turcotte & Emmerman 1983) but can be achieved by convective thinning of the lithosphere (Mareschal 1983, Yuen & Fleitout 1985) as a result of the gravitational instability inherent in the density contrast between mantle lithosphere and asthenosphere. Alternatively, Bird (1979) has proposed that this thinning could occur by delamination of the mantle lithosphere as it is torn away from the crust by a wedge of asthenosphere. Bird & Baumgardner (1981) have shown that a steady state convective regime can be established in the wedge, and therefore suggested that delamination, once triggered, would proceed at a constant rate unless stopped by the breaking of the delaminating mantle lithosphere or by a plate boundary interaction. On the other hand, Houseman *et al.* (1981) have proposed that the thickened lithosphere in convergent belts would either detach and sink as a blob into the mantle or be absorbed by downgoing convection. They modelled the interaction of a thickened thermal boundary layer with a convecting mantle and showed that mantle lithosphere would return rapidly to its normal thickness as the thickened mantle lithosphere is removed by enhanced convection.

We have attempted to determine the flow induced in the mantle by the sinking of delaminating mantle lithosphere in order to determine whether the delamination would propagate, and the nature of the resulting crustal stresses. Determining the mechanical evolution of a gravitationally unstable lithosphere is a complex problem because the crust and mantle materials have differing and strongly temperature-dependent rheologies. We must assume that the evolution of the system can be adequately described by fluid flow equations corresponding to a single (viscous) rheology. We model the mantle lithosphere as a fluid with no flexural rigidity. Creep is a thermally activated process and experimental data show that for olivine, creep would dominate the long time-scale behaviour at temperatures over 1000 K (Carter 1976). We would expect continental mantle lithosphere to behave very differently from oceanic lithosphere, which near the surface is cooler, and therefore behaves more elastically. The crust is modelled as a highly viscous fluid with a viscosity two orders of magnitude greater than that of the asthenosphere. This is clearly unrealistic, and inadequate to model the evolution of crustal thickness, but should allow us to estimate the general nature of crustal stresses produced by mantle flow. The higher viscosity results in a much longer stress relaxation time for the crust than for mantle lithosphere and asthenosphere. Permanent deformations take place on a time scale of the order of 1 Myr and, on that time-scale, the higher viscosity does not mimic an elastic rheology.

The equations and method of solution are summarised in the Appendix, and the values of physical parameters for each model are given in Table 1. The models are two-dimensional, and the region has been made as large as possible to reduce the effect of the boundary conditions. The feasible size of the modelled region is limited by the need to have as fine a grid as possible, and by practical restrictions on computing time. For the four models discussed here, the width of the region is 1200 km, and the depths are 450 km (Model 1), 550 km (Model 2) and 600 km (Models 3 and 4). The

TABLE 1. *Values of physical parameters used in Models 1–4*

		Crust	Mantle Lithosphere	Asthenosphere
Normal thickness (km)		30	80	
Average density (g/cm^3)		2.8	3.2	3.18
Average viscosity (Pa.s)	no. 1	$10^{22.48}$	10^{22}	$10^{21.30}$
	no. 2	10^{23}	$10^{22.48}$	$10^{21.78}$
	no. 3	10^{23}	10^{22}	$10^{21.50}$
	no. 4	10^{23}	10^{22}	$10^{21.50}$
Heat production (W/m^3)		10^{-7}	10^{-8}	0
Thermal diffusivity: 10^{-6} m^2/s				
Activation energy: 8×10^3 kJ/mole/K				

boundary conditions introduce several assumptions into the models: (1) the flow velocity vanishes at the lower boundary, (2) the stress vanishes on the free (upper) surface, (3) there is no flow across the lateral (side) boundaries. This third condition is particularly undesirable and has some unavoidable influence on the size and shape of the convection cells. In addition, because the mantle lithosphere cannot move across the lateral boundaries, the sinking of the slab in the collision zone will induce stretching between the collision zone and the lateral boundary. Similarly, the fixed end of the mantle lithospheric slab has the effect of hinging any delamination of the mantle lithosphere which may occur.

In the first model developed here (Model 1), the initial condition is one in which a mantle lithosphere of constant thickness is already delaminating. The finite element grids for the initial configuration and after 0.87 Myr are shown in Figs 5(a) and 5(b) respectively. The flow velocities after 0.82 and 0.87 Myr are shown in Figs 6(a) and 6(b). Although changes in crustal thickness are hardly discernible in the models due to the high crustal viscosity, zones of crustal extension and contraction can be predicted from calculation of strain rate, and by inspection of the two grids. There is a sharp transition in the crust between a central zone of contraction and two lateral zones of extension. Although a thin wedge of asthenosphere was introduced between crust and mantle (Figs 5 and 6) to permit the propagation of the delamination process, the mantle lithosphere is stretched and necked without appreciable growth of the asthenospheric wedge.

In the second model (Model 2), the initial condition is one of a delaminating thickened lithosphere. The finite element grids for the starting condition and after 1.76 Myr are shown in Figs 7(a) and 7(b), respectively. The flow velocities after 1.55 and 1.76 Myr are shown in Figs 8(a) and 8(b). The thickened lithosphere deforms and sinks like a drop in the asthenosphere rather than maintain its original shape. Increasing the viscosity of the mantle lithosphere does not significantly affect the style of deformation, although the deformation rate is decreased proportionally. As in the first model, no appreciable growth of the thin initial asthenospheric wedge between crust and mantle lithosphere is observed, suggesting that delamination would not propagate. Sharply defined zones of crustal extension and contraction develop (Fig. 8).

The numerical results from these two models suggest that: (1) the thickened mantle lithosphere would sink rapidly into the mantle without appreciable migration of the delamination front, (2) the process will induce sharp contrasts in crustal stress with a zone of contraction flanked by zones of extension, (3) the sinking slab deforms and eventually breaks away from the rest of the lithosphere.

We have investigated several initial lithospheric geometries, in which the delamination process is not underway, to determine the effect of the lithospheric geometry on the evolution of a collision suture and the conditions under which the delamination could take place. Models where the crust and mantle lithosphere were symmetrically and uniformily thickened did not readily evolve to a state of delamination. The stresses induced by crustal thickening counterbalanced the compressional stresses induced by the thickened mantle lithosphere. In order for delamination to proceed, it was necessary to have relatively attenuated crustal thickening, or an offset between crustal and mantle lithospheric thickening. The third model (Model 3) has an initial geometry with mantle lithospheric thickening but no crustal thickening (Fig. 9a). The perturbation of the lithospheric/asthenospheric boundary induces mantle flow (Fig. 9b), and the mantle lithosphere is pulled down into

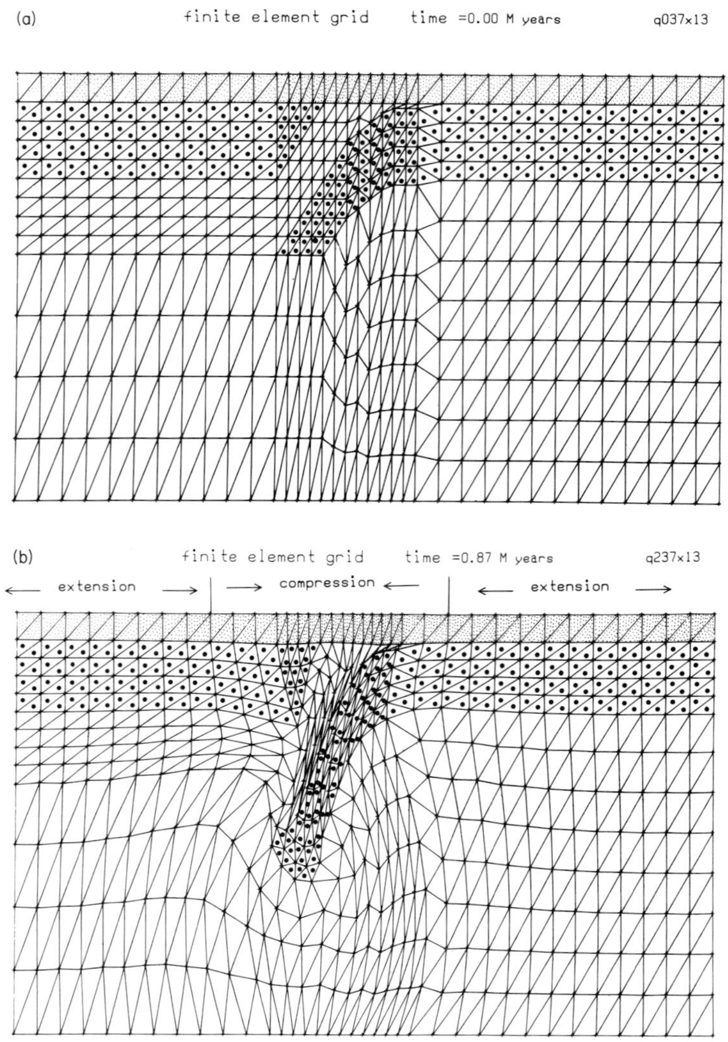

FIG. 5. Model 1: The finite element grid for the three-layer system comprising crust, mantle lithosphere and asthenosphere. (a) In the initial condition, and (b) after 0.87 Myr.

the asthenosphere. The mantle flow generates, as expected, a central region of crustal compression flanked by two regions of crustal extension. When crustal thickening is introduced into this model, crustal compressional in the central region is reduced because it is now opposed by the tension induced by a thickened crust. The balance between crustal and mantle lithospheric thickening will determine the state of stress in the collision zone crust.

In the last model (Model 4), the crustal thickening is displaced (offset) relative to the mantle lithospheric thickening (Fig. 10a), and the induced mantle flow is asymmetric (Fig. 10b). Delamination occurs as the mantle lithosphere is pulled down into the underlying asthenosphere. The resulting crustal stresses (Fig. 11) are strongly tensional W (left) of the collision suture. As the rheology used in these models is inappropriate for the crust, the computed stress magnitudes are not realistic, but the calculations give the general pattern of principal stress directions. As the model is allowed to evolve further (Fig. 10c), the central part of the col-

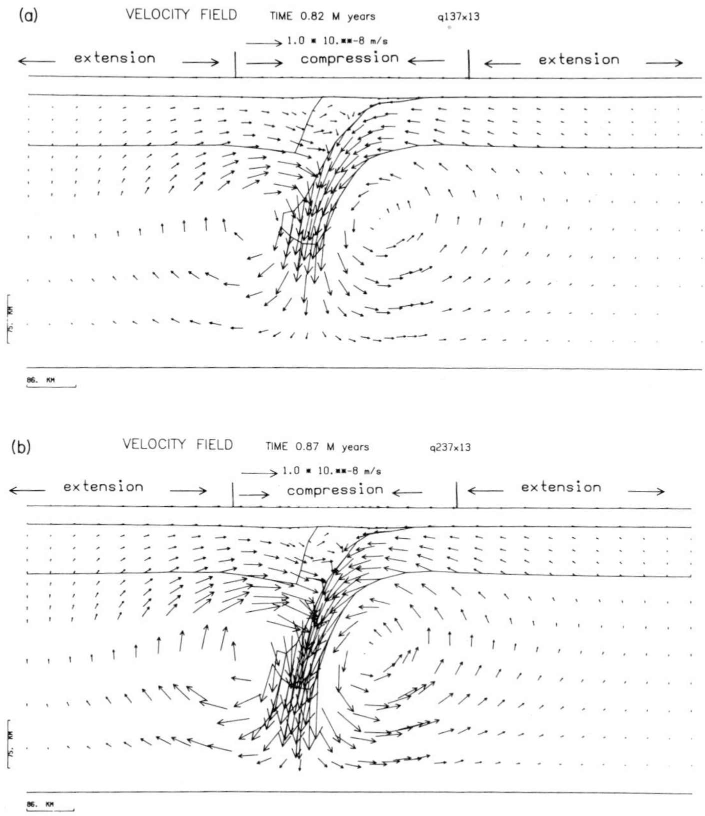

FIG. 6. Model 1: The flow pattern after, (a) 0.82, and (b) 0.87 Myr.

lision suture continues to be strongly compressional with an apparent eastward vergence. The tensional stress in the region E (right) of the suture is lower than that to the W, and the zones of tension and compression migrate eastward.

Discussion

Models 1 and 2 do not show any appreciable migration of the delamination front, in contrast to the predictions of Bird & Baumgardner (1981) who modelled the convection within the asthenospheric wedge and assumed a moving reference frame (i.e. that the hinge of the delaminating mantle lithosphere is moving). The establishment of a steady-state convective regime in the wedge led the authors to conclude that the migration of mantle lithospheric delamination is a viable process. The scale of our models is such that small-scale convection within the asthenospheric wedge, which may be important for migration of the mantle lithospheric delamination, cannot be resolved. The use of a much finer grid in the asthenospheric wedge did not succeed in demonstrating that the delamination propagates, nor did changes in the initial geometry. However, no firm conclusion can be drawn because the non-migration of the delamination may be an artefact of the lateral boundary conditions which maintain the configuration of the convection cells, and thus inhibit migration of the delamination front.

These numerical models show that flow induced by mantle lithospheric delamination will give rise to a region of crustal contraction flanked by two regions of extension. This juxtaposition of crustal extension and contraction during the evolution of collision zones has been predicted by many workers including McKenzie (1978) and Fleitout & Froidevaux (1982). In the

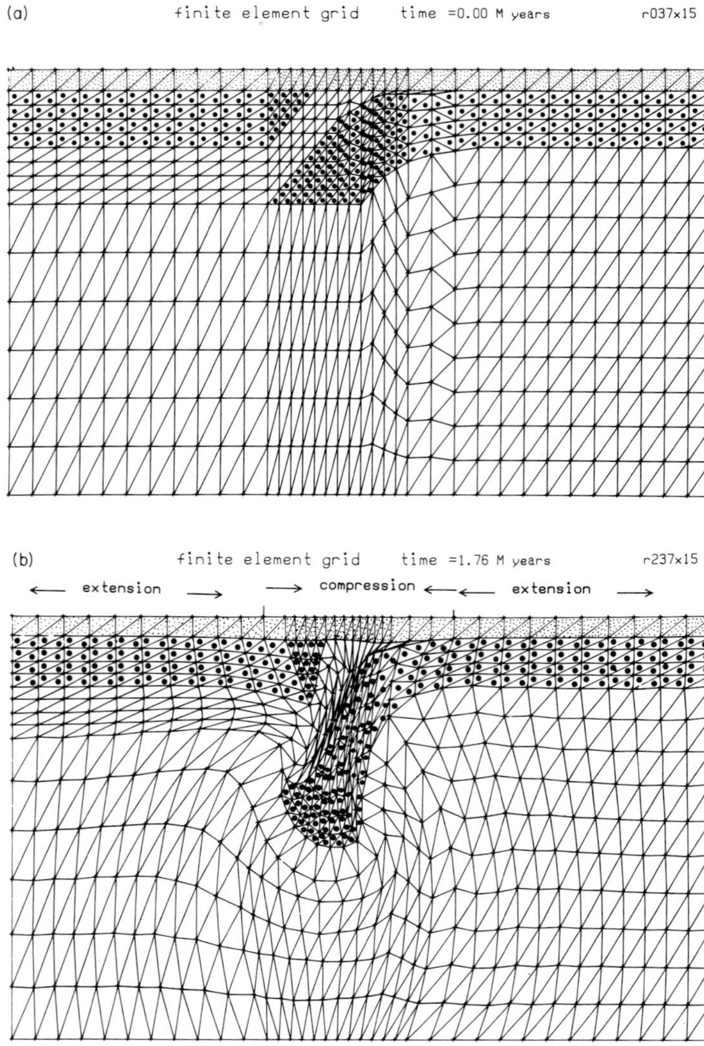

FIG. 7. Model 2: The finite element grid. (a) In the initial condition, and (b) after 1.76 Myr.

present study, the region of contraction corresponds to the Sicilian/Calabrian/Southern Apennine thrust belt and the regions of extension correspond to the Tyrrhenian and Ionian basins. As discussed above, the extension in the Tyrrhenian postdates the onset of collision between the Sardinia/Calabria arc and the continental margin of Adria. The Ionian Sea is characterized by basinal sedimentation from the Mesozoic, but appears to have undergone accelerated subsidence from early Tortonian time and is characterized by Neogene to Quaternary volcanism (Finetti 1985, Makris *et al.* 1986). The very thin basement (8 km), high Bouguer anomaly and low heat flow in the central part of the Ionian Sea has led most workers to consider that this basement is Mesozoic in age, and oceanic in nature (Hinz 1974, Finetti 1985, Makris *et al.* 1986). It appears to have a thick Mesozoic and Cenozoic sedimentary cover, and may be a part of Africa affected by oceanic rifting and continental margin subsidence. Recent seismic records have been interpreted to indicate that the subsidence of the Ionian basin relative to the Straits of Sicily (Fig. 1), with the formation of the Malta

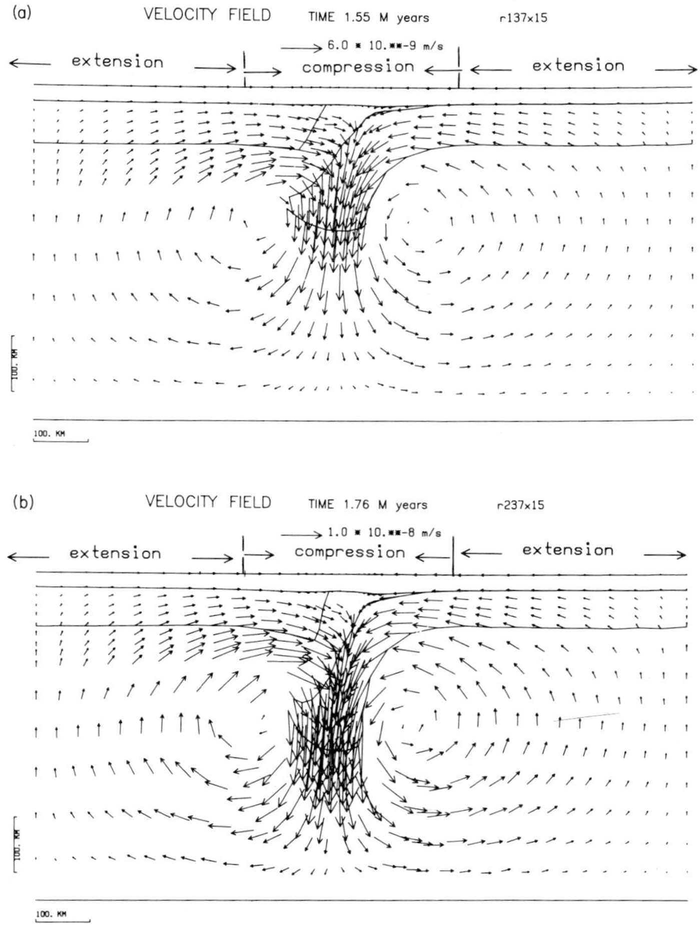

FIG. 8. Model 2: The flow pattern after, (a) 1.55 Ma, and (b) 1.76 Myr.

escarpment, occurred largely since early Tortonian time (Makris *et al.* 1986). Therefore, although part of the subsidence of the Tyrrhenian and Ionian Seas is contemporaneous, Tyrrhenian extension evolved much more rapidly than that in the Ionian Sea. Clearly, the symmetric extension either side of the region of contraction, shown in Models 1–3, did not occur. We suggest that the higher rate of extension for the Tyrrhenian Sea may be the result of the influence of initial lithospheric thickness variations in the convergent zone.

Models 3 and 4 indicate that asymmetric (offset) lithospheric thickening is required to generate asymmetric mantle flow (cf. Figs 9 and 10). The amount of offset in crustal and mantle lithospheric thickening was found to be non-critical in determining the general flow pattern. In each situation, the flow pattern and style of deformation were found to be resilient to change in physical parameters, although these parameters affect the rate of deformation. The asymmetric mantle flow (Fig. 10) gives rise to an asymmetry in crustal stress with a region of high tensional stress on the side of the collision suture characterised by thickened crust (Figs 10 and 11). The models suggest that the triggering of the delamination process, and the enhanced extension on the Tyrrhenian side of the Calabrian thrust belt, can be explained by asymmetric lithospheric thickening in the collision zone. Asymmetric lithospheric thickening

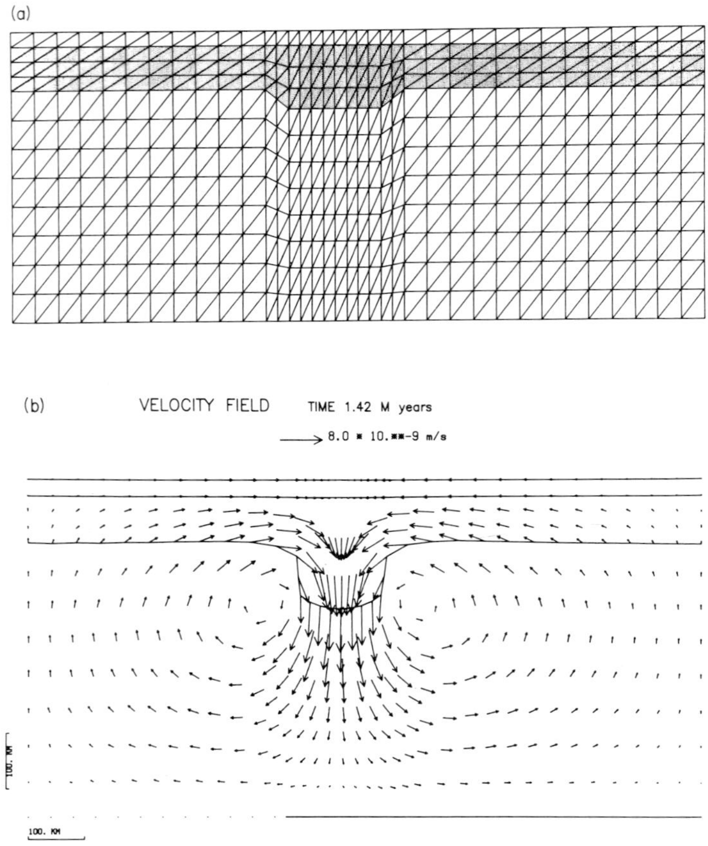

FIG. 9. Model 3: (a) The starting geometry is one of thickened mantle lithosphere with no crustal thickening. The zone of mantle lithospheric thickening is 240 km wide, and the thickening is from 120 to 160 km in the thickened zone. (b) The model is allowed to evolve from this starting geometry for 1.42 Myr, after which a symmetric flow velocity pattern is established.

is conceivable in this part of the collision belt. The collision was between a young island arc (Calabria/Sardinia) and an early Jurassic passive margin (Adria). The island arc was active until the onset of collision (16 Ma) and had been affected by late Cretaceous/Eocene 'Alpine' orogeny. We may expect, therefore, that the island arc had thick crust but relatively thin mantle lithosphere. On the other hand, the Jurassic passive margin which is characterized by carbonate platforms and intervening pelagic basins (with minor terrigenous influx) may have had a thin crust and a well developed mantle lithosphere. The collision of the two contrasting lithospheres may have produced asymmetric lithospheric thickening.

The numerical models demonstrate that the lithospheric geometry in the collision zone is critical to the subsequent evolution of the collision suture. The importance of density heterogeneities and thickness variations in the lithosphere as a source of crustal stress has been recognized for many years (e.g. Ramberg 1968, Artyushkov 1973), as has the importance of the Alpine 'verschluckungszone' (mantle lithospheric root) (Laubscher 1971). Surface wave data in the Swiss Alps indicates a thick mantle lithospheric root (Mueller 1983, Mueller & Panza 1986) which may contribute to maintaining the collision belt in a state of compression, and may account for the radial distribution of maximum principal stress directions in the western and central Alps (see Pavoni 1975). In the case of Tyrrhenian Sea, we propose that asymmetric lithospheric thickening gave rise to the rapid evolution of the collision zone into a

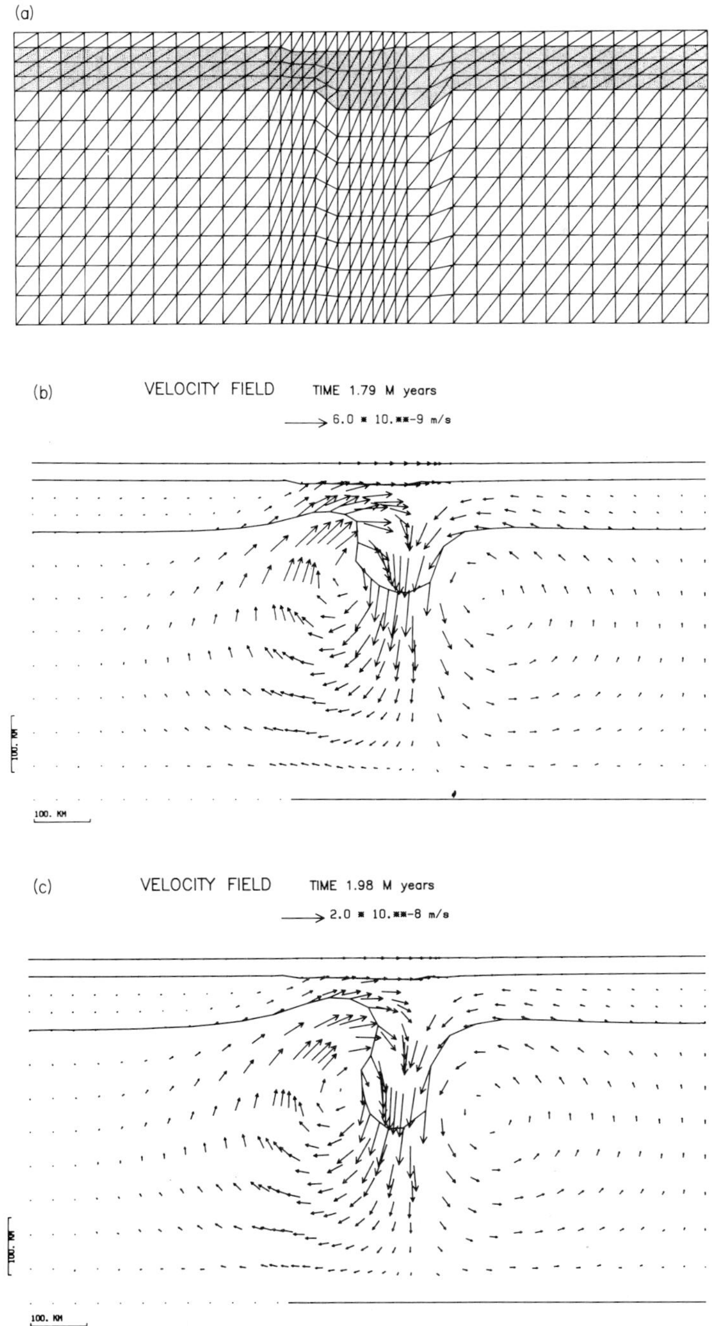

FIG. 10. Model 4: (a) The starting geometry is one of a thickened mantle lithosphere offset from a thickened crust. The thickened crust is offset to the W (left) by 60 km relative to the thickened mantle lithosphere. (b) The model is evolved for 1.79 Myr, and (c) for a further 190 kyr to 1.98 Myr.

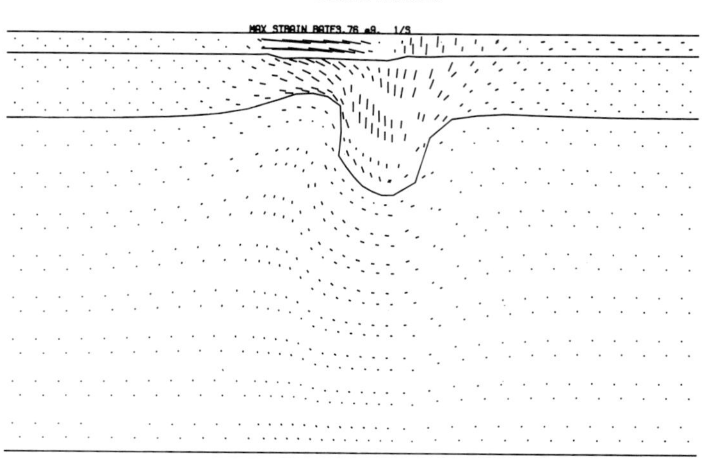

FIG. 11. Stress pattern corresponding to the flow velocities in Fig. 10(b). The bars indicate the direction and magnitude of the maximum tensional stress (maximum compression is perpendicular to the bars). A region of tension is flanked to the right (E) by a region of compression. This juxtaposition is analagous to the Tyrrhenian Sea and surrounding southern Apennine and Sicilian fold and thrust belt.

zone of extension flanked on one side by an outwardly vergent thrust belt. This mechanism can account for the relatively minor contemporaneous extension in the Ionian Sea and for the eastward migration both of Tyrrhenian extension and E-vergent thrusting. The same conditions may have occurred elsewhere, and they may explain the extension and contemporaneous compression associated with other Mediterranean 'episutural' basins. The location of these basins along the Alpine collision belt may coincide with segments of the collision suture characterized by asymmetric lithospheric thickening.

ACKNOWLEDGEMENTS: We acknowledge lively and constructive debate with Alberto Malinverno, Bruno D'Argenio and John Oldow. This work was partly supported by a NSERC operating grant to J. C. M. NSERC also provided computer time on the CRAY supercomputer at Dorval.

References

AMODIO-MORELLI M., BONARDI G., COLONNA V. *et al.* 1976. L'arco Calabro–Peloritano nell'orogene Appenninico–Maghrebide. *Memorie della Societa Geologica Italiana*, **17**, 1–60.

ARTYUSHKOV, E. V. 1973. Stresses in the lithosphere caused by crustal thickness inhomogeneities. *Journal of Geophysical Research*, **78**, 7675–708.

BARBERI F., INNOCENTI F., FERRARA G., KELLER J. & VILLARI L. 1974. Evolution of Eolian arc volcanism (southern Tyrrhenian Sea). *Earth & Planetary Science Letters*, **21**, 269–76.

BELLON, H., COULON, C & EDEL, J. B., Le deplacement de la Sardaigne. Synthese des donnees geochronologiques, magmatiques et paleomagnetiques. *Bulletin de la Societé Geologique de la France*, (7) XIX, 825–31.

BERGGREN W. A., KENT D. V. & VAN COUVERING J. A. 1985. The Neogene: Part 2. Neogene geochronology and chronostratigraphy. *In*: SHELLING N. J. (ed). *The Chronology of the Geologic Record.* Memoir of the Geological Society of London, No. 10, 211–60.

BIJU-DUVAL B., DERCOURT J. & LE PICHON, X. 1976. From the Tethys ocean to the Mediterranean seas: a plate tectonic model of the evolution of the western Alpine system. *In*: BIJU-DUVAL, B. & MONTADERT L. (eds). *International Symposium on the Structural History of the Mediterranean Basins. Split (Yugoslavia).* Editions Technip, Paris. 143–64.

BIRD, P. 1979. Continental delamination and the Colorado Plateau. *Journal of Geophysical Re-*

search, **84**, 7561–71.

—— & BAUMGARDNER, J. 1981. Steady propagation of delamination events. *Journal of Geophysical Research*, **86**, 4891–903.

BOCCALETTI, M., GUAZZONE, G. & MANETTI P. 1974. Evoluzione Paleografica e Geodinamica del Mediterraneo i Bacini Marginali. *Memorie della Societa Geologica Italiana*, **13(2)**, 1–39.

BONARDI, G., GIUNTA, G., PERRONE, V., RUSSO, M., ZUPPETTA, A. & CIAMPO, G. 1980. Osservazioni sull'evoluzione dell'arco Calabro-Peloritano nel Miocene inferiore. La formazione di Stilo-Capo d'Orlando. *Bulletin della Societa Geologica Italiana*, **99**, 365–93.

BOTT, M. H. P. 1971. Evolution of young continental margins and formation of shelf basins. *Tectonophysics*, **11**, 319–27.

BURRUS, J. 1984. Contribution to a geodynamic synthesis of the Provençal basin (north-western Mediterranean). *Marine Geology*, **55**, 247–69.

CARTER, N. L. 1976. Steady state flow of rocks. *Reviews of Geophysics and Space Physics*, **14**, 301–60.

CATALANO, R., CHANNELL J. E. T., D'ARGENIO, B. & NAPOLEONE, G. 1976. Mesozoic paleogeography of the southern Appennines and Sicily – problems of paleotectonics and paleomagnetism. *Memorie della Societa Geologica Italiana*, **15**, 95–118.

—— & D'ARGENIO, B. 1978. Facies analysis and paleogeography in western Sicily. *Geologica Roma*, **17**, 145–59.

CHANNELL, J. E. T. 1986. Paleomagnetism and continental collision in the Alpine Belt and the formation of latetectonic extensional basins. *In*: COWARD, M. P. & RIES, A. C. (eds). *Collision Tectonics*. Special Publication of the Geological Society of London, No. 19. 261–84.

——, D'ARGENIO, B. & HORVATH, F. 1979. Adria, the African promontory, in Mesozoic Mediterranean Palaeogeography. *Earth Science Reviews*, **15**, 213–92.

——, CATALANO, R. & D'ARGENIO B. 1980. Paleomagnetism and deformation of the Mesozoic continental margin in Sicily. *Tectonophysics*, **61**, 391–407.

——, OLDOW, J., CATALANO, R. & D'ARGENIO, B. 1986. Paleomagnetically deduced rotations in the west Sicilian thrust belt. *Geological Society of America Bulletin* (Abs. with Programs), **18(6)**, p. 562.

COCCO, E., CRAVERO, E., ORTOLANI, F., PESCATORE, T. S., RUSSO, M., TORRE, M. & COPPOLA, L. 1974. Le unita Irpine nell'area a Nord di Monte Marzano, Appenino Meridionale. *Memorie della Societa Geologica Italiana*, **13**, 607–54.

—— & PESCATORE, T. S. 1975. Facies pattern of the southern Apennines flysch troughs. *In*: SQUYRES, C. (ed). *Geology of Italy, Earth Science Society of Libya*. 289–303.

D'ARGENIO, B. 1976. Le piattaforme carbonatiche Periadriatiche una rassegna di problemi nel quadro geodinamico Mesozoico dell'area Mediterranea. *Memorie della Societa Geologica Italiana*, **13(2)**, 1–28.

DERCOURT, J., *et al*. 1986. Geologic evolution of the Tethys belt from the Atlantic to the Tamirs since the Lias. *Tectonophysics*, **123**, 241–315.

DEWEY J. F. 1980. Episodicity, sequence and style at convergent plate boundaries, *In*: STRANGEWAY, D. W. (ed). *The Continental Crust and its Mineral Deposits*, Geological Society of Canada Special Paper, **20**, 553–73.

FINETTI, I. 1985. Structure and evolution of the central Mediterranean (Pelagian and Ionian Seas). *In*: STANLEY, D. J. & WEZEL, F. -C. (eds). *Geological Evolution of the Mediterranean Basin*, 215–30. Springer-Verlag, Berlin.

—— & MORELLI, C. 1973. Geophysical exploration of the Mediterranean Sea. *Bolletino Geofisica, Teoretica e Applicata*, **15**, 263–340.

FLEITOUT, L. & FROIDEVAUX, C. 1982. Tectonics and topography for a lithosphere containing density heterogeneities. *Tectonics*, **1(1)**, 21–56.

GASPARINI, C., IANNACCONE, G., SCANDONE, P. & SCARPA, R. 1982. Seismotectonics of the Calabrian Arc. *Tectonophysics*, **84**, 267–86.

HINZ, K. 1974. Results of seismic refraction and reflection measurements in the Tyrrhenian Sea. *Geologisches Jahrbuch* 33–65.

HORVATH, F. & BERCKHEMER, H. 1982. Mediterranean back arc basins. *In*: BERCKHEMER, H. & HSÜ, K. (eds). *Alpine-Mediterranean Geodynamics*, 141–74. American Geophysical Union, Washington DC.

HOUSEMAN, G. A., MCKENZIE, D. P. & MOLNAR P. 1981. Convective instability of a thickened boundary layer and its relevance for the thermal evolution of continental convergent belts. *Journal of Geophysical Research*, **86**, 6115–32.

HUEBNER, K. 1975. *The Finite Element Method for Engineers*. Wiley (Interscience), New York.

KLIGFIELD, R., HUNZIKER, J., DALLMEYER, R. D. & SCHAMEL, S. 1986. Dating of deformation phases using K–Ar and $^{40}Ar/^{39}Ar$ techniques: results from the Northern Apennines. *Journal of Structural Geology*, **8(7)**, 781–98.

LAUBSCHER, H. P. 1971. The large-scale kinematics of the western Alps and the northern Apennines and its palinspastic implications. *American Journal of Science*, **271**, 193–226.

Leg 107 Scientific Drilling Party. 1986. A case study of a newly-rifted oceanic basin: preliminary results from ODP Leg 107 in the Tyrrhenian Sea. *Geotimes*, **31:8**, 11–14.

MAKRIS, J., NICOLICH, R. & WEIGEL, W. 1986. A seismic study in the western Ionian Sea. *Annales Geophysicae*, **6**, 665–78.

MALINVERNO, A. & RYAN, W. B. F. 1986. Extension in the Tyrrhenian Sea and shortening in the Apennines as a result of arc migration driven by sinking of the lithosphere. *Tectonics*, **5(2)**, 227–46.

MARESCHAL, J. C. 1983. Mechanisms of uplift preceding rifting. *Tectonophysics*, **94**, 51–66.

—— & WEST, G. F. 1980. A model for Archean tectonism, Part II, numerical models of vertical tectonism in greenstone belts. *Canadian Journal of Earth Sciences*, **17**, 60–71.

McKenzie, D. 1978. Active tectonics of the Alpine–Himalayan belt: the Aegean Sea and surrounding regions. *Geophysical Journal of the Royal Astronomical Society*, **55**, 217–54.

Morelli, C. 1975. Geophysics of the Mediterranean, *Newslett. C.I.M. Monaco Spec. Iss.*, **7**, 27–111.

Moussat, E., Rehault, J. P., Fabbri, A. & Mascle, G. 1985. Evolution geologique de la Mer Tyrrhenienne. *Comptes Rendus del' Academie des Sciences, Paris,* 301, Serie II, **7**, 491–96.

Mueller, St. 1983. Deep structure and recent dynamics in the Alps. *In*: Hsü, K. J. (ed). *Mountain Building Processes*, 181–99. Academic Press, London.

Mueller, S. & Panza, G. F. 1986. Evidence of a deep-reaching lithospheric root under the Alpine Arc. *In*: Wezel, F. C. (ed.) *The Origin of Arcs*, 93–113. Elsevier, Amsterdam.

Pescatore, T. 1977. The Irpinids: a model of tectonically controlled fan and base of slope sedimentation in southern Italy. *In*: Stanley, J. & Kelling, G. (eds). *Sedimentation in Submarine Canyons, Fans and Trenches*, 325–39. Dowsden, Hutchinson & Ross, New York.

Ramberg, H. 1968. Instability of layered systems in the field of gravity. I and II. *Physics of Earth & Planetary Interiors*, **1**, 427–74.

Rehault, J.-P, Boillot, G. & Mauffret, A. 1985. The western Mediterranean Basin, *In*: Stanley, D. J. & Wezel, F. -C. (eds). *Geological Evolution of the Mediterranean Basin*, 101–30. Springer-Verlag, Berlin.

——, Moussat, E. & Fabbri, A. 1987. Structural evolution of the Tyrrhenian back-arc basin. *Marine Geology*, **74**, 123–50.

Reutter, K. -J., Griese, P. & Closs, H. 1980. Lithospheric split in the descending plate: observations from the northern Apennines. *Tectonophysics*, **64**, T1–T9.

Savelli, C., Beccaluva, L., Deriu, M., Macciotta, G. & Maccioni L. 1979. K–Ar geochronology of the Tertiary 'calc-alkalic' volcanism of Sardinia (Italy). *Journal of Volcanology & Geothermal Research*, **5**, 257–69.

Scandone, P. 1979. Origin of the Tyrrhenian Sea and Calabrian Arc. *Bolletino della Societa Geologica Italiana*, **98**, 27–34.

Selli, R. 1985. Tectonic evolution of the Tyrrhenian Sea. *In*: Stanley, D. J. & Wezel, F. -C. (eds). *Geological Evolution of the Mediterranean Basin*, 131–52. Springer-Verlag, Berlin.

Strang, G. & Fix, G. 1973. *Analysis of the Finite Element Method*. Prentice-Hall, Englewood Cliffs, New Jersey.

Toksoz, M. N. & Hsui, A. T. 1978. Numerical studies of back-arc convection and the formation of marginal basins. *Tectonophysics*, **50**, 177–96.

Turcotte, D. L. & Emerman, S. H. 1983. Mechanisms of active and passive rifting. *Tectonophysics*, **94**, 39–50.

Van Bemmelen, R. W. 1972. Driving forces of Mediterranean orogeny (Tyrrhenian test-case). *Geologie en Mijnbouw*, **51**, 548–73.

Wezel, F. C. 1970. Numidian Flysch: an Oligocene–early Miocene continental rise deposit off the African platform. *Nature*, **228**, p. 275.

—— 1985. Structural features and basin tectonics of the Tyrrhenian Sea. *In*: Stanley, D. J. & Wezel, F. -C. (eds). *Geological Evolution of the Mediterranean Basin*, 153–94. Springer-Verlag, Berlin.

Yuen, D. & Fleitout, L. 1985. Thinning of the lithosphere by small scale convective destabilization. *Nature*, **313**, 125–8.

J. E. T. Channell, Department of Geology, University of Florida, Gainesville, FL 32611, USA.

J. C. Mareschal, Départment des Sciences de la Terre, Université du Québec à Montréal, Montréal PQ, H3C 3P8, Canada.

Appendix: The basic equations and the numerical method

The mechanical evolution of the lithosphere–asthenosphere system is determined by the equations of mass and momentum conservation. For an incompressible material, the continuity (mass conservation) equation takes the form:

$$\nabla.\bar{v} = 0$$

The equation of motion (momentum conservation) can be written as:

$$\rho\frac{d\bar{v}}{dt} = \nabla T + \rho\bar{g}$$

where T is the stress tensor, $\bar{g}$ is the acceleration of gravity, and $\bar{v}$ is the velocity.

It is necessary to add to this equation a constitutive equation to describe the rheology of the material. Although experimental studies on the rheology of rocks indicate that their behaviour under stress is best described by a non-linear creep law (Carter 1976), a linear constitutive equation was used in order to keep the computer time within reasonable limits.

For a newtonian fluid, the stress-rate of strain relationship is given by:

$$T = -PI + 2\mu D$$

where P is the pressure, I is the unit tensor, and D is the strain rate:

$$D = \tfrac{1}{2}\{\nabla\bar{v} + \nabla\bar{v}^{r}\}$$

The kinematic viscosity varies with temperatures as:

$$\mu = \mu_0 \exp\{\alpha/T\}$$

Introducing the constitutive equation into the momentum conservation equation yields the Navier–Stokes equation:

$$\rho\frac{d\bar{v}}{dt} = \mu\nabla^2\bar{v} + \rho\bar{g} - \nabla P$$

In most geological situations, the Reynolds number (i.e. the ratio of inertial to viscous forces) is extremely low, and the inertial term in the momentum equation can be neglected. The Navier–Stokes equation then becomes:

$$\mu\nabla^2\bar{v} + \rho\bar{g} - \nabla P = 0$$

The thermal evolution is determined by the heat equation:

$$\frac{dT}{dt} = \frac{\delta T}{\delta t} + \bar{v}\nabla T = \varkappa\nabla^2 T + \frac{S}{\rho C}$$

where T is temperature, C is the thermal capacity, S is the distribution of heat sources and $\bar{v}$ is velocity, $\varkappa$ is thermal diffusivity.

Although the thermal problem is solved and there is a formal dependance of the rheology on the temperature, the time-scales involved are much too short for the temperature to change significantly.

The numerical method used to solve the partial differential equations describing the flow and the temperature is the finite element method (e.g. Strang & Fix 1973, Huebner 1975).

The principle of the method is the following. In order to solve the equation:

$$O f = g \text{ in a domain } \Omega$$

where O is a differential operator, g is a known function, and f is the unknown function, subject to boundary conditions. f can be expanded in terms of a set of known functions:

$$f = \sum_{k=1}^{N} f_k \alpha_k$$

where $\{\alpha_k,\ k = 1, \ldots N\}$ is the set of known functions.

The coefficients of the expansion must be determined in such a way that:

$$O \sum_{k=1}^{N} f_k \alpha_k = g$$

is satisfied in some sense.

In the finite element method, the coefficients are determined by projecting the equations over the base functions. This gives the N conditions:

$$\sum_{k=1}^{N} f_k \int \alpha_l O \alpha_k d\tau = \int g\alpha_l d\tau \; ; j = 1, \ldots . N.$$

Since the α_l are known

$$O_{lk} = \int \alpha_l O \alpha_k \, d\tau$$

$$g_l = \int g\alpha_l \, d\tau$$

can be determined, and a system of N equations in N unknown is obtained.

$$\sum_{k=1}^{N} O_{lk} f_k = g_l$$

The set of functions is chosen to reduce the number of operations and the storage requirements involved in solving the linear system of equations. For instance, it is convenient to divide the domain into triangles or rectangles and to use functions that are polynomials within each subdomain and vanish outside. The flow equations have been solved using velocity and pressure rather than the stream function. The equations yield:

$$\sum_{l=1}^{N} u_l \int \int \nu \{\nabla \alpha_k . \nabla \alpha_l\} \, d\tau - \oint \nu \alpha_k \frac{\delta \alpha_l}{\delta n} \, ds$$

$$= \frac{1}{\rho} \sum_{l=1}^{N} P_l \int \int \alpha_k \frac{\delta \alpha_l}{\delta x} \, d\tau$$

$$\sum_{l=1}^{N} v_l \int \int \nu \{\nabla \alpha_k . \nabla \alpha_l\} \, d\tau - \oint \nu \alpha_k \frac{\delta \alpha_l}{\delta n} \, ds$$

$$= \frac{1}{\rho} \sum_{l=1}^{N} P_l \int \int \alpha_k \frac{\delta \alpha_l}{\delta y} \, d\tau + g \int \int \alpha_k d\tau$$

$$\sum_{l=1}^{N} u_j \int \int \alpha_k \frac{\delta \alpha_l}{\delta x} \, d\tau + \sum_{l=1}^{N} v_l \int \int \alpha_k \frac{\delta \alpha_l}{\delta y} \, d\tau$$

where ν is kinematic viscosity: $\nu = \mu/\rho$.

The method used and the tests of the program were described more completely by Mareschal & West (1980).

Deep-reaching geodynamic processes in the Alps

St. Mueller

SUMMARY: The Alps are a deep-reaching crust–mantle structure in Europe situated at the northernmost tip of the Adriatic promontory of the African plate. In that region the contact against the Eurasian plate is presently characterized by a NW–SE oriented compressional stress regime. A detailed regional analysis of seismic surface-wave dispersion and of P-wave travel time residuals ('seismic tomography') as well as long-range seismic refraction measurements and the interpretation of 'stripped' residual gravity anomalies have revealed a rather anomalous structure within the uppermost mantle under the Alpine arc. As a result of the massive continent–continent collision, parts of the lithosphere have been delaminated – a process which has led to 'flaking' in the upper crust, combined with a thickening of the entire crust and the formation of a pronounced relatively cold, dense and slowly subsiding 'lithospheric root' beneath the mountain chain. In this still continuing plate collision the lower parts of the lithosphere apparently have penetrated into the asthenosphere to a depth of 130 to 200 km in a steep zone of 'subfluence' ('Verschluckung'). On this scale the presently measurable uplift of the Alpine chain (about 0.1 cm per year) is a secondary effect due to isostatic rebound of less dense crustal masses which previously had been forced to greater depths. A configuration of this type generates regionally a compressive dynamics of its own on which – within a wider framework – rotational processes may be superimposed.

For a long time it has been the chief aim of the scientists working in the field of tectonics to find a satisfactory theory of orogenesis. There exists now an impressive list of publications through which the thoughts on the causes of orogeny can be traced historically. More than a century ago it was Suess (1875) who stated very clearly that the problems associated with the origin of mountain ranges could be solved only by studying in detail every chain with respect to its stratigraphy and structure. He furthermore pointed out that it was only possible to elucidate the mutual relationships of different mountain systems by employing all available geological methods and not by geometric speculation.

In the following decades several opposing schools of thought developed: the first one, the so-called school of 'fixists', adopted an elaborate classification of the Earth's major tectonic features in terms of cratons, geosynclines, orogens, etc., originally within a rigid framework of a contracting globe. In the view of the 'fixists', mountain building was primarily the result of asthenospheric and related lithospheric diapirism with thrusting and folding being explained either by gravity gliding or else by gravity spreading. Adherents of an expanding Earth view mountain building as the 'fixists' do, but invoke an essentially ensialic process. They assume that the oceanic lithosphere is not formed by 'oceanization', but rather by accretion and spreading along mid-oceanic ridges which are governed by the processes associated with an expanding Earth.

The second school followed a purely uniformitarian approach to tectonics which interpreted past tectonic events and fossil tectonic features in terms of the present tectonics of the Earth. These 'mobilists' opposed the fixism of the 'contractionists' and adopted the continental drift theory developed by Wegener (1912) which laid the foundations for the theory of plate tectonics (Wilson 1965). In the framework of global plate tectonics mountains are the product of collision and subduction processes at converging plate boundaries. Thrust and fold belts which consist of excess sediments and slivers of the underlying crystalline crust are of compressional origin. During the process of convergence sedimentary sequences have apparently been scraped off the basement and thin slices of crystalline rocks must have been delaminated and are effectively decoupled from the subducting lithospheric slabs.

Geological observations in mountain belts permit only a very limited insight into their deep structure and the processes that may have formed it. The brilliant early intuitions of Ampferer (1906), Argand (1916, 1924), Staub (1928) and other Alpine geologists, therefore, deserve our highest admiration. Their proposed models which were mostly rough sketches have led to a completely new picture of the deep-reaching structure in orogenic systems.

Ampferer (1906) had come to the conclusion that the forelands of mountain belts appeared to underthrust the orogens on a large scale thus causing the observed compression. Later he

From: Coward, M. P., Dietrich, D. & Park, R. G. (eds), 1989, *Alpine Tectonics*,
Geological Society Special Publication No. 45, pp. 303–328.

postulated a series of subcrustal currents which converge and subside at the site of an orogen, dragging down with them the crustal rocks of the overlying region to give rise to compression (cf. Fig. 13a). The disappearance of crustal units beneath the growing orogen was subsequently called 'subfluence' ('Verschluckung' in German).

Argand (1916 and 1924) explained the Mesozoic–Cenozoic tectonics of Eurasia as a consequence of the collision between African (i.e. Gondwana-related) elements and Eurasia. The mountain ranges that resulted directly from the compressional squeeze-out of the Tethys contents were called 'geosynclinal chains'. Within the Eurasian foreland the collision manifested itself in widespread deformational effects producing spectacular uplifts which paralleled the collision front. The horizontal pressure caused by the collision led to plastic folding of the entire crust in the form of 'basement folds' accompanied by thrusting of crustal slices that served to thicken the continental crust.

In all these various models presented at that time, however, the basic concept of plate tectonics–namely that the Earth's surface is covered by a finite number of quasi-rigid plates which move relative to each other, has not been mentioned once. It took another 45 years until a first serious attempt was made by Laubscher (1969 and 1970) to explain Alpine mountain building within the framework of plate tectonics. Based primarily on geophysical results obtained in the Alps and quantitative considerations of mass balance which included not only the crust but also the mantle part of the lithosphere he proposed a 'subfluence' model (Fig. 13b) following the original ideas of Ampferer (1906). In addition Laubscher (1969 and 1970) suggested detachment phenomena in the upper crust which allowed the horizontal transport of continental basement nappes. Bally (1981) took up this train of thought and proposed from balanced cross sections and their palinspastic reconstruction that a sizable but limited amount of continental lithosphere must have been subducted. He termed this process, which appeared to be associated with preceding or synchronous basement remobilization, Ampferer (or A-type) subduction in contrast to Benioff (or B-type) subduction which involves oceanic lithosphere. Again it was emphasized that decoupling and ductility contrasts within the continental crust and within the overlying sedimentary sequences were responsible for varying structural styles in mountain ranges.

The ongoing geodynamic processes in the Mediterranean–Alpine region are primarily governed by the relative movement of the African lithospheric plate against the Eurasian plate, resulting in a predominantly compressional regime along the zone of contact (see e.g. Mueller 1982a and b). Several sets of new geophysical data indicate that the Alpine arc is a deep-reaching continent–continent collision structure situated at the northernmost tip of the Adriatic promontory, a spur-like microplate which has moved with the African plate since the early Mesozoic thereby eventually creating the spectacular mountain range of the Alps. Seismological and gravimetric data have shown that the continuing collision process must have led to a considerable thickening of the crust under the Alps accompanied by the formation of a cold, dense and slowly subsiding 'lithospheric root' which substantially amplifies the regional compressive regime in the Alpine region (Mueller 1984).

Plate-tectonic framework

In the simple model of global plate tectonics, most of the Earth's seismicity is interpreted as being a direct consequence of the relative movements of large lithospheric plates (see, e.g. Minster & Jordan 1978). When studying ongoing geodynamic phenomena the plate boundaries and marginal zones between the different plates are therefore regions of particular interest. Mountain building processes are associated with convergent plate boundaries, i.e. subduction zones which normally penetrate into the Earth's mantle. In the present context only the plate boundary between Eurasia and Africa which extends from the Azores triple junction eastward into the Mediterranean–Alpine region is considered. The tectonic processes along this boundary zone can be understood as the result of active sea floor spreading in the Atlantic Ocean.

A higher spreading rate of the ocean floor in the south Atlantic (~4 cm per year) as compared to the central Atlantic (~3 cm per year) causes a gradual counterclockwise rotation of the African plate which presently in the western Mediterranean–Alpine area results in a roughly northward directed push against Eurasia, which in turn leads to a lithospheric shortening of 1 to 3 cm per year in the intervening boundary zone (see, e.g. Smith 1982). With a northwest–southeastward oriented spreading in the north Atlantic (at a rate of about 2 cm per year) the zone of plate contact is, therefore, to a large extent under compression (cf. Fig. 1b) thus providing the framework for mountain building

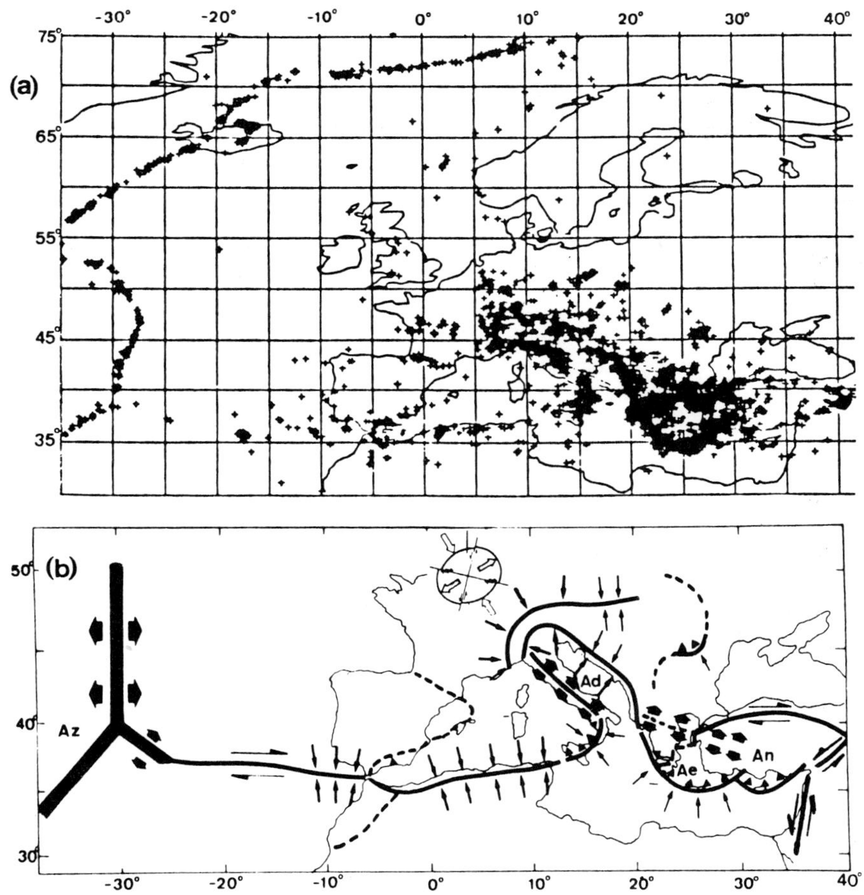

FIG. 1. (a) The plate boundaries between North America, Africa and Eurasia as outlined by the recent seismicity (after Waniek *et al.* 1982). The + signs represent the epicentres (1970–1980) of nearly 7000 earthquakes with magnitudes < 5.0. They clearly indicate the shape of the 'Adriatic promontory' of the African plate which extends all the way into the Alps. (b) Generalized plate boundaries and seismotectonic stress pattern in the eastern Atlantic as well as in the Mediterranean and Alpine region (after Udias 1982). Az = Azores triple junction; Ad = Adriatic promontory. Microplates: Ae = Aegean, An = Anatolian. Also indicated is a simplified seismotectonic scheme for central and northwestern Europe (after Ahorner 1975).

processes accompanied by active rifting in old preconditioned zones of lithospheric weakness (Mueller 1982b).

The recent seismic activity in the eastern Atlantic and in the Mediterranean–Alpine regions as summarized for the period 1970–1980 (Waniek *et al.* 1982) is depicted in Fig. 1a. It outlines very clearly the plate boundaries between North America and Africa against Eurasia; in particular it impressively illustrates the size of the Aegean–Anatolian microplate as well as the dimensions of the 'Adriatic promontory' of the African plate (Channell & Horváth 1976)–sometimes also called the Apulian microplate–which with its northernmost tip extends all the way into the Alpine arc. There are a few more sets of geophysical and geological data, such as the heat flow map of Europe (Cermak & Hurtig 1979) and the facies distribution map outlining the northern (European) and southern (African) margin of the Mesozoic Tethys (Channell *et al.* 1979), which confirm that an African 'spur' reaches all the way to the Alps. In addition Fig. 1a also demonstrates the existence of wide-spread 'intraplate' seismicity which must be ascribed to regional tectonic features on a smaller scale.

Earthquake mechanism studies are a powerful technique to provide information about the stresses acting in a particular region. Care must, however, be exercised in inferring regional stress patterns from isolated focal mechanism solutions. Only a sufficiently large number of well-determined solutions will be adequate to elucidate the seismotectonic stress field. Several types of regional stress distributions dominate the zone of contact between the African and Eurasian plates as can be seen in Fig. 1b (after Udias 1982). All of them are manifestations of the compressional stress regime due to the counterclockwise rotation of the African plate. The differential spreading along the Mid-Atlantic Ridge results in a right-lateral movement of Eurasia against Africa along the Azores–Gibraltar fracture zone. Latent subduction (Spakman 1986a) accompanied by mountain building can be traced from the Alboran Sea (Betic Cordillera) along the North African coast (Tell and Atlas Mountains) and around the Adriatic promontory (Alps and Dinarides), including the active Tyrrhenian and Hellenic arcs.

A tensional stress regime prevails all along the Apennine peninsula and in the back-arc basin of the Aegean Sea. The primary stress field due to the plate collision also radiates into central and northwestern Europe (cf. Ahorner 1975) as indicated in the small inset of Fig. 1b. It can be clearly seen that the Adriatic promontory of the African plate acts as an oblique 'indenter' (cf. Tapponnier & Molnar 1976) which has penetrated far into the Eurasian domain thus providing an explanation of why the collision structure of the Alps is in this particular location.

The lithosphere–asthenosphere system in the European–Mediterranean region

In the European–Mediterranean area a fairly large number of long-period seismograph stations has been recording teleseismic events during the past three decades. Suitable station combinations have permitted the measurement of fundamental mode phase velocities of Rayleigh waves from various azimuths in a wide period range across this seismograph network. Since the observed phase velocity dispersion is caused primarily by the geometric and physical properties of the lithosphere–asthenosphere system, this structure-dependent dispersion can be used to determine the average thickness and shear-wave velocity of the lower lithosphere or 'lid', once the crustal structure is known. Most of the European–Mediterranean region is sufficiently covered by phase velocity 'profiles', which provide the basis for a uniform 'hedgehog' inversion of the dispersion observations. In order to apply this method it must be assumed that a representative regionalization of the phase velocity dispersion for seismic surface waves can be achieved from the available observational data (Panza *et al.* 1980). In this procedure a systematic search in a multidimensional parameter space allows models to be determined that are compatible with a given range of observational uncertainty.

The results of this comprehensive inversion procedure are summarized in Fig. 2a, which shows a schematic map of the lithosphere–asthenosphere system for most parts of the European–Mediterranean area (Panza 1985). When interpreting this map, it is important to realize that it represents only an approximate solution to the inverse problem and is subject to inherent uncertainties (e.g. 15 to 20 km in lithospheric thickness) due to the limited resolution of long-period surface waves. Significant deviations from the 'normal' lithospheric thickness of 90 to 100 km in central and western Europe are, for instance, found in the Tyrrhenian and Balearic basins of the western Mediterranean Sea where the lithosphere thins to only about 30 km.

Another striking lithospheric feature which is conspicuously delineated in Fig. 2a, is the 'Central European Rift System'. It reaches from the North Sea to the western Alps with an average lithospheric thickness of about 50 km

FIG. 2. (a) Map of lithospheric thickness (in km) in the European–Mediterranean region, deduced from the regional dispersion analysis of seismic surface waves (after Panza 1985). Representative shear-wave velocities are given for the lower lithosphere or 'lid' (row of upper numbers) and for the asthenosphere (row of lower numbers). The three shaded areas indicate the postulated presence of 'lithospheric roots' to depths of about 200 km, while the thick dashed lines define the area of intermediate and deep focus earthquakes in the Tyrrhenian Sea. (b) Cross section through the lithosphere–asthenosphere system from the North Sea (A) to southern Tunisia (A′) along the 'European Geotraverse' (EGT) with representative shear-wave velocities (V_s, in km/s) for that profile. The vertical 'fence' hachures indicate the uncertainty in depth of the upper and lower bounds of the asthenosphere. Beneath the southern Alps the lithosphere penetrates into the asthenosphere forming a 'lithospheric root' which is shown by the black triangular arrow.

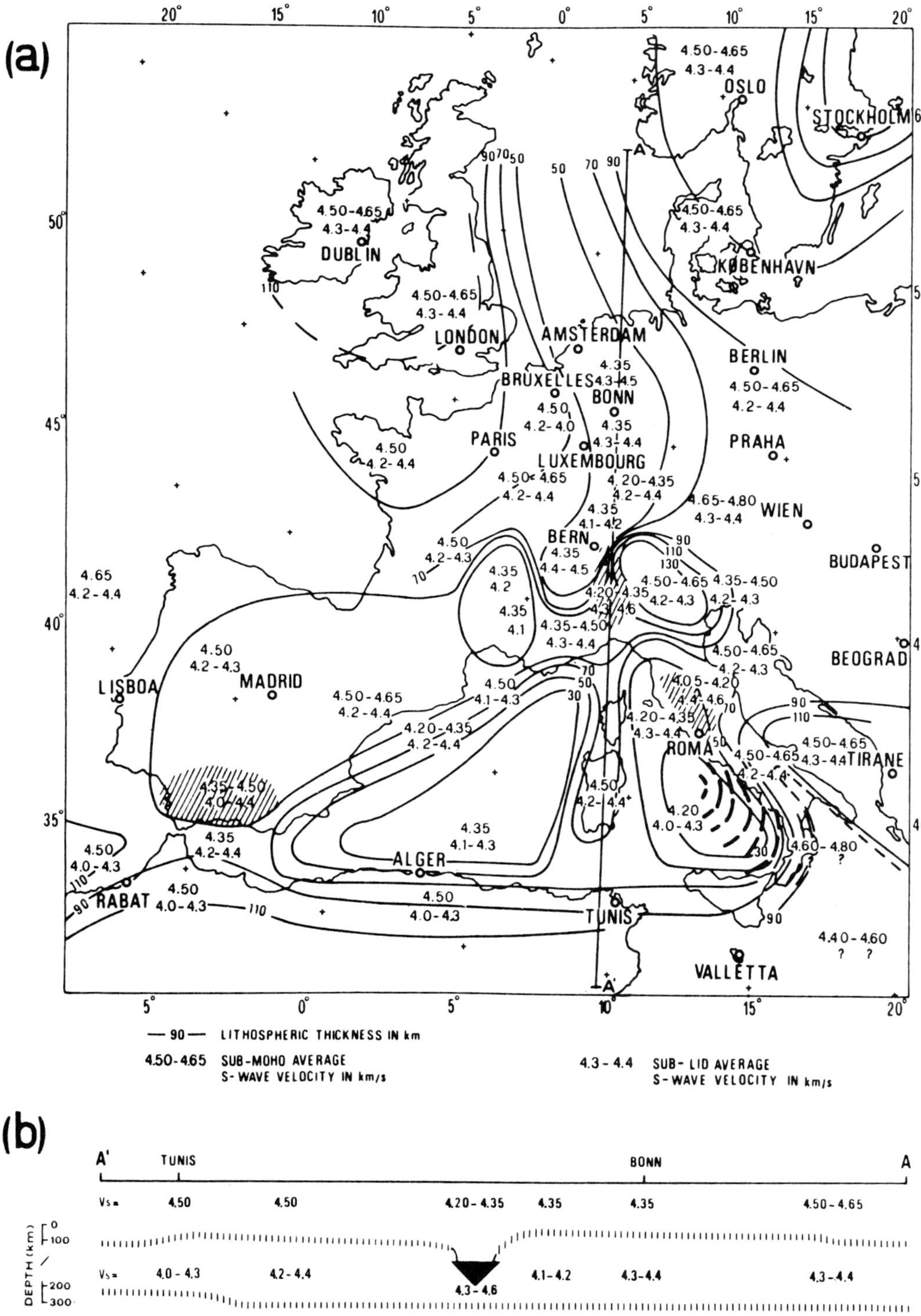
(a)
OSLO
STOCKHOLM
DUBLIN
KØBENHAVN
LONDON
AMSTERDAM
BERLIN
BRUXELLES
BONN
PARIS
LUXEMBOURG
PRAHA
WIEN
BERN
BUDAPEST
LISBOA
MADRID
BEOGRAD
ROMA
TIRANE
ALGER
RABAT
TUNIS
VALLETTA
— 90 — LITHOSPHERIC THICKNESS IN km
4.50-4.65 SUB-MOHO AVERAGE S-WAVE VELOCITY IN km/s
4.3 - 4.4 SUB-LID AVERAGE S-WAVE VELOCITY IN km/s
(b)
A' TUNIS BONN A
Vs = 4.50 4.50 4.20 - 4.35 4.35 4.35 4.50 - 4.65
Vs = 4.0 - 4.3 4.2 - 4.4 4.3 - 4.6 4.1 - 4.2 4.3 - 4.4 4.3 - 4.4
DEPTH (km) 0 100 200 300

and markedly lowered shear-wave velocities. Also noteworthy is the continuous structure which extends from North Africa through the Ionian Sea into the Adriatic Sea. There the thick lithosphere seems to have a clear African affinity in accordance with the concept of an Adriatic promontory of the African plate (Channell & Horváth 1976). Particularly interesting is the region of the Alps where the lithospheric thickness increases to 130 km and attains in places even greater values. These so-called 'lithospheric roots' are found under mountain belts in zones of continent–continent collision, such as the Betic Cordillera in southern Spain, the geothermally active Tuscany region in central Italy, and beneath the Alps (see the shaded areas in Fig. 2a). A north–south cross section (A–A′) through the lithosphere–asthenosphere system from the eastern part of the North Sea to southern Tunisia showing the shear-wave velocity distribution in the uppermost mantle, is displayed in Fig. 2b. It coincides with the so-called 'European Geotraverse' (Mueller & Banda 1983, Galson & Mueller 1987). The change in lithospheric structure from the margin of the Fennoscandian Shield across the 'Central European Rift System' to the neighbouring Alps, the island chain of Corsica and Sardinia, and finally to the North African craton can be clearly seen.

The conspicuous thickening of the Alpine crust and the asymmetric shape of the crust–mantle boundary (Mueller *et al.* 1980) are an indication of a much deeper-reaching structural anomaly beneath the Alps involving the entire lithosphere–asthenosphere system (cf. Figs 4 and 11). The detailed regional analysis of seismic surface-wave dispersion and P-wave travel time residuals ('acoustic tomography') in combination with seismic refraction and reflection observations in the northern Alpine foreland and along as well as across the strike of the Alps have revealed a rather anomalous distribution of P- and S-wave velocities with depth.

Outline of the geologic–tectonic units of the Alps

The Alps may schematically be subdivided into four major palaeogeographic and structural domains (Fig. 3a, after Frey *et al.* 1974):

(1) The Helvetic–Dauphiné domain

It consists of an old metamorphic basement complex which includes numerous Hercynian granites. The Argentera-, Pelvoux-, Belledonne-, Mont Blanc- and Aiguilles Rouges- as well as the Aar- and Gotthard-Massifs are the surface expression of this basement, which is not exposed in the eastern Alps. Recently doubts have been raised if the Gotthard-Massif and the Mont Chétif-'Massif'–a small basement complex south of the Mont Blanc–are really part of this domain (Trümpy 1980). The Helvetic and Ultrahelvetic nappes, that is the Permo–Mesozoic to Palaeogene covers, which have in part slipped outward, form a broad unit at the front of the orogen in the western and central Alps. The Helvetic domain can be traced continuously to the eastern Alps, where it becomes quite narrow.

(2) The Penninic domain

This highly complex domain comprises a pre-Triassic poly-metamorphic crystalline basement and a Mesozoic partially metamorphic cover. This cover is in some places overthrust by crystalline basement slices. Numerous tectonic units of Penninic basement are distinguished and marked in Fig. 3a (in the western Alps: the Bernhard–Brianconnais unit, the Monte Rosa–Gran Paradiso–Dora Maira unit, the Lepontine unit, the Penninic nappes of the Grisons – and in the eastern Alps: the central gneiss domes of the Tauern window, which are intruded by large bodies of Hercynian granites).

The numerous Mesozoic units of the Penninic domain can be differentiated by their lithologic composition, their depositional environment and by their different grade of metamorphism.

In the western and central Alps the following palaeogeographic zones can be distinguished: the Valais zone, the Subbriançonnais and Briançonnais zones and the Piédmont zone. In the Valais and Piédmont zones the crystalline basement is largely missing although by a 'substitution of the cover', for instance, the 'schistes lustrés' of the Piédmont zone are thrust over the middle Penninic crystalline massifs of the Gran Paradiso and Dora Maira. Furthermore it should be noted that the external parts of the Cretaceous Helminthoid flysch nappe presumably originate from the Piédmont zone.

It is thought that in the eastern Alps the Rhenodanubic flysch zone is derived from the Penninic domain. Furthermore a Mesozoic cover of the Penninic domain is found in the Engadine window, in the window of the 'Hohe Tauern' (see Fig. 3a) as well as in the Rechnitz–Bernstein window further to the east.

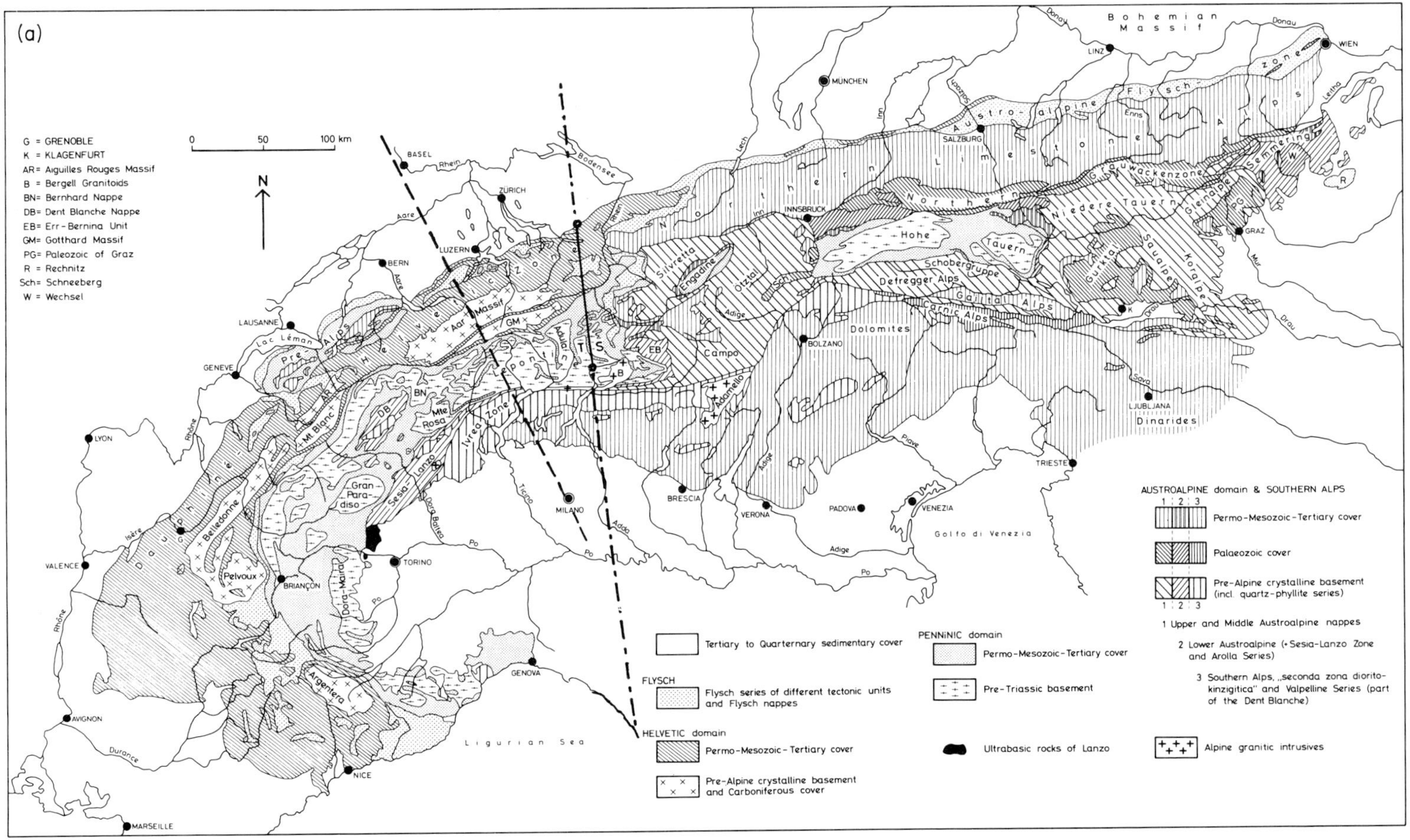

FIG. 3. (a) Sketch map of the geologic–tectonic units in the Alps (after Frey *et al.* 1974). The position of the lithospheric transect in the transition region between the central and eastern Alps which is discussed in this paper is indicated by a dash-dotted line. Along part of that profile (solid line) deep seismic reflection measurements have recently been carried out through the Helvetic and Penninic domains of the Alps in eastern Switzerland (cf. Finckh *et al.* 1987). In addition the position of the 'Swiss Geotraverse' from the southern Rhinegraben near Basel to Chiasso/Como and Milano in northern Italy is shown by a dashed line (cf. Rybach *et al.* 1980, Mueller 1982a). The Penninic nappes Suretta (S), Tambo (T) and Adula in the Lepontine Alps are also marked (cf. Fig. 9).

FIG. 3. (b) Topographic relief map of eastern Switzerland with the exact position of the 1986 deep seismic reflection profile (NFP 20). The line has a length of about 110 km (see Fig. 9) and extends from Wildhaus (W) in Toggenburg to Sovrana (southernmost shotpoint 17) in the Val Madris. Other shotpoints and locations: 1 = Säntis (cf. Figs 5 and 6); 4 = Sevelen (cf. Fig. 7); 10 = Canova (cf. Fig. 8); S = Sargans (between shotpoints 5 and 6); V = Vättis (cf. Fig. 9) near shotpoint 8, which is situated in an exposed 'window' of the Aar Massif. The Rhine–Rhône Line (RRL in Figs 4, 9 and 10) is traversed at Tamins (near shotpoint 9).

(3) The Austro–Alpine domain

The Austro–Alpine units occur mainly in the eastern Alps. There the Middle and Upper Austro–Alpine units form a huge overriding plate. The Middle Austro–Alpine is composed of extensive pre-Alpine metamorphic basement, which today forms the Silvretta–Oetztal–Campo mass to the west, the crystalline rocks south of the Tauern window as well as large nappe structures further to the east. The Upper Austro–Alpine thrust sheet consists of the Palaeozoic 'Grauwacken zone' and the very large Mesozoic sequence of the northern limestone Alps, the 'Drauzug' and some minor complexes. The Lower Austro–Alpine nappes comprise the following units: Semmering–Wechsel, the

'Radstätter Tauern' and the corresponding units in the transition region from the eastern Alps to the western Alps. In the western Alps the Lower Austro–Alpine domain is represented by the Dent Blanche and Sesia–Lanzo nappe system.

(4) The southern Alps

The southern Alps are separated from the other Alpine complexes by a major tectonic feature, the so-called 'Periadriatic Lineament', which in various parts carries different names, such as the 'Insubric Line' (cf. Figs 4 and 14a), the Giudicaria Line, the Pustertal/Gailtal Line and the Alpine–Dinaric Line.

The polymetamorphic crystalline basement of the southern Alps is overlain discordantly by unmetamorphosed Upper Carboniferous sediments, which are intruded by Permian granites and associated extrusives. The sedimentary sequence ranges from Carboniferous to Tertiary.

The spatial position of a lithospheric cross section which will be presented later in this paper (Fig. 15) is marked by a dash-dotted line in Fig. 3a. Along part of this profile (solid line) deep seismic reflection measurements have recently been carried out through the Helvetic and Penninic domains of the Alps in eastern Switzerland (Finckh *et al.* 1987). In the map of Fig. 3b the exact location of this deep reflection profile is shown. It reaches from Wildhaus (position W) in Toggenburg to Sovrana (position 17) at the southern end of the Val Madris close to the Swiss–Italian border northeast of Chiavenna. The dashed line in Fig. 3a (Basel–Milano) indicates the location of the so-called 'Swiss Geotraverse' (Rybach *et al.* 1980) which cuts across the Alps from the Rhinegraben in the north to the Po Plain in the south.

Results of deep seismic sounding experiments

Deep seismic sounding measurements have been carried out in the Alps since 1956 and are still going on. A number of important results have so far been obtained which can be summarized as follows (Miller *et al.* 1982): The Alpine crust is characterized by an asymmetric crust–mantle boundary (cf. M = Mohorovičić Discontinuity in Fig. 4) which reaches its greatest depth of some 55 km south of the central region and rises rather steeply towards the inner arc side of the Alps. The internal structure of the crust differs markedly between the eastern and the western Alps (see Fig. 3a) with more pronounced lateral velocity variations existing in the central and western Alps, which might be an indication that delamination and crustal stack-interfingering were the dominant processes forming the present-day crust, since crustal shortening was greater in the central und western Alps than in the eastern Alps.

Long-range seismic probing of the lithosphere

About 15 years ago it became apparent that specially designed experiments were necessary in order to gain more detailed knowledge on crustal and upper mantle structure sufficiently accurate for the development of geodynamic models. To this end international cooperative measurements along the 'Alpine Longitudinal Profile 1975 (ALP 75)' were carried out between Aix-les-Bains in France and the Pannonian Basin in Hungary. This long-range experiment, besides probing detailed crustal structure in the axial region of the Alps, also yielded information on the structure of the lower lithosphere beneath the Alps.

Long-range seismic refraction measurements in the northern Alpine foreland (Ansorge *et al.* 1979) and along the strike of the Alps (Miller *et al.* 1982) revealed the existence of a layer with a markedly high P-wave velocity of 8.4–8.6 km/s in the uppermost mantle. North of the Alps this layer lies at a depth of about 40 km and has a thickness of the order of 5 to 10 km (see Fig. 4) and lies between the crust–mantle boundary (M) and the relatively shallow base of the lithosphere located at depths of 50 to 55 km as determined from seismic surface-wave dispersion (Panza *et al.* 1980). As the Alps are approached this layer of high P-wave velocity (8.5 km/s) is found at somewhat greater depth (60 to 70 km). Following the axis of the Alpine chain from west to east, refraction seismic evidence (Miller *et al.* 1982) shows that roughly between the Mont Blanc Massif in the west and the eastern end of the Tauern window the mean P-wave velocity from a depth of 60 km down to at least 100 km amounts to about 8.5 km/s. In contrast to the marginal regions of the Alps in the west and east, where in the same depth range of the uppermost mantle a distinct layering has been found, no such structure could be detected beneath the innermost part along the axis of the Alps. The dimensions of this anomalous 'block' of high-velocity material are estimated to be about 500 km in length (west–east) and 100 to 150 km in width (north–south) reaching to a depth of at least 100 km. It is not yet clear if the marginal high P-wave velocity layer is connected directly with the thick central 'block' under the Alps.

Seismic refraction and reflection observations

A comprehensive study of the deep structure and recent dynamics in the central portion of the Alps in Switzerland was carried out between 1974 and 1980 as part of the International Geodynamics Project. The geophysical results obtained were presented in the form of a crust–mantle cross section for the 'Swiss Geotraverse' which transects the main tectonic and geologic units of the northern Alpine foreland and the Alps (Rybach *et al.* 1980). It extends from the southern Rhinegraben near Basel across the tabular and folded Jura mountains through the Molasse basin, the Helvetic nappes, the Aar and Gotthard Massifs, the Penninic nappes of the Lepontine and the southern Alps to the Tertiary sediments of the Po Plain as indicated in Fig. 3a.

By combining all available seismic refraction and reflection data along the 'Swiss Geotraverse' a representative crustal cross section could be constructed (Mueller *et al.* 1980) which is shown in Fig. 4. The crustal structure proposed appears quite complex, but is well documented for the northern foreland of the Alps, the Lepontine area in the Penninic domain and the southern Alps, while the details of the transition between these three major structural units, i.e. the extensions to depth of the Rhine–Rhône Line (RRL) and of the Insubric Line (IL), are still somewhat uncertain.

Starting in the northwest close to the pronounced mantle upwarp under the southern Rhinegraben the entire crust dips down towards the southeast beneath the thickening wedge of the Molasse basin. Only minor internal variations of seismic velocities and thicknesses of the crustal layers seem to occur with the lower crustal layers and the crust–mantle boundary (M discontinuity or 'Moho') running roughly parallel to the contact between the Mesozoic formations and the basement. Heavy bars in Fig. 4 mark the depth to the M discontinuity as it has been determined by a standard evaluation of the wide-angle reflection observations (f^2, x^2 method). Also indicated in Fig. 4 is the high P-velocity layer (8.5 km/s) in the lower lithosphere discussed in the previous section and the lithosphere–asthenosphere boundary (cf. Fig. 11) based on seismic surface-wave analysis (Panza & Mueller 1979).

A low-velocity zone–with a velocity reduction of three to ten per cent–separates the upper part of the crystalline basement with P-wave velocities of about 6 km/s from the middle crust under the northern Alpine foreland between Basel and Luzern (Fig. 4). The following segment of the 'Swiss Geotraverse' to the southeast, i.e. the crustal structure beneath the

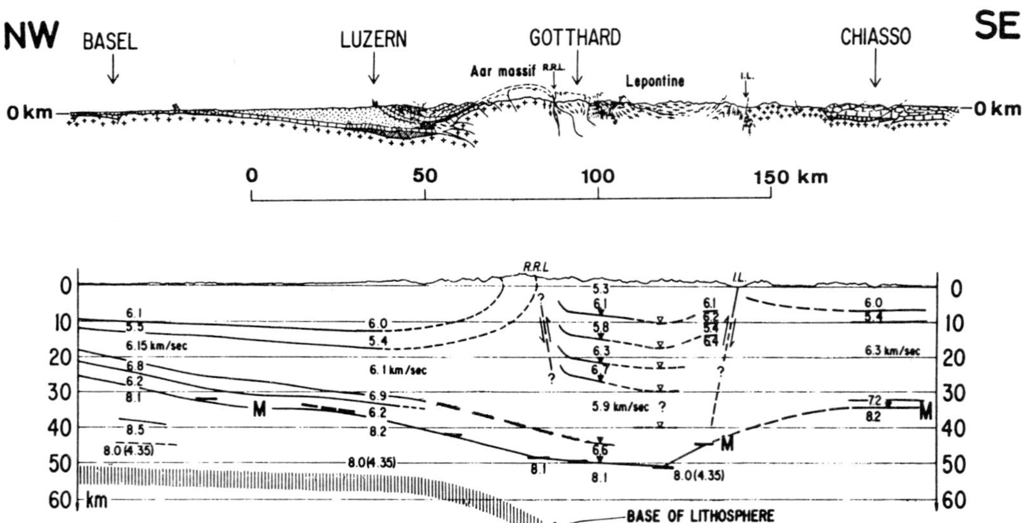

FIG. 4. Representative lithospheric cross section along the 'Swiss Geotraverse' from Basel to Chiasso/Como (cf. Mueller 1982a). P-wave velocities in km/s. RRL = Rhine–Rhône Line; IL = Insubric Line; M = Crust–mantle boundary (Mohorovičić Discontinuity). The base of the lithosphere ('fence' hachures, cf. Fig. 11) has been determined from a regional analysis of seismic surface-wave dispersion (S-wave velocities given in brackets).

Helvetic domain and the Aar Massif, seems to be rather complex and is not yet fully understood. In the axial core of the Aar Massif the surface geology with the granitic outcrops and the relatively low P_g–velocities there suggest that in this area the upper crust has been sheared off near or at the base of the low-velocity zone (cf. Fig. 10). The compressional shortening of the crust combined with dilatancy effects and the involvement of fluids with elevated pore pressures have led to a delamination or 'decoupling' of the upper layers in such a way that they could be bent upward in a flake-type manner as indicated in Figs 4 and 10.

It should be noted that the backward thrust of the geologic units observed near the surface seems to be restricted only to the upper portion of an otherwise relatively 'homogeneous' crust without pronounced velocity variations down to a depth of approximately 15 to 20 km under the central part of the Aar Massif. The grade of metamorphism observed in that region of the Alps (Frey *et al.* 1980) suggests that the material of the exposed Aar Massif once resided at depths of about 10 to 15 km. If the crustal flake in Figs 4 and 10 is bent back into its presumed original position it would fall into the appropriate depth range.

Basically the same structure model is – but now mirror-imaged – found beneath the southern Alps (Fig. 4). Again the upper crust is 'decoupled' from the underlying middle and lower crust with the 'Moho' dipping down towards the north. Delaminated structures of that type have very descriptively been termed as being due to 'crocodile tectonics'.

In contrast to this type of structure a very different crustal structure is found under the Lepontine area of the Penninic domain in the central Alps (Fig. 4). There is evidence that the lower crust of the northern foreland extends all the way under the Penninic block of the Lepontine Alps (Mueller *et al.* 1980). Whether in this region two separate crustal slabs are superimposed on each other or – more likely – a stack of several crustal slices is shoved into each other like a deck of cards cannot yet be decided. This complex crustal structure which is characterized by an overall low average P-wave velocity beneath the central Alps is corroborated by the presence of highly metamorphosed units at the surface which must have been uplifted by about 15 to 20 km (Frey *et al.* 1980).

The type of crustal structure determined from seismic refraction measurements in the Alps is strongly supported by recent results of continuous deep seismic reflection profiling in eastern Switzerland along a 110 km long traverse (see Fig. 3b) from Wildhaus (W) in Toggenburg to Sovrana (17) at the southern end of Val Madris north of the Engadine Line in the Val Bregaglia (close to the Swiss–Italian border NE of Chiavenna). This project is part of the Research Programme 20 of the Swiss National Science Foundation entitled 'Deep Geological Structure of Switzerland' (ref. Finckh *et al.* 1987). In addition to continuous 'Vibroseis' sounding, explosive charges (100–300 kg of dynamite) were fired and recorded along the profile by the geophone array (240 groups) with a total spread length of 19.2 km.

A few selected examples of the preliminary seismic reflection data obtained from explosive sources will be presented here and compared to the seismic refraction results (cf. Fig. 4). In Fig. 5 a seismogram section is shown which was recorded between Wildhaus (W) and Sargans (S), i.e. in the distance range of about 8 to 27 km from shotpoint 1 'Säntis' (see Figs 3b and 9). The first band of reflections behind the first arrivals corresponds to the Triassic marker of the northern Alpine foreland above the rather transparent upper crystalline crust. Quite conspicuous is the fairly broad dominant reflection band between about 12 and 13 seconds two-way time (TWT). It corresponds to the crust–mantle transition zone ('Moho') which dips towards the centre of the Alps and can be traced continuously all the way into the Penninic domain close to Thusis (cf. Figs 3b and 9). It actually consists of two separate reflection bands which are termed X (after Klemperer *et al.* 1986) and M (for 'Moho'). There also seem to be some faint traces of an intracrustal reflector at two-way times of 8 to 9 seconds.

To illuminate details of the 'Moho' reflection band in Fig. 5 an enlargement of single traces is displayed for the two-way time interval 11 to 14 seconds in Fig. 6. Two discrete reflection bands at 11.8 seconds (X band) and 12.6 seconds (M band) can be discerned which were not clearly resolved in Fig. 5. The X band in Fig. 6 appears more pronounced and can be correlated continuously across the geophone spread shown. A similar 'Moho' pattern is seen in the enlarged seismogram section of Fig. 7 for shotpoint 4 'Sevelen' (see Figs 3b and 9), where the two 'Moho' reflection bands now come in somewhat later – still 0.8 seconds apart – but the lower reflection band (M) appears this time to be more emphasized. Additional clear reflections can be traced across the section at about 9.2 sec two-way time; they would in the classical notation correspond to the 'Conrad Discontinuity' (C). Finally, an example of an enlarged observation in the Penninic domain is depicted

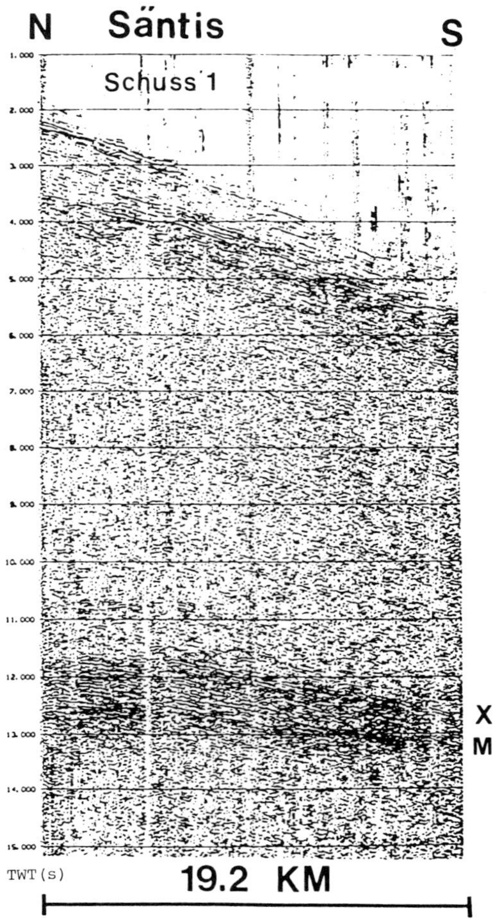

FIG. 5. Reflection seismogram section (after Finckh *et al.* 1987) recorded between Wildhaus (Toggenburg) and Sargans (Rhine valley) in the distance range of about 8 to 27 km from shotpoint 1 'Säntis' (for position see Figs 3b and 9). For a discussion see text. TWT = two-way time in seconds. The crust-mantle transition is characterized by two distinct reflection 'bands' (X and M).

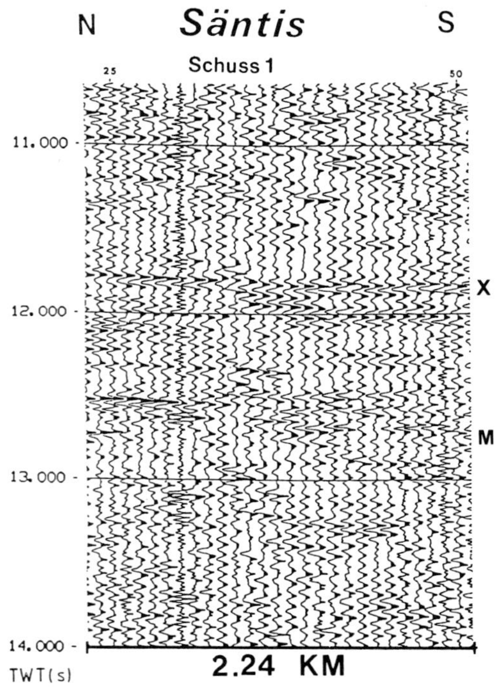

FIG. 6. Details of the two reflection 'bands' (X and M) in Fig. 5 illustrated by an enlargement of single traces in the two-way time (TWT) interval of 11 to 14 seconds (after Finckh *et al.* 1987). For a discussion see text.

in Fig. 8 (shotpoint 10 'Canova'; see Figs 3b and 9). It also exhibits a weak C reflection and the two 'Moho' reflection bands, about 0.8 seconds apart, again as in Fig. 6 with the upper X band more clearly defined, but with signs of rather abrupt vertical offsets.

The twofold band character of the 'Moho' reflections is most likely due to a lamination, i.e. an alternate sequence of thin high- and low-velocity layers, which seems to be typical of the relatively abrupt transition from the lower crust to the uppermost mantle. Speculations on the origin of the X and M reflection bands characterizing the crust–mantle transition have been published by Klemperer *et al.* (1986).

All the explosion–seismic reflection data for the traverse through eastern Switzerland have been evaluated and are presented in the form of hand-migrated line drawings in Fig. 9. The positions of the shotpoints (1 through 17) are marked. In the northern part of the section the Triassic marker can clearly be seen between shotpoints 1 and 6. The middle and lower crust with sporadic 'Conrad' reflections and the two 'Moho' reflection bands is 'decoupled' from the upper crust in the depth range from 10 to 20 km (cf. Fig. 4). This lower crustal slab dips towards the south and seems to move intact down into the deep-reaching subduction zone (cf. Figs 11, 13b and 14a). It can be traced well beyond the Rhine–Rhône Line (RRL in Fig. 4) all the way under the Penninic nappes near Thusis (Fig. 9).

The same deep-reflection pattern that has been identified in the Helvetic and the northern Penninic domains in eastern Switzerland (Fig. 9) has also been found in the region of the

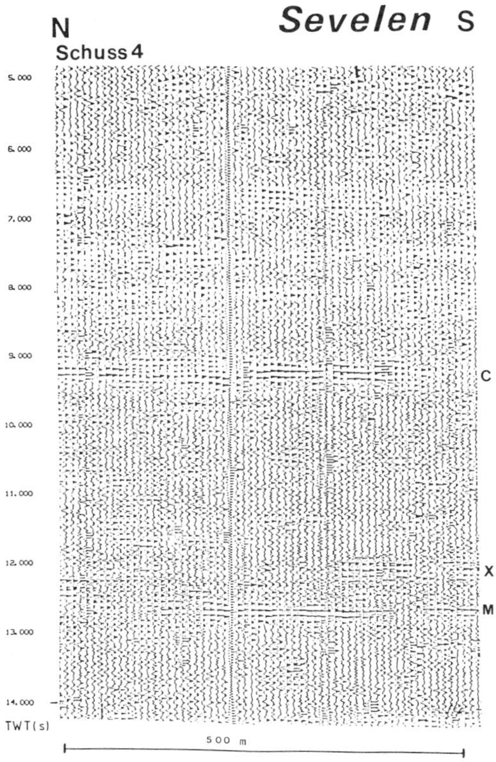

FIG. 7. Enlarged reflection seismogram section for shotpoint 4 'Sevelen' (for position see Figs 3b and 9). In addition to the two reflection 'bands' (X and M) between 12 and 13 seconds TWT another clear reflection 'band' can be traced at about 9.2 seconds TWT, which corresponds to the so-called 'Conrad Discontinuity' (C). For a discussion see text.

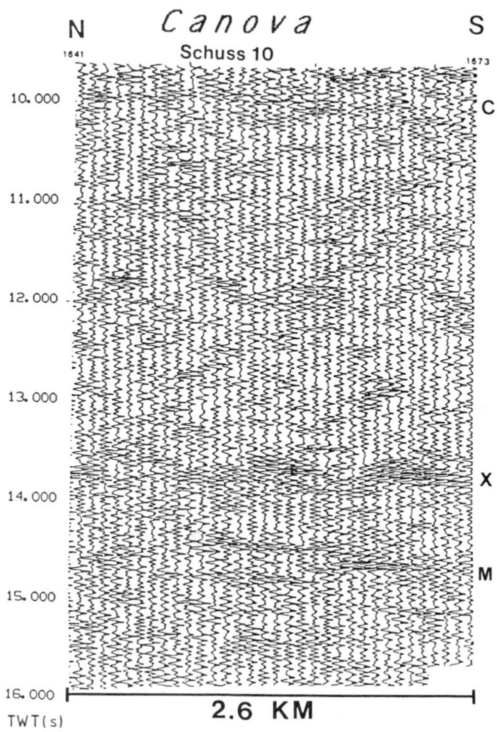

FIG. 8. Details of the two reflection 'bands' (X and M) in the lower crust as recorded from shotpoint 10 'Canova' (for position see Figs 3b and 9). The two separate reflection 'bands' about 0.8 seconds apart, can clearly be discerned (after Finckh *et al.* 1987). In addition a somewhat diffuse C reflection band can be identified at about 10 seconds TWT.

western transect through the Swiss Alps. These new results will be published soon. A corresponding pattern has also been shown for the French–Italian ECORS/CROP profile (Bayer *et al.* 1987) which roughly runs from Geneva to Torino across the Gran Paradiso. There the Belledonne Massif plays the same role as the back-thrust Aar Massif, and the Vanoise region is the exact counterpart of the northern Penninic domain in Switzerland.

Further to the south no near-vertical reflections from the lower crust and 'Moho' have been observed although the seismic refraction results show clear evidence of wide-angle reflections from the crust–mantle boundary at depths of 50 to 52 km (cf. Fig. 4). This apparent discrepancy may be explained by the different degree of resolution of the two seismic methods. Reflection seismology sees more details of the deep-reaching complex 'collision suture' (cf. Figs 13b and 14a & b) but fails for steeply dipping layers, while the refraction method with its longer wavelengths integrates over the local inhomogeneities of the highly deformed crust–mantle boundary.

At the southern end of the profile the 'Moho' rises gently towards the Engadine Line (EL in Fig. 9) in the Val Bregaglia. It should be noted that the steeply dipping reflectors in the southernmost part of the section have so far not been properly migrated; they actually come from a much more southern position. The flat reflectors in the upper crust which lie at depths of about 15 km between shotpoints 13 and 17 must be attributed to the well-known pile of Penninic nappes (Suretta, Tambo, Adula). Their northern front (PF) can be traced close to the surface at the Rhine–Rhône Line (RRL) near Tamins. A more detailed preliminary interpretation of the data displayed in Fig. 9 has been published recently (Finckh *et al.* 1987).

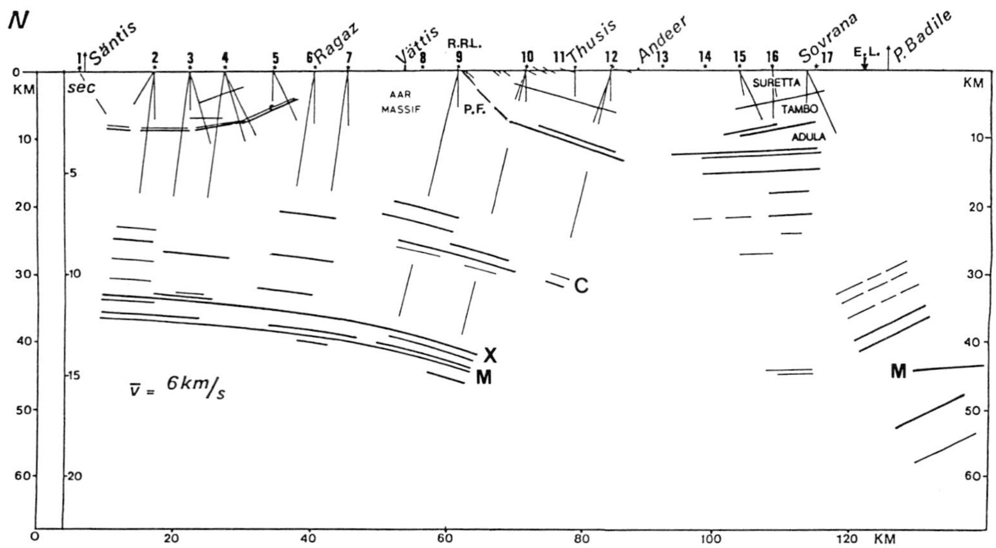

FIG. 9. Preliminary interpretation of the explosion–seismic reflection data for the traverse through eastern Switzerland (cf. Fig. 3b). The most conspicuous reflection 'bands' are represented by hand-migrated line drawings (after Finckh *et al.* 1987). In the north the lower and middle crust with the C, X and M reflectors appear to be 'decoupled' from the upper crust and seem to move intact down into the steeply southward dipping subduction zone (cf. Figs 11, 13b and 14a). In the upper crust the Aar Massif which is exposed in a 'window' near Vättis separates the Helvetic domain with its distinct Triassic marker (cf. Fig. 5) from the Penninic domain with the Suretta, Tambo and Adula nappes (cf. Fig. 3a). The 'Penninic Front' (PF) outcrops at the Rhine–Rhône Line (RRL) near Tamins (shotpoint 9). The steeply northward dipping reflectors in the southernmost part of the section have not been properly migrated. There is a distinct gap in the reflection bands (C, X and M) delineating the lower crust between shotpoints 10 and 12 in the north and shotpoints 16 and 17 in the south which has not yet been explained. When approaching the Engadine Line (EL) the M-discontinuity seems to rise gently towards the south. It should be noted that the crust-mantle boundary (M) as derived from seismic refraction measurements appears to be continuous. For a discussion see text.

Published results obtained from commercial seismic reflection measurements (BEB 1979) show a much more complex structure of the Mesozoic and Tertiary sediments and their interaction with the northern front of the Aar Massif than has been assumed previously. As pointed out above, the Hercynian central granitic core of the Aar Massif is most likely to be directly connected with the laccolithic zone of granitic intrusions which compose the sialic low-velocity zone in the upper crust of tectonically active areas (Mueller 1977) and which here has been pushed upwards in a flake-type manner (see Figs 4 and 10).

The sketch in Fig. 10 explains both the newly obtained data from the reflection measurements in eastern Switzerland as well as the seismic refraction observations in the Aar Massif. The 'delamination' model proposed previously for the development of the Aar Massif (Mueller *et al.* 1976), i.e. the uplift of an upper crustal flake sheared off near or at the base of the low-velocity zone in the middle crust, is compatible with the structure of the upper crystalline basement exposed at the surface and that derived from seismic reflection measurements (Fig. 9). The schematic model shown in Fig. 10 implies that the outer (northern) part of the Aar Massif, i.e. the gneissic envelope of the granitic intrusive masses, has been back-thrust and slices of upper basement rocks ride on the sedimentary zone of contact which eventually peters out towards the southeast. Additional evidence of the thrust character of the northern boundary of the Aar Massif is also provided by the focal mechanisms of earthquakes in that region which contain a clear component of thrust motion (Ahorner *et al.* 1972).

Evidence of a high-velocity 'lithospheric root' beneath the Alps

On a large scale the existence of a high-velocity 'block' within the upper mantle under the

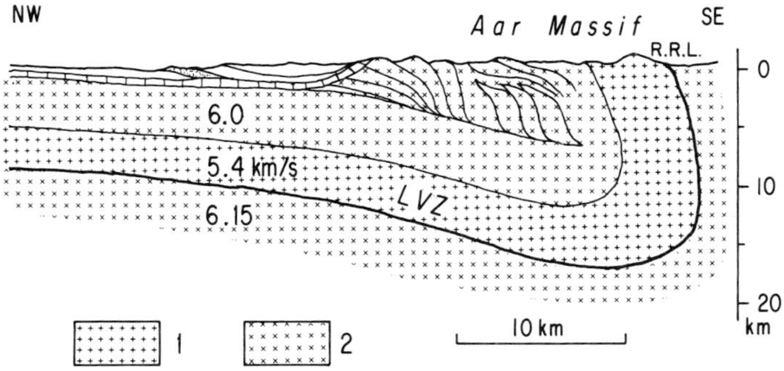

FIG. 10. Structural sketch map of the mechanical behaviour of the upper crust beneath the northern part of the Swiss Alps. Zero depth designates here the mean elevation above sea level. 1 = Quasi-continuous low-velocity layer consisting of a laccolithic zone of granitic intrusions into the cyrstalline basement (cf. Mueller 1977). 2 = Upper crystalline basement consisting of metamorphic rocks, such as gneisses, schists and migmatites. RRL = Rhine–Rhône Line. In this model the granitic, flake-type central core of the Aar Massif (cf. Figs 3a and 4) is embedded in gneissic envelopes and linked with the low-velocity zone (LVZ) (5.4 km/s) under the northern foreland of the Alps (after Mueller *et al.* 1980, modified). It is assumed that the base of the low-velocity zone (LVZ) serves as a 'decoupling horizon' (zone of weakness) which permits 'delamination', flaking, back-folding of the metamorphosed central granitic core and back-thrusting of the northern gneissic envelope. For a discussion see the text, also the remarks in conjunction with Fig. 4.

Alps has been deduced from the dispersion of Rayleigh waves (Panza & Mueller 1979). Fig. 11 shows a schematic cross section of the lithosphere–asthenosphere system along a profile from Basel to Milano which roughly coincides with the 'Swiss Geotraverse' (Rybach *et al.* 1980). The hatched areas in this figure mark the range of uncertainty for the bottom of the crust (M), of the lithosphere and of the asthenosphere. The centre part is characterized by a high-velocity 'block' of subducted lower lithosphere penetrating into the asthenosphere to depths of 150 to 220 km (Panza & Mueller 1979). This lithospheric 'block' of higher velocity could correspond to two slabs of lower lithosphere subducted to the south and to the north, respectively, during the plate collision process ('subfluence' or, in German, 'Verschluckung') forming the Alps (Ampferer 1906).

A crust–mantle model of this type not only provides a reasonable solution for the deposition of the excess lithospheric material (of about 300 to 400 km in width) which must have been subducted during the shortening of the Alpine lithosphere, but it also gives an explanation for the apparently non-existing difference in the P-wave travel time residuals of teleseismic events observed at seismograph stations in the northern Alpine foreland and in the Alps (Baer 1980). Figure 12 demonstrates that the proposed model with a high-velocity 'block' in the uppermost mantle under the Alps will provide the required compensation for the extra time delay caused by the much thicker 'slower' crust beneath the central portion of the Alps. It should be noted that the zone of near-vertical subduction or 'subfluence' is not symmetrical with respect to the central zone of the Alps, but appears to be somewhat displaced southeastwards (see Fig. 11), i.e. towards the inner side of the Alpine arc.

This new geophysical model of the crust–mantle system in the Alps (Mueller & Panza 1986) is in full accord with geologic–tectonic schemes which have been suggested earlier. Ever since Eocene/Oligocene times, i.e. for the last 40 Ma, the continuing collision of the African plate with the Eurasian plate has led to a considerable shortening of the lithosphere in that zone of contact. The geophysical evidence presented so far indicates that after the closure and subduction of the south Penninic Ocean relatively dense and cool lithospheric material has been pushed almost vertically into the upper mantle forming a 'Verschluckungszone', (as suggested by Ampferer 1906) which reaches to depths of about 200 km.

An anomalous mantle zone under the Alps had already been postulated more than 55 years ago (Kraus 1931). The sketch in Fig. 13a illustrates the schematic model of a 'double orogen' that was envisaged with lithospheric material 'flowing down into the mantle' from both sides. In the foreland, extensional structures associated with rifting and volcanism may thus be created. As a consequence of such deep-

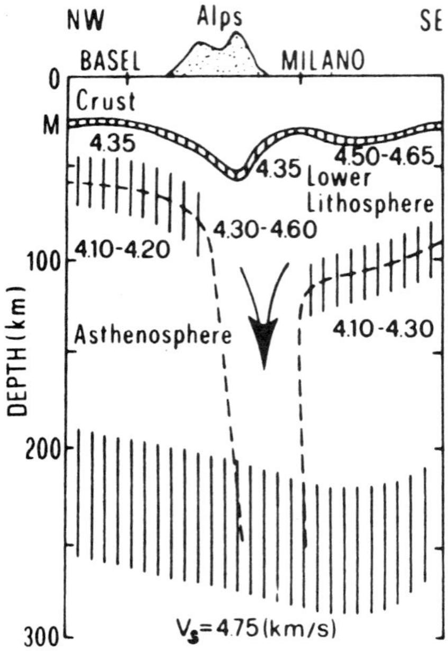

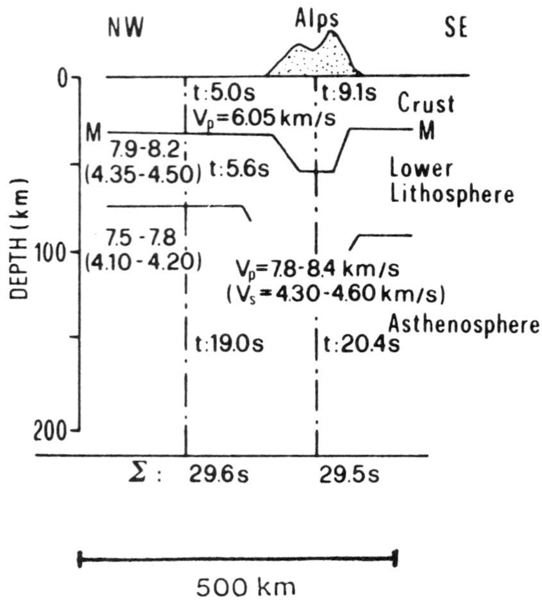

FIG. 11. (left) Crust–mantle cross section (cf. Fig. 3a) from the southern Rhinegraben (near Basel) to the Po Plain (southeast of Milano) derived from the simultaneous inversion of available surface-wave dispersion data (after Panza and Mueller 1979). The numbers represent shear-wave velocities V_s in km/s. Hatched areas indicate the range of uncertainty of the base of the crust (M), of the lithosphere and of the asthenosphere. In the collision process the lower lithosphere has penetrated into the asthenosphere thus forming a nearly vertical zone of 'subfluence' ('Verschluckung').

FIG. 12. (right) Schematic crust–mantle cross section through the central Alps along the 'Swiss Geotraverse' (compare with Fig. 11; after Panza and Mueller 1979). The one-way travel times (t) of P-waves propagating vertically upwards from the base of the asthenosphere to the surface corroborate the hypothesis that the time delay caused by the Alpine low-velocity crustal root must be compensated by the underlying high-velocity 'lithospheric root' which is penetrating into the asthenosphere (after Baer 1980). In the model a ratio V_p/V_s of 1.82 has been assumed.

reaching lithospheric subduction, slivers of upper crustal material ('flakes') were peeled off and piled up, thus creating the complex nappe edifice of the Alpine mountains as shown in Fig. 13b (Laubscher 1974). Deeper parts of the crust were shoved into each other (Mueller *et al.* 1980) and in this manner formed the less dense crustal root with a poorly defined M discontinuity (Laubscher 1974) above the much wider, denser and deeper-reaching bivergent 'lithospheric root' beneath the Alps (Fig. 13b) which most easily explains where the excess lithospheric material must have gone during this plate collision process.

Modelling of the 'lithospheric root'

The tectonics of the Alps are usually displayed in the form of two-dimensional cross sections. As the arc of the western Alps has a radius of only about 100 km, which is of the same order as the thickness of the lithosphere, it is necessary to view the actual situation in all three dimensions (Laubscher 1985). The pile of nappes which has been mapped on the surface consists of rocks derived from sediments or the uppermost 5 km, more or less of crystalline basement (Fig. 14a). It is apparent that the surficial masses were delaminated (or 'decoupled' in one or several low-velocity zones) from the main body of the crust and lower lithosphere which must have been subducted (Mueller *et al.* 1980). Since upper crust of both the African and Eurasian plates is involved in the Alpine nappe edifice, both African and Eurasian lithosphere may have been subducted. The most plausible solution would be by double or 'bivergent' subductions as illustrated in Fig. 14b (Laubscher 1985).

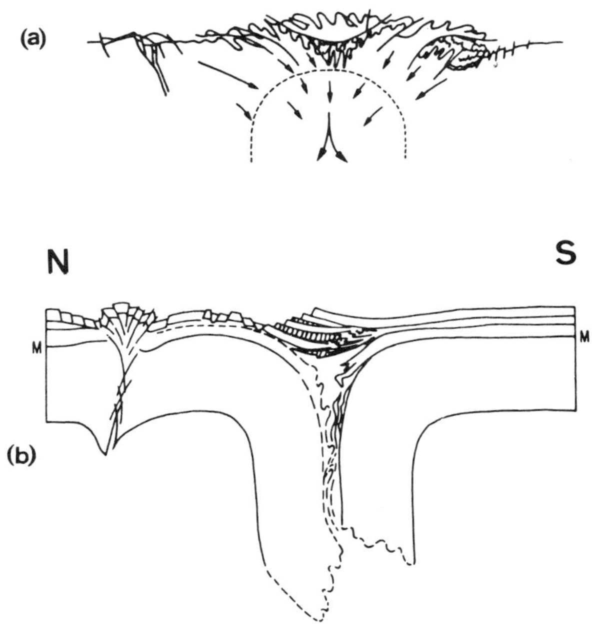

FIG. 13. Geologic–tectonic concepts of the deep structure beneath the Alps: (a) Schematic model of the Alps as a symmetrical 'double orogen' with lithospheric material flowing downwards into the mantle (after Kraus 1931). This model basically corresponds to a nearly vertical zone of 'subfluence'. (b) Schematic cross section through the crust–mantle system of the Alps (after Laubscher 1974). Caused by the plate collision process the subducting northern and southern lithospheres seem to form a nearly vertical 'bivergent zone of subfluence' in the Alpine region which penetrates deeply into the asthenosphere.

A rather detailed model of lithospheric thickness in the Alps has been derived recently based on the investigation of representative average P-wave travel time residuals (Babuška *et al.* 1985). The residuals were corrected for crustal structure near the stations using deep seismic sounding results. In order to minimize the effects due to different source regions and deep mantle paths a normalization was introduced for each event in which the mean residual at 15 chosen 'base stations' was subtracted from each residual. The majority of the variations in the relative residuals could be explained by variations of lithospheric thickness. Two zones with a rather thick lithosphere were found beneath the Alps, one in the western Alps (see Fig. 14c) and another one in the eastern Alps. They both show a NNE trend and seem to dip steeply towards the southeast reaching to depths of 180–200 km. The similarity of this result compared to the three-dimensional 'corner model' of the lithosphere–asthenosphere system in Fig. 14b (Laubscher 1985) is quite striking and the dimensions involved are in very good agreement with the other geophysical data sets presented in this paper.

Between the shallow and deep lithospheric masses there exists a depth range of profound disharmony which makes it virtually impossible to combine the visual surface geology (Fig. 14a) with the deep structures deduced by geophysical methods (Figs 14b & c). Smoothing out the surficial nappes in the western Alps proves shortening both in a north–south and in an east–west direction. Consequently, keeping the Eurasian plate (E) in a fixed position, the Adriatic promontory of the African plate (A) must have moved relatively northwestwards (Fig. 14b). The motion at the plate boundary must, therefore, have had a compressive and a strike-slip component which at depth looks quite different from what can be mapped at the surface.

Field evidence supports the concept of 'flake tectonics' requiring zones of weakness within the lithosphere which are identical with depth ranges of lowered seismic velocities (Mueller 1977). Three depth levels have been identified where slivers of crustal material can be sheared off rather easily: (1) at the top of the crystalline basement beneath the sedimentary cover ('Jura' type), (2) at or near the lower boundary of the

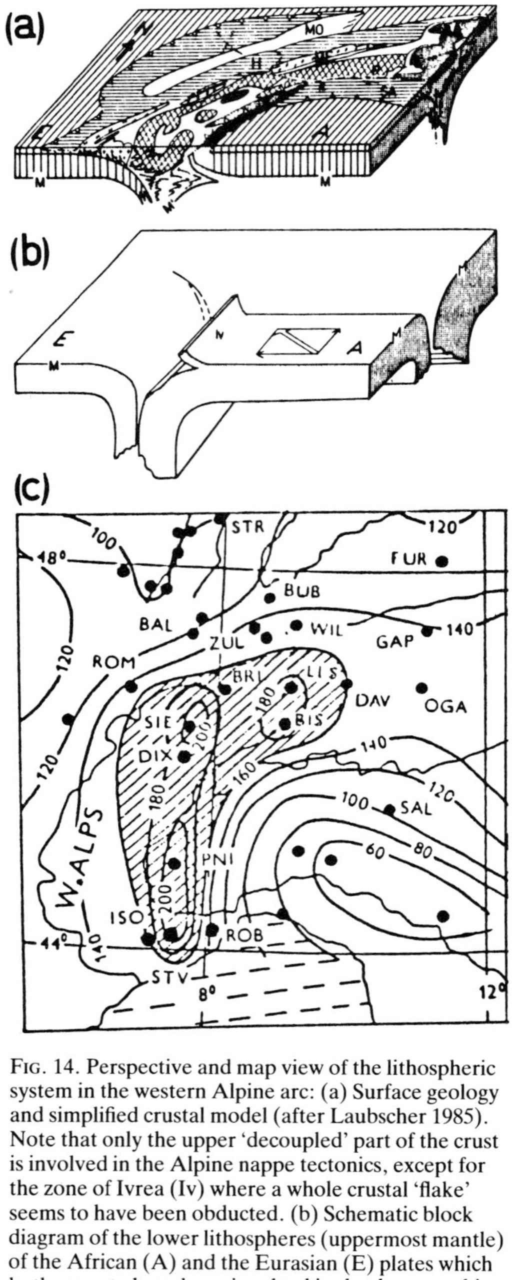

FIG. 14. Perspective and map view of the lithospheric system in the western Alpine arc: (a) Surface geology and simplified crustal model (after Laubscher 1985). Note that only the upper 'decoupled' part of the crust is involved in the Alpine nappe tectonics, except for the zone of Ivrea (Iv) where a whole crustal 'flake' seems to have been obducted. (b) Schematic block diagram of the lower lithospheres (uppermost mantle) of the African (A) and the Eurasian (E) plates which both seem to have been involved in the deep-reaching collision process (after Laubscher 1985). A double (or 'bivergent') subduction zone appears to be the most reasonable model to explain the geophysical observations. (c) Map of lithospheric thickness as derived from representative average P-wave travel time residuals (after Babuška *et al.* 1985). The region of thickest lithosphere (> 160 km) is hatched. Full circles denote the seismic stations where the observations have been made.

upper crustal (or 'sialic') low-velocity zone ('Central Massif' type, cf. Figs 4 and 10), and (3) immediately above the M-discontinuity ('Ivrea' type). The Ivrea body (Iv in Figs 14a & b) is a prime example where the obducted 'flake' consists of lower crustal material and even thin slices of upper mantle rocks.

Deep structure, gravity anomalies and dynamic processes

Guided by the original geometric model of Argand (1916 and 1924) and based on the results of the early seismic investigations in the central Alps (Mueller *et al.* 1976), Hsü (1979) designed some palinspastic reconstructions of crustal structure under the Swiss Alps for three different periods of time. The reconstructions depict the proposed deformation of the crustal structure which took place after continental collision in Eocene time assuming underthrusting of successive crustal wedges and décollement over thrusts of the sedimentary cover in the Helvetic domain. It was postulated that there has been no subduction of lithosphere under the Alps during the past 40 Ma, but it may have taken place during an earlier geologic epoch.

Along the same line an even simpler approach to this problem was followed by Karner & Watts (1983) who tried to reproduce the Bouguer gravity anomaly observed along the Swiss Geotraverse (Rybach *et al.* 1980) by superposing an 'obducted' southern crustal slab of presumably Adriatic–African affinity on top of an underriding crustal plate which corresponds to the upper part of the Eurasian lithosphere. The chosen two-dimensional geometry and the densities assigned to the various geological formations are rather arbitrary and do not correspond to the seismic results. It has long been recognized that the maximum crustal thickness is reached north of the Insubric Line and the location of the greatest depth to the crust–mantle boundary does not correspond to the location of the peak elevation in the central and western Alps. This asymmetry is due to a deep-reaching lithospheric subduction zone which has a slight southward-directed dip. The density distribution for the gravity model in the Penninic domain with high velocities and low densities in the obducted slab over low velocities and higher densities in the underriding crustal plate (with an additional denser subcrustal load attached to it) bears no resemblance to what the presently available sets of geophysical and geological data tell us. It is, therefore, obviously not sufficient to model the lithospheric structure in the Alps by simply considering elastic flexure of two superposed crustal slabs which are assumed to be homogeneous and not to interact with each other.

A synthesis of the available new seismic and gravimetric information provides a unique data set, which has permitted a quantitative modelling of the lithospheric structure along a north–south profile across the central Alps (Schwendener 1984, Schwendener & Mueller 1985). Any reasonable model will have to satisfy the following three boundary conditions for the centre of the Alps: (1) a maximum crustal thickness of some 50 km, (2) a minimum Bouguer gravity anomaly of −180 mgal, (3) an average topographic relief of some 2500 m. Combining these conditions leads either to an anomalously low crustal density or to the postulate of a deep-seated density anomaly in the upper mantle. In order to resolve this problem by modelling it has been crucial to rely on a good-quality seismic data base (see, e.g. Mueller *et al.* 1980) and to include the northern and southern Alpine forelands in the modelling process.

In a first step, the observed Bouguer gravity anomalies (BG in Fig. 15b) were reduced with three-dimensional models for all known near-surface density anomalies, such as the sediments of the northern Molasse Basin, the Quaternary sediments of the Alpine valleys and the sediments of the Po Plain. Seismic reflection and refraction data as well as borehole logs and geoelectric soundings were used to define the geometry and density distributions of these near-surface structures. A three-dimensional computer model of a simplified three-layer crust and the underlying upper mantle was then constructed. In order to find a realistic density distribution a least-squares inversion technique was applied to this crust–mantle model and the reduced Bouguer anomalies. Bodies with negative density contrasts were suggested to exist in the middle crust beneath the northern margin of the Alps and near the surface in the northern Apennines as indicated by the gravity, seismic, or geological data.

As a result of this stepwise 'stripping' procedure a smooth positive residual gravity anomaly of about 500 km in width has been obtained with a maximum amplitude close to +80 mgal (RG in Fig. 15b) which is situated in the region of the Po Plain (cf. Fig. 15a). The good coverage with high-quality seismic data, on which the construction of this gravimetric model was based, precludes an intracrustal source for this positive residual anomaly. It must, therefore, originate in the upper mantle in concordance with the findings from seismic surface-wave studies (Figs 2 and 11) and travel time investigations (Fig. 12). The shape of the anomalous body in the upper mantle was originally chosen according to the temperature anomalies (maximum difference of −400° to −500°C compared to the ambient temperature) calculated with models of the Alpine orogeny (Kissling *et al.* 1983, Werner & Kissling 1985). It should be noted, however, that the resulting positive density contrast was too large to be explained solely by thermal contraction due to the negative geothermal anomaly as suggested by the kinematic models. The presence of high-density (and associated high-velocity) material must, therefore, be ascribed to subducted lithosphere which has penetrated into the asthenosphere beneath the southern Alps and the Po Plain as suggested earlier (Panza & Mueller 1979).

So far the dynamical aspects of the 'lithospheric root' model have not been considered. They have recently been investigated for a two-dimensional geodynamic model along a profile from southwestern Germany across the eastern Swiss Alps, the Po Plain, the northern Apennines to the western Mediterranean Sea (Werner 1985). As an additional constraint in the model calculations the observed uplift and subsidence rates along this profile had to be satisfied. Since only average crust–mantle structures are taken into account, a two-dimensional treatment seemed to be justified. The method of calculation is based on the Navier–Stokes equation applied to a horizontally layered viscous halfspace in which the lateral density variation is assumed to behave as a harmonic function (Fleitout & Froidevaux 1982, Werner 1985).

Two sets of input parameters have to be extracted from observations or plausible assumptions have to be made: (1) the 'load' represented by the density distribution, and (2) the viscosity which is dependent on depth (i.e. on pressure and temperature). Three types of 'load' along the profile (Fig. 15) have to be considered: (a) the topographic 'load', including sedimentary basins and sea water, (b) the varying crustal thickness (Mueller *et al.* 1980, Miller *et al.* 1982), (c) the 'lithospheric root', characterized by relatively high seismic wave velocities (Panza & Mueller 1979, Baer 1980), whose increased density has been interpreted as being primarily due to a negative temperature anomaly (Kissling *et al.* 1983, Werner & Kissling 1985).

The density distribution of the model (see

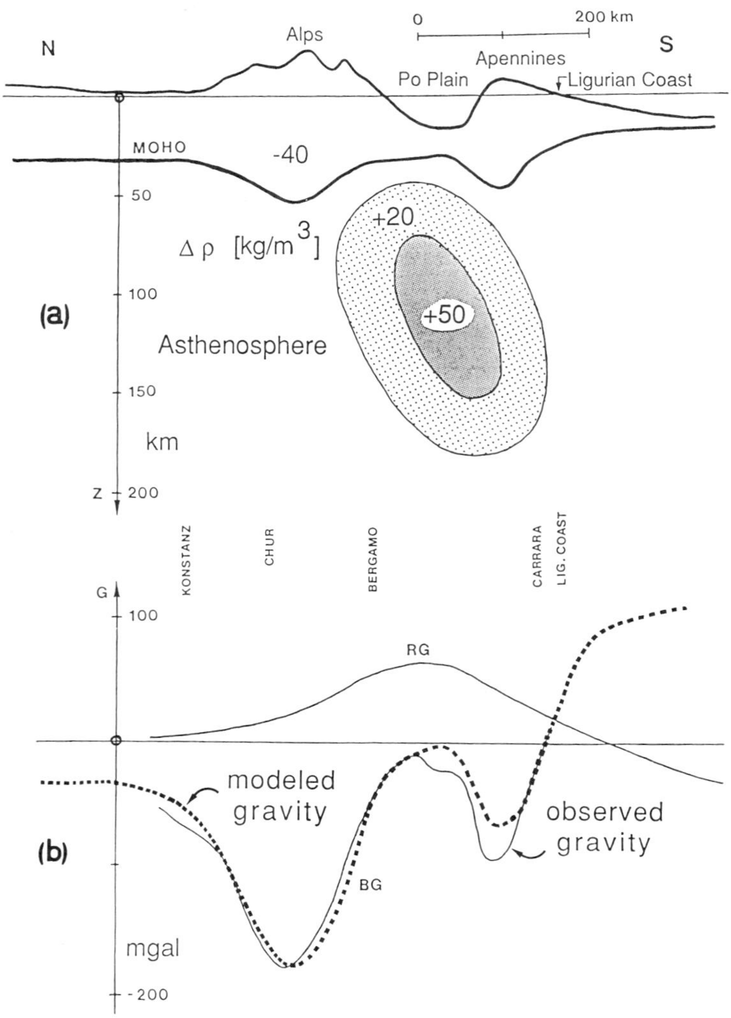

FIG. 15. Geodynamic model cross section (cf. Fig. 3a) across the Alps, the Po Plain and the northern Apennines (after Werner 1985). (a) Model structure consisting of the topographic 'load', the varying crustal thickness and the 'lithospheric root' with increased density. (b) Residual gravity effect (RG) due to the mantle anomaly sketched in Fig. 15a. The dotted curve BG shows the combined theoretical gravity effect caused by the mantle anomaly and the density jump at the crust–mantle boundary (MOHO) as compared to the observed gravity anomalies (solid line BG) after near-surface density disturbances have been eliminated (after Schwendener & Mueller 1985).

Fig. 15a) – without the topographic 'load' – must reproduce the observed Bouguer gravity anomalies (Schwendener 1984) along the profile (BG in Fig. 15b). As is to be expected, these anomalies exhibit pronounced negative values under the Alps and the Apennines and they become strongly positive in the western Mediterranean Sea. The residual gravity anomaly (RG in Fig. 15b) is the positive gravity anomaly which remains after the elimination of all crustal effects, i.e. it is caused by the 'lithospheric root' of increased density residing in the upper mantle beneath the southern Alps and the Po Plain. As a result of the modelling calculations relative uplift rates of about 0.1 cm per year are obtained in the Alps which agree with geodetic observations (Gubler *et al.* 1981). In the northern Alpine foreland and in the western Mediterranean Sea subsidence is prevailing which seems to be quite realistic, while the foundering of the Po Plain could not yet properly be reproduced. At greater depths the model in Fig. 15a postulates the 'lithospheric root' to sink into the asthenosphere with a slight southward-directed dip at a present rate of roughly 0.1 cm per year.

Seismic tomography of the upper mantle

The data so far presented for the existence of a 'lithospheric root' beneath the southern Alps and the Po Plain unfortunately do not allow the geometry of the subducted lithospheric slab to be outlined. It has only been through the advent of seismic body wave tomography (Aki *et al.* 1977) that the structure of the upper mantle beneath the Mediterranean–Alpine region could be illuminated more in detail (Spakman 1986a & b). 'Acoustic tomography' is an inversion method which estimates the P-wave velocity in the interior of the Earth from the travel times of compressional waves observed at the Earth's surface. In the linearized approach this method inverts for the deviation of the P-wave velocity field from a radially symmetric reference model of the Earth using the difference between the observed and calculated travel time, i.e. the so-called travel time residual or 'delay time' (cf. Figs 12 and 14c). For the purpose of effective computation a part of the Earth's interior is divided into a number of cells that are penetrated by seismic rays from teleseismic events. In every cell the mean seismic velocity anomaly is estimated by linear inversion of the 'delay times' observed at the Earth's surface. Modern computing facilities enable inversion of several ten thousands of unknown model parameters using hundred thousands of observational data (see, e.g. Nolet 1985).

Using the tomographic inversion of about 550 000 delay times coming from approximately 30 000 regional seismic events observed at 937 stations the large-scale P-wave velocity structure of the upper mantle beneath the European–Mediterranean region could be estimated (Spakman 1986b). Preliminary results of this analysis yielded a three-dimensional mantle structure of the central Mediterranean–Alpine area which is presented in the form of 5 parallel north–south oriented cross sections, each 2000 km long (Figs 16 and 17). All five profiles transect the northernmost tip of the Adriatic promontory of the African plate. The cross section in the middle (labelled C) corresponds to the 'European Geotraverse' (EGT). While the

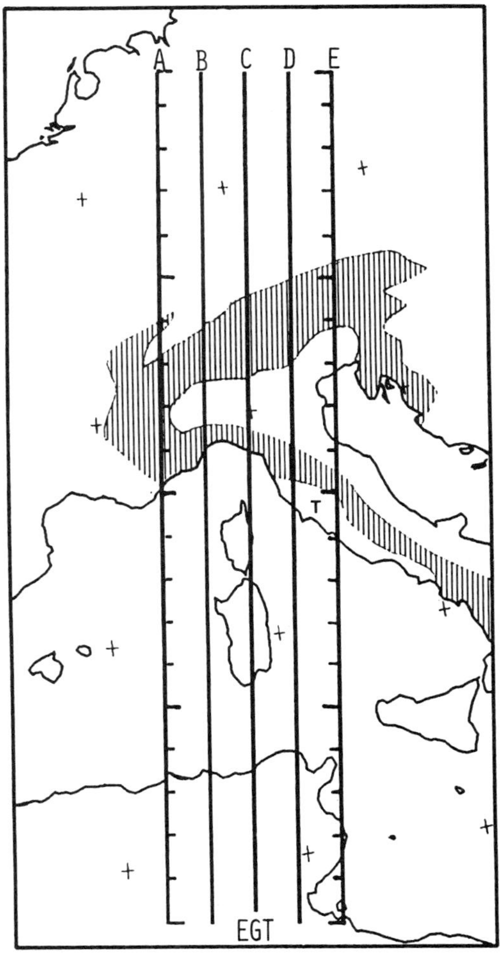

FIG. 16. Schematic map of the Alpine–Mediterranean area with the position of five cross sections (A . . . E), each 2000 km in length, which have been constructed by P-wave tomography (after Spakman 1986b). The cross section in the centre (C) corresponds to the 'European Geotraverse' (EGT). T = location of the north–central Italian upper mantle anomaly in Tuscany (cf. Panza *et al.* 1982).

central depth range between about 100 and 250 km in each section is characterized by a pronounced low-velocity zone within the upper mantle which is ascribed to the asthenosphere, significant variations in the thickness and properties of the lithosphere above are evident (cf. Fig. 2).

In profiles A and B the high-velocity 'lithospheric root' beneath the Alps dipping to the south can be clearly seen reaching its greatest depth (140 and 180 km, respectively) in the distance range from 600 to 800 km (measured from the northern end of the lines in Fig. 16).

The northern Apennines and parts of the Gulf of Genova are also underlain by a high-velocity lithosphere which extends to depths of 70–90 km (cf. Fig. 2a). For section C (EGT) a pseudo-symmetrical (bivergent?) 'lithospheric root' can be identified in the same distance range which is in agreement with the models proposed in Figs 11, 13b and 14b, but this configuration seems to be the exception rather than the rule. In the eastern Alps (cross sections D and E) two separate subducted slabs are visible which would indicate that after a first subduction phase– extending to a depth of nearly 130 km – this process terminated (~20 Ma ago?), a 'roll-back' occurred and a second subduction zone became active about 200 km further to the north, again reaching to depths of approximately 130 km. This rather unexpected new finding explains a number of subduction-related phenomena close to the northern margin of the eastern Alps which so far have erroneously been ascribed to a northward-directed subduction zone.

It is also worthwhile to mention the huge low-velocity body associated with the 'Central European Rift System' (cf. Fig. 2a) which in the sections A and B goes down to depths of a least 300 km. In sections D and E it can be seen that the asthenosphere comes close to the surface in the Tuscany region (T in Fig. 16) of central Italy (cf. Fig. 2a). In the same location a pronounced upper mantle anomaly characterized by low S-wave velocities was found several years ago from regional seismic surface-wave studies (Panza *et al.* 1982). In the light of the mantle sections D and E in Fig. 17 it is suggested here that both observations can be explained by back-arc spreading in the inner part of the Alpine arc. This extensional regime apparently extends under the northern Apennines and into the southern Po Plain.

The structure of the upper mantle below a depth of about 280 km is associated with higher P-wave velocities as shown in cross section E of Fig. 17. This depth range is normally termed the 'mantle mesosphere'. Clearly visible in all five cross sections is the deep-reaching plate boundary between Eurasia and Africa about 200 km from the southern end of the profiles which is about 150 to 200 km inland from the North African coast (see Fig. 16), i.e. south of the Tell and Atlas Mountains. Quite conspicuous is the subducted African slab in cross section A which has penetrated into the upper mantle to a depth of nearly 600 km. This rather large subducted slab can be traced further to the west into the Alboran Sea (Spakman 1986a) and probably has taken up most of the lithospheric shortening in the western Mediterranean region.

The stress regime in the Alps

In the context of the simple plate-tectonic model presented so far it has for some time been difficult to explain well-documented observations published 12 years ago (Pavoni 1975) which exhibit striking coincidence of the directions of maximum crustal shortening and the trajectories of maximum horizontal compressive stress in the arc-like western Alps and Jura Mountains. The orientation of the trajectories of maximum horizontal shortening (perpendicular to the fold axes) as deduced from a kinematic analysis of the Pliocene fold structures in the Alpine arc and its foreland exhibit the same pattern as the trajectories of maximum horizontal pressure (P axes) determined from the focal mechanisms of twenty earthquakes in the western Alps. Only a mechanism specifically characteristic for the Alpine lithospheric structure can provide an explanation for the cause and effect of these peculiar observations (Mueller 1984).

As pointed out above the collision of two continental plate margins will not only lead to a thickening of the crust, but will also produce a cold, dense and slowly subsiding 'lithospheric root' beneath a mountain chain (see Figs 13b and 14) which will induce strong regional compressive stresses able to maintain the mountain building process without requiring sustaining forces transmitted from far away (Fleitout & Froidevaux 1982). This positive feedback process is inherently unstable; whenever it reaches a climax episodic phases of strong compression will be observed accompanied by uplift (see e.g. Kissling *et al.* 1983).

Palaeomagnetic evidence indicates that the Iberian peninsula, the island chain of Corsica and Sardinia as well as the Apennine peninsula have all undergone rotations in a counterclockwise direction to a varying degree (Lowrie 1980). There is reason to believe that at least for Italy such a slow rotation is still in progress. Focal mechanism studies of earthquakes in the central Alps (Pavoni 1980) show predominantly left-lateral movement with the axes of maximum horizontal compression perpendicular to the Alpine arc. It is suggested here that these earthquake mechanisms are reactions of the rigid upper crust in the Alps in response to the continuing counterclockwise rotation of the Apennine peninsula.

Conclusions

From the presently available geophysical data it can be stated that the Alps are a deep-reaching

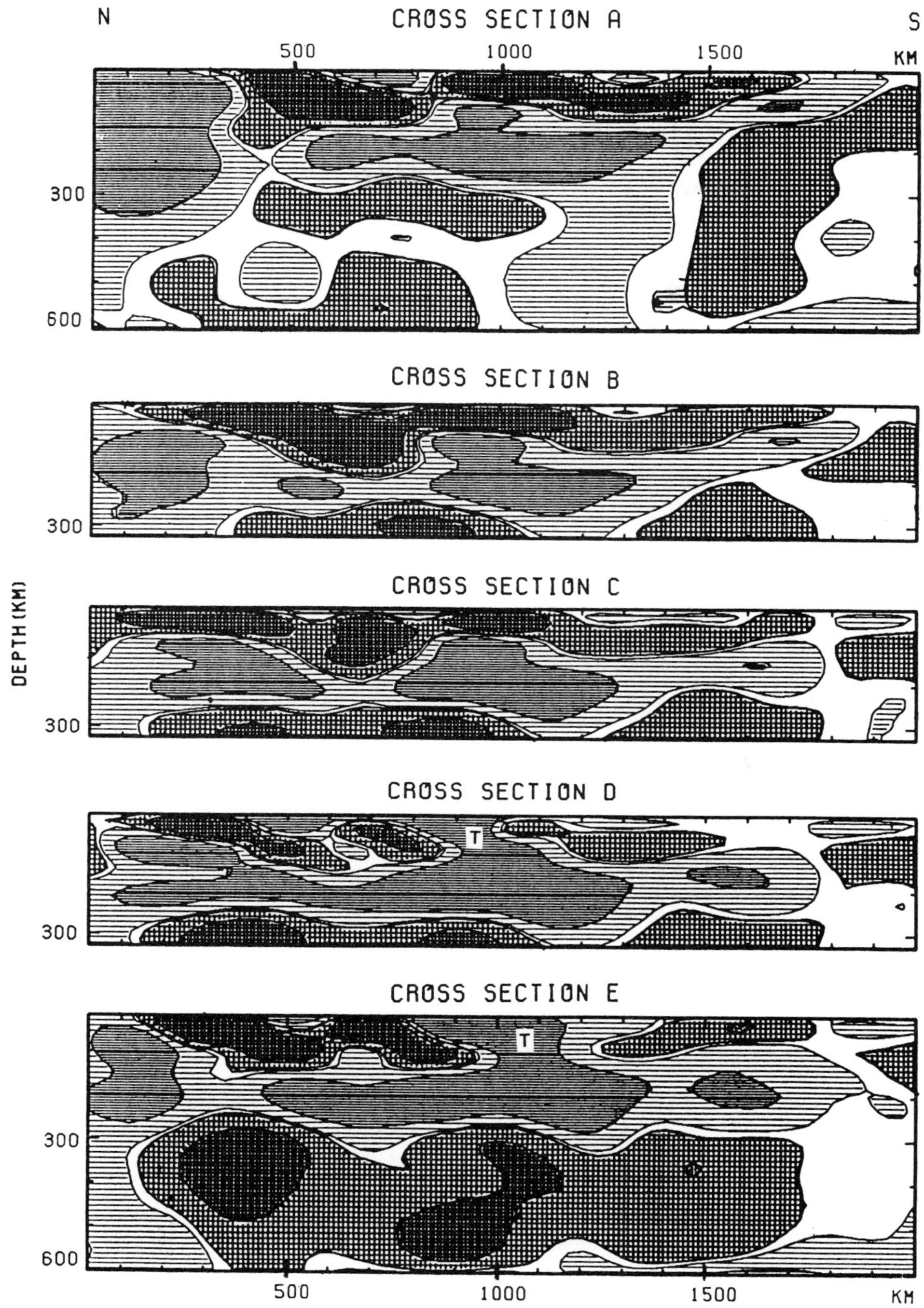

FIG. 17. Upper mantle cross sections (A . . . E) through the Alps (after Spakman 1986b) whose geographic position is indicated in Fig. 16. The shading indicates mantle P-wave velocity anomalies relative to the mean Jeffreys–Bullen velocity–depth model. Close horizontal shading: less than − 1.0%; wide horizontal shading: −1.0 to −0.1%; widely cross-hatched shading: +0.1 to +1.0%; closely cross-hatched shading: greater than +1.0%. For a detailed discussion see text. Due to the compressed depth scale details of crustal structure are, of course, not resolved in the cross sections. The depth range between about 100 and 250 km is characterized by the pronounced low velocities associated with the asthenosphere, while significant variations in the thickness and properties of the lithosphere above are evident. Note the difference in character of the southward dipping fast 'lithospheric root' beneath the western Alps (cross sections A and B) as compared to the quasi-symmetrical (bivergent?) 'lithospheric root' under the 'European Geotraverse' (cross section C) and the eastern Alps (cross sections D and E), where two separate subducted slabs can clearly be discerned. In the lower two sections the slow Tuscany (T) anomaly (for location see Fig. 16) is also distinctly visible; it could be interpreted as being due to back-arc spreading which caused a regional plume of asthenospheric material to rise close to the surface.

crust–mantle structure in Europe situated at the northernmost tip of the Adriatic promontory of the African plate. The contact between the African and the Eurasian plates in that region is presently characterized by a NW–SE directed compressional stress regime. A detailed regional analysis of seismic surface-wave dispersion and of P-wave travel time residuals ('acoustic tomography') as well as long-range seismic refraction measurements and the interpretation of 'stripped' residual gravity anomalies have revealed a rather anomalous structure within the uppermost mantle under the Alpine arc. As a result of the massive continent–continent collision parts of the lithosphere have been delaminated–a process which has led to 'flaking' in the upper crust, combined with a thickening of the entire crust and the formation of a pronounced relatively cold, dense and slowly subsiding 'lithospheric root' beneath the southern Alps and the Po Plain. In this still ongoing plate collision the lower parts of the lithosphere apparently have penetrated into the asthenosphere to a depth of 130 to 200 km at the inner side of the Alpine arc. On this scale the continuing uplift of the Alpine mountain chain (by about 0.1 cm per year) is a secondary effect mainly due to isostatic rebound of less dense crustal masses which previously had been forced to greater depths. A lithospheric configuration of this type generates regionally a compressive dynamics of its own on which – within a wider framework – rotational processes may be superimposed.

ACKNOWLEDGEMENTS: Thanks are due to a large number of individuals and institutions who now for over 30 years have cooperated unselfishly within several international research projects and whose contributions have made it possible to assemble a very comprehensive data set about the deep structure of the Alps. The author wishes to acknowledge the generous financial assistance of the Swiss National Science Foundation, which in particular supports the National Research Programme 20 'Deep Geological Structure of Switzerland'. Sincere thanks go to the author's colleagues and collaborators at the Institut für Geophysik der ETH Zürich for many stimulating discussions. The author is also indebted to the Swiss Association of Petroleum Geologists and Engineers for granting permission to reproduce several figures from their 'Bulletin' (Vol. **54**, Nr 124, April 1987).
Contribution No. 560, ETH Geophysics, Zurich, Switzerland.

References

AHORNER, L. 1975. Present-day stress field and seismotectonic block movements along major fault zones in central Europe. *Tectonophysics*, **29**, 233–49.

——, MURAWSKI, H. & SCHNEIDER, G. 1972. Seismotektonische Traverse von der Nordsee bis zum Apennin. *Geologische Rundschau*, **61**, 915–42.

AKI, K., CHRISTOFFERSON, A. & HUSEBYE, E. S. 1977. Determination of the three-dimensional seismic structure of the lithosphere. *Journal of Geophysical Research*, **82**, 277–96.

AMPFERER, O. 1906. Über das Bewegungsbild von Faltengebirgen. *Jahrbuch der kaiserlichen und königlichen Geologischen Reichsanstalt (Wien)*, **56**, 539–622.

ANSORGE, J., BONJER, K. -P. & EMTER, D. 1979. Structure of the uppermost mantle from long-range seismic observations in southern Germany and the Rhinegraben area. *Tectonophysics*, **56**, 31–48.

ARGAND, E. 1916. Sur l'arc des Alpes occidentales. *Eclogae geologicae Helvetiae*, **14**, 145–91.

—— 1924. La tectonique de l'Asie. *Comptes Rendus de la Congrès Géologique Internationale en Belgique* (1922), **13**, 171–372.

BABUŠKA, V., PLOMEROVA, J., SILENY, J. & BAER, M. 1985. Deep structure of the lithosphere and seismicity of the Alps. *Proceedings of the 3rd International Symposium on the Analysis of Seismicity and Seismic Risk* (Liblice Castle, Czechoslovakia), June 17–22.

BAER, M. 1980. Relative travel time residuals for teleseismic events at the new Swiss seismic station network. *Annales de Géophysique*, **36**, 119–26.

BALLY, A. W. 1981. Thoughts on the tectonics of folded belts. *In*: MCCLAY, K. R. & PRICE, N. J. (eds). *Thrust and Nappe Tectonics*. Special Publication of the Geological Society of London. No. 9, 13–32.

BAYER, R., CAZES, M., DAL PIAZ, G. V., DAMOTTE, B., ELTER, G., GOSSO, G., HIRN, A., LANZA, R., LOMBARDO, B., MUGNIER, J. -L., NICOLAS, A., NICOLICH, R., POLINO, R., ROURE, F., SACCHI, R., SCARASCIA, S., TABACCO, I., TAPPONNIER, P., TARDY, M., TAYLOR, M., THOUVENOT, F., TORREILLES, G. & VILLIEN, A. 1987. Premiers résultats de la traversée des Alpes occidentales par sismique reflexion verticale (Programme ECORS–CROP). *Comptes-Rendus de l'Académie des Sciences de Paris*, **305**, II, 1461–70.

BEB Gewerkschaften Brigitta und Elwerath Betriebsführungsgesellschaft m. b. H. 1979. Auf der Suche nach Erdgas und Erdöl in der Schweiz (Firmen-Prospekt) Hannover, B. R. D.

CERMAK, V. & HURTIG, E. 1979. Heat flow map of Europe. *In*: CERMAK, V. & RYBACH, L. (eds). *Terrestrial Heat Flow in Europe*. Springer-Verlag, Berlin.

CHANNELL, J., D'ARGENIO, B. & HORVATH, F. 1979. Adria, the African promontory. *In*: Mesozoic Mediterranean paleogeography, *Earth Science Reviews*, **15**, 213–92.

—— & HORVATH, F. 1976. The African–Adriatic promontory as a paleogeographic premise for Alpine orogeny and plate movements in the Carpatho–Balkan region. *Tectonophysics*, **35**, 71–101.

FINCKH, P., FREI, W., FREEMAN, R., HEITZMANN, P., LEHNER, P., MUELLER, ST., PFIFFNER, A. & VALASEK, P. 1987. Nationales Forschungsprogramm 20 'Geologische Tiefenstruktur der Schweiz' – Problemstellung und erste Resultate. *Bulletin der Vereinigung Schweizerischer Petroleum-Geologen und-Ingenieure*, **54**, 59–74.

FLEITOUT, L. & FROIDEVAUX, C. 1982. Tectonics and topography for a lithosphere containing density heterogeneities. *Tectonics*, **1**, 21–56.

FREY, M., HUNZIKER, J. C., FRANK, W., BOCQUET, J., DAL PIAZ, G. V., JAEGER, E. & NIGGLI, E. 1974. Alpine metamorphism of the Alps. *Schweizerische Mineralogische und Petrographische Mitteilungen*, **54**, 247–90.

——, BUCHER, K., FRANK, E. & MULLIS, J. 1980. Alpine metamorphism along the Geotraverse Basel–Chiasso: a review. *Eclogae geologicae Helvetiae*, **73**, 527–46.

GALSON, D. A. & MUELLER, ST. 1987. The European Geotraverse (EGT) Project: A progress report. *In*: FUCHS, K. & FROIDEVAUX, C. (eds). *Composition, Structure and Dynamics of the Lithosphere–Asthenosphere System*. Geodynamics Series of the American Geophysical Union, Washington, **16**, 253–72.

GUBLER, E., KAHLE, H.-G., KLINGELÉ, E., MUELLER, ST. & OLIVIER, R. 1981. Recent crustal movements in Switzerland and their geophysical interpretation. *Tectonophysics*, **71**, 125–52.

HSÜ, K. J. 1979. Thin-skinned plate tectonics during Neoalpine orogenesis. *American Journal of Science*, **279**, 353–66.

KARNER, G. D. & WATTS, A. B. 1983. Gravity anomalies and flexure of the lithosphere at mountain ranges. *Journal of Geophysical Research*, **88** (B 12), 10449–77.

KISSLING, E., MUELLER, ST. & WERNER, D. 1983. Gravity anomalies, seismic structure and geothermal history of the central Alps. *Annales Geophysicae*, **1**, 37–46.

KLEMPERER, S. L., HAUGE, T. A., HAUSER, E. C., OLIVER, J. E. & POTTER, C. J. 1986. The Moho in the northern Basin and Range province, Nevada, along the COCORP 40° N seismic-reflection transect. *Geological Society of America Bulletin*, **97**, 603–18.

KRAUS, E. C. 1931. Die Alpen als Doppelorogen. *Geologische Rundschau*, **22**, 65–78.

LAUBSCHER, H. P. 1969. Mountain building. *Tectonophysics*, **7**, 551–63.

—— 1970. Bewegung und Wärme in der alpinen Orogenese. *Schweizerische Mineralogische und Petrographische Mitteilungen*, **50**, 565–96.

—— 1974. The tectonics of subduction in the Alpine system. *Memorie della Società Geologica Italiana*, Supplemento 2 al volume XIII, 275–83.

—— 1985. The tectonics of the western and southern Alps: Correlation between surface observations and deep structure. *In*: GALSON, D. A. & MUELLER, ST. (eds). *Second EGT Workshop – The Southern Segment*, 93–101. European Science Foundation, Strasbourg, France.

LOWRIE, W. 1980. A paleomagnetic overview of the Alpine system. *Mémoire du B.R.G.M. (Orléans, France)*, **115**, 316–30.

MILLER, H., MUELLER, ST. & PERRIER, G. 1982. Structure and dynamics of the Alps—a geophysical inventory. *In*: BERCKHEMER, H. & HSÜ, K. J. (eds). *Alpine–Mediterranean Geodynamics*. Geodynamics Series of the American Geophysical Union, Washington, **7**, 175–203.

MINSTER, J. B. & JORDAN, T. H. 1978. Present-day plate motions. *Journal of Geophysical Research*, **83**, (B11), 5331–54.

MUELLER, ST. 1977. A new model of the continental crust. *In*: HEACOCK, J. G. (ed). *The Earth's Crust*. Geophysical Monograph Series of the American Geophysical Union, Washington, **20**, 289–317.

—— 1982a. Deep structure and recent dynamics in the Alps. *In*: HSÜ, K. J. (ed). *Mountain Building Processes*, 189–99. Academic Press, London.

—— 1982b. Geodynamic aspects of the Mediterranean–Alpine region. *Revista de Geofisica (Madrid)*, **38**, 103–11.

—— 1984. Dynamic processes in the Alpine arc. *Annales Geophysicae*, **2**, 161–4.

——, EGLOFF, R. & ANSORGE, J. 1976. Struktur des tieferen Untergrundes entlang der Schweizer Geotraverse. *Schweizerische Mineralogische und Petrographische Mitteilungen*, **56**, 685–92.

——, ANSORGE, J., EGLOFF, R. & KISSLING, E. 1980. A crustal cross section along the Swiss Geotraverse from the Rhinegraben to the Po Plain. *Eclogae geologicae Helvetiae*, **73**, 463–85.

—— & BANDA, E. 1983. The European Geotraverse Project. *Terra Cognita*, **3**, 291–4.

—— & PANZA, G. F. 1986. Evidence of a deep-reaching lithospheric root under the Alpine arc. *In*: WEZEL, F.-C. (ed). *The Origin of Arcs*. Developments in Geotectonics, **21**, 93–113. Elsevier, Amsterdam.

NOLET, G. 1985. Solving or resolving inadequate and noisy tomographic systems. *Journal of Computational Physics*, **61**, 463–82.

PANZA, G. F. 1985. Lateral variations in the lithosphere in correspondence of the southern segment of EGT. *In*: GALSON, D. A. & MUELLER, ST. (eds). *Second EGT Workshop – The Southern Segment*, 47–51. European Science Foundation, Strasbourg, France.

—— & MUELLER, ST. 1979. The plate boundary between Eurasia and Africa in the Alpine area. *Memorie di Scienze Geologiche (Padova)*, **33**, 43–50.

——, MUELLER, ST. & CALCAGNILE, G. 1980. The gross features of the lithosphere–asthenosphere system in Europe from seismic surface waves and body waves. *Pure and Applied Geophysics*, **118**, 1209–13.

——, ——, CALCAGNILE, G. & KNOPOFF, L. 1982. Delineation of the north–central Italian upper mantle anomaly. *Nature*, **296**, 238–9.

PAVONI, N. 1975. Zur Seismotektonik des Westalpenbogens. *Vermessung, Photogrammetrie, Kul-*

turtechnik, **3/4**, 185–87.

Pavoni, N. 1980. Crustal stresses inferred from fault-plane solutions of earthquakes and neotectonic deformation in Switzerland. *Rock Mechanics*, Supplement **9**, 63–8.

Rybach, L., Mueller, St., Milnes, A., Ansorge, J., Bernoulli, D. & Frey, M. 1980. The Swiss Geotraverse Basel–Chiasso: a review. *Eclogae geologicae Helvetiae*, **73**, 437–62.

Schwendener, H. 1984. Ein gravimetrisches Krusten–Mantel–Modell für ein Profil vom nördlichen Alpenvorland bis an die Ligurische Küste. *Geodätisch–geophysikalische Arbeiten in der Schweiz*, Kümmerly & Frei (Bern), Band 36, 160 pp.

—— & Mueller St. 1985. New evidence for a density anomaly in the upper mantle below the Southern Alps. *In*: Galson, D. A. & Mueller St. (eds). *Second EGT Workshop–The Southern Segment*. European Science Foundation (Strasbourg, France), 115–20.

Smith, D. G. (ed). 1982. *The Cambridge Encyclopedia of Earth Sciences*. Cambridge University Press, 496 pp.

Spakman, W. 1986a. Subduction beneath Eurasia in connection with the Mesozoic Tethys. *Geologie en Mijnbouw*, **65**, 145–53.

—— 1986b. The upper mantle structure in the central European–Mediterranean region. *In*: Freeman, R., Mueller, St. & Giese, P. (eds) *Third EGT Workshop–The Central Segment*. European Science Foundation (Strasbourg, France), 215–21.

Staub, R. 1928. *Der Bewegungsmechanismus der Erde, dargelegt am Bau der irdischen Gebirgssysteme*. Gebrüder Bornträger, Berlin.

Suess, E. 1875. *Die Entstehung der Alpen*. W. Braumüller, Wien.

Tapponnier, P. & Molnar, P. 1976. Slip-line field theory and large-scale continental tectonics. *Nature*, **264**, 319–24.

Trümpy, R. 1980. *Geology of Switzerland – a Guide Book*. Wepf & Co., Basel–New York, 334 pp.

Udias, A. 1982. Seismicity and seismotectonic stress field in the Alpine–Mediterranean region. *In*: Berckhemer, H. & Hsü, K. J. (eds). *Alpine–Mediterranean Geodynamics*. Geodynamics Series of the American Geophysical Union Washington, **7**, 75–82.

Waniek, L., Van Gils, J. M. & Ritsema, A. R. 1982. European Seismological Commission–Activity Report 1980–1982. K.N.M.I. De Bilt (The Netherlands), 90 pp.

Wegener, A. 1912. Die Entstehung der Kontinente. *Geologische Rundschau* **3**, 276–92.

Werner, D. 1985. A two-dimensional geodynamic model for the southern segment of the EGT. *In*: Galson, D. A. & Mueller, St. (eds). *Second EGT Workshop–The Southern Segment*. European Science Foundation (Strasbourg, France), 65–9.

—— & Kissling, E. 1985. Gravity anomalies and dynamics of the Swiss Alps. *Tectonophysics*, **117**, 97–108.

Wilson, J. T. 1965. A new class of faults and their bearing on continental drift. *Nature*, **207**, 343–7.

St. Mueller, Institute of Geophysics, Swiss Federal Institute of Technology, ETH-Hönggerberg, CH-8093 Zürich, Switzerland.

Indentation and rotation in the western Alpine arc

P. Vialon, P. Rochette & G. Ménard

SUMMARY: The overall structural aspects of the external western Alpine arc and foreland may be consequences of NW–SE compression ahead of crustal thrusting on the northern front and of mainly sinistral wrenching along its southwestern lateral border. However such a model assumes a simple straight-line displacement which, while possibly valid on the continental scale, cannot be considered for the external arc since a model of linear translation and rotation around a distant pole is not compatible with the structural organization of the western arc of the Alps. The constant tangential stretching parallel to the curved front, the fan-ring shortening and shear directions on décollement levels, the complex nappe transport trajectories and the widespread dextral wrench-faulting around the arc, can be better explained by local rotation of a block belonging to the European margin, dragged along the western sidewall of a large indenter moving northwestwards. This laterally displaced block produces an anticlockwise ring shear about a pole of rotation within it. Consequently, the Alpine arc cannot be understood in terms of plane-strain models along local 'transport' directions. The deformation process corresponds instead to lateral rotation–expulsion of a deformed unit belonging to the European Plate margin, jammed against its stable foreland.

The general tectonic framework

The French–Italian arc of the Alps, extending over a distance of 300 km, forms the western end of an E–W trending mountain range some 1000 km long. This arc can be imagined as part of the European Plate margin deformed along the western sidewall of a large 'Apulian' indenter moving and thrusting northwestwards (Argand 1924, Goguel 1963, Tapponnier 1977). As a consequence of this continental indentation, the curvature of the external zone between Switzerland and the Mediterranean Sea may be due to movements of the crystalline basement pushing away the surrounding cover. At this scale, and assuming compatibility with the Apulian movement as a first approximation, the displaced units of the External Crystalline Massifs may act essentially by NW–SE compression along the frontal alignment of the northern branch (Aar – Mt Blanc – Belledonne), with the Swiss molasse basin running along and in front of the deformed cover of this branch. After the sudden change in orientation of the crystalline massifs near Grenoble, the Pelvoux–Argentera branch would be in a position approximately parallel to the direction of displacement (Fig. 1) (Ménard 1979, 1980).

The arrangement of the cover-basement interface and other geophysical data (Ménard 1980, Ménard & Thouvenot 1984) would seem to suggest, as would the main map pattern of the External Crystalline Massifs, that the crystalline basement is an angular-shaped block with its limits depending on the attitude of Hercynian faults and especially of the so-called Cévennes faults (Vialon 1974), hidden under the deformed sedimentary cover. This previously faulted basement of the Alps is assumed to be rather similar to the nearby Massif Central, from which it inherits the geometry of extension structures and sedimentary distribution in the Alpine ocean (Arnaud *et al.* 1978, Barféty *et al.* 1979, Boillot *et al.* 1984). In the eastern part of the French Alps, a thickened crust is found with a low seismic velocity zone (LVZ) at a depth between 10 and 20 km under the External Crystalline Massifs. This can be interpreted in terms of stacking by crustal-scale thrusting, the lower part beneath the LVZ being the bent basement of the overlapped foreland (Fig. 2) (Ménard & Thouvenot 1984).

Thus, it seems possible that the basal framework, which governs the tectonic evolution of the arc area, corresponds to the pattern of the old basement faults, especially those trending N50 (Cévennes trend) and N140, which were reactivated several times. The previously faulted crystalline basement was reactived by normal faulting, flattening at depth on a delamination level, during rifting of the passive European margin of the Alpine ocean (Boillot *et al.* 1984). Closure of the ocean by northwestward movements carried this region into collision. The shortening occurred on the previous discontinuities but in a reverse sense (Gillcrist *et al.* 1987). The northwestward slip acted on the flat, deep crustal delamination planes and on the Cévennes fault, transforming them into reverse

From: Coward, M. P., Dietrich, D. & Park, R. G. (eds), 1989, *Alpine Tectonics*, Geological Society Special Publication No. 45, pp. 329–338.

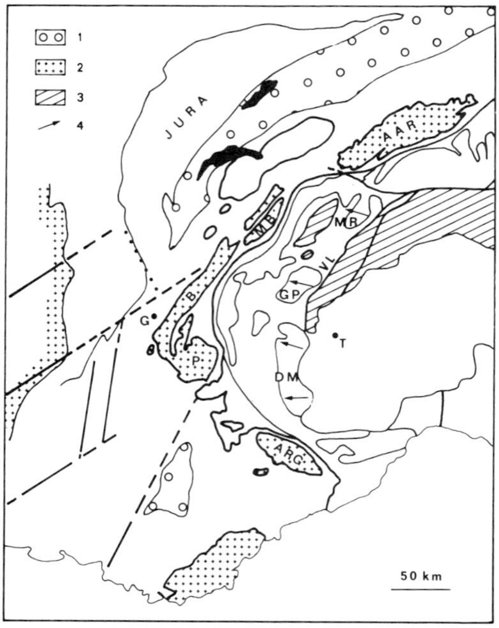

FIG. 1. General geological setting of the western Alpine arc and foreland. 1 = Molasse basin in front of the main NE–SW trending chain and in front of the northwestwards advancing nappes. 2 = Crystalline basement showing an angular outcrop pattern (near Grenoble = G) between Mt Blanc (= MB), Belledonne (= B), Pelvoux (= P) and Argentera (= ARG) Massifs. This outcrop pattern depends on the orientation of the Hercynian faults. The orientation of the MB–B alignment seems to deviate from the Aar trend and the main Hercynian faults (Cévennes trends) given for the western foreland of the Alps. 3 = Austro–Alpine domain thrusting over the Pennine units (in white), with the outlines of the Monte Rosa (MR), Gran Paradiso (GP) and Dora Maira (DM) internal massifs where the transport lineations (4) are shown. VL = Viu-Locana suture. T = Torino. Data from Lacassin (1987) and Malavielle *et al.* (1984).

faults, and leading to crustal thickening by thrusting in the eastern part of the external arc. An accretionary wedge developed in front of the crustal thrusts on the bent foreland where, under the load of advancing nappes, a further frontal molasse basin was formed (Fig. 2). To the southwest, where the NW–SE shortening and thrusting of crustal slices did not occur, a 'transform zone' appeared to absorb and transfer the displacement away from the stable western foreland by re-using the NW–SE discontinuities. A true NW–SE trending lateral ramp or a large vertical shear zone would have been able to accommodate the resulting sinistral wrenching.

As a consequence of this oversimplified mechanism, which assumes a transport direction always in a NW–SE straight line, shortening predominates in front of the Aar–Mt Blanc–Belledonne branch of the Crystalline Massifs. Both the shortening and the stretching directions lie in a vertical finite-strain XZ plane which also contains the NW–SE displacement direction. Along the axes of the NE–SW trending folds, only slight extension is to be expected if at all. The NW–SE vertical plane, which is a plane strain section, is therefore the best plane for drawing balanced cross-sections through the Alps. This is generally accepted (Boyer & Elliott 1982, Butler *et al.* 1986) and Fig. 2, constructed at a crustal scale, has been drawn according to this theory.

On the other hand, in the assumed lateral wrenching zone, a lesser degree of shortening than in the frontal area is probable, and the deformation includes significant stretching parallel to the fold axes.

Many observations fit quite well with this idealized structural pattern. However, several observations do not fit the pattern. For example a rectilinear translation associated with a distant pole of rotation, as assumed by a general movement either in a NW–SE (as above) or other direction (Laubscher 1971, Beach 1981, Butler *et al.* 1986, Ricou & Siddans 1986), is highly questionable in the western arc of the Alps, since transport and shortening seem to be divergent around this arc. These fan directions might be explained by final overflowing of the superficial cover occurring as gravitational spreading and sliding. However, the fold and thrust geometries and vergences and the nappe displacements, including the basement nappes and the structures of the internal metamorphic zones, are generally thought to be due to movements perpendicular to the curved structural strikes. This would mean that the transport directions are radial and converge into the same central zone situated in the Po plain. The question arising in this case concerns the origin of the translated material.

Reality of movement directions

Throughout the entire western arc of the Alps it is thus difficult or impossible to assume a simple straight-line translation model and a centrifugal expulsion model. Furthermore, the problem arises of whether to take into account, together with the westwards movements, the movement associated with a northwestwards collision.

In the internal zones, between Monte Rosa,

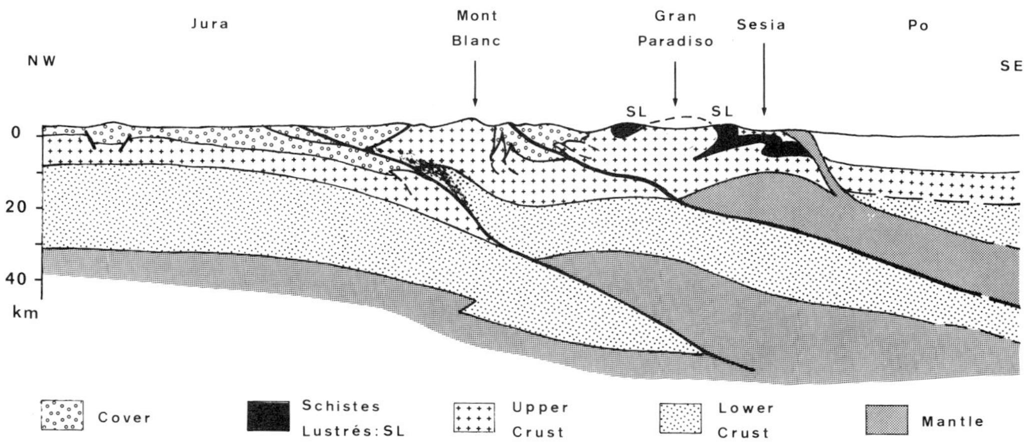

FIG. 2. Crustal scale geological balanced cross-section between the Jura and the Po plain, assuming that the NW–SE section plane is a plane strain section. The bent basement of the northwestern foreland is overlapped by the advancing Alpine nappes, and the basement in front of the nappes forms the External Crystalline Massifs. The thrusts initiate on deep mantle slices, whose shape and emplacement have been deduced from gravimetric and seismic data.

south of the Simplon Line (Mancktelow 1985, Lacassin 1987, Mancel & Merle 1987), and the southern part of the Dora Maira Internal Crystalline Massifs (Vialon 1966, Malavieille *et al.* 1984), the lineations and folds interpreted to be parallel to the transport directions are essentially E–W. They result of deformation coeval with the metamorphic events beginning in the Middle Cretaceous and proceeding through the Eocene continental collision, around 40 Ma. This E–W displacement pattern is continuous through the Schistes Lustrés and Briançonnais zones as far as the external parts of the arc and the External Crystalline Massifs. However, there, the structures become progressively divergent, as shown both by the curved directions of folds and thrusts and by the lineations and traces of nappe displacement (Fig. 3), and also appear to be increasingly recent towards the external zones.

For example, in the southern part of the arc, between and in front of the Pelvoux and Argentera, in the Embrunais–Ubaye Helminthoides flysch, the Cretaceous displacement is found to be northwestward, then becoming westward and even southwestward in the Oligocene (Merle & Brun 1984, Choukroune *et al*, 1986). In the foreland during the Neogene, the Digne nappe was displaced SW and southwards (Gidon & Pairis 1986), so that its western border acts as a dextral N–S lateral ramp (Fig. 3).

In the same way, in the northern part of the arc, behind the Mt Blanc massif, the internal E–W lineations become NW and further to the North, the Helvetic nappes move along well-documented complicated trajectories (Ramsay & Huber 1983, Dietrich & Durney 1986). During the earlier overthrust shear, the displacement was northward, then northwestward and more recently westward. This final displacement seems to be associated with large-scale dextral transpressional movements (Burkhard 1986, Dietrich pers. comm.), which can explain the deviated nappe displacement trajectories. Between the Simplon area and the Rhône valley, up to the Aiguilles Rouges Massif, these late movements affect a wide deformation zone, partially obliterating earlier orientations (Steck 1984).

In front of the External Crystalline Massifs, the displacements in the Subalpine zone are believed to have been concentrated on several ductile levels acting during the Miocene, and mainly on the Callovo–Oxfordian 'Terres Noires'. It is possible to follow deformation evolution along the arc in such shear levels by studying the magnetic susceptibility of rocks and their anisotropy (Rochette & Lamarche 1986, Lamarche 1987, Lamarche & Rochette 1987). In the Mt Blanc–N Belledonne area, considerable deformation has occurred connected with metamorphism; the transport direction is NW. At the same levels, deformation progressively diminishes southwards, the metamorphism disappears and the transport

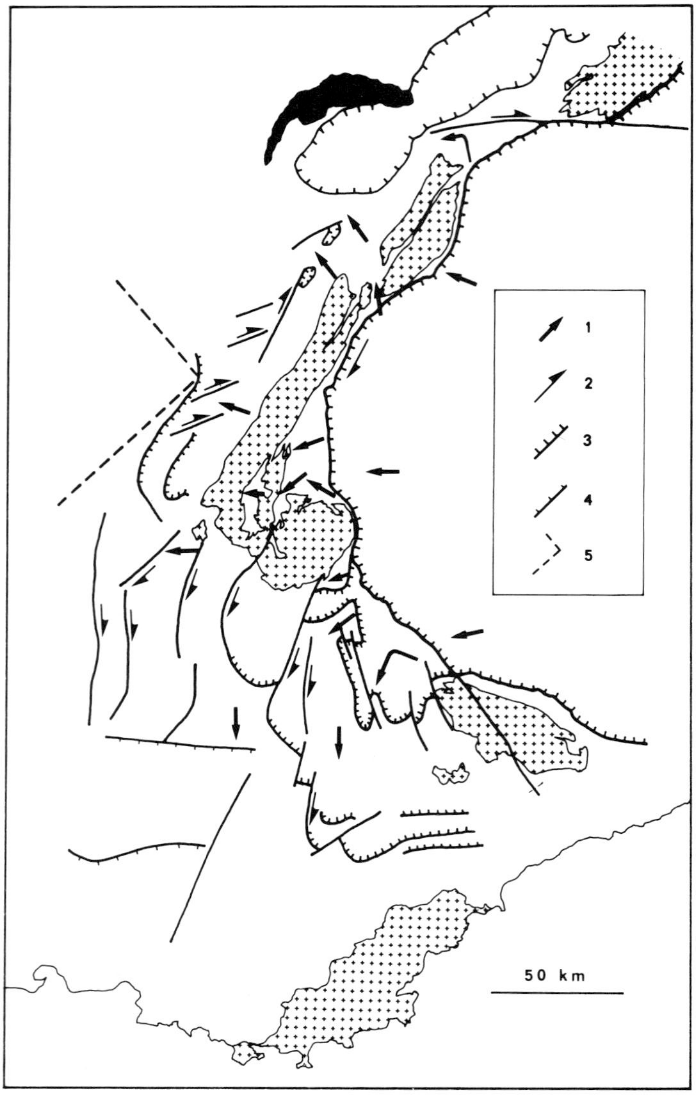

FIG. 3. The main displacement directions in the external zone, shown by the transport lineations of nappes and the stretching lineations in the sheared ductile levels (= 1) and by the principal dextral strike slip faults (= 2). 3 = Main thrust limits. 4 = Thrust between the Provençal and Subalpine domains. 5 = Margin of the fixed platform of the Bas Dauphiné foreland. Data from 1/50 000 geological maps of SE France (published by the BRGM) and by Burkhard (1986), Choukroune *et al.* (1986), Dietrich & Durney (1986), Gidon & Pairis (1986), Lacassin (1987), Lamarche (1987), Malavieille *et al.* (1984), Merle & Brun (1984), Pijolat *et al.* (1981), Ramsay & Huber (1983).

direction becomes E–W (and even WSW according to the present state of knowledge) (Fig. 4, and Aubourg 1987, pers. comm.).

There is therefore a change in displacement directions in time and space along the curvature of the western arc. The oldest directions are in the internal zones with a present orientation that is fairly regularly E–W: these are the so-called transversal lineations of the chain. The more recent lineations become divergent in the outer zones. The pattern of movement directions in the Subalpine zone is roughly helicitic (see Fig. 5) from the NW–SE trend in the northern Helvetic area, to the N–S trend near

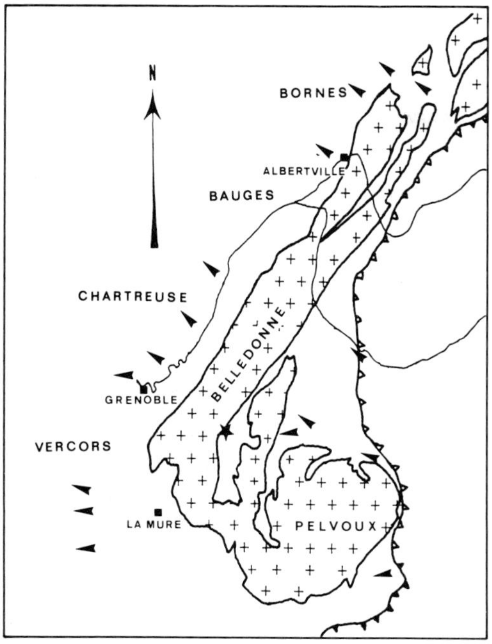

FIG. 4. Map of magnetic lineations in the sedimentary cover (Jurassic black shales and slates of the Dauphinois zone and Eocene flysch) of the External Crystalline Massifs. Arrows indicate the preferred orientation (star when vertical) of magnetic minerals (elongated magnetite grains and phyllosilicates). This orientation is either parallel to the S_0/S_1 intersection in weakly deformed rocks, or to a ductile stretching parallel to the displacement in the sheared levels as shown here. The data are the results of measurements carried out by Rochette (1983), Lamarche (1987) and Aubourg (unpublished report, Grenoble University, 1987).

the Provence zone. Similarly, there is significant stretching around and ahead of the External Crystalline Massifs, even along the previously mentioned frontal northern branch. This stretching is parallel to the 'en échelon' fold axes, this arrangement being a normal complement of the helicitic arrangement of the movement trends. The axial stretching, approximately parallel to the curved chain, is represented by either mineral orientation (Pijolat *et al.* 1981) or transverse tensile fractures, depending on the lithology. However, the main map features expressing the stretching are the generalised dextral strike-slip faulting (including the above-mentioned lateral sinistral wrenching zone), cutting obliquely through the Subalpine folds (Fig. 3).

It would therefore seem that the external zone is governed by recent (since the Miocene?) outwards thrusting and general dextral shear working along an approximately semi-circular boundary between the relatively stable foreland and the internal zone. This internal part appears to rotate anticlockwise about a pole within it. The simpler model of a straight northwestward translation of the inner zones, as described above, is nonetheless still partially valid. Indeed, there are differences of deformation between the northern Subalpine zone and the southwestern zone. The shortening in the north is more pronounced and linked to metamorphism. The northwestwards thrusting and frontal consequences (bent foreland, molasse basin, etc) are confirmed by the positions of the main structural units and of the subsiding basin. Geophysical data (Perrier & Vialon 1980) also confirm this situation and show that the crustal thickening by existing basement delamination in the northern area is absent in the southwestern zone. Thus, the displaced units must have rotated during the general northwestwards translation, or by successive increments, while the deformation and thrusts were progressing outwards.

The ring shear model

If the northwestwards translation is ignored and consideration is given only to the curved dextral wrenching inferred from the structures of the external Alpine arc, the results of the simple shear experiments of Mandl *et al.* (1977) in a ring shear apparatus are interesting. Radial deviated compression directions are observed in a vertical circular shear band obtained between a fixed outside wall and a coaxial inner rotating cylinder (Fig. 5). These directions, always situated at 45° to the circular border planes, are comparable to the helicitic fan of finite displacement and shortening trends shown by the external Alpine structures. This similarity reinforces the idea that the internal displaced zones of the Alps act against the foreland in an anticlockwise rotation as in Mandl's experimental ring shear device or in Emmons' model (1969) of a curved strike-slip displacement.

The rotation revealed by the structures of the outer zones of the arc probably began with the Tertiary. It may still be active today because the present stress field, drawn from the focal mechanisms of recent seismic activity (Fig. 6) (Fréchet 1978, Ménard & Fréchet 1988), is described in the western Alps as a fan of compression directions quite similar to those obtained in the experimental device.

The rotating unit transports the kinematic

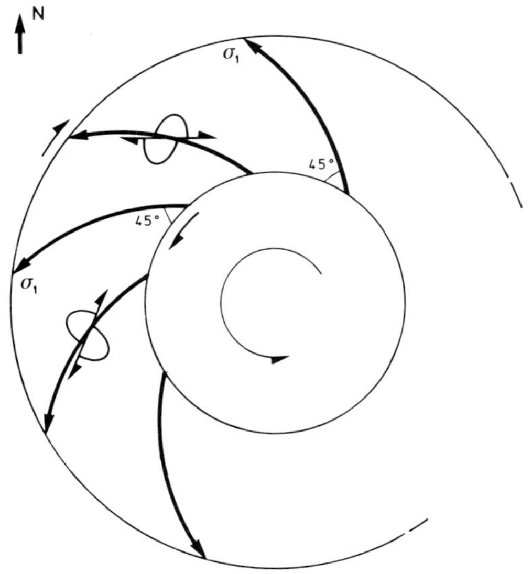

FIG. 5. Ring shear model of Mandl *et al.* (1977). The radially deviating compression directions are observed in the dextral vertical circular simple shear band between a fixed outer limit and a coaxial inner anticlockwise-rotating cylinder. The finite strain ellipses are shown with the inferred dextral strike-slip faults, which are the main characteristic map features in the western Alpine arc. Compare with Fig. 6, and Gidon (1974).

indicators acquired during previous stages of rotation. The Cretaceous transport lineations, currently E–W in the internal zones, diverging forwards and becoming increasingly recent, are typical examples. The recent rotation of this inner part of the arc, which still belongs to the European margin, has to be subtracted to obtain the true transport direction from the orientation of these lineations.

Structural interpretation

The resulting displacement pattern around the Alpine arc has been described as if the arc was a continuous medium. In fact, the behaviour of the different units and the movement directions depends on the nature of the formations involved, on the orientations of their boundaries and on previous discontinuities. The crystalline basement of the External Crystalline Massifs can be considered as rather rigid and its borders angular shaped, governed by the orientation of the old Hercynian faults. A rough idea of the external shape of the internal rotating unit acting on the surrounding soft sedimentary cover can therefore be obtained.

In the foreland north of Grenoble the Isère basement fault, parallel to the Cévennes trend of the nearby Massif Central, provides an external boundary. It forms an outer zone which stops and moulds the Subalpine folds and thrusts as well as the southern Jura folds. This outer platform can be likened to a fixed buttress in the foreland and it is between this platform and the External Crystalline Massifs that the ring shearing is thought to take place. Gratier *et al.* (this volume) argue that the cover of the External Crystalline Massifs in the Grenoble area is deformed and displaced by an anticlockwise rotation between the outer fixed limit and the inner crustal thrusts.

On a larger scale, the entire rotated unit and the mechanism of its displacement can be approximately defined (Fig. 7). The unit could be roughly a quadrangle in which the palaeomagnetic directions are deviated consistently relative to stable Europe (Westphal 1973, Schott 1976, Bogdanoff & Schott 1977). The external boundaries would be the Mt Blanc–Belledonne and Pelvoux–Argentera alignments, but excluding the Argentera Massif itself, where the palaeomagnetic orientation is identical to that of stable Europe (this does not necessarily mean that the Argentera Massif would represent a fixed point since it may have been translated without rotation). The northern boundary would be the complex Helvetic dextral transpressional zone. The internal boundary would lie between Simplon and Argentera, along the Viu–Locana suture and the edge of the Po Plain (Fig. 1). It is important to underline that the sedimentary cover of the foreland is considerably displaced and deformed over its basement in front of the External Crystalline Massifs. The positions of both the western molasse basin and of the Jura are governed by the localization and shape of these massifs and their sedimentary cover. They also coincide with the only zones where evaporite horizons are present at the basement/cover interface (Vann *et al.* 1986), horizons which provide favourable conditions for the localization of displacements and deformation.

The restored position of such a unit before rotation can only be approximate. Indeed the retro-rotation must be added to a translation (which must take into account the dominant northwestwards thrusting already described) between the relative 'fixed' situation of Argentera and the Helvetic dextral transpressional zone. The initial Hercynian orientation of the Belledonne–Mt Blanc alignment may have been parallel to the Cévennes trend. This seems probable because the border faults

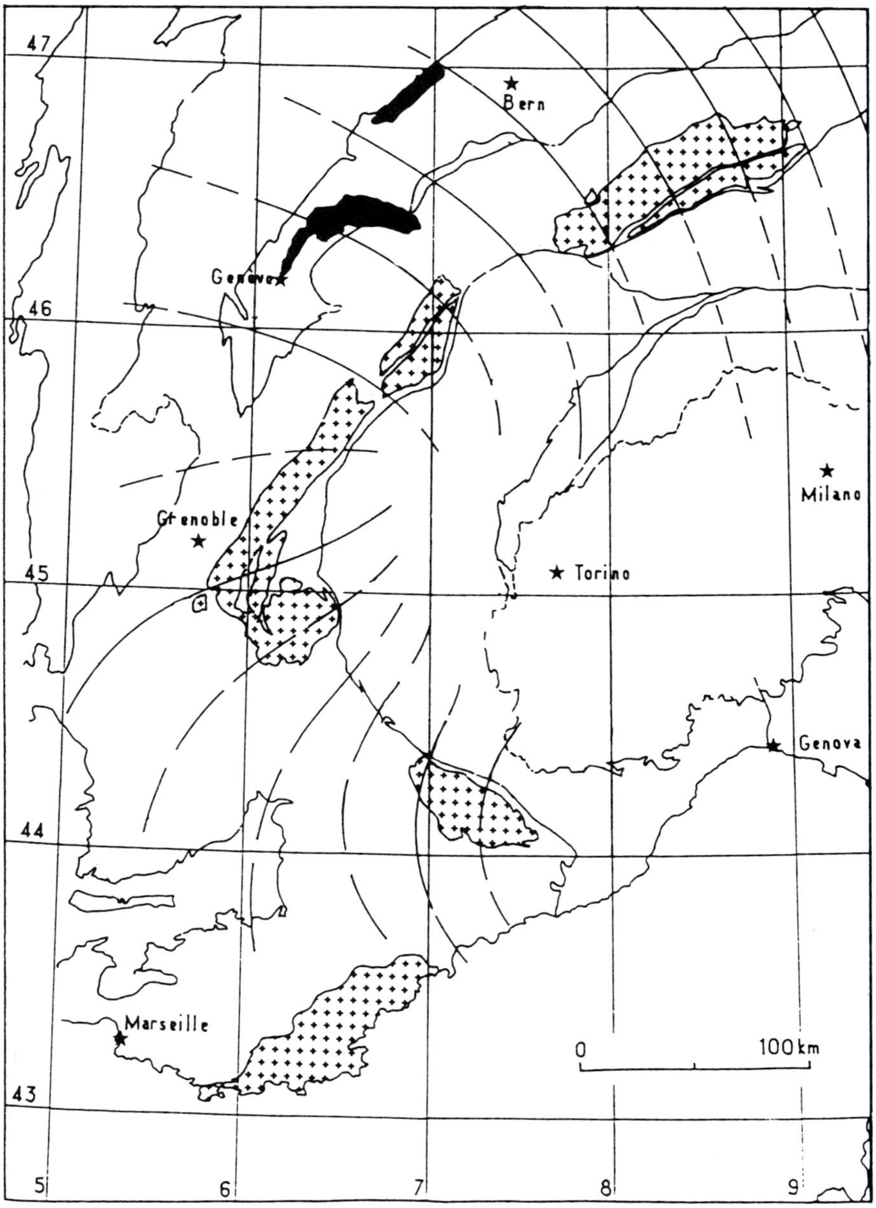

FIG. 6. Present stress field in the western Alpine arc drawn from the focal mechanisms of recent seismic activity (from Fréchet 1978, and Ménard & Fréchet 1989). Divergent lines are the orientations of P axes of earthquake mechanisms (see also Pavoni 1986). Compare with Fig. 5 and with the whirling model of Gidon (1974).

of the crystalline massifs are the same old basement faults which have been reactivated several times in both the Alps and in the Cévennes. By acting inversely along the Helvetic transpressional zone and by replacing the crystalline massifs in this assumed previous position, the restored orientation becomes the double-lined corner shown in Fig. 7(a) or the dash-lined quadrangle in Fig. 7(b). This replaces the palaeomagnetic direction of the Triassic rocks of the southern Pelvoux parallel to the stable Europe direction (C restored to C', Fig. 7(a)). The palaeomagnetic reference sites of Dourmillouse (= C), and Ponsonnière (= B), are

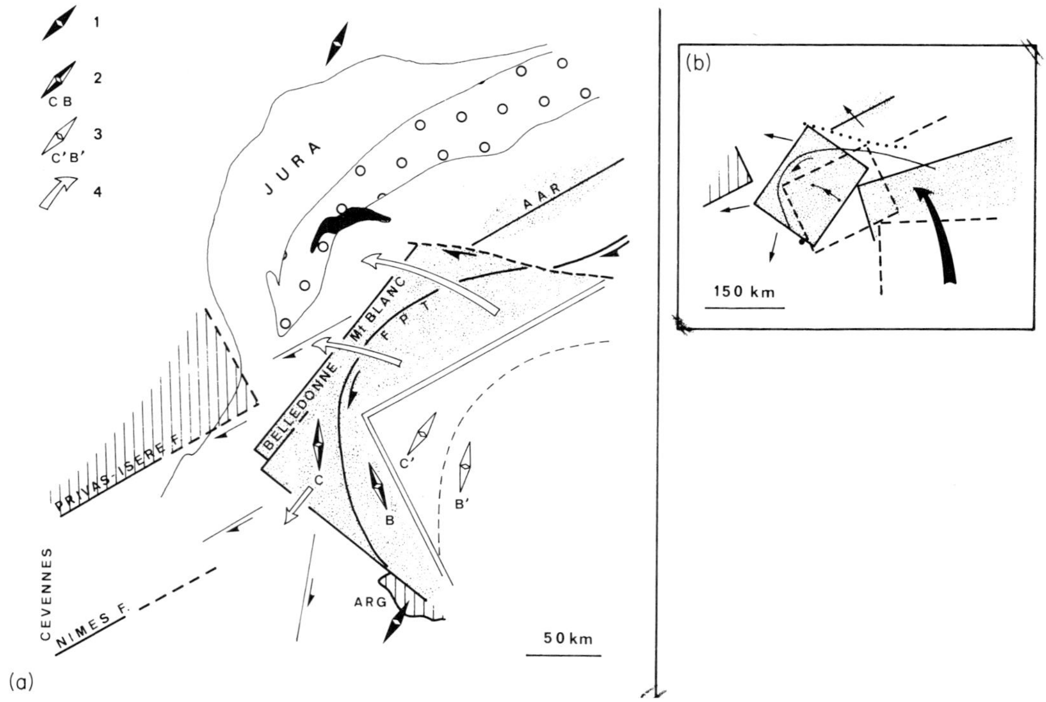

FIG. 7. Interactions of translation and rotation of the Mt Blanc–Belledonne–Pelvoux block at the western sidewall of the Apulian indenter.
(a) The shaded part indicates the present position of displaced blocks carrying the previously-thrusted internal units limited by the frontal Pennine thrust (= FPT). In the western foreland of the Alps and Jura the stable angular buttress determined by the Isère fault proceeding from the Cévennes trend is vertically hatched as well as the 'fixed' Argentera Massif (= ARG).

(1) Palaeomagnetic directions of Permo–Triassic rocks in stable Europe: data from Vosges and Argentera.
(2) Present palaeomagnetic orientations in Triassic rocks of the adhering cover of the Crystalline basement (= C, in the Southern Pelvoux, locality Dourmillouse) and in Permo–Triassic rocks of the Briançonnais zone (= B, North of Briançon, locality Ponsonnière).
(3) Palaeomagnetic orientations after retro-rotation of the crystalline alignment according to the Cévennes trend (double lined corner). The orientation of the crystalline basement C' reverts to the stable Europe orientation. The Briançonnais B' needs an additional retro-rotation along the FPT to be restored to the stable Europe direction.
(4) Displacement of the Mt Blanc–Belledonne block from its initial position (double lined corner), leading to the fan-shaped orientation of the dextral strike-slip faults in the foreland. The translation-rotation acts about the 'fixed' point of Argentera and along the dextral strike-slip Simplon–Rhône zone at the southern part of the Aar Massif.

(b) Large-scale sketch map showing the possible lateral expulsion-translation of the western block along the sidewall of the Apulian indenter, the motion of which is shown by a large black arrow. Present positions of the units, shaded; assumed initial positions, dash lined. The Mt Blanc–Belledonne–Pelvoux block is idealized as a perfect quadrangle in spite of deformations. Fixed blocks, vertically hatched. Black point, position of Argentera. Small arrows, divergent external movements.

used because these sites lie on the inside of the crystalline and Briançonnais units and are probably not affected by the border deformation of these units. However this retro-rotation is insufficient for restoring the palaeomagnetic direction of the Permo–Triassic rocks of the Briançonnais zone (B to B', Fig. 7(a)) into the stable Europe orientation. One explanation is that the Frontal Pennine Thrust acts as a strike-slip fault after its northwestward thrusting and before the external displacements. Therefore, successive increments of rotation of previously detached units are needed while thrusts were progressing outwards.

By the same retro-movements and retro-rotations the oldest internal transversal lineations of the chain could be restored to their initial orientation. This would mean early northwards displacements, probably associated with the well-known large displacement of the Apulian indenter and of nappes located in the main northern trend of the Alps, in front of which the molasse basin will subsequently be emplaced. However, some slight divergence of this reconstituted direction can be expected. This divergence may be interpreted either as representing fan-shaped stress trajectories in front of pushing forces due to the collision of plates (Pavoni 1961, 1986, Tapponnier 1977) or as a progressive change in the collision direction due to the well-documented rotation of Italy (Lowrie & Alvarez 1975, Soffel 1975).

Finally, as already similarly proposed by Goguel (1963), the northwestwards displacement of the assumed, but well-accepted Apulian indenter may be responsible for lateral expulsion and translation of a block previously dissected when part of the European margin. This laterally displaced and deformed unit, dragged in an anticlockwise rotation (Fig. 7(b)) and jammed against the stable foreland while rotating about an internal pole during its northwestwards translation, may have developed a dextral ring shear on its external border, leading to a divergent pattern of shortening and thrusting in the western Alpine arc. Thus, the translation of the Apulian mesoplate induced deformations and displacements depending on local boundary conditions of the previously weakened southern edge of the European Plate.

ACKNOWLEDGEMENTS: The authors would like to thank P. Cobbold, D. Dietrich, J. P. Gratier and L. Jenatton for discussions, criticisms and many suggestions that helped to improve this paper.

References

ARGAND, E. 1924. La tectonique de l'asie. *Congrés géologique international XIII, Belgique 1922*, **1**, 171–372.

ARNAUD, H., BARFETY, J. C., GIDON, M. & PAIRIS, J. L. 1978. A propos du rhegmatisme des zones externes alpines au S de Grenoble. *Comptes-Rendus Académie des Sciences–Paris, D*, **286**, 1335–8.

BARFETY, J. C., GIDON, M., LEMOINE, M. & MOUTERDE, R. 1979. Tectonique synsédimentaire liasique dans les massifs cristallins de la zone externe des Alpes occidentales françaises: la faille du Col d'Ornon. *Comptes-Rendus Académie des Sciences–Paris D*, **289**, 1207–10.

BEACH, A. 1981. Thrust tectonics and cover-basement relations on the northern margin of the Pelvoux massif, French Alps. *Eclogae Geologicae Helvetiae*, **74**, 471–79.

BOGDANOFF, S. & SCHOTT, J. J. 1977. Etude paléomagnétique et analyse tectonique dans les schistes rouges permiens du S de l'Argentera. *Bulletin de la Société Géologique de France*, **19**, 909–16.

BOILLOT, G., MONTADERT, L., LEMOINE, M. & BIJU-DUVAL, B. 1984. *Les marges continentales actuelles et fossiles autour de la France*. Masson.

BOYER, S. E. & ELLIOTT, D. 1982. Thrust systems. *American Association of Petroleum Geologists Bulletin*, **66**, 1196–230.

BURKHARD, M. 1986. Déformation des calcaires de l'Helvétique de la Suisse occidentale (Phénomènes, mécanismes et interprétations tectoniques). *Revue de Géologie dynamique et de géogéographie physique*, **27**, 281–301.

BUTLER, R. W. H., MATTHEWS, S. J. & PARISH, M. 1986. The NW external Alpine Thrust Belt and its implications for the geometry of the western Alpine Orogen. *In*: COWARD, M. & RIES, A. C. (eds), *Collision Tectonics*. Special Publication of the Geological Society No. **19**, 245–60.

CHOUKROUNE, P., BALLEVRE, M., COBBOLD, P., GAUTHIER, Y., MERLE, O. & VUICHARD, J. P. 1986. Deformation and motion in the western alpine Arc. *Tectonics*, **5**, 215–26.

DIETRICH, D. & DURNEY, D. W. 1986. Change of direction of overthrust shear in the Helvetic nappes of western Switzerland. *Journal of Structural Geology*, **8**, 389–98.

EMMONS, R. C. 1969. Strike-slip rupture patterns in sand models. *Tectonophysics*, **7**, 71–87.

FRECHET, J. 1978. *Sismicité du SE de la France et une nouvelle méthode de zonage sismique*. Unpublished 3e cycle thesis, University of Grenoble.

GIDON, M. 1974. L'arc alpin a-t-il une origine tourbillonaire? *Comptes-Rendus Académie des Sciences Paris D*, **278**, 21–4.

—— & PAIRIS, J. L. 1986. La nappe de Digne (Chaines subalpines méridionales): origine, déplacement et signification régionale. *Comptes-Rendus Académie des Sciences de Paris*, **303**, serie II, no. 10, 981–4.

GILLCRIST, R., COWARD, M. & MUGNIER, J. L. 1987. Structural inversion and its controls: examples from the Alpine foreland and the French Alps. *Geodinamica Acta (Paris)*, **1**, 5–34.

GOGUEL, J. 1963. L'interprétation de l'arc des Alpes occidentales. *Bulletin de la Société Géologique de France*, **5**, 20–33.

LACASSIN, R. 1987. Kinematics of ductile shearing from outcrop to crustal scale in the Monte Rosa nappe, western Alps. *Tectonics*, **6**, 69–88.

LAMARCHE, G. 1987. *Analyse structurale et fabrique magnétique. L'exemple des calcschistes et des flyschs de la zone dauphinoise (Alpes françaises)*. Unpublished 3e cycle thesis, University of Grenoble.

—— & Rochette, P. 1987. Microstructural analysis and origin of lineations in the magnetic fabric of some alpine slate. *Tectonophysics*, **139**, 285–93.

Laubscher, H. 1971. The large scale kinematics of the western Alps and the northern Apennines and its palinspastic implications. *American Journal of Science*, **271**, 193–226.

Lowrie, W. & Alvarez, W. 1975. Paleomagnetic evidence for rotation of the Italian peninsula. *Journal of Geophysical Research*, **80**, 1579–92.

Malavielle, J., Lacassin, R. & Mattauer, M. 1984. Signification tectonique des linéations d'allongement dans les Alpes occidentales. *Bulletin de la Société Géologique de France*, **27**, 895–906.

Mancel, P. & Merle, O. 1987. Kinematics of the northern part of the Simplon Line (central Alps) *Tectonophysics*, **135**, 265–75.

Mancktelow, N. 1985. The Simplon Line: a major displacement zone in the western Lepontine Alps. *Eclogae Geologicae Helvetiae*, **78**, 73–96.

Mandl, G., de Jong, L. N. J. & Maltha, A. 1977. Shear zones in granular material. *Rock Mechanics*, **9**, 95–114.

Menard, G. 1979. *Relations entre structures profondes et structures superficielles dans le SE de la France. Essais d'utilisation des données géophysiques*. Unpublished 3e cycle thesis, University of Grenoble.

—— 1980. Profondeur du socle antétriasique dans le SE de la France. *Comptes Rendus Académie des Sciences de Paris, D*, **280**, 299–302.

—— & Thouvenot, F. 1984. Ecaillage de la lithosphère européenne sous les Alpes occidentales: arguments gravimétriques et sismiques liés à l'anomalie d'Ivrea *Bulletin de la Société Géologique de France*, **26**, 875–84.

—— & Frechet, J. 1989. Mécanismes au foyer de seismes des Alpes occidentales et modèle de déformation actuelle de la chaine. *Earth Planetary Science Letter*, (in press).

Merle, O. & Brun, J. P. 1984. The curved translation path of the Parpaillon nappe. (French Alps). *Journal of Structural Geology*, **6**, 711–9.

Pavoni, N. 1961. Faltung durch Horizontalverschiebung. *Eclogae Geologicae Helvetiae*, **54**, 515–34.,

—— 1986. Regularities in the pattern of major fault zones of the earth and the origin of arcs. *In*: Wezel, F. C. (ed.) *The origin of arcs. Developments in Geotectonics*, **21**, 63–78.

Perrier, G. & Vialon, P. 1980. Les connaissances géophysiques sur le SE de la France. Implications géodynamiques. *26e Congrès Géologique international Paris*, Excursions guide, 13–20.

Pijolat, B., Gay, M., Gratier, J. P. & Vialon, P. 1981. Les variations des valeurs de la déformation dans un système de plis par cisaillement. *Revue de Géologie Dynamique et de Géographie Physique*, **23**, 195–201.

Ramsay, J. G. & Huber, M. 1983. *The techniques of modern Structural Geology. Vol. I: Strain analysis*, pp. 276–8. Academic Press, London.

Ricou, L. E. & Siddans, A. W. B. 1986. Collision tectonics in the Western Alps. *In*: Coward, M. & Ries, A. C. (eds). *Collision Tectonics*, Special Publication of the Geological Society no. 19, 229–44.

Rochette, P. 1983. *Propriétés magnétiques et déformations dans les roches sédimentaires alpines*. Unpublished 3e cycle thesis, University of Grenobles.

—— & Lamarche, G. 1986. Evolution des propriétés magnétiques lors des transformations minérales dans les roches: example du Jurassique dauphinois (Alpes françaises). *Bulletin de Minéralogie*, **109**, 687–96.

Schott, J. J. 1976. *Etude paléomagnétique des sédiments permiens de l'Argentera*. Unpublished 3e cycle thesis, University of Strasbourg.

Soffel, H. 1975. The paleomagnetism of age dated Tertiary volcanites of the Monti Lessini (northern Italy) and its implication to the rotation of northern Italy. *Journal of Geophysics*, **41**, 385–400.

Steck, A. 1984. Structures et déformations tertiaires dans les Alpes centrales. *Eclogae Geologicae Helvetiae*, **77**, 55–100.

Tapponnier, P. 1977. Evolution tectonique du système alpin en Méditerranée: poinçonnement et écrasement rigide-plastique. *Bulletin de la Société Géologique de France*, **19**, 437–60.

Vann, I. R., Graham, R. M. & Hayward, A. B. 1986. The structure of mountain fronts. *Journal of Structural Geology*, **8**, 215–27.

Vialon, P. 1966. *Etude géologique du massif cristallin Dora Maira Alpes cottiennes internes. Italie*. Doctorat Etat thesis, University of Grenoble.

—— 1974. Les déformations synschisteuses superposées en Dauphiné. *Bulletin Suisse de Minéralogie et Pétrographie*, **54**, 663–90.

Westphal, R. 1973. Etudes paléomagnétiques de quelques formations permiennes et triasiques dans les Alpes occidentales (France). *Tectonophysics*, **17**, 323–335.

P. Vialon, P. Rochette & G. Ménard, Université Joseph Fourier, LGIT–IRIGM, BP 53X, F38041 Grenoble Cedex, France.

Plate-scale kinematics and compatibility of crustal shear zones in the Alps

R. Lacassin

SUMMARY: Crustal-scale thrusts of the western Alps are characterized by W to WNW directed movements of several hundreds of kilometres. In the internal zones, prominent lineations mark these slip directions. In the central Alps, coeval right-lateral strike slip (Insubric fault) and N–NNW vergent thrusting (Helvetic and more external thrusts) occurred during the upper Tertiary, on crustal-scale faults. It is proposed that WNW directed convergence between NW Apulia and Europe was accommodated mainly by the combination of movements on three first-order faults: in the central Alps, this oblique convergence is partitioned in two components (thrust and strike slip) at a deep fault bifurcation; in the western Alps, it accounts for the observed thrusts. This direction of convergence in the western and central Alps is markedly different from the coeval relative motion of Africa and Europe (N to NNW directed convergence). It implies either specific movements of Apulia with respect to Africa, or the lateral extrusion of NW Apulia during the indentation of the African promontory during the collision.

Looking at mountain belts, one striking tectonic problem is to understand the relationship between displacements on crustal-scale faults and the bulk plate convergence. The latter, constrained by oceanic reconstructions, constitutes the kinematic boundary conditions for the mountain belt system. On the other hand, the kinematics of large faults, with offsets of 10 to 100 kilometres at least, is determined either by balanced cross sections, or by microtectonic analyses of faults, shear zones and high ductile strains. This latter approach emphasizes the regional compatibility of sample- to outcrop-scale analyses of deformation mechanisms, to demonstrate that stretching lineations, associated with clear shear criteria in highly deformed rocks, lie close to the slip direction in a crustal-scale shear zone (e.g. Escher & Watterson 1974, Mattauer 1975, Nicolas & Boudier 1975, Lacassin 1987). Large-scale studies of stretching lineations associated with large deformations thus allow us to determine the directions of transport on large ductile thrusts (e.g. Malavieille *et al.* 1984). On the other hand, the restoration of balanced cross-sections (Hossack 1979, Boyer & Elliott 1982) often gives good estimates of shortening but generally few constraints on movement directions; furthermore these estimates seem to be more accurate in the external zones than in imbricates of basement thrust sheets because of the absence of internal deformation and the existence of easily measured bedding.

Some authors have attempted to compare directly the slip directions marked by the lineations to the plate convergence vector (Shackleton & Ries 1984, Baird & Dewey 1986). Shackleton & Ries (1984) considered the lineation directions to be parallel to the convergence, their variations being signs of changes, in time or space, of the relative plate motion. This assumption should be valid for a linear mountain belt with one major fault only, but the stretching lineations measured in the field are generally not parallel to the plate convergence even in the most simple chains: for instance in the Himalayan Main Central Thrust the lineations show a diverging radial pattern along the slightly arcuate belt (Brunel 1986). Besides, the convergence in many collision zones is partitioned into several crustal-scale coeval faults with mainly thrust or strike-slip displacements (Mattauer 1986). Attempts to reconstruct the bulk kinematics of this system of faults are thus possible using detailed tectonic studies on each of these major faults. In the ideal case, the 3-D addition of the slip vectors measured on each fault (direction, sense and amount or rate of displacement) will give a good estimate of the relative convergence.

Laubscher (1971, 1975) and Tapponnier (1977) attempted to put geological constraints on plate tectonics and, following Laubscher, this approach is now attempted in the Alps with the aim of discussing the bulk geometry and mechanics of the collision. So far, the Alps are probably the best studied mountain belt. They present a strongly arcuate structure (e.g. Argand 1916, Goguel 1963), which formed as a result of the bulk N-directed convergence between Africa and Europe. The slip directions observed on the crustal-scale faults vary along

From: Coward, M. P., Dietrich, D. & Park, R. G. (eds), 1989, *Alpine Tectonics*, Geological Society Special Publication No. 45, pp. 339–352.

the arc and are often different from the convergence vector (Fig. 1); for instance in the western Alps, characterized by large W-vergent thrusts and minor E-vergent back-thrusts, the shortening direction is at a high angle to the convergence (Lacassin 1987). This paper will summarize published and some new tectonic results in the western and central Alps, and will discuss in conclusion a model involving the partitioning of the bulk convergence into several fault zones and, at a larger scale, the squashing of the African promontory (Apulia) during the NNW-directed convergence.

Boundary conditions

Progressively more accurate reconstructions of the Atlantic opening (e.g. Pitman & Talwani 1972, Biju-Duval *et al.* 1977, Olivet *et al.* 1984, Klitgord & Shouten 1985, Savostin *et al.* 1986) have been used to calculate the relative movements between Africa and Europe (e.g. Dewey *et al.* 1973, Biju-Duval *et al.* 1977, Olivet et al. 1984, Savostin et al. 1986, Le Pichon *et al.* in press). The first reconstructions (Pitman & Talwani 1972, Biju-Duval *et al.* 1977) suggested complex relative movements between Africa and Europe, with an important dextral relative motion during the late Cretaceous and Paleocene. Tapponnier (1977) remarked that such complex movements do not easily fit the geological observations and also run against the ideas about the mechanics of a large collision zone; he proposed more smooth plate trajectories, with a sinistral motion of Africa with respect to Europe during the Jurassic and Cretaceous, progressively followed by a N-directed convergence. Recent reconstructions

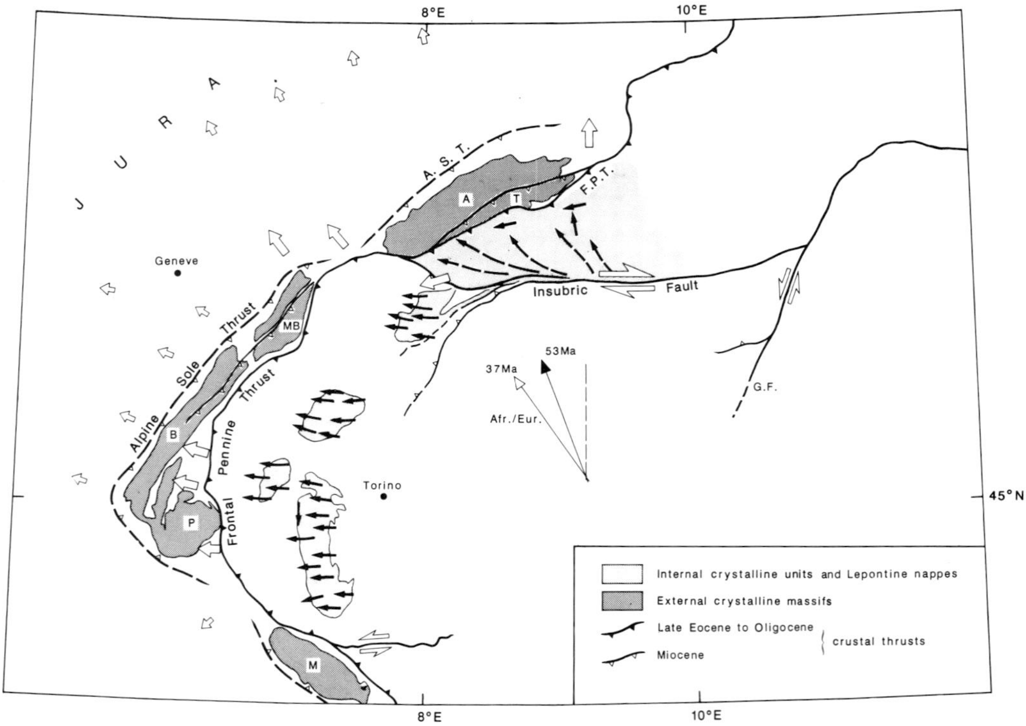

FIG. 1. Map of crustal-scale structures in the western and central Alps. Upper Tertiary crustal faults: frontal Pennine thrust (FPT, dark triangles), Alpine sole thrust (AST, open triangles), Insubric strike-slip fault (IF), Giudicarian fault (GF). Slip directions on crustal shear zones mainly of Eocene and Oigocene age are marked by lineations (after Malavieille *et al.* 1984, and Choukroune *et al.* 1986), and are indicated for the Internal Crystalline Massifs and Lepontine nappes (light shaded). Large open arrows represent Oligocene and Miocene slip directions on external thrusts as marked by lineations; small open arrows are shortening directions accounting for external folds and thrusts. Motion of Africa with respect to Europe is figured for the period between 53 and 37 Ma (Black arrow) and at 37 Ma (open arrow), calculated for a point 50 km S of Milano using Olivet *et al.*'s (1984) reconstruction. External Crystalline Massifs are shaded (A.: Aar; T.: Tavetsch-Gotthard; M.B.: Mont Blanc; B.: Belledone; P.: Pelvoux; M.: Argentera-Mercantour).

(Olivet *et al.* 1984, Klitgord & Shouten 1985, Savostin *et al.* 1986) support these conclusions. In these models, the dextral relative motion suggested by previous models has disappeared. Between 180 Ma and 80 Ma, they show a bulk left-lateral motion of Africa relative to Europe. Since 80 Ma, N- to NNW-directed convergence of Africa with respect to Europe is well documented by the models based on Atlantic data (Olivet *et al.* 1984, Klitgord & Shouten 1985). The model of Savostin *et al.* (1986), based for this period on Arctic ocean data, suggests a minor episode of dextral relative movement between 35 and 20 Ma, during an otherwise similar history. According to Savostin *et al.* (1986) and Le Pichon *et al.* (in press), the differences between the 'Atlantic' and 'Arctic' reconstitutions should account for relative motion between western Europe and the whole of Eurasia during the Oligocene.

The rate of relative motion, calculated for a point at the Northern tip of the Apulian promontory (Savostin *et al.* 1986), has remained close to 1 cm/yr since 80 Ma, suggesting 800 km of N- to NNW-directed convergence between Africa and Europe, only a part of which is absorbed in the Alps.

The specific motions of Apulia relative to Europe and Africa are poorly known. According to the review by Westphal *et al.* (1986), palaeomagnetic studies in stable Apulia (e.g. channell & Tarling 1975, Vandenberg 1983, Marton & Veljovic 1983) suggest that Apulia was locked to Africa from the Triassic to the Lower–Middle Cretaceous; it further suffered an anticlockwise rotation of 20° during the Middle or Upper Cretaceous and then moved again with Africa.

E–W directed thrusting in the western Alps

The western Alps are characterized by several crustal-scale thrusts due to the obduction of ophiolite and 'Schistes Lustrés' nappes, and to the stacking of former continental margins (Mattauer & Tapponnier 1978, Malavieille *et al.* 1984, Gillet *et al.* 1986). These thrusts are progressively more recent towards the W (Debelmas & Kerckhove 1980): late Cretaceous to late Eocene thrusts (W-vergent) and back-thrusts (E-vergent) in internal zones, Oligocene and Miocene W-directed thrusts in more external zones (clear on the recent ECORS profile, Bayer *et al.* 1987), the latter being situated under the external crystalline massifs (Fig. 1, Ménard 1979, Ménard & Thouvenot 1984).

Internal zones

The internal zones show high ductile strains with prominent stretching lineations (Fig. 2a), often parallel to the axes of isoclinal folds and probably sheath-folds. Extensive studies (Malavieille *et al.* 1984, Choukroune *et al.* 1986) have shown that these lineations trend on the average WNW–ESE to E–W (Fig. 1). In the Internal Crystalline Massifs they vary from N110°E in the Monte Rosa nappe (Lacassin 1987), N90°E in the Gran Paradiso (e.g. Bertrand 1968, Vearncombe 1980) and Ambin massif (e.g. Gay 1972, Malavieille & Etchecopar 1981, Malavieille 1982) to N90°/N80°E in the Dora-Maira massif (Vialon 1966). Clear shear criteria such as meter-scale shear zones, asymmetric pressure shadows, tails on deformed porphyroclasts (Fig. 2b) and quartz c-axes fabrics are generally associated with these lineations. This intense non-coaxial deformation has thus been interpreted as the result of large progressive shearing in a direction close to the lineation trend (Malavieille *et al.* 1984, Choukroune *et al.* 1986). Minimum shear strains (γ) are often of the order of 20, with displacements of several tens of kilometres at least (Lacassin & Mattauer 1985). Metamorphic and radiometric data suggest that most of the lineations of the western internal Alps are Eocene in age and formed in crustal-scale W- or E-verging thrusts during continental collision (Malavieille *et al.* 1984, Choukroune *et al.* 1986).

In other internal units like the 'Schistes Lustrés' (Caron 1977) or the Vanoise massif (Ellenberger 1958), the deformations are apparently more complex than in the Internal Crystalline Massifs, the observed intricate geometry probably resulting from mechanical instabilities which favour multi-phase folding rather than shear zones in well layered sedimentary rocks. Despite this complexity it is generally possible to demonstrate SE–NW to E–W directed thrusting in the Vanoise Massif (Platt & Lister 1985) and E–W directed thrusting in the Southern Briançonnais and 'Schistes Lustrés' nappes (Tricart 1980, 1984).

External zones

In the external zones of the western Alps, W- to WNW-vergent thrusts, mainly Oligocene and Miocene in age (Debelmas & Kerckhove 1980, Tricart 1984), are probably deep-rooted between two crustal-scale thrust zones (Fig. 1, Beach 1981a, Malavieille *et al.* 1984, Ménard & Thouvenot 1984, Butler 1986): the frontal Pennine thrust (FPT), and the Alpine sole

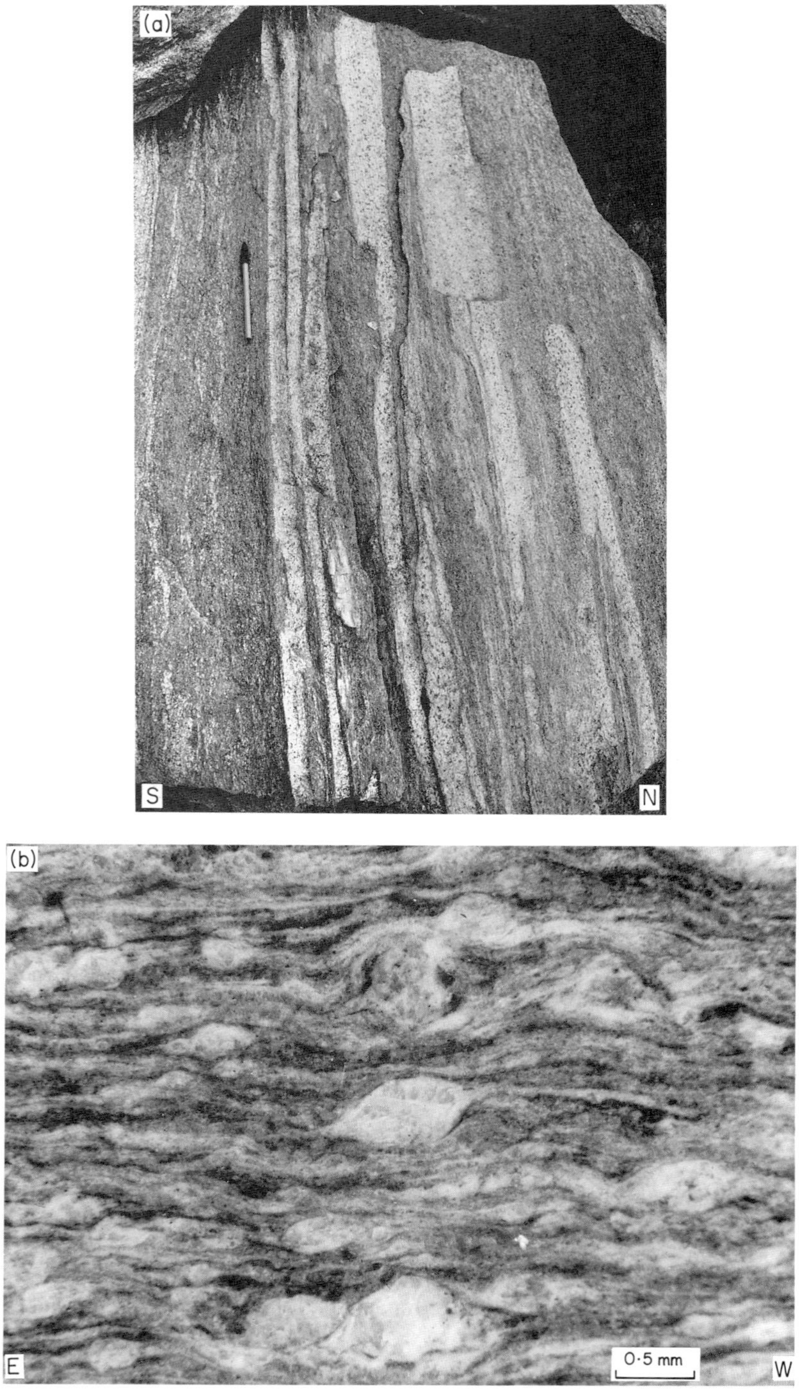

FIG. 2. Characteristic of deformation in the Internal Crystalline Massifs: (a): mineral and stretching lineations parallel to the fold axes of aplitic veins (Portjengrat, Monte Rosa nappe, Lacassin 1987); (b) tails on porphyroclasts indicating W-vergent shear (W of Mattmark Lake, Monte Rosa nappe, Lacassin 1987).

thrust (AST) beneath the External Crystalline Massifs.

These thrust movements and related E–W shortening have produced significant ductile strains in the Dauphinois Liassic shales, N of Pelvoux (Gratier & Vialon 1980, Gamond 1980, Beach 1981c): close to the frontal Pennine thrust, stretching lineations associated with shear criteria suggest important westwards shearing (Fig. 3); further W, subvertical cleavage and extensive folding imply strong E–W shortening of the former Bourg d'Oisan half-graben (Gratier & Vialon 1980). South of Pelvoux, the Champsaur Nummulitic flysch is affected by W- to WSW-vergent folds in front of the Briançonnais thrust, with E–W stretching lineations in the basal décollement zone (Plotto 1977, Tricart 1980). In the 'Chaînes Subalpines' (Vercors, Chartreuse, Bauges) and Southern Jura, the geometry of fold axes and thrusts (e.g. Doudoux *et al.* 1982, Mugnier *et al.* 1987) suggest WNW-directed thrusting on the Alpine sole thrust.

Several attempts to restore parts of, or the whole external thrust system have led to estimates of the Oligocene–Miocene shortening. In the Chartreuse massif, Mugnier *et al.* (1987) estimated 14 to 25 km of shortening between the Belledonne external crystalline massif and the foreland. Beach's (1981b) cross sections of the Dauphinois zone suggest 70 km of shortening between the External Crystalline Massifs and the frontal Pennine thrust (FPT). Butler's cross section in the same area (1986) implies shortening of the order of 180 km from the FPT to the foreland. Further N, in the Mont Blanc area, Butler (1985) estimated 126 km of shortening between the Belledonne massif and the FPT, and 385 km of late Eocene to Miocene shortening across the western Alps (Butler 1986), this latter value accounting for both external and some internal structures (Briançonnais thrusts and backthrusts). Despite some problems regarding Butler's restorations (see Platt 1984) which might result in overestimates of shortening (for instance in estimating the shortening in basement imbricates where bedding planes are unknown), 100 to 150 km of minimum shortening seems probable for the external structures, 200 to 300 km including the late Eocene internal structures. Besides, these values accord with the data of the recent ECORS profile (Bayer *et al.* 1987), showing for instance 30 to 50 km of Moho offset on the Alpine sole thrust.

Right lateral displacements in the central Alps

The large W- to WNW-directed convergence in the western Alps requires right lateral displacements in the central Alps for kinematic compatibility. So far, oblique shortening is supported by field observations in the Swiss Alps, showing the combination of thrust and strike-slip movements.

Indeed, along the Insubric fault, a large zone is affected by late, mainly strike-slip deformations (Lacassin 1986). North of Lago di Como, the Pennine nappe foliations are subvertical with down-dip lineations (Wenk 1955); a few kilometres N of the Insubric fault (site a, Fig. 4), this fabric is cross-cut by ductile to brittle shear zones (Fig. 5a) with mainly dextral movements; close to the Insubric fault (sites b and c, Fig. 4), a wide zone of ductile to brittle deformation includes low T° mylonites, cataclasites and gouges, with many right-lateral shear criteria (Fig. 5b). This deformation affects the Bergell intrusive body which shows C–S microstructures (Berthé *et al.* 1979) in the vicinity of the Insubric fault (sites c and d, Fig. 4, and Fig. 5c). At the regional scale, the horned shape of the Bergell Massif (Fig. 4), with a 48 km-long tail, is probably due to dextral shearing, as well as the later right lateral, N100° to N130° trending faults (site e, Fig. 4) which offset this tail (Fumasoli 1968). Further east, the same type of microstructures are observed for example at the Tonale Pass (site f, Fig. 4, and Fig. 5d). Further W, dextral shearing is also observed (Lacassin 1986), for example in the eastern part of the Monte Rosa nappe near Locarno (site g, Fig. 4, and Fig. 5e) and at Villadossola (site h, Fig. 4).

This deformation occurred on the Insubric fault between 30 Ma and the present time. This accords with the age of the Bergell intrusion (Gulson & Krogh 1973), with the observation of deformation structures syn- to post-Lepontine metamorphism, and with radiometric ages on biotites from the Insubric fault zone (21 Ma K–Ar and Rb–Sr, Armstrong *et al.* 1966; 11.6 Ma Ar^{39}–Ar^{40}, Monié 1985). An offset of 40 to 50 km is required to explain the elongated tail of the Bergell massif, but should be considered as a minimum estimate since the widest cataclasite and gouge zones along the Insubric fault can be found further S.

Thrusts and deformations in the external zones of the Swiss Alps (Helvetic nappes, external thrusts in the molasse basin and Jura) occurred during Oligocene to Pliocene time

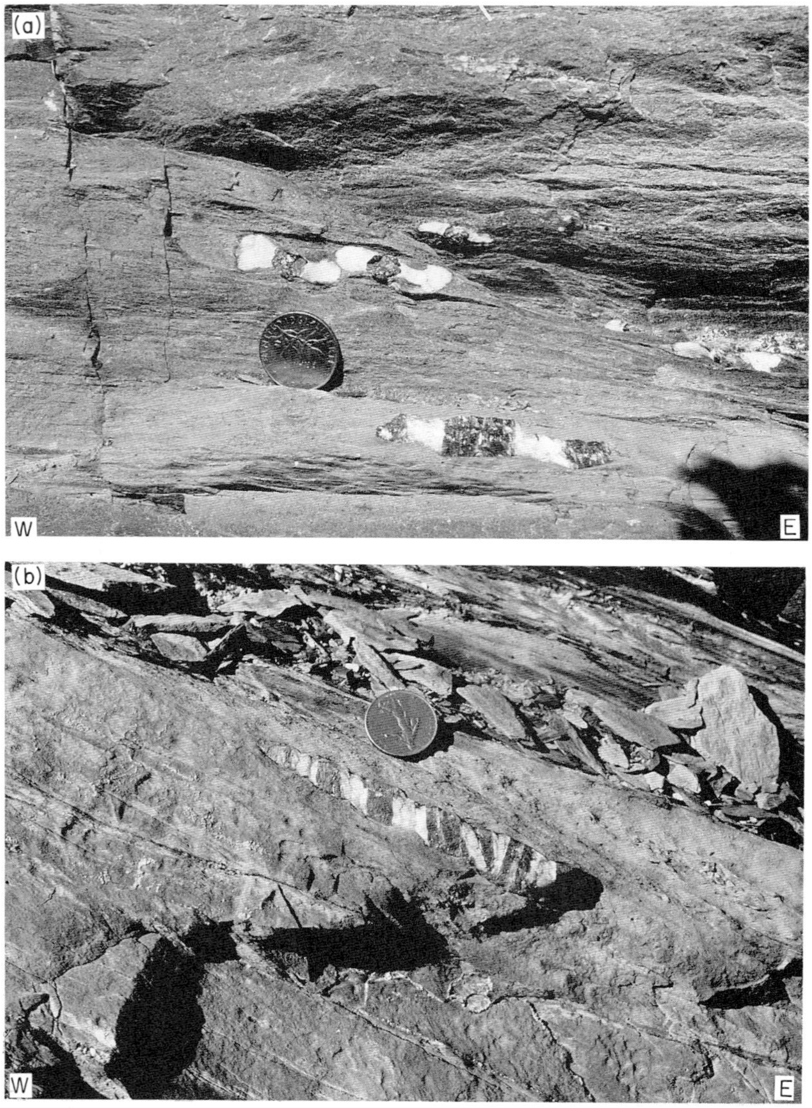

FIG. 3. Ductile deformation in Liassic shales of the Dauphinois zone: (a) asymmetric pressure-shadow crystallizations on pyrite, indicating shear to the W (Malavieille & Etchecopar 1987), along the road from La Grave to the Lautaret pass, N of Pelvoux; (b) stretched belemnite parallel to lineation, note rotation of fragments in agreement with a westwards sense of shear, near Les Hières, NW of La Grave.

(Trümpy 1969, 1973, Milnes & Pfiffner 1980), and are therefore coeval with the right lateral displacements observed further S. Transport directions, marked by lineations, are towards the N in the eastern Helvetic nappes (Schmid 1975, Pfiffner 1981). In the western nappes, stretching lineations trend on average N–S except in the highly sheared, inverted limb of the Morcles–Doldenhorn nappe (Ramsay 1981, Dietrich & Song 1984, Dietrich & Durney 1986). Dietrich & Durney (1986) emphasize an anticlockwise rotation of the shear directions, the major NW–directed shear, marked by the lineations in the inverted limb, being overprinted by a WNW- to W-directed shear recorded by the calcite microfabrics (Dietrich & Song 1984). In the Aar Massif, NNW–SSE-directed shortening occurred on conjugate shear zones (Choukroune & Gapais 1983). The most external thrusts in the Molasse basin are NNW-

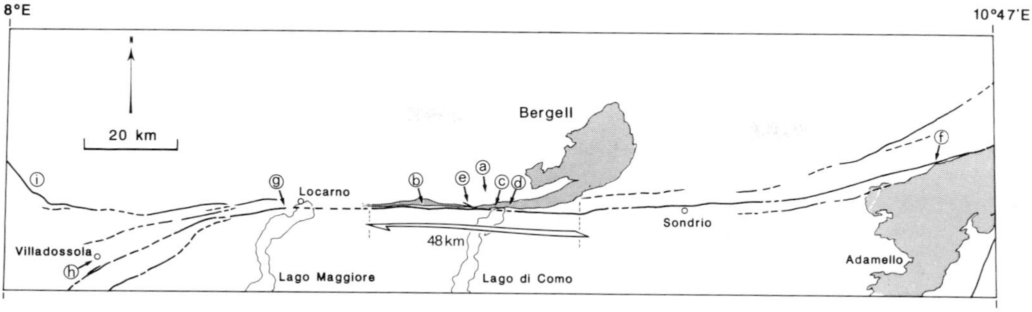

FIG. 4. Late Tertiary structures along the Insubric fault. Faults are traced from *Landsat* image analysis. Sites discussed in text and the location of photographs of Fig. 5 are indicated by circled letters. The 48 km long tail of the Bergell intrusive body is interpreted as the result of dextral shearing. Oligocene intrusives (Bergell, Adamello) are shaded.

vergent on average as in the Northern Jura, where 15 to 20 km of shortening occurred during the Pliocene (Laubscher 1971). According to Hsü (1979), three crustal-scale structures account for these deformations: the frontal Pennine thrust, a major thrust between Aar and Tavetsch, and the late Alpine sole thrust beneath the Aar massif. Hsü (1979) estimated 100 km of crustal shortening on these thrusts from 40 Ma to the present.

The late Eocene to Oligocene deformations observed in the internal Pennine nappes form a complex strain pattern with combined or superposed N- to NW-vergent, and W- to WSW-vergent ductile shearing (e.g. Steck 1984, Cobbold *et al.* 1985, Merle 1987).

Discussion

The kinematics of Tertiary crustal-scale faults have been summarized in Table 1, from the microtectonic data and estimates of shortening discussed above. First, note that coeval displacements occurred in the central and western Alps with several slip directions. At least for the well constrained upper Tertiary deformations, a valuable approach could be to consider a simple system of three major, first-order faults that account for the convergence between NW Apulia and Europe (Fig. 6a): the western and central Alps frontal thrusts with slip to the WNW and to the NNW respectively (frontal Pennine thrust during the Oligocene, present Alpine sole thrust during the Miocene), and the Insubric right lateral strike-slip fault. A rigid block kinematic analysis of this fault system (Fig. 6b) implies a dextral offset of the order of 110 km on the Insubric fault zone, and 65 km of shortening in the central Alps, accounting for 150 km of convergence and shortening in the western Alps during the late Tertiary. This estimate assumes continuous and coeval movements on each fault for the determination of finite displacements using the velocity diagram of Fig. 6(b). Such a geometry and interaction of crustal faults might explain the large-scale Tertiary kinematics of this part of the Alps: W-vergent thrusts in the western Alps and NW–SE shortening in the central Alps. Nevertheless this oversimplified model cannot account for the movements on important second order faults and the related penetrative deformation, and it cannot explain all the small scale observations.

The above fault geometry implies a connection between the Insubric fault and the frontal Pennine thrust: the rigid block model of Fig. 6 postulates a simple triple junction between the three major faults. In the Alps the junction between the Insubric fault and the FPT is represented by a wide and complex relay zone. During the late Tertiary, the dextral movements are probably transformed on several faults, as the Simplon fault zone (site i, Fig. 4), which is characterized by ductile to brittle, N110W-vergent normal faulting (Mancktelow 1985). Along the WSW–ESE directed segment of the FPT between Aar and Mont Blanc, a component of dextral shear is suggested by the anticlockwise rotation of shear directions in the western Helvetic nappes (Dietrich & Durney 1986). During the late Eocene and Oligocene, the more penetrative deformations in the Pennine nappes probably account for the transformation of the dextral strike-slip component in the central Alps towards the external parts of the Alps (Steck, 1984, Choukroune *et al.* 1986).

The NW–SE convergence in the central Alps

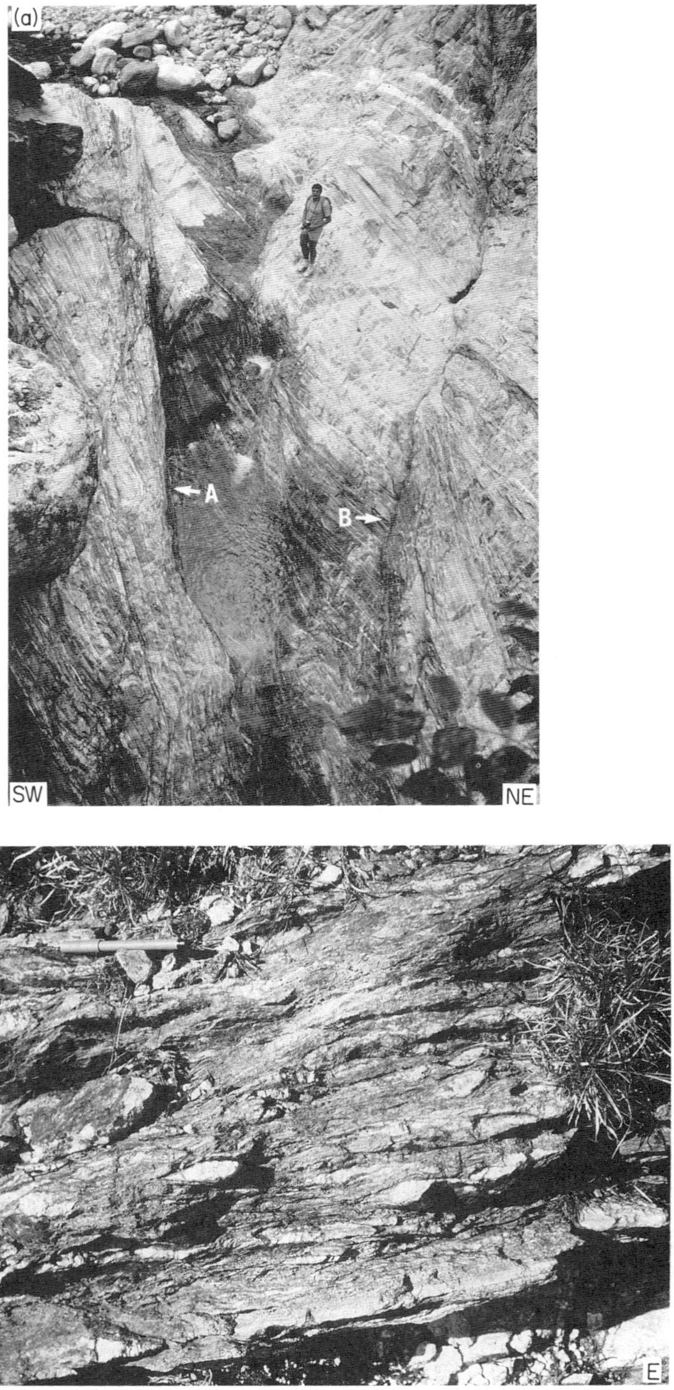

FIG. 5. Characteristic of deformation along the Insubric fault, indicating dextral shear movement. Sites are indicated on Fig. 4. (a): Ductile (A), and brittle (B) dextral shear-zones cross-cutting Pennine nappe foliations at site a (A and B trend respectively N140°E and N150°E); (b): Dextral shearing in micaschists of the Tonale series at site b (near Passo San Jorio); (c): Bergell tonalite with C/S microstructures at site c (arrows outline one C plane); (d) C/S microstructures in micaschists at site f (Passo di Tonale); (e) Quartz vein affected by dextral shearing at site g (Lacassin 1986).

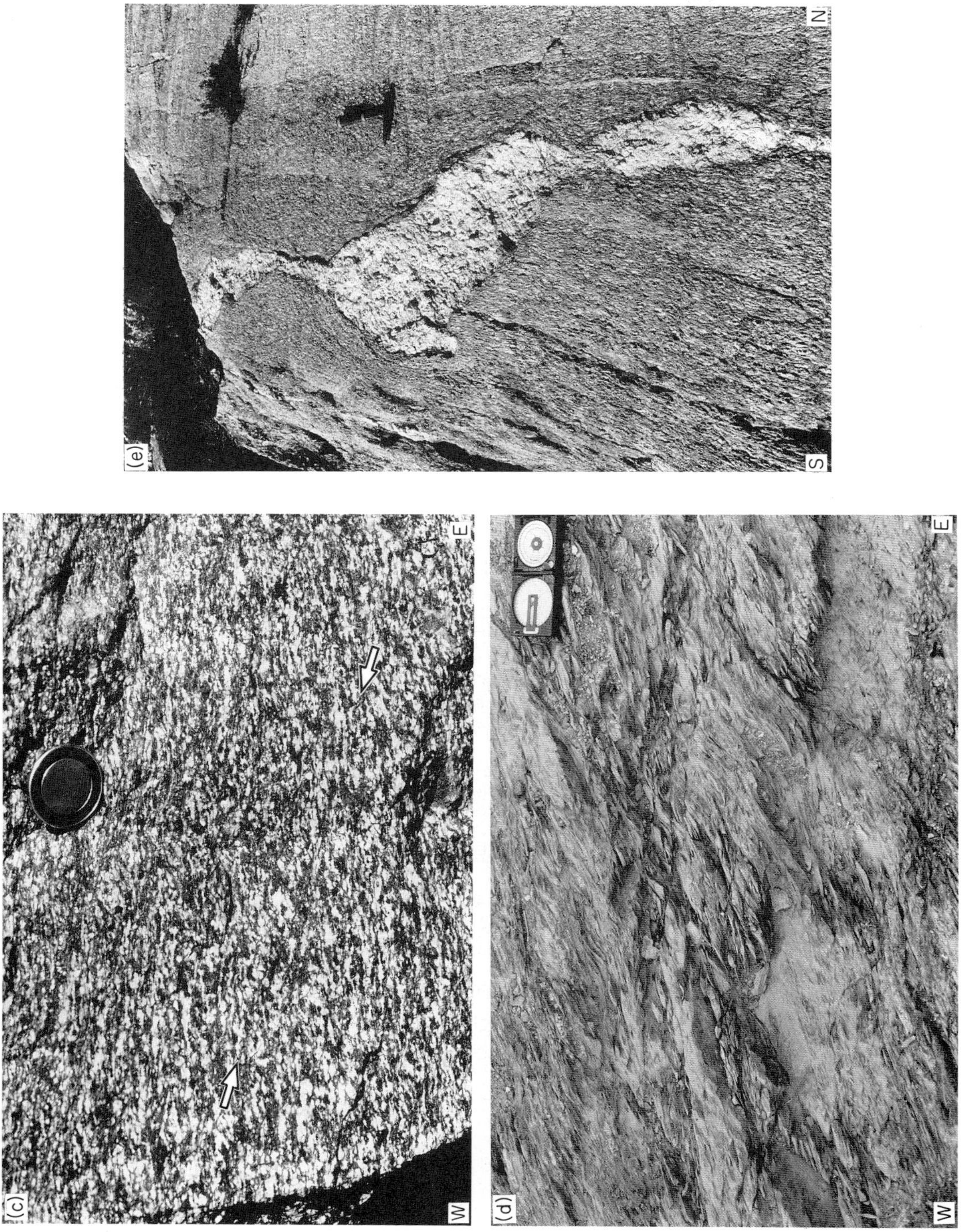
(e)
S
N
(c)
W
E
(d)
W
E

TABLE 1. *Present data on the kinematics of the deformation of the central and western Alps, see references in the text. (FPT: frontal Pennine thrust, ICM: Internal Crystalline Massifs).*

	Slip direction	Amount	Age (Ma)
Western Alps	N120°E to N80°E (thrust)		60–0 (mainly 60–30 Ma in internal zones)
External zones	N120°E to N80°E (thrust)	≥ 100 to 150 km (between FPT and foreland) ≥ 200 to 300 km (between ICM and foreland)	35–0 40–0
Central Alps	Combination of N150°E to N80°E (thrusts) N90°E (strike-slip)		40–0
Insubric fault	N90°E dextral	≫ 50 km	30–0
External thrusts	N150°E to N180°E	≈ 100 km (between FPT and foreland) Jura ≈ 15 km Helvetic ≈ 75 km	40–0 5–0 4–12

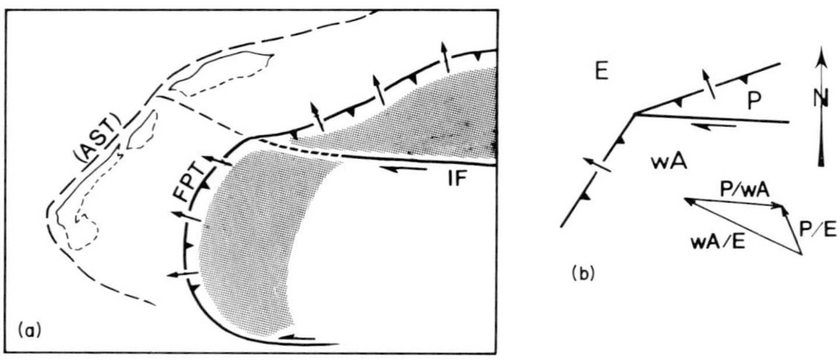

FIG. 6. Simplified sketch of crustal-scale fault kinematics during late Tertiary. FPT: frontal Pennine thrusts, IF: Insubric fault, AST: trace of future Alpine sole thrust, internal zones are shaded. Hinterland-facing borders of External Crystalline Massifs are dashed as they result from stacking of units situated between FPT and AST. On (b), rigid block kinematics is assumed between the three major faults; wA: west Apulia, P: internal zones of central Alps; E: Europe.

may be accounted for by a fault bifurcation geometry (Fig. 7): a deep oblique convergence on a lithospheric subduction plane (Mattauer 1986) is partitioned along the bifurcation line in two faults, with mainly strike-slip and thrust movements respectively. As in Fig. 6, this partitioning may be represented by a velocity diagram (Fig. 7a) in which the addition of the strike-slip and thrust components accounts for the convergence. Such a model of convergence partitioning at a fault bifurcation should apply to most oblique collisions (e.g. Reed *et al.* 1970, Mattauer & Collot 1986, Van Den Driessche 1986).

On a larger scale, the kinematics of the western and central Alps implies a W- to WNW-directed relative convergence between Apulia and Europe at least since the Eocene, as already suggested by Laubscher (1971). This direction of convergence is markedly different from the relative motion of Africa with respect to Europe during the same period. According to Savostin *et al.* (1986), a W-directed convergence could have occurred between 35 and 20 Ma only, if

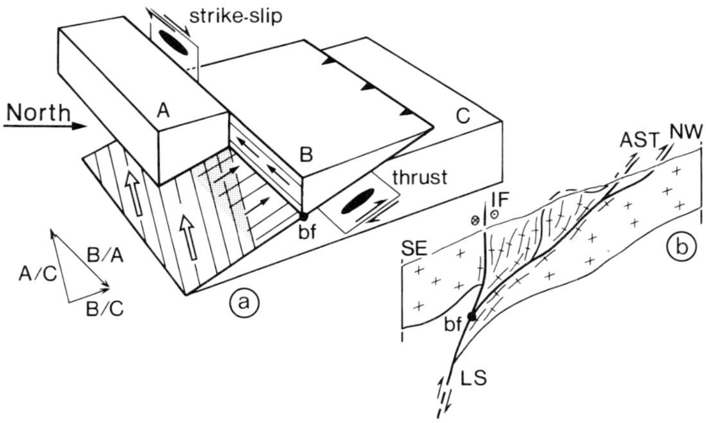

FIG. 7. Fault bifurcation model (a): deep oblique convergence (open arrows) between A and C is partitioned at fault bifurcation (bf) into pure strike-slip and thrust components (black arrows). Black ellipses represent stretching lineations in crustal shear-zones. The shaded zone shows superposition of the two 'phases' (*sensu stricto*) of deformation. (b): for analogy, simplified crustal-scale cross-section of the central Alps after Mattauer (1986). LS: oblique lithospheric subduction, IF: Insubric strike-slip fault; AST: Alpine sole thrust; elongated crosses indicate deformed continental crust.

the Arctic ocean reconstructions apply to the Alpine foreland (see above), but cannot explain the whole strain pattern of the western Alps. Furthermore, other reconstructions (Olivet *et al.* 1984, Klitgord & Shouten 1985) only suggest NNW-directed convergence between Africa and Europe. This discrepancy between large-scale convergence and Alpine kinematics could be explained by a specific relative motion of Apulia with respect to Europe. Yet current palaeomagnetic data do not support large movements between Apulia and Africa during the Tertiary (Westphal *et al.* 1986), but good data are too scarce for a definitive conclusion. Another interpretation assumes that this westward movement of Apulia with respect to Europe would characterize the displacement of only a small part of Apulia. Following Tapponnier's model (1977), the northwestern part of Apulia has been extruded towards the west by the squashing of Apulia during the NNW-directed convergence. This model requires a left-lateral strike-slip fault at the SE border of the extruded block which could be the Giudicarian fault, since this fault shows evidence of sinistral movements.

In conclusion, the model presented here (Figs 6 and 7) accounts quite well for data which have come to light through modern tectonic analyses in the western and central Alps. Nevertheless, much more detailed observations are required both in the area described here and in adjacent ones (eastern Alps, Appennines), and new palaeomagnetic data are needed to discuss more precisely the large-scale kinematics of Apulia relative to Europe.

ACKNOWLEDGMENTS: This work has been supported by a DGRST grant, UA 266 of CNRS, and the ATP Géodynamique of CNRS-INAG, during my 3eme cycle cursus at the Laboratoire de Géologie Structurale in Montpellier, and by the ATP Sismogenèse–plis et failles and the programme DBT of CNRS-INSU and UA 1093 of CNRS since 1986. Special thanks go to J. Malavieille, M. Mattauer and P. Tapponnier for discussions. This is IPGP contribution no. 997.

References

ARGAND, E. 1916. Sur l'arc des Alpes Occidentales. *Eclogae Geologicae Helveticae*, **14**, 145–91.

ARMSTRONG, R. L., JÄGER, E. & EBERHARDT, P. 1966. A comparison of K–Ar ages on Alpine biotites. *Earth and Planetary Science Letters*, **1**, 13–19.

BAIRD, A. W. & DEWEY, J. F. 1986. Structural evolution in thrust belts and relative plate motion: the upper Pennine Piémont zone of the internal Alps, southwest Switzerland and northern Italy. *Tectonics*, **5**, 375–87.

BAYER, R., CAZES, M., DAL PIAZ, G. V., DAMOTTE, B., ELTER, G., GOSSO, G., HIRN, A., LANZA, R., LOMBARDO, B., MUGNIER, J. L., NICOLAS, A., NICHOLICH, R., POLINO, R., ROURE, F., SACCHI,

R., Scarascia, S., Tabacco, I., Tapponnier, P., Tardy, M., Taylor, M., Thouvenot, F., Toreilles, G. & Villien, A. 1987. Premiers résultats de la traversée des Alpes occidentales par sismique rélexion verticale (Programme ECORS–CROP). *Comptes-Rendus des scéances de l'Académie des Sciences Paris*, Série 2, **305**, 1461–70.

Beach, A. 1981a. Thrust tectonics and crustal shortening in the external French Alps based on a seismic cross-section. *Tectonophysics*, **79**, T1–T6.

—— 1981b. Thrust structures in the eastern Dauphinois zone (French Alps), north of the Pelvoux Massif. *Journal of Structural Geology*, **3**, 299–308.

—— 1981c. Some observations on the development of thrust faults in the Ultradauphinois zone, French Alps. *In*: McClay, K. R. & Price, N. D. (eds). *Thrust and Nappe Tectonics*, Special Publication of the Geological Society, London, No. 9, 329–34.

Berthe, D., Choukroune, P. & Jegouzo, P. 1979. Orthogneiss, mylonite and non-coaxial deformation of granites: the example of the south Armorican shear zone. *Journal of Structural Geology*, **1**, 31–42.

Bertrand, J. M. 1968. Etude structurale du versant occidental du Massif de Grand-Paradis. *Travaux du Laboratoire de Geologie de la Faculté de Sciences*, Grenoble, **44**, 55–87.

Biju-Duval, B., Dercourt, J. & Le Pichon, X. 1977. From the Tethys Ocean to the Mediterranean Sea: a plate tectonic model of the evolution of the western alpine systems. *In*: Biju-Duval, B. & Montadert (eds). *International Symposium on the structural history of the Mediterranean basins*, Split, 134–64.

Boyer, S. E. & Elliott, D. 1982. Thrust systems. *American Association of Petroleum Geologists Bulletin*, **66**, 1196–230.

Brunel, M. 1986. Ductile thrusting in the Himalayas: shear sense criteria and stretching lineations. *Tectonics*, **5**, 247–65.

Butler, R. W. H. 1985. The restoration of thrust systems and displacement continuity around the Mont Blanc massif, NW external Alpine thrust belt. *Journal of Structural Geology*, **7**, 569–82.

—— 1986. Thrust tectonics, deep structure and crustal subduction in the Alps and Himalayas. *Journal of the Geological Society of London*, **143**, 857–73.

Caron, J. M. 1977. Lithostratigraphie et tectonique des Schistes Lustrés dans les Alpes cottiennes septentrionales et en Corse orientales. *Sciences Géologique*, **48**, 326 pp.

Channel, J. E. T. & Tarling, D. H. 1975. Paleomagnetism and the rotation of Italy. *Earth and Planetary Science Letters*, **25**, 177–88.

Choukroune, P. & Gapais, D. 1983. Strain pattern in the Aar Granite (central Alps): orthogneiss developed by bulk inhomogeneous flattening. *Journal of Structural Geology*, **5**, 411–18.

——, Ballevre, M., Cobbold, P., Gautier, Y., Merle, O. & Vuichard, J. P. 1986. Deformation and motion in the western alpine arc. *Tectonics*, **5**, 215–26.

Cobbold, P. R., Merle, O. & Barbotin, E. 1985. The fan like strain pattern in the Lepontine Alps and its history. *Terra Cognita*, **5**, 253.

Debelmas, J. & Kerckhove, C. 1980. Les Alpes franco–italiennes. *Géologie Alpine*, **56**, 21–58.

Dewey, J. F., Pitman, W. C., Ryan, W. B. F. & Bonnin, J. 1973. Plate tectonics and the evolution of the Alpine system, *Geological Society of America Bulletin*, **84**, 3137–80.

Dietrich, D. & Song, H. 1984. Calcite fabrics in a natural shear environment, the Helvetic nappes of western Switzerland. *Journal of Structural Geology*, **6**, 19–32.

—— & Durney, D. W. 1986. Change in overthrust shear in the Helvetic nappes of western Switzerland. *Journal of Structural Geology*, **8**, 389–98.

Doudoux, B., Mercier, B. & Tardy, H. 1982. Une interprétation nouvelle de la structure des massifs subalpins savoyards (Alpes occidentales): nappes de charriage oligocènes et déformations superposées. *Comptes-Rendus des scéances de l'Académie des Sciences Paris*, Série 2, **295**, 63–8.

Ellenberger, J. 1958. Etude géologique du pays de Vanoise. *Mémoire du Service de la Carte Géologique de France*, 561 pp.

Escher, A. & Watterson, J. 1974. Stretching fabrics, folds and crustal shortening. *Tectonophysics*, **39**, 121–39.

Fumasoli, M. W. 1968. Geologie des Gebietes nördlich und südlich der Jorio Tonale Linie im Western von Gravedona. *Mitteilungen aus dem geologischen Institut der ETH Zürich*, **194**, 230 pp.

Gamond, J. F. 1980. Direction de déplacement et linéation: cas de la couvérture sédimentaire dauphinoise orientale. *Bulletin de la Société Géologique de France*, **22**, 429–36.

Gay, M. 1972. Le massif d'Ambin et son cadre de Schistes Lustrés (Alpes franco–italiennes). Evolution structurale. *Archives des Sciences*, Genève, **25**, 165–214.

Gillet, P., Choukroune, P., Ballevre, M. & Davy, P. 1986. Thickening history of the western Alps. *Earth and Planetary Science Letters*, **78**, 44–52.

Goguel, J. 1963. L'interprétation de l'arc des Alpes occidentales, *Bulletin de la Société Géologique de France*, **5**, 20–33.

Gratier, J. P. & Vialon, P. 1980. Deformation pattern in a heterogeneous material: folded and cleaved sedimentary cover immediately overlying a crystalline basement (Oisans, French Alps). *Tectonophysics*, **65**, 151–80.

Gulson, B. L. & Krogh, T. E. 1973. Old lead components in the young Bergell Massif, south-east Swiss Alps. *Contributions to Mineralogy and Petrology*, **40**, 239–52.

Hossack, J. R. 1979. The use of balanced cross-sections in the calculation of orogenic contraction: a review. *Journal of the Geological Society of London*, **136**, 705–11.

Hsü, K. J. 1979. Thin-skinned plate tectonics during neo-alpine orogenesis. *American Journal of Science*, **279**, 353–66.

Klitgord, K. D. & Shouten, H. 1985. Plate kinematics of the central Atlantic. *In*: Tucholke, B. E. & Vogt, P. R. (eds). *The geology of North America: the western Atlantic region*, Geological Society of America, DNAG Series, **1**.

Lacassin, R. 1986. Cisaillement progressif dextre dans la partie W de la faille insubrienne (Alpes centrales). *Comptes-Rendus des séances de l'Académie des Sciences Paris*, Série 2, **303**, 229–34.

—— 1987. Kinematics of ductile shearing from outcrop to crustal scale in the Monte Rosa nappe, western Alps. *Tectonics*, **6**, 69–88.

—— & Mattauer, M. 1985. Kilometre scale sheath fold at Mattmark and implications for transport direction in the Alps. *Nature*, **316**, 739–42.

Laubscher, H. P. 1971. The large-scale kinematics of the western Alps and the northern Apennines and its palinspastic implications. *American Journal of Science*, **271**, 193–226.

—— 1975. Plate boundaries and microplates in alpine history. *American Journal of Science*, **275**, 865–76.

Le Pichon, X., Bergerat, F. & Roulet, M. J. in press. Plate kinematics leading to Alpine belt formation: a new analysis. *Geological Society of America Bulletin*, special volume, in the honour of J. Rodgers.

Malavielle, J. 1982. *Etude microtectonique de la déformation ductile dans de grands chevauchements crustaux. Exemple des Alpes franco–italiennes et de la Corse*. Unpublished thesis, University of Montpellier.

—— & Etchecopar, A. 1981. Ductile shear deformation of quartzite in an Alpine crustal thrust (Ambin massif). *Tectonophysics*, **78**, 65–71.

——, Lacassin, R. & Mattauer, M. 1984. Signification tectonique des linéations d'allongement dans les Alpes occidentales. *Bulletin de la Société Géologique de France*, **26**, 895–906.

Mancktelow, N. 1985. The Simplon line: a major displacement zone in the western Lepontine Alps. *Eclogae Geologicae Helveticae*, **78**, 73–96.

Marton, E. & Veljovic, D. 1983. Paleomagnetism of the Istria peninsula, Yugoslavia, *Tectonophysics*, **91**, 73–87.

Mattauer, M. 1975. Sur le mécanisme de formation de la schistosité dans l'Himalaya. *Earth and Planetary Science Letters*, **28**, 144–54.

—— 1986. Intracontinental subduction, crust mantle décollement and crustal stacking wedge in the Himalayan and other collision belts. *In*: Coward, M. P. & Ries, A. C. (eds). *Collision Tectonics*. Special Publication of the Geological Society of London No. 19, 37–50.

—— & Tapponnier, P. 1978. Tectonique des plaques et tectonique intracontinentale dans les Alpes franco–italiennes. *Comptes-Rendus des scéances de l'Académie des Sciences Paris*, **287**, 899–902.

—— & Collot, B. 1986. Continental subduction, thrusting and strike-slip faulting in the Canadian Cordillera. *Bulletin de la Société Géologique de France*, **2**, 899–909.

Menard, G. 1979. *Relations entre structures profondes et superficielles dans le SE de la France. Essai d'utilisation de données géophysiques*. Unpublished thesis, University of Grenoble.

—— & Thouvenot, F. 1984. Ecaillage de la lithosphère européenne sous les Alpes occidentales: arguments gravimétriques et sismiques liés à l'anomalie d'Ivrea. *Bulletin de la Société Géologique de France*, **26**, 875–84.

Merle, O. 1987. Histoire de la déformation dans les Alpes Lépontines occidentales. *Bulletin de la Sociéte Géologique de France*, **2**, 183–90.

Milnes, A. G. & Pfiffner, O. A. 1980. Tectonic evolution of the central Alps in the cross section St Gallen–Como. *Eclogae Geologicae Helveticae*, **73**, 619–33.

Monie, P. 1985. La méthode $^{39}Ar^{40}Ar$ appliquée au métamorphisme alpin dans le massif du Mont Rosa (Alpes occidentales); chronologie detaillée depuis 110 Ma. *Eclogae Geologicae Helveticae*, **78**, 487–516.

Mugnier, J. L., Arpin, R. & Thouvenot, F. 1987. Coupes équilibrées à travers le massif de la Chartreuse. *Geodinamica Acta*, **1**, 125–37.

Nicolas, A. & Boudier, F. 1975. Kinematic interpretation of folds in alpine-type peridotites. *Tectonophysics*, **25**, 233–60.

Olivet, J. L., Bonnin, J., Beuzart, P. & Auzende, J. M. 1984. Cinématique de l'Atlantique nord et central. *Publication Centre National pour l'Exploitation des Océans*, **54**, 108 pp.

Pfiffner, O. A. 1981. Fold-and-thrust tectonics in the Helvetic nappes (E. Switzerland). *In*: McClay, K. R. & Price N. D. (eds). *Thrust and Nappe Tectonics*, Special Publication of the Geological Society of London No. 9, 319–27.

Pittman, W. C. & Talwani, M. 1972. Sea-floor spreading in the north Atlantic. *Geological Society of America Bulletin*, **83**, 619–32.

Platt, J. P. 1984. Balanced cross-sections and their implications for the deep structure of the northwest Alps: discussion. *Journal of Structural Geology*, **6**, 603–06.

—— & Lister, G. S. 1985. Structural history of high-pressure metamorphic rocks in the southern Vanoise massif, French Alps, and their relation to Alpine tectonic events. *Journal of Structural Geology*, **7**, 19–35.

Plotto, P. 1977. *Structures et déformations des Grés du Champsaur au SE du massif du Pelvoux*. Unpublished thesis, University of Grenoble.

Ramsay, J. G. 1981. Tectonics of the Helvetic nappes. *In*: McClay, K. R. & Price, N. D. (eds). *Thrust and Nappe Tectonics*, Special Publication of the Geological Society of London No. 9, 293–309.

Reed, J. C., Bryant, B. & Myers, W. B. 1970. The Brevard zone: a reinterpretation. *In*: Fischer, O. W., Petitjohn, F. I. Reed, J. C. & Weaver, K. M. (eds). *Studies of Appalachian Geology: Central and Southern*, 261–69. J. Wiley.

Savostin, L. A., Sibuet, J. C., Zonenshain, L. P., Le Pichon, X. & Roulet, M. J. 1986. Kinematic evolution of the Tethys belt from the Atlantic ocean to the Pamir since the Triassic. *Tectonophysics*, **123**, 1–35.

SHACKLETON, R. M. & RIES, A. C. 1984. The relation between regionally consistent stretching lineations and plate motions. *Journal of Structural Geology*, **6**, 111–17.

SCHMID, S. M. 1975. The Glarus overthrust: field evidence and mechanical model. *Eclogae Geologicae Helveticae*, **68**, 247–80.

STECK, A. 1984. Structures de déformation tertiaire dans les Alpes centrales. *Eclogae Geologicae Helveticae*, **77**, 55–100.

TAPPONNIER, P. 1977. Evolution tectonique du système alpin en Méditerranée: Poinçonnement et écrasement rigide plastique. *Bulletin de la Société Géologique de France*, **19**, 437–60.

TRICART, P. 1980. *Tectoniques superposées dans les Alpes occidentales, au sud du Pelvoux: évolution structurale d'une chaîne de collision*. Unpublished thesis, University of Strasbourg.

—— 1984. From passive margin to continental collision: a tectonic scenario for the western Alps. *American Journal of Science*, **284**, 97–120.

TRÜMPY, R. 1969. Die Helvetischen Decken der Ostschweiz: Versuch einer palinspatischen Korrelation und Ansätze zu einer kinematischen Analyse. *Eclogae Geologicae Helveticae*, **62**, 105–38.

—— 1973. L'évolution de l'orogénèse dans les Alpes centrales: interprétation des données stratigraphiques et tectoniques. *Eclogae Geologicae Helveticae*, **66**, 1–10.

VANDENBERG, J. 1983. Reappraisal of paleomagnetic data from Gargano (South Italy). *Tectonophysics*, **98**, 29–41.

VAN DEN DRIESSCHE, J. 1986. Cinématique de la déformation ductile dans la cordillère canadienne: relations chevauchements-décrochements. *Bulletin de la Société Géologique de France*, **2**, 911–20.

VEARNCOMBE, J. R. 1980. *The geology of the Gran Paradiso region of the western Alps*. Unpublished thesis, University of Swansea.

VIALON, P. 1966. *Etude géologique du massif cristallin Dora-Maira, Alpes cottiennes internes, Italie*. Unpublished thesis, University of Grenoble.

WENK, E. 1955. Eine Strukturkarte der Tessinalpen. *Schweizerische Mineralogische und Petrographische Mitteilungen*, **35**, 311–19.

WESTPHAL, M., BAZHENOV, M. K., LAUER, J. P., PECHERSKY, D. M. & SIBUET, J. C. 1986. Paleomagnetic implications on the evolution of the Tethys belt from the Atlantic ocean to the Pamir since the Triassic. *Tectonophysics*, **123**, 37–82.

R. LACASSIN, Laboratoire de Tectonique, Mécanique de la Lithosphère; Institut de Physique du Globe de Paris; 4 pl. Jussieu, 75252 Paris Cedex 05, France.

Alpine thermal evolution in the central and the western Alps

J. C. Hunziker, J. Desmons & G. Martinotti

SUMMARY: In the western and central Alps, high-pressure assemblages dated as Eoalpine are distributed in the Pennine, Piemonte and Austro–Alpine units. The starting point of metamorphism must have been at low temperatures as both pre-Alpine metamorphic basement and Mesozoic sediments have been involved. In those parts of the Sesia–Lanzo zone where no later Tertiary high temperature overprint modified or obliterated the early Alpine history, the Eoalpine peak of metamorphic pressure was reached at about 110 Ma, the temperature peak at 85 Ma and a cooling down to 300°C at 60 Ma. The radiometric evidence gives insight into the Cretaceous orogenic events which also include penetrative deformation, nappe thrusting and uplift. The Pennine and Austro–Alpine units are both considered to have been located south of the Jurassic oceanic gap and to have formed parts of the fragmented northern margin of Gondwana.

The Tertiary Mesoalpine and Neoalpine metamorphic events in the western Alps are separated in time by 32–30 Ma magmatic intrusions. In the central Alps both events are distinguished on the basis of an additional schistosity connected with the second event. The Neoalpine ages are considered as resulting from the Mesoalpine thermal conditions outlasting the Mesoalpine time and, in the Simplon region, as recrystallization ages related to late mylonitization.

The convergence rates during the Cretaceous orogenesis were one order of magnitude greater than during the Tertiary orogenesis. Sedimentation in some Pennine units continued through Eoalpine times until Palaeogene; this is interpreted as resulting from the imbrication of blocks of different evolution in Tertiary times.

Since the beginning of this century (Argand 1916), the Alps have been known as one of the best examples of Stille's (1924) theory of episodical changes. In the well known controversy between Stille and Gilluly (see Trümpy 1973), i.e. periodical catastrophes versus continuous orogenic deformation, radiometric data and the plate tectonic theory are often cited as conclusive points (Hsü this volume). Nevertheless, be it atavistic conservationism or the simple fact that, as it is frequently the case in such discussions, the truth may be on both sides, the papers of the present volume show vividly that this discussion is still subconsciously going on. On one side we are continuously confronted with new information from structural geology according to which, depending on the preference, a *de jure* continuous evolution can be explained either as successive deformation phases numbered from D_1 up to possibly D_9, or as a unique continuous rotational deformation. On the other side, geochronological data are classified into several age groups, perhaps only in an attempt to gain a better overview, but possibly also on account of their being interpreted as pointing to succeeding phases.

Since Trümpy's (1973) paper the central and western Alps have been described as having evolved during three episodes: Eoalpine (Cretaceous, 140–60 Ma), Mesoalpine (Eocene–Oligocene, 45–30 Ma) and Neoalpine (Oligocene–Miocene, 30–10 Ma), separated by periods of tranquillity.

Admittedly, plate tectonic motions as estimated for different times and/or different places, show differences in rate of around one order of magnitude, as well as differences in direction. On the other hand, geochronological data, notably because of lack of proper compilations, are often uncritically taken where convenient, or simply omitted. All this tends to obscure the picture, especially when mean rock deformation rates are calculated over long periods, and contributes to uncertainty regarding one or another tectonothermal reconstruction.

For the new edition of the metamorphic map of the Alps three maps containing all available radiometric data are being compiled. These maps have so far been published only for the eastern Alps (Frank *et al.* 1986). For the central and western Alps data extracted from over 120 papers are plotted on three maps at the scale of 1:500 000 by the first two authors of the present paper (Desmons & Hunziker 1987). Map 1 comprises Rb–Sr, Sm–Nd and K–Ar whole rock data, as well as zircon and monazite U–Pb results, and shows mainly the pre-Alpine story. Map 2 shows the Rb–Sr mineral data and map 3 the K–Ar and $^{39}Ar–^{40}Ar$ mineral data. They will be published together with the new metamorphic map.

From: Coward, M. P., Dietrich, D. & Park, R. G. (eds), 1989, *Alpine Tectonics*, Geological Society Special Publication No. 45, pp. 353–367.

The aim of this paper is to show that the thermal evolution has been stepwise and to discuss the three stages of the thermometamorphic evolution in relation to Alpine geology.

Eoalpine period

As has been known for a long time, in the western Alps the Alpine metamorphic cycle started with a high-pressure phase. Mineral parageneses dated as Eoalpine are found in the following areas from the Pennine and Austro–Alpine crystalline basement and from the Piemonte zone (internal Pennine cover) (Fig. 1, nos 8 to 1):

(8) In the Dora—Maira Massif the association of pyrope, kyanite, talc, phengite and quartz with coesite relics within pyrope garnet grains, interpreted as Alpine, suggests ultra-high pressures of the order of 28 kbar and temperatures as high as 700–750°C (Chopin 1984). Phengite lying in a foliation younger than the garnet yielded an age of around 110 Ma (Goffé & Chopin 1986). Ages down to 60 Ma from various gneisses in the massif (Vialette & Vialon 1964, Hunziker unpublished data) can be interpreted as cooling ages.

(7) In the Gran Paradiso Massif pressures have been estimated as 8–15 kbar at temperatures of 400–500°C. The oldest Alpine phengite ages are 110 Ma; again cooling ages are down to 60 Ma (Chopin 1979, Chopin & Maluski 1980, Dal Piaz & Lombardo 1986, Goffé & Chopin 1986, Hurford pers. comm.).

(6) Some of the most spectacular occurrences of high-pressure rocks are from the Sesia-Lanzo zone where not only basic rocks have been transformed into eclogites: high pressure conditions are also revealed by granitoid rocks bearing jadeite, quartz and garnet (Dal Piaz *et al.* 1972, Compagnoni *et al.* 1977); late mylonites as well as slickensides show oriented glaucophane and omphacite and late cross-cutting quartz veins exhibit omphacite, garnet, zoisite and rutile.

Geochronological data from the internal part of the Sesia–Lanzo zone point to a pressure peak at around 110 Ma and to a temperature peak around 85 Ma, followed by a cooling phase down to 300°C at ~ 60 Ma (Oberhänsli *et al.* 1985). The Mesoalpine event in this internal part of the Sesia–Lanzo zone reached temperatures of around 250°C only, allowing for brittle deformation during the Tertiary, as revealed by fission track evidence (Hurford & Hunziker 1985). Fission track zircon ages are of 35 Ma and fission track apatite ages of ~ 25 Ma indicate the drop of temperature below approximately 100°C at the Oligocene–Miocene boundary. In other words, in this segment of the Alpine belt only one Alpine orogenic and metamorphic phase can be detected (apart from the 32–30 Ma magmatic event) and no evidence for a further metamorphic evolution is to be found. The only later orogenic activity in this zone was uplift and erosion, contributing to the thick lower Tertiary infill of the Ligurian basin of the Po Plain.

(5) In the Briançon–Bernhard nappe Alpine blueschist conditions in the crystalline basement have been estimated at ≥8 to ≥10 kbar and from 200 to 360–470°C (Gay 1972, Bocquet [Desmons] 1974, Desmons 1977, Saliot *et al.* 1980, Wust and Baehni 1986, Escher *et al.* 1987). A few 3 T phengites have been dated as ~ 100–65 Ma (Bocquet [Desmons] *et al.* 1974, Desmons & Hunziker unpublished data).

(4) In the Monte Rosa Massif eclogite and white schist parageneses have been described, revealing P–T conditions of 10–16 kbar at 400–500°C. Ages of 100 and 110 Ma have been obtained for the phengite (Wetzel 1972, Hunziker 1974, Chopin & Monié 1984, Monié 1985).

(3) In the Piemonte ophiolites described by Bearth (1962), Oberhänsli (1982) suggested P–T conditions of 10–16 kbar at 400–700°C for the eclogitic pillow lavas from Zermatt and Meyer (1983) estimated maximum pressures of ≤ 20 kbar at about 600°C for the Allalin gabbro. Hunziker (1974) dated the eclogitic pillow lavas at 100–80 Ma.

In the Queyras region of SE France *P–T* conditions in jadeite–glaucophane– and/or lawsonite bearing metaophiolites have been variously estimated at >10–11 kbar and 250–300°C (Bocquet [Desmons] 1974; Desmons 1977) or 7–10 kbar and 350–400°C (Saliot *et al.* 1980). Ages from 120 to 70 Ma have been obtained (Bocquet [Desmons] *et al.* 1974; Delaloye & Desmons 1976).

(2) In the central Adula nappe small boudins and lenses of basic composition have been described within younger barrovian type amphibolite facies country rocks (Heinrich 1982, 1983, 1986). The basic rocks show eclogite and blueschist parageneses indicating P–T conditions between 12 and 22

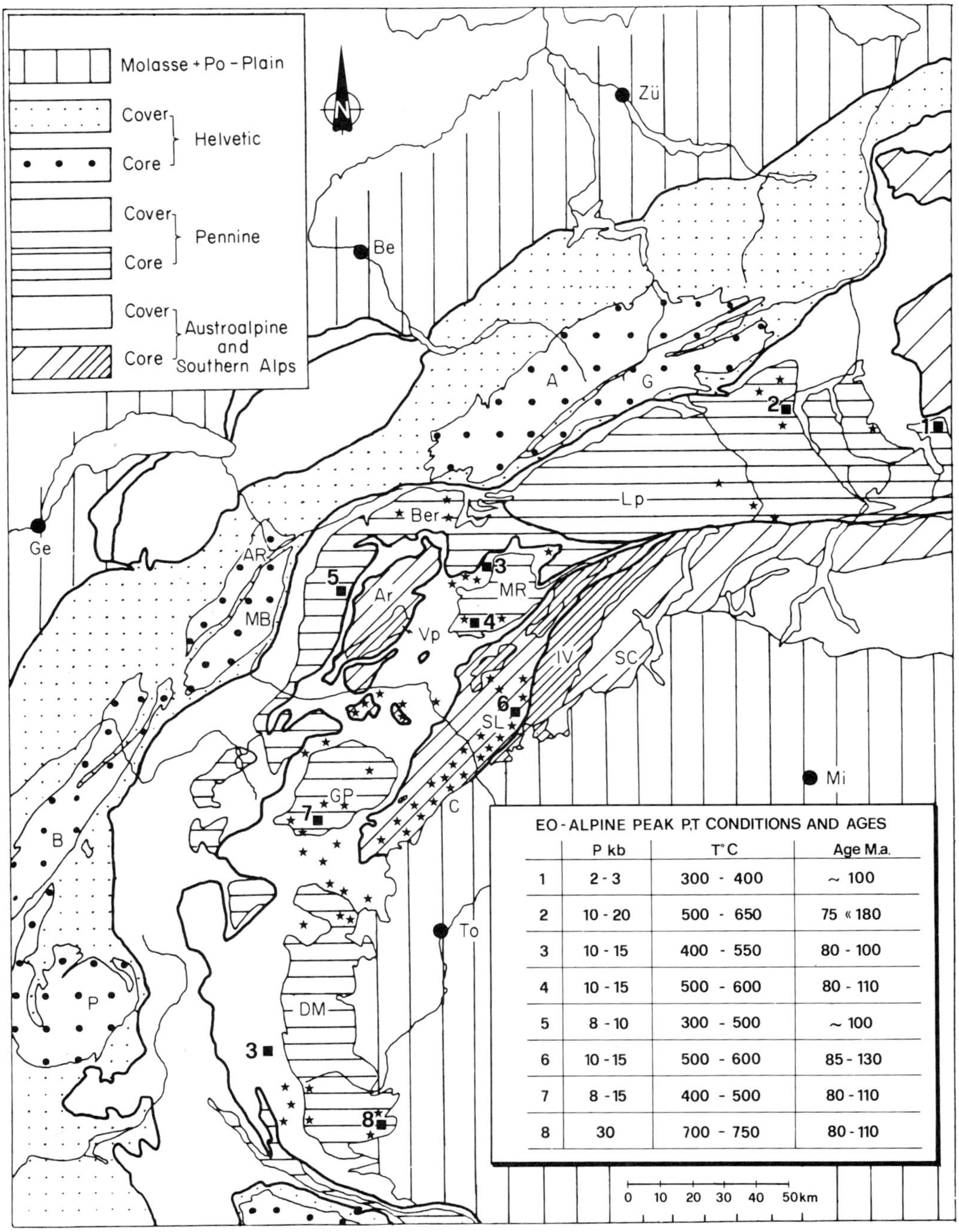

EO-ALPINE PEAK P,T CONDITIONS AND AGES

	P kb	T°C	Age M.a.
1	2 - 3	300 - 400	~ 100
2	10 - 20	500 - 650	75 « 180
3	10 - 15	400 - 550	80 - 100
4	10 - 15	500 - 600	80 - 110
5	8 - 10	300 - 500	~ 100
6	10 - 15	500 - 600	85 - 130
7	8 - 15	400 - 500	80 - 110
8	30	700 - 750	80 - 110

FIG. 1. Tectonic sketch map of the central and western Alps. The numbers refer to the locations of rock units in high-pressure facies radiometrically dated as Eoalpine (see text). Peak $P-T$ conditions and ages are shown in the inset table. Stars indicate localities of eclogites. Abbreviations: A = Aar Massif; Ar = Arolla series; AR = Aiguilles Rouges Massif; B = Belledonne Massif; Be = Berne; Ber = northern Bernhard; C = Canavese; DM = Dora–Maira Massif; G = Gotthard Massif; Ge = Geneva; GP = Gran Paradiso Massif; IV: Ivrea–Verbano zone; Lp = Lepontine area; MB = Mont Blanc Massif; Mi = Milan; MR = Monte Rosa Massif; P = Pelvoux Massif; SC = Strona–Ceneri zone; SL = Sesia–Lanzo zone; To = Turin; Vp = Valpelline nappe; Zü = Zürich.

kbar at 500–650°C. These parageneses are of Alpine age according to Muralt (1986), but the possible time span so far remains considerable, between 76 and 180 Ma. Similar P–T values were also derived from metapelites by Löw (1985, 1986). In the adjacent Mesozoic Mesocco zone glaucophanitic pillow lavas have been described (Gansser 1937, Oberhänsli 1978).

(1) In the Austro–Alpine Platta nappe blue amphiboles have been dated at around 100 Ma (Deutsch 1983, Deutsch & Steiger 1985). The $P-T$ conditions were estimated at 2–3 kbar and ~ 300–400°C (Oberhänsli 1986).

The distribution of Eoalpine ages from high pressure rocks throughout the western Alps sets severe problems for palaeogeographic reconstructions. According to the classical reconstruction (sketch A in Fig. 2) a southern plate, to which the Austro–Alpine and south Alpine units belonged, collided in Alpine times with a northern plate, the European plate, i.e. the Pennine and Helvetic domains, the Piemonte oceanic basin lying between both (e.g. Debelmas & Lemoine 1970, Debelmas & Kerckhove 1980).

The peculiar situation in the western Alps is caused by the fact that the continental crust undisputedly underwent Eoalpine high pressures in the Lower to Middle Cretaceous. The high pressure parageneses, according to classical palaeogeographic reconstructions, developed on both sides (Pennine and Austro–Alpine) of the Piemonte basin in its classically assumed position. Moreover, the high pressure parageneses in the continental crust were coeval with both the sedimentary deposition on the basin margins (Lemoine *et al.* 1984, Marthaler 1984) and the eclogitization of the oceanic crust (Hunziker 1974) (Fig. 1). We are therefore confronted with the fact that during the Cretaceous high pressure event, both plate margins, the northern rim of the southern plate and the southern rim of the northern or European plate, as well as the oceanic crust between both, all underwent pre collisional high pressure metamorphism at approximately the same time, during the Lower to Middle Cretaceous, and this while coeval sedimentation was taking place on the margins of the same basin. This seems most unlikely (see Hunziker & Martinotti 1984) and shows that all the so far presented tectonic models go lame, including that of Dal Piaz *et al.* (1972). These facts are only reconcilable if one of the following three assumptions is accepted:

(1) Two continuous subduction or thickening zones were present during Lower to Middle Cretaceous on both sides of the oceanic basin, an idea discussed by Hsü (this volume) (sketch A in Fig. 2). Evidence for the S-dipping subduction zone has been partially preserved, as revealed by geophysical data (Finckh *et al.* 1987), but no evidence of a N-dipping subduction zone has been found. This together with increasing evidence of bilateral vergence of the Alpine orogenesis (Cassano *et al.* 1986) does not support this assumption. However, it is not contradicted by the third proposal hereafter (sketch C in Fig. 2).

(2) It has been suggested that the Austro–Alpine Sesia–Lanzo zone formed part of the European plate (Aubouin *et al.* 1977, Mattauer *et al.* 1987) (sketch B in Fig. 2). This assumption does not take account of the fact that the Sesia–Lanzo zone for lithological reasons cannot be detached from the Ivrea–Verbano zone, that is, from the southern Alps which clearly belong to the southern plate (Compagnoni *et al.* 1977, Martinotti & Hunziker 1984).

(3) Another quite intriguing alternative, which has been independently proposed by both Hunziker & Martinotti (1984) and Desmons (1986, see also Radelli & Desmons 1987), is the idea of a different palaeogeographic setting, with the main suture located between the Pennine and the Helvetic domains. Thus the Pennine domain would form part of the southern plate and all the high-pressure rocks be on one side of the oceanic basin. In addition to the evidence from the high-pressure evolution outlined above, other lines of evidence support this reconstitution. The lithological similarities that exist between Pennine rocks and S Alpine rocks of the Strona–Ceneri zone are striking and the distinction of Pennine from Austro–Alpine units in the transition zone of the eastern Alps has always been a delicate question (Trümpy 1975). Moreover, characteristics of the sedimentary facies of the Triassic are more compatible with a connection of the Briançonnais with both the Austro–Alpine and S Alpine Alpine Triassic realm, than with a connection with the German-type Triassic as exposed in the Helvetic domain and further northwards. Palaeomagnetic arguments are still weak, but point to similar conclusions, stable European pole positions occurring only from the Helvetic domain northwards (Heller 1980, Westphal *et al.* 1986). Sketches C and D in Fig. 2 show the recon-

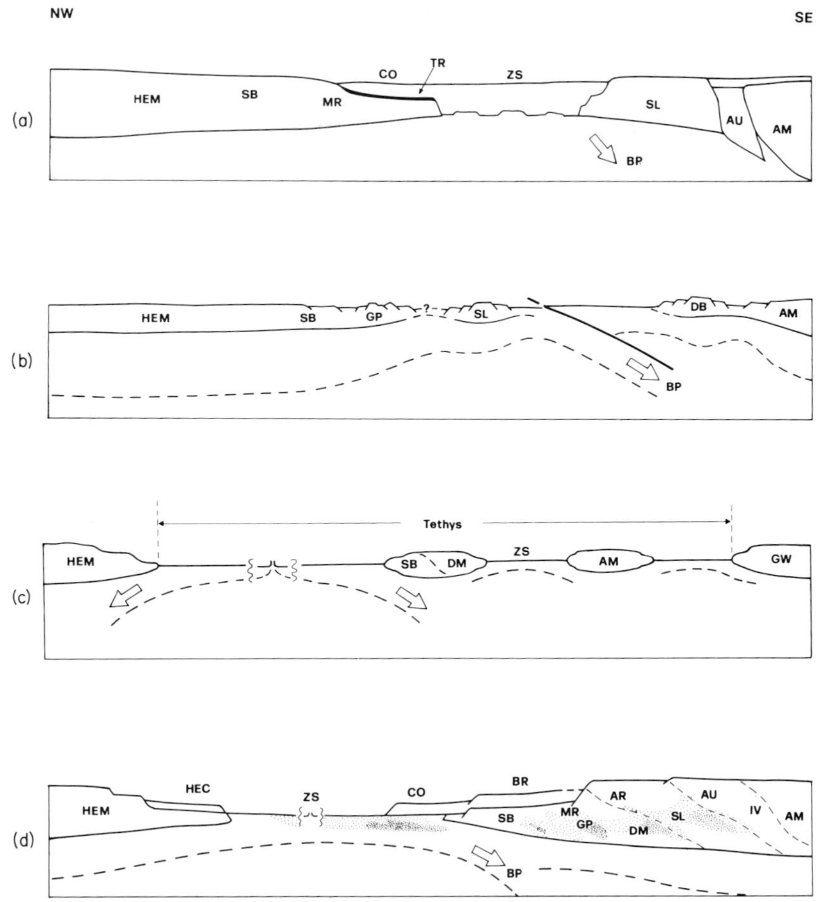

FIG. 2. Four possible palaeogeographic configurations at the beginning of the Alpine cycle (note the difference in scale of sketch C): A, The classical reconstruction, e.g. Debelmas & Lemoine (1970); Dal Piaz *et al.* (1972); Hsü (this volume); B, Mattauer *et al.* (1987); C, According to Radelli & Desmons (1987); D, Hunziker & Martinotti (1984). Abbreviations: *Helvetic*: HEM = Helvetic external Massifs; HEC: Helvetic cover; *Pennine*: SB = Bernhard–Briançon basement nappes; BR = Bernhard–Briançon cover, Mesozoic to Tertiary; MR = Monte Rosa nappe; GP Gran Paradiso basement; DM: Dora–Maira basement; TR: Pennine Triassic cover; ZS: Pennine Zermatt–Saas zone, Mesozoic oceanic sequence; CO: Pennine Combin zone, Mesozoic margin sequence of the oceanic basin; *Austro–Alpine and S-Alpine*: SL = Sesia–Lanzo zone; DB = Dent Blanche nappe; AR = Arolla unit; AM = southern Alps; IV: Ivrea–Verbano zone; AU- Upper Austro–Alpine units, Valpelline and II DK. GW = Gondwana. BP = Future Benioff plane. The dotted areas in sketch (d) show the relative intensity of the succeeding Eoalpine high pressure metamorphism.

structions that follow from the proposals of, respectively, Radelli & Desmons (1987) and Hunziker & Martinotti (1984).

According to Radelli & Desmons (1987) the Piemonte ophiolites would be derived from a back-arc basin marginal to the southern plate, situated behind thinned continental fragments (Pennine, Austro–Alpine, etc.) stripped off from Gondwana during the Triassic–Jurassic (sketch C in Fig. 2). In this model the crust generated at the mid-Tethyan ridge itself would be entirely consumed by subduction in a Benioff zone located at the site of the Valais trench, i.e. the Sion–Courmayeur line and its extensions (Fig. 3). The ocean-type crust of the marginal basin was partly consumed, partly emplaced on the Pennine fragments during the Mid-Cretaceous orogenesis. The tectonic imbrication of both Pennine and Austro–Alpine fragments as well as ophiolites and margin sediments was also an effect of the Middle Cretaceous orogenesis during which Gondwana collided with its formerly stripped-off fragments. Thus, the

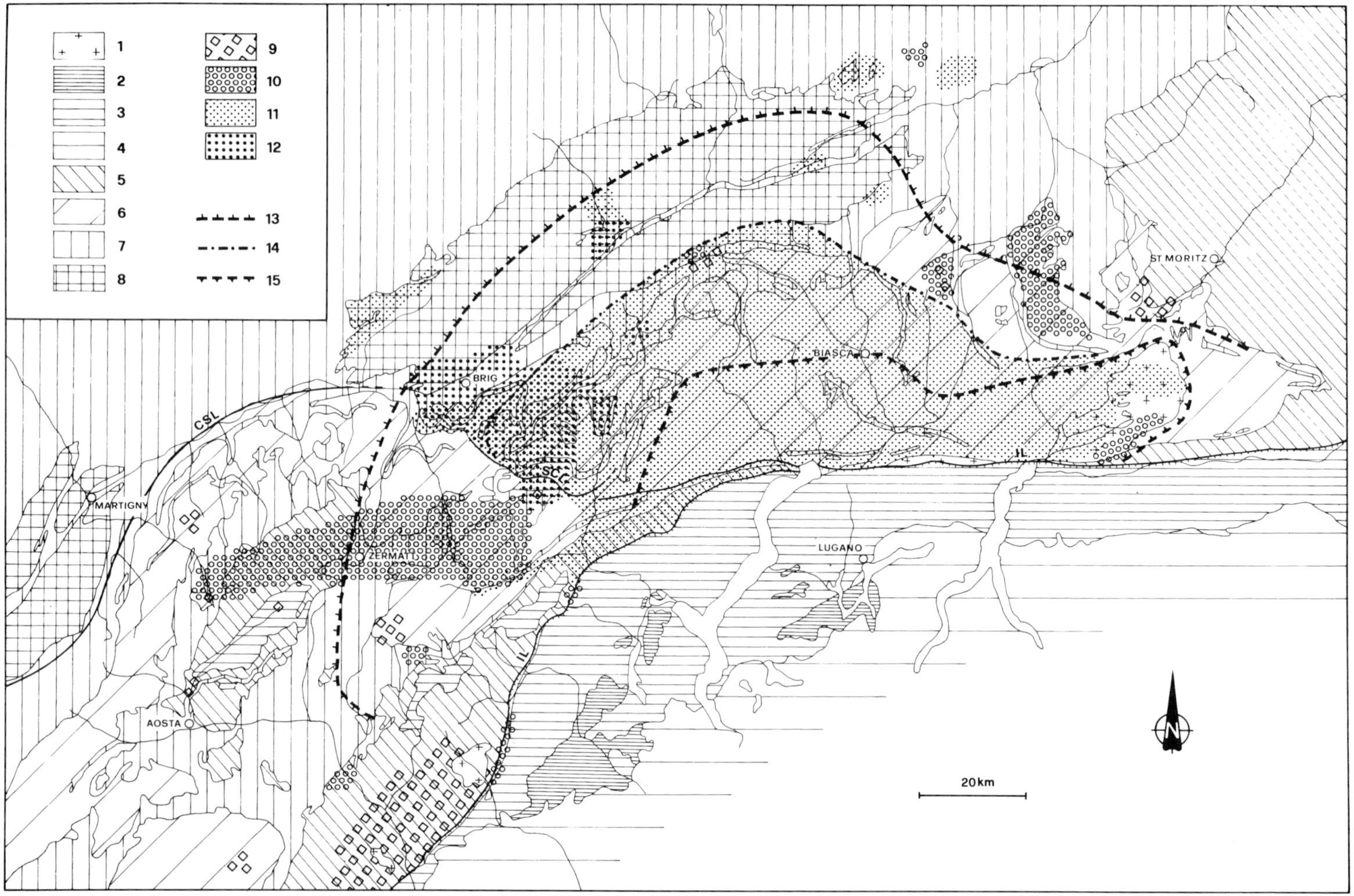

FIG. 3. Distribution map of Alpine K–Ar mineral ages in the central and northern western Alps (only ascertained and clear-cut published data have been taken into account). Tectonic units and abbreviations: 1 = Tertiary granitoids: Biella, Traversella, Bergell; 2 = Permian magmatics of the southern Alps: Baveno granite suite and S–Alpine Permian volcanics, predominantly rhyolites; 3 = S–Alpine basement, including the Ivrea zone, the Valpelline series and the II DK zone; 4 = S–Alpine sedimentary cover, Mesozoic and younger; 5 = Austro–Alpine nappes; 6 = Pennine basement and pre-Triassic sediments; 7 = Pennine Mesozoic and Neogene sediments, Helvetic and Canavese zones, Permian to Tertiary in age; 8 = external Helvetic basement massifs; 9 = field of Eoalpine mineral age data: 140–60 Ma; 10 = field of Mesoalpine mineral age data; 45–30 Ma; 11 = field of Neoalpine mineral ages of 30–15 Ma; 12 = field of Neoalpine mineral ages of 15–8 Ma; 13: stilpnomelane- out boundary; 14 = Alpine staurolite-in boundary; 15 = diopside-calcite-in boundary; IL = Insubric Line; SCL = Simplon Line; CSL = Sion-

presently observed intercalation of calcschists related to trenches (Piemonte, Valais, Mesocco, Arosa, etc.) between various Pennine and Austro–Alpine basement fragments, need not be interpreted as reflecting a primary palaeogeographic arrangement, but can be a consequence of Cretaceous collisional events, later followed by the Tertiary thrusting and backthrusting.

The initial Eoalpine pressure conditions, ranging from 8 to ≥28 kbar according to currently accepted petrological data, would require crustal depths of 25 to 100 km. We are aware that such depths may seem geologically unrealistic, but we shall provisionally accept them in the following discussions. The oceanic basin was according to commonly accepted reconstructions not very large (from 700 to ~1500 km (Trümpy 1960, and in Debelmas *et al.* 1983). Such a width applies only to Radelli & Desmons' (1987) marginal basin, not to the Tethyan ocean proper. In order to attain the required depths, in the case of a narrow ocean basin it is not very plausible to call upon several contemporaneous subduction zones. Hunziker & Martinotti (1984) preferred to relate the Eoalpine high pressure metamorphism to a unique subduction zone in connection with crustal thickening of the northern rim of the southern plate. However, in the view of the Piemonte ophiolites derived from a marginal basin, its crust may have been consumed through manifold subduction and this would easily allow for slightly different $P-T$ regimes within the high-pressure rocks. According to our working hypothesis of the suture lying at the Sion–Courmayeur line, all basement sequences showing Eoalpine high-pressure assemblages are related to the southern plate.

The first evidences of Eoalpine crustal thickening in the western Alps point to as far back as 130 Ma (Hunziker 1970, 1974, Steinitz & Jäger 1981, Oberhänsli *et al.* 1985). The Sesia zone had cooled below 300°C by the end of the Cretaceous, at around 60 Ma. If we take this cooling as indicating the end of the Eoalpine times, it suggests a total of about 70 Ma for the whole Eoalpine event. Already a slow plate motion of the order of 1–2 cm/a leads to 700–1400 km total convergence, in fairly good agreement with estimations of Trümpy (1960) and Hsü (this volume).

As noted above, the internal parts of the Sesia–Lanzo zone are the only segment of the central and western Alps where the Eoalpine metamorphic overprint was the last Alpine metamorphic effect recorded. After cooling below 300°C at about 60 Ma a brittle compressional deformation phase took place around 35 Ma (Hurford & Hunziker 1985), followed by the deposition of the Oligocene volcaniclastic sequence and dyke intrusion around 30 Ma (Scheuring *et al.* 1974, Cassano *et al.* 1986). These facts indicate that, for the internal part of the Sesia–Lanzo zone, we can adequately use the term orogenesis for the Eoalpine phenomena. It comprised an initial high pressure metamorphism, different subsequent deformations, cooling and uplift of a mountain belt.

According to Radelli & Desmons' (1987) proposal, the Cretaceous orogenesis, as mentioned above, affected all the Pennine, Austro–Alpine, etc., fragments detached from Gondwana. Pushed to the N by the Afro–Antarctic expansion site, Africa collided with its fragments in the Middle Cretaceous, finally closing the marginal basin. Flysch sedimentation in Briançon and other Pennine cover sequences are another record of this Cretaceous orogenesis. Flysch shows a decrease in age from Cretaceous in the internal parts of the Alpine chain to Eocene in the more external parts of the Briançonnais or even Lower Oligocene in the Helvetic domain (Debelmas 1970, Wildi 1987).

Tertiary metamorphism

A further widespread group of Alpine radiometric data covers the time span between 45 and 8 Ma, obviously corresponding to quite different geodynamic processes.

A first group of ages between 45 and 30 Ma can be distinguished. They predated the calcalkaline cross-cutting magmatism of the Traversella, Biella and Bergell intrusions and the trachy-andesitic volcanics, both of approximately 32–30 Ma. They also predate the onset of the first molasse sediments both N in the German and Swiss plateau molasse and S in the Tertiary basin of the Po Plain, which started in the Tongrian stage of the Lower Oligocene (Gasser 1968, Cassano *et al.* 1986).

A second group of ages can be identified, ranging between 30 and 8 Ma. Following Trümpy (1973) we adopt the terminology of Mesoalpine for the first group and of Neoalpine for the younger group of ages. The Tertiary metamorphic overprint affected rocks from both the northern and the southern plate, together with the emplaced ophiolites. This can be taken to indicate that by this time all the units had already collided and were close to each other.

In the western Alps the high-pressure Eoalpine terrains show their proper peculiar deformational style and pattern (Gosso *et al.*

1979, Williams & Compagnoni 1983). In the external Sesia–Lanzo zone a transition between Eoalpine and Mesoalpine ages has been recorded (Hunziker 1974). Where the mica ages have been reset by the Mesoalpine metamorphism, a new steeply dipping schistosity can be detected, with isoclinal folds, e.g. in the Gneiss Minuti of the external Sesia–Lanzo zone. This age transition goes hand in hand with the recrystallization of the Eoalpine high-pressure phengite 3T polytype into a lower-pressure phengite 2M polytype. In places the geochronological rejuvenation took place earlier than the polytype change, thus leaving 3T phengites as relics of the earlier high-pressure phase in a greenschist lithology, as for instance in the Briançonnais basement of the Bernhard nappe (Frey *et al.* 1983). In the internal western Alps, the Mesoalpine event is bracketed by the 32–30 Ma old unmetamorphosed calc-alkaline and trachy-andesitic magmatism. New fission track evidence (Hurford pers. comm.) also confirms this fact. We are confronted with cooling below 250°C in the internal parts of the belt in the Lower Oligocene and cooling below around 100°C in the Lower Miocene (Hurford & Hunziker 1985).

In the fifties Wenk (1956) introduced the term Lepontine for the structural style of the Tertiary Barrovian-type amphibolite facies metamorphism of the Ossola–Ticino area. This term was later extended to the greenschist facies overprint of the western Alps (Bearth 1962) and subsequently used for the Tertiary cooling ages in the central Alps (Jäger *et al.* 1967, Hunziker 1970, Jäger 1970, Köppel & Grünenfelder 1975, Deutsch & Steiger 1985). This led to the dispute concerning the age of the peak of this metamorphic phase.

On Fig. 3 we have synthesized the available K–Ar age information on micas and amphiboles. We see quite clearly that E and W of the Ossola–Ticino region a Mesoalpine field of mineral ages can be identified, approaching the staurolite isograd from lower grades. Inside the staurolite field, however, practically all the mineral ages are Neoalpine.

Following the staurolite isograd from both Mesoalpine fields, i.e. the Monte Rosa and Adula, to the N, we find a new schistosity S of the Gotthard Massif, towards the Lucomagno area (Chadwick 1965, Frey 1969, Probst 1980, Löw 1986). Two distinct tectonic areas can be recognized, one in the E and W, with staurolite in equilibrium with the Mesoalpine schistosity and with Mesoalpine white mica ages of 35–40 Ma, and the other S of the Gotthard Massif (Lucomagno region) where staurolite is in equilibrium with Neoalpine, 25 Ma old white micas and a new schistosity. It has therefore been proposed (Hunziker & Martinotti 1984, Desmons & Hunziker 1987) to drop the term Lepontine metamorphic phase and replace it by Mesoalpine and Neoalpine phases, which describe more adequately what has occurred. The Neoalpine event is mainly recorded at the present level of erosion within the staurolite isograd of the Ossola–Ticino area as a consequence of a late uplift (Hurford 1986, Schmid *et al.* 1987, 1988).

Outside of this high-grade central Ossola–Ticino field, Neoalpine mineral ages in the central Alps are restricted to late shear zones in the Aar and Gotthard Massifs, to the Helvetic nappes, and to mylonites of the Simplon area and of the Insubric Line system. In the western Alps they have also been found in the internal Penninic and the Helvetic zones (Bocquet [Desmons] *et al.* 1974). Among these cases, the Simplon Line mylonites have a special significance (Mancktelow 1985, Steck 1984, 1987). The Neoalpine ages, as already pointed out, are younger than 30 Ma. Most of the mica ages within the staurolite isograd fall into the time span 30–15 Ma, while in the Simplon area we find an age range of 15 to 8 Ma (Fig. 3). This field extends from the Simplon line east- and northwards into the Pennines, but also comprises the southern rim of the Gotthard and Aar Massifs. These youngest ages are predominantly found in the Mesozoic schists (Frank 1979), which exhibit a more plastic deformation at around 300°C than the more rigid crystalline basement rocks which, according to Mancktelow (1985), only show mylonite textures close to the planes of maximum displacement. The coincidence of the mylonitic deformation and recrystallization with the young mica ages leads to the interpretation that the Neoalpine micas were completely reset by a younger ductile deformation and recrystallization around 300°C.

On the Maggia profile through the Ossola–Ticino region Hurford (1986) conclusively showed that the Neoalpine mica ages represent cooling ages from peak temperatures still prevailing at the beginning of the Neoalpine time and were related to backfolding of the root zone. In view of the slow heat transfer of rocks, we therefore have to conclude that there is no Neoalpine metamorphic phase in the Ossola–Ticino region, but rather that the thermal conditions outlasting the Mesoalpine metamorphism are responsible for the Neoalpine reequilibration.

As already stated, the Mesoalpine event in

the western Alps involved nappe emplacement, deformation and metamorphism as well as uplift and cooling, and the whole event terminated with the deposition of the Lower Oligocene molasse sediments which contain Alpine clastics. In other words, with the Mesoalpine we again find a genuine orogenesis, but completely contrasting in geodynamic setting and style with the previous Eoalpine phase.

The Mesoalpine ages have been related to the metamorphic effects of the piling up due to the Tertiary collision phase, an adequate time span being allowed for the building up of the required thermal conditions. An alternative explanation of the Mesoalpine metamorphic phase is that of the thermal dome proposed by Wenk (1962), which accords well with a model proposed by Radelli (1986). This author suggests that the mantle upwelling that generated the Tethys spreading, underlying the mid-ocean ridge, did not die out with the collision event and disappearance of the ocean domain. Although capped by continental crust and momentarily inactive at the surface, it kept on living. As soon as the collision had caused the southern plate to dock along the European one and, thus, to come to a relative halt, its thermal effect would heat the overlying pile. This was all the easier because the pile was thicker due to Cretaceous thrusting and, thus, warmed up by radioactivity and preserved from ready heat loss by its thickness. This heating resulted in the Mesoalpine metamorphic phase which, strangely enough, is of the same ~ 38 Ma age all over the Alps. The same cause eventually produced the 'Periadriatic' magmatism. Again other ideas are exposed by Merle *et al.* (this volume) and by Schmid *et al.* (this volume).

The characterisation of three distinct tectono-metamorphic phases does not imply that all measured radiometric ages fall into one of the three clearly defined groups. As we have already seen, the Neoalpine ages are, in part, a consequence of a delaying of the Mesoalpine uplift. An analogous mechanism (in addition to Ar inheritance) may be taken into account to explain several ages that lie between Eo- and Mesoalpine, i.e. between 60 and 45–30 Ma. If a rock sequence remained at a depth of around 10 km after the Eoalpine cycle, the micas remained partly open systems, as shown by Bocquet [Desmons] *et al.* (1974). In this model one has also to take into account the fact that the underplating had brought cool material below the rock sequence which was rather thin and that no deformation occurred precisely during the Palaeocene period. Otherwise the rock sequence considered would have experienced recrystallization under greenschist facies conditions.

From the Dora–Maira through the Piemonte to the Briançon units, Goffé & Chopin (1986) postulated a discontinuously westwards- migrating high-pressure front decreasing westwards in age and intensity and evolving from ~ 110 to 35–40 Ma. The mechanism assumed by these authors is that of initial deep subduction followed by westwards migrating intracontinental overthrusts. A process similar to that proposed above, with rock bodies momentarily stationary at depth, might explain the spread of the Cretaceous high-pressure mineral ages in both the Dora–Maira and Piemonte zones. However, considering the Cretaceous ages yielded by phengites from the jadeite-bearing Briançon crystalline basement, we think that another model for the evolution is required, which also takes into account the distribution of the obviously Tertiary lawsonite, carpholite and crossite in the Briançon cover and Piemonte calcschists, indicating lower pressures between 3 and 6 kbar.

Discussion

So far we have described the different stages in the tectono-thermal evolution of the western and central Alps and discussed tectonic models. Now let us examine closely their relationship with the more general framework of Alpine geology. Petrological and isotopic data generally concern deep-seated geological processes, whereas geological observations relate to more superficial phenomena. The interrelation of both lines of observation is not always *a priori* clear.

Geochronological and petrological data concerning the metamorphic phase of the Eoalpine orogenesis set constraints on the Cretaceous tectono-thermal evolution, but also present severe problems. According to generally accepted models the high pressure event is connected with subduction and/or crustal thickening followed by rapid uplift. In addition to ophiolites and margin sequences, considerable amounts of continental crust in the Alps have also been affected by the very high to high pressures. During the last fifteen years increasing amounts of continental crust have been interpreted as being involved in Eoalpine orogenesis, from only the upper Pennine crystalline massifs and the Sesia–Lanzo zone in the early seventies (Dal Piaz *et al.* 1972, Bocquet [Desmons] 1974) to the lower Pennine nappes in more recent years (Heinrich 1982, 1986, Escher *et al.* 1987).

Although the incorporated Pennine and

Austro–Alpine crustal fragments may have experienced granulite and/or eclogite facies conditions in early Palaeozoic or earlier times, they all represent rocks that in late Variscan times were at upper crustal levels. On analogy with the Mesozoic history of the Ivrea–Verbano zone, Oberhänsli *et al.* (1985) proposed for the Sesia–Lanzo zone a rather cool early Mesozoic starting point corresponding to a depth of around 10 km. For the Monte Rosa, Gran Paradiso, Dora-Maira and Briançonnais Pennine elements, the same proposal can be put forward. At the present level of erosion only the higher parts of this continental crust have so far been exposed. For the lower parts of the continental crust involved in the Eoalpine crustal thickening, higher starting temperatures have to be assumed. This would create different $P-T$ paths during crustal thickening, possibly leading to high-pressure amphibolite or even granulite or eclogite facies assemblages (B-type eclogites according to Coleman *et al.* 1965), as exposed, for example, in some Norwegian localities (Austrheim & Griffin 1985, Griffin *et al.* 1985, Austrheim 1987). To assume low-temperature starting points leads to the possibility of inheritance of older, in this case pre-Alpine, isotopic as well as geochemical systems, as the rocks reached and maintained blocking temperatures in pre-Alpine times. On the other hand, should the Eoalpine cycle have started at elevated temperatures, above the blocking temperatures of geochronological systems, the only Alpine ages would be preserved, as the previous memory would have been wiped out during the high temperature event.

Detrital blue amphibole and lawsonite have been reported from various syn- to postorogenic sediments, of Mid-Cretaceous to Tertiary age (Niggli 1970, Bocquet [Desmons] 1966, 1974, Mange & Oberhänsli 1982, Wildi, 1985, Winkler & Bernoulli 1986, Decker *et al.* 1987). The presence of detrital blue amphiboles in south-Pennine flysch, deposited at around 85 Ma, sets a younger time limit for the crystallisation of blue amphibole during the Eoalpine event. Radelli & Desmons' (1987) geological reconstitution discussed earlier takes into account the fact that before 85 Ma, while some portions of the Alps were still undergoing subduction and/or crustal thickening, other portions obviously had already reached the erosional level and others still constituted sedimentary basins. Furthermore, some cover sequences show sedimentation up to around lower Upper Cretaceous in the Piemonte margins (Lemoine *et al.* 1984) and up to the Palaeogene in the Briançonnais, but with sedimentary gaps (Ellenberger 1958). In the internal parts of the Alpine chain the pressure peak according to geochronological data was already reached before 100 Ma, that is during the Early Cretaceous (see Fig. 1). The temperature maximum was attained in the Sesia–Lanzo zone at around 85 Ma, i.e., during Cenomanian times. The subsequent temperature decrease is possibly related to the beginning of uplift and denudation in this region, while in portions that now are more externally located, the high-pressure generating environment still persisted.

In the Briançon–Bernhard zone, there is a metamorphic jump between both cover and basement, the latter containing jadeite and glaucophane as well as 3T polytypes of white K–micas (Frey *et al.* 1983). Reliable Eoalpine radiometric ages have been measured in the Vanoise and Grand–Saint–Bernard areas (Bocquet [Desmons] *et al.* 1974, Hunziker 1974, Desmons & Hunziker unpubl. data). At first sight the coincidence between Eoalpine metamorphism in the Briançon basement and sedimentation that continued up to the Palaeogene in the whole Briançon–Bernhard cover sequence, seems to be inconceivable without admitting either (1) a broadly allochthonous cover, or (2) an imbrication of Eoalpine metamorphosed Briançon basement portions with portions of other Briançon basement plus sedimentary cover that both escaped Eoalpine metamorphism. The second hypothesis supposes, for example, that the external Vanoise cover was deposited on the zone Houillère or directly on the Sapey, Ruitor or similar basement, which would imply a considerable, but so far little documented, backthrusting.

The first clearly cross-cutting magmatic activity in the western Alps is the calc-alkaline magmatism of 32–30 Ma located on the eastern fringe of the Sesia zone. From Eoalpine times traces of magmatism are extremely scarce. Local remnants of Maastrichtian volcanics, apparently of alkaline composition, have recently been found in the Briançon cover (Deville 1987). The lack of Eoalpine magmatism has so far been explained by a continuation of the compression throughout the Cretaceous and Mesoalpine times. A hypothesis might be that the abundant andesitic volcani-clastic content of the Taveyannaz graywackes, whose deposition is thought to go back to Upper Eocene times (Sawatzki 1975, Stalder 1979), corresponds to the volcanic activity following the Cretaceous orogenic phase. At the present erosion level it is the only possible relic of such a magmatism.

Other considerations may be brought into the discussion. First, basic to intermediate volcanics of up to lower Upper Cretaceous age are documented in the calcschists of the Combin or related Piémont units; they differ from the ophiolites by their younger age and by their lithological association: the Upper Cretaceous volcanics are only associated with volcanogenic detritus and/or olistoliths and do not form an ophiolite sequence as do the Upper Jurassic Alpine ophiolites. Owing to possible syn- to post-sedimentary element mobility their geochemical affinities are not yet well defined. Is this magmatism a last effect of the thermal disturbance within the fragmented northern margin of Gondwana? Second, there is the question of magmatism related to the Tethys subduction in the Benioff zone under the European Plate (as proposed in sketch C in Fig. 2). Here, we may make a comparison with the western margin of America, where, as emphasized by Radelli (1986a, b), there is no continuous or regular magmatic activity, but a pulsating and local one, in places up to more than 60 Ma later than the onset of the subduction. Thus, other factors play a role in the generation of the calc-alkaline magmatism (Radelli 1988). In the Alpine belt also, we see a Cretaceous magmatism well developed in the Carpathians and further E in the Pontides, but not in other parts. More investigations and thought are needed on these topics.

The Tertiary Meso- and Neoalpine metamorphisms and collisional deformations differed in convergence rates from the Eoalpine metamorphism and associated plate convergence and consumption, by around one order of magnitude. Whereas for the Eoalpine orogenesis rates of the order of cm/a are widely accepted (e.g. Hsü this volume), evidence has been presented by Pieri & Groppi (1981) and Cassano *et al.* (1986) that during the Tertiary cycle the convergence rates associated with nappe formation reached the order of mm/a only. This drastic velocity difference in the Alps implies that first order changes occurred after the Cretaceous Eoalpine orogenesis, which led to the consumption of at least 700 km of oceanic lithosphere, as well as to imbrication and early nappe thrusting. (A good example of this Eoalpine imbrication is the Valpelline unit lying on top of the Arolla in the Dent Blanche nappe). In contrast to the consumption of more than 700 km of oceanic crust and the Eoalpine nappe emplacement, the slower Tertiary movements only led to refinements of the pre-existing picture, implying the further imbrication of the Mesozoic sediments between the different basement nappes (e.g., the Mesocco zone between the Adula and Tambo basement units; the profile of the Zermatt–Simplon area as reconstructed by Milnes *et al.* 1981), considerable gravity nappe tectonics (e.g. the Schistes Lustrés nappes onto the Briançon sequence and the Helminthoid flysch nappes), as well as the backfolding of Mesozoic and pre-Mesozoic units. With respect to syn- to post-orogenic sedimentation these two cycles, the Cretaceous and the Tertiary, show marked differences as well: the Eoalpine orogenesis mainly produced flysch sedimentation, with very little molasse, whereas the Tertiary phases are predominantly recorded by immediately succeeding molasse sedimentation with only local flysch sequences (Debelmas 1970, Wildi 1987).

Acknowledgements: The manuscript benefitted by fruitful discussions with M. Frey, P. Heitzmann, R. Oberhänsli, L. Radelli, St. Schmid and A. Zingg. The reviews of K. Brodie and M. Frey, as well as the correction of our continental English by K. Brodie and E. H. Brown, greatly improved the paper.

References

Argand, E. 1916. Sur l'arc des Alpes occidentales. *Eclogae geologicae Helvetiae*, **14**, 145–91.

Aubouin, J., Blanchet, R., Labesse, B. & Wozniak, J. 1977. Alpes occidentales et Alpes orientales: la zone du Canavese existe-t-elle? *Société géologique de France, Comptes-Rendus sommaires*, 155–8.

Austrheim, H. 1987. Eclogitization of lower crustal granulites by fluid migration through shear zones. *Earth and Planetary Science Letters*, **81**, 221–32.

—— & Griffin, W. L. 1985. Shear deformation and eclogite form tion within granulite-facies anorthosites of the Bergen Arcs, western Norway. *Chemical Geology*, **50**, 267–81.

Bearth, P. 1962. Versuch einer Gliederung alpinmetamorpher Serien der Westalpen. *Schweizerische mineralogische und petrographische Mitteilungen*, **42**, 127–37.

Bocquet [Desmons], J. 1966. Le delta miocène de Voreppe. Etude des faciès conglomératiques du Miocène des environs de Grenoble. *Travaux du Laboratoire de Géologie de l'Université de Grenoble*, **42**, 53–75.

—— 1974. *Etudes minéralogiques et pétrologiques sur les métamorphismes d'âge alpin dans les Alpes françaises.* Unpublished Doctorat ès Sciences thesis, University of Grenoble.

——, DELALOYE, M., HUNZIKER, J. C. & KRUMMENACHER, D. 1974. K–Ar and Rb–Sr dating of blue amphiboles, micas and associated minerals from the western Alps. *Contributions to Mineralogy and Petrology*, **47**, 7–26.

CASSANO, E., ANELLI, E., FICHERA, R. & CAPELLI, V. 1986. Pianura padana. Interpretazione integrata di dati geofisici e geologici. *Atti 73. Congressso della Societá Geologica Italiana*, Roma.

CHADWICK, B. 1965. *The structural and metamorphic geology of the Lukmanier region, Ticino-Grisons, Switzerland*. Unpublished PhD thesis, University of London.

CHOPIN, C. 1979. *De la Vanoise au massif du Grand Paradis*. Unpublished 3e cycle thesis, University of Paris VI.

—— 1984. Coesite and pure pyrope in high-grade blueschists of the western Alps: a first record and some consequences. *Contributions to Mineralogy and Petrology*, **86**, 107–18.

—— & MALUSKI, H. 1980. ^{40}Ar–^{39}Ar dating of high pressure metamorphic micas from the Gran Paradiso area (western Alps): evidence against the blocking temperature concept.*Contributions to Mineralogy and Petrology*, **74**, 109–22.

—— & MONIÉ, P. 1984. A unique magnesiochloritoid-bearing, high-pressure assemblage from the Monte Rosa, western Alps: petrologic and ^{40}Ar–^{39}Ar radiometric study. *Contributions to Mineralogy and Petrology*, **87**, 388–398.

COLEMAN, R. G., LEE, D. E., BEATTY, S. B. & BRANNOCK, W. W. 1965. Eclogites and eclogites: their differences and similarities. *Bulletin of the Geological Society of America*, **76**, 483–508.

COMPAGNONI, R., DAL, PIAZ, G. V., HUNZIKER, J. C., GOSSO, G., LOMBARDO, B. & WILLIAMS, P. F. 1977. The Sesia–Lanzo zone, a slice of continental crust with alpine high pressure–low temperature assemblages in the western Italian Alps. *Rendiconti della Societá Italiana di Mineralogia e Petrologia*, **33**, 281–334.

DAL PIAZ, G. V., HUNZIKER, J. C. & MARTINOTTI, G. 1972. La zona Sesia–Lanzo e l'evoluzione tettonico-metamorfica delle Alpi nordoccidentali interne. *Memorie della Societá Geologica Italiana*, **11**, 433–60.

—— & LOMBARDO, B. 1986. Early Alpine eclogite metamorphism in the Pennine Monte Rosa–Gran Paradiso basement nappes of the north-western Alps. *Geological Society of America Memoir*, **164**, 249–65.

DEBELMAS, J. 1970. La position des formations de type flysch dans le dévelopement orogénique des chaînes méditerranéennes. *Bulletin de la Société géologique de France* (7), **12**, 595–98.

—— & LEMOINE, M. 1970. The western Alps: palaeogeography and structure. *Earth-Science Reviews*, **6**, 221–56.

—— & KERCKHOVE, Cl. 1980. Les Alpes franco–italiennes. *Géologie Alpine*, **56**, 21–58.

DECKER, K., FAUPL, P. & MÜLLER, A. 1987. Synorogenic sedimentation in the northern Calcareous Alps during the early Cretaceous. *In*: FLÜGEL, H. W. & FAUPL, P. (eds). *Geodynamics of the Eastern Alps*. Deuticke, Vienna, 126–141.

DELALOYE, M. & DESMONS, J. 1976. K–Ar radiometric age determination of white micas from the Piémont zone, French–Italian Alps. *Contributions to Mineralogy and Petrology*, **57**, 297–303.

DESMONS, J. 1977. Mineralogical and petrological investigations of Alpine metamorphism in the internal French western Alps. *American Journal of Science*, **277**, 1045–66.

—— 1986. The Alpine metamorphisms and their environments in the western Alps: unsolved problems. *Schweizerische Mineralogische und Petrographische Mitteilungen*, **66**, 29–40.

—— & HUNZIKER, J. C. 1987. Maps of geochronological data from the central and western Alps. *Terra Cognita*, **7**, p 92.

DEUTSCH A. 1983. Datierungen an Alkaliamphibolen und Stilpnomelan aus der südlichen Platta-Decke (Graubünden). *Eclogae Geologicae Helvetiae*, **76**, 295–308.

—— & STEIGER, R. H. 1985. Hornblende K–Ar ages and the climax of Tertiary metamorphism in the Lepontine Alps (south–central Switzerland): an old problem reassessed. *Earth and Planetary Science Letters*, **72**, 175–89.

DEVILLE, E. 1987. *Etude géologique en Vanoise orientale (Alpes occidentales françaises, Savoie)*. Unpublished doctoral thesis, University of Savoy.

ELLENBERGER, F. 1958. *Etude géologique du pays de Vanoise*. Mémoire explicatif de la Carte géologique de France.

ESCHER, A., MASSON, H. & STECK, A. 1987. *Coupes géologiques des Alpes occidentales suisses*. Landeshydrologie und -geologie, geologische Berichte Nr. 2, Bundesamt für Umweltschutz, Bern.

FINCKH, P., FREI, W., FREEMAN, P., HEITZMANN, P., LEHNER, P., MUELLER, ST., PFIFFNER, A. & VALASEK, P. 1987. Nationales Forschungsprogram 20: Geologische Tiefenstruktur der Schweiz—Ausführung, Problemstellung und erste Resultate. *Bulletin der Vereinigung schweizerischer Petroleum Geologen und Ingenieuren*, **54**, Nr. 124, 59–74.

FRANK, E. 1979. *Metamorphose mesozoischer Gesteine im Querprofil Brig-Verampio: mineralogisch-petrographische und isotopengeologische Untersuchungen*. Unpublished thesis, University of Bern.

FRANK, W., KRALIK, M., SCHARBERT, S. & THÖNI, M. 1986. Geochronological data from the eastern Alps. *In*: FLÜGEL, H. W. & FAUPL, P. (eds). *Geodynamics of the eastern Alps, Deuticke, Vienna. 272–81.*

FREY, M. 1969. Die Metamorphose der Keupers vom Tafeljura bis zum Lukmanier-Gebiet. *Beiträge zur geologischen Karte der Schweiz*, neue Folge, **137**.

FREY, M., HUNZIKER, J. C., JÄGER, E. & STERN, W. B. 1983. Regional distribution of white K-mica polymorphs and their phengite content in the central Alps. *Contributions to Mineralogy and Petrology*, **83**, 185–97.

GANSSER, A. 1937. Der Nordrand der Tambodecke.

Geologische und petrographische Untersuchungen zwischen San Bernardino und Splügenpass. *Schweizerische mineralogische und petrographische Mitteilungen*, **17**, 291–523.

GASSER, U. 1968. Die innere Zone der subalpinen Molasse des Entlebuchs (Kt. Zürich): Geologie und Sedimentologie. *Eclogae geologicae Helvetiae*, **61**, 229–319.

GAY, M. 1972. Le massif d'Ambin et son cadre de Schites Lustrés (Alpes franco-italiennes). Evolution métamorphique. *Archives des Sciences, Genève*, **25**, 5–100.

GOFFÉ, B. & CHOPIN, C. 1986. High-pressure metamorphism in the western Alps: zoneography of metapelites, chronology and consequences. *Schweizerische Mineralogische und Petrographische Mitteilungen*, **66**, 41–52.

GOSSO, G., DAL PIAZ, G. V., PIOVANO, V. & POLINO, R. 1979. High pressure emplacement of early-Alpine nappes, postnappe deformations and structural levels. *Memorie di Science Geologiche*, **32**, 1–15.

GRIFFIN, W. L., AUSTRHEIM, H., BRASTAD, K., BRYHNI, I., KRILL, A. G., KROGH, E. J., MØRK, M.B.E., QUALE, H. & TØRUDBAKKEN, B. 1985 High-pressure metamorphism in the Scandinavian Caledonides. *In*: GEE, D. G. & STURT, B. A. (eds). *The Caledonide Orogen–Scandinavia and related areas*, 783–801. Wiley, London.

HEINRICH, C. A. 1982. Kyanite–eclogite to amphibolite facies evolution of hydrous mafic and pelitic rocks, Adula nappe, central Alps. *Contributions to Mineralogy and Petrology* **81**, 30–8.

—— 1983. *Die regionale Hochdruckmetamorphose der Aduladecke, Zentralalpen, Schweiz*. Unpublished Thesis, University of Zürich.

—— 1986. Eclogite facies regional metamorphism of hydrous mafic rocks in the Central Alpine Adula Nappe. *Journal of Petrology*, **27**, 123–54.

HELLER, F. 1980. Paleomagnetic evidence for late Alpine rotation of the Lepontine area. *Eclogae Geologicae Helvetiae*, **73**, 607–318.

HUNZIKER, J. C. 1970. Polymetamorphism in the Monte Rosa, western Alps. *Eclogae Geologicae Helvetiae*, **63**, 151–61.

—— 1974. Rb–Sr and K–Ar age determination and the Alpine tectonic history of the western Alps. *Memorie degli Istituti di Geologia e Mineralogia dell'Universitá di Padova*, **31**.

—— & MARTINOTTI, G. 1984. Geochronology and evolution of the western Alps. *Memorie della Societá geologica italiana*, **29**, 43–56.

HURFORD, A. J. 1986. Cooling and uplift patterns in the Lepontine Alps, south-central Switzerland and an age of vertical movement in the Insubric fault line. *Contributions to Mineralogy and Petrology*, **92**, 413–27.

—— & HUNZIKER, J. C. 1985. Alpine cooling history of the Monte Mucrone eclogites (Sesia–Lanzo zone): fission track evidence. *Schweizerische Mineralogische und Petrographische Mitteilungen*, **65**, 325–34.

JÄGER, E. 1970. Rb–Sr systems in different degrees of metamorphism. *Eclogae Geologicae Helvetiae*, **63**, 163–72.

——, NIGGLI, E. & WENK, E. 1967. Rb–Sr Altersbestimmungen an Glimmern der Zentralalpen. *Beiträge zur geologischen Karte der Schweiz*, **134**.

KÖPPEL, V. & GRUNENFELDER, M. 1975. Concordant U–Pb ages of monazite and xenotime from the central Alps and the timing of the temperature Alpine metamorphism, a preliminary report. *Schweizerische mineralogische und petrographische Mitteilungen*, **55**, 129–32.

LEMOINE, M., MARTHALER, M., CARON, M., SARTORI, M., AMAUDRIC DU CHAFFAUT, S., DUMONT, T., ESCHER, A., MASSON, H., POLINO, R. & TRICART, P. 1984. Découverte de foraminifères planctoniques du Crétacé supérieur dans les Schistes lustrés du Queyras (Alpes Occidentales). Conséquences paléogéographiques. *Comptes-Rendus de l'Académie des Sciences de Paris* (II), **299**, 727–32.

LÖW, S. 1985. *Die tectono-metamorphe Entwicklung der nördlichen Adula-Decke, Zentralalpen, Schweiz*. Unpublished thesis, University of Basel.

—— 1986. Ein tektono-metamorphes Entwicklungsmodell der nördlichen Adula–Decke. *Schweizerische mineralogische und petrographische Mitteilungen*, **66**, 129–34.

MANCKTELOW, N. 1985. The Simplon Line: a major displacement zone in the western Lepontine Alps. *Eclogae Geologicae Helvetiae*, **78**, 73–96.

MANGE-RAJETZKY, M. & OBERHÄNSLI, R. 1982 Detrital lawsonite and blue sodic amphibole in the Molasse of Savoy, France and their significance in assessing Alpine evolution. *Schweizerische mineralogische und petrographische Mitteilungen*, **62**, 415–36.

MARTHALER, M. 1984. Géologie des unités penniques entre le val d'Anniviers et le val de Tourtemagne (Valais, Suisse). *Eclogae Geologicae Helvetiae*, **77**, 395–448.

MARTINOTTI, G. & HUNZIKER, J. C. 1984. The Austro–Alpine system in the western Alps: a review. *Memorie della Societá Geologica Italiana*, **29**, 233–50.

MATTAUER, M., MALAVIEILLE, J. & MONIÉ, P. 1987. Une coupe lithosphérique des Alpes occidentales dans l'hypothèse où Sezia (*sic*) n'est pas d'origine sud-alpine. *Comptes-Rendus de l'Académie des Sciences de Paris* (II), **304**, 43–48.

MEYER, J. 1983. *Mineralogie und Petrologie des Allalingabbros*. Unpublished thesis, University of Basel.

MILNES, A. G., GRELLER, M. & MÜLLER, R. 1981. Sequence and style of major post-nappe structures, Simplon–Pennine Alps. *Journal of Structural Geology*, **3**, 411–20.

MONIÉ, P. 1985. La méthode $^{39}Ar-^{40}Ar$ appliquée au métamorphisme alpin dans le Massif du Mont Rose (Alpes occidentales) Chronologie détaillée depuis 110 Ma. *Eclogae Geologicae Helvetiae*, **78**, 487–516.

MURALT, R. 1986. *Mineralogisch-geologische Untersuchungen in der Adula–Decke am Nordhang des San Bernardino Passes (Graubünden, Schweiz)*. Unpublished diploma thesis, University of Bern.

NIGGLI, E. 1970. Alpine Metamorphose und alpine Gebirgsbildung. *Fortschritte der Mineralogie*, **47**, 16–26.

OBERHÄNSLI, R. 1978. Chemische Untersuchungen an Glaukophan-führenden basischen Gesteinen aus den Bündnerschiefern Graubündens. *Schweizerische mineralogische und petrographische Mitteilungen*, **59**, 139–56.

—— 1982. The P–T history of some pillow-lavas from Zermatt. *Ofioliti*, **7**, 431–6.

—— 1986. Blue amphiboles in metamorphosed Mesozoic mafic rocks from the central Alps. *Geological Society of America Memoir*, **164**, 239–247.

——, HUNZIKER, J. C., MARTINOTTI, G. & STERN, W. B. 1985. Geochemistry, geochronology and petrology of Monte Mucrone: an example of Eoalpine eclogitization of Permian granitoids in the Sesia–Lanzo zone, western Alps, Italy. *Chemical Geology*, **52**, 165–84.

PIERI, M. & GROPPI, G. 1981. Subsurface geological structure of the Po Plain, Italy. *Programma finalizzato di Geodinamica*, Centro Nazionale della Ricerca, Publication No. 144.

PROBST, PH. 1980. Die Bündnerschiefer des nördlichen Penninikum zwischen Valser Tal und Passo di San Giacomo. *Beiträge zur geologischen Karte der Schweiz*, neue Folge **153**,

RADELLI, L. 1986a. An essay on the Southern Basin and Range. *Bolletino del Departamento de Geologia Uni-Son* (Mexico), **3**, 1, 51–146.

—— 1986b. On geologic inquiry. *Bolletino del Departmento de Geologia Uni-Son* (Mexico), **3**, 2, 31–54.

—— 1988. Arcs and magmatism. *Bolletino del Departamento de Geologia Uni-Son* (Mexico), **4**, 1.

—— & DESMONS, J. 1987. Pennique, Téthys et orogenèse crétacé moyen dans les Alpes. *Comptes Rendus de l'Académie des Sciences de Paris* (II), **305**, 1375–78.

SALIOT, P., DAL PIAZ, G. V. & FREY, M. 1980. Métamorphisme de haute pression dans les Alpes franco–italo–suisses. 26th International Geological Congress *Géologie Alpine*, **56**, 203–35.

SAWATZKI, G. 1975. *Etude géologique et minéralogique des flyschs à grauwackes volcaniques du synclinal de Thônes (Haute-Savoie, France)*. Unpublished thesis, University of Genève.

SCHEURING, B., AHRENDT, H., HUNZIKER, J. C. & ZINGG, A. 1974. Palaeobotanical and geochronological evidence for the Alpine age of the metamorphism in the Sesia zone. *Geologische Rundschau*, **63**, 305–326.

SCHMID, S. M., ZINGG, A. & HANDY, M. 1987. The kinematics of movements along the Insubric Line and the emplacement of the Ivrea zone. *Tectonophysics*, **135**, 47–66.

STALDER, P. J. 1979. Organic and inorganic metamorphism in the Taveyannaz sandstone of the Swiss Alps and equivalent sandstones in France and Italy. *Journal of Sedimentary Petrology*, **49**, 463–82.

STECK A. 1984. Structures de déformations tertiaires dans les Alpes centrales. *Eclogae Geologicae Helvetiae*, **77**, 55–100.

—— 1987. Le Massif du Simplon–Réflexions sur la cinématique des nappes de gneiss. *Schweizerische Mineralogische und Petrographische Mitteilungen* **67**, 27–45.

STEINITZ, G. & JÄGER, E. 1981. Rb–Sr and K–Ar studies on rocks from the Suretta nappe; eastern Switzerland. *Schweizerische mineralogische und petrographische Mitteilungen*, **61**, 121–31.

STILLE, H. 1924. *Grundfragen der vergleichenden Tektonik*. Berlin, Borntraeger.

TRÜMPY, R. 1960. Palaeotectonic evolution of the central and western Alps. *Bulletin of the Geological Society of America*, **71**, 843–908.

—— 1973. The timing of orogenic events in the central Alps. *In*: DEJONG, K. A. (ed). *Gravity and Tectonics*, Wiley, London.

—— 1975. Penninic–Austro–Alpine boundary in the Swiss Alps: a presumed former continental margin and its problems. *American Journal of Science*, **275-A**, 209–38.

VIALETTE, Y. & VIALON, P. 1964. Etude géochronologique de quelques micas des formations du massif Dora–Maira, Alpes cottiennes piémontaises. *Annales de la Faculté des Sciences de l'Université de Clermont-Ferrand*, **25**, 91–9.

WENK, E. 1956. Die lepontinische Gneissregion und die junge Granite der Valle della Mera. *Eclogae geologicae Helvetiae*, **49**, 139–52.

—— 1962. Plagioklas als Index mineral in den Zentralalpen. *Schweizerische mineralogische und petrographische Mitteilungen*, **42**, 139–52.

WESTPHAL, M., BAZHENOV, M. L., LAUER, J. P., PECHERSKY, D. M. & SIBUE, T. J. C. 1986. Paleomagnetic implications on the evolution of the Tethys belt from the Atlantic ocean to the Pamirs since the Triassic. *Tectonophysics*, **123**, 37–82.

WETZEL, R. 1972. Zur Petrographie und Mineralogie der Furgg zone (Monte Rosa–Decke). *Schweizerische Mineralogische und Petrographische Mitteilungen*, **52**, 161–236.

WILDI, W. 1985. Heavy mineral distribution and dispersal pattern in Penninic and Ligurian Flysch basin (Alps, northern Apennines). *Giornale di Geologia* (30), **47**, 77–99.

—— 1987. Les régions sources du matériel terrigène dans les flyschs alpins. *Géologie Alpine*, Mémoire hors série **13**, 379–88.

WILLIAMS, P. F. & COMPAGNONI, R. 1983. Deformation and metamorphism in the Bard area of the Sesia–Lanzo zone, western Alps, during subduction and uplift. *Journal of Metamorphic Geology*, **1**, 119–40.

WINKLER, W. & BERNOULLI, D. 1986. Detrital high-pressure/low temperature minerals in a late Turonian flysch sequence of the eastern Alps (western Austria):implications for early Alpine tectonics. *Geology*, **14**, 598–601.

WUST, G. H. & BAEHNI, L. A. 1986. The distinctive tectonometamorphic evolution of two basement complexes belonging to the Grand–Saint–Bernard nappe (Val de Bagnes, Valais). *Schweizerische Mineralogische und Petrographische Mitteilungen*, **66**, 53–71.

J. C. HUNZIKER & J. DESMONS, Institut de Minéralogie UNIL BFSH-2, CH 1015 Lausanne CRPG, BP 20, F-54 501 Vandoelivre-les-Nancy cedex, France.

G. MARTINOTTI, Dipartimento di Scienze delta Terra, University of Torino, via Accademia delle Scienze 5, 1-10123 Torino, Italy.

Unravelling the thermo-tectonic evolution of the Alps: a contribution from fission track analysis and mica dating

A. J. Hurford, M. Flisch & E. Jäger

SUMMARY: Fission track analysis of apatite and zircon, and K–Ar and Rb–Sr dating of white mica and biotite provide time and temperature constraints for direct dating of deformation phases and in establishing cooling and uplift rates and patterns. In the western Alps, timing of the Sesia–Lanzo zone Eoalpine metamorphic temperature peak is confirmed between 110 and 85 Ma, possibly preceded by the pressure maximum. The slowly cooling Sesia zone reached 200–250°C by ~33 Ma and ~100°C by ~25 Ma, with fast uplift during the late Oligocene to Miocene Insubric phase; cooling in the adjacent Gran Paradiso Massif occurred slightly later, with fission track zircon and apatite ages of 30±2 and ~22 Ma respectively. No evidence to support an inverted metamorphism of the Gran Paradiso was found. In the Silvretta nappe (E Alps) three different uplift styles and periods are distinguished: uparching between 110 and 35 Ma, homogeneous uplift during 35 and 2 Ma and recent eastward tilting. The data imply that updoming of the Engadine Window occurred before ~31 Ma. In the central Lepontine Alps, a uniform late Neogene to recent mean uplift rate of ~0.5 mm/a succeeded early Miocene uplift at ~2.2 mm/a in the southern steep belt (root zone), the latter resulting from a major phase of backthrusting of the central Alps along the Insubric Line. Clear-cut time-temperature paths are not found for the Suretta Nappe (central Alps): fission track data delimit the Miocene cooling, but speculative interpretation of phengite-mica and K–Ar amphibole ages suggests that the main deformation (possibly Ferrera phase) occurred between 55 and 36 Ma. Schematic models are advanced to describe the asynchronous thermal evolution of different parts of the central Lepontine Alps.

Background and methodology

In attempting to unravel the complex history of the Alpine chain, with the interaction of the African and Eurasian tectonic plates and the intervening Tethyan ocean, crustal shortening, polyphase deformation, polyphase metamorphism and uplift, it is essential, in addition to discriminating areas by tectonic style and metamorphic grade, to establish a time framework within which the orogenic events may be placed. Radiometric age determination techniques provide unique data fundamental to establishing such a chronology. Occasionally, a metamorphic event or deformation phase can be dated directly. More commonly the isotopic systems in minerals and rocks record specific times in the thermal history of a rock and, when deciphered, these data can be utilised to define cooling and uplift rates and styles. In particular, the ability of the fission track dating technique to provide time data for the low-temperature history of a rock has promoted widespread use of the method in many active mountain belts globally, including the Himalayas (Zeitler *et al.* 1982), the Rockies (Naeser 1979), the Andes (Kohn *et al.* 1984), the Transantarctic mountains (Fitzgerald *et al.* 1986, 1987), Tibet (Lewis, unpublished data) and the New Zealand Alps (White & Green 1986). It is the purpose of this contribution to review the use which may be made of fission track analysis together with K–Ar and Rb–Sr dating of micas in the Alps, to provide time and temperature data for the reconstruction of cooling patterns and the differentiation of uplift movement; examples of recently published work from the western, central and eastern Alps are presented as illustration.

The concepts and methodologies of the K–Ar and Rb–Sr methods have been comprehensively described (for example see Purdy & Jäger (1976) and Flisch (1986a) for K–Ar details; see Jäger (1979) for description of Rb–Sr determinations). Normally Rb–Sr mica ages are corrected using the $^{87}Sr/^{86}Sr$ ratio measured for the whole-rock, and are thus effectively 2-point mica–whole-rock isochrons; this correction assumes that a sample was heated sufficiently during a metamorphic event to permit strontium homogenisation between minerals within the total rock system. Strictly, in a slowly-cooled rock, correction of measured Rb–Sr biotite ages should use the $^{87}Sr/^{86}Sr$ ratio of the whole-rock minus the white mica phase, since the white mica is closed to strontium before the biotite closure temperature is reached. Since 1977, most K–Ar and Rb–Sr ages have been calculated with IUGS recom-

From: COWARD, M. P., DIETRICH, D. & PARK, R. G. (eds), 1989, *Alpine Tectonics*, Geological Society Special Publication No. 45, pp. 369–398.

mended constants (Steiger & Jäger 1977); ages published earlier may require recalculation to permit direct comparison.

A brief outline of the more novel fission track dating method is presented here, more detailed description of the systematics of the method having been provided by several authors (e.g. Fleischer *et al.* 1975, Hurford 1986a). Spontaneous nuclear fission of the more abundant isotope of uranium, ^{238}U results in the passage of large, highly energetic fission fragments through the lattice of the host material leaving in their wake a damaged zone of disrupted ions. These fission tracks are stored in the host, provided that the host material remains below certain critical temperature thresholds, an important factor reconsidered below. Upon analysis, the tracks are revealed by chemically etching an internal face polished through the sample, the weaker fission-damaged zone being preferentially attacked (Fig. 1). The etched tracks, typically 10 to 20 μm in length, are observed and counted using high power optical microscopy. The number of spontaneous tracks depends upon the time during which tracks have been accumulating, the rate at which fission occurs, the amount of uranium in the sample and the length of the tracks, the latter governing the probability of intersection with the viewed surface. Uranium content is conveniently assessed by irradiating the sample with thermal (low energy) neutrons which induces fission in a fraction of ^{235}U atoms, producing a second induced track density which may be similarly revealed and counted. The sequences of procedural techniques most commonly used are schematically represented in Fig. 2. The most useful minerals for fission track analysis are apatite, zircon and sphene, having uranium contents between ~10 and 1000 ppm. Other minerals such as epidote, garnet and muscovite are generally uranium-poor and represent a quantitatively insignificant proportion of all fission track analyses.

System calibration in fission track dating has been a contentious and confusing subject (see Hurford & Green 1982, 1983). The lack of a uniquely quantified ^{238}U spontaneous fission decay constant and a multiplicity of techniques and conventions for assessing neutron fluence, demands evaluation by each worker of their chosen parameters by analysis of materials regarded as age standards. An alternative approach circumvents absolute evaluation of decay constant and neutron fluence, relying instead upon a direct comparison with age standard analyses using a proportionality factor *zeta* (Fleischer & Hart 1972, Hurford & Green 1983,

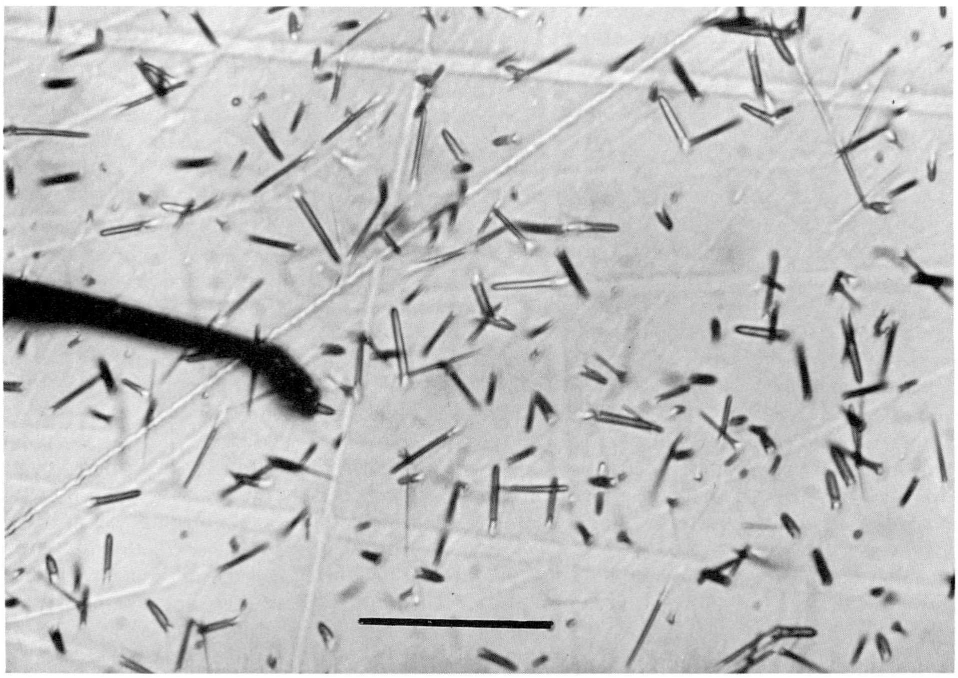

FIG. 1. Photomicrograph of etched ^{238}U spontaneous fission tracks in apatite; scale bar is ~20 μm.

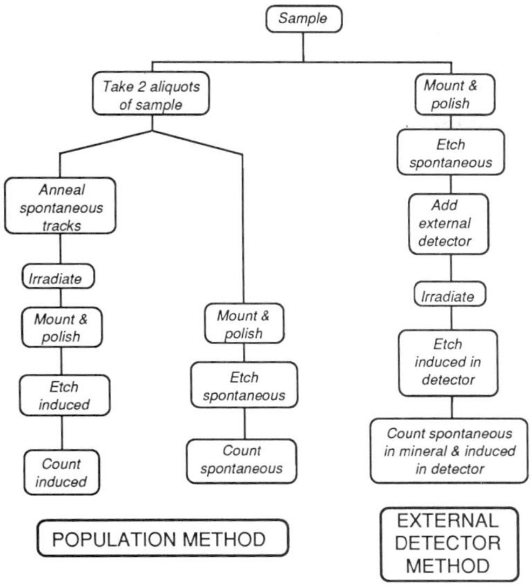

FIG. 2. Schematic flow-chart of the most used fission track analytical procedures.

Green 1985). All published or reported fission track results should be critically examined for details of calibration and supporting analyses of standards.

Tools for interpretation

The concept of blocking or closure temperature represents the keystone for discussion of radiometric data and thermal history. Utilized by Armstrong (1966) and Harper (1967) as a means of interpreting mineral-age patterns in regionally metamorphosed terrains, the most widespread use of the concept has been in the Alps (Jäger *et al.* 1967, Purdy & Jäger 1976, and others). Dodson (1973) offered the simple definition of closure temperature as being 'the temperature of a system at the time represented by its apparent age'. In essence the closure temperature of a mineral-isotopic system represents a critical threshold above which the radiogenic daughter product is lost and below which the system is 'blocked' against thermal disturbance and radiogenic products are accumulated. Clearly closure cannot be instantaneous and there exists a transitional temperature range in which the daughter product is partially lost and partially retained. The rate of cooling will also have some effect upon the closure temperature, although for the apatite fission track system annealing studies show that an order of magnitude increase in time reduces the temperature necessary to achieve a certain annealing effect by only about 10°C (Wagner & Reimer 1972, Green *et al.* 1988a).

Estimations of closure temperature have been made from laboratory experiment and from geological evidence. For fission tracks in apatite, extrapolation of the results of early laboratory annealing studies to geological time provided an estimate of ~100°C for the closure temperature, below which tracks are effectively retained (Wagner 1968, Naeser & Faul 1969). In a refinement of this approach, Gleadow *et al.* (1983) and Green *et al.* (1986) have demonstrated from very detailed laboratory annealing experiments and drill-hole sample studies that apatites can provide information about palaeotemperatures between 20 and 130°C. Note that in contrast to the theoretical modelling of closure temperatures (Dodson & McClelland-Brown 1985 and references therein), Laslett *et al.* (1987) and Green *et al.* (1988b) demonstrate the lack of evidence for describing track annealing in apatite by first-order kinetics, preferring instead a model which allows for differences in activation energy for annealing, dependent both upon chemical composition of the apatite and orientation of the track, annealing being noticeably anisotropic. Extrapolation to geological time of laboratory annealing experiments on zircon suggests effective closure temperatures >300°C (Fleischer *et al.* 1965, Krishnaswami *et al.* 1974) in stark contrast to geological evidence which favours a lower temperature between 175 and 250°C (Gleadow & Brooks 1979, Harrison *et al.* 1979, Zeitler *et al.* 1982, Hurford 1986b). Although no direct evidence is available to explain this difference, it is conceivable that the presence of fluid phases, pressure and the chemistry of the zircon itself may be contributory factors. From the geological arguments, a reasonable estimate of closure temperature for tracks in zircon is 200–250°C.

Effective closure temperatures for the retention of radiogenic argon and strontium in micas have been the subject of some debate. Jäger *et al.* (1967) and Purdy & Jäger (1976) derived a value of 300±50°C from the coincidence of reset Hercynian K–Ar and Rb–Sr biotite ages with the stilpnomelane metamorphic isograd in the Aar Massif, to the north of the central Lepontine Alps. The similarity of K–Ar and Rb–Sr biotite ages in slowly cooled terrains (Siegenthaler 1984, Dempster 1985) supports the idea of a common closure temperature. By a similar comparison of the Alpine resetting of pre-Alpine muscovite Rb–Sr ages with the staurolite-chloritoid transition, Purdy & Jäger (1976) proposed a closure temperature of

500±50°C for strontium in muscovite, whilst 350°C for argon in muscovite represents their extrapolation between measured biotite and Rb–Sr muscovite ages. The approach assumes *a priori* that the opening and closing temperatures of mineral-isotope systems are identical, a factor which is far from self-evident. Some support for the Alpine biotite closure temperature of 300±50°C may be gained from a value of >300°C from contact metamorphic studies (Hanson & Gast 1967), from 300–345°C given by a tectonothermal argument of Dallmeyer (1978), and from 280±40°C, resulting from argon diffusion studies by Harrison & McDougall (1980a). Alternative values include ~225°C from the extreme extrapolation of limited borehole data (Turner & Forbes 1976), and 350–400°C from argon diffusion experiments (Berger & York 1981). In addition, 400°C (or even higher) has been proposed by Verschure *et al.* (1980) as a consequence of a 870 Ma Sveconorwegian age retained by a brown biotite despite the growth of younger green Caledonian biotite. Arnold & Jäger (1965) have reported pre-Alpine Rb–Sr and K–Ar biotite ages in a 1 m ultrabasic inclusion within a quartz diorite lens in the greenschist facies Alpine chloritoid zone, in the tunnel Nalps–Curnera (Gotthard Massif), both being enclosed in a biotite-sillimanite gneiss. The diorite gave an Alpine Rb–Sr biotite age of 22.7±2.0 Ma and a pre-Alpine K–Ar biotite age of 363±15 Ma; the gneiss gave a similar pre-Alpine K–Ar biotite age, but a Rb–Sr age of 54±8 Ma. Arnold & Jäger explain the argon ages in terms of excess argon, but the ultrabasic pre-Alpine Rb–Sr age is accounted for by an absence of fluid phases, which inhibited strontium exchange. Clearly the presence or absence of fluid phases may have a critical bearing on closure temperature.

In certain very specific circumstances, petrofabric analysis may reveal the growth of a particular mineral to be synchronous with deformation; under the supposition that the mineral has grown below its closure temperature, the measured age should yield the time of the tectonic phase. Steiger (1964) and Deutsch & Steiger (1985) have described Alpine biotites and hornblendes which may so define the age of deformation; see also the discussions below on phengite formation in the Sesia zone of the western Alps and the ages of phengite and hornblende in the Suretta Nappe of the central Alps.

Since mineral-isotopic systems close at different temperatures, analysis of co-existing systems yields a series of time-temperature coordinates which permit construction of first approximation cooling paths for a sample. Figure 3 illustrates such a cooling curve constructed from mica and fission track age data obtained for a single 30 kg orthogneiss sample from the central Penninic Alps. The sequence of measured ages: Rb–Sr muscovite > K–Ar muscovite > Rb–Sr biotite ≈ K–Ar biotite > FT zircon > FT apatite is consistent with the assumed effective closure temperature for the different mineral and isotope systems and has been reported in studies from many different areas (Purdy & Jäger 1976, Wagner *et al.* 1977, Harrison & McDougall 1980a, Dempster 1985, Hurford 1986b). Regional comparison of the styles of cooling curves and rates of cooling provides a useful key to differentiating the tectonic history of structurally distinct units (see Wagner *et al.* 1977).

The blocking temperature concept and the resulting cooling ages have not been without criticism: see, e.g. Chopin & Maluski (1980). It should however be noted that Jäger and her coworkers erected the model *specifically* for the Lepontine Penninic Alps and for those samples which, lying within well-defined metamorphic isograds, demonstrated marked regional trends of ages interpreted as cooling stages subsequent to a mid-Tertiary regional metamorphism. No satisfactory overall alternative explanation of these Lepontine ages has been proffered. Note however that under other circumstances complete rejuvenation may not occur although the closure temperature of a mineral-isotope system is reached – see the discussion below on the Austro–Alpine Silvretta nappe.

An alternative approach uses apatite fission track data to yield uplift rates without the need to assume a value for the geotherm. For a given intrusive or basement body, apatite age invariably increases with altitude due to the earlier passage of the upper sample below the closure temperature for the retention of tracks. Thus for a limited lateral distance, the difference in elevation divided by the difference in age provides a direct measure of uplift. Such an approach was adopted by Wagner *et al.* (1979) in determining the uplift rate of the Bergell intrusives (Fig. 4), a study which revealed a significant decrease of the uplift rate with time. The studies of Wagner & Reimer (1972) and Wagner *et al.* (1977 and 1979) signified the highly important first application of fission track analysis to tectonic studies in the Alps. [Note that although apatite ages obtained by Hurford (1986b) agree within analytical error with those reported by Wagner *et al.* (1977), differences in system calibration do make absolute comparison of values between different workers and labora-

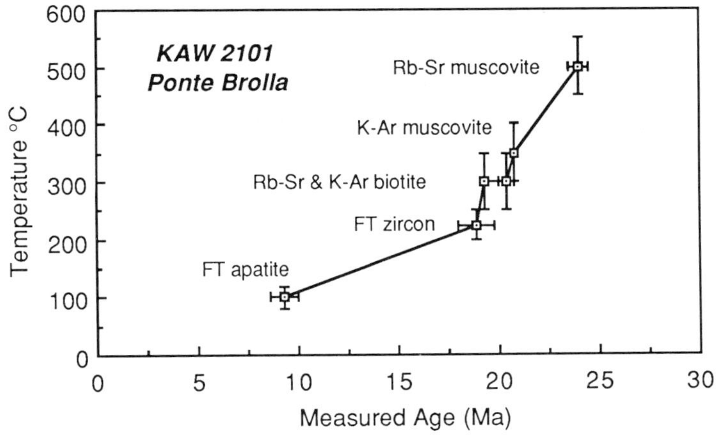

FIG. 3. Cooling curve constructed for a single sample 30 kg orthogneiss sample from Ponte Brolla in the Maggia Valley, central Alps from the analysis of different minerals using different isotopic systems.

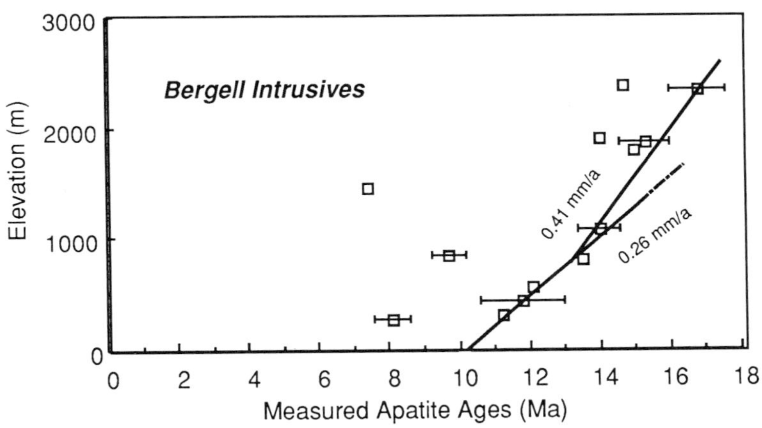

FIG. 4. An example from the Bergell Intrusion of calculation of uplift rate from the increase of fission track apatite age with altitude; the break in slope indicates a decreasing uplift rate with time (redrawn from Wagner *et al.* 1979).

tories difficult. All radiometric ages *must* be considered within the uncertainties set by the analyst.]

As Parrish (1983, 1985) has noted, the approximations of uplift rate provided by these procedures are dependent upon critical isotherms remaining horizontal and uninfluenced by surface topography and variable thermal conductivity. The isotherms should also reside at constant depth relative to the surface; this requires both that the isotherms should not relax and sink, for example following tectonism, and that the rate of uplift should equal the rate of erosion. Comparison of model-derived uplift rates and measured 'apparent' uplift rates for the Coast Mountains in British Columbia show agreement within error of model and measured values where uplift is 0.3 mm/a or less, but that where rates are greater than 0.5 mm/a, uplift has a significant effect upon the movement of geotherms and geothermal gradient.

An additional tool for the interpretation of fission track data is the distribution of track length in apatite, which is diagnostic of the style of cooling and will reveal more complex thermal histories. The sensitivity of the apatite fission track system to elevated temperatures is reflected in reductions both in the length of spontaneous tracks and in their density (and hence age). An awareness that the knowledge of track length was critically important to the interpretation of age was noted at an early stage (see, e.g. Wagner & Storzer 1970). After initial controversial attempts to 'correct' fission track apatite ages, the concept has developed in two directions: measurement of confined and projected track lengths – see Fig. 5. Horizontal confined tracks (Bhandari *et al.* 1971) lie parallel to the viewed surface and although totally enclosed within the crystal, intercept other outcropping tracks or fractures which facilitate the passage of etchant to the confined track. Measurement of confined tracks, although subject to certain bias (Laslett *et al.* 1984), yields the most precise estimate of true etchable track length (see Gleadow *et al.* 1986 and references therein). It is essential to grasp that each track has been formed at a different stage in the history of the sample and thus has experienced a different proportion of that total history. The distribution of confined lengths therefore records the integrated thermal history of that sample. Clear differences have been distinguished (Gleadow *et al.* 1986) between apatites from undisturbed, fast-cooled volcanic rocks, where mean track lengths are 14–15.6 μm with standard deviations of 0.8–1.2 μm, and from slowly cooled and undisturbed basement, where the mean track lengths are typically 12.5–13.5 μm with standard deviations of 1.3–1.7 μm (see Fig. 6). Partial annealing will cause shortening of some or all of these tracks, depending upon when annealing takes place and its intensity, resulting in a broadening of the distribution. Moore *et al.* (1986), working in southeastern Australia, and Green (1986) in a study in northern England have each demonstrated that the transition from unaffected slow-cooling ages, *via* mixed, partially reset ages, to total annealing is mirrored by significant changes in the track length distribution. Those tracks formed before and during the annealing period are progressively shortened until, in those samples where the age is totally reset, they are eliminated; the population of longer tracks represents those formed since the elevated temperatures waned. In such case studies, the slowly-cooled, mixed and reset natures of the measured apatite ages can only be discriminated by confined track length measurement. A measured apatite age alone can, in most cases, be interpreted in several ways; only with the help of confined track length measurements can an age be interpreted both uniquely and with confidence.

With these tools of data interpretation at our disposal, we consider five areas of the Alps for which new fission track and mica data have been acquired: the Sesia–Lanzo zone of the western Alps, the Gran Paradiso Massif, the Austro–Alpine Silvretta nappe of the eastern Alps, the Lepontine region and the Suretta nappe of the central Alps (Fig. 7). We conclude by examining the contribution which these new data make to our understanding of the timing, 'peak' temperature age, and duration of the central Alpine mid-Tertiary metamorphism.

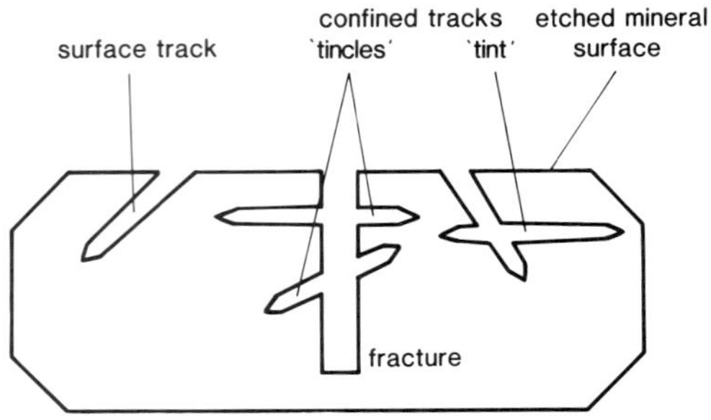

FIG. 5. Stylised section through an apatite crystal to illustrate the difference between confined and projected fission track lengths (after Gleadow *et al.* 1983).

The Sesia–Lanzo zone of the western Alps

The Sesia–Lanzo zone, the most internal unit of the western Alps, is an elongated body some 150 by 25 km, stretching from Locarno in the northeast to near Turin in the southwest (see Figs 7 and 8). Together with the Dent Blanche nappe system, the Sesia zone represents the uppermost tectonic unit in the western Alps, comparable with the Austro–Alpine nappes of the eastern Alps. The zone is bounded to the southeast by the Insubric fault line which separates the western Alps with Alpine metamorphism and deformation from the southern Alpine Ivrea zone, itself consisting of pre-Alpine high grade paragneisses, basic and ultrabasic rocks, virtually unaffected by metamorphism during Alpine times (Zingg 1983). To the northwest, the Sesia zone nappes lie above the Schistes Lustrés and metaophiolites of the underlying Pennine Piémont zone, a former oceanic trough lying between the European and African continents. Three major rock complexes in the Sesia zone have been distinguished on grounds of lithological difference and metamorphic evolution: the Sesia Eclogitic Mica Schist, the Gneiss Minuti Complex and the Second Dioritic Kinzigitic complex (Dal Piaz *et al.* 1972, Compagnoni 1977). These complexes can be subdivided into two elements – the Upper and

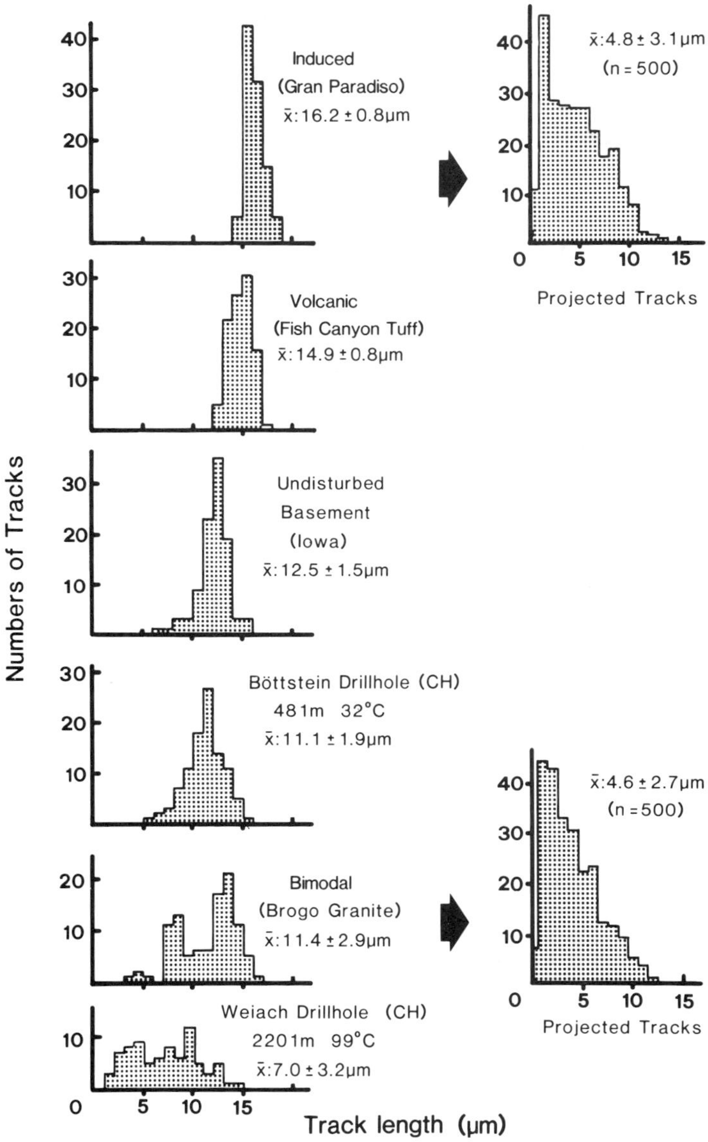

FIG. 6. Examples of typical patterns of confined fission track length in apatite reflecting different thermal histories (as measured in Berne by AJH, 100 tracks measured for each determination). *Projected* track length distributions for two samples are similar, masking the very different thermal histories revealed by the *confined* track lengths. Böttstein and Weiach cores made available courtesy of NAGRA.

Lower Sesia elements – according to their degree of Alpine overprint (Carraro *et al.* 1970, Dal Piaz *et al.* 1971, Compagnoni *et al.* 1977); the two elements may also be separated by a major thrust plane.

The Upper Sesia element consists of a number of Klippen from the Second Dioritic Kinzigitic zone, with pre-Alpine kinzigites, amphibolites, marbles and ultramafics having affinities to Ivrea lithologies perfectly preserved despite the Alpine orogenesis (Compagnoni *et al.* 1977). Alpine metamorphism of two phases can be observed: relics of polyphase high pressure metamorphism are of very localized occurrence. More developed, but again localised, is a later polyphase greenschist-facies episode, accompanied by multiple deformation producing in places complete textural reworking. In contrast,

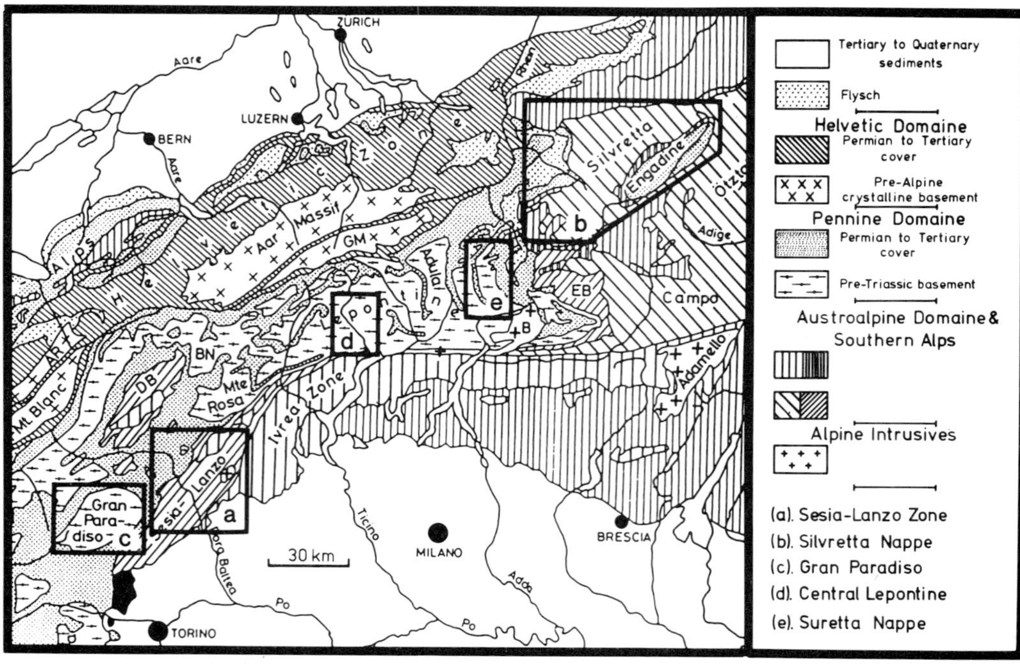

FIG. 7. Extract from a geological sketch map of the Alps (Frey *et al.* 1974) showing location of the 5 studied areas, a to e. DB = Dent Blanche Nappe; BN = Bernhard Nappe; GM = Gotthard Massif; B = Bergell Intrusion; EB = Err-Bernina Nappe. See Frey *et al.* for more detailed explanation of the geologic and tectonic units.

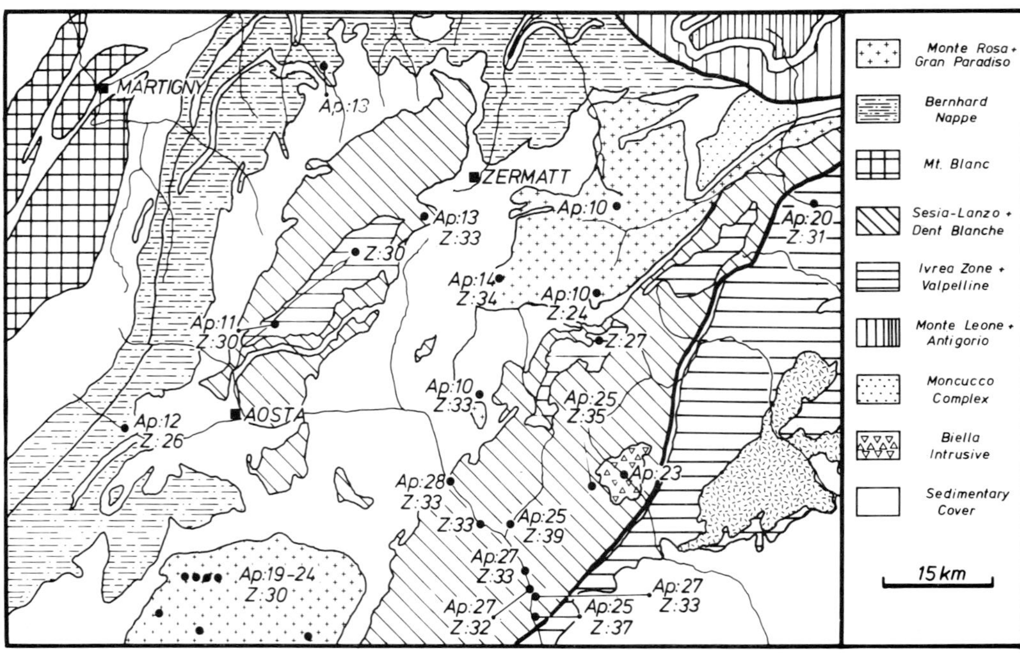

FIG. 8. Sample location map for western Alps, Sesia–Lanzo zone, showing measured fission track ages: Z = zircon, Ap = apatite. The Simplon and Insubric fault lines are shown as heavy lines trending NW–SE and NE–SW respectively.

the Lower Sesia element, comprising the Eclogitic Mica schists and Gneiss Minuti complexes, is characterized by the presence of abundant pre-Alpine granitoids and by more pervasive Alpine deformation and metamorphism. Compagnoni *et al.* (1977) view the Eclogitic Mica schists and Gneiss Minuti complexes as having a common pre-Alpine history, becoming distinguishable only as a consequence of Alpine events. However, Passchier *et al.* (1981) report that these two complexes are separated in the Aosta Valley by a thrust which, together with the accompanying jump in metamorphic grade, argues for a further subdivision of the Sesia zone into middle and lower elements.

The Eclogitic Mica schists are related to an early Alpine high pressure metamorphism which predated the overthrusting of the Austro–Alpine nappes onto the Piémont zone. The Gneiss Minuti complex of metagranitoids, augengneisses and paraschists is typified by greenchist facies assemblages. Although less intensively studied, the complex is divisible into a northeastern sector with early Alpine non-eclogitic assemblages, a NW margin of the zone with mid-Tertiary greenschist facies conditions and an overprinting sector where high pressure low-temperature minerals are overprinted by greenschist facies (Dal Piaz *et al.* 1972).

The Sesia zone is generally thought to represent the Alpine subduction of pre-Alpine basement rocks from the northwestern margin of the southern plate, the subduction providing sufficient depth for high pressure metamorphism: a minimum lithostatic load of 50 to 60 km is necessary to produce the pressures of 15–17 kb and temperatures of 500–600°C estimated by Compagnoni (1977, see also discussion by Stöckhert *et al.* 1986). A difficulty which confronts such modelling is the need to uplift and unroof the crustal fragments whilst preserving the newly-formed high pressure assemblages. Various models advocating low thermal regimes and very rapid uplift, both vertical and oblique, have been reviewed by Compagnoni *et al.* (1977). Rubie (1984) proffered the concept of prolonged subduction of cool oceanic lithosphere below the already subducted Sesia unit which, together with underplating maintained a sufficiently low temperature to prevent metamorphic overprinting during uplift. Hunziker *et al.* (this volume) consider alternative tectonic models for the evolution of the Sesia zone.

The chronology of the Alpine metamorphic phases in the Sesia has been based largely upon K–Ar white mica analyses (Hunziker 1974, Scheuring *et al.* 1974) which yield early Alpine ages of between 60 and 90 Ma. Detailed analysis of white micas of different grain size fractions from the Lower Aosta Valley (Stöckhert *et al.* 1986) indicated that only ~80 Ma ages from coarse and slightly deformed phengites could be accepted as essentially undisturbed cooling ages. The observed scatter is attributable to deformation with consequent redistribution (not necessarily loss) of radiogenic argon. Systematically lower ages for fine-grained phengites are interpreted as mixtures of fragmented high-pressure white micas with newly crystallised phengites, grown at the expense of high pressure minerals. The age of the early Alpine, Cretaceous high pressure metamorphism is bracketed by a Rb–Sr whole-rock isochron age of 129±15 Ma for the Monte Mucrone eclogite and a Rb–Sr isochron age of 114±1 Ma for high pressure minerals from the same site (Oberhänsli *et al.* 1985). Hunziker (1974) reported K–Ar white mica ages of ~38 Ma for the external Sesia zone where greenschist facies conditions had completely reset the isotopic systems. The similarity of age with the central Pennine metamorphism – the Lepontine Phase – (Hunziker 1969, Jäger 1973) has supported a contemporary origin concept. Little has been known of the history of the Sesia–Lanzo zone since mid-Tertiary times until the recent application of fission track analysis to zircon and apatite from 14 Sesia samples and 4 Dent Blanche samples, and to samples from the adjacent Penninic and Ivrea zones (Hurford & Hunziker 1985, Hurford, Hunziker & Stöckhert work in progress). Figure 8 shows the location of analysed samples together with their measured fission track ages. Both zircon and apatite fission track ages from the lower Sesia element demonstrate reasonable consistency: 11 zircon ages range between 32.2±2.3 Ma and 39.7±2.7 Ma, with a weighted mean of 33.8±1.8 Ma (2σ). 10 individual apatite ages range between 23.8±1.1 and 28.6±4.8 Ma, with a weighted mean of 25.2±1.6 Ma (2σ).

Figure 9 shows a schematic and speculative cooling curve for the Sesia–Lanzo zone, which combines these fission track results with the other limited age data which are available. The postulated pathway starts at the Jurassic–Cretaceous boundary with the assemblage K-feldspar + plagioclase + biotite + quartz; subduction or crustal thickening leads to the high pressure paragenesis jadeite + garnet + clinozoisite + quartz for which stable isotope and partition coefficient studies yield temperatures in excess of ~500°C, timed between 130 and 100 Ma (Desmons & O'Neill 1978, Oberhänsli *et al.* 1985). These latter authors argue that a

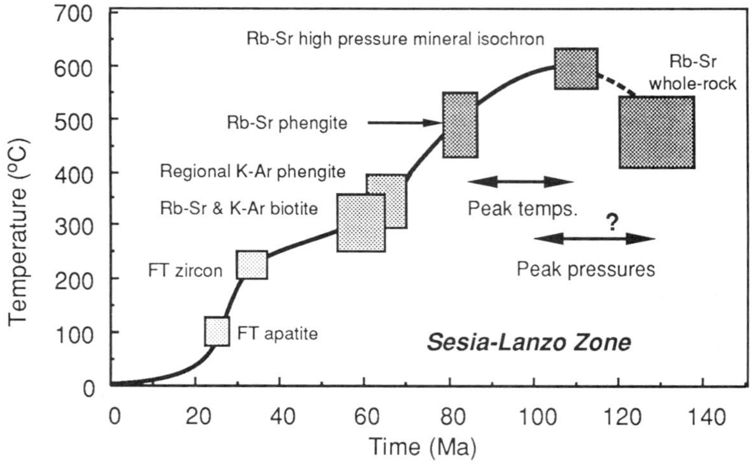

FIG. 9. Speculative time–temperature path for Sesia–Lanzo zone derived from measured radiometric ages.

temperature increase, coincident with a pressure decrease, yields the paragenesis chloromelanite + garnet + albite + glaucophane + phengite + quartz, indicating temperatures of ~600°C. The timing of the temperature peak can be estimated to be between 110 and 85 Ma: a clinozoisite-phengite Rb–Sr mineral isochron gives an age 85±1 Ma (Oberhänsli *et al.* 1985), whilst other Rb–Sr results on high pressure phengites yield broadly similar or slightly younger ages (Stöckhert, Hurford, Schaltegger & Jäger, work in progress). As discussed above, the regional scatter of Rb–Sr and K–Ar mica ages provides a broad pattern of slow cooling through the latest Cretaceous and earliest Palaeocene. Fission track zircon data indicate a cooling of the Sesia to 200–250°C by ~33 Ma whilst the mean apatite age of 25.2±1.6 Ma requires cooling through ~100°C by late Oligocene. This age–temperature pattern, somewhat imprecise in the older stages, suggests progressive cooling subsequent to the climax of the early Alpine high pressure metamorphism, possibly disrupted by reheating and/or tectonism. A relatively quiescent period during the early Palaeogene, with cooling at only ~3°C/Ma, offers some support for the 'Restoration Phase' of Trümpy (1960).

This cooling pattern differs markedly from those for the central Alps as given in Fig. 17. A further contrast with many apatite uplift studies (e.g. Wagner *et al.* 1977) is that within analytical uncertainty, the ages measured in the Sesia–Lanzo zone are indistinguishable, no correlation of apatite age with sample elevation being apparent despite altitudes ranging from 300 to ~2000 m. The most probable interpretation of such a lack of age-altitude dependency is to suppose a fast rate of uplift during which samples at different elevations were raised rapidly through the apatite closure temperature at times which cannot be resolved separately by fission track analysis. The magnitude of analytical uncertainty can be used as a measure of resolution of ages at different elevations and thus provide limits for plausible, detectable rates of uplift. Simple modelling shows that uplift at 1 mm/a maintained for a vertical distance of ~2 km in the Sesia zone requires an uplift time of 2 Ma, a time period not resolvable in this apatite data set where the mean individual 1σ analytical error is ±1.9 Ma. If such a rapid uplift hypothesis were true, 1 km/Ma would represent the minimum uplift rate operative during the passage of present-day outcrop samples through the ~100°C isotherm in Oligocene times. Fortunately a second parameter of apatite fission track analysis offers confirmation of rapid cooling of Sesia apatites. Figure 10 compares typical confined fission track length distributions measured for Sesia–Lanzo, central Penninic, western Alpine Penninic and Ivrea samples. Patterns typical of slow-cooling have been measured for all Penninic and Ivrea samples, with mean track lengths between 12.5 and 13.5 μm and standard deviations between 1.3 and 1.7 μm. Patterns for the Sesia samples are significantly different: mean lengths vary between 13.8 and 14 μm with tighter distributions, standard deviations being from 1.2 to 1.6 μm. Comparison of Figs 6 and 10 shows the Sesia track length patterns almost approximate to

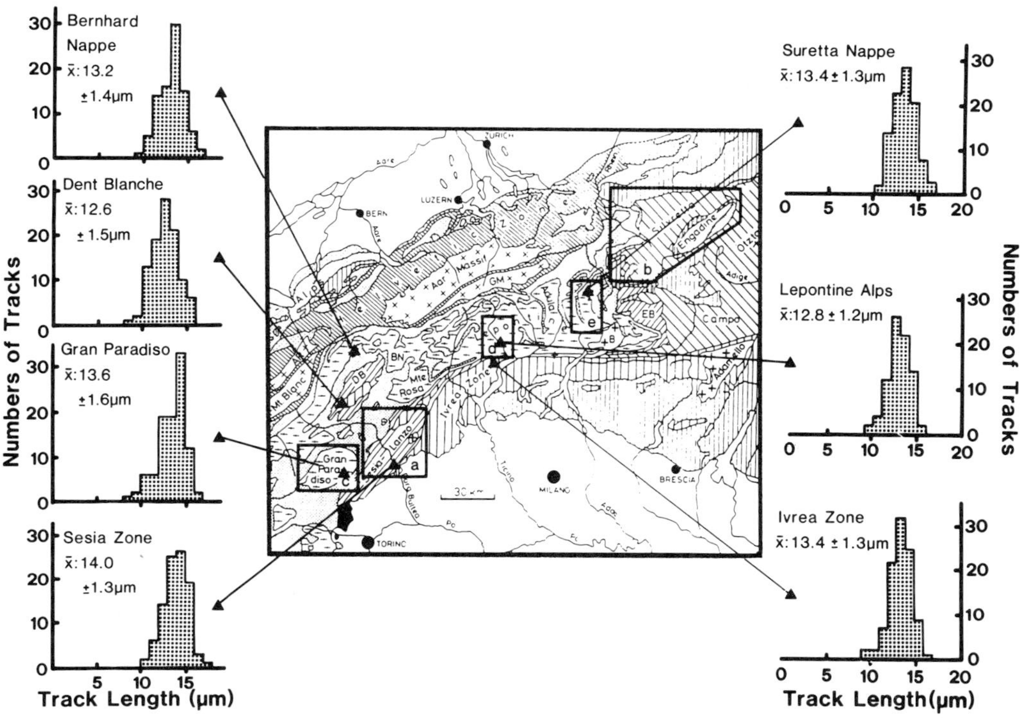

FIG. 10. Comparison of apatite confined track length patterns measured for different Alpine tectonic units; the longer mean length is typical of the Sesia samples and reflects a more rapid cooling. (100 lengths measured for each analysis). Base map, similar to Fig. 7, from Frey *et al*. (1974).

volcanic type distributions, indicating rapid cooling and substantiating the thesis of fast uplift of the Sesia–Lanzo zone.

Juxtaposition of the Sesia–Lanzo zone with polyphase Alpine metamorphism, against the Ivrea zone where Alpine metamorphic effects are absent, demonstrates that the Sesia zone has been uplifted, most probably in part along the intervening Insubric fault line. In broadest terms, the late uplift of the Sesia Zone is bracketed in time by the fission track apatite and zircon ages, i.e. between ~25 and ~33 Ma. Although the evidence of the track length data provides confidence for interpretation of the ~25 Ma apatite ages as the time of rapid *cooling*, this age represents only a lower limit for the timing of fast *uplift*. To define this time more precisely, the cautions regarding the position and movement of isotherms cited above must be duly considered: the calculated minimum uplift rate of 1 mm/a for the Sesia zone exceeds Parrish's threshold of 0.5 mm/a, implying a dragging upward or folding of the isotherms, and a consequent temporary increase in the geothermal gradient. The interval between this uplift with its accompanied erosion and the ensuing rapid cooling at ~25 Ma consequent upon the relaxation of the geotherms, cannot be estimated with certainty, but possibly may not exceed a few Ma. By assuming, from this apatite fission track evidence, that rapid uplift and cooling of the Sesia zone occurred in Chattian times, a broad contemporaneity may be seen with the major uplift to the north of the Insubric Line, demonstrated in the Lepontine Penninic Alps by Hurford (1986b – see below). Although the cause of the two events may be the same, viz. crustal shortening by indentation from the southern tectonic plate, the affect in the Sesia is minor when compared to the major backthrusting of the central Alps onto the southern Alps (Schmid *et al.* 1987). Since present-day outcropping Sesia rocks yield zircon fission track ages of ~33 Ma, an estimated burial of ~7–8 km at that time results from the first approximation of a normal geothermal gradient of 30°C/km. Apatite ages of ~25 Ma provide the next limitation of 3–4 km burial (with a gradient of 30°C/km), or significantly less with an uplifted and condensed sequence of isotherms. Accordingly, the difference in burial permits a first-order estimate of ~5 km uplift for the Sesia zone during the Oligocene–Miocene Insubric Phase (see Schmid *et al.* this volume).

Gran Paradiso Massif

Apatite and zircon fission track ages have been measured for a profile of 8 orthogneiss samples from the Hercynian basement of the Gran Paradiso Massif of the western Alps (Fig. 7). Preliminary results indicate a uniform zircon fission track age of ~30±2 Ma for samples from altitudes ranging between 750 and 4060 m above sea level. Apatite ages from the same samples show a very modest dependency upon elevation, with ages ranging from ~20 Ma at 750 m to 24.5 Ma at 4060 m (Hurford & Hunziker, in press). Whilst these apatite results lie broadly within the scatter of data from a previous study, the present zircon results are at total variance with the inverted age profile reported by Carpena (1985) and interpreted by her as inverted metamorphism resulting from the emplacement of the Dent Blanche nappe over the Gran Paradiso Massif ~40 Ma b.p. The reason for this discrepancy of zircon results is at present not known. We interpret the new data as cooling ages, both zircon and apatite being slightly younger than the results found in Sesia–Lanzo zone. The moderate apatite gradient with altitude suggests an uplift rate of 0.75 mm/a; apatite confined track length patterns are typical of the undisturbed basement type (Fig. 6). The data suggest an Oligocene–Miocene cooling and uplift history somewhat similar to that found in the Sesia–Lanzo zone and contradict Carpena's model of inverted metamorphism.

The Silvretta nappe of the eastern Alps

The upper Austro–Alpine Silvretta nappe (Fig. 11), one of the highest tectonic units of the Alps, is a basement unit lying in the border region between Austria and Switzerland. The Palaeozoic polymetamorphic crystalline core consists mainly of biotite plagioclase paragneisses, granitic two-mica or biotite orthogneisses and amphibolites. It is overlain in the west by autochthonous and para-authochthonous Triassic sediments. Both the crystalline core and the overlying Triassic sediments have undergone a weak Alpine diagenetic to greenschist metamorphism. The different Alpine overprint indicates a differential Alpine uplift history. The climax of the Alpine metamorphism has been reached during the Eoalpine phase (Trümpy 1960, Thöni 1981, Flisch 1986b). Recent investigations into the geology, tectonics, petrography and isotope geology of the Silvretta area can be found in Grauert (1969), Frapolli (1975), Maggetti (1986), Maggetti & Galetti (1984), Maggetti *et al.* (1987), Flisch (1986b, 1987) and Laubscher (this volume).

Fission track apatite analyses of 11 samples have permitted reconstruction of the Mid-Tertiary to recent uplift history of the Silvretta nappe (Flisch 1986b). Apatite ages measured for the nappe show a strong dependency upon sample elevation (Fig. 12) with the N–S profile AA′ of Fig. 13 revealing a good linear array indicating an uplift rate of ≈0.1 mm/a. In contrast, five samples fall off this line, one sample B to the west of the N–S profile AA′ yielding an age apparently too young, samples C–F lying to the east, ages apparently too high for their altitudes. Although samples B, c, D, E and F lie at approximately the same elevation (1600±250 m), they reveal a distinct trend from west to east of increasing apatite age (12.7 to 26.2 Ma). A possible interpretation of this trend (Flisch 1986b) assumes that, subsequent to an uniform uplift starting in the mid-Oligocene during which the closure temperature for apatite was passed, the entire Silvretta nappe was tilted toward the east during late Pliocene to recent times; as a result, at the same present-day altitude samples with younger ages are found in the west having been uplifted from deeper levels, whilst in the east older ages signify the relative depression of higher samples. This model can be supported by the following reasoning. A NE–SW trending horizontal reference contour or isopleth GG′ at a constant elevation of 1720 m can be constructed to pass through sample point D at that altitude and vertically above sample b (Fig. 13). Since the uplift rate of profile AA′ which contains sample point b is well defined, the corresponding age at 1720 m is precisely calculated as 16.9 Ma. A horizontal distance between samples D and b on profile GG′ of 18.2 km relates to an age difference of 6.8 Ma and shows a linear age increase eastwards of ≈0.37 Ma/km. Assuming a uniform eastward tilting of the nappe, this rate of age increase can be used to estimate the age of a sample anywhere on the 1720 m reference contour. Lines (thin dashed lines in Fig. 13) parallel to the N–S profile AA′ (and parallel to the strike of the assumed eastward dip) through sample points B, C, E and F intersect the 1720 m reference contour, permitting estimation of the

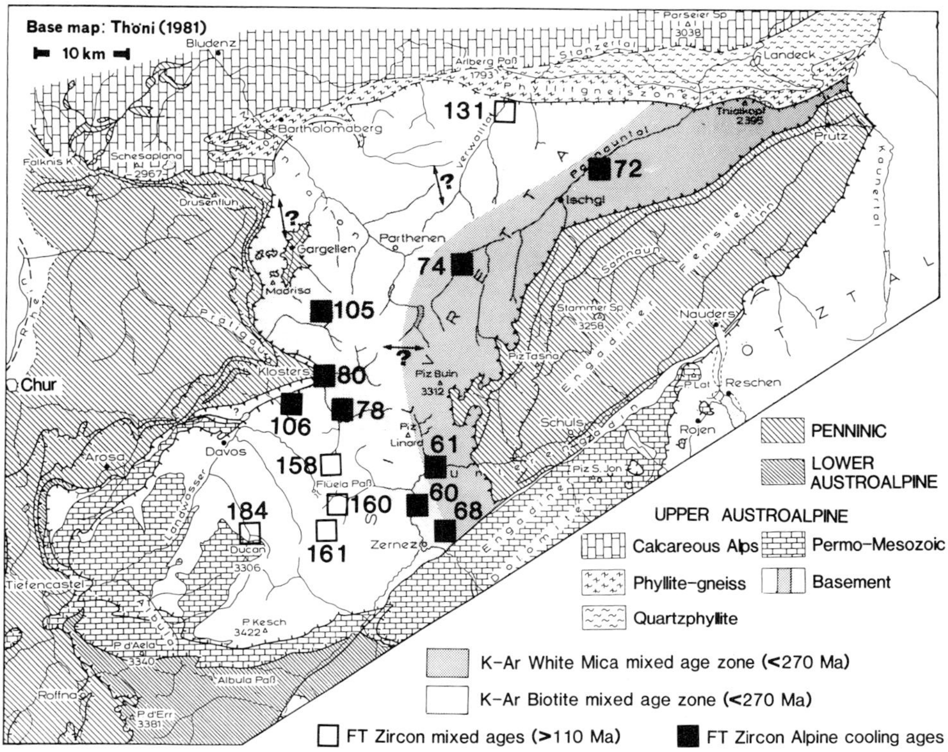

FIG. 11. Distribution of K–Ar white mica and biotite ages and fission track zircon ages in the Silvretta (from Flisch 1986b). K–Ar data from Thöni (1981); Kreczy (1981); Krähenbühl (1984); Flisch (1987). Arrows marked (?) indicate uncertain boundaries because only a few data are available. Shading indicates outcrop of the deeper parts of the Silvretta basement.

corresponding ages at the intersection points. This provides a second estimate of age vs altitude in different slices through the Silvretta (the starred points in Fig. 12). Joining the measured ages for the sample points B, C, E and F to their corresponding estimated ages along strike at 1720 m yields uplift rates of 0.1 to 0.12 mm/a for the Oligocene-Miocene, identical to the rate defined more precisely by the N–S section (Fig. 12).

Further support for such a model of a late easterly dip of the Silvretta is provided by recent geodetic measurements which show a young rapid uplift of 1.7 mm/a at Chur with a radial decrease in uplift rate away from Chur (Gubler 1976) resulting, for example, in a recent uplift rate at Davos of 1.2 mm/a. These recent rates differ dramatically from the Oligocene-Miocene uplift rate given by samples from profile AA′. A constraint on the duration of the present high uplift rates may be provided by assuming that the present-day zero apatite fission track age (at a temperature of ≈120°C) is at a depth of 1800 m below sea level (with a geothermal gradient of 30°C/km). Assuming that the recent uplift is not decreasing, the intersection of the lines representing the present estimated uplift rate of 1.0 mm/a for sample point e (near Klosters – see Gubler 1976) and the apatite determined uplift of 0.1 mm/a (profile AA′) gives a minimum age of 2.1 Ma b.p. for the start of the accelerated uplift (Fig. 12). To accommodate the model of an easterly dipping Silvretta nappe, an uplift rate of ≈0.4 mm/a operative during the last 2 Ma is required, for example, at point F. An alternative explanation of the easterly dip requires the start of the accelerated uplift to become successively younger from west to east, but maintaining an approximately constant uplift rate. The distribution of Neoalpine jointing (Krähenbühl 1984) in the central Silvretta nappe revealed a prominent N–S direction, parallel to sample profile AA′ and this jointing could relate to the strike direction of the proposed easterly dip.

In summary, the altitude and regional depen-

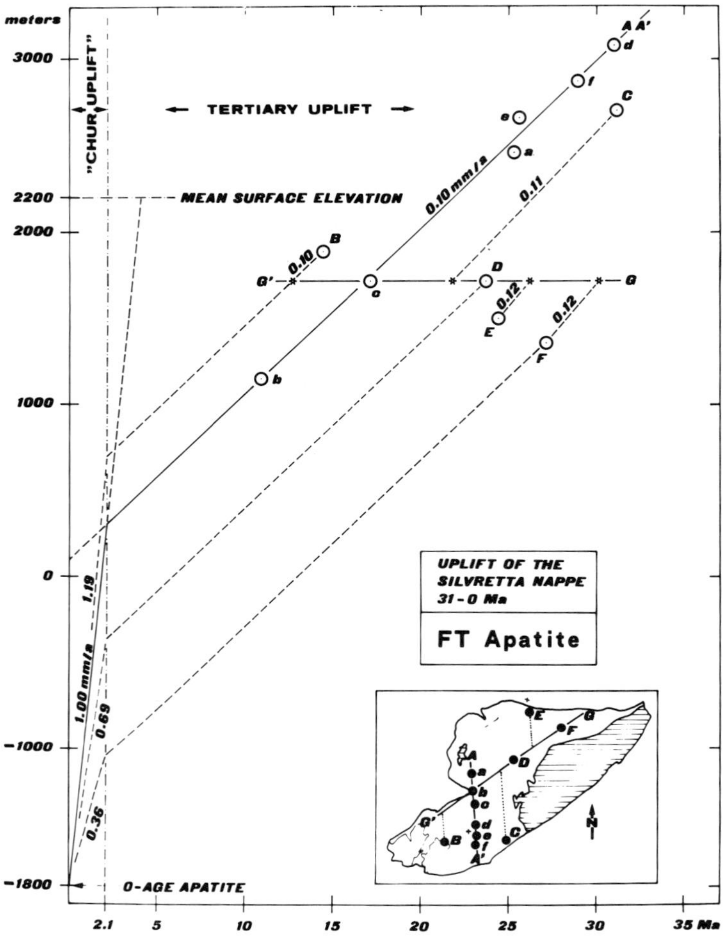

FIG. 12. Uplift paths for the Silvretta nappe from 31 Ma to present based upon fission track apatite age determinations (Flisch 1986b). Circled points are counted apatite fission track samples placed in their respective time-altitude positions. Starred points are ages calculated on the basis of a 1720 m isopleth relative to their geographical positions (see caption of Fig. 13). The numbers on the uplift lines in the 'Chur Uplift field' represent the uplift rates required for the last 2.1 Ma to explain the eastward tilting of the Silvretta nappe. The 1.19 mm/a and 1.0 mm/a uplift rates for sample B and profile AA′ respectively are of the same order as recent uplift rates reported by Gubler (1976).

dency of the fission track apatite age distribution indicates a homogeneous uplift from mid-Oligocene to Pliocene (0.1 mm/a) and a differential, eastward tilting uplift during Quaternary times.

Fission track zircon analyses of 14 samples show ages ranging between 184 and 60 Ma (Fig. 11). The ages of the samples from the northern and the southwestern regions are older than 130 Ma and, since they do not obviously correlate with a geological event, are therefore interpreted as Hercynian–Alpine mixed ages (open squares in Fig. 11). This implies that the closure temperature of the fission track zircon system estimated at 225±25°C has only been exceeded in the central parts of the Silvretta nappe during Alpine time (black squares in Fig. 11), resulting in cooling ages on the retrograde path of the Alpine metamorphism. The zircon ages also show that today's outcropping Silvretta nappe has cooled to 225±25°C before 60 Ma b.p. The two highest fission track zircon ages from the region where the Alpine temperature was higher than the zircon closure temperature (stilpnomelane zone, K–Ar biotite mixed age zone) are 106±7 Ma and 105±8 Ma (1σ) which implies an earlier climax of the Alpine metamorphism. Various lines of evidence contribute to defining

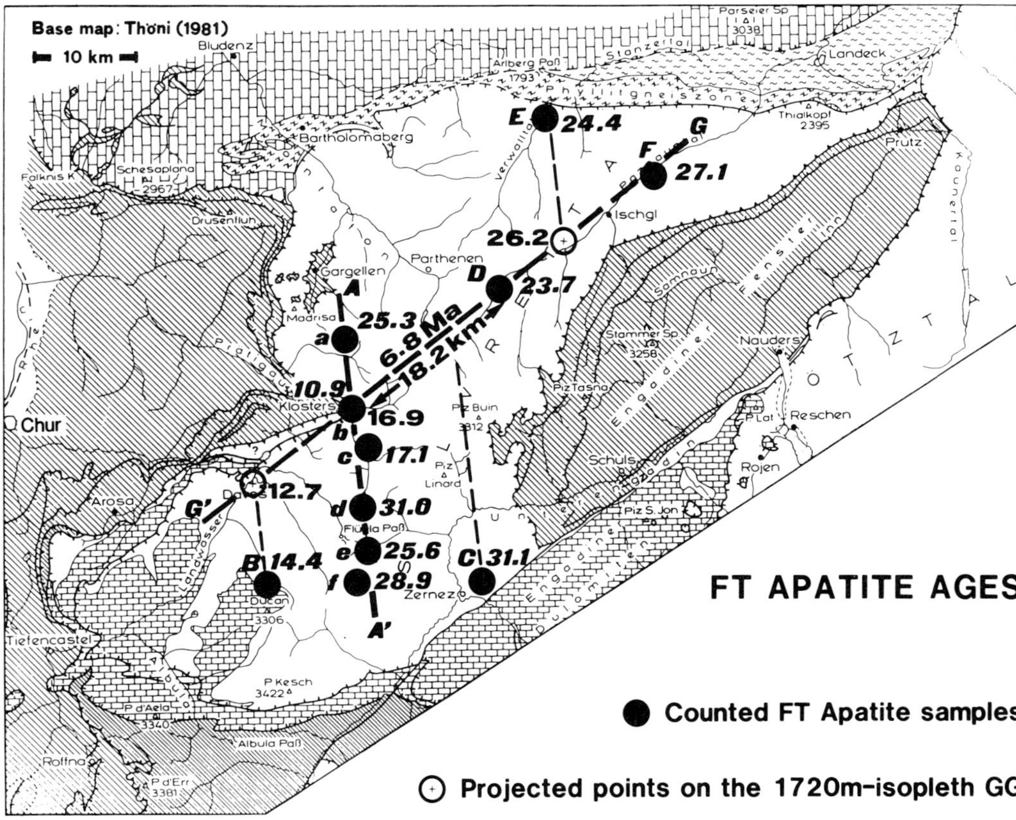

FIG. 13. Fission track apatite ages in the Silvretta nappe (Flisch 1986b). Black circles are sample localities with measured ages shown in italics. Open circles: two examples of sample points projected parallel to profile AA′ onto the 1720 m reference line (isopleth); the corresponding ages (normal numbers) calculated using an age increase of 6.8 Ma over 18.2 km from W to E (between sample points b and D). The calculated ages for the projected points are shown in Fig. 12 as stars. Note that the age of 16.9 Ma for locality b represents an age corrected vertically to lie on the 1720 m isopleth.

the timing of the Eoalpine metamorphic climax: in the western Alps ages between ≈80 and 129 Ma provide broad time limits, with some evidence for a peak at 85–110 Ma (see above). In the eastern Alps, Satir (1975) deduced an age of 130–110 Ma from mica formation ages for the Eoalpine Schneeberg metamorphism of the Ötztal crystalline basement. Thöni (1985 and refs. therein) preferred a climactic age of between 100 and 85 Ma, based upon numerous mica cooling ages and small-scale isochrons on rocks from upper Austro–Alpine nappes lying to the west of the Tauern window: the Ötztal and Campo nappes and S-charl units. From the study of Flisch (1986b), the age of the climax of Alpine metamorphism within the Silvretta nappe can be tentatively placed at 110 Ma, although to ensure that the climax is older than 105–106 Ma, further fission track zircon analyses are needed. A minimum age of 90 Ma may be concluded confidently for the metamorphic climax within the Silvretta nappe, (an age within the 2σ error of the fission track zircon ages), although in other upper Austro–Alpine units the metamorphic peak may have occurred later.

The closure temperature concept requires that rejuvenation of a certain mineral-isotope system commences when its closure temperature is reached on the prograde metamorphic path of a subsequent event. In most cases simply reaching the closure temperature of a mineral is insufficient to cause complete rejuvenation of the isotopic system, Thöni (1982) suggesting that for micas an additional temperature of 70–100°C is required. This hypothesis is supported in the eastern and northeastern Silvretta by the co-existence of K–Ar biotite and white mica mixed ages indicating that although temperatures were sufficient to open the white mica system (i.e. >350°C), they were insufficient to completely reset the biotite systems (Flisch 1986b), which require a temperature of 300±50°C to open.

From the patterns of Rb–Sr and K–Ar mica and fission track zircon ages, a clear regional variation of palaeotemperature with time may be deciphered. To establish such Rb–Sr and K–Ar mica age patterns in the Silvretta, results have been used from the studies of Grauert (1969), Kreczy (1981), Thöni (1981), Krähenbühl (1984) and Flisch (1986b). Throughout the nappe (with the exception of one sample from the margin of the Engadine window, Thöni 1981), Rb–Sr white mica ages have been interpreted as unaffected Hercynian cooling ages, indicating that a temperature of 500°C has not been reached anywhere in the Silvretta nappe during Alpine times. In those mineral-isotope systems having lower closure temperatures, an increase of Alpine influence from W to E in the central part of the Silvretta nappe is progressively revealed (Fig. 11). In the northern, western and southern parts Hercynian K–Ar white mica cooling ages are still found, in sharp contrast to the lower ages of between 270 and 110 Ma from the northeastern and eastern parts of the nappe, showing that in these latter regions temperatures of 350°C have been reached during Alpine time. This Alpine rejuvenation is more markedly shown by mineral-isotope systems with lower closure temperatures: first, by the the patterns of K–Ar biotite ages, where values of 270–110 Ma from the entire central part of the nappe show clear Alpine influence, and second, by the even more sensitive fission track zircon system, where in the central part of the nappe only Eoalpine ages <110 Ma are found (see Fig. 11), the south-western and northern margins showing Hercynian-Alpine mixed ages (Flisch 1986b), indicating that temperatures did not exceed 225±25°C during Alpine times in these regions.

Consideration of these patterns of measured mica and fission track zircon ages for samples in different parts of the Silvretta nappe permit a regional estimation of the degree of Alpine influence, and thus estimation of the temperature experienced during the Eoalpine metamorphism at any one sample point. Further temperature constraints are provided by both the regional distribution of quartz recrystallisation and Alpine formed stilpnomelane. Thus for each sample point, assuming a geothermal gradient of 30°C/km, the total uplift since the Eoalpine metamorphic climax may be calculated. In the legend of Fig. 14, the first number in the brackets gives the estimated temperature in °C for the different sample localities and the second number the estimated total uplift in km referred to a mean elevation of 2200 m (see Wagner *et al.* 1977). These patterns reveal a central uparching of the Silvretta nappe (Fig. 11), with an east-west strike and a westerly dip. The difference in uplift between the central and marginal Silvretta is a minimum of 5 km.

Figure 14 presents a composite uplift history for the Silvretta from ≈110 Ma to the present. Figure 14(a) shows, for each sample locality, how a point at the today's mean surface elevation of 2200 m has been uplifted during the last ≈110 Ma. Figure 14(b) shows the uplift evolution from 110 Ma to 35 Ma which leads to the 0 m altitude plane 35 Ma ago. The points which represent age values <110 Ma and >60 Ma on the uplift lines a, (b,X), c, (C,Z), D, F and B and the points with age values =110 Ma on the uplift lines d, e, f, B and E are derived from fission track zircon analyses. Three uplift periods can be distinguished: *updoming, homogeneous uplift* and *eastwards tilting*. The uplift younger than 35 Ma (recent to early Oligocene) is quite well-defined, because the recording of the homogeneous uplift is independent of the prevailing geothermal gradient, and for the calculation of the eastwards tilting (during the last 2 Ma) the present-day measured geothermal gradient of ≈30°C/km can be used with confidence. The absolute amount of uplift and the altitude position of the resulting points in the updoming-field (≈35 Ma to ≈110 Ma) are strongly dependent on the geothermal gradient – a value of 30°C/km has been used in calculations. The only restriction is the argument that, with a geothermal gradient of more than 36°C/km, sample B should give an Alpine cooling fission-track zircon age and not a Hercynian–Alpine mixed age. There are no strong arguments for or against a geothermal gradient of less than 30°C/km. The transition from the homogenous uplift to the updoming has been estimated at 35 Ma. This transition must be, in each case, younger than 43 Ma, otherwise the amount of homogeneous and tilting uplift would already be higher than the total estimated overburden for sample B. In this case sample B should show a fission track zircon cooling age ≤43 Ma and not a Hercynian–Alpine mixed age. The transition must also be older than the oldest fission track apatite age of 31 Ma (sample d). The estimation of 35 Ma for the transition implies a maximum error of +800 m and −400 m on the amount of the early Oligocene to recent uplift. As can be seen in Fig. 14(b), a part of the differential

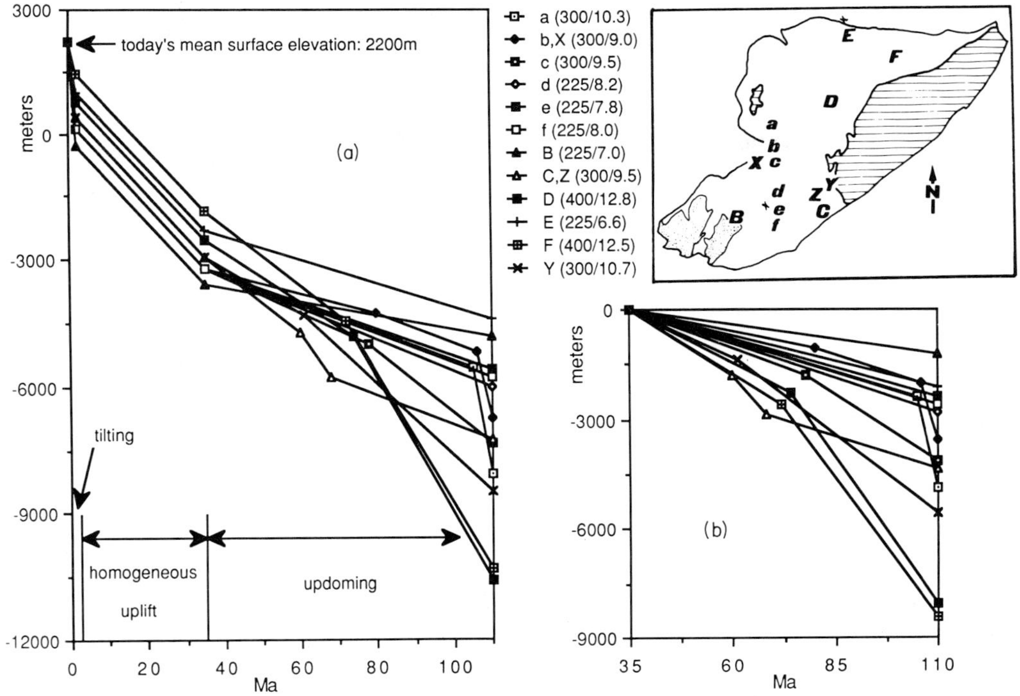

FIG. 14. Compilation of uplift history for the Silvretta nappe from Mid-Cretaceous until the present. In the legend the first number in brackets gives estimated temperature in °C for different sample localities, the second number the estimated total uplift in km referred to a mean elevation of 2200 m. (a) Shows the uplift patterns for points which today would lie on the mean surface elevation of 2200 m at the different sample localities. Three uplift periods can be distinguished: updoming (uparching or upwarping), homogeneous uplift and eastward tilting. (b) Shows the uplift from 110 to 35 Ma of points which lay at 0 m above sea level 35 Ma ago. It can be seen that a certain part of the differential uplift is older than 80 Ma and much is older than 60 Ma.

uplift of different sample localities is significantly older than 80 Ma, and much is older than 60 Ma, as revealed by the spread of points derived from the zircon analyses. This shows that much of demonstrable differential uplift must be older than the updoming of the Engadine Window which is itself younger than 60 Ma (see below). Although the exact uplift mechanism in the time interval 100–80 Ma can only be surmised, the following represents a conceivable scenario: at ~110 Ma, the start of the compression of the upper Austro–Alpine environment with consequent uparching and generation of topographic highs, results in the cooling recorded by Mid-Cretaceous zircon ages, and contemporaneous deposition of flysch-like Triazza Formation (Furrer 1985). This implies that a certain part of the total demonstrable differential uplift could be older than the shearing of the Silvretta nappe from its continental basement. At 100–80 Ma, increasing crustal shortening leads to shearing of the major Austro–Alpine nappes (e.g. northern calcareous Alps, the Ötztal, Campo–Languard and Silvretta nappes). Continued compression resulted in the piling-up of the Austro–Alpine nappes, with the consequent increase of temperature to a metamorphic maximum in the lower units (e.g. Sesvenna crystalline rocks and Engadine dolomites). This model presumes that the Alpine metamorphic conditions in the Silvretta nappe can only be explained by the differential uplift, the normal sedimentary overburden of ≈4 km (Furrer 1985) and internal shearing within the nappe itself. Such internal shearing produced large folds in the overlying sediments and faults in the basement rocks (e.g. Sertig–Flüela fault) of western Silvretta, resulting in a thickening of the whole sedimentary pile, and a temperature increase sufficient to explain the observed high-diagenetic metamorphism of the sediments of the Ducan or Landwasser trough of the southern Silvretta.

In published studies (e.g. Pirkl 1980) the

updoming of the Engadine Window is placed in the Miocene. Such an updoming should be reflected in the apatite age pattern since this records the uplift history since ≈30 Ma: no such evidence is found, indicating that the updoming of the Engadine Window must have occurred earlier than mid-Oligocene, although it must be younger than the youngest sediments of upper Palaeocene to lower Eocene age, found in the Window (Rudolf 1982, Oberhauser 1983).

Central Lepontine alps

In one study along 40 km of the Swiss Geotraverse, 19 samples, chiefly Penninic orthogneisses, were collected through the Maggia valley, Canton Ticino, south-central Switzerland (Hurford 1986b, Fig. 15). The line of study transects the mid-Tertiary culmination of the Lepontine Alps, with its classical mid-Tertiary Barrovian metamorphism revealed as a series of roughly concentric isograd surfaces which cross-cut nappe boundaries (Bearth 1958, Niggli 1960, Wenk 1962, Frey *et al.* 1980 and references therein). Definition of the timing of the metamorphic peak has been controversial and we return to this below. The southern end of the Maggia profile is cut by the Insubric fault line, a major lineament which separates the central Alpine highly metamorphosed rocks from the southern Alps, relatively unmetamorphosed during Alpine times.

Figure 16 presents the measured isotopic ages for the Maggia Valley plotted along a NNW–SSE profile. Because of the altitude dependency of the apatite ages (see above), they are recalculated to a common elevation of 350 m, the mean height of the valley floor, to permit direct comparison. The samples north of the Insubric Line may be divided tectonically into three groups: within error limits, the northern half of the profile, with sub-horizontal gneisses, shows reasonable consistency of zircon fission track and K–Ar and Rb–Sr mica ages, but gradually increasing apatite fission track ages. South of Maggia–Moghegno to Ponte Brolla, a transition zone is entered where effects of Neo-Alpine deformation are detectable and measured ages for all minerals show gradual increases. In a third area, adjacent to the Centovalli Line (a brittle branch of the Insubric Line), the southern steep belt, with vertical and sub-vertical gneisses, is entered and there is a strong grouping of zircon and mica ages between 19 and 24 Ma. For the single Ivrea zone sample south of the Insubric Line, apatite gives an age of 15 Ma, zircon of 48 Ma, whilst micas yield pre-Alpine ages. Analysis of apatite confined track lengths revealed typical slow-cooling values for all samples, irrespective of elevation or tectonic position, with a mean length of 12.7 μm and a mean standard deviation of 1.6 μm (cf. Fig. 6). Such length patterns indicate a similarity of cooling style for all parts of this central Alps profile during the late Neogene.

Figure 17 shows composite cooling paths for samples from the three areas north of the Insubric Line. Approximately similar cooling patterns are exhibited by the three areas, each with seemingly continuous cooling since the late Oligocene – early Miocene, at overall rates of 20–26°C/Ma. In each area, a period of rapid cooling is followed by considerably slower cooling at a rate of 11–17°C/Ma from the late Miocene to the present. Interpretation of cooling in terms of uplift necessitates selecting a value for the geothermal gradient and assuming its constancy: use of a normal gradient value of 30°C/km to calculate the following 'apparent' uplift rates gains some support from model calculations by Werner (1980). Late Miocene to recent cooling in the Steep Belt has a mean rate of 14°C/Ma which corresponds to an uplift rate of 0.47 mm/a, probably resulting from isostatic adjustment (Neugebauer *et al.* 1980). Maintained over 19 Ma, this rate produces an uplift of the Lepontine Alps of some 9 km. The more accelerated period of cooling occurs progressively earlier toward the S: in the southernmost area between 24 and 19 Ma, in the northernmost area between 19.5 and 16 Ma. For the steep belt with a mean cooling rate of 65°C from 23 to 19 Ma, an apparent uplift rate of ~2.2 mm/a is derived; maintained over 4 Ma, the minimum time period of fast cooling, a minimum further uplift of the central Lepontine Alps of 9 km is found.

Within the southern steep belt area, the cooling curves are not obviously affected by the Centrovalli fault, indicating that the vertical displacements along this branch of the Insubric Line are minor. Figures 16 and 17 also emphasize the tremendous difference in thermal history shown by the single Ivrea zone sample. Calculation of uplift for this Ivrea sample over a similar time period of 23 Ma, using a mean cooling rate of ~7°C/Ma results in an overall uplift of the Ivrea zone by 5–6 km. This gives a net difference in uplift of the central Lepontine Alps relative to the southern Alps of 12–13 km since early Miocene. Comparison of metamorphic grade of present day juxtaposed central and southern Alpine rocks (Frey *et al.* 1980) indicate maximum temperatures of 650 and 200°C respectively: persisting with a constant

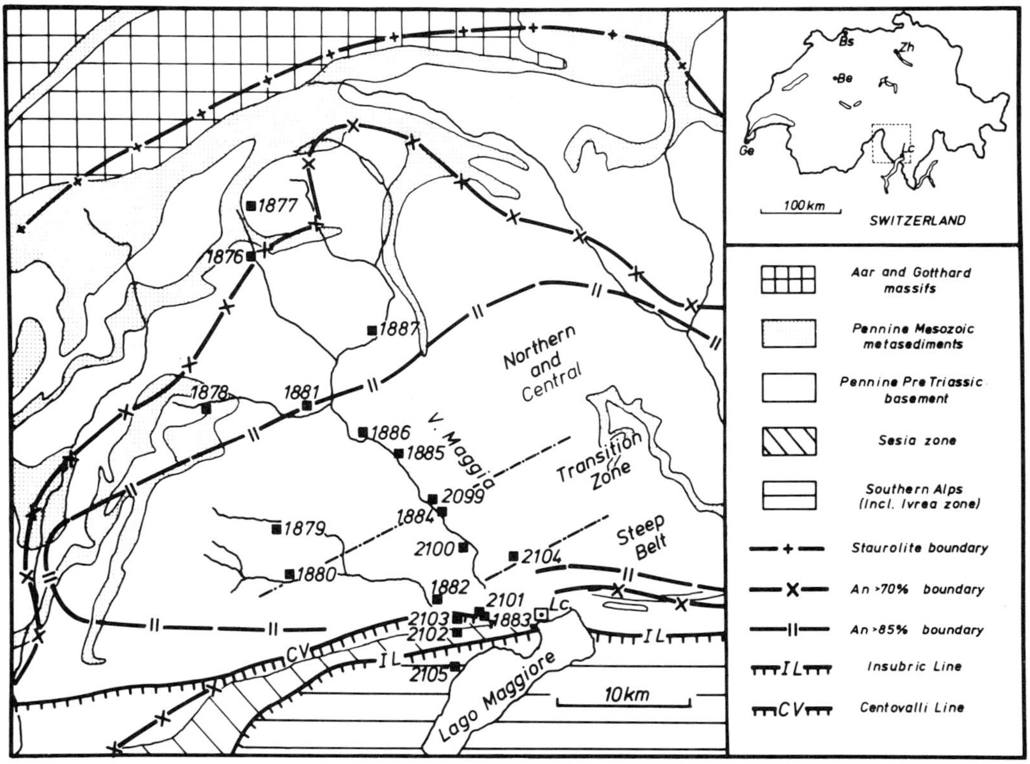

FIG. 15. Sample location map for the central Lepontine Maggia study (Hurford 1986b). Lc: Locarno; Be: Berne: Bs: Basle; Zh: Zürich; Ge: Geneva. Base map is from Wagner *et al.* (1977) with metamorphic isograds from Jäger *et al.* (1967).

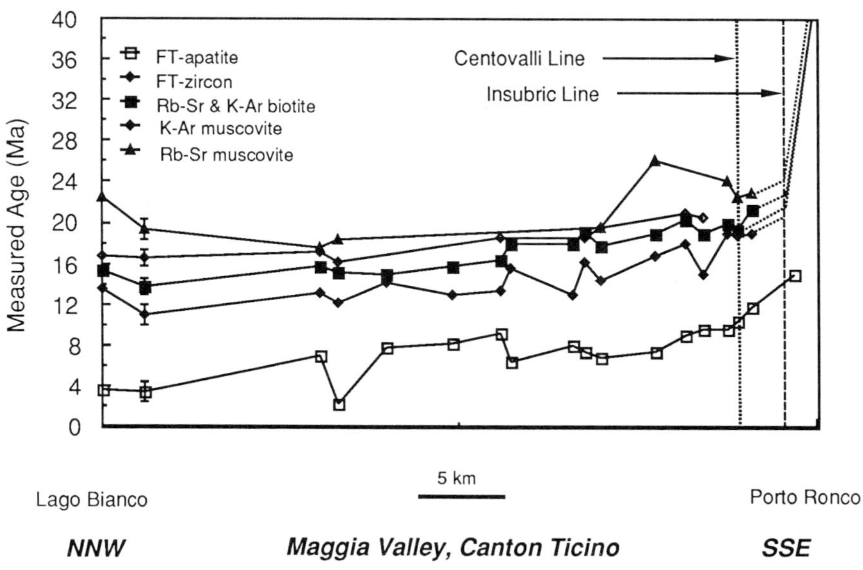

FIG. 16. Graphical representation of measured mineral ages along the Maggia Valley profile, central Alps. Apatite ages are recalculated to a common altitude of 350 m. The shown 2σ error bars are typical of each age (redrawn from Hurford 1986b).

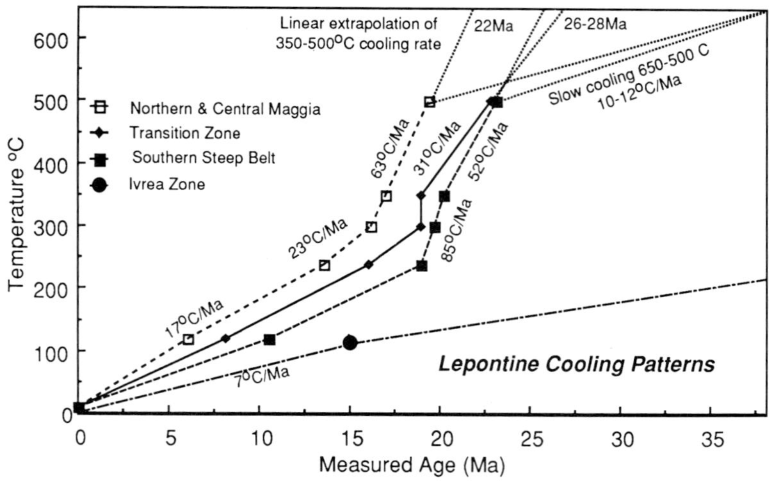

FIG. 17. Composite cooling paths for the three areas N of the Insubric Line in the Maggia Profile, and low temperature cooling path of the Ivrea zone sample. (Redrawn from Hurford 1986b).

and normal geothermal gradient indicates a differential uplift of the central Alps by more than 15 km. Additional uplift of the central Alps can be accommodated in the present model by extrapolating the cooling paths in Fig. 17 to higher temperatures and earlier times than the artificial limit imposed by the 500°C Rb–Sr muscovite chronometer: for example 1 Ma linear back extrapolation of the annealing curves provides an additional uplift of the central Alps of ~2 km. The structural analysis of Schmid *et al.* (1987) revealed that the intervening Insubric mylonite belt accommodates a backthrusting motion of the central Alps over the southern Alps. The coincidence of the southern limit of Pennine rapid cooling with the Insubric mylonites provides strong evidence for a correlation between the rapid cooling and backthrusting during active Neoalpine crustal shortening. Since rapid cooling must postdate the associated uplift, 23 Ma represents a minimum age for the onset of backthrusting, the 500°C chronometer again providing an artificial older time limit. As parts of the Bergell batholith are affected by the schistosity within the southern steep belt (Heitzmann 1975, Vogler & Voll 1981), the backthrusting must postdate the intrusion of the Bergell Massif. From the measured Maggia ages, and assuming that the intrusion age of the Bergell Massif is around 30 Ma (Gulson 1973, Deutsch & Steiger 1985), the age of backthrusting of the central Alps may be bracketed between 25 to 30 Ma. Both the relation of the steep belt schistosity to the Novate intrusion and the age of intrusion itself are problematical, the Rb–Sr whole-rock isochron reported by Gulson (1973) being proportional to a grossly imprecise age of 25±80 Ma – an age often quoted without the uncertainty. Recalculation of the data with IUGS constants and York least squares regression analysis yields a Rb–Sr whole-rock age of 30±13 Ma for the Novate intrusion, still an imprecise age which does not help to define an age for the backthrusting. The drift northward in the Maggia toward younger ages of rapid cooling is interpreted in terms of a time lag between a fixed time of rapid uplift due to backthrusting and the time of rapid cooling.

The Suretta nappe of the central Alps

The studies in the western, eastern and central Alps described above show how the combination of Rb–Sr and K–Ar dating of micas with fission track dating of apatite and zircon results in detailed time-temperature pathways, essential for constraining the timing of tectonic and metamorphic evolution. One area in the central Alps illustrates that such clear-cut cooling patterns are not always forthcoming, imposing limits on interpretation.

Traversing eastward from the Maggia area, the easternmost part of the central Alps is entered, formed by the Tambo and Suretta middle Pennine nappes which lie at the margin of the mid-Tertiary Barrovian metamorphism centred in the Lepontine. The Suretta nappe comprises a pre-Alpine core, enveloped in a Mesozoic parautochthonous cover of quartzites and carbonates (Grünenfelder 1956, Milnes & Schmutz 1978) (Fig. 18). The northern frontal part of the nappe core is today formed by the Roffna gneiss, a granitic gneiss which, from zircon Pb–Pb analyses (Hanson *et al.* 1969), was probably intruded as a granite porphyry ~350 Ma b.p. The present-day southern part of the core is an heterogeneous crystalline mass, the Timun complex, which may have originated from pelitic sediments and tuffs during a Palaeozoic amphibolite-facies metamorphism (Zurflüh 1961, Wenk 1974). The Suretta nappe is affected throughout by a low-grade greenschist-facies Alpine overprint, being traversed by the stilpnomelane isograd, but with most of the nappe lying outside the staurolite isograd. Milnes & Schmutz (1978) have differentiated four deformation phases of thrusting and folding in the general area of the Suretta nappe.

Rb–Sr whole-rock and white mica analyses by Steinitz & Jäger (1981) extended the initial studies of Hanson *et al.* (1969). A combination of the results obtained for the frontal part of the nappe produced a Rb–Sr whole-rock isochron age of 118±7.8 Ma, with an initial $^{87}Sr/^{86}Sr$ isotopic ratio of 0.7288±0.0016. Assuming an open system for this northern part of the Suretta nappe, this age may provide the timing of an Eoalpine metamorphic event, possibly synchronous with an early-Alpine deformation phase (Milnes & Schmutz 1978). A similar Rb– Sr whole-rock evolution diagram for the southern part of the Suretta nappe (Steinitz & Jäger 1981) clearly shows that these samples do not possess the early Alpine open system found in the north, but contain an older, possibly Hercynian system, partially disturbed during Alpine times. Rb–Sr analysis of phengite from the same rocks in the N and S of the Suretta nappe revealed a very different story: seven of the 11 measured phengite–whole-rock ages lie in the range 35–40 Ma. Since these analysed samples have not experienced temperatures in excess of 500–550°C during Alpine times (lying outside the staurolite isograd of the Lepontine metamorphic aureole) these ages were interpreted by Steinitz & Jäger (1981) as phengite formation ages, assuming small scale strontium equilibration at that time. The four Rb–Sr phengite–whole-rock ages lying outside the 'Lepontine' age range at 22.8±0.7, 30.7±9.7, 31.1±5.4 and 45.5±2.7 Ma are variously interpreted by Steinitz & Jäger as cooling ages, or in disequilibrium with the whole-rock.

Because of the distinctive penetrative schistosity in which the fine, newly crystallized white mica is orientated, the indication is that the 35–40 Ma phengite ages (box a in Fig. 19) are not only formation ages, but also date a deformation phase; Steinitz & Jäger proposed this to be the Niemet phase of Milnes & Schmutz. However, since these phengites lie within a schistosity identified as the main foliation of the Roffna gneiss, it may be postulated from the literature (Milnes & Schmutz 1978, Milnes & Pfiffner 1980) that the deformation belongs instead to the Ferrera phase.

In a further study, Flisch & Sonsuk (work in progress) have turned their attention to blue amphibole from within the sedimentary cover of the Suretta nappe at Cröt, Avers (Fig. 18). From petrographic evidence, the amphibole needles lie within the main schistosity and are aligned parallel to a stretching lineation. Moreover, the needles are themselves broken and stretched parallel to this lineation, which seems to indicate that they grew during the Ferrera deformation phase. K–Ar analysis of different grain-size fractions yielded ages ranging between 49±1 and 55±1 Ma (Fig. 19) which compare with a K–Ar phengite age of 41±0.5 Ma from the same rock. The blue amphibole argon ages could be interpreted as either the time of amphibole formation, contemporaneous with early Ferrera deformation movements, or as resulting from the incorporation of inherited argon within the amphiboles, in which case 55 to 49 Ma would be maximum ages for both crystallisation and deformation. It is well documented that amphiboles are especially prone to contain high concentrations of excess ^{40}Ar (see e.g. Zeitler & Wijbrans, 1985, 86, Harrison & McDougall 1980b). The customary way of detecting the presence of excess argon is to plot K–Ar analytical results on a $^{40}Ar/^{36}Ar$ *vs* $^{40}K/^{36}Ar$ diagram (Flisch 1986a): the Suretta amphibole data yield a $^{40}Ar/^{36}Ar$ intercept ratio of 322, significantly in excess of the atmospheric argon ratio of 295.5 and indicating the presence of excess or inherited ^{40}Ar in these samples (Flisch & Sonsuk, work in progress). Accordingly, these amphibole results must be considered as providing a maximum age for both crystallisation and the deformation in which the needles are aligned, probably the Ferrera deformation phase.

K–Ar phengite ages of 38.4±1.6 to 49.6 Ma measured by Steinitz & Jäger (1981) for the Suretta (plus one age of 165±6 Ma) were similarly interpreted as resulting from inherited ^{40}Ar since, in all but one case, K–Ar ages were higher than those found by Rb–Sr analysis (the reverse of the usual sequence – see above). The exception, a quartzite sample KAW 1380

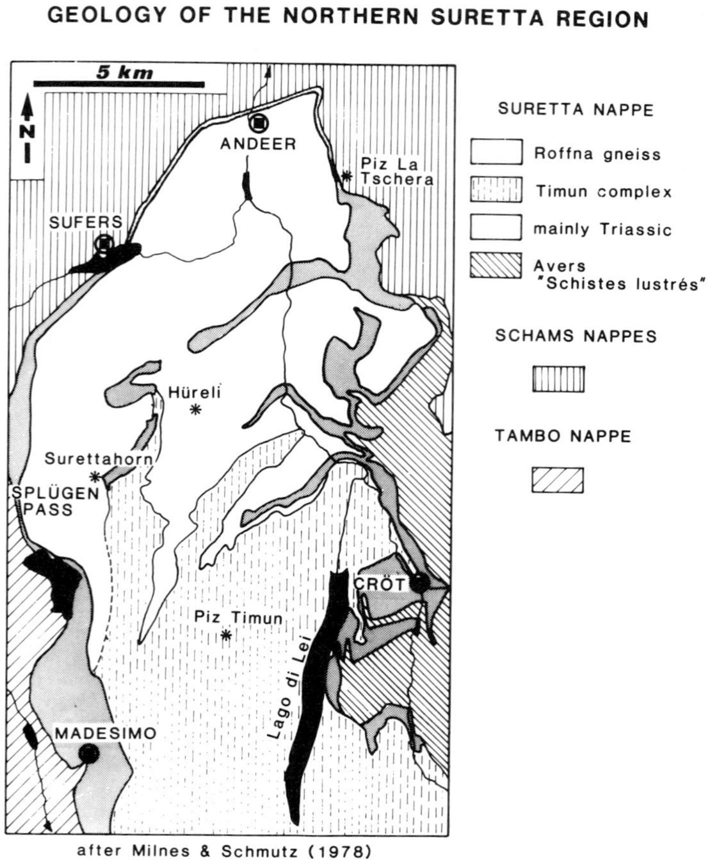

FIG. 18. Geological sketch map of Suretta nappe after Milnes & Schmutz (1978).

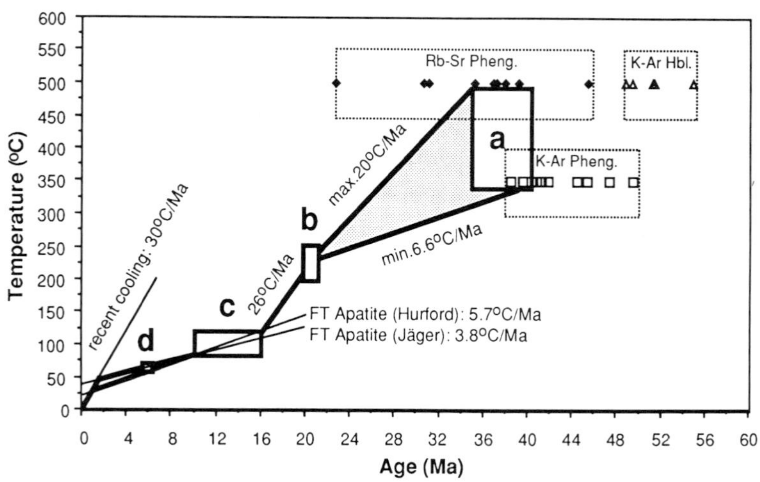

FIG. 19. Postulated time–temperature paths for the frontal Suretta nappe; see text for further explanation.

which yielded a Rb–Sr phengite – whole-rock age of 39.3±2.9 and a K–Ar age of 38.4±1.6 Ma, was deemed free of argon overpressure problems (no other potassium phase being present) and thus probably represented a system chemically independent of the crystalline core during mid-Tertiary metamorphism.

The above discussion begins to point to phengite and amphibole formation being Palaeogene in age, synchronous with a deformation, possibly of the Ferrera phase. A potential conflict arises here. If the Eoalpine Rb–Sr isochron age of Steinitz & Jäger (1981) is also interpreted as recrystallisation contemporaneous with a penetrative foliation this, conceivably, could also be considered as the Ferrera deformation, which would then be dated at 118±7.8 Ma. In such a case, both the phengite and amphibole results must be considered as cooling ages, or (conceivably) formation ages during a second event which had no disturbing effect upon the whole-rock systems. Recently Onstott & Peacock (1987) have proposed a closure temperature of 450–550°C for retention of argon in amphibole. The possibility that the apparently well-defined Eoalpine Rb–Sr isochron is an artefact must also be considered: Marquer (1987) has argued from unexpected Eoalpine Rb–Sr whole-rock ages found in the Aar Massif, that progressive deformation of a granitic body can result in partial rejuvenation of the Rb–Sr whole-rock systems yielding seemingly useful, but meaningless isochrons.

Consideration of the available Rb–Sr and K–Ar radiometric data together with the uncertainties discussed above surrounding their interpretation, permits the imprecise time-temperature pattern of Fig. 19 to be drawn for the Suretta. Within the wide spread of data, we consider that box a represents a reasonable estimate of the timing of phengite formation, with the blue amphiboles crystallising at the same time, or earlier, at <55 Ma ago. Quite clearly this initial age data needs supplementing, particularly by the analysis of samples upon which the effect of the deformation phases is precisely known.

In an attempt to define further the Alpine evolution of the Suretta nappe, especially during the lower temperature cooling and uplift stages, fission track ages on apatite and zircon have recently been measured (Jäger, Wagner & Hurford, work in progress). Apatite and zircon were separated from 11 samples, roughly constituting two N–S trending profiles. Four fission track Suretta zircon ages group tightly around 20–21 Ma (box b in Fig. 19), whilst a single zircon from the Tambo nappe gives 18.1±1.2 Ma, marginally different within error. The substantial age difference of 18 Ma between the zircon and white mica results signifies a slow overall cooling rate of between 6.6. and 20°C/Ma for the Oligocene to earliest Miocene interval, in stark contrast to the cooling and uplift patterns reported for the western and Lepontine Alps during this time period. The difference in measured zircon and apatite fission track data indicate a cooling of 26°C/Ma for the Burdigalian. A distinct break of slope indicating a pronounced deceleration of cooling from Langhian times onward occurs at box c, as defined by the apatite uplift path. The apatites were dated independently in two separate studies; most concentrates possessed exceedingly low and grossly inhomogeneous uranium contents and all crystals contained high densities of inclusions and defects, making analysis relatively imprecise. Resulting apatite ages ranged between 9.9±1.1 and 15.7±0.9 Ma. Both analysts found a clear increase of age with increasing elevation, the regressed slopes of the two data sets corresponding to uplift rates of 0.13 and 0.2 mm/a respectively (corresponding to cooling rates of 3.8 and 5.7°C/Ma, assuming a geothermal gradient of 30°C/Ma) which, viewed within the analytical uncertainties and extremely complicated samples, represents reasonable agreement. This defines a fairly precise segment of the time-temperature path in the Suretta and is included in Fig. 19 together with an estimated timing of the passage of the sample through ~65°C (box d), derived from a consideration of the relative proportions of long (>10 μm) projected track lengths (see Fig. 6) in the spontaneous and induced track apatite aliquots (see Wagner, in press). Measurement of confined spontaneous track lengths (e.g. Fig. 10) revealed typical slow-cooling patterns, commensurate with the calculated slow uplift rates. The cooling history of the Suretta is completed in Fig. 19 by inclusion of the geodetically measured recent uplift of ~1.0 mm/a.

The timing of the Mid-Tertiary metamorphism

Timing of the Mid-Tertiary metamorphism responsible for the marked concentric zonation of mineral isograds in the Lepontine central Alps has long been a contentious subject. Published debate centres upon whether metamorphism was mono- or polyphase, with a climax at ~38 Ma (Jäger 1973) or ~23–29 Ma (Deutsch & Steiger 1985 and references therein). Seemingly, a tacit assumption in discussion has been the ubiquity of the peak temperature and unique nature of the subsequent cooling throughout the region, both considerations appearing unreasonable.

The case for a Lepontine metamorphism in the central and western Alps has been argued (Jäger 1973) as a consequence of numerous Rb–Sr phengite ages around 35–38 Ma, measured for rocks in the Monte Rosa nappe of the western Alps (Hunziker 1969, 1970, Bigioggero *et al.* 1981), in samples from N, E and W of the Lepontine high-grade Barrovian metamorphic area (Jäger 1970) and, as described above, from the Suretta nappe in the eastern part of the central Alps (Steinitz & Jäger 1981). Since these ages are found in samples lying outside the Mesoalpine staurolite isograd but within the chloritoid zone, these temperature limitations indicate that phengite crystallized near the maximum temperature of the Mesoalpine metamorphism and that subsequent temperatures were insufficient to rejuvenate the Rb–Sr system. As a consequence, 38–35 Ma has been interpreted as dating the peak of the Lepontine metamorphism, after which continuous cooling occurred.

The K–Ar analysis of structurally orientated Alpine hornblendes from the southern and central Lepontine Alps (Steiger 1964, Deutsch & Steiger 1985) produced a scatter of results between 23.4 and 31.9 Ma, with an actinolite age of 57.1±2.8 Ma; 23–29 Ma was proposed as the age of the peak of high-grade amphibolite facies metamorphism. Such a timing is consistent with U–Pb monazite ages of Köppel & Grünenfelder (1975, 1978) who reported 22 Ma in the central and northern Lepontine Alps, and ~27 Ma in the S, just N of the Insubric Line. Deutsch & Steiger (1985) admit the significance of the prominent belt of ~38 Ma Rb–Sr white mica ages around the Lepontine Alps aureole to be unclear, but venture the opinion that they signify an earlier Mesoalpine phase of deformation and/or metamorphism.

Given the different depths of burial of the present-day outcropping rocks, the dynamics of cooling and the demonstrable deformation during the Mid-Tertiary, asynchronous differential heating and cooling of different parts of the central Lepontine Alps must have been inevitable. Figure 20 advances two possible schematic time–temperature paths to describe the Mid-Tertiary Alpine metamorphism, providing two thermal end-members for each model representing the present-day outcropping central high-grade and marginal low-grade areas.

In the central high-grade areas, curve (**i**) in Model A supposes that early Palaeogene deep burial beneath the Pennine nappe pile produced maximum metamorphic temperatures of ~650°C (Frey *et al.* 1980 and refs therein) between 45 and 35 Ma which were either maintained or which waned very slowly. In earliest Chattian times, the rate of cooling dramatically increased most probably as a consequence of the backthrusting of the central onto the southern Alps. If the rapid cooling to 500±50°C by ~24 Ma and 225~25°C by ~20 Ma (Fig. 17) is extrapolated linearly to ~650°C at earlier times, an age of 25–27 Ma may be tentatively proposed for the onset of cooling, with uplift and backthrusting preceding the fast cooling. In such a model, U–Pb monazite ages of ~27 Ma from near the Insubric Line may be accommodated as cooling ages, perhaps subsequent to complete lead loss during amphibolite-facies conditions at 35–45 Ma. Support for the continued burial of the Lepontine during the early Oligocene (curve (i) of model A) must be envisaged by analysis of deformation in the Helvetic zone and molasse deposition in the north, where the Oligocene corresponds to increasing burial of the external massifs. In the more marginal parts of the Lepontine Alpine metamorphic aureole, path (ii) shows that peak temperatures of less than 500°C are reached at the same time as the central area, 45–35 Ma. The preservation of Rb–Sr phengite formation ages of 38±2 Ma indicate that subsequently temperatures declined slowly, until the effect of backthrusting and rapid cooling was experienced.

Scenario B considers two episodes of Alpine metamorphism: an earlier, widespread upper greenschist to lower amphibolite facies event is still recorded in the peripheral areas by the phengite formation ages ~38 Ma (path iv in Fig. 20). A second metamorphism at 29–23 Ma was limited to the central high-grade areas (path iii), synchronous with Neoalpine deformation (see e.g. Deutsch & Steiger 1985) such as the backthrusting of the central Alps and the steep belt formation (Schmid *et al.* this volume). The heat source for a second metamorphic episode between 29 and 23 Ma is not immediately apparent.

From due consideration of available radiometric data, our preference is for model A, which approximates to the Lepontine Metamorphism of Jäger (1973), a term used originally to describe specifically the timing of metamorphic peak temperatures, during the interval 35–40 Ma. In attempting to unravel the thermotectonic history of the central Alps, we consider it to be more valuable to use the term Lepontine Metamorphism to describe the whole Mesoalpine to Neoalpine metamorphic evolution within the

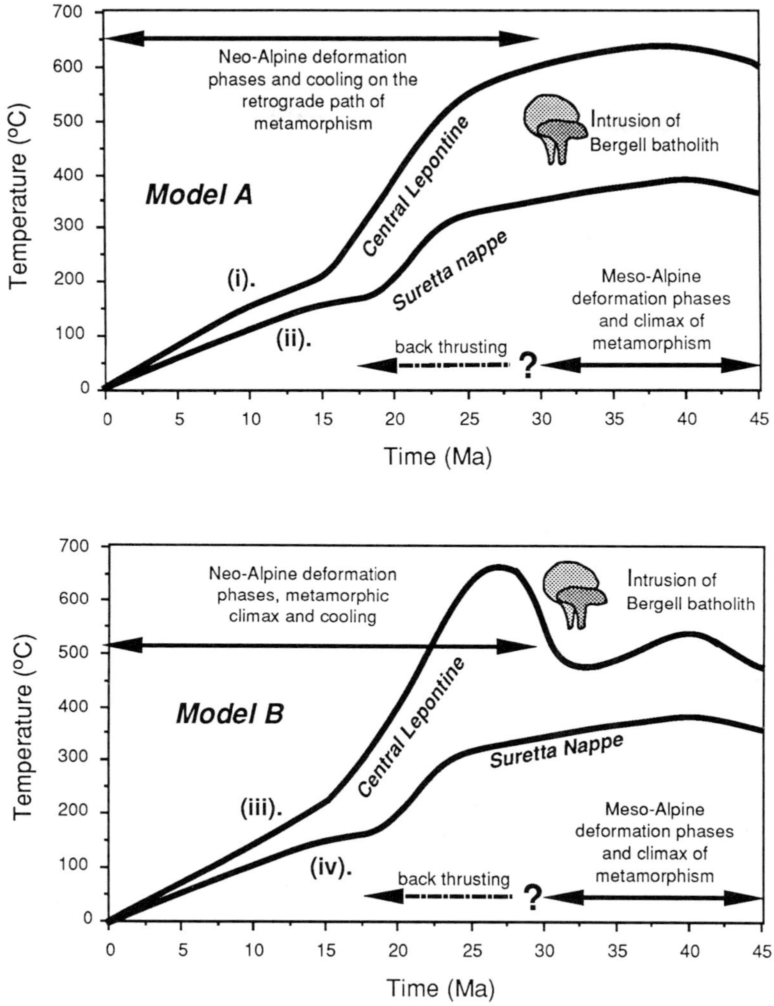

FIG. 20. Two schematic models for time and temperature during the evolution of the Lepontine metamorphism in the central Alps; in each case, the two curves represent thermal end–members.

Lepontine Alps and adjoining areas. In the Lepontine metamorphism temperature maxima were reached in the interval late Eocene to early Oligocene (Mesoalpine), with subsequent cooling patterns on the retrograde paths strongly contrasting according to horizontal and/or vertical position, generating mica and fission track cooling ages, or formation ages which reflect mineral growth below their closure temperatures during several important Neoalpine deformation phases.

Conclusions

Fission track analysis, and Rb–Sr and K–Ar dating of micas can provide unique information: under favoured circumstances, analysis of syntectonically grown minerals can yield directly the age of deformation; more usually, assessment of time and temperature stages in the history of a sample permits-reconstruction of the cooling and uplift patterns.

For the Sesia–Lanzo zone of the western Alps the high-pressure Eoalpine metamorphism is timed at 110–85 Ma, with the pressure maximum probably preceding the temperature peak. A broad pattern of slow-cooling through the late Cretaceous and earliest Tertiary reaches ~225°C by 33 Ma and ~100°C by 25 Ma. Similarity of apatite ages at different elevations and mean confined track lengths of ~14 μm indicate late Oligocene fast cooling, conceivably a function of Insubric phase tectonism. Zircon and apatite fission track ages for the Penninic Gran Paradiso indicate a monotonic cooling of ~10°C/Ma during the Miocene to recent. Similarity of zircon ages from samples at different elevations contests a previous model of inverted metamorphism resulting from the Dent Blanche nappe emplacement.

In the eastern Alps, fission track and mica data reveal a Mid-Cretaceous to recent uplift history of the Silvretta nappe which contains three totally contrasting styles: between 110 and 35 Ma the Silvretta was upwarped (or uparched) along an E–W strike with a westerly dip, with a minimum differential uplift of 5 km. From 35 to ~2 Ma, the Silvretta was uplifted relatively uniformly, whilst in very recent times rapid uplift centred on Chur has produced an eastwards tilt. Previously postulated Miocene updoming of the Engadine window is not reflected in the apatite data and probably occurred prior to ~30 Ma.

For the central Lepontine Alps, a fairly uniform late Neogene to recent mean uplift rate of ~0.5 mm/a confirms previous studies. Cooling from higher temperatures occurred earlier in the south where uplift rates of ~2.2 mm/a in the southern steep belt (or root zone) indicate >9 km of early Miocene uplift of the northern Pennine block, probably by backthrusting along the Insubric fault line. In the Suretta nappe, at the eastern edge of the central Alps, fission track ages broadly similar to the Lepontine are found, indicating a slow overall cooling rate from Miocene to recent. Interpretation of the older phengite-mica and K–Ar hornblende ages is more speculative, but could bracket the age of the main deformation (equivalent to the Ferrera phase?) between 55 and 36 Ma.

Consideration of different models for the asynchronous thermal history of the central Lepontine Alps favours a pattern where maximum temperatures were reached in the Mesoalpine (late Eocene to early Oligocene), with subsequent cooling paths strongly contrasting according to horizontal or vertical position.

Differential cooling and uplift rates, as demonstrated in the above case studies usually reveal tectonic activity, with radiometric ages offering chronological and temperature constraints for structural modelling. Application of dating, and petrofabric and petrographic techniques to samples which have been well-defined structurally in the field, clearly yields the most valuable means of deciphering complex metamorphic and deformation histories. Nevertheless, as demonstrated by early studies in the Alps, much information can be gleaned in a reconnaissance survey from the analysis of a few well-chosen samples.

ACKNOWLEDGEMENTS: We are grateful to our colleagues in Berne for their logistical support of all aspects of isotope analysis and for their discussion of the various topics included in this paper. Constructive criticism of an early version of this manuscript by Martin Casey, Dorothee Dietrich, Johannes Hunziker, Adrian Pfiffner, Stefan Schmid, Bernhard Stöckhert and André Zingg has substantially improved the end-product. Our research is generously supported by the Schweizerischer Nationalfonds zur Förderung der wissenschaftlichen Forschung.

References

ARMSTRONG, R. L. 1966. K–Ar dating of plutonic and volcanic rocks in orogenic belts. *In*: SCHAEFFER, O. A. & ZÄHRINGER, J. (eds). *Potassium-Argon Dating*, 177–33, Springer-Verlag, Berlin.

ARNOLD, A. & JÄGER, E. 1965. Rb–Sr Altersbestimmungen an Glimmern im Grenzbereich zwischen voralpinen Alterswerte un alpiner Verjüngung der biotite. *Eclogae geologicae Helvetiae*, **58**, 369–90.

BEARTH, P. 1958. Über einen Wechsel der Mineralfazies in der Wurzelzone des Penninikums. *Schweizerische Mineralogische und Petrographische Mitteilungen*, **38**, 363–73.

BERGER, G. W. & YORK, D. 1981. Geothermometry from $^{40}Ar-^{39}Ar$ dating experiments. *Geochimica et Cosmochimica Acta*, **45**, 795–811.

BHANDARI, N., BHAT, S. G., LAL, D., RAJAGOPALAN, G., TAMHANE, A. S. & VENKATAVARADAN, V. S. 1971. Fission fragment tracks in apatite: recordable track lengths. *Earth and Planetary Science Letters*, **13**, 191–9.

BIGIOGGERO, B., BORIANI, A., COLOMBO, A. & TUNESI, A. 1981. Età e carattere petrochimico degli ortogneiss della zona Moncucco–Orselina nell'area Ossolana. *Rendiconti della Società Italiana di Mineralogia e Petrologia*, **38**, 207–18.

CARPENA, J. 1985. Tectonic interpretation of an inverse gradient of zircon fission track ages with respect to altitude: alpine thermal history of the Gran Paradiso basement. *Contributions to Mineralogy and Petrology*, **90**, 74–82.

CARRARO, F., DAL PIAZ, G. V. & SACCHI, R. 1970. Serie di Valpelline e II Zona Diorito Kinzigitica sono i relitti di un ricoprimento proveniente dalla zona Ivrea-Verbano. *Memorie della Società Geologica Italiana*, **9**, 197–224.

CHOPIN, C. & MALUSKI, H. 1980 $^{40}Ar-^{39}Ar$ dating of high pressure metamorphic micas from the Gran Paradiso area (western Alps): evidence against the blocking temperature concept. *Contributions to Mineralogy and Petroloy*, **74**, 109–22.

COMPAGNONI, R. 1977. The Sesia–Lanzo zone: high pressure–low temperature metamorphism in the Austro–Alpine continental margin. *Rendiconti della Società Italiana di Mineralogia e Petrologia*, **33**, 335–74.

——, DAL PIAZ, G. V., HUNZIKER, J. C., GOSSO, G., LOMBARDO, B. & WILLIAMS, P. F. 1977. The Sesia–Lanzo zone, a slice of continental crust with Alpine high pressure–low temperature assemblages in the western Italian Alps. *Rendiconti della Società Italiana di Mineralogia e Petrologia*, **33**, 281–334.

DALLMEYER, R. D. 1978. $^{40}Ar-^{39}Ar$ incremental release ages of hornblende and biotite across the Georgia Inner Piedmont: their bearing on late Palaeozoic–early Mesozoic tectonothermal history. *American Journal of Science*, **278**, 124–49.

DAL PIAZ, G. V., GOSSO, G. & MARTINOTTI, G. 1971. La II Zona Dioritico–Kinzigitica tra la Valsesia e la Valle d'Ayas (Alpi Occid.). *Memoire della Società Geologica Italiana*, **10**, 257–76.

——, HUNZIKER, J. C. & MARTINOTTI, G. 1972. La zona Sesia–Lanzo e l'evoluzione tettonico-metamorfica delle Alpi nordoccidentali interni. *Memoire della Società Geological Italiana*, **11**, 433–60.

DEMPSTER, T. J. 1985. Uplift patterns and orogenic evolution in the Scottish Dalradian. *Journal of the Geological Society of London*, **142**, 111–28.

DESMONS, J. & O'NEILL, J. R. 1978. Oxygen and hydrogen isotope compositions of associated rocks from the eastern Sesia zone (western Alps, Italy). *Contributions to Mineralogy and Petrology*, **71**, 13–22.

DEUTSCH, A. & STEIGER, R. H. 1985. Hornblende K–Ar ages and the climax of Tertiary metamorphism in the Lepontine Alps (south– central Switzerland): an old problem reassessed. *Earth and Planetary Science Letters*, **72**, 175–89.

DODSON, M. H. 1973. Closure temperature in cooling geochronological and petrological systems. *Contributions to Mineralogy and Petrology*, **40**, 259–74.

—— & MCCLELLAND-BROWN, E. 1985. Isotopic and palaeomagnetic evidence for rates of cooling, uplift and erosion. *In*: SNELLING, N. J. (ed). *The Chronology of the Geological Record*. Memoir of the Geological Society of London No. **10**, 315–25.

FITZGERALD, P. G., SANDIFORD, M., BARRETT, P. G. & GLEADOW, A. J. W. 1986/87. Asymmetric extension associated with uplift and subsidence in the Transantarctic Mountains and Ross Embayment. *Earth and Planetary Science Letters*, **81**, 67–78.

FLEISCHER, R. L., PRICE, P. B. & WALKER, R. M. 1965. Effects of temperature, pressure and ionisation on the formation and stability of fission tracks in minerals and glasses. *Journal of Geophysical Research*, **70**, 1497–502.

—— & HART, H. R. 1972. Fission track dating: techniques and problems. *In*: BISHOP, W. W., MILLER, J. A. & COLE, S. (eds). *Calibration of Hominid Evolution*, 135–70. Scottish Academic Press, Edinburgh.

——, PRICE, P. B. & WALKER, R. M. 1975. *Nuclear Tracks in Solids: Principles and Applications*. University of California Press, p. 605.

FLISCH, M. 1986(a). K–Ar dating of Quaternary samples. *In*: HURFORD, A. J., JÄGER, E. & TEN CATE, J. A. M. (eds). *Dating Young Sediments: Proceedings of the Workshop, Beijing, China, September 1985* 299–323 CCOP–UNESCO Technical Publication **16**.

——, 1986(b). Die Hebungsgeschichte der oberostalpinen Silvretta–Decke seit der mittleren Kreide. *Bulletin der Vereinigung schweizerische Petroleum–Geologen und Ingenieure*, **53**, 23–49.

—— 1987. *Geologische, petrographische und isotopengeologische Untersuchungen an Gesteinen des Silvretta-Kristallins*. Unpublished PhD thesis, University of Berne.

FRAPOLLI, G. 1975. Petrografia della regione del passo della Flüela. *Schweizerische Mineralogische und Petrographische Mitteilungen*, **55**, 307–64.

FREY, M., HUNZIKER, J. C., FRANK, W., BOCOUET, J., DAL PIAZ, G. V., JÄGER, E. & NIGGLI, E. 1974. Alpine metamorphism of the Alps–a review. *Schweizerische Mineralogische und Petrographische Mitteilungen*, **54**, 247–90.

FREY, M., BUCHER, K., FRANK, E. & MULLIS, J. 1980. Alpine metamorphism along the geotraverse Basel–Chiasso–a review. *Eclogae geologicae Helvetiae*, **73**, 527–46.

FURRER, H. 1985. Field workshop on Triassic and Jurassic sediments in the eastern Alps of Switzerland: a guide book. *Mitteilungen aus dem geologischen Institut ETH und Universität Zürich*, **248**.

GLEADOW, A. J. W. & BROOKS, C. K. 1979. Fission track dating, thermal histories and tectonics of igneous intrusions in east Greenland. *Contributions to Mineralogy and Petrology*, **71**, 45–60.

——, DUDDY, I. R. & LOVERING, J. F. 1983. Fission track analysis: a new tool for the evaluation of thermal histories and hydrocarbon potential. *Journal of the Australian Petroleum Exploration Association*, **23**, 93–102.

——, ——, GREEN, P. F. & LOVERING, J. F. 1986. Confined fission track lengths in apatite: a diagnostic tool for thermal history analysis. *Contributions to Mineralogy and Petrology*, **94**, 405–15.

GRAUERT, B. 1969. *Die Entwicklungsgeschichte des Sivrettakristallins auf Grund radiometrischer Altersbestimmungen*. Unpublished PhD thesis, University of Berne.

GREEN, P. F. 1985. Comparison of zeta calibration baselines for fission track dating of apatite, zircon and sphene. *Chemical Geology (Isotope Geoscience Section)*, **58**, 1–22.

—— 1986. On the thermo-tectonic evolution of northern England: evidence from fission track analysis. *Geological Magazine* **123**, 493–506.

——, DUDDY, I. R., GLEADOW, A. J. W., TINGATE, P. R. & LASLETT, G. M. 1986. Thermal annealing of fission tracks in apatite 1: a qualitative description. *Chemical Geology (Isotope Geoscience Section)*, **59**, 343–54.

——, ——, —— & LOVERING, J. F. 1988a. Apatite fission track analysis as a palaeotemperature indicator for hydrocarbon exploration. *In*: NAESER, N. D. (ed). Society of Economic Paleontologists and Mineralogists Special Publication (*in press*).

——, —— & LASLETT, G. M. 1988b. Can fission track annealing in apatite be described by first-order kinetics? *Earth and Planetary Science Letters* (*in press*).

GRÜNENFELDER, M. 1956. Petrographie des Roffnakristallins in Mittelbünden und seine Eisenvererzung. *Beiträge zur Geologischen Karte der Schweiz, Geotechnische Serie*, **35**.

GUBLER, E. 1976. Beitrag des Landesnivellements zur Bestimmung vertikaler Krustenbewegungen in der Gotthard Region. *Schweizerische Mineralogische und Petrographische Mitteilungen*, **56**, 675–8.

GULSON, B. L. 1973. Age relations in the Bergell region of the south-east Swiss Alps: with some geochemical comparisons. *Eclogae geologicae Helvetiae*, **66**, 293–313.

HANSON, G. N. & GAST, P. W. 1967. Kinetic studies in contact metamorphic zones. *Geochimica et Cosmochimica Acta*, **31** 1119–53.

——, GRÜNENFELDER, M. & SOPTRAYANOVA, G. 1969. The geochronology of a recrystallised tectonite in Switzerland–the Roffna gneiss. *Earth and Planetary Science Letters*, **5**, 413–22.

HARPER, C. T. 1967. The geological interpretation of potassium-argon ages of metamorphic rocks from the Scottish Caledonides. *Scottish Journal of Geology*, **3**, 46–66.

HARRISON, T. M., ARMSTRONG, R. L., NAESER, C. W. & HARAKAL, J. E. 1979. Geochronology and thermal history of the Coast Plutonic complex, near Prince Rupert, B. C. *Canadian Journal of Earth Sciences*, **16**, 400–10.

—— & MCDOUGALL, I. 1980a. Investigations of an intrusive contact, northwest Nelson, New Zealand I: Thermal, chronological and isotopic constraints. *Geochimica et Cosmochimica Acta*, **44**, 1985–2003.

—— & —— 1980b. Investigations of an intrusive contact, northwest Nelson, New Zealand II: Diffusion of radiogenic and excess ^{40}Ar in hornblende revealed by ^{40}Ar/^{39}Ar age spectrum analysis. *Geochimica et Cosmochimica Acta*, **44**, 2005–20.

HEITZMANN, P. 1975. Zur Metamorphose und Tektonik im südöstlichen Teil der Lepontinischen Alpen. *Schweizerische Mineralogische und Petrographische Mitteilungen*, **55**, 467–522.

HUNZIKER, J. C. 1969. Rb–Sr Altersbestimmungen aus den Walliser Alpen. Hellglimmer und Gesamtgesteinsalterwerte. *Eclogae geologicae Helvetiae*, **62**, 527–42.

—— 1970. Polymetamorphism in the Monte Rosa, western Alps. *Eclogae geologicae Helvetiae*, **63**, 151–61.

—— 1974. Rb–Sr and K–Ar age determination and the Alpine tectonic history of the western Alps. *Memorie degli Istituti Geologia e Mineralogia dell' Università di Padova*, **31**, p. 54.

HURFORD, A. J. 1986a. Application of the fission track dating method to young sediments: principles, methodology and examples. *In*: HURFORD, A. J., JÄGER, E. & TEN CATE, J. A. M. (eds). *Dating Young Sediments: Proceedings of the Workshop, Beijing, China, September 1985*, 199–233. CCOP–UNESCO Technical Publication **16**, Bangkok.

—— 1986b. Cooling and uplift patterns in the Lepontine Alps, south–central Switzerland, and an age of vertical movement on the Insubric fault line. *Contributions to Mineralogy and Petrology*, **93**, 413–27.

—— & GREEN, P. F. 1982. A users' guide to fission track dating calibration. *Earth and Planetary Science Letters*, **59**, 343–54.

—— & —— 1983. The zeta age calibration of fission track dating. *Isotope Geoscience*, **1**, 285–317.

—— & HUNZIKER, J. C. 1985. Alpine cooling history of the Monte Mucrone eclogites (Sesia– Lanzo zone): fission track evidence. *Schweizerische Mineralogische und Petrographische Mitteilungen*, **65**, 325–34.

——, —— & STÖCKHERT, B. (in press). Thermal history of the western Alps: a reconstruction. *Contributions to Mineralogy and Petrology*.

—— & —— (in press). A revised thermal history for the Gran Paradiso Massif. *Contributions to Mineralogy and Petrology*.

JÄGER, E. 1970. Rb/Sr systems in different degrees of metamorphism. *Eclogae geologicae Helvetiae*, **63**, 163–72.

—— 1973. Die alpine Orogenese im Lichte der radiometrischen Altersbestimmung. *Eclogae Geologicae Helvetiae*, **66**, 11–21.

—— 1979. The Rb–Sr method. *In*: JÄGER, E. & HUNZIKER, J. C. *Lectures in Isotope Geology*, 11–26, Springer-Verlag, Berlin.

——, NIGGLI, E. & WENK, E. 1967. Rb–Sr Altersbestimmungen an Glimmern der Zentralalpen. *Beiträge zur Geologischen Karte der Schweiz NF 134*, Schweizerischen Geologischen Kommission, Berne, p. 67.

KOHN, B. P., SHAGAM, R., BANKS, P. O. & BURKLEY, L. A. 1984. Mesozoic–Pleistocene fission track ages on rocks of the Venezuelan Andes and their tectonic implications. *Memoir of the Geological Society of America*, **162**, 365–84.

KÖPPEL, V. & GRÜNENFELDER, M. 1975. Concordant U–Pb ages of monazite and xenotime from the central Alps and the timing of high temperature Alpine metamorphism: a preliminary report. *Schweizerische Mineralogische und Petrographische Mitteilungen*, **55**, 129–32.

—— & —— 1978. The significance of monazite U–Pb ages: examples from the Lepontine area of the Swiss Alps. *In*: ZARTMAN, R. E. (ed). *Short papers of the 4th International Conference of Geochronology Cosmochronology and Isotope Geology*, 226–7. US Geological Survey Open File Report **78–701**.

KRÄHENBÜHL, R. 1984. *Petrographisch Geologische*

Untersuchungen in der Silvretta Masse vom Flüelawisshorn zum Gorihorn und Rosställispitz (Kanton Graubünden). Unpublished diploma thesis, University of Berne.

KRECZY, L. 1981. *Seriengliederung, Metamorphose und Altersbestimmung in der Region der Thialspitze SW Landeck, Tirol*. Unpublished Diploma Thesis, University of Freiburg-in-Breisgau.

KRISHNASWAMI, S., LAL, D., PRABHU, N. & MACDOUGALL, D. 1974. Characteristics of fission tracks in zircon: applications to geochronology and cosmology. *Earth and Planetary Science Letters*, **22**, 51–9.

LASLETT, G. M., GLEADOW, A. J. W. & DUDDY, I. R. 1984. The relationship between fission track length and track density in apatite. *Nuclear Tracks*, **9**, 29–38.

——, GREEN, P. F., DUDDY, I. R. & GLEADOW, A. J. W. 1987. Thermal annealing of fission tracks in apatite 2: a quantitative analysis. *Chemical Geology (Isotope Geoscience Section)*, **65**, 1–15.

MAGGETTI, M. 1986. Petrographische Kriterien zur Alterstellung des Hochdruckereignisses in der Silvrettamasse. *Fortschritte der Mineralogie Beiheft*, **1**, 64, 103

—— & GALETTI, G. 1984. Chemie und geotektonische Position von Metabasiten aus dem Südosten der Silvretta (Schweiz). *Schweizerische Mineralogische und Petrographische Mitteilungen*, **64**, 423–50.

——, —— & STOSCH, H. G. 1987. Eclogites from the Silvretta nappe (Switzerland): geochemical constraints on the nature and geotectonic setting of their protoliths. *Chemical Geology*, **64**, 319–34.

MARQUER, D. 1987. *Transfert de matière et déformation progressive des granitoïdes*. Memoires et Documents du Centre Armoricain d'Etude Structurale de Socles, University of Rennes, France.

MILNES, A. G. & SCHMUTZ, H. U. 1978. Structure and history of the Suretta nappe (Pennine zone, central Alps)–a field study. *Eclogae geologicae Helvetiae*, **71**, 19–33.

—— & PFIFFNER, O. A. 1980. Tectonic evolution of the central Alps in the cross section St Gallen–Como. *Eclogae geologicae Helvetiae*, **73**, 619–33.

MOORE, M. E., GLEADOW, A. J. W. & LOVERING, J. F. 1986. Thermal evolution of continental margins: new evidence from fission tracks in basement apatites from southeastern Australia. *Earth and Planetary Science Letters*, **78**, 255–70.

NAESER, C. W. 1979. Fission track dating and geologic annealing of fission tracks. *In*: JÄGER, E & HUNZIKER, J. C. *Lectures in Isotope Geology* 154–69. Springer-Verlag, Berlin.

—— & FAUL, H. 1969. Fission track annealing in apatite and sphene. *Journal of Geophysical Research*, **74**, 705–10.

NEUGEBAUER, H. J., BROTZ, R. & RYBACH, L. 1980. Recent crustal uplift and the present stress-field of the Alps along the Swiss Geotraverse Basel–Chiasso. *Eclogae geologicae Helvetiae*, **73**, 489–500.

NIGGLI, E. 1960. Mineral-zonen der alpinen Metamorphose in den Schweizer Alpen. *International Geological Congress Copenhagen*, 132–8.

OBERHÄNSLI, R., HUNZIKER, J. C., MARTINOTTI, G. & STERN, W. B. 1985. Geochemistry, geochronology and petrology of Monte Mucrone: an example of Eoalpine eclogitization of Permian granitoids in the Sesia–Lanzo zone, western Alps, Italy. *Chemical Geology (Isotope Geoscience Section)*, **52**, 165–84.

OBERHAUSER, R. 1983. Mikofossilfunde im Nordwestteil des Unterengadiner Fensters sowie im Verspalaflysch des Rätikon. *Jahrbuch der geologischen Bundesanstalt (Wien)*, **126**, 71–93.

ONSTOTT, T. C. & PEACOCK, M. W. 1987. Argon retentivity of hornblendes: a field experiment in a slowly cooled metamorphic terrane. *Geochimica et Cosmochimica Acta* **51**, 2891–903.

PARRISH, R. 1983. Cenozoic thermal evolution and tectonics of the Coast Mountains of British Columbia 1: Fission track dating, apparent uplift rates and patterns of uplift. *Tectonics*, **2**, 601–31.

—— 1985. Some cautions which should be exercised when interpreting fission track and other dates with regard to uplift rate calculations (abstract). *Nuclear Tracks*, **10**, 425.

PASSCHIER, C. W., URAI, J. L., VAN LOON, J. & WILLIAMS, P. F. 1981. Structural geology of the central Sesia–Lanzo zone. *Geologie en Mijnbouw*, **60**, 497–507.

PIRKL, H. 1980. Die westlichen Zentralalpen (vom der Silvretta zum Brenner). *In*: OBERHAUSER, R. (ed). *Der geologische Aufbau Österreichs*, 332–47. Springer-Verlag, Vienna, New York.

PURDY J. W. & JÄGER, E. 1976. K–Ar ages on rock-forming minerals from the central Alps. *Memorie degli Istituti di Geologia e Mineralogia dell' Università di Padova*, **30**, 30 pp.

RUBIE, D. C. 1984. A thermal-tectonic model for high pressure metamorphism and deformation in the Sesia zone, western Alps. *Journal of Geology*, **92**, 21–36.

RUDOLF, J. 1982. Tieferes Tertiär im oberen Fimbertal. *Neues Jahrbuch für Geologie und Paläontologie, Monatshefte*, **181–3.**

SATIR, M. 1975. Die Entwicklungsgeschichte der westlichen Hohen Tauern und der südlichen Ötztalmasse aufgrund radiometrischer Altersbestimmungen. *Memorie degli Istituti Geologia e Mineralogia dell'Università di Padova*, **30**, 84 pp.

SCHEURING, B., AHRENDT, H., HUNZIKER, J. C. & ZINGG, A. 1974. Palaeobotanical and geochronological evidence for the Alpine age of metamorphism in the Sesia zone. *Geologische Rundschau*, **63**, 305–26.

SCHMID, S. M., ZINGG, A & HANDY, M. 1987. The kinematics of movement along the Insubric Line and the emplacement of the Ivrea zone. *Tectonophysics*, **135**, 47–66.

SIEGENTHALER, R. 1984. *Alter und Geochemie von Glimmern und Feldspäten*. Unpublished PhD thesis, University of Berne.

STEIGER, R. H. 1964. Dating of orogenic phases in the central Alps by K–Ar ages of hornblende,

Journal of Geophysical Research, **69**, 5407–21.
—— & Jäger, E. 1977. Subcommission on geochronology: convention on the use of decay constants in geo- and cosmochronology. *Earth and Planetary Science Letters*, **36**, 359–62.
Steinitz, G. & Jäger, E. 1981 Rb–Sr and K–Ar studies on rocks from the Suretta nappe, eastern Switzerland. *Schweizerische Mineralogische und Petrographische Mitteilungen*, **61**, 121–31.
Stöckhert, B., Jäger, E. & Voll, G. 1986. K–Ar age determinations on phengite from the internal part of the Sesia zone, western Alps, Italy. *Contributions to Mineralogy and Petrology*, 92, 456–70.
Thöni, M. 1981. Degree and evolution of the Alpine metamorphism in the Austro–Alpine unit west of the Hohe Tauern in the light of K–Ar and Rb–Sr age determinations on micas. *Jahrbuch der geologischen Bundesanstalt (Austria)* **124**, 111–74.
—— 1982. Der Einfluss der kretazischen Metamorphose im Westabschnitt der ostalpinen Einheit: Interpretation geochronologischer Daten. *Mitteilungen der Gesellschaft der Geologie- und Bergbaustudenten in Österreich (Wien)*, **28**, 17–34.
—— 1985. *Die Rb–Sr Kleinbereichsisochronen-Methode: ein unverlässliches Instrument zur Datierung geologischer Ereignisse in polymetamorphen Gebeieten?* Unpublished Habilitationschrift, University of Vienna.
Trümpy, R. 1969. Palaeotectonic evolution of the central and western Alps. *Geological Society of America Bulletin*, **71**, 843–908.
—— 1980. *Geology of Switzerland–a guide book. Part A: an outline of the geology of Switzerland.* Wepf, Basel.
Turner, D. L. & Forbes, R. B. 1976. K–Ar studies in two deep basement drillholes: a new estimate of argon blocking temperature for biotite. *Transactions of the American Geophysical Union*, **57**, 353.
Verschure, R. H., Andriessen, P. A. M., Boelrijk, N. A. I. M., Hebeda, E. H., Maijer, C., Priem, H. N. A. & Verdurmen, E. A. Th. 1980. On the thermal stability of Rb/Sr and K/Ar biotite systems: evidence from coexisting Sveconorwegian (ca.870 Ma) and Caledonian (ca.400 Ma) biotites from SW Norway. *Contributions to Mineralogy and Petrology*, **74**, 245–52.
Vogler, W. S. & Voll, G. 1981. Deformation and metamorphism at the south margin of the Alps, East of Bellinzona, Switzerland. *Geologische Rundschau*, **70**, 1232–62.
Wagner, G. A. 1968. Fission track dating of apatites. *Earth and Planetary Science Letters*, **4**, 411–15.
—— (in press) Apatite fission track geochronothermometer to 60°C: projected length studies. *Chemical Geology (Isotope Geoscience Section)*.
—— & Storzer, D. S. 1970. Die Interpretation von Spaltspurenaltern am Beispiel von natürlichen Gläsern, Apatiten und Zirkonen. *Eclogae geologicae Helvetiae*, **63**, 335–44.
—— & Reimer, G. A. 1972. Fission track tectonics: the tectonic interpretation of fission track ages. *Earth and Planetary Science Letters*, **14**, 263–8.
——, Reimer, G. M. & Jäger E. 1977. Cooling ages derived by apatite fission track, mica Rb/Sr and K/Ar dating: the uplift and cooling history of the central Alps. *Memorie degli Istituti di Geologia e Mineralogia dell' Università di Padova*, **30**, 27 pp.
——, Miller, D. S. & Jäger, E. 1979. Fission track ages on apatite of Bergell rocks from central Alps and Bergell boulders in Oligocene sediments. *Earth and Planetary Science Letters*, **45**, 355–60.
Wenk, E. 1962. Plagioklas als Indexmineral in den Zentralalpen. *Schweizerische Mineralogische und Petrographische Mitteilungen*, **42**, 139–53.
Wenk, H. R. 1974. Two episodes of high grade metamorphism in the northern Bergell Alps. *Schweizerische Mineralogische und Petrographische Mitteilungen*, **54**, 555–65.
Werner 1980. Probleme der Geothermik in Bereich der Schweizer Zentralalpen. *Eclogae geologicae Helvetiae*, **73**, 513–25.
White, S. H. & Green, P. F. 1986. Tectonic development of the Alpine fault zone, New Zealand: a fission track study. *Geology*, **14**, 124–7.
Zeitler, P. K., Tahirkheli, R. A. K., Naeser, C. W. & Johnson, N. M. 1982. Unroofing history of a suture zone in the Himalaya of Pakistan by means of fission track annealing ages. *Earth and Planetary Science Letters*, **57**, 227–40.
—— & Wijbrans, J. R. 1985/86. A reassessment appraised: comment on 'Hornblende K–Ar ages and the climax of Tertiary metamorphism in the Lepontine Alps (south-central Switzerland): an old problem reassessed' by Alexander Deutsch and Rudolf H. Steiger. *Earth and Planetary Science Letters*, **76**, 390–92.
Zingg, A. 1983. The Ivrea and Strona–Ceneri zones (southern Alps, Ticino and N. Italy)–a review. *Schweizerische Mineralogische und Petrographische Mitteilungen*, **63**, 361–92.
Zurflüh, E. 1961. *Zur geologie des Monte Spluga,* Unpublished PhD thesis, ETH, Zürich.

A. J. Hurford, University of London Fission Track Research Group, Department of Geological Sciences, University College, Gower Street, London WC1E 6BT, UK.
M. Flisch & E. Jäger, Laboratory for Isotope Geology, University of Berne, Erlachstrasse 9a, CH-3012 Berne, Switzerland.

A review of palaeomagnetic and magnetic anisotropy results from the Alps

F. Heller, W. Lowrie & A. M. Hirt

SUMMARY: Numerous palaeomagnetic studies in the Alpine region have documented the effects of tectonic displacement and rotation. Regions in the southern Alps are characterized by palaeomagnetic directions which are more similar to directions predicted by the African path of apparent polar wander than by the Eurasian path. Attempts to define the extent of the region with this African affinity are found in the internal zone of the western Alps, and also in the Lepontine of southern Switzerland, yet directions with European affinity have been reported from S of the Insubric Line. Although the northern calcareous Alps are of southern Tethyan origin, palaeomagnetic directions reveal that these allochthonous units have undergone substantial rotations, in some places clockwise and in others anticlockwise, during emplacement. Several palaeomagnetic studies in the eastern and central Alps show rotated directions due to strong local tectonic effects. In the western Alps, European style directions are found in the undeformed rocks of the Dôme de Barrot and the external Dauphinois zone, whereas directions from the internal zone tend to have African affinity.

Deformation produces not only a deflection of the stable palaeomagnetic vector in a rock. If the deformation is penetrative, grain rotations falsify the palaeomagnetic vector. Preferential grain alignment caused by deformation produces a magnetic anisotropy which can be measured easily and quickly in many rocks. Analogously to the strain fabric, the magnetic fabric can be described by an anisotropy ellipsoid. Several studies in the Alps have confirmed that the principal axes of the magnetic anisotropy are aligned with the corresponding axes of the strain ellipsoid. In deformed Permian red beds in the Alpes Maritimes the anisotropy ellipsoid changes shape and orientation like the strain ellipsoid during progressive deformation. A good qualitative agreement between magnetic fabric and petrofabric is also found in Jurassic shales and slates from the Dauphinois zone of the western Alps, where magnetic lineations often correspond to the direction of nappe transport. In strongly deformed Permo–Triassic red beds from the Lower Glarus nappe complex in central Switzerland the quantitative relationship between strain and magnetic fabric was used to derive equivalent strain ellipsoids in rocks that had no conventional strain markers. In the eastern Alps, the sense of motion on the Insubric Line has been postulated from magnetic fabric observations. Magnetic anisotropy results have been used in attempts to correct palaeomagnetic vectors for the effect of deformation. This technique may be viable in weakly deformed rocks, but it fails in strongly deformed samples; the limits of its applicability have not yet been determined satisfactorily.

Palaeomagnetic results have provided important, sometimes provocative, information about the geodynamic history of the Mediterranean region. Caught between the colliding African and European Plates, the area developed in a complex manner which involved microplate rotations in some domains, as for example in Corsica and Sardinia. The regions involved directly in the collision suffered extensive overthrusting. In rocks from the allochthonous overthrust units the palaeomagnetic method is a useful way to detect rotations, by means of the accompanying deflection of the declination of the remanent magnetization.

The interpretation of palaeomagnetic results in the form of rotations or translations necessitates comparison with directions from autochthonous regions. In the case of the Alpine–Mediterranean region, reference directions are deduced from the African and European apparent polar wander (APW) paths. There are enough data from Eurasia to define an APW path, but the data from Africa are sparse. To increase the data available, Van der Voo & French (1974) reconstructed the positions of the circum-Atlantic continents for Pangaea at various stages of opening of the Atlantic ocean, and combined available palaeomagnetic data from different continents to compute optimized pole positions from the late Carboniferous to the present. Another method used to construct an artificial African APW path consists of stepwise rotating the well-defined North American APW path about the poles of rotation for the opening of the Atlantic (VandenBerg 1979). The reference paths for Eurasia and Africa are shown in Fig. 1.

A fundamental problem arises, because the

From: Coward, M. P., Dietrich, D. & Park, R. G. (eds), 1989, *Alpine Tectonics*, Geological Society Special Publication No. 45, pp. 399–420.

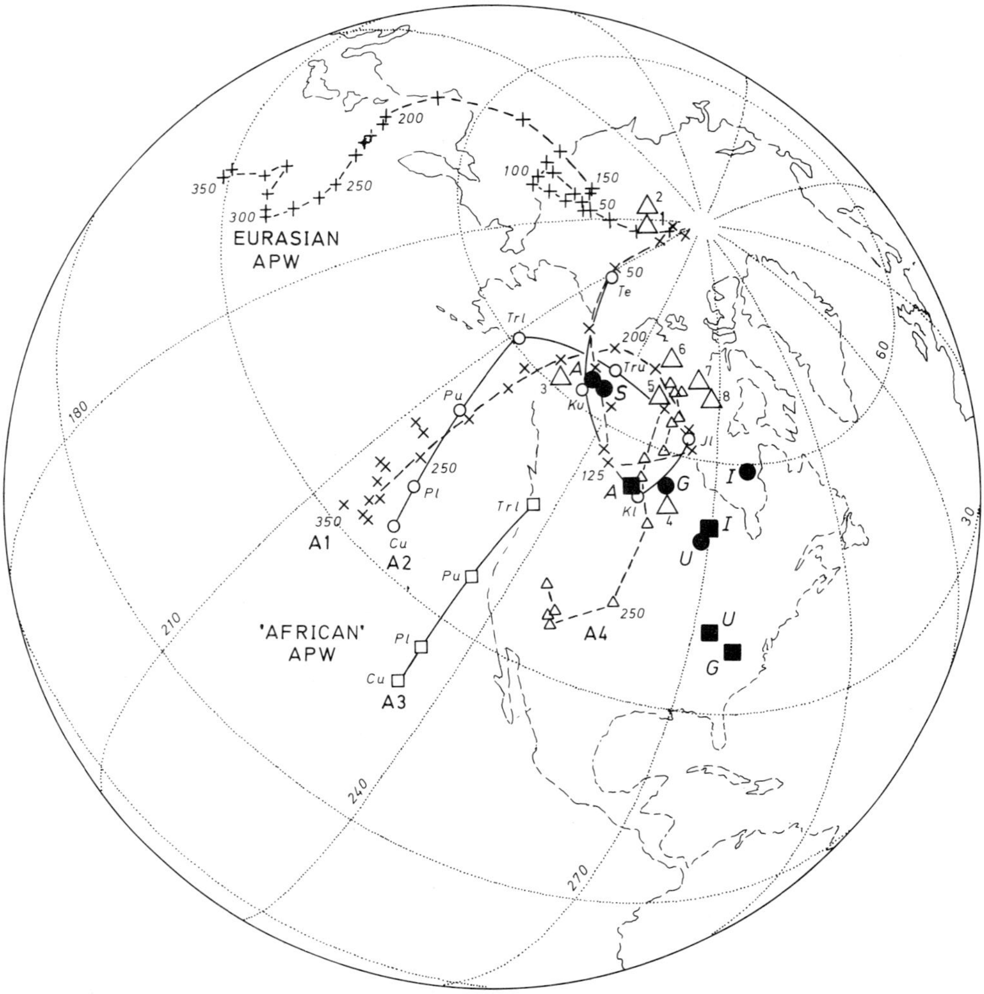

FIG. 1. Apparent polar wander (APW) reference paths for Eurasia (Irving 1977) and Africa, and palaeomagnetic pole positions measured on the Adriatic promontory. 'African' APW paths: (A1) obtained by rotating the North American APW path of Irving (1977) about the poles of opening of the Atlantic; (A2) assuming the Bullard fit of circum-Atlantic continents, and (A3) assuming an alternative fit (Van der Voo & French 1974); (A4) Gondwanaland (Irving 1977). Numbers along the APW refer to ages (in Ma) and letters denote geological time intervals using conventional nomenclature. Large open triangles mark pole positions calculated by Vanden Berg (1979) from African palaeomagnetic data for the following ages: (1) Miocene–Pliocene, (2) Eocene to Miocene, (3) Upper Cretaceous, (4) Lower Cretaceous, (5) Lower to Middle Jurassic, (6) Upper Triassic to Lower Jurassic, (7) Lower to middle Triassic, (8) Upper Permian to Lower Triassic. Upper and lower Cretaceous pole positions for sites on the Adriatic promontory (A = Southern Alps, G = Gargano, I = Istria, S = Sicily, U = Umbria) are shown by solid circles and squares, respectively (after Lowrie 1986).

reference paths for both Africa and Europe are often too imprecisely defined for the purposes of a tectonic study. The errors associated with palaeomagnetic pole positions are generally around 5° for a well-defined pole. Some parts of the African and European paths are much less precisely known, and have errors of more than 10°. The analytical methods in palaeomagnetic research have been gradually refined, and the identification of stable directions through application of progressive demagnetization techniques has become firmly established. However, a few early palaeomagnetic poles were determined without these more stringent laboratory methods and must now be considered suspect. Even a good palaeomagnetic study can rarely define a pole position to better than 5°. Thus the cumulative errors of observation and the reference curves limit the resolution with which translations and rotations can be determined.

Several comprehensive reviews have been written about palaeomagnetic results from the entire Alpine–Mediterranean region, including the Alpine chain (Zijderveld & Van der Voo 1973, Lowrie 1980, VandenBerg & Zijderveld 1982, Westphal *et al.* 1986, Channell 1986). The overview of palaeomagnetic results from the Alps given here enlarges the known picture with some recent studies. In addition, a large part of this review is a summary of magnetic fabric investigations in deformed rocks from the Alpine region, which until now have not been included in reviews of magnetic data.

The review consists of two parts. First, conventional palaeomagnetic results are reviewed from the southern Alps, eastern, central and western Alps. Then the results of magnetic fabric investigations in these areas are presented.

Palaeomagnetic results from the southern Alps

The Permian and Triassic rocks of the southern Alps were for many years the main focus of palaeomagnetic studies in Italy. Although the number of reference directions from stable Europe north of the Alps was then small, palaeomagnetic directions found in different regions of the southern Alps were consistently different from the European directions (Fig. 2) and suggested that the southern Alps were part of the Adriatic microplate (Dewey *et al.* 1973). The magnetization directions found in Permian porphyries near Bolzano (Fig. 2: Site 1) were steeper in inclination than expected, and the declinations were rotated counter-clockwise by more than 50° (Van Hilten 1960, 1962, Zijderveld *et al.* 1970). Palaeomagnetic poles calculated for Permian and Triassic redbeds and volcanics from the Vicentinian Alps (Fig. 2: Site 2) showed these same characteristics (de Boer 1963, 1965). An early, controversial interpretation was offered by Van Hilten (1964), who postulated that there had been a dextral megatectonic shear (which came to be known as the 'Tethys twist') between Europe and Africa in Jurassic time. As palaeomagnetic data became more abundant it became clear that the directions in the southern Alps were more similar to results from Africa than from stable Europe. However, in an evaluation of the palaeomagnetic data from the continents surrounding the Atlantic ocean, Van der Voo & French (1974) found that they were more consistent when an alternative reconstruction of the circum-Atlantic continents to the model of Bullard *et al.* (1965) was used. The better fitting model required a major dextral shear between Africa and Europe in the Triassic. Irving (1977), in an analysis of palaeomagnetic pole positions, inferred a reconstruction of the relative positions of Europe, Africa and North America, which required an even greater shear, several thousand kilometres in length.

The African affinity of palaeomagnetic directions from the southern Alps has been documented in many individual studies and summarized in earlier reviews. The palaeomagnetic pole positions from Permian porphyritic rocks from the Lugano basin (Fig. 2: Sites 3, 4; Van Hilten & Zijderveld 1966, Heiniger 1979), from Permian and Triassic rocks from the Dolomites and the Lombardy Alps (Fig. 2: Sites 5, 6, Zijderveld & De Jong 1969, Manzoni 1970, Heiniger 1979), and from Permo–Triassic red sandstones and volcanics from the eastern margin of the Adriatic plate near Ljubljana (Fig. 2: Site 7, Soffel *et al.* 1983) are fairly consistent. However, they lie far from the European apparent polar wander path, and are quite close to the African curve, even though this is in part rather poorly known (Fig. 1).

Apart from studies of Permian and Triassic volcanics and red beds, palaeomagnetic results from the southern Alps have been obtained in Cretaceous and Late Jurassic limestones and some Tertiary basalts. In three regions in the southern Dolomites, Channell & Tarling (1975) reported directions from 17 sites (Fig. 2: Site 8) which were concordant with values predicted from the African rather than the European apparent polar wander path, and implied that little rotation had taken place between the units. VandenBerg & Wonders (1976, 1980) found that palaeomagnetic poles for Upper Cretaceous limestones from the Vicentinian Alps were not significantly different from the African path. The African association of the palaeomagnetic directions in the southern Alps also has been observed in several magnetostratigraphic studies in limestones from the Belluno basin (Fig. 2: Site 9, Channell *et al.* 1979a, Lowrie *et al.* 1980, Channell & Medizza, 1981).

The agreement is restricted to sites which have not suffered severe tectonic displacement, and is better for the Belluno basin than for the Lombardian basin farther to the W. For

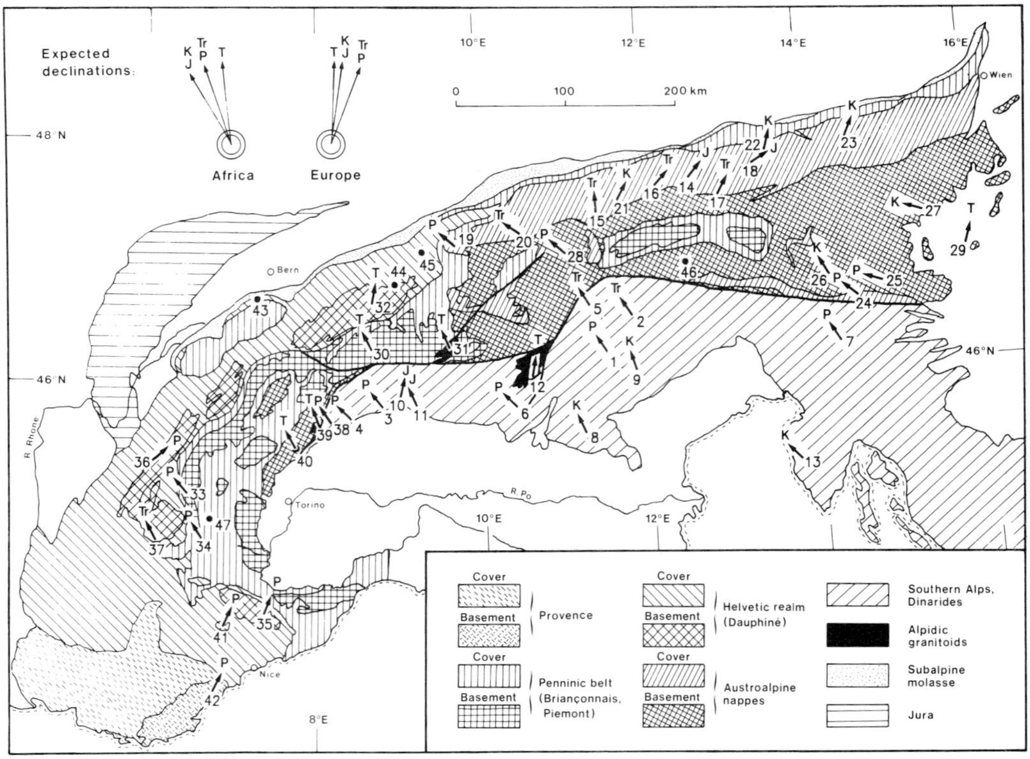

FIG. 2. Declinations of palaeomagnetic vectors at sites in different tectonic units of the Alpine domain. The northwesterly directions with an 'African' affinity are clearly distinct from the north-northeasterly 'Eurasian' directions. Letters associated with each direction denote the age of the formation studied, using conventional nomenclature. Numbers identify sites referred to in the text. Geological map after Lemoine (1978).

example, Lower Jurassic limestones from the Breggia gorge in southern Switzerland (Fig. 2: Site 10) had a clearly defined polarity pattern but showed anomalous clockwise rotated directions; the effects of a major drag fault could not be corrected by simple rigid body rotation about a horizontal axis (Horner & Heller 1983). However, the mean direction of the nearby Lower Jurassic section of Alpe Turati (Fig. 2: Site 11) was rotated anticlockwise in close agreement with the expected African pattern. Permo-Triassic beds which were metamorphosed in the contact aureole of the Eocene to Oligocene intrusions of the Adamello massif (Fig. 2: Site 12) have characteristic magnetization components with apparent European affinity (Kipfer & Heller 1988). Evidently, this area at the northern margin of the southern Alps underwent anticlockwise rotation before the emplacement of the Adamello intrusions.

The African association of palaeomagnetic data from the southern Alps is taken as supporting evidence that the Periadriatic region acted as a promontory to the African Plate during the Mesozoic evolution of the Mediterranean and was an instrumental factor in the Alpine orogeny (Channell & Horvath 1976, Channell *et al.* 1979b, Channell 1986). Other parts of the autochthonous Adriatic promontory are found in Gargano, Italy, and Istria, Yugoslavia. Cretaceous palaeomagnetic data from weakly magnetized, autochthonous limestones in Istria (Fig. 2: Site 13, Marton & Veljovic 1983) are rather badly scattered but broadly support the purported role of Istria as part of the Adriatic promontory in the geodynamic evolution. However, the declinations of Cretaceous limestones from Istria and Gargano are rotated by about 17° anticlockwise relative to coeval directions from the southern Alps (Lowrie 1986), which may imply that these regions are not autochthonous, or that the Adriatic promontory became decoupled from the African plate in the late Tertiary (VandenBerg 1979).

The eastern Alps

The northern calcareous Alps

Most of the eastern Alpine palaeomagnetic data come from the northern calcareous Alps, which consist of a series of large décollement nappes of variable lateral extent and thickness. They rest allochthonously on top of Penninic basement and were transported into their present position since the late Eocene (Gwinner 1971).

The earliest palaeomagnetic results from the northern calcareous Alps (Hargraves & Fischer 1959) were obtained with methods customary at the time, which did not include the use of modern demagnetization techniques. Consistent directions of the natural remanent magnetization were observed in early Jurassic red limestones and Middle Jurassic maroon radiolarites from a section in the Staufen–Höllengebirge thrust sheet (Fig. 2: Site 14). The formation mean directions were interpreted as being in close agreement with the values predicted from the European polar wander path available at that time. Subsequently, the quantity of data has increased, the quality has improved and the reference polar wander paths have become more securely established. Comparison with the recent European polar wander curve of Westphal *et al.* (1986) indicates a clockwise rotation of about 20° of the Jurassic sites in the Northern Calcareous Alps.

Permo–Triassic (Scythian) red sandstones from different units in the Inntal and Staufen–Höllengebirge nappes have been investigated by Soffel (1979). The characteristic remanent magnetization directions (Fig. 2: Sites 15, 16) were isolated by progressive alternating field or thermal cleaning. Comparison with directions calculated for the sampling locations from the European polar wander path showed that the inclinations were too steep for rocks of that age. The tectonically undeformed rocks have evidently been overprinted with a later remagnetization, perhaps during the late Cretaceous or early to Middle Tertiary. Supposing an Alpine age of remagnetization, both slight clockwise and anticlockwise rotations were interpreted for these areas. A third sandstone unit from Leogang in the Werfen nappe (Fig. 2: Site 17) yielded an early Triassic direction with almost no rotation relative to stable Europe (Soffel 1979).

The Liassic red nodular limestones of the Osterhorn group (Staufen–Höllengebirge nappe) in the eastern northern calcareous Alps were found to carry a stable Jurassic magnetization (Fig. 2: Site 18), which showed a large clockwise rotation of about 45° relative to the Eurasian plate (Mauritsch & Frisch 1978). On the other hand south-easterly declinations were measured in the Permian quartz porphyries of the Lechtal nappe near St Anton (Fig. 2: Site 19), which suggest that the northern calcareous Alps here in the W may have undergone an anticlockwise rotation by 60° since the Permian (Soffel 1975). Similar rotations have been reported at several sites in the northern Hauptdolomit near Bludenz (Fig. 2: Site 20, Becke 1980).

The youngest primary palaeomagnetic directions in the northern calcareous Alps have been reported from flyschoid sediments of the Upper Cretaceous/Lower Tertiary Gosau formation (Becke & Mauritsch 1985). Hemipelagic limestones yielded coherent patterns of normal and reversed magnetization, especially across the well developed Cretaceous/Tertiary boundary section at Elendgraben near Gosau (Preisinger *et al.* 1986). The mean direction of magnetization of three sections near Brandenberg, Gosau and Weisswasser (Fig. 2: Sites 21–23, Mauritsch (1986), which are separated by a distance of about 200 km, indicates a late clockwise rotation of 10°–15° with respect to stable Europe.

The interpretation of the palaeomagnetic results from the northern calcareous Alps is somewhat controversial. Mauritsch & Frisch (1980) view this region as a former part of the Adriatic promontory, which disintegrated during the Alpine orogeny in the late Cretaceous or early Tertiary into three major units, corresponding to the Adriatic block, the south Alpine block and the Austro–Alpine–Carpathian block. Each block rotated differently with respect to Eurasia. The mainly clockwise rotation in the eastern part and the anticlockwise rotation in the western part are explained by deformation of a sedimentary basin, which was originally V-shaped southwards, into the present rectilinear body of the northern calcareous Alps (Becke & Mauritsch 1985). According to Soffel (1979), the pattern of rotations does not reflect any systematic bending, but can be explained by local effects which may be typical of a specific nappe or part of it. This applies to the three different directions found in the three regions investigated in the Staufen–Höllengebirge nappe. VandenBerg & Zijderveld (1982) tentatively speculated that the clockwise rotated declinations were suggestive of a European rather than African origin. However, this proposal disregards geological evidence for a southern Tethyan origin of the northern calcareous Alps.

A definitive explanation for these rotations can not yet be made, because the available palaeomagnetic data are too scarce. More high quality studies are necessary to document and characterize clearly and in detail the significance of the components of natural remanence measured. The age of the magnetization must be established more conclusively, the structural corrections must be appropriate for the structural complexity of the region, and the possible effects of deformation on the remanent magnetization must be evaluated. The northern calcareous Alps are, in this respect, an interesting field for comprehensive palaeomagnetic and tectonic studies.

The Austro-Alpine nappes

Palaeomagnetic data from the Upper Austro-Alpine nappes to the south of the northern calcareous Alps show anticlockwise rotations (Agnoli *et al.* 1984). Permian sediments at Christofberg and St Paul (Figure 2: Sites 24, 25) have been rotated by about 80°; the late Cretaceous Gosau sediments at Wietersdorf and near Graz (Fig. 2: Sites 26, 27) have both normal and reversed magnetizations, rotated by about 50° and 70° anticlockwise, respectively. These results suggest a two-phase emplacement of these nappes, similar to the history of the northern calcareous Alps. Diabase intrusions near to, but north of the Insubric Line (Fig. 2: Site 28) yielded consistent directions with shallow inclinations, as expected for late Palaeozoic rocks (Förster *et al.* 1975, Marton 1987). The directions suggest an anticlockwise rotation of about 50° relative to Europe but no rotation relative to the southern Alps and Africa. These results agree with the geological evidence that the Austro-Alpine Nappes belong to the Apulian margin. This observation has to be taken into account in interpretations of the geological nature of the Periadriatic Line and the movements which occurred along it. All the tectonic movements ended before the emplacement in the Miocene of the Styrian trachyandesites (Fig. 2: Site 29), in which the palaeomagnetic directions show no indications of tectonic displacement (Pohl & Soffel 1982).

Palaeomagnetic results from the central Alps

The Lepontine area in the central Alps underwent high grade metamorphism between late Eocene and early Miocene, caused by deep burial during the formation of the Alpine nappes. At the same time granitoid bodies such as the Bergell intrusion were emplaced. The metamorphic zoning is simple and the end of the metamorphic events postdates the tectonic deformation.

Several palaeomagnetic studies have been carried out on the metamorphic rocks (mainly gneisses) of the Lepontine area, and on the Bergell granite (Fig. 2: Sites 30, 31, Heller 1971, 1973, 1977, 1980). A steeply N-dipping magnetic foliation observed in the Bergell granite deflected the declinations, and necessitated a correction. Magnetic anisotropy, characterized by a horizontal magnetic foliation without a preferred lineation, was noticed in the Val Antigorio and Val Leventina sites, but had no detectable influence on the declinations. The directions of magnetization in the Penninic nappes in the Lepontine area are rotated anticlockwise by 27° with respect to contemporaneous directions from stable Europe (Fig. 3). This rotation, which is very similar to the late Tertiary rotation of the Adriatic promontory (VandenBerg 1979), was ascribed to a left-lateral rotation of the Lepontine area caused by dextral shear motions along the Insubric Line and possibly between the already formed Alpine chain and stable Europe. Declinations at two sites in the Aar massif further to the N (Fig. 2: Site 32) were compatible with contemporaneous Miocene directions in stable Europe, although the inclinations were anomalously steep.

Palaeomagnetic directions in Permo–Triassic red beds of the Lower Glarus nappe complex of the Helvetic zone in eastern Switzerland were found to be carried by hematite (Hirt *et al.* 1986). These rocks possess a strong anisotropy, with a magnetic foliation close to the slaty cleavage and a N–S lineation. This implies that the ferromagnetic grains were realigned during the Alpine deformation. The characteristic directions showed large between-site scatter, but the site mean directions were streaked along a great circle. As a possible interpretation for this distribution of directions, it was suggested that they were the result of penetrative simple shear deformation associated with the nappe-internal deformation.

Complicated remanences are found in the Blegi oolites of Jurassic age in the autochthonous sedimentary cover of the Aar Massif. The risk of mineralogical alteration during heating in air necessitated thermal demagnetization in a nitrogen atmosphere. Palaeomagnetic directions from haematitic oolites have N to NE stable end-point directions. In magnetite-chamosite oolites palaeomagnetic directions with steep inclinations (70°–80°) and northerly declinations are found, most probably carried

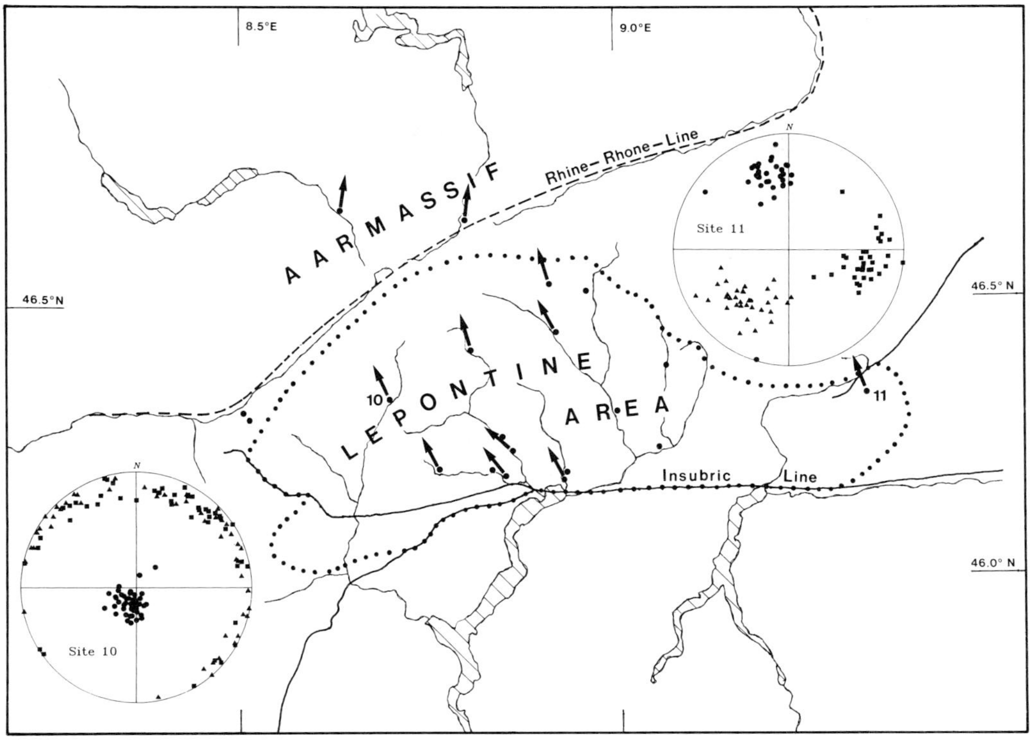

FIG. 3. Directions of magnetization in the Lepontine area are rotated anticlockwise by 27° with respect to stable Europe. Typical European Late Tertiary directions are found in the Aar Massif to the N of the region (from Heller, 1980). Insets show lower hemisphere equal area plots of the the principal axes of magnetic susceptibility in the Val Antigorio (site 10) and in the Bergell intrusion (site 11). Symbols for individual susceptibility axes: squares, k_{max}; triangles, k_{int}; circles, k_{min}.

by pyrrhotite and magnetized during the Alpine deformation. Late Tertiary directions to be expected for this region of the central Alps calculated from the European apparent polar wander path (Irving 1977) should dip to the north at around 60°. Magnetic fabric studies in the Blegi oolite at Färnigen give an extension direction dipping at 70–80° northwards. The observed remanent magnetization is possibly deflected by the anisotropy towards this direction. Alternatively, as suggested also by the two sites of Heller (1980) from the Aar Massif, the steep inclinations may be due to a more regional tilting toward the N related to uplift of the massif.

The Helvetic nappes are affected by low-grade metamorphism which varies in intensity with position in the nappe and with lithology. Kligfield & Channell (1981) investigated the magnetic properties of limestones of Middle and late Jurassic, and Cretaceous age from the Helvetic nappes at sites in which the deformation was accommodated by pressure solution. Fold tests, which rotate the folded layers together with the magnetization vectors back into the original horizontal attitude, showed that the magnetization component isolated by alternating field and thermal cleaning postdates the Neogene folding in these limestones. The dominant magnetization was a viscous component carried by very fine pyrrhotite. The widespread remagnetization of these limestones meant that they could not be used for tectonic interpretations.

Viscous remanent magnetization predominates also in fine-grained turbiditic sandstones from folded beds in the Schlieren flysch (Channell *et al.* 1980). The most stable magnetization components, which reside in single domain magnetite grains so fine sized as to be near to the superparamagnetic threshold, are greatly dispersed after tilt correction. This failed fold test indicates that the remanent magnetization was acquired after Alpine displacement of the flysch units.

Palaeomagnetic results from the western Alps

The Dauphinois and Briançonnais zones

The western Alps are divided into an external (Dauphinois) zone and an internal zone, made up of the Briançonnais and Piémontais zones. The parautochthonous external, Dauphinois zone which is pierced by crystalline massifs, has been displaced by a few tens of kilometres with respect to stable Europe (Vialon *et al.* this volume, Gratier *et al.* this volume). The internal zone is more disturbed and displaced. Palaeomagnetic investigations have been carried out on Permian and Triassic rocks from both the Dauphinois and Brainçonnais zones. Stable palaeomagnetic directions in Permian rhyolitic ignimbrites from the valleys of Ponsonniere and Guil in the Briançonnais zone were rotated anticlockwise by about 30° relative to the European directions (Fig. 2: Site 33, 34), and more closely resembled the 'African' directions found in the southern Alps (Roche & Westphal 1969, Van der Voo & Zijderveld 1969, Westphal 1973). The Guil ignimbrite data were doubted by the authors of these studies because the correction for bedding tilt was uncertain. Henry (1976) has maintained that the Permian volcanics from the Guil valley are too deformed for palaeomagnetic study.

In the Dauphinois zone, Permian directions from undisturbed outcrops in the Argentera massif (Fig. 2: Site 35) agree with those of stable Europe (Bogdanoff & Schott 1977). Samples taken close to the western edge of the Belledonne massif (Fig. 2: Site 36) also had palaeomagnetic directions near to those of stable Europe, but Triassic rocks from the Pelvoux massif (Fig. 2: Site 37) had rotated directions similar to those found in the Briançonnais zone (Westphal 1973, Henry 1980).

The natural remanent magnetizations of Jurassic schists from the sedimentary cover on the east side of the external crystalline massifs of Belledonne and les Grandes Rousses were found to be carried by pyrrhotite, formed after the main Alpine deformation (Rochette & Lamarche 1986). The synmetamorphic palaeomagnetic directions showed the effects of later deformation in the Dauphinois zone, including tilting by up to 20° and possible anticlockwise rotation (Lamarche *et al.* 1988). The results favored a tectonic origin for basement uplift by means of a crustal thrust with a flat-and-ramp geometry.

The Ivrea–Verbano and Sesia–Lanzo zones

The mafic and ultramafic rocks of the Ivrea–Verbano zone suffered amphibolite to granulite facies metamorphism more than 160 Ma ago, and possess a remanent magnetization acquired before the Jurassic (Heller & Schmid 1974). Stable Permo–Carboniferous directions in the Valle Sessera region are rotated about 50° anticlockwise relative to European normal directions (Fig. 2: Site 38). The directions resemble those measured in the Bolzano porphyries (Zijderveld *et al.* 1970). After correction for tectonic tilting, the directions measured in an Oligocene andesite dyke (Fig. 2: Site 39) agreed with those found in the Bergell granite (Heller 1973). Thus the directions measured in the Ivrea–Verbano zone have an African affinity, like those observed in the southern Alps.

The latest metamorphism experienced by the polymetamorphic rocks in the Sesia–Lanzo zone, the innermost structural unit of the western Alps, was the Alpine phase. Postmetamorphic andesitic and lamprophyric dykes intruding the metamorphics are of Oligocene age. Magnetization directions from these dykes formed two groups, corresponding to the inner, southeastern and outer, northwestern parts of the Sesia–Lanzo zone, respectively (Lanza 1977, 1979). The inner group were believed to be tectonically displaced, and after a correcting rotation of 40° about a horizontal axis were brought into agreement with the results from the outer group. The direction of the stable component of magnetization in Oligocene intrusives of the Traversella massif (Fig. 2: Site 40) was close to the directions measured in these dykes (Lanza 1984). After tilt corrections, the magnetization directions from these sites in the Sesia–Lanzo zone were also close to the directions from the Bergell massif (Heller 1973).

The Alpes Maritimes

The undeformed rocks of the Dôme de Barrot were involved in a palaeomagnetic investigation (Van den Ende 1977), which showed that the stable directions conformed well with expected Permian directions for this region (Fig. 2: Site 41). Henry (1973) found that the minimum susceptibility axes in these undeformed rocks are sub-normal to bedding, demonstrating that compaction in these rocks is more important than eventual tectonic foliation. The unknown effects of this compaction on a pre-existing remanent magnetization caused Henry (1973) to doubt the validity of the palaeomagnetic directions. The effects of compaction on the remanent directions are dependent on when the magnetizations were acquired, which is often a difficult question to answer unequivocally. Late Permian results from the Esterel Massif on the Mediterranean coast (Zijderveld 1975) gave the same directions as found in the Dôme de Barrot (Fig. 2: Site 42).

Magnetic fabric and strain

Deformation has two main effects on palaeomagnetic vectors. The first is simple tilting, which can be corrected for by a rigid body rotation. The second effect arises when rocks suffer penetrative deformation and the palaeomagnetic vector is deflected by grain rotations. This produces a magnetic anisotropy due to the preferential alignment of grains of ferromagnetic or paramagnetic minerals. The most common method of analysis (and the one on which most of the investigations reported here are based) is by the anisotropy of magnetic susceptibility (AMS). More complex techniques are needed to separate magnetic subfabrics in rocks where more than one family of minerals is anisotropic.

The magnetic anisotropy can be described analogously to strain. The geometric representation of magnetic fabric results is by use of a magnetic anisotropy ellipsoid. The attitude of this ellipsoid is fully determined by the directions of the principal axes. The principal susceptibilities ($k_{max}>k_{int}>k_{min}$) are used to describe the shape of the magnetic anisotropy ellipsoid, in the same way as the principal strains ($X>Y>Z$) describe the shape of the strain ellipsoid. A large number of samples are measured from each site, and an average ellipsoid is calculated. The mean direction of the minimum susceptibility axes defines the normal to a magnetic foliation plane, and the mean direction of the maximum susceptibility axes defines a magnetic lineation.

The pre-Alps

Three significant anisotropy studies have been made in sedimentary rocks from the pre-Alps. One of the first investigations of the relation between magnetic fabric and local structural patterns was performed on slates, limestones, sandstones and grit from small-scale folds near the summit of the Niessen nappe (Fuller 1964). The coarser grained sediments possessed an oblate susceptibility magnitude ellipsoid in which the k_{min} axes were normal to the bedding plane. The limestones and slates, on the other hand, had an oblate fabric with the magnetic foliation plane parallel to the axial planar cleavage in the small scale folds. No lineation could be defined in these samples. Pyrrhotite was found to be responsible for the observed magnetic anisotropy in the limestones. The anisotropy of pyrrhotite has a magnetocrystalline origin, with high susceptibility in the crystal basal plane and a low susceptibility parallel to the crystal c-axis. Fuller (1964) speculated that the fabric in the limestone slate samples resulted either directly or indirectly from the stress environment at the time of the formation of cleavage. Since the magnetic anisotropy of the underlying coarse-grained sediments was controlled by bedding compaction, the magnetic fabric of the different rock types appears to be sensitive to the local small-scale stress environment, rather than the regional stress.

Channel *et al.* (1979) investigated sediments from outcrops of the Gurnigel nappe between Lake Thun and Lake Geneva (Fig. 2: Site 43). This 1200–1600 m thick sequence of alternating sandstones and shales is from Maastrichtian to Middle Eocene in age, and is thought to have been deposited in the S Penninic or Ligurian realm. Four sites were sampled in the undeformed flysch at Fayaux, Falli Hölli, Zollhaus and Bärgli. The susceptibility magnitude ellipsoid shapes were predominantly oblate with the minimum axes of susceptibility normal to the bedding plane. This fabric is associated with magnetite grains which occur along the (010) twin planes of albite (Channell *et al.* 1980). Because the clusters of magnetite grains tend to be oriented parallel to the maximum elongation axes of the host albite, the magnetic fabric of the flysch is due to the orientation of albite and other detrital grains by gravitational and hydrodynamic forces.

A weak alignment of the k_{max} axes was present at Bärgli, Falli Hölli and Fayaux. At Fayaux this lineation was parallel to the flow direction during deposition of the flysch, as determined from flute casts. The magnetic susceptibility of these samples was very weakly anisotropic; the maximum and intermediate axes were often indistinguishable. In spite of the weak lineations, the magnetic fabric in this flysch has sedimentary characteristics, and could be useful in studying palaeo-flow directions.

The central Alps

The relationship between magnetic fabric and deformation has been observed in several studies in the central Helvetic Alps. A quantitative correlation between the strain and the magnetic anisotropy has been made in the deformed Jurassic oolitic limestones from the sedimentary cover of the Aar massif (Kligfield *et al.* 1982) and from Permo–Triassic red beds from the Helvetic nappes in eastern Switzerland (Hirt 1986). Qualitative comparisons between magnetic fabric and deformation have been made in granitic rocks of the Bergell massif (Heller 1973) and in granites and granitic gneisses of the Maggia nappe in the Lepontine Alps (Hirt & Lowrie 1985).

Sedimentary cover of the Aar Massif

A thin, ferruginous Jurassic oolitic limestone from the autochthonous cover of the Aar Massif was recognized by Heim (1919) as a potential strain indicator. Baker (1964) found a qualitative correspondence between the directions of the principal axes of the strain ellipsoids, as determined from the oolite shapes, and the directions of the principal axes of magnetic susceptibility anisotropy ellipsoids. Kligfield *et al.* (1982) confirmed this spatial relationship in anticlinal and synclinal folds in the Maderanertal (Fig. 2: Site 44). In specimens from both folds the k_{min} axes were nearly normal to the cleavage and to the XY plane of the finite strain ellipsoid (X>Y>Z). The k_{max} axes were parallel to the stretching direction (X), or to the mineral elongation within the cleavage (Fig. 4). The k_{int} axes were generally parallel to the regional strike.

Oolites with two different magnetic mineralogies have been described in these rocks. In hematitic oolites, hematite is the only magnetic mineral present, as platelets coating the deformed ooids. In chamositic oolites, magnetite is the main magnetic mineral, although pyrrhotite and hematite may also be present. Euhedral, individually undeformed magnetite grains, which crystallized after the oolites were deformed, string together along the curved ooid boundaries, giving rise to a mimetic shape anisotropy. In both types of oolite the distribution of magnetic minerals most likely to be responsible for the magnetic fabric conforms to the shape of the deformed ooids, so that the magnetic fabric also shows this deformation.

A quantitative correlation was established between the finite strain and the magnetic anisotropy in the hematite-rich layers. This correlation allowed Kligfield *et al.* (1982) to derive a large number of equivalent strain values from the magnetic fabric measurements around the two folds. With this dense network of strain data, the relationship of strain variation to the geometry of the fold can be examined in detail.

Red beds of the Lower Glarus nappe complex

A quantitative correlation between finite strain magnetic fabric was made for the Permian Verrucano and Triassic Quartenschiefer formations of the Lower Glarus nappe complex in the central Helvetics (Hirt 1986). At 12 sites from the Murgtal and Flumserberg areas (Fig. 2: Site 45) the finite strain ellipsoid was determined from reduction spots within the red beds. A good correlation was found between the magnitudes and directions of the principal axes of the strain and magnetic anisotropy ellipsoids. Using this correlation, an equivalent strain ellipsoid was derived from magnetic anisotropy measurements for 15 additional sites in the Lower Glarus nappe complex at which reduction spots or other strain markers are not present (Fig. 5). In the eastern part of the nappe complex, the k_{max} axes describe a NW extension lineation, in contrast to the N–S lineation found from strain observations and magnetic anisotropy measurements in other parts.

The Bergell massif

A comparison of the magnetic fabric with macroscopic structures in the porphyritic granites, granite aplites and inclusions of the Bergell massif in the Val d' Albigna (Fig. 2: Site 31) was made by Heller (1973). Magnetite was found to be responsible for the low-field susceptibility anisotropy. The shape of the susceptibility magnitude ellipsoids in the porphyritic granites and aplites gradually changed from oblate on the northern margin of the intrusion to prolate in the center (Fig. 6a). The magnetic lineation in the Albigna valley was very consistent and coincided with the linear structure defined by the long axes of alkali feldspar phenocrysts. The foliation of the oblate magnetic anisotropy ellipsoids agrees well with the orientation of biotite (001) planes (Fig. 6b). At the extreme northern margin of the massif, the planar magnetic fabric agrees well with the orientation of platy alkali-feldspar crystals. Only in the central area of the intrusion, where the susceptibility ellipsoids are prolate, is there a poor agreement between the biotite fabric and magnetic fabric. Here the magnetic lineation parallels the intersection line of two planar biotite fabrics.

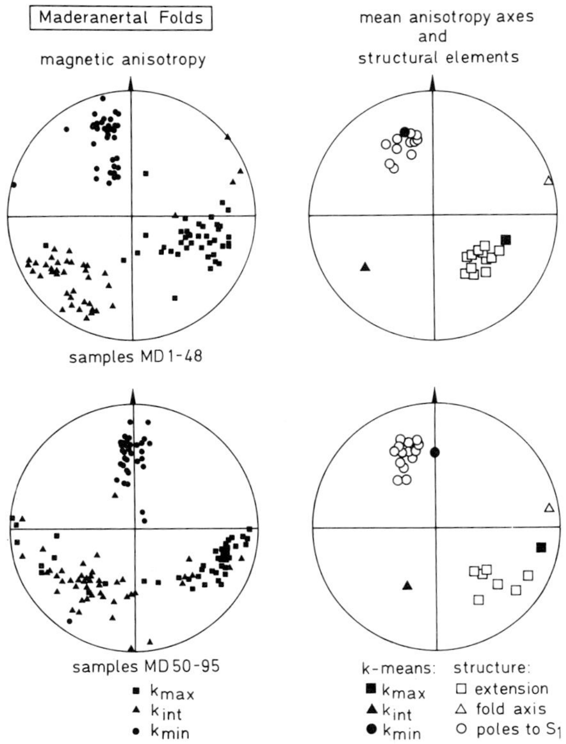

FIG. 4. Comparison of the directions of the principal susceptibility axes in deformed oolites (left) and structural elements (right) for Maderanertal folds. The directions are plotted on lower hemisphere, equal area projections. S_1 is the Calanda phase cleavage (from Kligfield *et al.* 1982).

The Lepontine Alps

Three main phases of deformation identified in the northern region of the Lepontine Alps can be related to: (i) nappe emplacement; (ii) post-nappe recumbent folding; (iii) late Alpine folding, with formation of backfolds and crenulation, respectively (Ramsay & Allison 1979, Huber *et al.* 1980). Shear zones are associated with the first two phases of deformation. The main thermal metamorphic event reached its peak (amphibolite facies) between the second and third deformation stages (Huber *et al.* 1980). The metamorphism outlasted the third deformation stage, so that most minerals of these basement gneisses were recrystallized (Simpson 1981).

The same style of magnetic fabric was found in the granitic gneisses and in diorite and aplite dykes from this region. The axes of magnetic anisotropy ellipsoids for individual samples formed three distinct groups; the k_{min} axes were normal to the schistosity plane, within which the k_{max} axes were parallel to the mineral elongation (Hirt and Lowrie 1985). The petrographic and magnetic fabrics are believed to have originated during recrystallization after the major deformation stages. Analysis of high-field torque measurements indicated that the paramagnetic minerals (predominantly biotite) and ferromagnetic minerals (magnetite and pyrrhotite) had similar anisotropic distributions.

The eastern Alps

The Periadriatic Line

Several studies have used the anisotropy of magnetic susceptibility to determine the sense of motion on the Periadriatic Line in south Tyrol. Rathore & Becke (1980) found a strong magnetic fabric parallel to the line, but limited to a narrow zone on either side of it. No effects of shearing could be detected at distances greater than 2 km from the line. Rathore & Heinz (1980) noted that the anisotropy ellipsoids

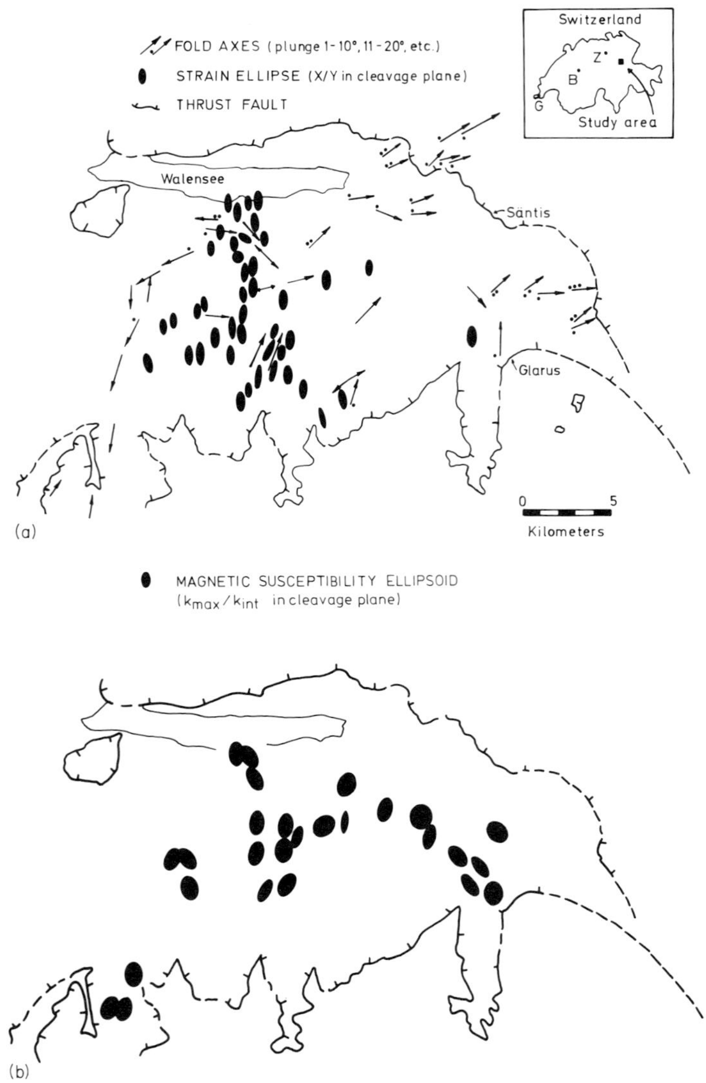

FIG. 5. (a) Generalized map of the Lower Glarus nappe complex (after Pfiffner 1981) with horizontal projections of strain ellipsoids from Huber (1964), Schielly (1964), Ryf (1965), Kühn (1966), Markus (1967), Richter (1968) and Pfiffner (1981). (b) Horizontal projections of strain ellipsoids calculated from the correlation of strain with magnetic fabric (after Hirt 1986).

did not change in orientation across the shear zone, and concluded that relative rotations N and S of the line did not occur.

Because the flattening of oblate anisotropy ellipsoids of tectonic origin indicates compressional strain, Rathore (1985) reasoned that the k_{min} axes should point in the direction of the shear zone. The magnetic foliation planes in sites located along the Periadriatic Line are steeply dipping. Rathore & Becke (1980) resolved the declinations of the k_{min} axes of these sites into components normal and parallel to the shear zone. Analysing the directions parallel to the Periadriatic Line, they found predominantly dextral motion along the line with small regions of sinistral motion. A similar study on the Judicarian Line segment of the Periadriatic Line (Rathore 1980) indicated sinistral motion along the NE side of the line and regions with sinistral and dextral motion along the SW side of the line. According to these magnetic fabric analyses, the dominant motion along the entire line was sinistral.

Rathore (1985) cautioned that this analysis of motion from the anisotropy ellipsoids is simplistic and can only be viewed as an empirical evaluation. The Periadriatic Line is not straight, as assumed by the analysis, and any deviation of the true shear zone from rectilinearity influences the resolution of the k_{min} directions into components normal and parallel to the line. The significance of magnetic lineations was not evaluated. No detailed investigations were made in these studies to identify the minerals responsible for the magnetic anisotropy. It is critical to the analysis to know when the minerals causing the anisotropy were formed, since this places limits on the timing and extent of the motion. As in the northern Lepontine Alps, recrystallization has left a magnetic fabric which is not related to the development of the shear zones, but only to the final stage of deformation.

The Riesenferner massif

Large tonalite and granodiorite bodies have intruded the metamorphic Austro-Alpine nappes between the Insubric Line and the Hohe Tauern window. Henry (1975) found well grouped magnetic lineations in the Riesenferner tonalite (Fig. 2: Site 46); similar petrofabric lineations are not visible in the field. With the aid of supporting evidence from microtectonic studies, the lineation was linked to late Alpine deformations. The structural analysis on the basis of magnetic anisotropy data suggested that the massif underwent a second phase of E–W compression, nearly perpendicular to the late Alpine compression.

The western Alps

The Alpes Maritimes

The red Permian mudstones of the Alpes Maritimes exhibit a wide range of strains as a result of the Alpine deformation. The spectrum extends from the tectonically undeformed state to penetrative deformation with well developed slaty cleavage. In the SW part of the region the undeformed rocks of the Dôme de Barrot show abundant sedimentary structures. The intensity of deformation increases northeastwards and slaty cleavage appears.

The progressive deformation in the red shales, mudstones and slates of the Alpes Maritimes is expressed in deformed reduction spots. These serve as finite strain indicators and provide a quantitative model for the progressive deformation (Graham 1978). The magnetic fabric mirrors the strain pattern (Kligfield *et al.* 1981). The magnetic anisotropy ellipsoid is initially oblate, flattened in the bedding, corresponding to the effects of compaction only. Superposition of tectonic shortening produces prolate susceptibility magnitude ellipsoids with long axes parallel to pencil-cleavage lineations and bedding-cleavage intersections. Further tectonic shortening results in oblate ellipsoids with short axes normal to cleavage, and eventually with long axes parallel to extension lineations within the cleavage plane (Fig. 7). A quantitative correlation between natural strain and magnetic anisotropy was established for these rocks (Kligfield *et al.* 1981, 1983). The AMS-strain correlation was used by Siddans *et al.* (1984) to extend the strain data base by converting the more abundant magnetic fabric measurements to equivalent strain values. The combination of AMS data with structural geological field evidence provided a more detailed evaluation of the regional tectonic history of the Alpes Maritimes.

The Dauphinois zone

In the Dauphinois zone, the effects of Alpine deformation are evident in the progressive development of planar and linear fabrics. A good qualitative correspondence exists between magnetic fabric and petrofabric in Jurassic shales and slates. It follows the SW to NE deformation gradient, with fabrics that ranged in type from sedimentary to highly deformed (Rochette & Vialon 1984). The deformation history was characterized by pressure solution at an early stage, followed by grain rotation and finally recrystallization at higher stages of deformation. The magnetic susceptibility magnitude ellipsoid exhibited the same progressive change of shape as observed in the Alpes Maritimes, with oblate initial ellipsoids changing to prolate and back to oblate during more extreme deformation. At very high strains the AMS ellipsoid was prolate with long axis parallel to the mineral extension lineation within the cleavage.

Further investigations of the magnetic fabrics in these deformed Jurassic calcareous black shales, schists and slates confirmed and enhanced the good association with deformation pattern (Lamarche & Rochette 1987a). The origin of lineations in the magnetic fabric was investigated with microstructural analysis.

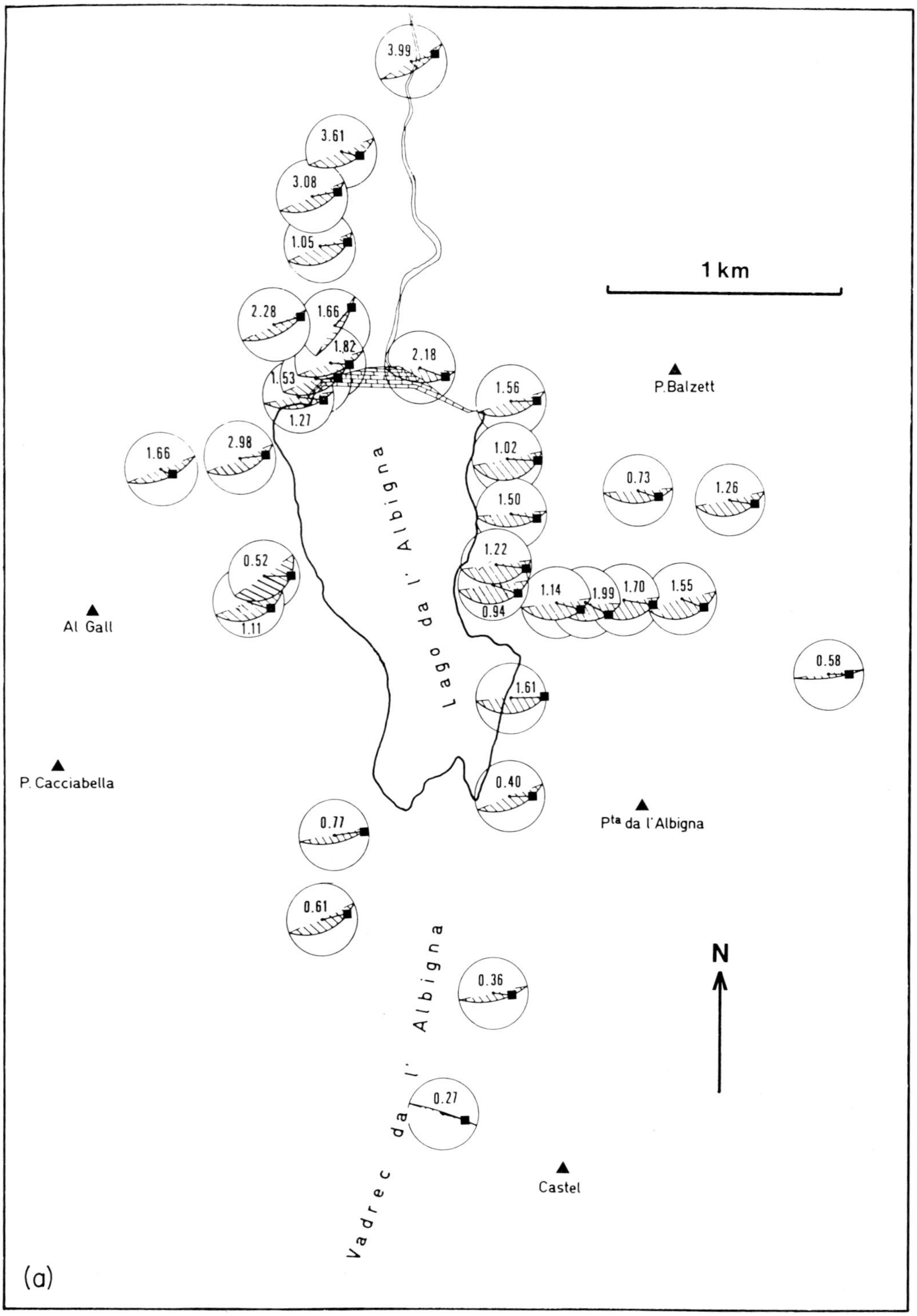
1 km
P. Balzett
Lago da l' Albigna
Al Gall
P. Cacciabella
Pta da l'Albigna
Vadrec da l' Albigna
Castel
N
(a)

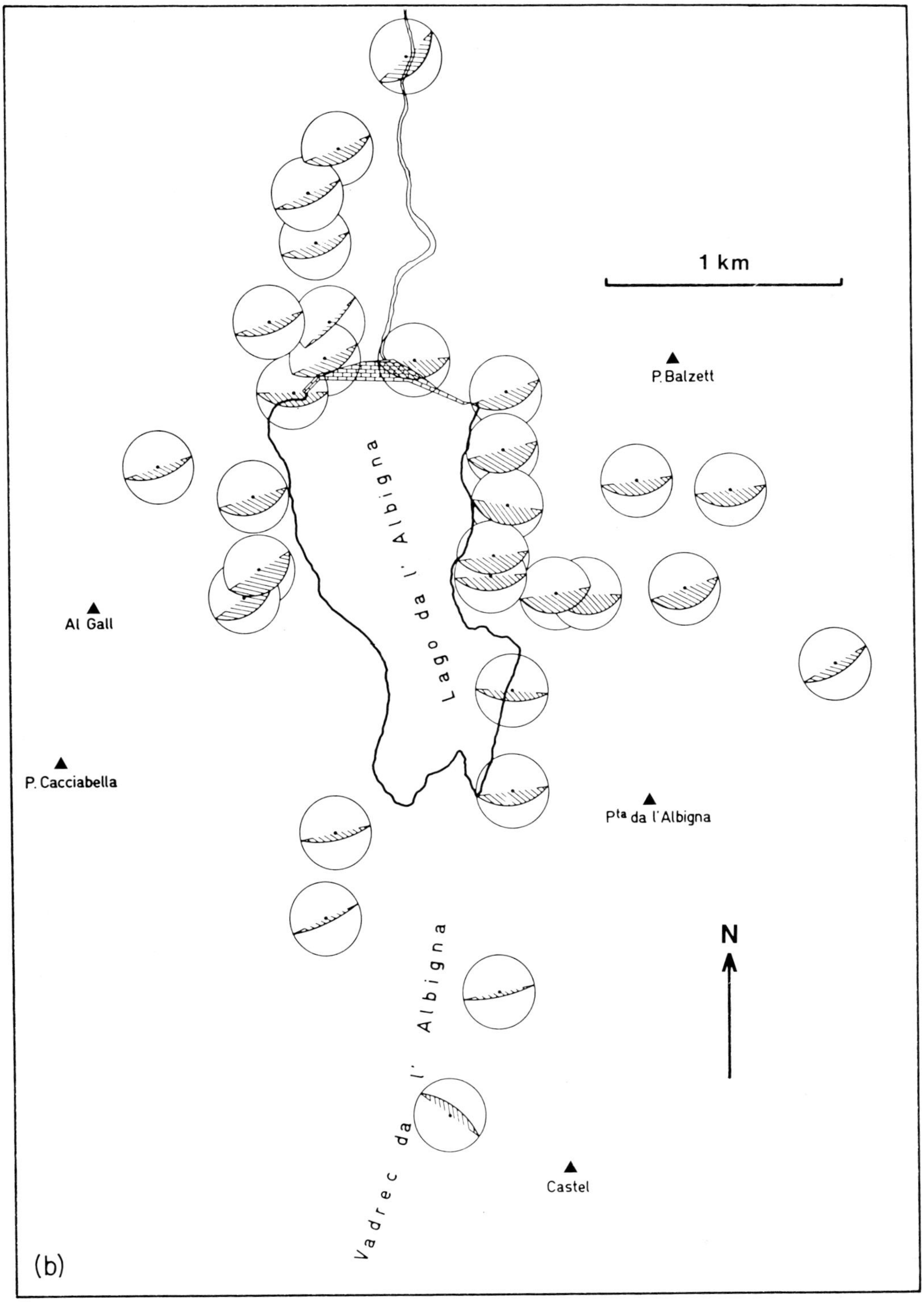

FIG. 6. Lower hemisphere, equal-area projections of (a) the magnetic foliation planes, and (b) the macroscopic biotite, planar fabric of the porphyritic granites from the Albigna valley in the northern part of the Bergell intrusion. In (a) squares represent the position of the k_{max} axis in each magnetic foliation plane. The number at each site denotes an oblateness quotient (OB); OB = magnetic foliation/lineation; OB > 1 for an oblate ellipsoid, OB < 1 for a prolate ellipsoid (from Heller 1973).

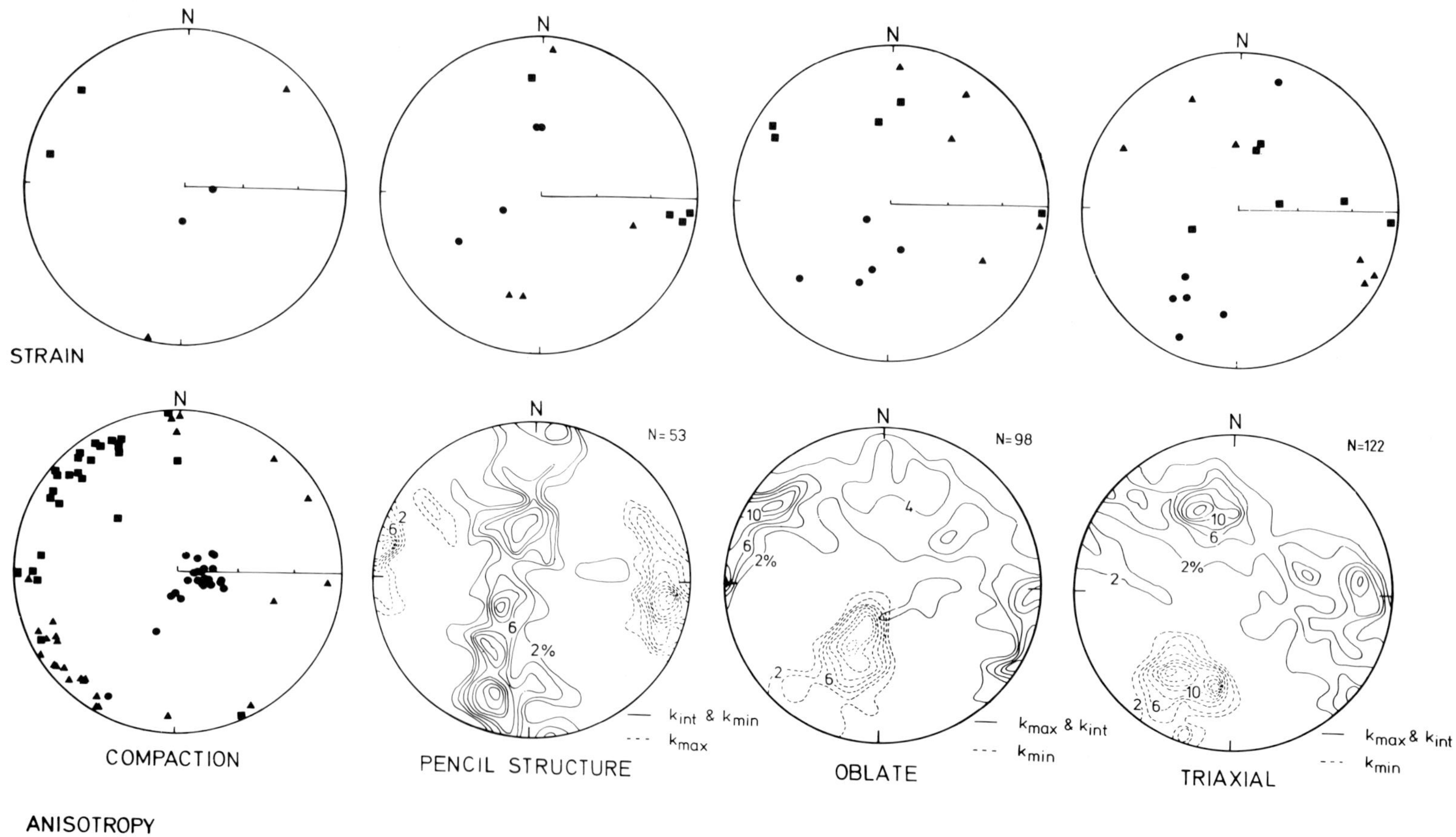

FIG. 7. Comparison of the directions of the principal axes of natural strain (X>Y>Z) and magnetic susceptibility anisotropy ($k_{max}>k_{int}>k_{min}$) for the deformed redbeds of the Alpes Maritimes (Kligfield *et al.* 1981, 1983). Symbols for individual axes: squares, X or k_{max}; triangles, Y or k_{int}; circles, Z or k_{min}.

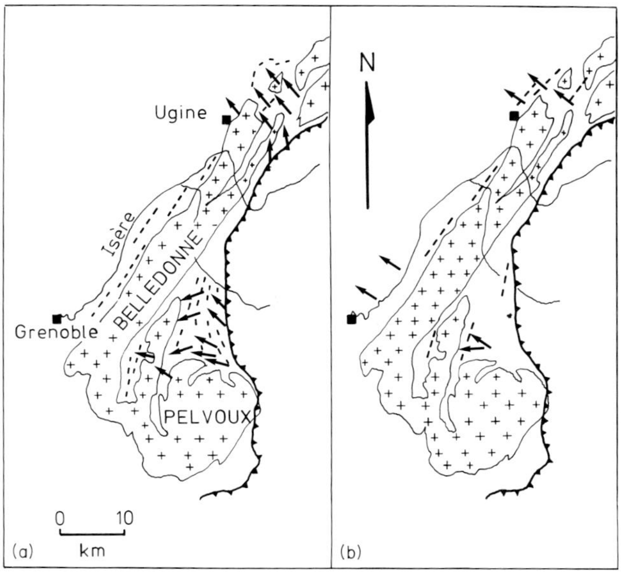

FIG. 8. Structural lineation features in the Dauphinois zone of the French Alps (Lamarche & Rochette 1987a). Arrows indicate the transport directions, and dotted lines the intersection lineations L_1 of bedding and slaty cleavage. Data determined (a) from field measurements, (b) from studies of the magnetic anisotropy.

The magnetic lineation is parallel to one of two structural lineations in the rocks: (a) the stretching lineation which indicates transport direction, or (b) the intersection of two planar fabrics. In some sites a magnetic lineation was found which had no visible microstructural expression but which provided additional information for interpretation of the kinematic deformation. AMS ellipsoids were also found to correlate well with the structural features in Eocene flysch in the Ultradauphinois of the French Alps (Lamarche & Rochette 1987b). Deformation and AMS ellipsoids were flattened in the cleavage plane. Magnetic lineations indicated the transport directions, or more rarely the intersections of more than one planar fabric (Fig. 8).

In the AMS studies of rocks from the Dauphinois and Ultradauphinois zones in the French Alps particular attention was given to understanding the origin of the magnetic anisotropy by means of careful evaluation of the rock magnetic mineralogy. Low-field AMS results are generally attributed to the ferromagnetic grains (magnetite, hematite or pyrrhotite) in a rock, but the investigations of magnetic mineralogy in the Dauphinois shales and slates emphasized the importance of phyllosilicates as carriers of the magnetic fabric (Rochette & Vialon 1984, Rochette & Lamarche 1986).

The Montgenèvre massif

The relationship between magnetic anisotropy and petrofabric has been studied in some gabbros in the ophiolites of the Montgenèvre massif (Fig. 2: Site 47, Wagner *et al.* 1981). Although broken into a series of tectonic slices by the Alpine orogeny, the massif has suffered only weak Alpine metamorphism. The anisotropy of magnetic susceptibility was attributed to paramagnetic minerals such as hornblende, formed during ocean floor metamorphism. The relationship between magnetic fabric and petrofabric showed the progressively increasing effects of the deformation, which was implied to have had a simple shear mechanism.

Discussion

Numerous palaeomagnetic studies in the Alpine region have shown both local and large scale effects of Alpine deformation. Palaeomagnetists usually avoid sampling rocks that have been heavily tectonized, because of the unknown effects the deformation can have on the palaeomagnetic vector. Even when internal deformation is negligible, the simple tilt correction by a rigid body rotation about the strike line is usually an inadequate correction for tectonic displacement. Depending on the nature of the fold and the attitude of the fold axis, more extensive correction by at least another rigid body rotation is often needed. However, successive rotations are non-commutative, and because the sequence in which they must be made is not always known uniquely, a certain amount of uncertainty remains. Many palaeomagnetic studies reported in this review were only subjected to a single simple tilt correction. Nevertheless, the results show some remarkable consistencies. Large areas in the Alpine domain have coherent palaeomagnetic directions that clearly relate to either the Eurasian or African apparent polar wander paths (Figs 1 and 2).

The pioneering studies of the use of anisotropy of magnetic susceptibility as a petrofabric indicator were made in deformed sedimentary rocks in the Appalachian mountain belt (Graham 1954, 1966). As illustrated by the many case examples described here, studies of the magnetic fabric of deformed rocks in the Alps have further documented the excellent correspondence between the directions of principal strain and principal susceptibilities. Research has shown that the shape of the susceptibility magnitude ellipsoid changes with progressive deformation in the same way as the strain ellipsoid. Quantitative relationships between the principal axes of susceptibility and strain have been used to derive large bodies of equivalent strain data from magnetic fabric measurements. The relationship cannot be linear over a large range of strain, because the maximum magnetic anisotropy is limited; the magnetic anisotropy remains finite, even for infinite strain. The correct way to correlate the corresponding principal axes of these ellipsoids has not been established definitively, but the potential of providing a petrofabric indicator where other strain markers are absent makes this a promising field for further research (Graham 1954).

Attempts have been made to correct the remanent magnetizations in deformed beds of the Alpes Maritimes rocks for the effects of strain (Kligfield *et al.* 1983, Cogné & Perroud 1985, Lowrie *et al.* 1986). The validity of this type of correction has important implications for palaeomagnetism, because it would enable the interpretation of palaeomagnetic studies in deformed rocks which hitherto had to be avoided. Attempts at unstraining palaeomagnetic vectors have assumed the March model of ductile deformation. For the wide range of deformation in the Alpes Maritimes this is an oversimplification. Extreme deformation is often accompanied by recrystallization, which requires a change of the rheological model. Moreover, the relationship between strain and magnetic anisotropy is nonlinear. For very high strains the magnetic anisotropy tends asymptotically to a limiting value. Kligfield *et al.* (1983) showed that unstraining with the March model gives reasonable results in rocks which have suffered only low deformation, but cautioned that such corrections in the Alpes Maritimes rocks are probably inapplicable beyond the pencil-structure stage of deformation. Cogné & Perroud (1985, 1987) contest this conservative attitude, but ignore the physical and geological limitations of the March model of deformation. Indiscriminate use of this model to unstrain palaeomagnetic vectors in highly strained rocks can not be justified.

In any palaeomagnetic or magnetic fabric study it is important to understand the magnetic mineralogy of the rocks studied. This is especially true in the Alps. Due to the complex tectonic and metamorphic history, the original magnetic minerals may have been altered. The remanent magnetization may be carried by a generation of magnetic minerals formed, or remagnetized, during the metamorphism. The magnetic fabric may postdate the structural fabric, or represent a late phase of deformation only. In each case it is essential to know what magnetic minerals carry the magnetic signature. Extension of palaeomagnetic or magnetic fabric results to tectonic interpretations is of limited value unless the rock magnetism is clearly understood.

References

AGNOLI, F., MAURITSCH, H. & REISINGER, J. 1984. Paläomagnetische Arbeiten im Berichtsjahr 1983/84. *Jahresbericht 1983, Hochschulschwerpunkt S15 Graz*, 7–15.

BAKER, D. W. 1964. *Structural petrology of the Windgällen fold*. Unpublished PhD thesis, University of Zürich.

BECKE, M. 1980. *Geologisch-tektonische und paläomagnetische Untersuchungen im Mieminger Gebirge (Nördl. Kalkalpen, Tirol)*. Unpublished PhD thesis, University of Innsbruck.

—— & MAURITSCH, H. J. 1985. Die Entwicklung der Nördlichen Kalkalpen aus paläomagnetischer Sicht. *Archiv für Lagerstättenforschung der geologischen Bundesanstalt*, **6**, 113–16.

BOGDANOFF, S. & SCHOTT, J. J. 1977. Etude paléomagnétique et analyse tectonique dans les schistes rouges permiens du Sud de l'Argentera. *Bulletin de la Société géologique de France*, **19**, 909–16.

BULLARD, E. C., EVERETT, J. E. & SMITH, A. G. 1965. The fit of the continents around the Atlantic. *In*: *A Symposium on Continental Drift. Philosophical Transactions of the Royal Society, London*, **258A**, 41–51.

CHANNELL, J. E. T. 1986. Palaeomagnetism and continental collision in the Alpine belt and the formation of Late-tectonic extensional basins. *In*: COWARD, M. P. & RIES, A. C. (eds). *Collision Tectonics*, Geological Society Special Publication No. 19, pp. 261–84.

—— & HORVATH, F. 1976. The African/Adriatic promontory as a palaeogeographical premise for Alpine orogeny and plate movements in the Carpatho-Balkan region. *Tectonophysics*, **35**, 71–101.

—— & TARLING, D. H. 1975. Palaeomagnetism and the rotation of Italy. *Earth and Planetary Science Letters*, **25**, 177–88.

——, LOWRIE, W. & MEDIZZA, F. 1979a. Middle and early Cretaceous magnetic stratigraphy from the Cismon section, northern Italy. *Earth and Planetary Science Letters*, **42**, 153–66.

——, D'ARGENIO, B. & HORVATH, F. 1979b. Adria, the African promontory, in Mesozoic Mediterranean palaeogeography. *Earth Science Reviews*, **15**, 213–92.

——, HELLER, F. & VAN STUIJVENBERG, J. 1979. Magnetic susceptibility anisotropy as an indicator of sedimentary fabric in the Gurnigel flysch. *Eclogae geologicae Helvetiae*, **72**, 781–7.

——, FREEMAN, R. & HELLER, F. 1980. Magnetic properties of Swiss flysch. *Journal of Geophysics*, **48**, 60–66.

—— & MEDIZZA, F. 1981. Upper Cretaceous and Palaeogene magnetic stratigraphy and biostratigraphy from the Venetian (southern) Alps. *Earth and Planetary Science Letters*, **55**, 419–32.

COGNE, J. P. & PERROUD, H. 1985. Strain removal applied to paleomagnetic direction in an orogenic belt: the Permian red slates of the Alpes Maritimes, France. *Earth and Planetary Science Letters*, **72**, 125–40.

—— & —— 1987. Unstraining paleomagnetic vectors: the current state of debate. *Eos*, **68**, 705–12.

DE BOER, J. 1963. Geology of the Vicentinian Alps with special reference to their paleomagnetic history. *Geologica Ultraiectina*, **11**, 178 pp.

—— 1965. Paleomagnetic indications of megatectonic movements in the Tethys. *Journal of Geophysical Research*, **70**, 931–44.

DEWEY, J. F., PITMAN III, W. C., RYAN, W. B. F. & BONNIN, J. 1973. Plate tectonics and the evolution of the Alpine system. *Geological Society of America Bulletin*, **84**, 3137–80.

FÖRSTER, H., SOFFEL, H. & ZINSSER, H. 1975. Palaeomagnetism of rocks from the eastern Alps from north and south of the Insubrian line. *Neues Jahrbuch für Geologie und Paläontologie Abhandlungen*, **149**, 112–27.

FULLER, M. D. 1964. On the magnetic fabric of certain rocks. *Journal of Geology*, **72**, 368–76.

GRAHAM, J. W. 1954. Magnetic susceptibility anisotropy, an unexploited petrofabric element. *Bulletin of the Geological Society of America*, **65**, 1257–8.

—— 1966. Significance of magnetic anisotropy in Appalachian sedimentary rocks. *In*: STEINHART, J. S. & SMITH, T. J. (eds). *The Earth Beneath the Continents*, Geophysical Monograph 10, American Geophysical Union, Washington, 627–48.

GRAHAM, R. 1978. Quantitative deformation studies in the Permian rocks of the Alpes–Maritimes. *In*: *Proceedings of Symposium in Honour of Prof. J. Goguel, Bureau de Recherches Géologiques et Minères*, 219–238.

GWINNER, M. P. 1971. *Geologie der Alpen*. Schweizerbart'sche Verlagsbuchhandlung, Stuttgart, 477 pp.

HARGRAVES, R. B. & FISCHER, A. G. 1959. Remanent magnetism in Jurassic red limestones and radiolarites from the Alps. *Geophysical Journal*, **2**, 34–41.

HEIM, A. 1919. *Geologie der Schweiz* (Bd. II/I), Tauchnitz, Leipzig.

HEINIGER, C. 1979. Palaeomagnetic and rock magnetic properties of the Permian volcanics in the western southern Alps. *Journal of Geophysics*, **46**, 397–411.

HELLER, F. 1971. Remanent magnetization of the Bergell granite. *Zeitschrift für Geophysik*, **37**, 557–71.

—— 1973. Magnetic anisotropy of granitic rocks of the Bergell Massif (Switzerland). *Earth and Planetary Science Letters*, **20**, 180–8.

—— 1977. Palaeomagnetic data from the western Lepontine area. *Schweizerische Mineralogische und Petrographische Mitteilungen*, **57**, 135–43.

—— 1980. Palaeomagnetic evidence for late Alpine rotation of the Lepontine area. *Eclogae geologicae Helvetiae*, **73**, 607–18.

—— & SCHMID, R. 1974. Paläomagnetische Untersuchungen in der Zone Ivrea–Verbano (Prov. Novara, Norditalien): Vorläufige Ergebnisse. *Schweizerische Mineralogische und Petrographische Mitteilungen*, **54**, 229–42.

HENRY, B. 1973. Studies of microtectonics, anisotropy of magnetic susceptibility, and palaeomagnetism of the Permian Dôme de Barrot (France). Palaeomagnetic and paleosedimentological implications. *Tectonophysics*, **17**, 61–72.

—— 1975. Microtectonique et anisotropie de susceptibilité magnétique du massif tonalitique des Riesenferner-Vedrette di Ries (Frontière Italo-Autrichienne). *Tectonophysics*, **27**, 155–65.

—— 1976. Relations entre contraintes tectoniques et propriétés magnétiques des roches volcaniques permiennes de la vallée du Guil (Briançonnais, Alpes Françaises). *Pure and Applied Geophysics*, **114**, 685–700.

—— 1980. *Contribution à l'étude des propriétés magnétiques de roches magmatiques des Alpes.*

Conséquences structurales, régionales et générales. Unpublished doctoral thesis, University of Paris.

HIRT, A. M. 1986. *Paleomagnetic and magnetic anisotropy techniques applied to tectonically deformed regions*. Unpublished PhD thesis, Swiss Federal Institute of Technology, Zürich.

—— & LOWRIE, W. 1985. Analysis of magnetic anisotropy in the Maggia Nappe, Lepontine Alps, Switzerland (abstract). *Abstract Volume 1: International Association of Geomagnetism and Aeronomy, Prague, Czechoslovakia*, 5–7 August, 1985, p. 211.

——, —— & PFIFFNER, O. A. 1986. A paleomagnetic study of tectonically deformed red beds of the Lower Glarus nappe complex, eastern Switzerland. *Tectonics*, **5**, 723–31.

HORNER, F. & HELLER, F. 1983. Lower Jurassic magnetostratigraphy at the Breggia Gorge (Ticino, Switzerland) and Alpe Turati (Como, Italy). *Geophysical Journal of the Royal Astronomical Society*, **73**, 705–18.

HUBER, M., RAMSAY, J. & SIMPSON, C. 1980. Deformation in the Maggia and Antigorio nappes, Lepontine Alps. *Eclogae Geologicae Helvetiae*, **73**, 593–606.

HUBER, R. 1964. Etude géologique du massif du Gufelstock avec stratigraphie du Verrucano. *Mitteilungen der geologischen Institute, ETH Zürich und Universität Zürich: neue Folge*, **23**.

IRVING, E. 1977. Drift of the major continental blocks since the Devonian. *Nature*, **270**, 304–9.

KIPFER, R. & HELLER, F. 1988. Palaeomagnetism of Permian red beds in the contact aureole of the Tertiary Adamello massif (northern Italy). *Physics of the Earth and Planetary Interiors*, **52**, 365–75.

KLIGFIELD, R. & CHANNELL, J. E. T. 1981. Widespread remagnetization of Helvetic limestones. *Journal of Geophysical Research*, **86**, 1888–1900.

——, OWENS, W. H. & LOWRIE, W. 1981. Magnetic susceptibility anisotropy, strain and progressive deformation in Permian sediments from the Maritime Alps (France). *Earth and Planetary Science Letters*, **55**, 181–9.

——, LOWRIE, W. & PFIFFNER, O. A. 1982. Magnetic properties of deformed oolites from the Swiss Alps: the correlation of magnetic anisotropy and strain. *Eclogae geologicae Helvetiae*, **75**, 127–57.

——, ——, HIRT, A. M. & SIDDANS, A. W. B. 1983. Effect of progressive deformation on remanent magnetization of Permian redbeds from the Alpes Maritimes (France). *Tectonophysics*, **97**, 59–85.

KÜHN, H. 1966. *Observations géologiques dans la région du Spitzmeilen*. Unpublished Diploma thesis, Swiss Federal Institute of Technology.

LAMARCHE, G. & ROCHETTE, P. 1987a. Microstructural analysis and origin of lineations in the magnetic fabric of some Alpine slates. *Tectonophysics*, **139**, 285–93.

—— & —— 1987b. La fabrique magnétique du flysch dauphinois (Alpes françaises): origine et application quantitative. *Geodinamica Acta*, **1**, 101–11.

——, MENARD, G. & ROCHETTE, P. 1988. Données paléomagnétiques sur le basculement tardif de la zone dauphinoise interne (Alpes occidentales). *Comptes-Rendues Académie de Science de Paris*, in press.

LANZA, R. 1977. Palaeomagnetic data from the andesitic and lamprophyre dikes of the Sezia–Lanzo zone (western Alps). *Schweizerische Mineralogische und Petrographische Mitteilungen*, **57**, 281–90.

—— 1979. Palaeomagnetic data on the andesitic cover of the Sezia–Lanzo zone (western Alps). *Geologische Rundschau*, **68**, 83–92.

—— 1984. Palaeomagnetism in the Traversella Massif. *Bollettino Geofisica Teorica ed Applicata*, **26**, 115–24.

LEMOINE, M. 1978. *Geological Atlas of Alpine Europe and Adjoining Alpine Areas*. Elsevier, Amsterdam, 584 pp.

LOWRIE, W. 1980. A paleomagnetic overview of the Alpine system. *Bureau de Recherches Géologiques et Minères, Mémoire*, **115**, 316–30.

—— 1986. Paleomagnetism and the Adriatic promontory: a reappraisal. *Tectonics*, **5**, 797–807.

——, CHANNELL, J. E. T. & ALVAREZ, W. 1980. A review of magnetic stratigraphy investigations in Cretaceous pelagic rocks. *Journal of Geophysical Research*, **85**, 3597–605.

——, HIRT, A. M. & KLIGFIELD, R. 1986. Effects of tectonic deformation on the remanent magnetization of rocks. *Tectonics*, **5**, 713–22.

MANZONI, M. 1970. Palaeomagnetic data of Middle and Upper Triassic age from the Dolomites (eastern Alps, Italy). *Tectonophysics*, **10**, 411–24.

MARKUS, J. H. 1967. Geologische Untersuchungen in den Flumserbergen (St Galler Oberland). *Mitteilungen der geologischen Institute, ETH Zürich und Universität Zürich: neue Folge*, **71**.

MARTON, E. 1987. Palaeomagnetism and tectonics in the Mediterranean region. *Journal of Geodynamics*, **7**, 33–57.

—— & VELJOVIC, D. 1983. Paleomagnetism of the Istria peninsula, Yugoslavia. *Tectonophysics*, **91**, 73–87.

MAURITSCH, H. J. 1986. Der Stand der paläomagnetischen Forschung in den Ostalpen. *Leobner Hefte zur Angewandten Geophysik*, **1**, 141–60.

—— & FRISCH, W. 1978. Palaeomagnetic data from the central part of the northern calcareous Alps, Austria. *Journal of Geophysics*, **44**, 623–37.

—— & —— 1980. Paleomagnetic results from the eastern Alps and their comparison with data from the southern Alps and the Carpathians. *Mitteilungen der österreichischen geologischen Gesellschaft*, **73**, 5–13.

PFIFFNER, O. A. 1981. Fold and thrust tectonics in the Helvetic nappes (E Switzerland). *In*: MCCLAY, K. R. & PRICE, N. J. (eds). *Thrust and Nappe Tectonics* Special Publication of the Geological Society of London No. 9. 319–27.

POHL, J. & SOFFEL, H. 1982. Palaeomagnetism of Tertiary volcanics of Styria (Austria). *Geologisches Jahrbuch*, **D52**, 137–47.

PREISINGER, A., ZOBETZ, E., GRATZ, A. J.,

LAHODYNSKY, R., BECKE, M., MAURITSCH, H. J., EDER, G., GRASS, F., RÖGL, F., STRADNER, H. & SURENIAN, R. 1986. The Cretaceous/Tertiary boundary in the Gosau basin, Austria. *Nature*, **332**, 794–9.

RAMSAY, J. G. & ALLISON, I. 1979. Structural analysis of shear zones in an alpinised Hercynian granite. *Schweizerische mineralogische und petrographische Mitteilungen*, **59**, 251–79.

RATHORE, J. S. 1980. Evidence for sinistral movements along the Judicarian Line drawn from a study of magnetic fabrics in the region of Mt Croce and Asten (south Tyrol). *Geologische Rundschau*, **69**, 678–94.

—— 1985. Some magnetic fabric characteristics of sheared zones. *Journal of Geodynamics*, **2**, 291–301.

—— & BECKE, M. 1980. Magnetic fabric analysis in the Gail valley (Carinthia, Austria) for the determination of the sense of movements along this region of the Periadriatic Line. *Tectonophysics*, **69**, 349–68.

—— & HEINZ, H. 1980. The application of magnetic susceptibility anisotropy analyses to the study of tectonic events on the Periadriatic Line. *Mitteilungen der österreichischen geologischen Gesellschaft*, **71/72**, 275–90.

RICHTER, H. 1968. Die Geologie der Guschagruppe im St. Galler Oberland. *Mitteilungen der Geologischen Institute, ETH Zürich and Universität Zürich: neue Folge*, **99**.

ROCHE, A. & WESTPHAL, M. 1969. Sur la direction d'aimantation des roches volcaniques permiennes de la vallée du Guil. *Comptes-Rendus Sommaires des Séances de la Société Géologique de France*, **7**, 239.

ROCHETTE, P. & VIALON, P. 1984. Development of planar and linear fabrics in Dauphinois shales and slates (French Alps) studied by magnetic anisotropy and its mineralogical control. *Journal of Structural Geology*, **6**, 33–8.

—— & LAMARCHE, G. 1986. Evolution des propriétés magnétiques lors des transformations minérales dans les roches: exemple du Jurassique Dauphinois (Alpes françaises). *Bulletin de Minéralogie*, **109**, 687–96.

RYF, W. H. 1965. Geologische Untersuchungen im Murgtal (St Galler Oberland). *Mitteilungen der geologischen Institute, ETH Zürich und Universität Zürich: neue Folge*, **30**.

SCHIELLY, H. P. 1964. Geologische Untersuchungen im Deckengebiet des westlichen Freiberges (Kt. Glarus). *Mitteilungen der geologischen Institute, ETH Zürich und Universität Zürich: neue Folge*, **44**,

SIDDANS, A. W. B., HENRY, B., KLIGFIELD, R., LOWRIE, W., HIRT, A. M. & PERCEVAULT, M. N. 1984. Finite strain patterns and their significance in Permian rocks of the Alpes Maritimes (France). *Journal of Structural Geology*, **6**, 339–68.

SIMPSON, C. 1981. Ductile shear zones: a mechanism of rock deformation in the orthogneisses of the Maggia nappe, Ticino. *Mitteilungen der geologischen Institute, ETH Zürich und Universität Zürich: neue Folge*, **236**.

SOFFEL, H. 1975. The palaeomagnetism of the Permian effusives near St. Anton, Vorarlberg (Austria) and the anticlockwise rotation of the northern calcareous Alps through 60°. *Neues Jahrbuch für Geologie und Paläontologie Monatshefte*, **6**, 375–82.

—— 1979. Palaeomagnetism of Permo–Triassic red sandstones from the northern calcareous Alps. *Journal of Geophysics*, **45**, 447–50.

——, POHL, W. & BUSER, S. 1983. Paleomagnetism of Permo–Triassic rocks from Northern Slovenia, Yugoslavia, and the eastern margin of the Adriatic plate. *Tectonophysics*, **91**, 301–20.

VANDENBERG, J. 1979. Palaeomagnetic data from the western Mediterranean: a review. *Geologie en Mijnbouw*, **58**, 161–74.

—— & WONDERS, A. A. H. 1976. Paleomagnetic evidence of large fault displacement around the Po basin. *Tectonophysics*, **33**, 301–20.

—— & —— 1980. Paleomagnetism of late Mesozoic pelagic limestones from the southern Alps. *Journal of Geophysical Research*, **85**, 3623–27.

—— & ZIJDERVELD, H. 1982. Paleomagnetism in the Mediterranean area. *Alpine Mediterranean Geodynamics, AGU Geodynamics Series*, **7**, 83–112.

VAN DEN ENDE, C. 1977. *Palaeomagnetism of Permian red beds of the Dôme de Barrot (S France)*. PhD thesis, State University of Utrecht, 171 pp.

VAN DER VOO, R. & ZIJDERVELD, J. D. A. 1969. Paleomagnetism in the western Mediterranean Area. *Verhandelingen Kon. Ned. Geol. Mijnbouwk, Genoot. Geol. Ser.* **26**, 121–38.

—— & FRENCH, R. B. 1974. Apparent polar wandering for the Atlantic-bordering continents: late Carboniferous to Eocene. *Earth Science Reviews*, **10**, 99–119.

VAN HILTEN, D. 1960. Geology and Permian paleomagnetism of the Val-di-Non area. *Geologica Ultraiectina*, **5**, 95 pp.

—— 1962. Presentation of paleomagnetic data, polar wandering and continental drift. *American Journal of Science*, **260**, 401–26.

—— 1964. Evaluation of some geotectonic hypotheses by paleomagnetism. *Tectonophysics*, **1**, 3–71.

—— & ZIJDERVELD, J. D. A. 1966. The magnetism of the Permian porphyrites near Lugano (northern Italy/Switzerland). *Tectonophysics*, **3**, 429–66.

WAGNER, J. -J., HEDLEY, I. G., STEEN, D., TINKLER, C. & VUAGNAT, M. 1981. Magnetic anisotropy and fabric of some progressively deformed ophiolitic gabbros. *Journal of Geophysical Research*, **86**, 307–15.

WESTPHAL, M. 1973. Etudes paléomagnétiques de quelques formations permiennes et triasiques dans les Alpes Occidentales (France). *Tectonophysics*, **17**, 323–35.

——, BAZHENOV, M. L., LAUER, J. P., PECHERSKY, D. M. & SIBUET, J. C. 1986. Paleomagnetic implications on the evolution of the Tethys belt from the Atlantic ocean to the Pamirs since the Triassic. *Tectonophysics*, **123**, 37–82

Zijderveld, J. D. A. 1975. *Paleomagnetism of the Esterel*. Unpublished PhD thesis, State University of Utrecht.

—— & De Jong, K. A. 1969. Paleomagnetism of some late Paleozoic and Triassic rocks from the eastern Lombardic Alps, Italy. *Geologie en Mijnbouw*, **48**, 559–64.

——, Hazeu, M. & Van Der Voo R. 1970. Shear in the Tethys and the Permian paleomagnetism in the southern Alps, including new results. *Tectonophysics*, **10**, 639–61.

—— & Van Der Voo, R. 1973. Paleomagnetism in the Mediterranean area. *In*: Tarling, D. H. & Runcorn, S. K. (eds). *Implications of Continental Drift to the Earth Sciences*, 133–61.

F. Heller, W. Lowrie & A. M. Hirt, Institut für Geophysik, ETH-Hönggerberg, 8093 Zürich, Switzerland.

Time and place in Alpine orogenesis – the Fermor Lecture

K. J. Hsü

SUMMARY: Alpine geologists have suggested that the main structures of the Alps have been formed by a late Eocene event, which lasted at most a few million years. Data on earlier and later deformations have been interpreted as the evidence of Eoalpıne and Neoalpine phases. The supposedly episodic nature of the Alpine orogenesis has been considered to be a fundamental contradiction of the seafloor spreading which predicts slow and gradual movement.

The purpose of this contemplation on time and place in Alpine orogenesis is to point out that the paradox is man-made: the interpretation of the episodic nature of orogeny is a relic of the defunct catastrophic theory of orogeny of Elie de Beaumont (1831). Divorced from that discredited assumption, the orogenic data in the Alps are consistent with a postulate of continuous orogenic deformation since the early Cretaceous.

The Alps is a collision-type of orogen. Using the late Eocene collision-event as a time marker, we can recognize two major stages of deformation: the Eoalpine subduction, and the Neoalpine post-collisional. The geological evidence is consistent with the postulate of a gradual build-up of the Alpine edifice, consequent upon the approach between two continental blocks since the early Cretaceous and continuing deformation after the late Eocene collision.

The Eoalpine subduction should have started 130 Ma ago, some 50 Ma before the approach of the African and European Plates. This and other geological evidence necessitates an assumption of relative movements among 3 plates to explain the Alpine orogenesis; the middle Mediterranean Plate (M-plate) had movements independent of Africa since 130 Ma BP. The kinematics of plate displacements in the Alpine–Mediterranean region have been reconstructed, using the data on Atlantic magnetic lineations to fix the positions of Africa and Europe, and geological evidence to guide the movement of the M-Plate. This kinematic analysis, harmonized with geological data, indicates that the Alpine Orogenesis resulted from the relative movement between the European- and M-Plates at rates of about 1 centimetre per year. There is no contradiction between Alpine geology and the plate-tectonic theory.

Catastrophism and uniformitarianism

I was here in February delivering a talk on the inevitability of catastrophic events in the history of life, during the Lyell Symposium and in this Lyell Auditorium of the Geological Society, named to honour Charles Lyell, the Uniformitarian and a former president of the society. It was a provocation necessitated by geological evidence. I am happy to say that I have come here today not to bury Lyell, but to praise him. With the firm establishment of the plate tectonic theory during the last decade, Lyellian uniformitarianism in interpreting the chronology of orogenesis has achieved its final triumph.

There was a sound method for determining the relative sequence, but no reliable method to determine the absolute duration, of geological time during the 19th century. Whereas in every other respect geologists were bound by observations, experiences and natural laws, the duration of time was the variable which they could dispose wholly freely. Whether past changes could be catastrophic, or whether apparent catastrophes were the artefacts of an imperfect geological record, was a matter of personal belief, or an article of faith. When alternative interpretations are not falsifiable by observational data, scientists tended to rely on their intuition to choose the more credible. Explanations based upon personal experiences are easily understood, but personal experiences are limited in scope because of the brevity of our life span.

Lyell's insistence that the causes of geological changes in the past never differed in kind or in energy from those we are now experiencing was a premise. This premise was easily accepted by the over-confident scientists of the late 19th century, who thought they had found the answer for everything. With the establishment, early in this century, of a time-scale by radiometric dating, we can examine the soundness of the Lyellian assumption by studying the variation of the energy released by geological processes during Earth history.

The energy release of a natural event varies

From: COWARD, M. P., DIETRICH, D. & PARK, R. G. (eds), 1989, *Alpine Tectonics*, Geological Society Special Publication No. 45, pp. 421–443.

from time to time: There are big storms; there are smaller ones. There are big earthquakes, there are smaller ones. There are big meteorites, there are small ones. There is an inversely linear log/log law, such that the frequency and magnitude of naturally occurring processes is inversely linear when those parameters are plotted on a log/log graph. In other words, catastrophic events seldom occur, but will be inevitable if enough time has elapsed (Hsü 1982).

That such a log/log relation exists was first noted by seismologists (Gutenberg & Richter 1954). Now that data on millions of earthquakes are available, the linear relationship is well established (Hsü 1987a). The same relation has been obtained for meteorite impacts: billions of impact craters have been identified on the Moon and the magnitude/frequency relation is also approximately log/log linear (Shoemaker 1966, 1983).

The energy release of any naturally occurring event is, however, finite. One can ask if the most energetic, or the least frequent, event is likely to happen in historical time of 10^3 years or in geological time of 10^9 years. The statistics on impact indicate that actualistic events give little indication of the destructive potential of those of the past. In our life time, we are witnessing events which cannot alter terrestrial environment; the largest impact in a life time should have an energy equivalent to the explosive energy of about 1 megaton of TNT. Yet impacts a billion times more energetic have happened several times in the history of life on Earth (Shoemaker 1983). There is good geological evidence that one such impact took place at the end of the Mesozoic Era and triggered an environmental crisis and a mass extinction (Alvarez *et al.* 1980, Hsü 1980). Lyell and Darwin erred when they refused to accept the evidence of environmental catastrophes, which were responsible for mass extinctions.

The statistics of earthquakes tells a different story. Few earthquakes can have a magnitude larger than 10, and probably none ever exceeded magnitude 12. The earthquakes taking place during a person's life time has thus a range of magnitudes close to the whole spectrum of possible energy release by earthquakes (Fig. 1). One cannot assume that the Earth has ever been hit by an earthquake a billion times more energetic than the largest one of the century. On this score, Lyell is correct, earthquakes of the past were of the same kind and of the same energy. Since earthquakes are an expression of mountain-building processes, Lyellian uniformitarianism is applicable to interpret orogenesis.

Time in Alpine orogenesis

The deformational structures of the Swiss Alps had not escaped a naturalist of the 18th century, Benedict de Saussure. 'The famed mineralogist describes the calcareous strata which compose the mountain of Axenberg, on the side of the Lake of Lucerne, as having from top to bottom of the mountain the form of the letter S compressed (*écrasée*).' (Playfair 1802, p. 222). Hutton correctly identified the inclined bedding as a structure of orogenic deformation, or angular elevation, and used unconformity as the evidence for an orogenic event. He further considered 'the highly inclined position, and the manifold inflexion of the strata (manifestations) of the violence with which they have been forced up from their original place' (Playfair 1802, p. 47).

Hutton erred when he considered the severity of deformation a sign of violence, but the idea of slow strain rate was apparently not comprehensible to 19th century naturalists. Elie de Beaumont (1831) postulated periodical catastrophes to explain the 'revolutionary' transgressions and regressions of the seas postulated by Cuvier. A corollary was the scheme of dividing geological formations by mountain-building events. The concept was simple and alluring. With the introduction of the geosynclinal concept by Hall, the periodicity idea became popular. Geological history, it was thought, was comparable to human history; there were long ages of tranquillity, alternating with comparatively brief epochs of revolution (Dana 1873). With their mechanistic love for order, earth history was regarded, by doctrinaire authorities around the turn of the 20th century, to have been marked by several critical episodes, during which there were rapid geographic changes which caused rapid biological and climatic changes (e.g. T. C. Chamberlin 1898, 1907).

How many such 'critical episodes' were there? I attempted a review when I wrote my master's thesis (Hsü 1950) and found that Le Conte, following Elie de Beaumont, started with 4 in 1875. Dana doubled the number in 1895. The number was halved by Haug in 1901, although Chamberlin and Salisbury in 1905 wanted again an increase to 6. After that it was a steady rise: Blackwelder proposed 11 in 1914, and Stille in 1924, just after the Great Inflation of Germany, listed 30 orogenic epochs of short duration and global importance. Bucher added a few more in 1933. Eventually we lost count.

Unconformities were the evidence for orogenic deformation. Bucher (1933), relying upon an analysis of a few local unconformities, reiter-

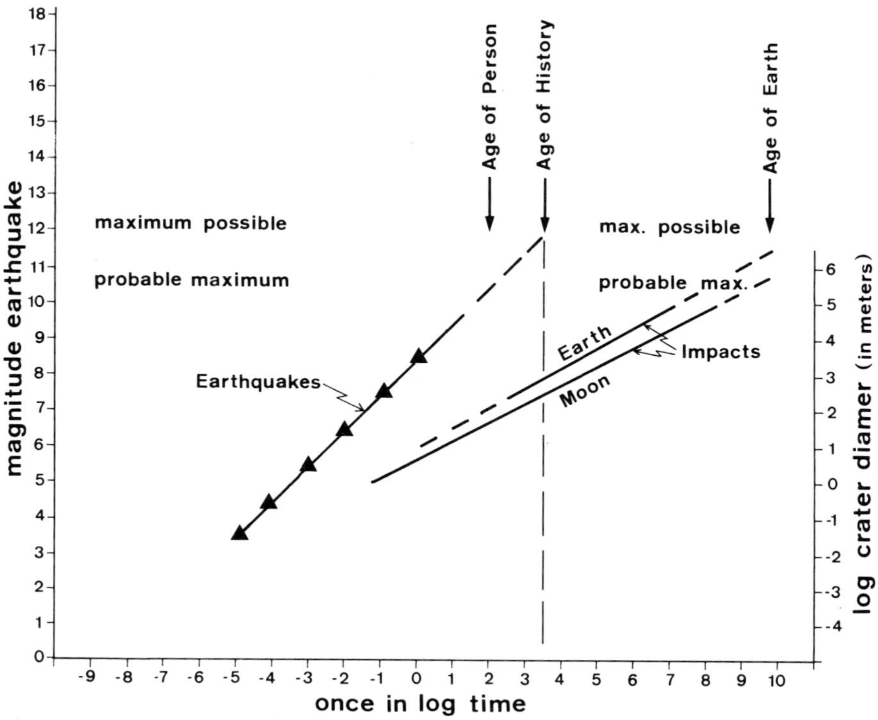

FIG. 1. An earthquake event with maximum magnitude is likely to take place within historical time, if not in a person's life time. Lyell was correct in assuming that mountain-building processes during the past had about the same energy as those of the present; substantive uniformitarianism is applicable to tectonic interpretations. A meteorite-impact event with maximum energy release took place only a few times in the earth history; those were the rare events postulated by catastrophists to explain mass extinctions.

ated the dogma of Elie de Beaumont, Dana and Stille, that orogenic movements are limited to relatively short epochs separated by much longer epochs of rest or of negative movement'.

The concept that orogenic movements are global revolutions of short duration dominated the thinking of geologists during the first half of this century. This concept was in conflict with the Lyellian uniformitarianism. When I went to the United States for my graduate studies, I had the chance to study with Ed Spieker and Jim Gilluly, two of the great enthusiasts of the Lyellian doctrine. Spieker used to tell us that the word 'revolution' should be expunged from the vocabulary of a geologist (see Spieker 1946). Gilluly's (1949) presidential address at the Geological Society of America on the 'Distribution of mountain-building in geologic time' is now a classic.

The controversy between Gilluly and Stille was well known to all geology students in our day, but the plate tectonic theory has rendered all such discussions superfluous. Orogenic movements are consequences of displacements of lithospheric plates, which have been moving at linear rates. There have been no global catastrophes of violence as Hutton or Elie de Beaumont once envisioned.

Although the new theory is generally accepted, old concept dies hard. Unconformity is still held up as the evidence of episodic diastrophism. True, nobody speaks of global catastrophes, but the term 'orogenic phases' is still current. When indications for deformation at various geological times are found, the evidence is interpreted as recurring 'phases', not as various manifestations of one, long continuous process. Therefore, Rudolf Trümpy could speak of a geological contradiction in his farewell address to the Swiss Federal Institute of Technology. While accepting the theory of steady displacement of lithospheric plates, Trümpy (1987) wrote:

> 'From the analyses of deformation processes in the Alps, one obtains the result that many tectonic processes, which affected large areas, took place in *surprisingly short* time, within the span of a few

million years.' (Freely translated by Hsü, italics Trümpy's.)

He added:

'Although it may have been a regional prejudice, because the 'classical' orogeneses, namely the Caledonian, Variscan, and Alpine, were concepts all originating from Europe, a *concentration of deformational phases* is indisputable.' (Freely translated by Hsü, italics Trümpy's.)

Trümpy concluded, therefore, that the geophysical postulate of gradual displacement of lithospheric plates is in conflict with the geological evidence of episodic orogeny.

This question should not be overlooked. The apparent conflict suggests that either one or both concepts are not exactly correct. The purpose of my talk is to examine the evidence for orogenic deformation in the Swiss Alps in an attempt to resolve the paradox.

Place in Alpine orogenesis

I used to be impressed by paleontologists who could name a new species on the basis of a couple of bone fragments. The Peking Man, *Pithecanthropus pekinensis*, was, for example, named on the basis of two teeth found as raw material for aphrodisiac in a Chinese pharmacy. The palaeontologists do not have magic, they have acquired their ability after painstaking studies of the anatomical structures of living and fossil organisms, after every bone has been named and described, its place in the body ordered, and its functional utility understood.

The progress of structural geology during the last few decades has gone so far that we should be able to reconstruct an entire orogenic system on the basis of a few pieces of cores obtained from deep wells drilling into the basement of a sedimentary basin. Trümpy bemoaned the fact that foreign scientists could now drive across Switzerland in a couple of days and think they understand the Alpine orogenesis; Swiss geologists had been working for generations to clarify the intricate relations among hundreds of nappes, slices, or 'schuppen' and they still had difficulty putting them back to their place of origin.

When we put a jigsaw puzzle together, it is much easier if the pieces are cut from a picture than if the pieces are cut from a plywood board. The picture is, in modern scientific terminology, the model. Alpine geologists have been working with a model, the *Deckenbau* model, developed by Marcel Bertrand, Eduard Suess, Albert Heim, Pierre Termier, Maurice Lugeon, Hans Schardt, Emil Argand, Rudolf Staub, and refined by the French and Swiss geologists of the post-war generation, among them, my venerated colleague Rudolf Trümpy. They have named every 'piece of the bone in this organism', and they have found out the functional utility of every part of the Alpine structure. After this achievement, it is no longer a difficult task to translate the Alpine literature into the language of the plate-tectonics theory. One could indeed drive across Switzerland in a few days to present an easily understood outline of the Alpine *Deckenbau*, without having to resort to the use of many terms which were indispensable during the reconstruction of the Alpine edifice. This is what I shall try to do in this talk.

Swiss Alps as a model for continental collision

An Alps formed by continental collision has been an established concept in Switzerland since the turn of the century, while Americans were still quarrelling about the physical impossibility of continental 'drift'. Prior to a collision, oceanic crust should have been underthrust, or 'subducted', only a trace of the ocean being found in the ophiolitic melanges of the suture zone. After collision, one of the two continental blocks has to be underthrust, the other overriding. In the structures of every collision-type of orogen, the three major elements are thus: (1) the overriding continent, (2) the ophiolitic melange, and (3) the underthrust continent.

Elements more or less corresponding to those three have long been recognized in the Swiss Alps (see Trümpy 1960). The crust of the overriding continent has been sliced into numerous wedges and overthrust northward. Those constitute the Austro–Alpine nappes of eastern Switzerland and their correlative, the Dent Blanche nappe (DB in Fig. 2), in the W. They are also called rigid-basement nappes, because they were displaced bodily; the 'Hercynian' basement and the overlying sedimentary cover have not been subjected to much strain or metamorphism during the Alpine orogenesis. The more 'interior', or more southerly, overriding wedges have, however, been more deeply buried and thus metamorphosed during the Alpine deformation; they are the Sesia (Se) and Canavese (Ca) zones. Those have brought up into contact with the rigid basement of the southern Alps by back-thrusting (Schmid this volume).

The ophiolitic melanges of the Swiss Alps are, as a rule, found directly under the Austro–Alpine nappes. They are known by various local names, e.g., Arosa Schuppen zone and Platta Decke in the E, and Saas–Zermatt zone in the W. The ophiolites have been labelled High Penninic because they are superposed tectonically above mobilized basement nappes of the Pennine Alps.

There are exceptions to the rule. The Margna nappe (MS) of eastern Switzerland, for example, is sandwiched between two ophiolitic melanges (see Trümpy 1980). Furthermore, ophiolitic rocks are present in the Misox zone between the Adula (Ad) and Tambo (Tb) nappes in eastern Switzerland, and submarine basalts are present in the Valais zone (V) in western Switzerland. Those occurrences do not necessarily contradict the plate-tectonic theory, but they do require a model more complicated than a simple collision of two blocks to explain the Alpine Orogenesis. I shall discuss this in a later section.

The margin of the northern continent has been thrust under the overriding Austro–Alpine nappes, with the ophiolitic melanges sandwiched between the underthrust and the *traineau écraseur*. The structures of the underthrust are the most complicated in the Alps. The three subdivisions are (1) the mobilized basement nappes (mB in Fig. 3), (2) the mobilized cover (mC), and (3) the *décollement* cover nappes (CT).

The best known mobilized basement nappes are the Bernhard (BD) and Monte Rosa (MR) nappes in western Switzerland, the Adula, Tambo and Suretta (Su) nappes in eastern Switzerland, and the Ticino nappes (A, L, ML, Ma, Sim) in southern Switzerland.

The mobilized sedimentary cover is the more metamorphosed *Schistes Lustrés* of the west and the less metamorphosed *Bündnerschiefer* of the east. Slightly metamorphosed cover in the northern Penninic region has been referred to as the pre-flysch formations of the Valais Belt (V) in the west, and as the Prätigau pre-flysch (P) in the east (Trümpy 1960).

Detached sedimentary cover has been piled up in the high calcareous Alps of Switzerland in the form of the Helvetic nappes. The mainly carbonate sequence was laid down on the southern margin of the Mesozoic and early Palaeogene Europe.

Cover nappes, present as klippes in the pre-Alps, include the Simmen (Si), Breccia (Br), and Médian (or Klippen, KL) Nappes. The Médian sediments were the cover of a basement which has become the Bernhard nappe; *Briançonnais* and *Sub-Briançonnais* are paleogeographical terms designating the sedimentation sites of the Médian and Bernhard rocks (Trümpy 1960). The Breccia nappe consists mainly of sedimentary breccias deposited at the foot of a steep slope of the Piémont Ocean, or the Italian–French–Swiss segment of the Tethys. The rocks of the Simmen nappe are mainly coarse clastics, with detritus of radiolarites and ophiolites.

The third group of cover nappes are the flysch and Wildflysch nappes. Their origin and the palaeogeographic significance will be discussed in a later section.

The part of the northern continent which has not been greatly displaced is traditionally referred to as autochthonous or parautochthonous. The Palaeozoic granitic basement is exposed in the Aar (AM) and Gotthard (GM) Massifs, and their sedimentary cover forms parautochthonous nappes (PA).

North of the Alpine Front are the molasse basin and the Jura Mountains. They are, strictly speaking, not a part of the Alps, but their deformation is a continuation of the process which has formed the Alps.

The ghost of Elie De Beaumont

The edifice of the Swiss Alps was, of course, not formed in one catastrophic upheaval. Trümpy (1973) suggested three orogenic phases: Palaeoalpine, Mesoalpine, and Neoalpine, with intervening stages of subsidence or of vertical uplift; the Austro–Alpine is Palaeoalpine, the Pennine Mesoalpine, and the Helvetic Neoalpine. He believed, however, that 'compressional deformations took place during distinct, relatively short periods; uplift and erosion followed only after a significant delay' (Trümpy 1973, p. 247).

This conclusion, in my opinion, is not justified by facts, but is a consequence of his assumption. Trümpy's judgment has been betrayed by the title of his paper: *The Timing of Orogenic* ***Events*** *in the central Alps*, which assumed that there have been events. Trümpy (p. 236) wrote:

> 'The main deformation of the western Alps occurred ... at the end of the Eocene and/or the beginning of the Oligocene. The time of this Mesoalpine paroxysm is very well established in the western Alps. It affected the whole Penninic belt, the Ultrapenninic units ... and probably also part of the Ultradauphinois belt ...
>
> The lower bracket is determined by the age of the youngest formations preserved

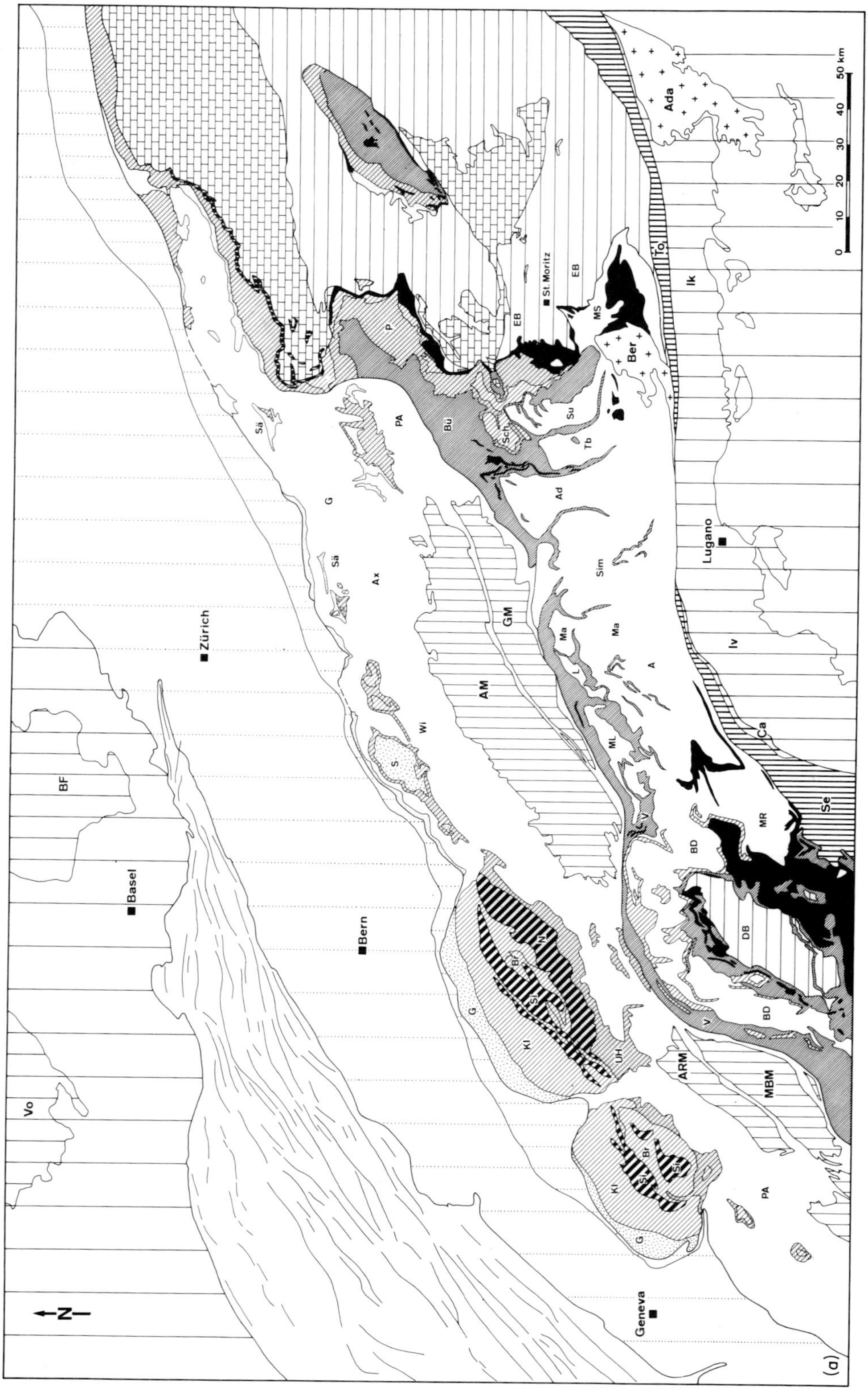
Basel
Zürich
Bern
Geneva
Lugano
St. Moritz
Ada
Ber
AM
GM
ARM
MBM
BF
Vo
0 10 20 30 40 50 km
(a)

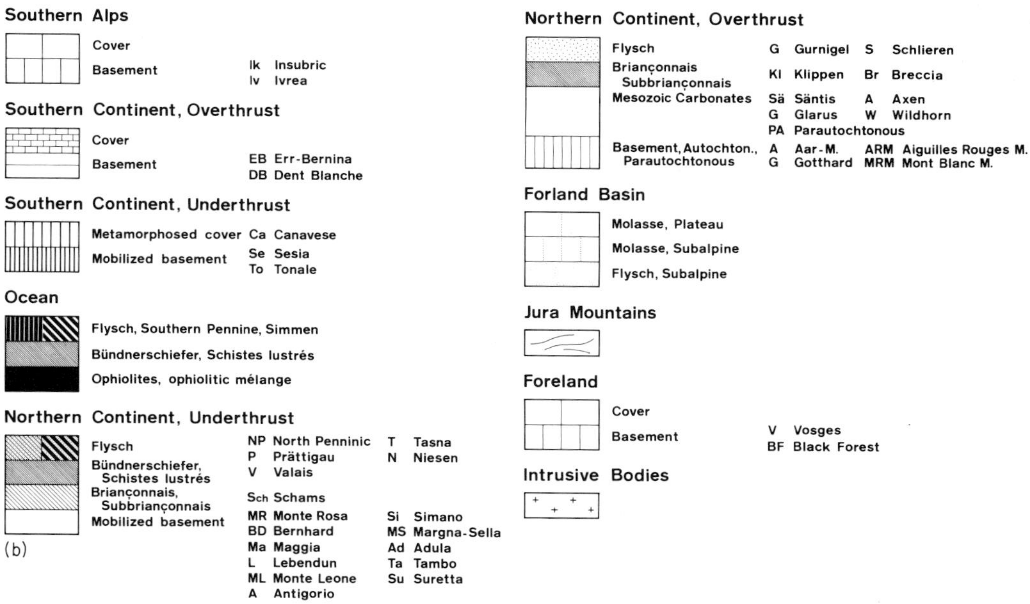

FIG. 2. (*Left, Key above.*) Tectonic map of the Swiss Alps. This map is revised from the Geological Map of Switzerland and Tectonic Map of Switzerland, published by the Swiss Geological Commission at the scale of 1:500 000. The traditional nappes are classified according to their position in inter-continental collision. They are tectonic units of the overriding block (mainly Austro–Alpine), in the suture zone (mainly ophiolites), and from the underying block (mainly Penninic and helvetic Nappes).

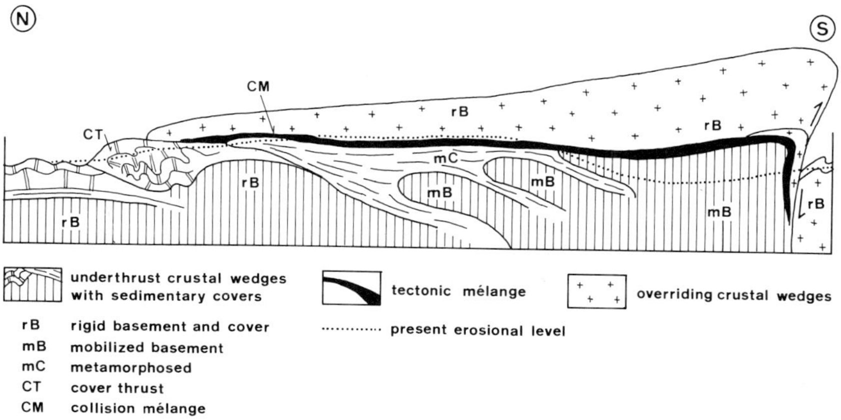

FIG. 3. Schematic section showing the Swiss Alps as a model for the collision-type of orogeny. Although the overriding block consists mainly of rigid basement nappes (Austro–Alpine), the more deeply buried part of the block, the Sesia zone, has been mobilized, and is now brought up by backthrusting to be in contact with the rigid basement of the southern Alps. The basement of the underthrust block is largely mobilized (Penninic), whereas the sedimentary cover on the margin was sheared off above décollement horizons to form cover nappes (Helvetic). The molasse is a foreland basin.

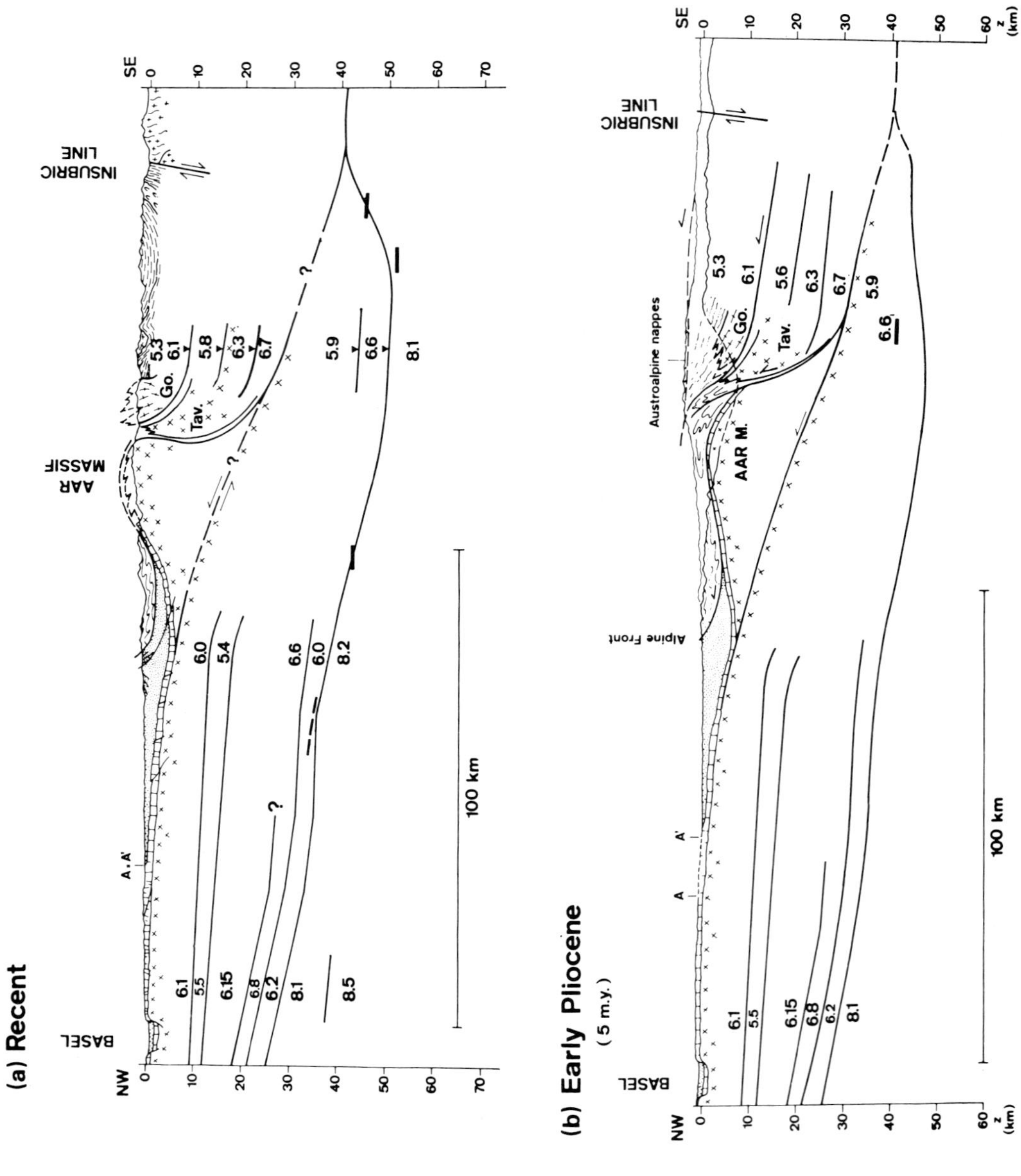
(a) Recent
NW
SE
BASEL
A, A'
AAR MASSIF
INSUBRIC LINE
Go.
Tav.
100 km
(b) Early Pliocene
(5 m.y.)
A
A'
Alpine Front
AAR M.
Austroalpine nappes
z (km)

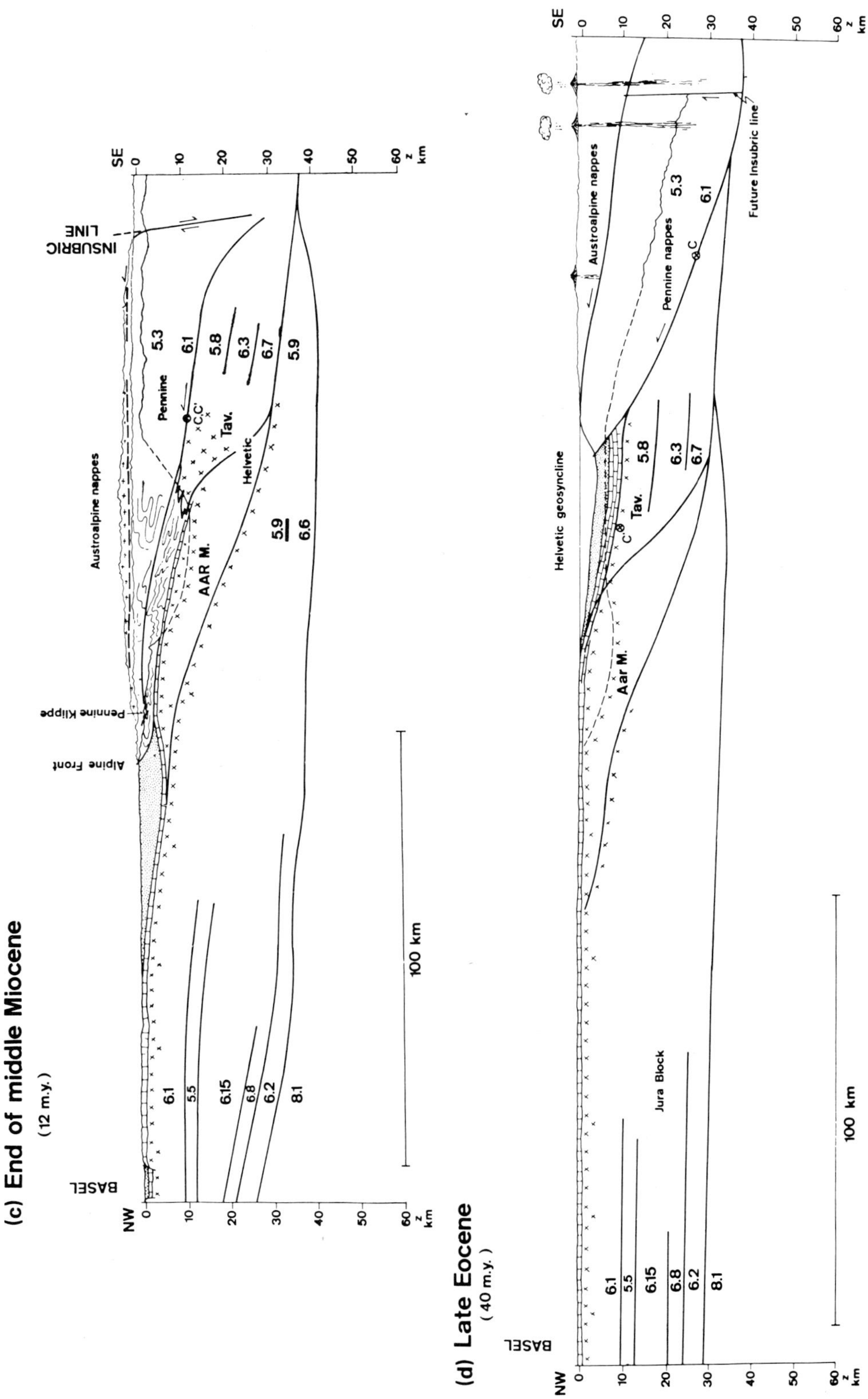

FIG. 4. Neoalpine Orogenesis (after Hsü 1979). The Neoalpine orogenesis is interpreted as a consequence of underthrusting of crustal wedges after the collision. The underthrusting took place along low-angle shear surfaces within the continental crust, which has been detached from the underlying mantle at the Moho.

in this belt. The flysch of the Briançonnais belt is certainly in part of middle Eocene (Lutetian) age and may comprise the early upper Eocene . . . The best upper bracket is furnished by the angular unconformity of molasse beds in Liguria . . . Conglomerates, deposited over a hilly landscape, truncate folded, thrust, and metamorphosed Penninic rocks. The conglomerates . . . are considered to belong to the lower Oligocene.'

The conclusion that a post-Middle Eocene and pre-early Oligocene event 'affected the whole Penninic belt' is based upon the concept of episodic orogeny of the Elie de Beaumont/Dana/Stille/Bucher tradition. Assuming that there is a time for everything, the geologic evidence was made into a drama of three acts: (1) A geosynclinal stage of sedimentation in the Penninic realm which ended with the Briançonnais flysch in Middle or late Eocene, (2) An episodic orogenic event which affected the *whole* Penninic belt, and (3) the deposition of post-orogenic molasse sediments above the deformed and eroded Penninic rocks.

I italicize the word 'whole' because this is the key word. The deformation did not have to affect the whole region all at once, the deformation could have happened at a different time at a different place. The Penninic rocks of Liguria could have been metamorphosed at an earlier age than the overthrusting of the Briançonnais flysch. Without the assumption of an episodic orogeny affecting the whole region, the pre-molasse unconformity in Liguria has little relevance to the timing of the post-flysch deformation in the pre-Alps.

Radiometric dates of Cretaceous Penninic deformation and metamorphism in Liguria are, in fact, not uncommon. Trümpy had known about those dates; but those, in his opinion, indicated a Cretaceous event – an earlier episode, prior to the 'late Eocene paroxysmal event which affected the whole of the Penninic Belt'. Even today the concept of an 'Eoalpine *event*' prevails in the geological literature (e.g., Hunziker & Martinotti 1984, Tollmann 1987).

The plate tectonic model and the Stille dogma are a geological contradiction, as Trümpy emphasized. There can be no compromise. One cannot pay lip service to the plate tectonic theory, while continuing to adhere to the traditional thinking of 'orogenic events', and of 'tectonic phases'. That there were orogenic events is a hypothesis, not an inescapable conclusion. That there was a paroxysmal late Eocene Mesoalpine event is circular reasoning, starting with a premise of episodic orogeny which is also the conclusion. Proceeding from a premise that orogenesis does not consist of episodic *events*, but is a long and continuous process, I see no conflict between Alpine geology and the new theory of global tectonics.

Time and its place in Alpine orogenesis

We all accept the conclusion that the Swiss Alps have been formed by continental collision, and that the collision was an event. But, in the scheme of plate tectonics, the Alps were not formed by the event, but by continuous deformation before and after the event. Using the time of collision as a reference datum, we can recognize two stages of compressional deformation to replace the three 'events' proposed by Trümpy. Prior to the collision, the deformation was continent-ocean interaction and should have involved the subduction of oceanic lithosphere. This stage is the Eoalpine deformation. The orogenesis after the collision resulted from continent–continent inter-action, when one continental block was thrust under another. Deeply buried continental basement and part of its sedimentary cover were metamorphosed and mobilized, another part of the cover was stripped off and pushed onto the foreland as cover nappes. Meanwhile, partial melting of the underthrust crustal wedges led to the origin of post-collision granites. Further crustal shortening caused the thin-skinned deformation of the sedimentary cover of the foreland. The deformations after the collision have been called Mesoalpine and Neoalpine (Trümpy 1973). Since there is no clearcut event marker to separate the two, I prefer to refer to all those as the Neoalpine orogenesis (Hsü 1979).

The geological evidence is most clearcut for the latest history. I shall, therefore, start from the Holocene and go back in time to relate the time and place of Alpine orogenesis:

(1) Plio–Quarternary Jura deformation. The *décollement* of the Jura Mountains has been held up as a model of thin-skinned cover deformation. Reconstructions give a crustal shortening of 10–20 km under the Jura since the end of the Miocene some 5 Ma ago (Laubscher 1971); at rate of about 0.2–0.4 cm yr^{-1}. The Jura deformation can be explained by assuming an underthrusting of the Jura basement. Relative to that was the continuous northward movement of a crustal wedge under the Aar Massif (Fig. 4a, b). The push detached the molasse

and Jura sedimentary sequences from their basement. While the molasse basin was bodily displaced, the Jura cover was thrown into disharmonic folds. At the same time, the advancing wedge was slightly warped, causing the uplift of the Aar Massif.

(2) Oligo–Miocene Helvetic, Ultrahelvetic, and pre-Alpine deformations. Trümpy suggested a Miocene phase of Helvetic and an Oligocene phase of pre-Alpine deformation. We all agree that the Helvetic nappes represent thin-skinned cover deformation under compression. The pre-Alpine nappes were once considered gravity-slide structures (Lugeon & Gagnebin 1941, Trümpy 1973). This hypothesis was popular during the 1950s when geophysicists thought overthrusting of thin slabs of weak sedimentary strata mechanically impossible by a 'push from behind' (Hubbert & Rubey 1959). In fact, 'push from behind' was a strawman, because knowledgeable Alpine geologists had always invoked the 'plunger action' of a *traineau écraseur* to explain the deformation of cover nappes (Termier 1903). The stress-transmittal presents no problem, if all the detachment thrusts, the Helvetic, the Ultrahelvetic, and the pre-Alpine nappes, have been formed by the shearing stress induced by overriding Austro–Alpine nappes. The ductile deformation of carbonate strata resulted in the Helvetic recumbent folds, and the brittle deformations of flysch sandstones formed the broken formations of the Ultrahelvetic Wildflysch. There is no need to postulate different 'phases', with different stress conditions, for the different cover nappes.

The Helvetic deformation, which was mainly Miocene, has involved a crustal shortening of about 100 km, at a rate of about 0.5 cm yr^{-1} since 25 Ma BP. Kinematic analysis of plate positions suggest a total of 200–250 km of crustal shortening during the 40 Ma since the late Eocene (Le Pichon 1968), when continental collision took place. From this, one may conclude that another 100 or 150 km of crust disappeared during the Oligocene, from under the flysch and pre-Alpine nappes. Summing up, I suggest that the Helvetic, flysch, and pre-Alpine nappes were formed steadily as their basement was being thrust under the Austro–Alpine nappes, at a rate of about 0.5–0.7 cm yr^{-1}, during the Oligocene and Miocene (Fig. 4c, d).

In an analysis of the statics and kinetics of the Glarus overthrust, its displacement rate was estimated by relating stress, strain rates and temperature of deformation. I concluded that the thrust moved at a rate not less than 0.2 cm and not more than 10 cm yr^{-1} (Hsü 1969). The present kinematic estimate of 0.5–0.7 cm yr^{-1} falls well within the range obtained previously on the basis of experimental rock deformation.

(3) Oligo–Miocene Penninic deformation. The sedimentary cover, including sediments as young as Middle Eocene or lower Upper Eocene, of the Briançonnais basement had been sheared off to form the Médian nappes before the basement was underthrust and metamorphosed. The *décollement* took place at about 45 or 40 Ma BP. Radiometric dates indicate that the main Pennine metamorphism started at about 38 Ma BP (Niggli 1970). Thus the rocks at depth were metamorphosed at about the same time as, or only shortly after, the sedimentary cover on top was pushed away by the advancing Austro–Alpine nappes.

Radiometric ages of metamorphism or of cooling (after uplift) range to as young as early late Miocene (12 Ma BP). Tectonic and structural analyses, such as those presented in this volume, have suggested that the Pennine deformation was long and involved, including several stages of varying directions of tectonic transport. The available geological data are thus quite consistent with the postulate of continuing deformation after the collision event. The plutonic alterations of the underthrust basement proceeded while the cover nappes advanced. The thickening of continental crust caused the eventual isostatic uplift of the piled-up crustal wedges, as well as the cooling of metamorphosed Penninic rocks when they were unroofed by erosion (Fig. 4c, d).

(4) Neoalpine granites. The Bergell granites (Ber.) constitute the only major batholith of the Swiss Alps. This body is intrusive on its northeastern border, but concordant to the Penninic country rocks on its southwestern border. Pebbles identified as having been derived from the Bergell are present in the Como molasse of Oligocene age. Radiometric dates of the granites, ranging from 30–18 Ma, have been considered cooling ages (see summary in Trümpy 1980).

The Bergell granites are post-collision granites. They owed their origin to partial melting of a subducted continental crustal wedge. The field evidence indicates intrusive emplacement during an earlier stage of Neoalpine Orogenesis, and also that the body was deformed under plutonic conditions upon the continuation of crustal shortening.

(5) Eoalpine subduction. Using late Eocene collision as an event marker, the earlier deformations in the Alpine realm should have been the subduction of oceanic crust. This corollary of the plate-tectonic theory has been appreciated by our colleagues in Austria, where evidence for Cretaceous deformation and metamorphism is unmistakable (Flügel *et al.* 1987, Frank 1987). Tollmann (1987) wrote:

> 'The subduction of the different parts of the base of the geosyncline and of the adjoining margin of the foreland and also particularly the thrusting of the nappes during the orogenesis did not take place at equal rates; instead distinct *phases* of higher activity alternated with more quiet epochs.' (Italics by Tollmann)

This is a pronouncement by an authority deeply rooted in the tradition of the theory of geosycline and of episodic orogenic phases. I know of no geological data in the Alps to contradict the assumption of an actualistic analogue through the comparison of Eoalpine subduction to the deformation in circum-Pacific regions.

Two circum-Pacific types of orogen have been distinguished: (1) the Andean type, dominated by igneous activities related to the partial melting of subducted plate, and (2) the Franciscan type, characterized by ophiolitic mélanges with exotic blocks of high-pressure metamorphics (Hsü 1972a). There was little andesitic volcanism or batholithic intrusion during the Eoalpine Orogenesis. This has been explained by the fact that the relative motion across the Tethyan plate-boundary has a large lateral component (Hsü 1972b). There is, however, much evidence for the Franciscan type of deformation.

The indications of Eoalpine deformations by subduction are the radiometric ages of up to 130 Ma for the eclogites and of 90–60 Ma for other high-pressure metamorphics of the Sesia zone (Dal Piaz, *et al.* 1972, Hunziker 1974), of 100–80 Ma for the glaucophane-bearing rocks in ophiolitic mélanges of the western Alps (Bocquet *et al.* 1974), and of 110–70 Ma for mélange rocks containing alkali amphiboles in the Platta nappe of eastern Switzerland (Deutsch 1983). Applying the Franciscan model, one can postulate that the high-pressure rocks formed at great depth were jacked up by under-plating, and raised to great heights by a thick pile of ophiolitic mélanges, which had been tucked under the Austro–Alpine basement. The mélanges in coastal mountains on the southern margin of the Cretaceous Tethys could then supply detritus for deep-sea turbidites.

This scenario is confirmed by the finding of clastic grains of ophiolitic rocks in Penninic flysch of Aptian age in Austria (Flügel *et al.* 1987), and of glaucophane and lawsonite detritus in the south Penninic Flysch of Turonian/Coniacian age in eastern Switzerland. These sediments of 110–85 Ma age were deposited not very long after the start of the Eoalpine metamorphism (Flügel *et al.* 1987, Winkler & Bernoulli 1986). In western Switzerland, the Simmen Nappe consists of slices of ophiolite, pelagic sediments, and of the Simmen flysch. The flysch conglomerates include clasts from ophiolitic mélange, as well as detritus from an Austro–Alpine provenance. The flysch sandstones, ranging from late Cenomanian to Santonian (95–80 Ma), are distinguished by the detritus derived from ophiolites, such as chrome spinel (Wildi 1985). The Simmen flysch had thus a palaeogeographic position equivalent to that of the south Penninic Flysch of the east; those turbidites were laid down in foredeep and/or on abyssal plain N of an Eoalpine subduction zone.

The presence of abundant glaucophane detritus in the Niesen flysch has been reported on the basis of second-hand information (Niggli 1970). If this is true, the Niesen flysch should have been deposited at a site adjacent to the Simmen. The report is, however, refuted by more recent studies which showed that the Niesen heavy-mineral suite consisted of zircon, tourmaline, sphene and apatite, indicative of derivation from terranes underlain by continental crust (Wildi 1985). If so, the palaeogeographic position of Niesen remains as much of a mystery as before. New evidence tends to support the classical interpretation of a northern Penninic origin for the Niesen (Homewood, pers. comm.).

The model and the real world

Assuming a simple two-plate collision model, the place for the northern continent should be beneath the ophiolite mélange, and that for the southern continent above. The timing for the nappe deformation involving the rocks on the margin of the northern continent (Middle and Lower Penninic) should have started after the collision, and that for the overlying tectonic units (Austro–Alpine) before the collision. These predictions have been proved false: ophiolite mélanges have been found above and below Pennine nappes, and radiometric dates of Eoalpine metamorphism are not uncommon in Pennine rocks. The simple model has thus to be modified to accommodate the apparently conflicting evidence.

No one has ever assumed that the Alpine history was simple. The presence of ophiolites at apparently more than one tectonic level within the Pennine belt and the Eoalpine ages of deformation have been explained by several hypotheses:

(1) The 'involuted-mélange' hypothesis. This hypothesis assumed the presence of one ophiolite mélange, derived from the subduction of one major ocean, the Piémont Trough. One can assume that the mélange has been folded and involuted, so that they appear to be be present at several tectonic levels.
(2) The multiple-collision hypothesis. One could assume one ocean, but also that there was a collision earlier than late Eocene. The Eoalpine mélange was subsequently folded and/or thrust during the Neoalpine orogenesis, so that it is seen at various tectonic levels within the Lower Austro–Alpine and Penninic nappe complexes.
(3) The multiple-subductions hypothesis. One could assume two Eoalpine subduction zones, at least during certain time-intervals, on both margins of the Piémont Ocean. According to this hypothesis, the main ophiolite mélanges of the Saas/Zermatt and Arosa zones were formed in a S-dipping zone, but the mélanges in the middle and lower Pennine complexes may have been resulted from subduction on the northern margin of the Tethys Ocean.
(4) The multiple-oceans hypothesis. One could assume that each ophiolite body represents the relic of one small ocean basin between islands or submarine swells within the Tethyan Ocean. One variant of this hypothesis assumes that the 'geosynclinal' sequence of the Valais belt, which includes locally submarine volcanics, was deposited in the Valais Trough, which was separated by a swell from the main Piémont Trough. The Briançonnais/Sub-Briançonnais rocks of the Bernhard Nappe in the west, and of the Schams and Falknis/ Sulzfluh Nappes in the E, represent deposition on on that submarine swell.

The four hypotheses are not mutually exclusive, because even the assumption of one major Piémont Trough does not exclude the possibility that small backarc basins were present. Each hypothesis is applicable to explain one or more of the local situations. Since the first has been well formulated in the Alpine literature (see Trümpy 1960), I shall discuss briefly the second, third, and fourth models.

A Cretaceous collision in eastern Switzerland has been predicted on the basis of a kinematic analysis of plate movements (Hsü 1979b). Such an earlier collision could explain the anomalous occurrences of ophiolite mélanges both above and below the Margna nappe (MS) in the Upper Engadine of eastern Switzerland. This nappe has a basement core, and a sedimentary cover of Lower Austro–Alpine facies. But it has been traditionally considered a Penninic nappe, because it underlies the main ophiolite mélange of the Platta nappe. Yet the Margna nappe has been superposed on another ophiolitic mélange, which overlies tectonically the Penninic Suretta Nappe (Trümpy 1980). One could assume an ocean north of, and another S of, a Margna Island (in addition to the north Penninic Valais Trough). Alternatively one could consider the sub-Margna mélange a Cretaceous suture and the supra-Margna mélange an Eocene suture.

The postulate of a Cretaceous collision in eastern Switzerland is supported by the timing and the nature of Eoalpine metamorphism. The basement rocks altered by the Alpine metamorphism in the Austro–Alpine nappes have radiometric ages of 125–80 Ma (Satir 1975, Thöni 1983). The metamorphism of the gneissic rocks of Suretta has an 118 Ma age (Steinitz & Jäger 1981). Furthermore the metamorphism is not the high-pressure type, characteristic of the Eoalpine subduction.

The puzzling geology of the Schams Nappes, discussed briefly by Trümpy (1980, pp. 70–71), could be clarified by the assumption of a Cretaceous collision. Work by Stephan Schmid and his students is coming up with the evidence for earlier deformation with a N–S fold axis. Perhaps those cover nappes were stripped off and carried westward after the Cretaceous collision, before they were folded again during the Neoalpine deformation.

When I first proposed late Cretaceous collision, a colleague remarked that the idea is contradicted by the evidence in the Lower

Engadine Window area, where the Austro–Alpine nappes and ophiolite mélanges were overthrust on top of a Penninic flysch of Eocene age. This remark reveals the deep root of tradition in modern-day geologists. The evidence at Lower Engadine indicates that post-collisional movement continued until after Eocene, but it does not tell us when the continents on both sides of the Tethys first collided.

There is little evidence for a Cretaceous collision in the western Alps, where such a postulate depends upon an interpretation of the Cretaceous metamorphic ages (Tricart 1984). The evidence can, however, be better explained by the multiple-ocean or multiple-subduction hypotheses.

Hunziker suggests (this volume) that the main subduction zone of Eoalpine deformation could be north of the Briançonnais swell (see also Hunziker & Martinotti 1984). He seems thus to echo the opinion of Milnes *et al.* (1981), who believed that the Zermatt/Saas and Antrona ophiolite mélanges of the Penninic belt are one and the same between the Bernhard and Monte Rosa nappes. But if we accept the interpretation that the Briançonnais was the southern margin of Europe, the subduction zone consuming the Tethys Ocean had to be south of the Briançonnais and had to dip to the north. There is clear evidence for a Cretaceous S-dipping subduction zone on the Austro–Alpine margin. This fact, however, does not exclude the possibility that the Tethys, like the Celebes Sea today, was at times bounded by subduction zones on both sides. The postulate of N-dipping subduction on the Cretaceous European margin will not only explain the evidence of high-pressure metamorphism in the Penninic realm (Desmons 1986), but could also be invoked in support of a hypothesis that the Briançonnais was a part of an island arc system.

The Briançonnais was considered, in the classical geosynclinal theory of orogeny, a geanticline between the Piémont Ocean and the Valais trough. Interpreted within the framework of plate tectonics, such a positive feature underlain by continental crust could be an island arc. I now suggest a scenario that the arc was ripped apart from the southern margin of Europe during the early Cretaceous, while the crust of the Piémont Ocean was subducted down a N-dipping Benioff zone. The Valais trough was a back-arc basin behind the Briançonnais–Falknis island arc, like the Aegean basin behind the Cretan Arc. Neotectonic studies of west Pacific arcs suggest that back-arc underthrusting is a common phenomenon. The relic back-arc basin of the South China Sea is, for example, being subducted along an E-dipping Benioff zone under the Philippine Arc. Applying this model to interpret Alpine geology, one could suppose that this back-arc basin was eliminated by arc–continent collision, when the Briançonnais Arc was, through 'back-arc collapse' during the early Tertiary, sutured onto the European margin, along a S-dipping Benioff zone. According to this interpretation, the N Penninic ophiolites are the relics of a Cretaceous basin behind an island arc.

The geology of the Pennine Alps is so complicated that much future work is needed for the final clarification of the Cretaceous history of the northern Tethyan margin.

Time and place of flysch sedimentation

The term *flysch* was originally introduced to designate a lithostratigraphical unit of Eocene age in the Alps. Practically all flysch formations include turbidites, but not all turbidites are flysch (Hsü 1970). Flysch has been called synorogenic sediments and molasse post-orogenic; but those designations assume that orogenic movements were episodic events. In the scheme of continuous Alpine deformation, both flysch and molasse are syn-orogenic foredeep sediments, the former in deep sea trenches and the latter in foreland basins.

The Upper Eocene Sub-Alpine and north Helvetic flysch of the Alps were deposited on a European margin that had been elastically bent down by the weight of the advancing Austro–Alpine nappes; they are post-collision. The south Penninic, Simmen, and Niesen flysch, as we have stated previously, were pre-collision trench or abyssal sediments. What are, then, the Schlieren (S) and Gurnigel (G) flysch?

Those were once called 'Ultrahelvetic' because exotic blocks of Mesozoic carbonates associated with those flysch formations have a sedimentary facies indicative of deposition just south of the Helvetic realm. This interpretation assumed a southerly source for the flysch detritus, which is not Helvetic in origin. This was the case for the Schlieren, but not for the Gurnigel Flysch, which was found to have been derived from a northerly source (Crowell 1955, Hsü 1960).

Sedimentological data thus do not necessarily argue against an Ultrahelvetic origin of the Schlieren Flysch. The abundant large foraminifera in this flysch formation may well have been derived from a carbonate shelf on the southern

Helvetic margin. I have, therefore, considered the Schlieren to be Ultrahelvetic, and the Gurnigel, north Penninic (Hsü 1960, Hsü & Schlanger 1971).

The directions of the Palaeocene/Eocene current-transport for the Schlieren were northerly and easterly, and those for the Gurnigel were southerly and easterly, and a pink-granite detritus (Habkern Type) is common to both. The two flysch basins were thus apparently separated by a chain of islands on which a pink granite cropped out (Hsü 1960) and this chain of islands could well be an island arc – the 'Habkern Arc'. Hsü & Schlanger (1971) compared the Palaeocene southern European margin to the Aegean margin today. The Schlieren, like the Cretan, was a backarc basin, the Gurnigel, like the Hellenic Trench, was a forearc deep, whereas the Briançonnais, like the Mediterranean Ridge, was a rising 'geanticline'. The Simmen strata, like those on the Nile Fan today, were abyssal and submarine fan deposits in a Mediterranean ocean (Fig. 5).

According to this interpretation, the Habkern Arc came into existence in latest Maastrichtian time. The Palaeocene Schlieren basin occupied the same position north of an island arc as the Cretaceous 'Valais trough' north of the 'Briançonnais swell'. The genesis of the Habkern Arc would again imply the existence of a N-dipping subduction zone (beneath the Gurnigel Trench), along which the 'Valais trough' was subducted. At the same time, this Cretaceous back-arc basin was being consumed by a S-dipping Benioff zone N of the Briançonnais Arc.

If there had been such an arc, there should have been volcanoes. Evidence for late Cretaceous and early Palaeogene volcanism at or near the European margin has indeed been found by sedimentary petrographers. Winkler (1981) noted two detrital populations within the Schlieren flysch: One was fed longitudinally from a pink-granite source-terrane to the west, whereas the other was fed in from a southern volcanic arc. If the volcanic detritus is Maastrichtian and/or Palaeocene, its presence as Schlieren component supports the postulate of a back-arc setting. The Gurnigel flysch, the forearc trench sediments, received no volcanic detritus.

Andesitic volcanism apparently continued as late as Eocene, because the Taveyanne sandstone consists largely of volcanic debris, which is believed to have been derived from volcanism related to a subduction zone underlying the Ultrahelvetics (Sawatzki 1975, Giraud 1983).

One final piece of the jigsaw puzzle in favour of N-dipping subduction is the sudden deepening of the Helvetic, or the European, margin during the Aptian. The Barremian–Aptian shelf sediments of the Helvetic were overlain by open marine detrital sediments (Albian/Cenomanian) and by pelagic oozes and marls (Cenomanian/Turonian). The subsidence could not be related to isostasy, it could be explained by assuming an elastic downbend of the lithosphere when a passive margin was changed to an active margin by subduction (Hsü 1982b).

The N-dipping subduction zone under the 'Habkern Arc' like that under the Cretan Arc today, may only have a limited extent. There is little evidence the Schlieren backarc basin extended far eastward beyond the Wägital in eastern Switzerland (Fig. 5).

I have gone into much detail to present my interpretation of the Gurnigel and Schlieren flysch. The scenario is based upon a plate tectonic model, and seems far too complicated. Many Swiss scientists prefer the interpretation of the Fribourg School, who placed the Gurnigel and Schlieren south of the Briançonnais (Homewood & Caron 1982, p. 161). They argued that 'the earlier Maastrichtian sequences of the otherwise 'sandy' Flysch closely resemble the Helminthoid flysch facies.' I do not accept this kind of reasoning because similarity does not necessarily imply proximity. Furthermore, if Schlieren or Gurnigel turbidites were deposited in the Piémont Ocean, they should contain ophiolite detritus, such as those of the south Penninic, or the Simmen flysch. Neither Schlieren nor Gurnigel has such detritus.

Seafloor spreading and Alpine orogenesis

Numerous attempts have been made to relate the Alpine deformation to the relative displaceme etween the European and African Plates (e.g. Hsü 1971, Smith 1971, Dewey *et al.* 1973, Bijü Duval *et al.* 1977, Smith & Woodcock 1982, Savostin *et al.* 1986). This displacement can be calculated on the basis of the data on the north Atlantic magnetic lineations. There is general agreement that the extensional Alpine deformation, from Triassic to Cretaceous, was an expression of sinistral displacement and the compressional deformation, from Cretaceous to Recent, of dextral displacement.

Although a two-plate model gives a first-order correlation between seafloor-spreading, plate-displacement, and Alpine orogenesis, closer analyses have revealed contradictions. The

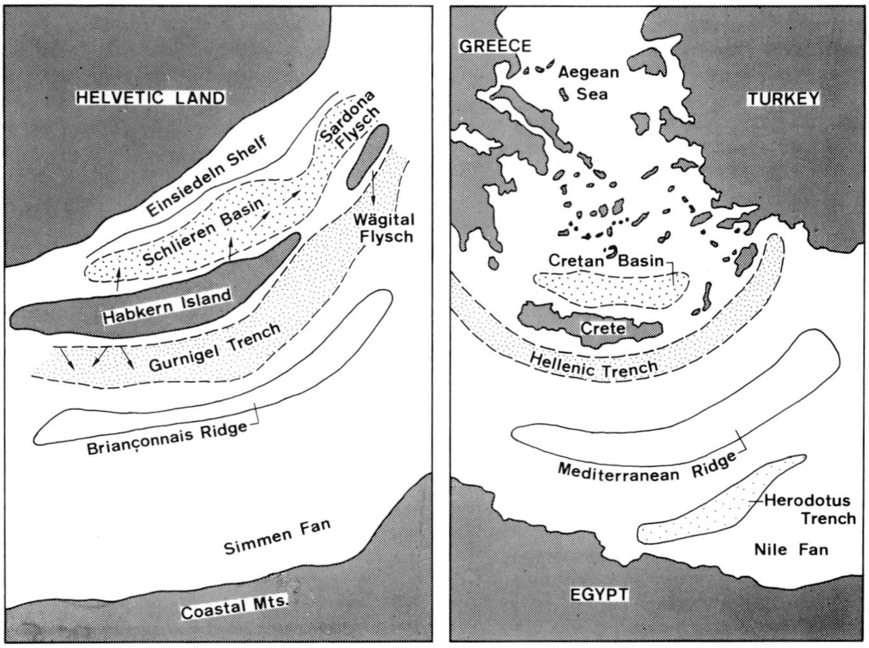

FIG. 5. Palaeocene palaeogeography of the western Alpine Tethys, compared to its actualistic analogue in the eastern Mediterranean. (After Hsü & Schlanger 1971). Two Palaeocene subduction zones in the Alps are assumed, one S-dipping under the Austroalpine coastal mountains, which were underlain by mélange terranes, and another N-dipping under the Gurnigel trench. In comparison, the present eastern Mediterranean has only one active, N-dipping, subduction zone under the Cretan Arc.

conventional schemes assume that Italy was a promontory of the African Plate. A dextral displacement since the late Cretaceous should result in seafloor spreading between the promontory and the Balkans (Fig. 6). No such Tertiary ocean existed. On the contrary, Italy and the Balkans have been compressed together.

A second problem which cannot be explained by the two-plate model is the start of the compression in the Alps. North Atlantic data predicted the onset at 81 Ma BP, or Santonian, when Europe, separating from North America, began to catch up with Africa (Dewey *et al.* 1973). Yet the initiation of Alpine subduction, indicated by the chronology of high-pressure metamorphism, should have been at 130 Ma BP, or Valanginian, some 50 Ma earlier. Another line of evidence, namely, the elastic down-bend of the Helvetic margin under the influence of a N-dipping subduction zone, placed the compression to an age not later than Aptian, or 110 Ma. The discrepancy between geophysical prediction and geological deduction is too large to be attributed to inaccuracies in dating. To resolve the dilemma, I proposed a 3-plate model in my *Society Lecture* to the European Geophysical Society (Hsü 1979b), and I have been experimenting with plate-movements ever since. Correlating geological and geophysical data, I found that Italy could have moved with Africa prior to the Cretaceous. The movements of the two, however, had to be different after the Valanginian Stage: while the Cretaceous movement of the African Plate was still sinistral relative to Europe until 80 Ma, the movement of the middle plate including Italy was dextral since 130 Ma.

A three-piece jigsaw puzzle

We have no need nor evidence to postulate, like Dewey *et al.* (1973), 15 or 20 micro-plates to interpret the Alpine orogenesis. Whereas a 2-plate model gives a first-order approximation, a 3-plate model provides second-order refinements. The third plate is the M-plate; M stands for middle, or for *Mediterranean*. This includes not only what has been commonly called the Apulian or Adriatic Block, but also the

Dinarides, the Hellenides, and parts of the western Mediterranean.

The M-plate proposed in this paper is underlain by continental lithosphere. The plate was separated from the European Plate by the Jurassic seafloor spreading and sutured onto the latter by the Cretaceous and Cenozoic Alpine orogenesis. The northern boundary of the M-plate is the southern boundary of the European Plate; it is defined by the suture between the two. The present Alpine suture extends eastward from the Betic Cordilleras in Spain. Discontinuous segments of the suture are identified in Sicily, Calabria and Corsica. In central Europe, the suture is manifested by the ophiolite mélanges of the Alps. Farther east, it is buried under the Austro–Alpine nappes and the southern Alps, only to crop out again in the Vardar zone of Yugoslavia and Greece.

The twists and turns of the suture in the western Mediterranean region reflect the post-collision fragmentation of the Alpine Orogen. The Balearic Islands, Sardinia, and Corsica were close to the shores of Spain and France in the Oligocene. They now constitute a remnant arc between the Balearic and Tyrrhenian basins, which owe their origin to back-arc seafloor spreading. The Sicily–Calabria basement nappes were part of a frontal arc that had been split off from eastern Sardinia before the arc collided with Italy in the Miocene (Argand 1924, Hsü 1971, 1977). For the sake of simplicity, we have omitted the intra-plate deformation from our consideration and placed the northern boundary of the M-Plate to coincide with the Oligocene position of the Alpine suture. The areas now underlain by the western Mediterranean basins are defined as a part of the M-plate.

The southern boundary of the M-plate is the northern boundary of the African Plate during the late Neogene. This boundary is defined by the subduction zone and transform fault south of the islands of Cyprus and of Crete in the eastern Mediterranean. It crosses central Sicily and extends westward under the flysch nappes of North Africa and from there to the Azore–Gibraltar fracture zone (Fig. 7).

To make the reconstructions presented in this paper, the positions of the African and European continents have been fixed on the basis of the data on magnetic lineation. Alan Smith kindly provided me with base maps on which the positions of the continents during seven time intervals since the Jurassic, namely, 0–20 Ma, 20–60 Ma, 60–80 Ma, 80–100 Ma, 100–120 Ma, 120–140 Ma, and 140–160 Ma, have been plotted. The position of the M-Plate, which is a micro-continent, should be shifted within the confines of the European and African continents.

The surface trace of the Alpine suture, shown by the solid line nM at 0 time in Fig. 6, is the northern margin of the M-Plate. The ancient southern margin of the European Plate, now buried under the nappes, is shown by a dashed line. The northern margin of the M-Plate should have been south of the dashed line before its collision with Europe.

Except for the Neogene reconstruction, the position of the M-Plate at time x is defined by moving its northern boundary away from the suture in a direction on the basis of the geological data until a best fit of its southern margin (sM at x Ma) with the northern coastline of Africa (nAc at x Ma) is obtained. There is a slight overlap between the M-Plate and western N Africa in several time slices of our reconstruction, but this overlap is an artefact caused by the simplifying assumption of plate rigidity; it should be absent or negligible if we could take into consideration Neogene intra-plate deformation, which has enlarged the outline of the western half of the M-plate. For this preliminary analysis, I have not made this correction.

0–20 Ma (Fig. 6a). The plate displacement between Europe and Africa in this interval has been largely absorbed by the consumption of the northern margin of the African Plate, the M-Plate having been sutured onto Europe. This subduction zone is well defined in the E by the Cyprus suture and by the Hellenic trench S of Crete. In the W, the suture extends from the Calabrian Arc, through central Sicily in the mélange (*'argille scagliose'*) country, and cuts across coastal North Africa under the flysch nappes.

The present northern boundary of the African Plate is the southern boundary of the M-Plate (onA=sM). The location of this line (onA) at 20 Ma is determined by its position relative to the African coastline at 20 Ma, which is fixed by an analysis of the Atlantic magnetic lineations. The area between the onA now and the onA at 20 Ma is the oceanic crust of the African Plate that has been consumed during the last 20 Ma.

Our reconstruction indicates a 250 km consumption of the Mediterranean seafloor during the last 20 Ma. at an average rate of about 1.25 cm yr^{-1}. A consumption of the African Plate at a rate of 1.4 cm yr^{-1} in the eastern Mediterranean has been estimated on the basis of deepsea drilling data (Hsü & Ryan 1973). The agreement is good.

The positioning of the southern boundary of the M-Plate fixes its northern boundary (nM).

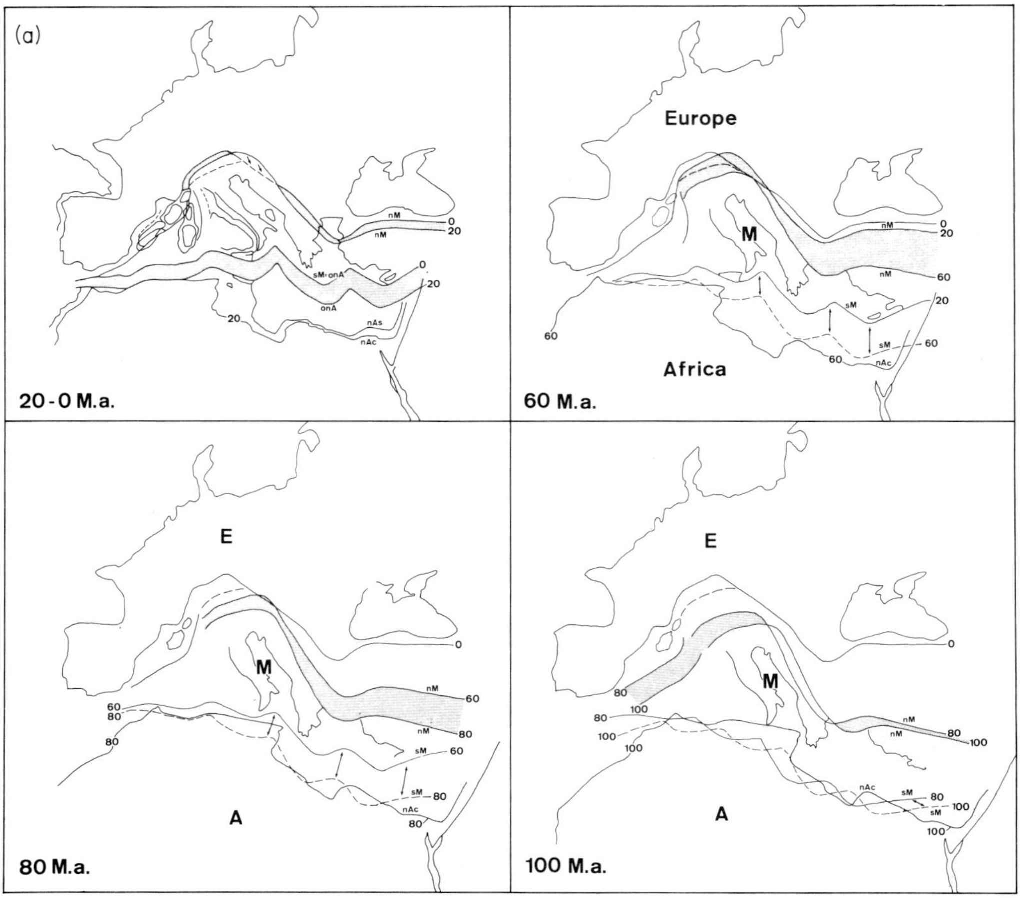

FIG. 6. A '3-plate Puzzle', showing the motions of Africa, M-Plate, and Europe since Middle Jurassic, 160 Ma. The northern and southern boundaries of the M-Plate are designated nM and sM, respectively, with the numbers (0, 20, 60, ...) indicating their positions at 0, 20, 60 ... Ma BP. The dashed line shows the edge of the European continent before continental collision. The coast line of North Africa is designated by nAc. The area underlain by oceanic crust which was consumed is shaded, and the arrows suggests the direction of seafloor spreading. See text for detailed explanations.

Overlap between nM and the southern boundary of Europe signifies compression and offset signifies extension. Our reconstruction shows an overlap in the Alpine Front, equivalent to a lithospheric shortening of 20 km at 0.1 cm yr^{-1} rate. The *décollement* deformations of the Helvetic Alps and the Jura Mountains were the structures produced during the interval, but the crustal shortening was about 100 km at 0.5 cm yr^{-1} rate (Hsü 1979). The shortening is greater than that predicted, but the discrepancy can be explained by the Neogene intra-plate deformation. The part of the M-Plate north of Balearic and Tyrrhenian basins may have been moved dozens of kilometres farther than the 20 km distance estimated on the basis of assuming plate-rigidity.

The offset of the northern boundary of the M-Plate in east-central Europe is also consistent with geological observations. The crust [illegible]er the Pannonian Basin has been extended [illegible] the Neogene. No oceanic crust is present und[illegible] the basin, but the basin subsidence should be an expression of the crustal thinning induced by extension.

20–60 Ma (Fig. 6a). This reconstruction is made by moving the M-Plate in a SSE direction until its southern boundary (sM) coincides more or less with the N African coastline (nAc) at 60 Ma; the position of the latter having been deter-

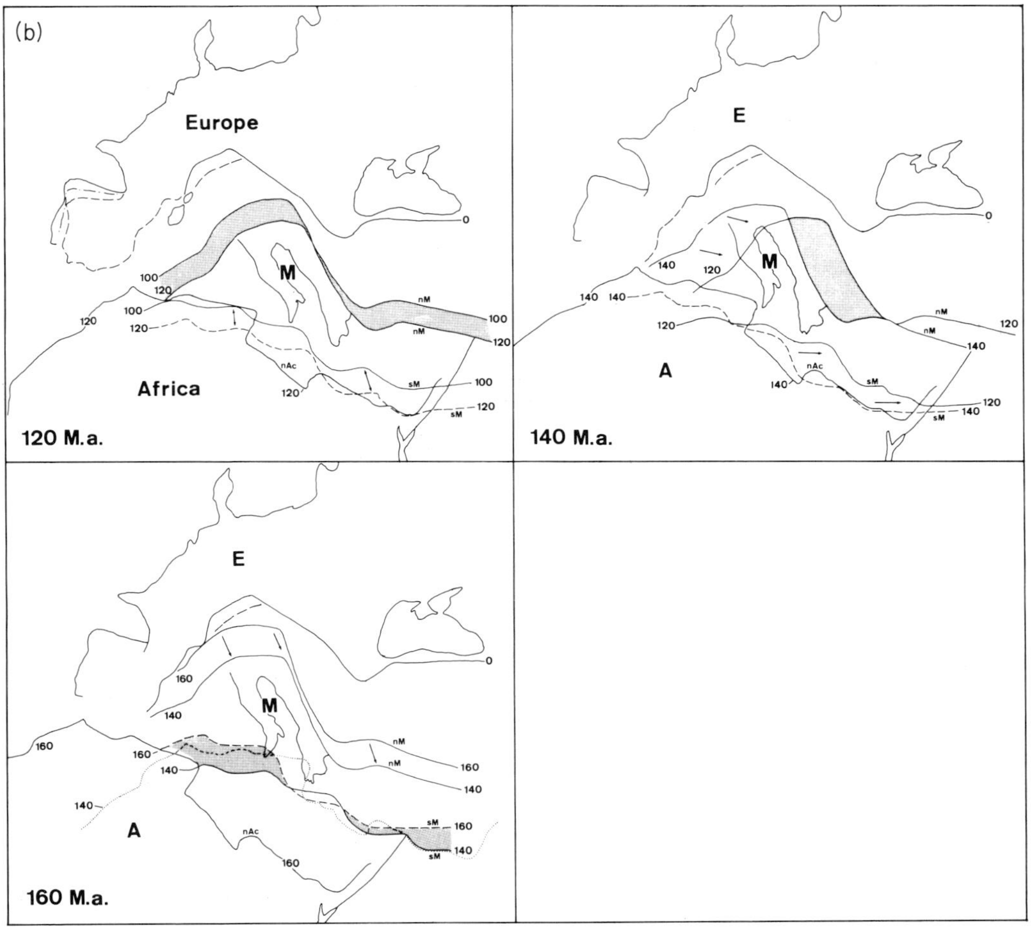

mined by the Atlantic magnetic lineations. The sketch shows that the M-plate moved 275 km to the NW during this 40 Ma interval at 0.7 cm yr^{-1} rate. This agrees with the geologically deduced rate of the Alpine deformation at 0.5–1 cm yr^{-1} (Hsü 1971, 1979).

The southern edge of Europe, shown by the dashed line in Fig. 6(a), is intermediate between the positions of the northern edge of the M-plate (nM) at 60 and 20 Ma respectively, indicating that the collision took place at 40 Ma, or in late Eocene, in good agreement with the geology. However, the collision was not synchronous everywhere; having taken place earlier in the E and later in the W. A detailed illustration of the movement of the northern boundary of the M-Plate relative to the southern margin of Europe is shown by Fig. 7.

The northward movement of the M-Plate caused a corresponding displacement of its southern margin (sM), resulting in the generation of the ocean crust in the eastern Mediterranean region; the compressional deformation of the Alps in the N was apparently related to the seafloor-spreading in the S.

60–80 Ma (Fig 6a). The reconstruction is based upon the same principle of moving the M-Plate in a southeasterly direction until the best fit with Africa is obtained. The displacement should have been 135 km in 20 Ma or 0.7 cm yr^{-1}. The rate is about the same as that during succeeding time interval (60–20 Ma), consistent with the postulate that the plate displacement was a consequence of seafloor-spreading at a linear rate.

The reconstruction suggests that collision had

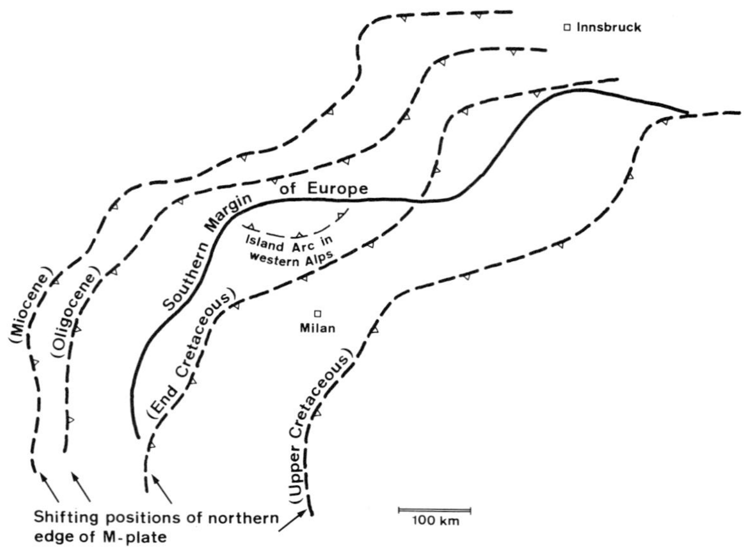

FIG. 7. The successive positions of the northern margin of the M-Plate relative to stable Europe. The configurations of the plate boundary and the changing directions of the M-plate motion led to an earlier collision in eastern Switzerland during the late Cretaceous.

not yet taken place at 80 Ma, confirming geological interpretations. The Upper Cretaceous South Penninic Flysch was deposited in a trench, when Tethyan ocean crust was being consumed by subduction during the Eoalpine deformation.

The Alpine Tethys underlain by ocean crust should have had a width of 250 km, fringed by a marginal sea some 150 km wide. The palaeogeography is thus comparable to that of the eastern Mediterranean today (Fig. 5); the postulated island arc in western Alps had a dimension similar to the Cretan Arc.

80–100 Ma (Fig. 6a). Moving M-Plate in an ESE direction until its southern boundary fits best with the northern boundary of Africa produces this reconstruction.

The assumption that the M-Plate moved dextrally relative to Europe is necessitated by the geological evidence that the Tethyan ocean between the two was not spreading, but being consumed during the Cretaceous Eoalpine deformation. The data on Atlantic magnetic lineations require, however, that Africa moved in sinistrally with respect to Europe, implying that Africa moved in a direction opposite to that of the M-Plate.

The displacement rate in the Alpine sector, calculated on the basis of this reconstruction, is again about 0.7 cm yr^{-1}; the linear rate of the Eoalpine subduction was apparently maintained since 100 Ma. While the Alps was being compressed, the Balkan region was extended; the lithospheric stretching in the Balkans should have resulted in crustal thinning. This prediction is in agreement with the geological observation that Upper Cretaceous 'miogeosynclinal' sediments of the Dinarides and Hellenides were laid down during this stage of crustal thinning and subsidence.

100–120 Ma (Fig. 6b). Moving the M-Plate to the SE until the best fit between the Africa and M-Plate is achieved gives this reconstruction. The displacement rate should be 250 km in 20 Ma, or 1.25 cm yr^{-1}, substantially more than the 0.7 cm yr^{-1} rate during the subsequent time intervals. I have no explanation for this.

The evidence of Eoalpine deformation excludes the possibility that the M-Plate moved as a promontory of the African Plate. The movement of the M-Plate with respect to Europe had to be dextral, while the African movement was sinistral.

The direction of the M-Plate movement is also constrained by the geological evidence of early Cretaceous deformation in the Balkans. A strike-slip motion of the M-Plate with respect to Balkan Europe is assumed in order to explain the deposition of Lower Cretaceous flysch formations of the Dinarides and Hellenides as the sediments in pull-apart basins.

120–140 Ma (Fig. 6b). Moving the M-Plate

eastward until the best fit with Africa is obtained gives this reconstruction; the position of Africa having been determined by the magnetic lineations of the Atlantic.

The kinematic analysis predicts extension in western Europe and compression in eastern Europe. This agrees with geological observations. In western Europe, the seafloor spreading, which started in Middle Jurassic, continued till earliest Cretaceous. In the Hellenides, there was orogenic deformation during late Jurassic and early Cretaceous, when subduction mélanges of the Vardar zone were overthrust onto the carbonate platform of the foreland (Smith & Moores 1974).

The Alpine Tethys Ocean should have reached its maximum extent during the earliest Cretaceous (130 Ma) before compression started. The area underlain by oceanic crust should have been 750 km. This figure is in agreement with an estimate of a 700 km crustal shorting in the Alps (see Trümpy 1960).

140–160 Ma (Fig. 6b). The movement of the M-Plate relative to Europe indicates a widening of oceanic realm, the Piémont Ocean, during the Jurassic. The displacement of some 300 km in 20 Ma gives a seafloor-spreading rate of 1.5 cm yr^{-1}.

The reconstruction shows that a narrow ocean between Europe and the M-Plate first was in existence before 160 Ma (Bathonian), as suggested by geology (see Kelts 1981). The oldest sediment in this ocean is the Callovian radiolarite, which locally overlies Alpine ophiolites (Kocher 1981).

The Swiss Alps as a model

It has been suggested that the Swiss Alps is a unique feature and should not be considered a model for other mountain ranges. The Swiss Alps is unique, but only in the sense that the rocks are well exposed and the geology is well understood. It is nevertheless a species in the family of collision-type mountains. Using the Alps as a model, I have been able to reconstruct the tectonic framework of other collision-type mountains, such as the Huanan Alps in China (Hsü *et al*. 1987).

In this contemplation on the time and place of Alpine orogenesis, I see no fundamental conflict between the geology of the Alps and plate tectonic theory. The apparent paradox has been man-made by those who have conditioned themselves too deeply to the Elie de Beaumont/Dana/Stille/Bucher theory of globally synchronous episodic orogeny.

ACKNOWLEDGEMENTS: This paper is a summary of my understanding of the Swiss Alps after 30 years of working off and on in the Alps, and after my Swiss friends taught me the ins and outs of the Alps. I am particularly grateful to Rudolf Trümpy, who seems to know every outcrop in, and remembers every detail about, our magnificent mountains. I have many fond memories of the trips we took together, and even of the arguments we have had. While we may not have arrived at the same conclusion on time and place in orogeny, this paper is nevertheless a testimony to his inspiration.

I have benefitted from discussions with Daniel Bernoulli, Stephan Schmid, Helmut Weissert, Johannes Hunziker, Celâl Şengör and many other friends. I am indebted to Alan Smith for having provided me with print-outs of his plate reconstructions. Rudolf Trümpy and Dorothee Dietrich read an earlier version of the manuscript, and their critiques led to its improvement. Finally, I would like to thank Mike Coward and the Geological Society for having invited me to give this Fermor Lecture. A first draft of the manuscript has been sitting in my desk drawer for 8 years. Without their urging, I may never have taken the time to tie the last loose ends together.

References

ALVAREZ, L., ALVAREZ, W., ASARO, F. & MICHEL, H. 1980. Extraterrestrial cause for the Cretaceous–Tertiary extinction. *Science*, **208**, 1095–108.

ARGAND, E. 1924. La tectonique de l'Asie. *13th Congress géologie internationella Bruxelles*. **1922**, 171–372.

BIJU-DUVAL, B., DERCOURT, J. & LE PICHON, X. 1977. From the Tethys Ocean to the Mediterranean Seas: a plate tectonic model of the evolution of the western Alpine system. *In*: BIJU-DUVAL, B. & MONTADERT, L. (eds). *Structural History of the Mediterranean Basins*, 143–183. Editions Technip, Paris.

BOCQUET, J., DELALOYE, M., HUNZIKER, J. C., KRUMMENACHER, D. 1974. K–Ar and Rb–Sr dating of blue amphiboles, micas and associated minerals from the western Alps. *Contributions to Mineralogy and Petrology*, **47**, 7–26.

BUCHER, W. H. 1933. *The Deformation of the Earth's Crust*. Princeton University Press, Princeton.

CHAMBERLIN, T. C. 1898. The ulterior basis of time divisions and the classification of geologic history. *Journal of Geology*, **6**, 449–62.

—— 1907. Diastrophism as ulterior basis of correlation. *Journal of Geology*, **17**, 685–93.

CROWELL, J. C. 1955. Directional-current structures from the pre-Alpine Flysch, Switzerland. *Geological Society of America Bulletin*, **66**, 1351–84.

Dal Piaz, G. V., Hunziker, J. C. & Martinotti, G. 1972. La zona Sesia–Lanzo e l'evoluzione tetonico-metamorfica delle Alpi nordoccidentali interne. *Memorie della Societa geologica Italiana*, **11**, 433–66.

Dana, R. D. 1973. On the origin of mountains. *American Journal of Science,* **2nd ser. 9**, 381–98.

Desmus, J. 1986. The Alpine metamorphisms and their environments in the western Alps: Unsolved problems. *Schweizerische Mineralogische und Petrographische Mitteilungen*, **66**, 29–40.

Deutsch, A. 1983. Datierung an Alkaliamphibolen und Stilpnomelan aus der südlichen Platta–Decke (Graubünden). *Eclogae Geologicae Helveticae*, **76**, 295–308.

Dewey, J. F., Pitman, W. C., Ryan, W. B. F. & Bonin, J. 1973. Plate tectonics and the evolution of the Alpine system. *Geological Society of America Bulletin*, **84**, 3137–80.

Elie de Beaumont, L. 1831. Researches on some of the revolutions on the surface of the Globe. *Philosophical Magazine*, **n.s. 10**, 241–64.

Flügel, H. W., Faupl, P. & Mauritsch, H. J. 1987. Implications on the Alpidic evolution of the eastern parts of eastern Alps. *In*: Flügel, H. W. & Faul, P. (eds). *Geodynamics of the Eastern Alps*, 407–14. Deuticke, Vienna.

Frank, W. 1987. Evolution of the Austro–Alpine elements in the Cretaceous. *In*: Flügel, H. W. & Faupl, P. (eds). *Geodynamics of the Eastern Alps*, 379–406. Deuticke, Vienna.

Gilluly, J., 1949. The distribution of mountain building in geologic time. *Geological Society of America Bulletin*, **60**, 561–90.

Giraud, J. D. 1983. *L'arc andösitique paléogène des Alpes occidentales: Pétrologie, volcanologie, geodynamique*. Thèse, Nice.

Gutenberg, B. & Richter, 1954. *Seismicity of the Earth and Associated Phenomena*. Princeton University Press.

Homewood, P. & Caron, C. 1982. Flysch of the western Alps. *In*: Hsü, K. J. (ed). *Mountain Building Processes*, 156–68. Academic Press, London.

Hsü, K. J. 1950. *Evidence for the dating of Paleozoic orogenies of eastern North America*. MA Thesis, Ohio State University, Columbus, Ohio.

—— 1960. Paleocurrent structures and paleogeography of the Ultrahelvetic Flysch basins, Switzerland. *Geological Society of America Bulletin*, **71**, 577–610.

—— 1969. A preliminary analysis of the statics and kinetics of the Glarus Overthrust. *Eclogae Geologicae Helveticae*, **62**, 143–54.

—— 1970. The meaning of the word flysch – a short historical search. *Geological Association of Canada, Special Paper*, **7**, 1–11.

—— 1971. Origin of Alps and western Mediterranean. *Nature*, **233**, 44–8.

—— 1972a. The concept of the geosyncline, yesterday and today. *Leicester Literary & Philosophical Society, Transactions*, **66**, 26–48.

—— 1972b. Alpine flysch in a Mediterranean setting. *International Geological Congress, 24th Session, Montreal, Canada, Proceedings*, Sec. **6**, 67–74.

—— 1977. Tectonic evolution of the Mediterranean basins. *In*: Nairn *et al.* (eds). *The Ocean Basins and Margins, IV Mediterranean*, 29–75. Plenum Press, New York.

—— 1979a. Thin-skinned plate tectonics during Neo-alpine orogenesis. *American Journal of Science*, **279**, 353–66.

—— 1979b. A kinematic analysis of plate-displacements in the Mediterranean region. *European Geophysical Society, 6th Annual Meeting Vienna. Society Lecture*.

—— 1980. Terrestrial catastrophe caused by cometary impact at the end of Cretaceous. *Nature*, **285**, 201–03.

—— 1982a. Actualistic catastrophism. *Sedimentology*, **30**, 3–9.

—— 1982b. Geosynclines in plate tectonic settings: Sediments in Mountains. *In*: Hsü K. J. (ed). *Mountain Building Processes*, 3–12. Academic Press, London.

—— 1982c. Alpine Mediterranean Geodynamics: Past, Present and Future. *American Geophysical Union Geodynamics Series*, **7**, 7–14.

—— 1987. Nuclear risk evaluation. *Nature*, **328**, 22.

—— & Schlanger, S. O. 1971. Ultrahelvetic flysch sedimentation and deformation related to plate tectonics. *Geological Society of America Bulletin*, **82**, 1207–18.

—— & Ryan, W. B. F. 1973. Deep-sea drilling in the Hellenic Trench. *Geological Society of Greece Bulletin*, **10**, 81–93.

—— Sun Shu & Li jiliang 1987. Mesozoic suturing in the Huanan Alps and the tectonic assembly of South China (in Chinese). *Scientia Sinica*, **1987,B**, 1107–115.

Hubbert, M. K. & Rubey, W. W. 1959. Role of fluid pressure in mechanics of overthrust faulting. *Geological Society of America Bulletin*, **70**, 115–66.

Hunziker, J. C. 1974. Rb–Sr and K–Ar age determinations and the Alpine tectonic history of the western Alps. *Memorie Istituto Geologicae Mineralogia della Universita di Padova*, **31**, 1–55.

—— & Martinotti, G. 1984. Geochronology and evolution of the western Alps: A review. *Memorie della Societa Geologica Italiana*, **29**, (in press).

Kelts, K. K. 1981. A comparison of some aspects of sedimentation and translational tectonics from the Gulf of California and the Mesozoic Tethys, Northern Penninic Margin. *Eclogae Geologicae Helvetiae*, **74**, 317–338.

Kocher, R. N. 1981. *Biochronostratigraphische Untersuchungen oberjurassischer radiolarienführender Gesteine, insbesondere der Südalpen*. Diss. ETH Zürich.

Laubscher, H. P. 1971. The large-scale kinematics of the western Alps and the northern Apennines and its palinspastic implications. *American Journal of Science*, **271**, 193–226.

Le Pichon, X. 1968. Seafloor spreading and continental drift. *Journal of Geophysical Research*, **73**, 3661–98.

Lugeon, M. & Gagnebin, F. 1941. Observations et

vues nouvelles sur la géologie des Préalpes romandes. *Laboratoire de Géologie de l'Universite de Lausanne Bulletin*, **72**, 1–90.

MILNES, A. G., GRELLER, M. & MULLER, R. 1981. Sequence and style of major post-nappe structures. *Journal of Structural Geology*, **3**, 411–20.

NIGGLI, E. 1970. Alpine Metamorphose und alpine Gebirgsbildung. *Fortschritte der Mineralogie*, **47**, 16–26.

PLAYFAIR, J. 1802. *Illustrations of the Huttonian Theory of the Earth*. Fascimile reprint by Dover, New York.

SATIR, M. 1975. Die Entwicklungsgeschichte der westlichen Hohen Tauern und der südlichen Oetztalmasse aufgrund radiometrischer Altersbestimmungen. *Memorie della Istituto Geologica er Mineralogia della Universita di Padova*, **30**, 84 pp.

SAVATSKI, G. G. 1975. *Etude géologique et minéralogique des flyschs à grauwackes volcaniques du synclinal de Thones (Hautes-Savoie, France). Grès de Taveyanne et grès du val dlliez*. Thesis, Univ. Genève.

SAVOSTIN, L., SIBUET, J. -C., ZONENSHAIN, L. P., LE PICHON, X. & ROULET, M. -J. 1986. Kinematic evolution of the Tethys belt from the Atlantic Ocean to the Pamirs since the Triassic. *Tectonophysics*, **123**, 1–35.

SHOEMAKER, E. M., 1966. Progress in the analysis of the fine structure and geology of the lunar surface from the Ranger VIII and IX photographs. *Jet Propulsion Laboratories Technical Report*, **32–800**, 249–337.

—— 1983. Asteroid and comet bombardment of the Earth. *Annual Review of Earth & Planetary Sciences*, **11**, 461–94.

SMITH, A. G. 1971. Alpine deformation and the oceanic areas of the Tethys, Mediterranean and Atlantic. *Geological Society of America Bulletin*, **82**, 2039–70.

—— & MOORES, E. M. 1974. Hellenides. *In*: SPENCER, A. M. (ed). *Mesozoic–Cenozoic Orogenic Belts*, 159–86. Scottish Academic Press, Edinburgh.

—— & WOODCOCK, N. H. 1982. Tectonic synthesis of the Alpine–Mediterranean region: A review. *American Geophysical Union Geodynamics Series*, **7**, 15–38.

SPIEKER, E. M. 1946. Late Mesozoic and early Cenozoic history of central Utah. *U.S. Geological Survey Professional Paper*, **205-D**.

STEINITZ, G. & JÄGER, E. 1981. Rb–Sr and K–Ar studies on rocks from the Suretta Nappe, eastern Switzerland. *Schweizerische Mineralogische und Petrographische Mitteilungen*, **61**, 121–31.

STILLE, H. 1924. *Grundfragen der vergleichenden Tektonik*. Bornträger, Berlin.

TERMIER, P. 1903. *Les montagnes entre Briançon et Vallouise*. Mémorie de la Service de Carte Geologique de la France, Paris, 182 pp.

THÖNI, M. 1983. The thermal climax of the early Alpine metamorphism in the Austro–Alpine thrust sheet. *Memorie della Istitut Geologicae Mineralogia della Universita di Padova*, **36**, 211–37.

TRICART, P. 1984. From passive margin to continental collision. *American Journal of Science*, **284**, 97–120.

TRÜMPY, R. 1960. Paleotectonic evolution of the central and western Alps. *Geological Society of America Bulletin*, **71**, 843–908.

—— 1973. The timing of orogenic events in the central Alps. *In*: DE JONG, K. A. & SCHOLTON, R. (eds). *Gravity and Tectonics*, 229–52. Wiley, New York.

—— 1980 *An outline of the geology of Switzerland*. Wept & Co., Basel, 104 pp.

—— 1987. Vom Sinn der Erdgeschichte. *Forschung u. Technik, neue zürcher Zeitung*, **8 April**, 65–6.

WILDI, W. 1985. Heavy mineral distribution and dispersal pattern in Penninic and Ligurian flysch basins (Alps, northern Apennines). *Giornale di Geologia, serie 3a*, **47**, no. 1–2, 77–100.

WINKLER, W. 1981. *Die Sedimentologie und Sedimentpetrographie des Schlierenflysches (Zentralschweiz)*. Diss. Fribourg.

—— & BERNOULLI, D. 1986. Detrital high-pressure/low-temperature minerals in a late Turonian flysch sequence of the eastern Alps (western Austria): Implications for early Alpine tectonics. *Geology*, **14**, 596–601.

K. J. HSÜ, Swiss Federal Institute of Technology, Zurich, Switzerland.

Note added in proof

North-dipping reflectors have been found between the Helvetic and Penninic units in eastern Switzerland (see Schw. Arbeitsgruppe f. Reflexionsseismile, Vorläufige Ergebuisse der Alpentraversen des NFP-20 'Geologische Fiefenstrukthr der Schweiz'). I consider this important discovery a verification of the 'N-dipping subduction zone under the Habkern Arc', as postulated on p. 435 of this article.

Index

References to figures appear in *italics* and those to tables appear in **bold**